DATA ANALYSIS FOR SCIENTISTS AND ENGINEERS

DATA ANALYSIS FOR SCIENTISTS AND ENGINEERS

STUART L. MEYER
Northwestern University

John Wiley & Sons, Inc., New York · London · Sydney · Toronto

To Jonathan, Eric, and David

"Now this is the peculiarity of scientific method, that when once it has become a habit of mind, that mind converts all facts whatsoever into science. The field of science is unlimited; its material is endless, every group of natural phenomena, every phase of social life, every stage of past or present development is material for science. *The unity of all science consists alone in its method, not in its material.* The man who classifies facts of any kind whatever, who sees their mutual relation and describes their consequences, is applying the scientific method and is a man of science. The facts may belong to the past history of mankind, to the social statistics of our great cities, to the atmosphere of the most distant stars, to the digestive organs of a worm, or to the life of a scarcely visible bacillus. It is not the facts themselves which form science, but the methods by which they are dealt with."

Karl Pearson, *The Grammar of Science*

Library of Congress Cataloging in Publication Data

Meyer, Stuart L 1937–
Data analysis for scientists and engineers.

Bibliography: p.
1. Mathematical statistics. 2. Probabilities.
I. Title.
QA276.M437 519.2 74–8873
ISBN 0-471-59995-6

Printed in the United States of America

10 9 8 7 6 5 4 3 2 1

PREFACE

This book evolved from a personal need: the need to have in one place, with a consistent style and notation, practically all that an experimental scientist *needs* to know to deal with data and *wants* to know to satisfy his or her intellectual curiosity.

For many years there has really been no discipline in which to teach data analysis *for experimental scientists.* The usual medium in science departments for imparting knowledge in this area has been the teaching laboratory. I believe that an appreciation of the means of extracting and evaluating information from experimental data is one of the basic goals of any laboratory course. Moreover, this is a skill and appreciation that should be developed as early as possible in a student's career, regardless of whether he or she intends to be an experimental or theoretical scientist. Therefore, although this book can be used in departments that have formal courses on the subject, I have assumed that the student's first acquaintance with data analysis will come through the science laboratory and that mature knowledge will come largely from self-study.

The early parts of the book are written to accompany a typical first-course science laboratory, and versions of Parts I and II have been used for several years at Northwestern University as a supplement to the beginning physics laboratory (and as a guide for the graduate teaching assistants). In addition, a version of Parts III and IV has been used in an integrated program of mathematics, physics, and chemistry for engineering students. Despite this history, the book is intended primarily for self-study and reference by the student, and much of the material goes beyond the usual content of a first course to provide the core of what is needed by a professional scientific investigator. Therefore, I have kept the discussion as complete and lucid as possible with many worked-out examples and cases that are of direct interest to the science student.

The wide availability of digital computer capability has made traditional, formerly cumbersome analysis methods more convenient and accurate and also has permitted the use of more sophisticated techniques that are not practical with slide rules, pencils, and paper alone. Some of these techniques are introduced here because they will become utilized more and more in the near future. I have not discussed programming or specific computer programs, since this material is readily available elsewhere. Also, the use of individual programs is often determined by the availability of library routines in a particular computer center, and it was more important to present the *methods* underlying applicable library programs instead of the programs themselves. Finally, the advances in technology have made available to *individuals* computing power that required a large facility only a decade ago. All of the numerical discussions and worked-out examples included here may be *followed* by the reader without *any* computational assistance and may be *duplicated* with a pocket calculator of the "electronic slide rule" type.

The level of this book requires only the rudiments of calculus as a preliminary. Although the early discussions are on a level that is understandable to the beginning student, the subject is developed so that it is fairly complete and self-contained, and it ends at a level of sophistication that is suitable for the professional scientist. This will enable the beginning student to "grow into" the book. The volume also may be used as a general reference and guide as the reader becomes more sophisticated in the design and analysis of experiments. Many discussions, formulas, tables, and graphs are included for convenience, completeness, and review. I hope that the book will find a useful place on the permanent bookshelf of the working scientist.

I wrote this work from the viewpoint of the *practitioner* of data analysis, that is, the scientist or engineer *using* the techniques, rather than from the viewpoint of the mathematician who develops new ones. Nevertheless, it is important for the practitioner to *understand* the bases of the techniques that he uses. For this reason, a fair amount of time is spent in developing the *ideas* of data analysis instead of merely presenting a series of recipes. Few problems encountered in the real world fit any one recipe exactly. It is important, therefore, to understand data-analysis concepts and methods thoroughly, and to have an appreciation for their spheres of usefulness.

The presentation is designed to be self-contained, eschewing the elegant obscurity of sophisticated methods of proof for the tedious clarity of more straightforward ones. All available sources have been utilized, and my major efforts are directed toward making the concepts palatable at the level of the reader and making them form a comprehensible, coherent, and useful whole. If I have erred, it has been on the side of too many explanation instead of too few, more steps in the mathematics rather than not enough steps; although this might please the experts less, it will help the students more. There is a famous explanation of why a classical work of J. Willard Gibbs was not widely known: "It is a little book which is little read because it is a little hard." I have tried to make this book easy to understand, even at the expense of greater than minimum length. However, there is sufficient depth in this single volume to reward careful and continued study.

Even the simplest problem often has subtleties that require more nuances of thought than the casual reader may consider at first glance. Some of the ramifications are not needed early in the book, and we sometimes return to the same problem again when it is appropriate to discuss other aspects of it. In addition, considerable material has been added because of my personal (possibly idiosyncratic) tastes. This material is usually set in special type or otherwise identified as not being absolutely essential to the main line of discussion.

I am pleased to acknowledge the support of the U.S. Army Research Office—Durham under Grant DA-ARO-D-31-124-71-G175 whose assistance facilitated the completion of the manuscript. I thank Publishers—Hall Syndicate and Cartoon Features for arranging permission to use cartoons and to Trendline Inc. for permission to use a graph. My sincere appreciation goes to Professors A. J. F. Siegert, Meyer Dwass, Robert Eisberg, and Hugh Young for reading parts of the manuscript and for helpful suggestions and to Mr. King L. Leung for able assistance with the calculation of the tables and graphs.

Stuart L. Meyer

Evanston, Illinois

CONTENTS

PART I
Introduction to Scientific Measurement

PART II

Introduction to Graphical Techniques and Curve Fitting

PART III
Probability

PART IV
Some Probability Distributions and Applications

PART V
Statistical Inference

APPENDICES

PART I

Introduction to Scientific Measurement

1

THE MEANING OF MEASUREMENT

The distinction is sometimes made between the "exact" sciences and the other sciences. In the first category are usually put the physical sciences such as physics and chemistry; in the second category are, broadly speaking, biology, psychology, etc. This terminology is unfortunate since it seems to imply that physical science makes statements that are true whereas the others do not. In fact, all scientists deal with the truth. The distinction is that the exact sciences can more easily measure or at least assign a value to the amount of truth in a scientific statement. It is preferable to call the physical sciences "quantitative" disciplines, since one can discuss in numerical fashion the amount of truth in any given statement of physical fact.

The alert student will note at this point that we have not yet defined truth. We must ask what it means to say that a measurement is "true" and what we mean when we say that a given statement about the physical world is "true." We make measurements in the physical sciences, and we perform experiments to test hypotheses about the physical world. There is no clear distinction between experiment and measurements, and the former of necessity involve the latter; that is, an experiment inescapably requires the measurement of *something.*

A Definition of Measurement

For the purposes of this discussion we shall consider that a measurement is the quantitative determination of the value of some *fixed* physical constant that is characteristic of a physical object or system or of a parameter that is needed for the description of a *reproducible* physical situation. Note carefully the adjectives *fixed* and *reproducible.* Both carry the necessary idea that the measurement can be repeated either tangibly or conceptually. The important point is that an *independent* determination of the value of the measured quantity can be made.

SIGNIFICANCE

Examples of quantities subject to measurement are legion: the velocity of light in vacuum; the mass of a proton; the length of an inch scale in centimeters. When the measurement is made, we assign a value to the result. If one looks in a handbook that is modern, one finds the following values: the velocity of light in vacuum is 2.997925×10^{10} cm/sec; the mass of the proton in MeV/c^2 (millions of electron volts divided by the square of the velocity of light) is 938.256 ± 0.005; and the length of an inch in centimeters is 2.54 *exactly*.

What do these numbers mean? It is intuitively easy to understand that one makes a measurement and gets a result. We have three results, however, and they appear to be in three different forms. The first number consists of seven digits with a decimal point as well as an exponent. By writing down seven digits, we are expressing a measure of the reliability of this number. We tacitly mean that the next to last digit is truly meaningful and that the last number is our best guess. We say that the number has seven significant figures. In this case the velocity of light is most likely to be halfway between the numbers 2.99792 and 2.99793. The use of the exponent is a standard way of separating the number of digits needed to specify the *magnitude* of a number from the number of digits needed to convey its *significance*.

The second number represents another way of conveying the significance of a numerical quantity. The mass of the proton is believed likely to be between the values 938.251 and 938.261 MeV. We shall discuss what "likely" means at a later time. At that point we shall be able to discuss whether two numbers that represent the same quantity are *consistent* with each other.

The last is the strangest one of all: 2.54 *exactly*. For one of the few times to be encountered, one's first impression is correct in this instance. This is a unique quantity known as a *defined constant*. This is the only kind of quantity that is exact. The metric and English systems of units arose independently, and for many years the conversion between them was a measured quantity like all the (undefined) quantities we shall encounter. If you find an old mathematics or physics book, you may even find the conversion constant written as 2.54001. (The U.S. Coast and Geodetic Survey still uses 2.54001.) Recently, however, it was decided to *fix* the inch as the distance that is equal to 2.54 of the units known as the centimeter. This was defined in terms of 1/100 of the distance between two ruled lines inscribed on a platinum-iridium bar kept under standard conditions in Sevres, France. Since 1960, the standard centimeter has been defined by international agreement in terms of the wavelength of the orange spectral line of the light emitted by the pure isotope krypton-86. The official centimeter is now defined to be 16,507.6373 wavelengths.

DIMENSIONS

The basic dimensions that we deal in are length (L), time (T), and mass (M). All else may be expressed in terms of these using various physical relations, since the two sides of any equation must have the same dimensions.

Thus, force F is defined in our basic set of dimensions by the relation between the acceleration produced on a mass and the force acting on it:

$$F = ma$$

We shall denote the equality of dimensions by $\stackrel{D}{=}$. Thus

$$F \stackrel{D}{=} MLT^{-2}$$

represents the dimensions of force.

The gravitational force equation tells us that the gravitational force between two masses M_1 and M_2 in isolation is proportional to the product of the masses and to the inverse of the square of the distance between them:

$$F = \frac{GM_1M_2}{r^2}$$

where G, the gravitational constant, has

dimensions

$$G \overset{D}{=} M^{-2}L^{+2}F = M^{-2}L^{+2}MLT^{-2}$$
$$\overset{D}{=} L^3M^{-1}T^{-2}$$

To determine the dimensions of any quantity we need only recall the equations relating the quantity of interest to quantities whose dimensions we know and treat the dimensions as ordinary algebraic symbols.

The arguments (x) of various functions such as $\exp(x)$ and $\sin(x)$ must be dimensionless.

UNITS

It is desirable to write all equations so that they are independent of the system of units. Since our systems of units are arbitrary, it would be absurd to do otherwise. It should be obvious that the left side of any equation must have the same units as the right side. This provides a convenient check on any equations we write down (a necessary but not sufficient condition for the correctness of the equation).

Except for dimensionless constants, all quantities used must be regarded in some system of units, for example, centimeter-gram-second (CGS) system, meter-kilogram-second (MKS) system, foot-pound-second (English) system, etc. As with dimensions, the units of the left-hand side of an equation must be the same as the right-hand side. The units should be continually checked through any calculation since errors arising from discrepancies in units are most common! All units used must be rationalized to be consistent, since combinations of mixed units are myriad and hence inconvenient.

For example, an English moat (it must be English) is to be filled from a reservoir. How many acre-feet of water must be drawn from the reservoir to fill the moat to a depth of 2.00 fathoms if the moat is 0.500 furlongs in length and 900 barleycorns wide?

$$\text{Volume } V = 900 \text{ barleycorn-furlong-fathoms}$$

We may always multiply anything by unity. We recognize that

$$1 \text{ barleycorn} = \frac{1}{3} \text{ inch}$$
$$1 \text{ fathom} = 6 \text{ feet}$$
$$1 \text{ furlong} = \frac{1}{8} \text{ mile}$$
$$1 \text{ acre} = 43{,}560 \text{ square feet}$$

Therefore

$$\begin{aligned} V &= 900 \text{ barleycorn} \frac{1}{3} \frac{\text{inch}}{\text{barleycorn}} \times \frac{1 \text{ foot}}{12 \text{ inch}} \\ &\quad \times \text{furlong} \frac{1}{8} \frac{\text{mile}}{\text{furlong}} 5280 \frac{\text{feet}}{\text{mile}} \\ &\quad \times \text{fathoms} \cdot 6 \frac{\text{feet}}{\text{fathom}} \\ &= (900)(1/3)(1/12)(1/8)(5280)(6) \text{ foot}^3 \\ &= 99{,}000 \text{ ft}^3 = 99{,}000 \text{ ft}^2 \frac{1 \text{ acre}}{43560 \text{ ft}^2} \\ &= 2.27\,27(\dot{2}\dot{7}) \text{ acre-ft} \end{aligned}$$

The advantages of the metric system should be obvious!

Units and standards of both the metric and the U.S. Customary Systems are discussed in detail in Appendix I.

Dimensional Analysis

Given a physical situation describable by physical variables $x_1, x_2, \ldots,$ we can sometimes deduce from dimensional analysis certain limitations on the form of any possible relationship among the variables. Dimensional analysis is not capable of completely determining the unknown functional relationship, but it can delimit the possibilities and, in some simple cases, it can give the complete relationship to within a constant of proportionality. Appendix II contains a discussion of this subject.

Testing Hypotheses

Physical hypotheses, often erroneously called "theories," must make testable predictions about the physical world to be meaningful. These predictions, to be testable, must be quantitative. That is, certain numerical quantities are predicted with a range of values. When measurements are made of these quantities and the results are compared with the predictions, we have performed an experiment to test the hypothesis.

The job of a scientist is to understand what is going on in the physical world. The function of the scientific investigator is not unlike that of the superrational detective of fiction. In the words of Sherlock Holmes as he explained his unraveling of a particularly knotty situation: "When we have eliminated the impossible, what remains is merely the improbable." One of the tasks of the experimental scientist is to determine the degree of improbability or, better, the *probability* of various occurrences.

The result of an experiment is quoted as a number together with a stated experimental uncertainty. As we have seen, this may be done with significant figures or with a statement that the result is "$x \pm \Delta x$ where Δx is the experimental uncertainty or "error." (We shall also see later that the result may be quoted "$x^{+\Delta x_1}_{-\Delta x_2}$.") We mean that we believe that the number we have measured has a high probability of lying within this range of values. Practically speaking, we believe that a valid but independent repetition of the experiment would more often than not yield a numerical result in this range. To be even more specific, if the error quoted is a "standard deviation," we believe *at this time* that, out of 100 repetitions of the measurement, over 68 will yield a result within the limit quoted. As we shall see, however, an additional measurement of comparable precision that is consistent with the existing number will result in a new "best value" and our belief *at that time* will be slightly different.

The prediction of the hypothesis under test is considered to be verified if the predicted range of values overlaps the range of measured value with its estimated uncertainty. The estimated uncertainty is very important. In general, the larger the experimental uncertainty, the wider the range of predictions that will be consistent with the result. A small uncertainty will enable an experimenter to attach more meaning to his result, since it will often enable him to eliminate some of several different hypotheses. *This is how progress in science is made.* Hypotheses are conjectured; they are tested by experiment; and some are rejected while others are provisionally accepted subject to further testing. Hypotheses that have successfully weathered all tests *so far conceived* are accorded the status of "principles" or "laws." The more tests a given principle has passed, the more firmly we believe in it, but even a hallowed principle must be rejected should it be inconsistent with a valid experimental result in the future.

2

THE CONDUCT OF AN EXPERIMENTAL INVESTIGATION

An experimental investigation may be considered to consist of three stages: the experiment is planned and designed; the measurements are performed and the data are taken; the data are analyzed, conclusions are drawn, and the results are reported in written form.

Design

What is involved in the design and planning of an experiment? First, the experiment must be based on an idea; it must have a point. That is, either a quantity is to be measured for its own sake or some hypothesis is to be tested in the sense of comparing its predictions with measured quantities. The designing and planning stage is probably the most important part of an experiment. Good experiments rest on good ideas; great experiments rest on great ideas. Good planning is a necessary although not sufficient condition for a good experiment.

However, the design of experiments is not something that can easily be taught. It requires experience and a mature viewpoint that only comes with experience. For this reason, it is not often part of one's early education in experimental methods. Usually, the first experiments in a typical laboratory course are designed by the instructor and laid out in general detail. The student should, however, try to appreciate the features of the design: what one is out to measure, why one does certain check measurements, etc. By understanding how to analyze experiments, one quickly learns what is important in their design. Very often the analysis of an experiment indicates that some quantity should have been measured that wasn't or that some other quantity should have been measured *better.* Often a bad experiment has as much *educational* value as a good one. If one can understand why a result is bad (i.e., is wrong or has an uncertainty that is larger than necessary), he is well on the way to being able to design and plan a *good* experiment. One must be able to walk before he can fly, and a knack for experimental design comes only after experience with performing and analyzing experiments.

There is an apt analogy with chess. Emmanuel Lasker, who was chess champion of the world for almost three decades, pointed out that, when instructing a novice chessplayer, one should start with the endgame position, then investigate the middle game tactics, and teach the chess openings last of all. Endgame situations leading directly to checkmate are obviously the most important positions in chess, and one has only to work backwards from there to determine how best to produce the desired result. Also, in learning about experimentation, if one can learn to appreciate what a successful final result is, then the factors that enter into successful experimental design become obvious. In our case, of course, the desired endgame position is to extract from an experimental situation the maximum possible information that bears on the initial scientific question.

Crucial Times During an Investigation

Furthermore, the most important times during an experimental study are the periods before and after the taking of data. The experiment should deal as directly as possible with the ideas one is investigating and minimize the influence of peripheral or background effects. The design of an experiment is directed to minimizing the uncertainties and maximizing the information content. This usually means getting a result with as small an error as possible. Since time, money, and effort are valuable commodities, the design should also direct the experimental measurements to the most efficient use of the resources available for the experiment.

The treatment of the data after the experimental observations are made is equally important. The only limitation on this importance is that one can at this stage only maximize what one has and extract as much information as the data *potentially* contain. A better design, on the other hand, could have produced a better experiment. It has been truly said that every experiment should be done twice because one knows how to do it the second time! If one looks to the data-analysis aspects at the beginning of the study, however, then one can determine how best to do the experiment the first time. In that sense data-analysis considerations enter the experiment at all stages and, certainly, an understanding of data-analysis principles is a *sine qua non* for a successful experiment.

3

THE SCIENTIFIC REPORT

As we have mentioned before, the end-game goal is the scientific report. One can claim to have performed a measurement, obtained a result, or completed an experiment, but this claim is generally accepted by the scientific community only after the results of the experimental investigation have been published. Although this is often a controversial issue, questions of priority in scientific accomplishment have historically been decided on the basis of who *published* first. Science historian Derek de Solla Price of Yale has written that science is not really science until it has been published in a professional journal.

Functions of the Scientific Report

Why is there this seemingly petty emphasis on the written word? The reason basically is that no result can be accepted until it has been described and laid out for the scrutiny and searching criticism of one's scientific colleagues. The situation is not unlike that in a court of law. An attorney presents evidence and elicits his client's version of the facts by asking questions of his witness. The evidence is not considered complete, however, until the attorney for the other side has had the opportunity to ask some pointed questions designed to probe for errors, misinterpretations, or distortions of fact. The judge may also ask questions to clarify the evidence presented and to determine its validity. It is clear that an advocate's case is strengthened greatly if his initial presentation is so clear, cogent, and compelling that it anticipates and answers in advance all possible questions or, at least, as many as possible.

When a scientist publishes the results of an investigation, he is presenting his case before the court of his peers. Most reputable scientific journals review papers before publication. This means that the editor submits the proposed publication to an expert in the field for his recommendations. This reviewer may ask questions, raise objections, or request clarification and, often, even rewriting of the paper. After the paper is finally published, it is available to all and sundry to read, digest, object to, and attack if they see fit. In the vernacular, when a scientist publishes a report, *he lays his reputation on the line.* Only when he is willing to do this are his results and conclusions to be believed and then, of course, they are believed or not on their own merits.

This makes clear the function of a scientific report: it is designed to present the results and conclusions as clearly as possible while describing all the procedures and assumptions of the investigation in enough detail to enable a scientific reader to make his independent assessment of the work or to repeat it. The report should answer all the questions another scientist could legitimately ask about the results and their believability.

It is not sufficient to say that one measured something unless the "how" is spelled out in detail (unless patently obvious). If pitfalls in the technique are known to exist, how were they avoided in this investigation? What precautions were taken? How were the results *and the uncertainties* in the results arrived at?

Even the conclusions reached by the author of the report are a consequence of his opinion. It is important to state enough of the facts of the experimental situation to enable the critical reader to reach his own conclusions and, at least, decide for himself if the author's judgment appears reasonable or unduly optimistic—even, occasionally, unduly conservative.

What is probably most crucial for the student is the ability to winnow the wheat (the information desired) from the chaff (the total yield of the experiment), which always involves some uncertainty. The important thing for the beginning scientist is to evaluate correctly the reliability of the result. It is then a matter of experience to design a new experiment with a better yield of information.

Components of the Scientific Report

In general, the scientific report should consist of four parts although the emphasis given to each part will vary greatly depending on the report and the kind of experiment to be reported. The four basic parts are the following: introduction, description of the technique, a discussion of the data, and a statement of the conclusions. The order will usually be as just presented, but not necessarily so. Just as the emphasis on the various parts is a matter of judgment for the author, so too is the order as well as the possibility of combining one or more of these parts or, even, subdividing them further. This arbitrary division is intended to serve only as a general guide to insure that the minimum presentation is made.

INTRODUCTION

The introduction includes the title and, occasionally, an abstract of the whole work. The introduction serves to state the aim and the general area of the report. The title should not be lengthy but should communicate some information concerning what is the subject of the report and specify the work as much as possible. The abstract that often appears after the title briefly describes the material and results of the report without presenting detailed arguments relating to the correctness of the results. The title should tell a prospective reader whether or not he is interested in reading the report. If the title does not immediately "turn him off," he may briefly scan the abstract to determine whether he wishes to read further. The names of the authors should appear in the vicinity of the title (and abstract, if included) to identify the work further and to provide the reader with the names of the responsible parties so that he may properly assign either credit or blame or may inquire further for more information. With the plethora of modern scientific publications, it is also necessary to use the title and authors' names for indexing purposes. Abstracts are also useful in this connection and publications exist that present only abstracts so that someone "searching the literature" can most easily determine what may be useful for him to read. The introduction may contain more background material starting at a more elementary level for a report directed at a student or lay audience than would be true for one aimed at a restricted group of highly specialized readers. In all cases the introduction should put the work to be reported in the context of other work and the current state of the art.

DESCRIPTION

The next important part of a scientific report is the description of the techniques employed *in this work.* This should include a discussion of what is novel or new about the procedure. The use of diagrams or photographs is often of great value in explaining the equipment and procedures used if not generally known.

This is also an appropriate place to discuss what the various difficulties of the technique are and how they have been surmounted or corrected for in the reported work. This is a most important part of the scientific report for it bears directly on the credibility of the work. Have all the difficulties been recognized? Have they been surmounted? What problems remain and how do they contribute to the uncertainty and reliability of the results?

DATA

The third part of the scientific report concerns the presentation of data. Here is the place to state what one can obtain from the technique and what one has been able to obtain. The data should be organized and significant features presented in a fashion that the intended reader will readily be able to digest. This presentation will frequently involve the use of tables or graphs. The old saying that a "picture is worth a thousand words" is often an understatement when applied to scientific research. The use of graphical techniques is so important that we defer detailed discussion to Part II. This section of the scientific report is the proper place for a discussion of salient features of the data that should be pointed out and emphasized.

DISCUSSION AND CONCLUSIONS

The fourth part of the scientific report contains a statement of the inferences that the authors wish to draw from their results as well as a justification and discussion of the arguments that lead to these conclusions. The certainty of the conclusions is an essential ingredient of this section and is tied to the reliability or uncertainty that attaches to the reduced data of the experiment as well as to the logic that connects the data with the conclusions. It is sometimes the case that the major inference that one can draw from an experiment is that *no* conclusion can be drawn from the data obtained.

NOTES

Formally, the last part of a scientific report is frequently a list of notes and references. If one wishes to utilize a result or argument that is already in the literature, it is sufficient only to cite the reference where the result or argument may be found. This is usually done by means of superscripted numbers that are then defined in a listing that usually appears at the end of the report. The definitions may also appear as footnotes. Parenthetical arguments that would otherwise interrupt the flow of the report may also be relegated to notes referenced by superscripted numbers.

4

PROCEDURE IN A LABORATORY AND THE LABORATORY NOTEBOOK

Naturally, the actual procedure to be followed during the data-taking part of the experiment varies greatly depending on the nature of the experiment. However, certain general truths obtain in all situations. The most important requirement of the actual laboratory phase of the work is thorough understanding of the goals and plan of the experiment. Preparation cannot be stressed too much as the greatest requirement of a successful experiment.

The ability to think on one's feet during an experiment is one of the talents that an instructional laboratory program is designed to foster. This ability, however, is meaningless unless the scientist (or student-scientist) has entered the laboratory with a clear idea of what he intends to do. The more difficulties that the scientist can anticipate in advance, the more likely is the success of the whole venture. Orderliness both in habits of thought about the experiment and in the actual performance of the physical manipulations required to make the observations and record the data are universal attributes of the professional experimenter.

The Laboratory Notebook

The recording of the data of the experiment is no less important than the setting up of the experimental equipment and the manipulations with it. One must always keep in mind that the scientist may often remember no more about the data-taking stage than is written in his notebook. All relevant information must be written down in easily intelligible form.

PERMANENCE AND COMPLETENESS

It is strongly recommended that the pages of the laboratory notebook be numbered and that they be bound so that none may be removed. The date, time, and name of the person taking data should be written as soon as the experimental work is commenced. Diagrams of the equipment, circuits, and mechanical structures must be noted. If particular instruments are used, their identifying numbers should be recorded and any peculiarities of operation remarked. If calibration procedures are required or if reliable calibrations exist, they should be recorded in the laboratory notebook or their location noted. Any information that *conceivably* could help in reconstructing what one has done or what happened in the laboratory should be recorded. An almost universal experience of experimental scientists is that even apparently irrelevant pieces of information have helped at one time or another in recovering data or correcting mistakes. If one is not sure that a temperature correction need be made, he should record the ambient temperature anyway. The sparks being generated by a machine across the corridor may not affect the experiment, but the situation should be noted in any case. Observations made can be ignored and data can be discarded, but there is no solution if a vital fact has not been recorded.

CORRECTIONS

It is also strongly recommended that the entries in the laboratory notebook be made in ink and that erasures not be made. If it is believed that incorrect entries have been made, the errors should be neatly crossed out with a fine line and corrections entered. If one has occasion to make entries on previous pages that were initially written at an earlier time, the date and time of the added entries should be noted as well as the name of the person making the additions. Mistakes are always a possibility. Not infrequently the mistake was in making the addition, and by avoiding erasures one has preserved the original, sometimes correct, entries. Often collaborating scientists may disagree about the validity of an entry. By signing and dating corrections, additions, or deletions (crossing out), the ultimate decision may be postponed until a thorough discussion is possible among the investigators concerned.

SUMMARIES

Summaries and preliminary data reductions should be made as soon as possible during the course of the data taking. This serves to uncover mistakes and generally helps to determine whether or not one is on the right track. If one discovers errors in procedure or malfunctions of equipment while yet in the laboratory, the fault is not permanent and may be rectified before leaving the laboratory. Constant checking of what one observes against what one expects is an integral part of the "thinking on one's feet" that is so essential to successful experimental work. Analysis while the situation is still fresh in the experimenter's mind is usually more fruitful of uncovering errors and deficiencies. Data grow stale and an experienced scientist often regards with a jaundiced eye an experiment reported too long after the data were actually taken. A successful experiment requires thorough understanding and control of the experimental situation. As the scientist's recollection of what happened dims with the passing of time, so do his chances of completely reconstructing and understanding what was happening.

5

EXPERIMENTAL ERRORS

It has been pointed out that experimental uncertainties always exist and that it is the scientist's prime function to minimize these. We must investigate the nature of experimental uncertainties in measurements. We have chosen to use as our standard the criterion of repeatability of measurements. When a given measurement is repeated, the resulting values do not usually agree *exactly.* The causes of the differences in the values must also enter into the values differing from the "true" value.

It should be stressed, perhaps, that one can never know the true or "exact" value of a quantity except for a defined constant. All one can determine is an estimate or *estimator* of the true value. We all have an intuitive notion that two identical and independent measurements of a quantity are better than one; three are better than two; and so on. If, in fact, we repeat the number of identical measurements N times and let N approach a very large number, the arithmetic average will approach a constant value. If the estimator approaches the "true" value, we say that the estimator is *consistent.* The ideal measurement produces an estimator that is consistent and without bias. We shall see that we may have estimators that are consistent because they converge to the true value as $N \to \infty$, but that are biased because they are not the best estimator of the true value for small N, being either too large or too small.

Random Errors

When a measurement is identically performed more than once and the results do not agree identically, we say that there has been a *random fluctuation.* A random or stochastic phenomenon is one characterized by the property that repeated occurrences of the phenomenon do not always lead to the same observed outcome. There does not exist deterministic regularity but, rather, repetition leads to different outcomes so that there exists statistical regularity. Because of these fluctuations, we say that measurements are subject to *random or statistical error.*

Here are examples of random errors.

1. Errors of judgment in estimating the fraction of the smallest division on an instrument scale. The observer's estimate may vary from time to time for a variety of (unpredictable) reasons. Such an error is truly random if the observer reads "high" about as often as he reads "low."
2. Variations due to fluctuating ambient conditions such as temperature changes, pressure changes, voltage changes, humidity changes, etc.
3. Variations in results due to disturbances not correlated with the experiment such as mechanical vibration from traffic outside the laboratory, electrical noise from nearby rotating machinery, or spurious signals picked up from another laboratory or experimental setup.
4. Differences due to deficiency in defining the quantity being measured. The height of a man is different when he awakens in the morning and when he retires at night after a working day spent compressing the cartilage between his vertebrae. His height is thus not an exact quantity, and even perfect measurements differ if made throughout the course of a day.
5. Intrinsically random processes exist such as the arrival of cosmic rays through a given detecting instrument. We are constantly being bombarded from outside our solar system by charged particles that arrive with a well-defined average rate. But, over a given interval of time, we cannot say when a given cosmic ray will arrive or even that one will arrive at all during the interval. The process is not deterministically regular. It is, however, statistically regular, and we can predict the *average* rate of arrival very well. Radioactive decay of an atom is another example of such a process.

Crudely speaking, measurements with random errors tend to be equally often higher and lower than the true value. This is *not* to say that "random errors cancel out." Our intuitive feeling, however, is that estimators subject only to random errors are consistent because they converge to the true value as $N \to \infty$.

Systematic Errors

Another class of error is *systematic error.* These errors are characterized by their deterministic nature. They are frequently constant. For example, if one measures the height of a man repeatedly (at the same time of day to avoid the error of definition) but forgets to remove his shoes, the heights recorded will be uniformly larger by the thickness of his shoes.

Here are other examples of systematic errors.

1. A biased error of judgment in reading instrument scales. If an observer always, or even most often, estimates the fraction of the smallest division high, there results a systematic error or bias toward high values.
2. A calibration error of the instrument itself will produce a bias in one direction.
3. Efficiency of observation is often a source of systematic error. If one is counting flights of birds to determine the relative numbers of blue birds and black birds, it may not be possible to count every bird that flies overhead, and the observer may miss more blue birds than black ones. This would, of course, distort the count. We shall later (Chapter 29) consider in detail an example of observational efficiency.
4. Variation in experimental conditions

has often introduced a systematic bias. If an instrument is calibrated at one temperature, but used at another, a systematic error will result.

5. Imperfect technique of myriad sorts results in systematic error.

Blunders

Before we discuss systematic and random errors more fully, it is probably of value to mention another frequent source of error that has been called *illegitimate error.* This results from outright blunders and mistakes. Errors of computation are in this class when the computational technique produces errors that are not negligible compared to the other errors in the experiment. It is highly improper, for example, to use a slide rule capable of maintaining three-significant-figure accuracy with data precise to five significant figures. The student should note that "slide-rule error" is not an acceptable remark in a laboratory report.

Analysis of Errors

Random errors are determinate by the procedures to be developed in this book. Errors are determinate when they may be evaluated by a logical procedure that is either experimental or theoretical. Systematic errors are often determinate since they may sometimes be evaluated by subsidiary experiments.

CORRECTIONS

It is frequently possible to remove some determinate errors by *corrections.* Thus, in one of our examples of systematic errors, knowledge of the thickness of the man's soles would enable the observer to correct his data to eliminate the bias. Random errors that are due to fluctuations of specific quantities may be removed by establishing the correlations of the fluctuations with the quantities and correcting the data. Thus, if temperature fluctuations cause variation in the length of a steel tape measure, one need only determine the temperature coefficient of expansion of the tape, note the temperature, and correct the length to some standard temperature. Careful consideration of an experiment and the performance of check experiments will often reveal the existence of systematic errors and frequently enable the experimenter to correct them.

Note that even correction procedures may involve some experimental uncertainty, and this uncertainty must be considered along with the others in the experiment.

PRECISION AND ACCURACY

If an experiment has small random errors, it is said to have high *precision.* If it has small systematic errors, it is said to have high *accuracy.* It is clearly the experimenter's aim to reduce both systematic and random errors as much as possible. However, there is little purpose to reducing the contribution of the random error to the total uncertainty if most of the uncertainty is contributed by the systematic error. The converse is also true. For economy of effort, one usually aims to achieve some sort of "match" between the sources of error, always bending his greatest effort to reducing the contribution of the larger of the two.

6

REJECTION OF DATA: CHAUVENET'S CRITERION AND ITS DANGERS

A few remarks are in order about one of the most difficult problems of experimental science: the question of mistakes in the data and the rejection of data.

When the measurement of a quantity is repeated several times, it often happens that one or more of the values differs from the others by relatively large amounts. There is no problem when these anomalous measurements can be directly traced to some systematic disturbance or fluctuation in the controlled conditions of the experiment. In this case, the values can be corrected for the effect of the known disturbance if it is determinate or the data may be rejected because of a known intercession of an extraneous influence. More difficult is the case where no cause for the anomalous values can be ascertained. The experimenter is often tempted to discard the anomalous values anyway on the grounds that some error in reading the instruments *must* have occurred. *This temptation must be resisted strongly.*

The first point to be made is that seemingly large fluctuations are often to be expected. When we discuss the normal distribution of error, we shall see that we may quantitatively determine how often the deviation from the mean of 1, 2, 3, or more "standard deviations" will occur. Thus, it is very often true that the seemingly anomalous values are perfectly acceptable. If the normal probability law (or the χ^2 criterion to be discussed later) indicates that the fluctuation is reasonable, obviously nothing is to be done and the data are certainly to be retained without change.

Now let us suppose that the deviation we are investigating has a very small chance of occurring. That is, we have computed the chance of obtaining one of our N values with a deviation from the mean as large as was observed, and the probability is calculated to be less than $1/N$. Because of random fluctuations in a series of N measurements, we may reasonably expect to observe the fluctuation with probability $1/N$ or a little less, but not very much less. It is a matter of fiat at what point one chooses to cut this; a widely used standard is Chauvenet's criterion, which states that if the probability of the value deviating from the mean by the observed amount is $1/2N$ or less, the data should be rejected. This means that the rejected data are not used in any way.

Without quibbling over the use of $1/2N$

as opposed to $1/3N$ or other numbers, we must point out several dangers. First, this criterion should never (if ever) be applied more than once. Obviously, one could reject the data with the maximum fluctuations, recompute the mean without the rejected data, and then reject data that differ from the new mean. This, of course, produces an unrealistically consistent set of data that is just as incorrect as an inconsistent set. The latter at least has the advantage of being an honest depiction of what happened in the laboratory.

A distinct danger in applying Chauvenet's or any other criterion for the rejection of data without determinate cause is that important effects may be "swept under the rug" in this way. In the history of experimental science, many "unexpected" or anomalous results proved to be very significant in themselves. When Lord Rutherford's students Geiger and Marsden first bombarded a gold foil with α-particles, a very small fraction of observed events indicated anomalously large scatterings. If these data had been rejected, it would have been left to someone else to discover the atomic nucleus. There are many other examples.

We should rather adopt the view that Chauvenet's criterion should be used to *flag suspicious situations* in the laboratory. When the deviation observed is larger than one can reasonably expect, this should serve as a stimulus to find out what happened. If it appears that nothing happened, then the data should generally be left as is unless the experimenter uses his judgment and experience to determine that it is more likely that the undetected systematic fluctuation occurred than that the effect is real. We cannot stress too strongly that judgment is involved here. The blind use of Chauvenet's criterion is a guarantee of never finding anything that was unanticipated at the beginning. We alluded earlier to the ability to think on one's feet. This means that the experimenter can shift the course of his investigation to pursue some new avenue if that seems interesting or fruitful.

When the fluctuations are large and unaccounted for, leaving the data in has at least the merit that the uncertainty that exists is flagged for the benefit of the scientific community (or whoever reads the report of the research). Rejecting data for the sake of appearing consistent is like covering warts; once it is discovered that they are there, it is worse than if they were in the open all the time. It is also dishonest. If your experiment is to be believed, it has to be accepted, warts and all.

7

THE PHILOSOPHY OF SAMPLING AND THE DEFINITION OF BASIC STATISTICAL CONCEPTS

It has been said that the process of analyzing data involves the study of populations and the study of variation. To understand this, we must first introduce the concept of *population.* Each measurement is an individual, and the repetition of the measurement generates an aggregate of results that we shall call the population. Indefinite repetition generates the (infinite) *parent population* or *universe.* Although we may never know the "true" value of a quantity—which is the object of our measurements—we tacitly seek to determine the features of the parent population of which our few measurements are but individuals.

Population Parameters

The almost intuitive directive to calculate the arithmetic mean shows our efforts to learn something about this population. It also contains the implicit assumption that the mean of an indefinitely repeated series of measurements (free from systematic error) will approach the "true" value we seek to determine. This, of course, requires the knowledge of an infinite population. What we attempt always to do is to determine as best we can the features of an infinite population from unbiased or *random sampling* of a finite population. Since we cannot attain exactitude in this way, our goal is to make the best estimate of the *parameters* of the infinite population together with some indication of the uncertainty of our estimate using our knowledge of the sample population. These estimates are called *statistics* and are calculated from our data.

CONSISTENCY AND BIAS

A statistic that converges to a fixed-value characteristic of the parent population is said to be a *consistent statistic.* For example, the mean obtained from a large sample will, when compared with other means drawn from other samples, agree more and more as the sample sizes increase. The value of the sample estimate converges to the value for the infinite population. We should distinguish at this point what is an *unbiased estimator.* This is a statistic that, when drawn from a sample of *any* size, represents on the average the corresponding quantity for the infinite population. Not all of our common statistics are unbiased estimators. Some are consistent statistics in that they converge to the value for the parent population as the sample size increases, but at any given (finite) sample size, their value has to be corrected to yield an unbiased best estimate of the value for the parent population. For n identical, independent measurements, we shall see later that both the sample mean and the sample variance are consistent statistics but only the sample mean is an unbiased estimator. To obtain a best estimate of the variance of the parent population, it is necessary to multiply the sample variance by the factor $n/(n-1)$.

Variation and Distributions

When we say we wish to get "better statistics," we mean that we wish to make more measurements to improve our estimates of the parameters of the parent population. In line with this goal, we note that the populations we study in experimental science show variation. To illustrate the opposite case, a population of *identical* individuals can be completely specified by describing one individual and noting the number of individuals. For the populations we deal with, the individuals are not identical, and we must specify not only the mean, but an indication of how the individuals vary from the mean. The study of variation requires the idea of a *frequency distribution.* We have a variable quantity called the *variate* and the frequency distribution specifies how often the variate takes on each of its possible values.

As we shall see later in more detail, frequency distributions may be *discrete* or *continuous,* depending on what values the variate may assume. In the former case the variate may take on only certain specified values. A simple example is the variate expressing the outcome of a coin toss. The possible outcomes are only twofold (dichotomous), being either heads or tails. The frequency distribution corresponding to this variate is discrete, involving only two values whose frequencies will always add up to the total number of trials. A variate that takes on any intermediate value in a given range requires a continuous frequency distribution. An example is the distribution of temperatures noted throughout a given time interval. Note that this may be *represented* by a discrete distribution if, for example, the observed temperatures are rounded off to the nearest integer degree Fahrenheit.

Frequency distributions are divided into two types, depending on how the distribution is mathematically expressed as a function of the variate. If it is expressed as the proportion of the population for which the variate is less than (or greater than) a given value, it is called an integral distribution function. Specifically, the *cumulative distribution function (c.d.f.)* describes the proportion of the population for which the variate is less than a given value. If one describes a continuous variate by expressing the (infinitesimal) proportion of the population for which the variate is within an (infinitesimal) interval of its range, we have a differential distribution function or *probability density function (p.d.f.).* Clearly the p.d.f. may be obtained mathematically by differentiating the c.d.f. Conversely, the c.d.f. is obtained by integrating the p.d.f. These distributions are discussed in more detail and with examples in Part III.

We must frequently study the simultaneous variation of two or more variates. Since the variation occurs independently or dependently, we say that we must study the *correlations* or the *covariation* of the population.

When we do an experiment, what we are after is not the result of a particular measurement, but some indication of what the population of all the possible experiments is. In fact, the calculation of mean value, variance, and standard error actually shows an effort to learn something about the population of these possibilities.

Functions of Statistics

One can regard the study of statistics as a threefold entity: (1) the study of populations to determine the properties of the aggregate population; (2) the study of the variations, that is, instead of just studying the aggregate characteristics of the population, we want also to study how the individual members of the population differ from each other, and how they vary from the aggregate or average values. One should tie this concept to the difference between emphasis on the mean of the sample and on the standard deviation of the sample. The former is a characteristic of the aggregate, and the latter is an indication of how the individual members differ from the mean of the population. (3) Another function of the experimenter is to reduce the bulk of his data to easily comprehensible form. We seek to express the meaning of any body of data by means of a comparatively few numerical values that express all the relevant information contained in the mass of data.

We shall use quantities called statistics. Statistics that address the function 1 are called measures of *central tendency*. Quantities that describe variation (function 2) are called measures of *dispersion*. We shall see later that statistics that are "sufficient" contain all the relevant information and may be used to satisfy function 3.

Definitions of Basic Statistical Concepts

To establish a basis for discussion, we must first of all introduce some simple statistical concepts.

MEAN OR ARITHMETIC AVERAGE: $\bar{x}$

If n independent measurements x_i of equal validity have been made where $i = 1, 2, \ldots, n$, the average or mean is calculated by summing the measured values and dividing by the total number n.

$$\bar{x} = \frac{1}{n} \sum_{i=1}^{n} x_i \tag{7.1}$$

For later use we may here distinguish our notation to specify that $\bar{f}$ denotes the *sample average* and $\langle f \rangle$ denotes the *population* average or *universe average* of the variate f.

MEDIAN

Of a set of n measurements, the median is the measured value that is as frequently exceeded as not. It is usually the value such that $n/2$ of the measurements have values less than the median and $n/2$ have values greater than the median. (See the example below for an exception.)

Formally, we define the median as follows: arrange the set of measurements in order of algebraic magnitude (including the sign) so that we have the sequence

$$x_1 \leq x_2 \leq \cdots \leq x_n$$

When n is odd, the median is the value x_k where $k = \frac{1}{2}(n + 1)$. When n is even, the median is not uniquely defined unless $x_l = x_{l+1}$ where $l = \frac{1}{2}n$, in which case the median is the common value. If the median is not uniquely defined in this way, we conventionally take the median to be $\frac{1}{2}(x_l + x_{l+1})$ and the median is thus not one of the actual measured values.

MODE

The mode is the most probable measured value. If n measurements have been made, the mode is the value that appeared most often. [*Note.* The mean, the median, and the mode are coincident only for a distribution of values (or a series of measurements) that is *symmetrical.*]

► **Example 7.1**

Consider the asymmetric distribution of numbers: 2, 3, 5, 6, 7, 8, 8, 9, 9, 9, 11, 11, 12. Here the mean value is 7.69, the mode is 9, and the median is 8. One may plot these (discrete) data as a frequency distribution (Figure 7.1). ◄

DEVIATION

This is the absolute value of the difference between a particular measurement and some value, usually the average, which is generally calculated from a set of measurements of the same quantity. It takes several forms.

1. Mean Deviation: a

The sum of the deviations divided by the number of observations. This gives a numerical guess concerning how far from the average the next measurement might be. If in a set of data the large deviations are regarded as no more significant than the smaller deviations, then the mean deviation is useful.

$$a = \frac{\Sigma|x - x_i|}{n} \tag{7.2}$$

2. Standard Deviation or Root-Mean-Square Deviation: σ

The square root of the average value of the square of the individual deviations. The square of this quantity is called the *variance*: σ^2. For n independent measurements of equal validity, the standard deviation is

$$\sigma = \sqrt{\frac{\Sigma(x_i - \bar{x})^2}{n}} = \sqrt{\overline{(x - \bar{x})^2}} \tag{7.3}$$

It weights the large deviations more heavily than the small, and it is generally the best estimate of truly random errors.

One may take the *average* of his experimental observations as the best estimate of the "true value", that is, the estimator of the parent population. To obtain the best estimate of the variance of the parent population, it is necessary to modify the sample variance.

$$\text{Estimate of } \sigma^2_{\text{parent}} = \frac{n}{n-1}\,\sigma^2_{\text{sample}} \tag{7.4}$$

This has the effect of replacing n by $n - 1$ in the definition for standard deviation for

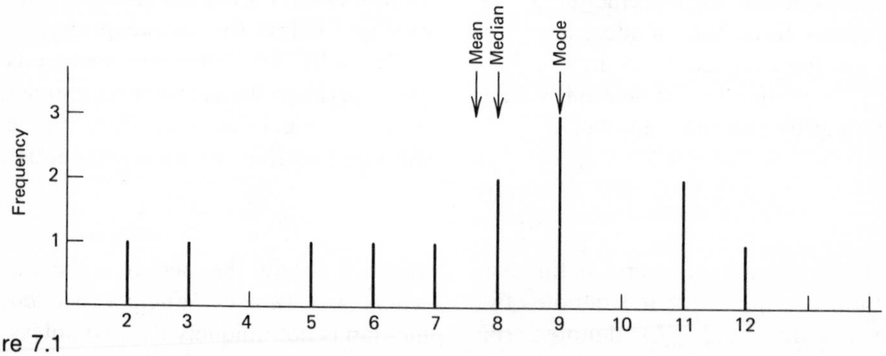

Figure 7.1
A plot of the frequencies of various numbers as in Example 7.1. The mean, median, and mode are indicated.

n observations:

$$s = \sqrt{\frac{\Sigma(x_i - \bar{x})^2}{n - 1}} \qquad (7.5)$$

Note that s^2 is now an unbiased estimator of the parent population variance, σ^2. In any case, either version of the sample variance is a consistent statistic, since each converges to the parent variance in the limit as $n \to \infty$. For most cases of interest, there is no significant difference between s and σ, and either may be used.

It is frequently more convenient to calculate the mean deviation than the standard deviation. It will be shown later that the ratio of standard deviation to mean deviation approaches a constant value as the number of observations gets large:

$$\lim_{n \to \infty} \frac{\sigma}{a} = 1.25 \qquad (7.6)$$

We state these relations without proof for the present and shall return to them later.

What can we make of these concepts for now?

Suppose one has made a series of identical, independent measurements $x_1, \ldots, x_i, \ldots, x_n$ of a quantity x:

The best guess or estimator of x is

$$\bar{x} = \frac{\Sigma x_i}{n}$$

The standard deviation is related to the average error of *each* measurement. If we make a frequency distribution of the values x_i, the standard deviation is related to the *shape* (especially the *width*) of the distribution.

3. Standard Deviation of the Mean

We have the intuitive belief that we should have more confidence in a result (i.e., assign a smaller error to the mean $\bar{x}$) if the values that are combined to obtain it are closely clustered. If we speak of the frequency distribution of x, we believe the mean is better determined, with a smaller error, according to the *narrowness* of the frequency distribution. We have seen, however, that the shape is determined by the average error per measurement, which is the same irrespective of the number of times we repeat the measurement. We have the further intuitive belief that our result should improve (i.e., $\bar{x}$ should have a smaller error) as the number of measurements increases.

The error to be assigned to the result is the *standard deviation of the mean*: $\Delta\bar{x}$

$$\Delta\bar{x} = \sqrt{\frac{s^2}{n}} = \sqrt{\frac{\Sigma(\bar{x} - x_i)^2}{n(n-1)}} \qquad (7.7)$$

This formula states that the error assigned to the mean value of a restricted number of observations decreases as the number of observations increases, but only as the square root. (See Chapter 8 for a discussion of this formula.) The error of the mean is directly proportional to the error of a single observation. Thus a reduction by a factor of 10 in the average error (standard deviation) of individual measurements produces the same improvement in the precision of the mean as do one hundred repetitions of the *series* of measurements; for example, suppose 25 measurements each of precision characterized by $\sigma_i = 0.05$ cm produce a mean value with $\sigma_{\bar{x}} = 0.01$. If an improved technique is used with individual error $\sigma_i = 0.005$, then a series of 25 measurements will yield a precision in the mean of $\sigma_{\bar{x}} = 0.001$. Without the improved technique, the measurement would have to be repeated not 25, but 2500 times to achieve the same precision on the mean. Moral: it always pays more to make a better measurement than to repeat an old one.

The following length measurements were made of an object where the measurements were made (in cm) *to the nearest half millimeter*: 10.7, 10.4, 10.1, 10.6(5), 10.5(5), 10.3, 10.1(5), 10.5, 10.4, 10.6, 10.8, 11.1, 10.9, 10.4(5), 10.5, 10.8(5), 10.5, 10.5, 10.7, 10.5(5), 10.7, 10.5, 10.9, 10.7, 10.6 being the first 25 measurements and 10.3(5), 10.5, 10.3(5), 10.7, 10.3, 10.7(5), 10.4(5), 10.5, 10.9, 10.4, 10.5, 10.3, 10.7, 10.2, 10.6(5),

10.5(5), 10.2(5), 10.7, 10.3, 10.8, 10.4, 10.5, 10.3, 10.7, 10.5, and 10.3 being another 26.

Rounding. Note that the figure in parentheses is not truly significant. The first problem that arises is that of reducing all the measurements to the same number of significant figures. This is the problem of *rounding*.

It is clear how to round numbers like 10.63 and 10.58. Each of these rounds to 10.6 since 10.63 is obviously closer to 10.6 than to 10.7. It is not so clear when one has numbers such as those above that end in 5. The number 10.55 is equally distant from 10.5 and 10.6. What is one to do? One could round to the nearest lower number. This is equivalent to merely dropping the last figure and is called *truncation*. Truncation clearly involves some error if the last figure is not 5. It also involves some error if the last figure is 5. One could, as a second alternative, round up to the next higher number. A third alternative is to round off according to some arbitrary but random rule. In fact, it is customary to round off numbers ending in 5 to the nearest *even* number.

Let us consider in Table 7.1 the numbers above that are in this category of ending in 5.

Table ***7.1***

Actual	Truncated	Rounded Up	Rounded to Even
10.65	10.6	10.7	10.6
10.55	10.5	10.6	10.6
10.15	10.1	10.2	10.2
10.45	10.4	10.5	10.4
10.85	10.8	10.9	10.8
10.55	10.5	10.6	10.6
10.35	10.3	10.4	10.4
10.35	10.3	10.4	10.4
10.75	10.7	10.8	10.8
10.45	10.4	10.5	10.4
10.65	10.6	10.7	10.6
10.55	10.5	10.6	10.6
10.25	10.2	10.3	10.2
136.55	135.9	137.2	136.6

It should be clear that the errors from rounding 5 are least if one rounds off to the nearest *even* number. (Obviously, the procedure is equally valid if one always rounded off to the nearest *odd* number.

Table ***7.2***[a]

Measured Value x	Number of Occurrences n	nx	$(x-\bar{x})$	$n\lvert x-\bar{x}\rvert$	$(x-\bar{x})^2$	$n(x-\bar{x})^2$	x^2	nx^2
10.1	1	10.1	-0.5	0.5	25×10^{-2}	25×10^{-2}	102.010	102.010
10.2	1	10.2	-0.4	0.4	16×10^{-2}	16×10^{-2}	104.040	104.040
10.3	1	10.3	-0.3	0.3	9.0×10^{-2}	9×10^{-2}	106.090	106.090
10.4	3	31.2	-0.2	0.6	4.0×10^{-2}	12×10^{-2}	108.160	324.480
10.5	5	52.5	-0.1	0.5	1.0×10^{-2}	5×10^{-2}	110.250	551.250
10.6	5	53.0	0.0	0.0	0.0	0	112.360	561.800
10.7	4	42.8	$+0.1$	0.4	1.0×10^{-2}	4×10^{-2}	114.490	457.960
10.8	2	21.6	$+0.2$	0.4	4.0×10^{-2}	8×10^{-2}	116.640	233.280
10.9	2	21.8	$+0.3$	0.6	9.0×10^{-2}	18×10^{-2}	118.810	237.620
11.0	0	0.0	$+0.4$	0.0	16×10^{-2}	0	121.000	0.000
11.1	1	11.1	$+0.5$	0.5	25×10^{-2}	25×10^{-2}	123.210	123.210
Σ	25	264.6		4.2	0.94	1.22		2801.740
$\frac{\Sigma}{25}$		10.58(4)		0.16(8)		.048(8)		112.0696

[a] Measurements in centimeters.

What is important is to be consistent.) Let us do this so that the individual measurements are in the same form (three significant digits).

We shall find it convenient to summarize our data in *tabular form*. We may consider the first 25 measurements to constitute determination 1 and the next 26 measurements to constitute (an independent) determination 2. We may also combine the data from (1) and (2) to make a determination of 51 measurements that we shall call (3). These are summarized in Tables 7.2, 7.3, and 7.4, respectively.

We first note that the mean values are quoted to four significant figures while the individual measurements are quoted to three. The last digits in the nx column are significant and so the total ($\Sigma\, nx$) should be quoted to 0.1 cm, requiring four signifi-

Table 7.3[a]

Measured Value x	Number of Occurrences n	nx	$(x - \bar{x})$	$n\lvert x - \bar{x}\rvert$	$(x - \bar{x})^2$	$n(x - \bar{x})^2$	x^2	nx^2
10.1	0							
10.2	2	20.4	−0.3	0.6	9.0×10^{-2}	18×10^{-2}	104.040	208.080
10.3	5	51.5	−0.2	1.0	4.0×10^{-2}	20×10^{-2}	106.090	530.450
10.4	5	52.0	−0.1	0.5	1.0×10^{-2}	5×10^{-2}	108.160	540.800
10.5	5	52.5	0.0	0.0	0.0	0	110.250	551.250
10.6	2	21.2	+0.1	0.2	1.0×10^{-2}	2×10^{-2}	112.360	224.720
10.7	4	42.8	+0.2	0.8	4.0×10^{-2}	16×10^{-2}	114.490	457.960
10.8	2	21.6	+0.3	0.6	9.0×10^{-2}	18×10^{-2}	116.640	233.280
10.9	1	10.9	+0.4	0.4	16.0×10^{-2}	16×10^{-2}	118.810	118.810
11.0	0	0.0	+0.5	0.0	25.0×10^{-2}	0	121.000	0.000
11.1	0	0.0	+0.6	0.0	36.0×10^{-2}	0	123.210	0.000
Σ	26	272.9		4.1		0.95		2865.350
$\frac{\Sigma}{26}$		10.49(6)		0.15(8)		0.036(5)		110.2058

[a] Measurements in centimeters.

Table 7.4[a]

Measured Value x	Number of Occurrences n	nx	$(x - \bar{x})$	$n\lvert x - \bar{x}\rvert$	$(x - \bar{x})^2$	$n(x - \bar{x})^2$	x^2	nx^2
10.1	1	10.1	−0.4	0.4	16.0×10^{-2}	16×10^{-2}	102.010	102.010
10.2	3	30.6	−0.3	0.9	9.0×10^{-2}	27×10^{-2}	104.040	312.120
10.3	6	61.8	−0.2	1.2	4.0×10^{-2}	24×10^{-2}	106.090	636.540
10.4	8	83.2	−0.1	0.8	1.0×10^{-2}	8×10^{-2}	108.160	865.280
10.5	10	105.0	0.0	0.0	0.0×10^{-2}	0×10^{-2}	110.250	1102.500
10.6	7	74.2	+0.1	0.7	1.0×10^{-2}	7×10^{-2}	112.360	786.520
10.7	8	85.6	+0.2	1.6	4.0×10^{-2}	32×10^{-2}	114.490	915.920
10.8	4	43.2	+0.3	1.2	9.0×10^{-2}	36×10^{-2}	116.640	466.560
10.9	3	32.7	+0.4	1.2	16.0×10^{-2}	48×10^{-2}	118.810	356.430
11.0	0		+0.5	0.0	25.0×10^{-2}	0	121.000	0.000
11.1	1	11.1	+0.6	0.6	36.0×10^{-2}	36×10^{-2}	123.210	123.210
Σ	51	537.5		8.6		2.34		5667.090
$\frac{\Sigma}{51}$		10.53(9)		0.16(9)		.0459		111.1194

[a] Measurements in centimeters.

cant figures. The denominator of the quotient $\Sigma\, nx/\Sigma\, n$ is arbitarily precise, and so the quotient should contain four significant figures. We have quoted all the means calculated to one significant figure more than the individual measurements. (Note that we sometimes carry along an extra figure in parenthesis to minimize rounding error in later calculating.) The ultimate justification, of course, is that *the average is more precise than any of the individual values and requires more significant figures.*

We may summarize the means, average deviations and standard deviations $s = \sqrt{\Sigma(x-\bar{x})^2/(n-1)}$ for the three determinations as in Table 7.5.

Table 7.5

Determination	n	$\bar{x}$	a	s
(1)	25	10.58(4)	0.16(8)	0.2(3)
(2)	26	10.49(6)	0.15(8)	0.1(9)
(1) + (2)	51	10.53(9)	0.16(9)	0.2(2)

We note that we lose a digit in having to calculate $(x_i - \bar{x})$ since $\bar{x}$ has one extra significant figure more than x_i that appears to be wasted if we have to subtract them.

Another Formula for the Variance. There is an alternative method of calculation that involves a most important formula for σ^2.

$$\sigma^2 = \frac{\Sigma(x-\bar{x})^2}{n} = \frac{\Sigma x^2}{n} - \frac{2\bar{x}\Sigma x}{n} + \frac{(\bar{x})^2\Sigma}{n}$$

Since

$$\frac{\Sigma x}{n} = \bar{x}$$

and

$$\frac{1}{n}\sum = \frac{n}{n} = 1$$

we get

$$\sigma^2 = \frac{\Sigma\, x^2}{n} - (\bar{x})^2$$

But $\Sigma x^2/n = \overline{x^2}$, the mean of x^2, and so we get the important formula

$$\sigma^2 = \overline{x^2} - (\bar{x})^2 \qquad (7.8)$$

We may now calculate σ^2 [and s^2 via (7.4)] using this formula having computed the necessary quantities in Tables 7.2, 7.3, and 7.4.

$$s_{(1)}{}^2 = \frac{25}{24}\{112.0696 - [10.58(4)]^2\}$$
$$= 0.0506 = [0.22(5)]^2$$

$$s_{(2)}{}^2 = \frac{26}{5}\{110.2058 - [10.49(6)]^2\}$$
$$= 0.0414 = [0.20(3)]^2$$

$$s_{(3)}{}^2 = \frac{51}{50}\{111.1194 - [10.53(92)]^2\}$$
$$= 0.0452(2) = [0.21(3)]^2$$

We see that the use of (7.8) rather than the definition (7.3) is both more useful and more efficient since the result is more precise.

We may now compute the ratio σ/a using s as our best estimate for σ_{parent}.

$$(\sigma/a)_{(1)} = \frac{0.22(5)}{0.16(8)} = 1.3(4)$$

$$(\sigma/a)_{(2)} = \frac{0.20(3)}{0.15(8)} = 1.2(8)$$

$$(\sigma/a)_{(3)} = \frac{0.21(3)}{0.16(9)} = 1.2(6)$$

This appears to be consistent with our prediction of 1.25 given that n is quite finite and that our determinations of σ and a are not good to three significant figures.

We further note that the average deviations agree quite well for the three determinations. This is correct, since a refers to an average for an individual measurement, and each of the three is made up of the same sort of individual measurements. The average deviation and the standard deviation are both related to the width of the distribution of individual measurements for each determination.

This is most easily seen from Figures 7.2, 7.3, and 7.4. Here we have plotted frequency *histograms* of the measurements. Superimposed on each plot is a theoretical

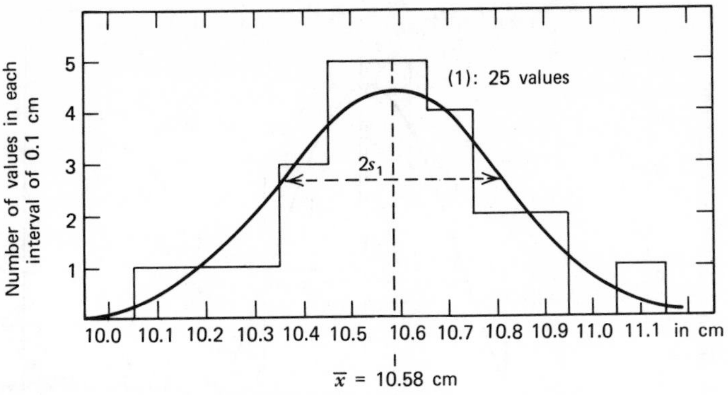

Figure 7.2
A histogram of 25 measurements from Table 7.2. The mean and variance are indicated. A suitably normalized gaussian curve with this mean and variance is superimposed.

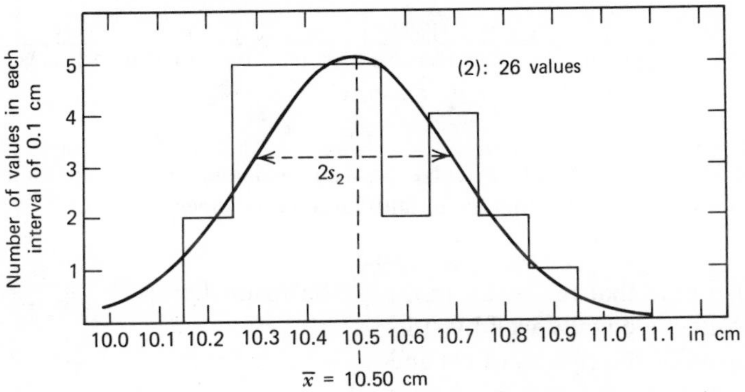

Figure 7.3
A histogram of 26 measurements from Table 7.3. The mean and variance are indicated. A suitably normalized gaussian curve with this mean and variance is superimposed.

gaussian curve that we shall discuss later. We see that the shapes of the three curves are very similar. What, then, have we gained from the extra 26 values in (3) compared with (1)?

The answer, of course, is that the mean is better determined in (3). Recalling our formula (7.7), we may express the various means along with the standard deviations *of the means.* ($S_{\bar{x}} \equiv \Delta \bar{x}$)

(1) $\bar{x} = 10.58 \pm 0.04$

(2) $\bar{x} = 10.50 \pm 0.04$

(3) $\bar{x} = 10.54 \pm 0.03$

We have chosen to express the standard deviations of the various means to only one significant figure. This raises the question of the *precision of the error* in the mean, the error of the error, if you will (we shall not inquire into the error in *this* quantity). It turns out that if we have obtained $\bar{x} \pm S_{\bar{x}}$ from a series of n independent measurements, then the standard deviation of $S_{\bar{x}}$ is obtained from the relation

$$\frac{2S_{\bar{x}}^4}{n-1} = \sigma^2(S_{\bar{x}}^2) \qquad (7.9a)$$

or

$$\sigma(S_{\bar{x}}) = \frac{S_{\bar{x}}}{\sqrt{2(n-1)}} \qquad (7.9b)$$

For the current example

$$S_{\bar{x}}^{(1)} = 0.04(5) \pm 0.01$$
$$S_{\bar{x}}^{(2)} = 0.04 \quad \pm 0.01$$
$$S_{\bar{x}}^{(3)} = 0.030 \quad \pm 0.004$$

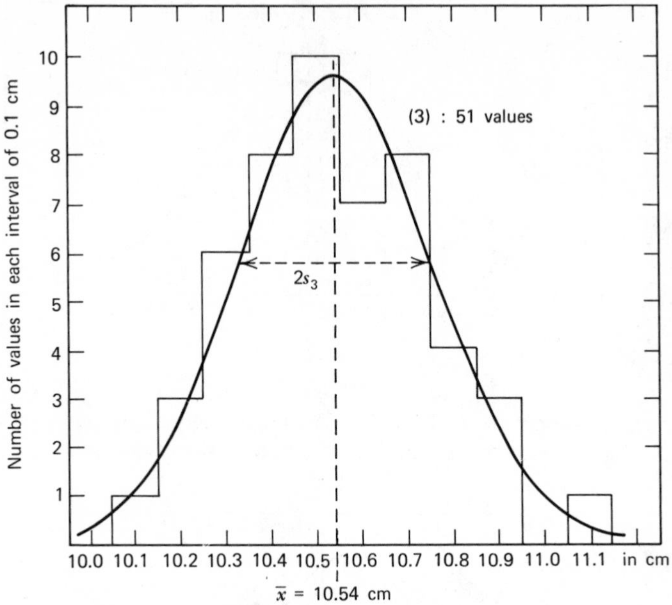

Figure 7.4
A histogram of 51 measurements from Table 7.4. The mean and variance are indicated. A suitably normalized gaussian curve with this mean and variance is superimposed.

We may also note that the result in (3) should be obtainable equally well by suitable combination of the results of (1) and (2), since 51 values are involved no matter how we split up the data. Ignoring the slight difference between 25 and 26, we note that the mean of (3) is, indeed, given by the average of the means of (1) and (2). The standard deviation is, however, not given by this simple relation. We shall return to this point shortly but, for the present, merely note that $\frac{1}{2}\sqrt{(0.04)^2 + (0.04)^2} = 0.03$. Here the two results $\bar{x}_{(1)}$ *and* $\bar{x}_{(2)}$ are of equal weight since the errors are the same. It thus appeals to "common sense" to average them to get the best combined value: $\bar{x}_{\text{best}} = \frac{1}{2}(\bar{x}_{(1)} + \bar{x}_{(2)})$ and equation (10.4d) will be seen to be relevant for the error in $\bar{x}_{\text{best}}$. A more important question is how to combine two or more independent measurements $\bar{x}_1, \bar{x}_2, \ldots,$ of the same quantity where the errors (or standard deviations, σ_i) are quite different. The weights should then be different. We shall consider the calculation of a "weighted average" in Chapter 10, but we may note the result here:

$$x_{\text{best}} = \frac{\sum_i x_i/\sigma_i^2}{\sum_i 1/\sigma_i^2} \tag{7.10}$$

$$\sigma_{x_{\text{best}}} = \left[\sum_i \frac{1}{\sigma_i^2}\right]^{-1/2} \tag{7.11}$$

OTHER MEASURES OF CENTRAL TENDENCY

To emphasize even at this early stage in our considerations that the arithmetic average is not universal in its utility, we consider two examples in which the arithmetic mean is not helpful.

Harmonic Mean

Consider a boat that goes 15 miles downstream at a speed of 5 miles per hour but returns upstream at a speed of only 3 miles per hour. The average speed is not $\frac{1}{2}(5+3) = 4$! The trip downstream took

$$\frac{15 \text{ mi}}{5 \text{ mi/hr}} = 3 \text{ hr}$$

while that upstream took

$$\frac{15\text{ mi}}{3\text{ mi/hr}} = 5\text{ hr}$$

The total trip of 30 miles took 8 hours so the true average speed is

$$\frac{30\text{ mi}}{8\text{ hr}} = 3.75\text{ miles per hour}$$

Note that $\frac{1}{2}(1/5 + 1/3) = 1/3.75$.

An average obtained in this way is called an harmonic mean (see Appendix A). Harmonic means are useful when considering *ratios* of things as data (e.g., miles *per* hour).

Geometric Mean

A rabbit breeder supplies rabbits to a laboratory and charges a price that is based on his costs, being 150% of his costs. Over the years it has been determined that the upkeep for one rabbit per month irrespective of age is A dollars. The breeder starts with a pair of rabbits at time $t = 0$. The rabbit population doubles every month for four months at the end of which time he has $2 \cdot 2^4 = 32$ rabbits. What is the average number of rabbits the breeder has had to keep for the four months?

The breeder argues that he started with 2 and ended with 32 rabbits yielding an average of 17 rabbits for 4 months. His cost basis is, therefore, $(17)(A)4 = 68A$.

The laboratory points out, however, that the geometric mean (see Appendix A) is more appropriate. This is $\sqrt{(2)(32)} = 8 = \sqrt[5]{2 \cdot 4 \cdot 8 \cdot 16 \cdot 32}$. If we note that the number of rabbits was successively 2, 4, 8, 16, 32 we see that the geometric mean represents the number at the midway point. In fact, we note that the number of "rabbit-months of upkeep" can be calculated out. During the fourth month there were 16 rabbits to keep up (assuming the doubling took place at the end of the period), during the third month there were 8, during the second only 4 rabbits and only 2 during the first month. The total is $(16 + 8 + 4 + 2) = 30$ rabbit months, which is an average of 7.5 for the four months. The geometric mean of 8 and the arithmetic mean of 17 should be compared with this figure.

Another example concerns a new stock on Wall Street that appreciated to 225% of its issue price by the end of the first year and went to 144% of that figure by the end of the second year. What is its average annual rate of appreciation?

The arithmetic average is $(225 + 144)/2 = 184.50$ whereas the geometric mean is

$$\sqrt{(144)(225)} = 180$$

The final price is $225 \times 144 = 324\%$ of the issue price. Now $(1.8)^2 = 3.24$, but $(1.845)^2 = 3.40$. Thus, the geometric mean is the correct measure.

We may generalize that the geometric mean is to be used when our data consist of rates of change or when the data are distributed exponentially.

8

DISCUSSION OF THE ANALYSIS OF SAMPLES

This section is for the intellectually curious and may be skipped at a first reading. The motivation is to provide a basis for formulas (7.4), (7.5), and (7.7).

We wish to use the result (7.8) to make plausible the statements in (7.4), (7.5), and (7.7). We may prove in fashion similar to the argument used to derive (7.8) that the following relation holds for the universe of population quantities:

$$\langle x^2 \rangle = \langle x \rangle^2 + \sigma^2_{\text{pop}} \tag{8.1}$$

Suppose we have a series of samples of independent measurements drawn from a universe with mean $\langle x \rangle$ and variance $\sigma^2 = \sigma^2_{\text{pop}}$.

$$\bar{x}_n = \frac{\Sigma x}{n} = \text{mean of a sample}$$

$$\begin{aligned}\langle \bar{x}_n \rangle &= \text{universe average of all sample means} \\ &= \left\langle \frac{1}{n} \sum x \right\rangle = \frac{1}{n} \sum \langle x \rangle = \frac{1}{n} n \langle x \rangle \\ &= \langle x \rangle\end{aligned}$$

This says that the average value of the mean of a sample of n observations drawn from the same population (identical measurements) is the same as the average value of each of the observations. This justifies our use of the sample average as the best estimate of the population average.

Now consider the variance of the universe of sample means (*Note.* This is not the variance of individual measurements but of their means):

$$\begin{aligned}\sigma^2_{\bar{x}_n} &= \langle (\bar{x}_n - \langle \bar{x}_n \rangle)^2 \rangle = \langle \bar{x}_n{}^2 \rangle - \langle \bar{x}_n \rangle^2 \\ &= \langle (\bar{x}_n)^2 \rangle - \langle x \rangle^2\end{aligned}$$

where

$$\begin{aligned}\langle (\bar{x}_n)^2 \rangle &= \left\langle \left(\frac{x_1 + \cdots + x_n}{n} \right)^2 \right\rangle \\ &= \frac{1}{n^2} \left\langle \left[\sum_{i=1}^{n} x_i^2 + \sum_{i \neq j} x_i x_j \right] \right\rangle\end{aligned}$$

We have previously specified that the measurements are independent, hence x_i and x_j are independent for $i \neq j$ and

$$\langle x_i x_j \rangle = \langle x_i \rangle \langle x_j \rangle = \langle x \rangle^2$$

From (8.1) we have $\langle x_i^2 \rangle = \sigma^2 + \langle x \rangle^2$ and so

$$\langle (\bar{x}_n)^2 \rangle = \frac{1}{n^2} \left[\sum_{i=1}^{n} \{\sigma^2 + \langle x \rangle^2\} + \sum_{i \neq j} \langle x \rangle^2 \right]$$

where the first sum has n terms and the second has $(n-1)n$ terms.

$$\begin{aligned}\therefore\ \langle (\bar{x}_n)^2 \rangle &= \frac{1}{n^2} [n\sigma^2 + n\langle x \rangle^2 + n(n-1)\langle x \rangle^2] \\ &= \frac{\sigma^2}{n} + \langle x \rangle^2 \\ &= \sigma^2_{\bar{x}_n} + \langle x \rangle^2\end{aligned}$$

Thus

$$\sigma_{\bar{x}}^2 = \frac{\sigma^2}{n}$$

This says that the variance of the means (of samples of n measurements) is equal to the variance of an individual observation divided by n. This is our statement (7.7) since it follows that the standard deviation of the mean of a sample of n is less by a factor $1/\sqrt{n}$ than that of an individual measurement.

Now consider the *sample* variance of individual measurements

$$\sigma_n^2 = \frac{\Sigma(x-\bar{x})^2}{n} = \frac{\Sigma x^2}{n} - (\bar{x})^2$$

Take the population average of all the sample variances

$$\begin{aligned}
\langle \sigma_n^2 \rangle &= \left\langle \frac{\Sigma x^2}{n} \right\rangle - \langle (\bar{x}_n)^2 \rangle \\
&= \frac{1}{n} \sum \langle x^2 \rangle - [\langle \bar{x}_n \rangle^2 + \sigma_{\bar{x}_n}^2] \\
&= \frac{1}{n} \sum [\langle x \rangle^2 + \sigma^2] - \left[\langle x \rangle^2 + \frac{\sigma^2}{n} \right] \\
&= \frac{1}{n} [n\langle x \rangle^2 + n\sigma^2] - \left[\langle x \rangle^2 + \frac{\sigma^2}{n} \right] \\
&= \sigma^2 - \frac{\sigma^2}{n} = \left(\frac{n-1}{n} \right) \sigma^2
\end{aligned}$$

Therefore, on the average, the sample variance is less than the population variance (of individual measurements) by a factor $(n-1)/n$. One may restate this to get the best estimate of the population variance in terms of the sample variance

$$\begin{aligned}
\sigma^2 &= \frac{n}{n-1} \langle \sigma_n^2 \rangle \approx \frac{n}{n-1} \frac{\Sigma(x-\bar{x})^2}{n} \\
&= \frac{\Sigma(x-\bar{x})^2}{n-1}
\end{aligned}$$

which are our statements (7.4) and (7.5).

A useful way to recall the $(n-1)$ in the denominator is to notice that a single measurement x_i suffices to yield a best value ($\bar{x} = x_i$) but no information on the distribution of values (or on the standard deviation of the mean). Hence, an expression for σ^2 in terms of the individual measurements should be indeterminate for $n = 1$. Relations (7.4) and (7.5) satisfy this condition. Also, two measurements suffice to determine the best value ($\bar{x}$) and the rms deviation ($S_{\bar{x}}$) but cannot yield information on the variance of $S_{\bar{x}}$.

9

DISCUSSION OF DISCRETE AND CONTINUOUS FREQUENCY DISTRIBUTIONS AND HISTOGRAMS

We have introduced the idea of frequency distributions in Chapter 7 with explicit examples of discrete distributions shown in Figure 7.1 and in the data Tables 7.2, 7.3, and 7.4. In the latter case, the measurements were actually drawn from a continuous distribution and quantized or made discrete by assigning the measurements to intervals. In the latter case, we have also the problem of comparing a frequency distribution with 51 entries (Table 7.4) with one containing a total of only 25 entries (Table 7.2). To facilitate such comparisons we introduce the idea of a normalized frequency distribution.

Normalized Frequency Distribution

We consider that the measurements x_i each occurred n_i times. If k different measurements are considered, we define the total number of measurements

$$N = \sum_{i=1}^{k} n_i = n_1 + n_2 + \cdots + n_k \quad (9.1)$$

The normalized or relative frequency of x_i is then defined to be

$$f_N(x_i) = \frac{n_i}{\Sigma n_i} = \frac{n_i}{N} \quad (9.2)$$

We note the normalization relation

$$\sum_{i=1}^{k} f_N(x_i) = \frac{\Sigma n_i}{N} = (1/N) \sum n_i = N/N = 1 \quad (9.3)$$

In these terms we may replot Figure 7.1 in Figure 9.1.

A plot of $f_N(x_i)$ versus x_i is the relative or normalized frequency distribution. We have defined this in terms of a total sample size N. In the spirit of Chapter 7, we are really interested in the limiting relative frequency distribution

$$\lim_{N\to\infty} f_N(x_i) = f(x_i) \quad (9.4)$$

or, the frequency distribution of the infinite population. This limiting frequency distribution represents our goal as stated in Chapter 7: to determine the features of the parent population or (infinite) universe of measurements.

Let us return to the continuous distribution. The length measurement x is a continuous variable since it could, in principle, take on any value. In Tables 7.2 to 7.4 we have listed the frequencies of measurements lying within various intervals in x. *Note that the x intervals in general need not be equal.*

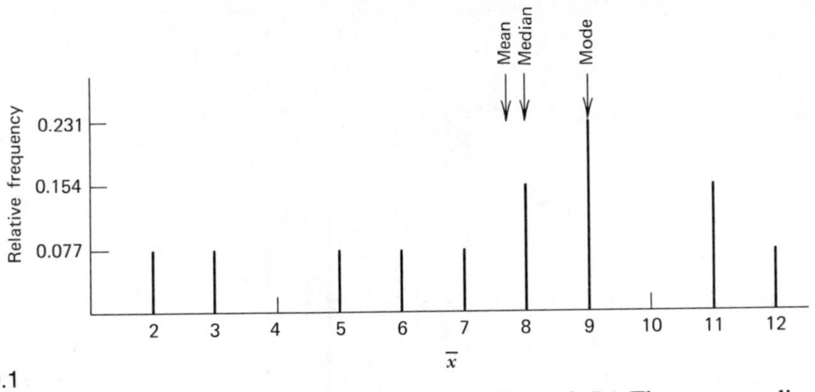

Figure 9.1
A plot of the relative frequencies of various numbers as in Example 7.1. The mean, median and mode are indicated.

We may define the relative frequency of measurements lying in any interval to be

$$\Phi_N(x_i) = \frac{n_i}{N} \quad \text{in the interval } x_{i-1} < x \leq x_i \tag{9.5}$$

where N is the total number of measurements and we have labeled the ordinate by the rightmost boundary of the interval. We could also, as we have done in Figures 7.2 to 7.4, label the ordinate by the center of the interval.

Normalized Frequency Histogram

We may now plot a *normalized frequency histogram.* On such a plot equal areas represent equal frequencies. If the data are divided into unequal ranges (x intervals), the areas must nevertheless correspond to the observed frequencies so that the areas standing on any interval of the baseline shall represent the actual frequency observed in that interval. We plot rectangles, with the heights proportional to the relative frequency per unit interval $f_N(x_i)$ and the area

$$(x_i - x_{i-1})f_N(x_i) = \Phi_N(x_i) = \frac{n_i}{N} \tag{9.6}$$

proportional to the relative frequency of obtaining a measurement in the interval $x_{i-1} < x \leq x_i$.

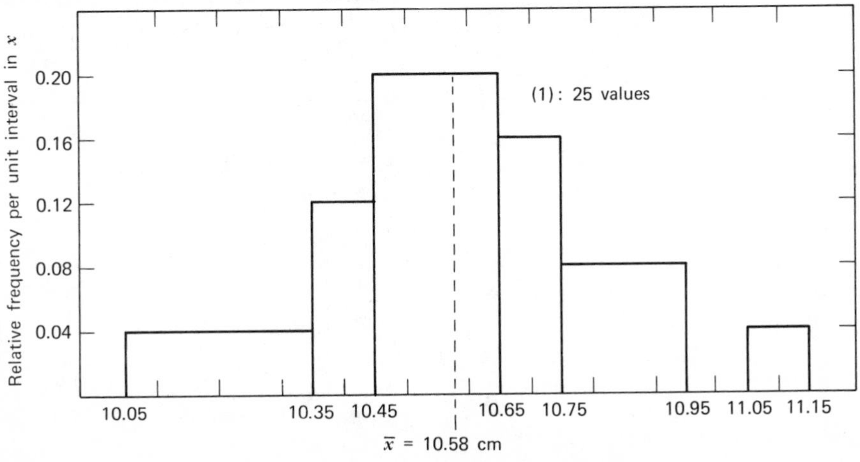

Figure 9.2
A *normalized* frequency histogram of 25 measurements from Table 7.2.

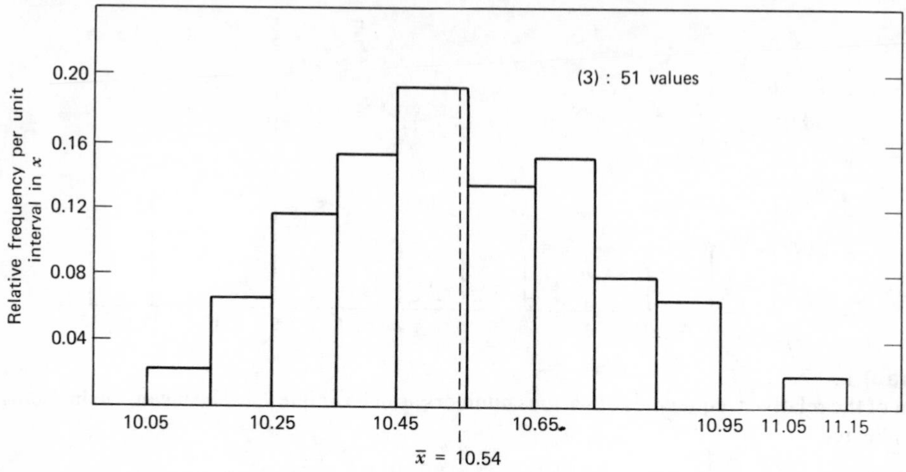

Figure 9.3
A *normalized* frequency histogram of 51 measurements from Table 7.4.

We see that $f_N(x)$ *is the relative frequency per unit interval in x.* To obtain the actual relative frequency $\Phi_N(x_i)$ for a particular interval i we must multiply $f_N(x_i)$ by the appropriate interval.

In these terms we may replot Figure 7.2 as a normalized frequency histogram in Figure 9.2.

We may now more easily compare the data of Table 7.2 with those of Table 7.4 shown in Figure 9.3.

The area A under the histograms is

$$\begin{aligned} A &= \sum_{i=1}^{k} (x_{i-1} - x_i) f_N(x_i) \\ &= \sum_{i=1}^{k} \frac{n_i}{N} \quad \text{by (9.6)} \\ &= \frac{1}{N} \sum n_i = 1 \end{aligned} \tag{9.7}$$

We may combine the discussion of discrete and continuum histograms by considering a smooth curve drawn through the plots. Such a curve $S_N(x)$ for the discrete distribution of Figure 9.1 is shown in Figure 9.4.

We have drawn $S_N(x)$ through the ordinates such that $S_N(x_i) = f_N(x_i)$.

We may also draw a smooth curve $S_N(x)$ through a histogram. Here, we choose $S_N(x)$ so that it corresponds closely to the ordinate at the midpoint of each interval

$$S_N(x^i) \approx f_N(x_i) \quad \text{where} \quad x^i = \frac{x_i + x_{i-1}}{2}$$

A more careful criterion chooses $S_N(x)$ so that the areas of each interval i

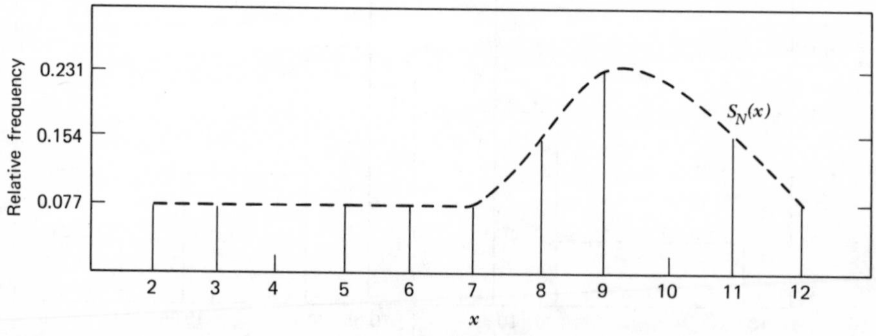

Figure 9.4
A smooth curve drawn through a normalized frequency histogram.

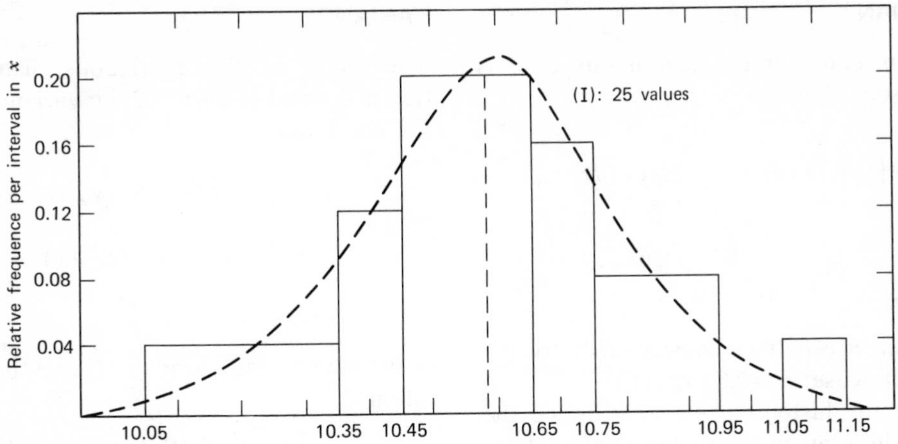

Figure 9.5
A smooth curve drawn through the normalized frequency histogram of Figure 9.2.

correspond

$$\int_{x_{i-1}}^{x_i} S_N(x)\,dx = (x_i - x_{i-1}) f_N(x_i) \tag{9.8}$$

This is the important definition and insures the normalization condition

$$\int_{-\infty}^{\infty} S_N(x)\,dx = \sum_{i=1}^{k} (x_i - x_{i-1}) f_N(x_i) = 1 \tag{9.9}$$

A smooth curve is drawn through the histogram of Figure 9.2 in Figure 9.5.

Note that it is not always possible to satisfy either the approximate criterion for $S_N(x)$ or the relation (9.8) *for each* interval. This is shown in Figure 9.5 where we have invoked the relation (9.8) for the *combined* interval 10.95 to 11.15.

Note that the curve drawn in Figure 9.5 is not too dissimilar from the Gaussian curve drawn in Figure 7.2. In the latter case we have used an explicit known form for the function $S_N(x)$ with appropriate normalization.

Definitions

We are now in a position to define the various statistical quantities introduced in Chapter 7 in terms of either the *frequency distribution function* (also called the *frequency function*), $f_N(X)$, for the discrete case or in terms of the *continuous frequency distribution function*, $S_N(X)$ (also called the *probability density function* or *p.d.f.*).

MODE

The *mode* or *most probable* value of x has the greatest frequency. As before, for the discrete distribution it is x_M corresponding to the largest $f_N(x_M)$ of all the $f_N(x_i)$. For the continuous function $S_N(x)$ it corresponds to x_M such that $S_N(x_M)$ is maximal.

For continuous and differentiable functions $S_N(x)$ the mode corresponds either to one of the extreme boundaries of the x interval (if it does not extend from $-\infty$ to $+\infty$) or to a local maximum such that

$$\left.\frac{dS_N(x)}{dx}\right|_{x=x_M} = 0$$

and

$$\left.\frac{d^2S_N(x)}{dx^2}\right|_{x=x_M} < 0$$

MEDIAN

The *median* for the continuous case is x_{med} such that

$$\int_{-\infty}^{x_{\text{med}}} S_N(x)\,dx = \int_{x_{\text{med}}}^{+\infty} S_N(x)\,dx = \frac{1}{2}$$

MEAN

The mean is defined as previously for the discrete case: $\bar{x} = \sum_{i=1}^{k} x_i f_N(x_i)$.

For the histogram we may define the mean in terms of the midpoint of each interval weighted by the relative frequency for the interval and summed over all the intervals.

$$\begin{aligned}\bar{x} &= \sum \frac{(x_i + x_{i-1})}{2}\Phi_N(x_i)\\ &= \sum \frac{(x_i + x_{i-1})}{2}(x_i - x_{i-1}) f_N(x_i)\\ &= \sum f_N(x_i)\Delta x_i (x_i)_{\text{avg}}\\ &= \sum_{i=1}^{k}\left\{\frac{x_i^2}{2} f_N(x_i) - \frac{(x_{i-1})^2}{2} f_N(x_i)\right\}\\ &= \tfrac{1}{2}\sum_{i=1}^{k} f_N(x_i)\{x_i^2 - x_{i-1}^2\} \qquad (9.10)\end{aligned}$$

For the continuous distribution the mean is defined to be

$$\bar{x} = \int_{-\infty}^{\infty} x S_N(x)\,dx \qquad (9.11)$$

MEAN OF A FUNCTION

We may generalize the mean value of an arbitrary function $\xi(x)$:

$$\langle \xi_N(x)\rangle = \overline{\xi(x)} = \int_{-\infty}^{\infty} \xi(x) S_N(x)\,dx$$

where the limiting distribution

$$\lim_{N\to\infty} S_N(x) = S(x)$$

for the infinite population if we interpret the result to be the universe average of $\xi(x)$.

$$\langle \xi(x)\rangle = \int_{-\infty}^{\infty} \xi(x) S(x)\,dx \qquad (9.12)$$

VARIANCE

The width of the continuous distribution is defined in terms of the variance *of the distribution*

$$\begin{aligned}\sigma^2 &= \int_{-\infty}^{\infty} (x - \bar{x})^2 S_N(x)\,dx \qquad (9.13)\\ &= \int_{-\infty}^{\infty} x^2 S_N(x)\,dx - (\bar{x})^2\end{aligned}$$

$$\sigma^2 = \overline{x^2} - \bar{x}^2 \qquad (9.14)$$

We have already used (9.11) and the relation

$$\overline{x^2} = \int_{-\infty}^{\infty} x^2 S_N(x)\,dx \qquad (9.15)$$

The relation (9.14) is general and holds for any distribution including the limiting distribution

$$\sigma^2 = \langle (x - \langle x\rangle)^2\rangle = \langle x^2\rangle - \langle x\rangle^2 \qquad (9.16)$$

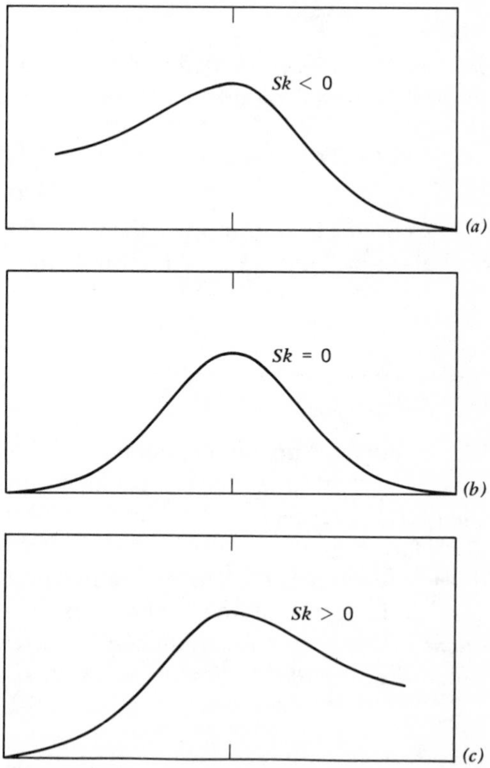

Figure 9.6
Distributions that tail to the left (*a*) have *negative* skewness; those which are symmetrical (*b*) have *zero* skewness; those which tail to the right (*c*) have *positive* skewness.

Skewness

For a symmetrical distribution, the mean, mode, and median coincide. Skewness, or deviation from symmetry, may be defined in terms of the noncoincidence of these "measures of central tendency." K. Pearson suggested measuring skewness by the relation

$$Sk \equiv \frac{(\text{mean} - \text{mode})}{\text{standard deviation}} \tag{9.17}$$

As shown in Figure 9.6*a*, distributions tailing to the left have negative skewness. Symmetrical distributions (e.g., Figure 9.6*b*) have zero skewness while distributions that tail to the right, as in Figure 9.6*c*, have positive skewness.

Sheppard's Correction for Grouping Data

When data whose underlying variate is continuous are grouped into intervals ("binned" or histogrammed) and the moments of the distribution (e.g., the mean and the variance) calculated as if the frequency function is discrete, some inaccuracy is introduced. This inaccuracy results because the data values are scattered throughout the interval whereas the grouping assigns them all to the center of the interval. Crudely speaking, the difficulty arises from two causes. First, if the interval is too wide the underlying curve may not be well approximated by a straight line over the interval. In this event, the mean of the area under the continuous curve may not coincide with the midpoint of the interval. Second, for higher moments such as the variance, raising deviations from a midpoint to powers means that values lying beyond the midpoint should add relatively more to the moment than values below the midpoint.

W. F. Sheppard showed that this effect can be corrected for values governed by most distributions (in particular, for normally distributed data). His analysis showed that the grouped mean usually needs no correction and that the grouped variance is simply corrected:

$$(\bar{x})_g = (\bar{x})_{\text{actual}} \tag{9.18}$$

and

$$(\sigma^2)_g = (\sigma^2)_{\text{actual}} + \frac{h^2}{12} \tag{9.19}$$

where h is the width of the interval.

For the benefit of the curious we present a heuristic discussion of (9.19).*

We note the definitions

$$N(\sigma^2)_g = \sum_{-\infty}^{\infty} x_i^2 \int_{x_i-h/2}^{x_i+h/2} f(x)\,dx$$

and

$$N(\sigma^2)_{\text{actual}} = \int_{-\infty}^{\infty} x^2 f(x)\,dx$$

where N is the total number of data points, and the x_i represent the midpoints of the intervals. The width of each interval is h.

We use the Taylor expansion about each x_i:

$$f(x_i+h) = f(x_i) + \frac{f'(x_i)}{1!}h + \frac{f''(x_i)}{2!}h^2 + \frac{f'''(x_i)}{3!}h^3 + \cdots$$

We write

$$N(\sigma^2)_g = \sum_{-\infty}^{\infty} x_i^2 \int_{-h/2}^{+h/2} f(x_i+u)\,du$$

$$= \sum_{-\infty}^{\infty} x_i^2 \int_{-h/2}^{h/2} \left[f(x_i) + \frac{f'(x_i)}{1!}u + \frac{f''(x_i)}{2!}u^2 + \cdots \right] du$$

and integrate term by term.

$$N\sigma_g^2 = \sum x_i^2 [f(x_i)u]_{-h/2}^{h/2} + \sum x_i^2 \left[\frac{f'(x_i)u^2}{(2)\cdot 1!}\right]_{-h/2}^{h/2} + \sum x_i^2 \left[\frac{f''(x_i)u^3}{3\cdot 2!}\right]_{-h/2}^{h/2} + \sum x_i^2 \left[\frac{f'''(x_i)u^4}{4\cdot 3!}\right]_{-h/2}^{h/2} + \cdots$$

*We rely on the analysis in Aitken, *Statistical Mathematics*, Oliver and Boyd, Edinburgh, 1957.

All the terms involving even powers of u will drop out leaving

$$N\sigma_g^2 = h \sum x_i^2 f(x_i) + h^3 \sum x_i^2 \frac{f''(x_i)}{2^2 \cdot 3!} + h^5 \sum x_i^2 \frac{f^{\mathrm{IV}}(x_i)}{2^4 \cdot 5!} + \cdots$$

The relation between an integral and a sum of equidistant values is given by the Euler–Maclaurin summation formula

$$\int_o^n f(x)\,dx = \tfrac{1}{2}f(0) + f(1) + \cdots + f(n-1) + \tfrac{1}{2}f(u) - b_2\{f'(n) - f'(0)\} - b_4\{f'''(n) - f'''(0)\} - \cdots - b_{2r}\{f^{(2r-1)}(n) - f^{(2r-1)}(0)\} - \sum_{m=0}^{n-1} \int_m^{m+1} P_{2r+1}(x-m) f^{(2r+1)}(x)\,dx$$

The feature of this is that a sum is expressed as an integral over the range plus correction terms that are linear in derivatives of odd order evaluated at the boundaries of the range. For most distributions of interest the derivatives do vanish at $\pm\infty$. For most functions used, if h is sufficiently small ($h < \sigma$) the sum may be replaced by an integral

$$N\sigma_g^2 = \int_{-\infty}^{\infty} x^2 f(x)\,dx + h^2 \int_{-\infty}^{\infty} \frac{x^2 f''(x)}{2^2 \cdot 3!}\,dx + h^4 \int_{-\infty}^{\infty} \frac{x^2 f^{\mathrm{IV}}(n)}{2^4 \cdot 5!}\,dx + \cdots$$

Integrating by parts we get

$$N\sigma_g^2 = N\sigma^2 + \frac{h^2}{24} 2(2-1) \int_{-\infty}^{\infty} f(x)\,dx$$

or

$$(\sigma^2)_{\text{actual}} = \sigma_g^2 - \frac{h^2}{12}$$

A similar analysis will show that no correction is required for odd moments.

In particular, $(\bar{x})_g = (\bar{x})_{\text{actual}}$.

10

PROPAGATION OF ERROR AND LEAST SQUARES

Suppose we consider a function $z = f(x, y, \ldots)$ of two or more measured quantities $x, y, \ldots$. The quantity z will be "in error" by an amount dz as a consequence of the errors in the measured quantities $x, y, \ldots$. The errors $dz, dx, dy, \ldots$, must be the same kind of quantity: either all average errors, all standard deviations, etc. We have previously emphasized the standard deviation as the most convenient and shall continue to do so in what follows.

General Case of Error Propagation

We may distinguish two cases. The errors dx, dy, for example, may be *independent*. In the case of independent errors we have the possibility of compensation. When the error in x causes z to be too large, the error in y may cause it to be too small. On the average, the total error in z will be algebraically less than the sum of the separate contributions of $dx, dy, \ldots$. There is no possibility of such compensation for *nonindependent* errors and these contributions do add algebraically according to the formulas of the calculus.

In this discussion we may use the variance to specify the error in the independent variables $x, y, \ldots$, recalling that the standard deviation is the square root of the variance. We may thus consider the $x, y, \ldots$ to have means $\langle x\rangle, \langle y\rangle, \ldots$ and variances $\sigma_x^2 = \langle(x - \langle x\rangle)^2\rangle = \langle(\Delta x)^2\rangle$, $\sigma_y^2 = \langle(y - \langle y\rangle)^2\rangle = \langle(\Delta y)^2\rangle$, etc. where $\Delta x = x - \langle x\rangle, \ldots$. We may also define quantities relating to the correlation or nonindependence of the errors. These quantities, called the *covariances*, are defined, for example,

$$\sigma_{xy} = \langle(x - \langle x\rangle)(y - \langle y\rangle)\rangle = \langle(\Delta x)(\Delta y)\rangle \tag{10.1}$$

We consider that the function varies slowly enough so that it may be represented by the first few terms in a Taylor series expansion of z about the point $P = (\langle x\rangle, \langle y\rangle, \ldots)$.

$$z \sim f(\langle x\rangle, \langle y\rangle, \ldots) + \left.\frac{\partial f}{\partial x}\right|_P (x - \langle x\rangle) + \left.\frac{\partial f}{\partial y}\right|_P (y - \langle y\rangle) + \cdots$$

where $\partial f/\partial x_i|_P$ is the partial derivative of f with respect to x_i evaluated at the point

$P = (\langle x\rangle, \langle y\rangle, \ldots)$. We may obtain $\langle z\rangle$ term by term noting that $\langle(x - \langle x\rangle)\rangle = 0 = \langle(y - \langle y\rangle)\rangle =$ etc.

$$\therefore \qquad \langle z\rangle \approx f(\langle x\rangle, \langle y\rangle, \ldots) \qquad (10.2)$$

This is the expected result that the best estimate of z is obtained by evaluating f at the best values for x, y, ...

Consider the variance of z.

$$(z - \langle z\rangle)^2 \approx \left(\frac{\partial f}{\partial x}\bigg|_P\right)^2 (x - \langle x\rangle)^2 + \left(\frac{\partial f}{\partial y}\bigg|_P\right)^2 (y - \langle y\rangle)^2 + \cdots + 2\left(\frac{\partial f}{\partial x}\bigg|_P\right)\left(\frac{\partial f}{\partial y}\bigg|_P\right)(x - \langle x\rangle)(y - \langle y\rangle) + \cdots$$

$$\sigma_z^2 = \langle(z - \langle z\rangle)\rangle^2 \approx \left(\frac{\partial f}{\partial x}\bigg|_P\right)^2 \sigma_x^2 + \left(\frac{\partial f}{\partial y}\bigg|_P\right)^2 \sigma_y^2 + 2\left(\frac{\partial f}{\partial x}\bigg|_P\right)\left(\frac{\partial f}{\partial y}\bigg|_P\right)\sigma_{xy} + \cdots$$

In general, if $z = f(x_1, \ldots, x_i, \ldots, x_n)$

$$\sigma_z^2 \approx \sum_{i=1}^{n} \left(\frac{\partial f}{\partial x_i}\bigg|_P\right)\left(\frac{\partial f}{\partial x_j}\bigg|_P\right) \times \langle(x_i - \langle x_i\rangle)(x_j - \langle x_j\rangle)\rangle$$

$$\sigma_z^2 \approx \sum_{i,j=1}^{n} \left(\frac{\partial f}{\partial x_i}\bigg|_P\right)\left(\frac{\partial f}{\partial x_j}\bigg|_P\right)\sigma_{x_jx_i} \qquad (10.3)$$

We note that this holds generally for the variables x, y, . . . , having independent errors and also correlated errors. The covariances are always symmetric ($\sigma_{x_ix_j} = \sigma_{x_jx_i}$) and are equal to zero for independent errors. (Note that $\sigma_{x_ix_i} = \sigma^2_{x_i\cdot}$)

To simplify discussion we shall not distinguish between sample average and population average but use the best value that is available from the samples. We shall use $\bar{x}$ for the best value of x and S_x = the best value for its standard deviation. The deviations will always be evaluated at the point $\bar{x}$, $\bar{y}$, etc., unless otherwise specified.

Independent Errors

Consider the simple case of two independent variables x, y, and $z = f(x, y)$. The errors in x and y are *independent* and *uncorrelated* ($S_{xy} = 0$):

$$S_z = \sqrt{\left(\frac{\partial f}{\partial x}\right)^2 S_x^2 + \left(\frac{\partial f}{\partial y}\right)^2 S_y^2} \qquad (10.4a)$$

This enables us to obtain several results of practical interest.

PRODUCT: $z = xy$

$$\frac{\partial f}{\partial x} = \bar{y} \qquad \frac{\partial f}{\partial y} = \bar{x}$$

$$S_z^2 = (\bar{y})^2 S_x^2 + (\bar{x})^2 S_y^2$$

If we divide by $\bar{z}$ and take the square root,

$$\frac{S_z}{\bar{z}} = \sqrt{\left(\frac{S_x}{\bar{x}}\right)^2 + \left(\frac{S_y}{\bar{y}}\right)^2} \qquad (10.4b)$$

(*Note.* The generalization (10.10) is needed for the case $\bar{y} = \bar{x} = 0$.) We may define the *fractional standard deviation in x_i or relative error* to be $\epsilon_{x_i} = S_{x_i}/\bar{x}_i$, the ratio of the standard deviation to the mean value. The relative error in the product of two independent measurements will be approximately the square root of the sum of their relative errors squared

$$\epsilon_z = \sqrt{\epsilon_x^2 + \epsilon_y^2} \qquad (10.4c)$$

We say that the relative errors ϵ_x and ϵ_y have been *added in quadrature*.

QUOTIENT: $z = x/y$

$$\frac{\partial f}{\partial x} = \frac{1}{\bar{y}} \qquad \frac{\partial f}{\partial y} = -\frac{\bar{x}}{(\bar{y})^2}$$

$$S_z^2 = \frac{1}{(\bar{y})^2} S_x^2 + \frac{\bar{x}^2}{(\bar{y})^4} S_y^2$$

and

$$\frac{S_z^2}{(\bar{z})^2} = \frac{S_x^2}{(\bar{x})^2} + \frac{S_y^2}{(\bar{y})^2}$$

which is the same relation as for the product of x and y.

SUM AND DIFFERENCE: $z = x \pm y$

$$\frac{\partial f}{\partial x} = 1 \qquad \frac{\partial f}{\partial y} = \pm 1$$

$$S_z^2 = S_x^2 + S_y^2 \qquad \text{or} \qquad S_z = \sqrt{S_x^2 + S_y^2} \tag{10.4d}$$

The absolute standard deviation of a quantity that is the sum or difference of measured quantities is the sum in quadrature of the absolute standard deviations of the measured quantities.

Note that a large *relative error* can result from the difference of two numbers that are nearly equal: $\epsilon_z = S_z/\bar{z}$ where $\bar{z}$ may be very small. If alternative procedures are possible, one should avoid taking the difference of two nearly equal numbers. It will often not be possible to avoid this kind of subtraction, and a large relative error remains.

► **Example 10.1**

Suppose one wishes to measure the voltage between two dynodes A and B of a photomultiplier tube.

I. We may measure $V_A = (2010 \pm 10)$ volts and $V_B = (1982 \pm 10)$ volts using a voltmeter capable of a relative error $\sim 10/2000 = \frac{1}{2}\%$. The voltage difference $V_A - V_B = 28$ volts with an error of $\sigma_{A-B} = \sqrt{10^2 + 10^2} = 14$ volts yielding $V_A - V_B = 28 \pm 14$ volts. $\epsilon_{A-B} = 50\%$.

II. We may measure the voltage difference directly with a voltmeter good to 10% and get 28 ± 3 volts. It would appear that there is usually a hard way to do business. Had you thought about the problem in advance, which way would you have done the measurement? ◄

PRODUCT OF FACTORS RAISED TO POWERS: $z = x^m y^n$

$$\frac{\partial f}{\partial x} = m(\bar{x})^{m-1}(\bar{y})^n \qquad \frac{\partial f}{\partial y} = n(\bar{x})^m(\bar{y})^{n-1}$$

It follows that

$$\epsilon_z = \frac{S_z}{\bar{z}} = \sqrt{m^2\left(\frac{S_x}{\bar{x}}\right)^2 + n^2\left(\frac{S_y}{\bar{y}}\right)^2}$$

$$= \sqrt{m^2\epsilon_x^2 + n^2\epsilon_y^2} \tag{10.4e}$$

A simple corollary of this result applies to a single quantity raised to a power: $z = x^n$

$$\epsilon_z = n\epsilon_x \tag{10.4f}$$

A trivial case that should be mentioned is the following.

CONSTANT MULTIPLIER: $z = Kx$

Here

$$S_z = KS_x \qquad \text{and} \qquad \epsilon_z = \epsilon_x \tag{10.4g}$$

Graphical Description of Error Propagation

For the general case of an arbitrary function of a single variate, $z = f(x)$, the propagation of uncertainty has a simple graphical description as shown in Figure 10.1.

The following basic relation holds if the function varies slowly in the vicinity of $\bar{x}$ (so that we take only first order terms in the Taylor series):

$$\Delta z = \left.\frac{df}{dx}\right|_{\bar{x}} \Delta x$$

and we may use σ_x for Δx getting σ_z for Δz.

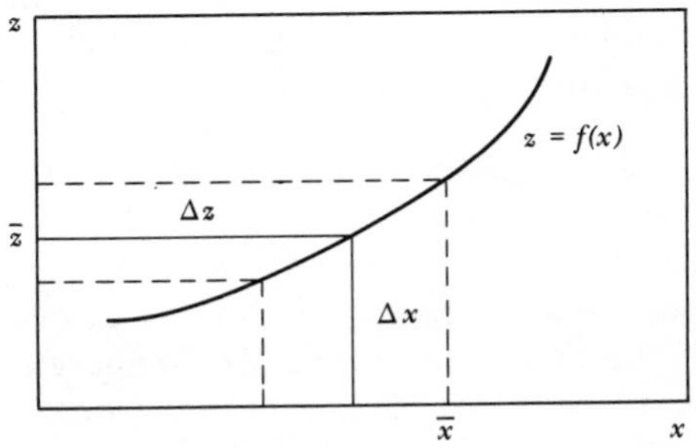

Figure 10.1
Curve of an arbitrary function of x. An uncertainty in x corresponds to an uncertainty in $z = f(x)$.

If the derivative $df/dx|_{\bar{x}}$ is large, a small deviation from $\bar{x}$ may produce a large deviation in z from $\bar{z}$. Here the derivative acts as an amplification factor.

If the function varies rapidly enough in the region within one or two standard deviations of $\bar{x}$, the linear approximation may not be good. Here a symmetric error in x propagates to an asymmetric error in z. An example is shown in Figure 10.2 for the function $z = 1/x$.

The arbitrary case is difficult to treat, and we shall restrict most of our discussion to regions sufficiently close to $\bar{x}$ to validate our approximations of linearity and symmetric errors. This is sufficiently general for most purposes. We point out the possibility, however, so that anomalous cases may be treated by special methods.

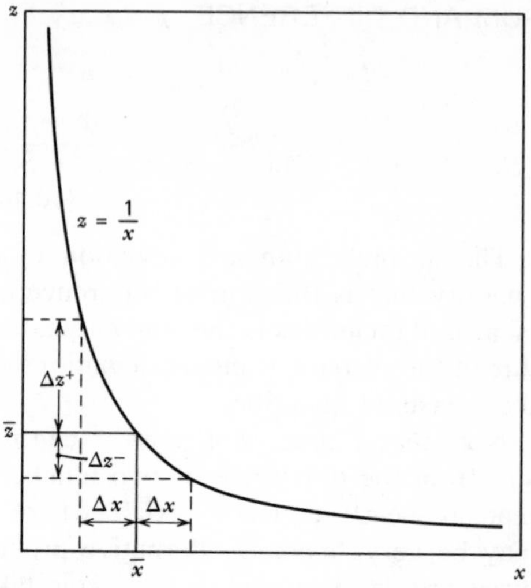

Figure 10.2
Curve of $z = 1/x$ versus x. A symmetrical uncertainty in x corresponds to an asymmetrical uncertainty in z.

Minimum Variance (Least-Squared Error)

We are now in a position to verify relations (7.10) and (7.11) that tell us how to obtain the "best" value of a quantity x from two or more independent measurements whose errors may be different. It is clear that one should weight a measurement more heavily if its error is small, but the exact weighting is not at once apparent.

Let us consider first the case of two independent measurements $(x_1 \pm \sigma_1,\ x_2 \pm \sigma_2)$ where we wish to obtain the "best value," $\bar{x}_{12} \pm \sigma_{12}$. We shall invoke what may be called the *principle of least squared error* to define the "best value", $\bar{x}_{12}$: $\bar{x}_{12}$ is that combination of x_1 and x_2 that yields the minimum σ_{12}^2.

We may make the reasonable assumption that

$$\bar{x}_{12} = ax_1 + (1-a)x_2$$

where a is the normalized statistical weight to be accorded to measurement x_1. We have just seen how to calculate σ_{12} for such an expression:

$$\sigma_{12}^2 = a^2\sigma_1^2 + (1-a)^2\sigma_2^2$$

We wish to minimize this quantity and so set

$$\frac{d\sigma_{12}^2}{da} = 2a\sigma_1^2 + 2(1-a)(-1)\sigma_2^2 = 0$$

This determines the normalized weights a and $(1-a)$ for us.

$$a = \frac{\sigma_2^2}{\sigma_1^2 + \sigma_2^2} = \frac{\dfrac{1}{\sigma_1^2}}{\dfrac{1}{\sigma_1^2} + \dfrac{1}{\sigma_2^2}}$$

$$1 - a = \frac{\sigma_1^2}{\sigma_1^2 + \sigma_2^2} = \frac{\dfrac{1}{\sigma_2^2}}{\dfrac{1}{\sigma_1^2} + \dfrac{1}{\sigma_2^2}}$$

Thus

$$\bar{x}_{12} = \frac{\dfrac{x_1}{\sigma_1^2} + \dfrac{x_2}{\sigma_2^2}}{\dfrac{1}{\sigma_1^2} + \dfrac{1}{\sigma_2^2}}$$

and

$$\sigma_{12}^2 = \left(\frac{1}{\sigma_1^2} + \frac{1}{\sigma_2^2}\right)^{-1}$$

This may be extended to a third measurement, $x_3 \pm \sigma_3$. We may now treat as two independent measurements $\bar{x}_{12} \pm \sigma_{12}$ and $x_3 \pm \sigma_3$. This immediately yields

$$\bar{x}_{123} = \frac{\dfrac{\bar{x}_{12}}{\sigma_{12}^2} + \dfrac{x_3}{\sigma_3^2}}{\dfrac{1}{\sigma_{12}^2} + \dfrac{1}{\sigma_3^2}} = \frac{\dfrac{x_1}{\sigma_1^2} + \dfrac{x_2}{\sigma_2^2} + \dfrac{x_3}{\sigma_3^2}}{\dfrac{1}{\sigma_1^2} + \dfrac{1}{\sigma_2^2} + \dfrac{1}{\sigma_3^2}}$$

with

$$\sigma_{123}^2 = \left(\frac{1}{\sigma_1^2} + \frac{1}{\sigma_2^2} + \frac{1}{\sigma_3^2}\right)^{-1}$$

(The mathematically sophisticated reader will recognize that we have gone through a "proof by induction.")

This generalizes to the weighted mean

$$\bar{x} = \frac{\sum_{i=1}^{n} \left(\frac{x_i}{\sigma_i^2}\right)}{\sum_{i=1}^{n} \left(\frac{1}{\sigma_i^2}\right)} \qquad (10.4h)$$

and

$$\sigma_{\bar{x}} = \left\{\sum \left(\frac{1}{\sigma_i^2}\right)\right\}^{-1/2} \qquad (10.4i)$$

for the general case of n independent measurements $x_i \pm \sigma_i$. These are just relations (7.10) and (7.11). We note that, as usual,

$$\bar{x} = \frac{1}{n}\sum_{i=1}^{n} x_i$$

the arithmetic average, for the case of n independent measurements of *equal* validity.

We shall see later on that this is just a special case of a general principle: the arithmetic average (or the weighted average, where appropriate) is the "least-squares best estimate" of the mean of the population of measurements.

Nonindependent or Correlated Errors

We have emphasized independent errors. If some of the errors are nonindependent, they may not be treated as if they are. The proper procedure is one of the following:

1. The problem may be restated in terms of variates that have independent errors.
2. The proper expressions for the derivatives using the "chain rule" of the differential calculus may be used.
3. The general formula (10.3) may be used if the covariances are known.

► **Example 10.2**

Consider $\epsilon = A/(A + B)$ where A and B, are independent measurements with mean and variance $\bar{A}$, σ_A^2 and $\bar{B}$, σ_B^2, respectively.

We may write this as $\epsilon = A/U$ where A and U are now non-independent since $U = A + B$.

The mean value of ϵ is obvious: $\bar{\epsilon} = \bar{A}/(\bar{A} + \bar{B})$. The calculation of interest is that of the variance of $\epsilon = \sigma_\epsilon^2$.

I. There is no problem if we express ϵ only in terms of independent quantities A and B

$$\sigma_\epsilon^2 = \left(\frac{\partial\epsilon}{\partial A}\right)^2 \sigma_A^2 + \left(\frac{\partial\epsilon}{\partial B}\right)^2 \sigma_B^2$$

where the derivatives are evaluated at $\bar{A}$, $\bar{B}$:

$$\frac{\partial\epsilon}{\partial A} = \frac{\bar{B}}{(\bar{A} + \bar{B})^2} \qquad \text{and} \qquad \frac{\partial\epsilon}{\partial B} = \frac{-\bar{A}}{(\bar{A} + \bar{B})^2}$$

$$\sigma_\epsilon^2 = \frac{(\bar{B})^2}{(\bar{A} + \bar{B})^4}\sigma_A^2 + \frac{(\bar{A})^2}{(\bar{A} + \bar{B})^4}\sigma_B^2 \qquad (10.5)$$

II. We consider $\epsilon = A/U$. As always $\bar{\epsilon} = \bar{A}/\bar{U} = \bar{A}/(\bar{A} + \bar{B})$.

$$\sigma_U^2 = \sigma_A^2 + \sigma_B^2 \quad \text{by (10.4d).}$$

By (10.3)

$$\sigma_\epsilon^2 = \left(\frac{\partial\epsilon}{\partial A}\right)^2 \sigma_A^2 + \left(\frac{\partial\epsilon}{\partial U}\right)^2 \sigma_U^2 + 2\left(\frac{\partial\epsilon}{\partial A}\right)\left(\frac{\partial\epsilon}{\partial U}\right)\sigma_{AU}$$

and, here, the covariance is not zero. We shall for readability use $\bar{A}$ and $\langle A\rangle$ interchangeably.

We may calculate the covariance

$$\begin{aligned}
\sigma_{AU} &= \langle(A - \bar{A})(U - \bar{U})\rangle \\
&= \langle(A - \bar{A})(A + B - \bar{A} - \bar{B})\rangle \\
&= \langle(A^2 + AB - A\bar{A} - A\bar{B} - \bar{A}A - \bar{A}B \\
&\quad + \bar{A}^2 + \bar{A}\bar{B})\rangle \\
&= \langle A^2\rangle + \langle AB\rangle - \bar{A}^2 - \bar{A}\bar{B} - \bar{A}^2 - \bar{A}\bar{B} \\
&\quad + \bar{A}^2 + \bar{A}\bar{B} \\
&= \langle A^2\rangle - \bar{A}^2
\end{aligned}$$

Since A and B are independent and $\langle AB\rangle = \bar{A}\bar{B}$

$$\sigma_{AU} = \sigma_A^2$$

$$\frac{\partial\epsilon}{\partial A} = \frac{1}{\bar{U}} = \frac{1}{\bar{A}+\bar{B}}$$

$$\frac{\partial\epsilon}{\partial U} = \frac{-\bar{A}}{(\bar{U})^2} = \frac{-\bar{A}}{(\bar{A}+\bar{B})^2}$$

$$\sigma_\epsilon^2 = \frac{1}{(\bar{A}+\bar{B})^2}\sigma_A^2 + \frac{(\bar{A})^2}{(\bar{A}+\bar{B})^4}(\sigma_A^2+\sigma_B^2) - \frac{2}{(\bar{A}+\bar{B})}\frac{\bar{A}}{(\bar{A}+\bar{B})^2}\sigma_A^2$$

$$= \left[\frac{(\bar{A}+\bar{B})^2+(\bar{A})^2-2\bar{A}(\bar{A}+\bar{B})}{(\bar{A}+B)^4}\right]\sigma_A^2 + \frac{(\bar{A})^2}{(\bar{A}+\bar{B})^4}\sigma_B^2$$

$$\sigma_\epsilon^2 = \frac{(\bar{B})^2}{(\bar{A}+\bar{B})^4}\sigma_A^2 + \frac{(\bar{A})^2}{(\bar{A}+\bar{B})^4}\sigma_B^2$$

This result is just (10.5), which is the same answer obtained in **I**. Note that the term involving the covariance may be positive or negative while the variances make only positive contributions to the sum. ◄

We may now evaluate our formulas for the propagation of error for the general case that the errors may be nonindependent or correlated.

The quantities given will be X and Y, that is, we start with their central values, their respective variances, and their covariance or correlated error: $\bar{X}$, $\bar{Y}$, S_x^2, S_y^2, and S_{xy}.

SUM AND DIFFERENCE: $Z = X \pm Y$

$$S_z^2 = \langle[(X+Y)-(\bar{X}\pm\bar{Y})]^2\rangle$$
$$= \langle X^2\rangle - (\bar{X})^2 + \langle Y^2\rangle - (\bar{Y})^2 \pm 2\{\langle XY\rangle - \bar{X}\bar{Y}\}$$

Note that

$$S_{xy} = \langle(X-\langle X\rangle)(Y-\langle Y\rangle)\rangle$$
$$= \langle XY\rangle - 2\langle X\rangle\langle Y\rangle + \langle X\rangle\langle Y\rangle$$
$$= \langle XY\rangle - \bar{X}\bar{Y}$$

Therefore

$$S_z^2 = S_x^2 + S_y^2 \pm 2S_{xy} \qquad (10.6a)$$

Note that the covariance term enters the sum and difference with different algebraic signs. The same result could have been obtained immediately from (10.3).

PRODUCT: $Z = XY$

Using (10.3), we obtain

$$\frac{S_z^2}{(\bar{Z})^2} = \frac{S_x^2}{(\bar{X})^2} + \frac{S_y^2}{(\bar{Y})^2} + 2\frac{S_{xy}}{(\bar{X})(\bar{Y})} \qquad (10.6b)$$

QUOTIENT: $Z = X/Y$

$$S_z^2 = \left(\frac{1}{\bar{Y}}\right)^2 S_x^2 + \left(\frac{-\bar{X}}{\bar{Y}^2}\right)^2 S_y^2 - 2\frac{\bar{X}}{\bar{Y}\bar{Y}^2}S_{xy}$$

$$\frac{Sz^2}{(\bar{Z})^2} = \frac{S_x^2}{(\bar{X})^2} + \frac{S_y^2}{(\bar{Y})^2} - 2\frac{S_{xy}}{\bar{X}\bar{Y}} \qquad (10.6c)$$

Note the similarity of this result to that for the product except for the sign of the covariance term.

► **Example 10.3**

We may return to the previous example $Z = A/U$ where $U = A + B$, and we are given as data $\bar{A}$, σ_A^2, $\bar{U} = \bar{A} + \bar{B}$, $\sigma_U^2 = \sigma_A^2 + \sigma_B^2$, and $\sigma_{AU} = \sigma_A^2$. By (10.6c)

$$\frac{S_z^2}{(\bar{Z})^2} = \frac{\sigma_A^2}{(\bar{A})^2} + \frac{\sigma_U^2}{(\bar{U})^2} - 2\frac{\sigma_{AU}}{\bar{A}\bar{U}}$$
$$= \frac{\sigma_A^2}{(\bar{A})^2} + \frac{\sigma_A^2+\sigma_B^2}{(\bar{A}+\bar{B})^2} - 2\frac{\sigma_A^2}{\bar{A}(\bar{A}+\bar{B})}$$
$$= \frac{\sigma_A^2(\bar{A}+\bar{B})^2+\sigma_A^2(\bar{A})^2-2\bar{A}(\bar{A}+\bar{B})\sigma_A^2}{(\bar{A})^2(\bar{A}+\bar{B})^2} + \frac{\sigma_B^2}{(\bar{A}+\bar{B})^2}$$

and

$$\frac{S_z^2}{(\bar{Z})^2} = \frac{\sigma_A^2(\bar{B})^2}{(\bar{A})^2(\bar{A}+\bar{B})^2} + \frac{\sigma_B^2}{(\bar{A}+\bar{B})^2}$$

Thus

$$S_z^2 = \frac{(\bar{B})^2}{(\bar{A}+\bar{B})^4}\sigma_A^2 + \frac{(\bar{A})^2}{(\bar{A}+\bar{B})^4}\sigma_B^2$$

which agrees with the previous result (10.5). ◄

PRODUCT OF FACTORS RAISED TO POWERS: $Z = X^m Y^n$

Again using (10.3) we get

$$S_z^2 = [m(\bar{X})^{m-1}(\bar{Y})^n]^2 S_x^2 + [n(\bar{X})^m(\bar{Y})^{n-1}]^2 S_y^2 + 2m(\bar{X})^{m-1}(\bar{Y})^n n(\bar{X})^m(\bar{Y})^{n-1} S_{xy}$$

and

$$\frac{S_z^2}{(\bar{Z})^2} = m^2 \frac{S_x^2}{(\bar{X})^2} + n^2 \frac{S_y^2}{(\bar{Y})^2} + 2\frac{mnS_{xy}}{(\bar{X}\bar{Y})} \tag{10.6d}$$

► **PROBLEM:**
Verify that this last equation yields the usual result (10.5) for $Z = A/U$ where $U = A + B$, A and B are independent and $\bar{A}$, $\bar{B}$, σ_A^2, σ_B^2 are given.◄

Covariances of Calculated Quantities

We have seen that we need only have the quantities $\bar{X}$, $\bar{Y}$, . . . , σ_x^2, σ_y^2, . . . , and σ_{xy}, . . . , to be able to propagate errors in the general case of nonindependence. We have still to calculate the covariances of the calculated quantities.

Suppose $Z = f(X, Y, \ldots)$ and $W = g(X, Y, \ldots)$. We need to obtain the covariance of Z and W for they are certainly correlated independent of the question of correlations among X, Y, etc.

We have (again using $\langle X \rangle = \bar{X}$ interchangeably)

$$\Delta Z = (Z - \langle Z \rangle) = \left.\frac{\partial f}{\partial x}\right|_P (X - \langle X \rangle) + \left.\frac{\partial f}{\partial y}\right|_P (Y - \langle Y \rangle) + \cdots$$

$$\Delta W = (W - \langle W \rangle) = \left.\frac{\partial g}{\partial x}\right|_P (X - \langle X \rangle) + \left.\frac{\partial g}{\partial y}\right|_P (Y - \langle Y \rangle) + \cdots$$

and thus obtain

$$\begin{aligned} \sigma_{ZW} = \sigma_{WZ} &= \overline{\Delta Z \Delta W} \\ &= \left.\frac{\partial f}{\partial x}\right|_P \left.\frac{\partial g}{\partial x}\right|_P \sigma_x^2 + \left.\frac{\partial f}{\partial y}\right|_P \left.\frac{\partial g}{\partial y}\right|_P \sigma_y^2 + \cdots \\ &\quad + \left\{ \left.\frac{\partial f}{\partial x}\right|_P \left.\frac{\partial g}{\partial y}\right|_P + \left.\frac{\partial f}{\partial y}\right|_P \left.\frac{\partial g}{\partial x}\right|_P \right\} \sigma_{xy} + \cdots \end{aligned} \tag{10.7}$$

► **PROBLEM:**
Suppose the area A and perimeter P of a rectangle of sides X and Y have been independently measured with the values $A \pm \Delta A = (6 \pm 0.6)\ \text{cm}^2$, $P \pm \Delta P = (10 \pm 1)$ cm, respectively (where the errors are standard deviations). Find the best values for X, Y, and their standard deviations and covariances. ◄

Generalization of (10.3)

The approximation (10.3), while usually adequate, may be extended somewhat *and must be* in certain circumstances. In particular, the relation (10.4b) is inadequate for the situation where $\bar{z} = \bar{x} = \bar{y} = 0$. Whenever one invokes a power series expansion, higher order terms must be included when it is possible for the lower order terms to vanish. To investigate this point we expand z about the point $P = (\langle x \rangle, \langle y \rangle, \ldots)$ as before but including more terms since we consider the possibility that

$$\left.\frac{\partial f}{\partial x}\right|_P = \left.\frac{\partial f}{\partial y}\right|_P = \cdots = 0$$

We write

$$\begin{aligned} z &\sim f(\langle x\rangle, \langle y\rangle, \ldots) \\ &+ \left.\frac{\partial f}{\partial x}\right|_P (x - \langle x\rangle) + \left.\frac{\partial f}{\partial y}\right|_P (y - \langle y\rangle) + \cdots \\ &+ \left.\frac{\partial^2 f}{\partial x\,\partial y}\right|_P (x - \langle x\rangle)(y - \langle y\rangle) + \cdots \\ &+ \frac{1}{2}\left.\frac{\partial^2 f}{\partial x^2}\right|_P (x - \langle x\rangle)^2 + \frac{1}{2}\left.\frac{\partial^2 f}{\partial y^2}\right|_P (y - \langle y\rangle)^2 + \cdots \end{aligned}$$

Form the squared quantity

$$\begin{aligned} (z - \langle z\rangle)^2 \approx &\left(\left.\frac{\partial f}{\partial x}\right|_P\right)^2 (x - \langle x\rangle)^2 \\ &+ \left(\left.\frac{\partial f}{\partial y}\right|_P\right)^2 (y - \langle y\rangle)^2 + \cdots \\ &+ 2\left(\left.\frac{\partial f}{\partial x}\right|_P\right)\left(\left.\frac{\partial f}{\partial y}\right|_P\right)(x - \langle x\rangle)(y - \langle y\rangle) + \cdots \\ &+ \frac{1}{2}\left(\left.\frac{\partial^2 f}{\partial x\,\partial y}\right|_P\right)\left(\left.\frac{\partial f}{\partial x}\right|_P\right)(x - \langle x\rangle)^2(y - \langle y\rangle) \\ &+ \frac{1}{2}\left(\left.\frac{\partial^2 f}{\partial x\,\partial y}\right|_P\right)\left(\left.\frac{\partial f}{\partial y}\right|_P\right)(y - \langle y\rangle)^2(x - \langle x\rangle) \\ &+ \cdots \end{aligned}$$

from which we may calculate the variance

$$\begin{aligned} \sigma_z^2 &= \langle (z - \langle z\rangle)^2\rangle \\ &= \left(\left.\frac{\partial f}{\partial x}\right|_P\right)^2 \sigma_x^2 + \left(\left.\frac{\partial f}{\partial y}\right|_P\right)^2 \sigma_y^2 + \cdots \\ &\quad + 2\left(\left.\frac{\partial f}{\partial x}\right|_P\right)\left(\left.\frac{\partial f}{\partial y}\right|_P\right)\sigma_{xy} + \cdots \\ &\quad + \left(\left.\frac{\partial^2 f}{\partial x\,\partial y}\right|_P\right)^2 \sigma_{x^2y^2} + \cdots \\ &\quad + \frac{1}{2}\left(\left.\frac{\partial^2 f}{\partial x\,\partial y}\right|_P\right)\left(\left.\frac{\partial f}{\partial x}\right|_P\right)\sigma_{x^2y} \\ &\quad + \frac{1}{2}\left(\left.\frac{\partial^2 f}{\partial x\,\partial y}\right|_P\right)\left(\left.\frac{\partial f}{\partial y}\right|_P\right)\sigma_{y^2x} + \cdots \end{aligned} \tag{10.8}$$

Equation (10.8) is the proper extension of (10.3) since all of the terms written may be either zero or nonzero for the general case of correlated errors where we introduce the notation

$$\begin{aligned} \sigma_{x^2y^2} &\equiv \langle (x - \langle x\rangle)^2(y - \langle y\rangle)^2\rangle \\ \sigma_{x^2y} &\equiv \langle (x - \langle x\rangle)^2(y - \langle y\rangle)\rangle \end{aligned} \tag{10.9}$$

etc.

For the case of uncorrelated or independent errors (10.8) simplifies since

$$\sigma_{xy} = 0 \text{ and}$$

$$\langle (x - \langle x\rangle)^2(y - \langle y\rangle)^2\rangle = \sigma_{x^2y^2} = \sigma_x^2\sigma_y^2$$

as well as

$$\begin{aligned} \sigma_{x^2y} &= \langle (x - \langle x\rangle)^2(y - \langle y\rangle)\rangle \\ &= \sigma_x^2\langle (y - \langle y\rangle)\rangle = 0 \end{aligned}$$

and

$$\begin{aligned} \sigma_{y^2x} &= \langle (y - \langle y\rangle)^2(x - \langle x\rangle)\rangle \\ &= \sigma_y^2\langle (x - \langle x\rangle)\rangle = 0 \end{aligned}$$

For errors in x and y being *independent* and *uncorrelated* we have

$$\begin{aligned} S_z^2 = &\left(\left.\frac{\partial f}{\partial x}\right|_P\right)^2 S_x^2 + \left(\left.\frac{\partial f}{\partial y}\right|_P\right)^2 S_y^2 \\ &+ \left(\left.\frac{\partial^2 f}{\partial x\,\partial y}\right|_P\right)^2 S_x^2 S_y^2 \end{aligned} \tag{10.10}$$

which is the proper generalization of (10.4a) and is valid for the situation where

$$\left(\left.\frac{\partial f}{\partial x}\right|_P\right) = \left(\left.\frac{\partial f}{\partial y}\right|_P\right) = 0$$

► **Example 10.4**

For the product, $z = xy$, (10.10) yields the results

$$S_z^2 = (\bar{y})^2 S_x^2 + (\bar{x})^2 S_y^2 + S_x^2 S_y^2 \tag{10.11}$$

This result is rigorous as can be seen from the following argument where X and Y are independent random variables.

By (9.16)

$$\begin{aligned} \sigma_{XY}^2 &= \langle (XY)^2\rangle - \langle XY\rangle^2 \\ &= \langle X^2\rangle\langle Y^2\rangle - \langle X\rangle^2\langle Y\rangle^2 \end{aligned}$$

since X and Y are independent.

Let us use the relations $\sigma_X^2 = \langle X^2\rangle - \langle X\rangle^2$ and $\sigma_Y^2 = \langle Y^2\rangle - \langle Y\rangle^2$. Thus,

$$\begin{aligned} \sigma_{XY}^2 &= \langle X^2\rangle\langle Y^2\rangle - [-\sigma_X^2 + \langle X^2\rangle] \\ &\quad \times [-\sigma_Y^2 + \langle Y^2\rangle] \\ &= \langle X^2\rangle\langle Y^2\rangle - [\sigma_X^2\sigma_Y^2 - \sigma_Y^2\langle X^2\rangle \\ &\quad - \sigma_X^2\langle Y^2\rangle + \langle X^2\rangle\langle Y^2\rangle] \\ &= -\sigma_X^2\sigma_Y^2 + \sigma_X^2\{\sigma_Y^2 + \langle Y\rangle^2\} \\ &\quad + \sigma_Y^2\{\sigma_X^2 + \langle X\rangle^2\} \end{aligned}$$

Therefore, we obtain the rigorous relation

$$\sigma_{XY}^2 = \sigma_X^2\langle Y\rangle^2 + \sigma_Y^2\langle X\rangle^2 + \sigma_X^2\sigma_Y^2 \tag{10.11a}$$

◄

We may extend the ideas of error propagation to the complex domain since one frequently finds it convenient to analyze problems in terms of complex variables.

Error Propagation with Complex Variables

The importance of the use of complex variables in physics and in computation is too obvious to bear emphasis here. It has become both necessary and convenient to fit experimental data to complex quantities. All too often, the correlations among fitted quantities is lost sight of when further calculations are made involving the fitted parameters. This is true for fitted complex quantities where too simplistic a view will lead one astray.

Let us consider that data have been fitted to yield values for $\mathbf{z}$ and that we are given $\mathbf{w} = f(\mathbf{z})$ where $\mathbf{w} = w_R + iw_I$ and $\mathbf{z} = z_R + iz_I$. We are further given central (fitted) values of $\mathbf{z}$ with associated errors: z_R, Δz_R, z_I, Δz_I, and $\overline{\Delta z_R\, \Delta z_I}$($\sqrt{\sigma^2}_{\text{Rez}} = \Delta z_R$, $\sqrt{\sigma^2}_{\text{Imz}} = \Delta z_I$, $\sigma_{\text{RezImz}} = \overline{\Delta z_R\, \Delta z_I}$). (These quantities are obtained from the error matrix, which is outputted by most fitting calculations).

Since $\mathbf{w} = f(\mathbf{z})$

$$\Delta\mathbf{w} = \frac{\partial f}{\partial \mathbf{z}}\Delta\mathbf{z} = \Delta w_R + i\Delta w_I$$
$$= \left(\text{Re}\frac{\partial f}{\partial \mathbf{z}} + i\ \text{Im}\frac{\partial f}{\partial \mathbf{z}}\right)(\Delta z_R + i\Delta z_I)$$

Equating real and imaginary parts, we get

$$\Delta w_R = \left(\text{Re}\frac{\partial f}{\partial \mathbf{z}}\right)(\Delta z_R) - \left(\text{Im}\frac{\partial f}{\partial \mathbf{z}}\right)(\Delta z_I)$$

and

$$\Delta w_I = \left(\text{Im}\frac{\partial f}{\partial \mathbf{z}}\right)(\Delta z_R) + \left(\text{Re}\frac{\partial f}{\partial \mathbf{z}}\right)(\Delta z_I)$$

We obtain the variances

$$\sigma^2_{\text{Rew}} = \overline{\Delta w_R^2} = \left(\text{Re}\frac{\partial f}{\partial \mathbf{z}}\right)^2 \sigma^2_{\text{Rez}} + \left(\text{Im}\frac{\partial f}{\partial \mathbf{z}}\right)^2 \sigma^2_{\text{Imz}} - 2\left(\text{Re}\frac{\partial f}{\partial \mathbf{z}}\right) \times \left(\text{Im}\frac{\partial f}{\partial \mathbf{z}}\right)\sigma_{\text{RezImz}} \qquad (10.12a)$$

and

$$\sigma^2_{\text{Imw}} = \overline{\Delta w_I^2} = \left(\text{Im}\frac{\partial f}{\partial \mathbf{z}}\right)^2 \sigma^2_{\text{Rez}} + \left(\text{Re}\frac{\partial f}{\partial \mathbf{z}}\right)^2 \sigma^2_{\text{Imz}} + 2\left(\text{Re}\frac{\partial f}{\partial \mathbf{z}}\right) \times \left(\text{Im}\frac{\partial f}{\partial \mathbf{z}}\right)\sigma_{\text{RezImz}} \qquad (10.12b)$$

and the covariance

$$\sigma_{\text{Rew Imw}} = \left(\text{Re}\frac{\partial f}{\partial \mathbf{z}}\right)\left(\text{Im}\frac{\partial f}{\partial \mathbf{z}}\right)[\sigma^2_{\text{Rez}} - \sigma^2_{\text{Imz}}] + \left[\left(\text{Re}\frac{\partial f}{\partial \mathbf{z}}\right)^2 - \left(\text{Im}\frac{\partial f}{\partial \mathbf{z}}\right)^2\right]\sigma_{\text{RezImz}} \qquad (10.12c)$$

These formulas are especially useful when the (complex) derivative can be obtained explicitly, either in closed form or as an ordered pair of numbers in a computer.

► **Example 10.5**

There is a well-known relation in nuclear physics between the (complex) scattering length $\mathbf{A}$ and the (complex) scattering phase shift $\boldsymbol{\delta}$: $k\mathbf{A} = \tan\boldsymbol{\delta}$ and $\mathbf{A} = A_R + iA_I$ and $\boldsymbol{\delta} = \delta_R + i\delta_I$ as usual. Suppose we are given $\bar{A}_R$, ΔA_R, $\bar{A}_I$, ΔA_I, and $\overline{\Delta A_R \Delta A_I} = \sigma_{\text{ReA ImA}}$. We wish to find the errors appropriate to $\boldsymbol{\delta}$.

Let $\mathbf{z} = k\mathbf{A}$ and $\mathbf{w} = \boldsymbol{\delta}$. Then $\mathbf{w} = \tan^{-1}\mathbf{z}$. We note that

$$\mathbf{z} = \tan\mathbf{w} = \frac{e^{i\mathbf{w}} - e^{-i\mathbf{w}}}{i(e^{i\mathbf{w}} + e^{-i\mathbf{w}})}$$

which yields the equation

$$i\mathbf{z}e^{2i\mathbf{w}} + i\mathbf{z} = e^{2i\mathbf{w}} - 1$$

which may be solved for

$$\mathbf{w} = +\frac{i}{2}\ln\left(\frac{1 - i\mathbf{z}}{1 + i\mathbf{z}}\right) = \frac{i}{2}\ln\left(\frac{i + \mathbf{z}}{i - \mathbf{z}}\right) = \tan^{-1}\mathbf{z}$$

We may now take the derivative

$$\frac{\partial\mathbf{w}}{\partial\mathbf{z}} = \frac{\partial}{\partial\mathbf{z}}(\tan^{-1}\mathbf{z}) = \frac{\partial}{\partial\mathbf{z}}\left\{\frac{i}{2}\ln\left(\frac{i + \mathbf{z}}{i - \mathbf{z}}\right)\right\} = \frac{1}{1 + \mathbf{z}^2}$$

Substituting, we get

$$\frac{\partial f}{\partial \mathbf{z}} = \frac{1}{1+(k\mathbf{A})^2} = \left(\mathrm{Re}\frac{\partial f}{\partial \mathbf{z}}\right) + i\left(\mathrm{Im}\frac{\partial f}{\partial \mathbf{z}}\right)$$

where

$$\mathrm{Re}\frac{\partial f}{\partial \mathbf{z}} = \frac{1+k^2(\bar{A}_R{}^2 - \bar{A}_I{}^2)}{[1+k^2(\bar{A}_R{}^2-\bar{A}_I^2)]^2 - 4k^4\bar{A}_R{}^2\bar{A}_I{}^2}$$

and

$$\mathrm{Im}\frac{\partial f}{\partial \mathbf{z}} = \frac{-2k^2\bar{A}_R\bar{A}_I}{[1+k^2(\bar{A}_R{}^2-\bar{A}_I^2)]^2 - 4k^4\bar{A}_R{}^2\bar{A}_I{}^2}$$

These may now be substituted into equations (10.12). ◄

PART II

Introduction to Graphical Techniques and Curve Fitting

Among the functions of graphical techniques is the display of experimental data. The large information content of pictures is aptly described by the old proverb that "a picture is worth a thousand words." Among the graphs that are used for display purposes primarily are the line, bar, pie, volumetric, and histogram forms of graph. See Chapter 11 for examples.

Another function of graphs is to serve as reassurance and a check of numerical analysis. Graphing facilitates the exposure of mistakes, for example, mistakes in calculations in tables of values may be most easily shown by plotting the results. This function of flagging errors is quite important in the use of graphs. Although it would be nice to have quantitative measures of fluctuations that are more than statistical, these criteria have intrinsic dangers, and the sixth sense or "nose" obtained with experience is equally valuable and possibly safer. (See the discussion of Chauvenet's criterion in Chapter 6.)

Graphing frequently serves as a useful technique to make a quick analysis of data, especially if one can utilize a straight-line format. There are dangers in this procedure that we will point out later concerning the fact that the proper statistical weights must be preserved.

"The smoothness of nature" is one of the features that make graphical techniques most useful. Data smoothing, *extrapolation* into regions not measured, and *interpolation* between points that have actually been measured all are procedures most easily done with graphical analysis. It usually turns out that graphical methods are faster although not as accurate as other numerical techniques.

Very often the idea of a graph is useful while the actual plotting of the graph is not necessary. Basically the graphical idea is to plot on some sort of coordinate paper the various values of one quantity corresponding to the values of another. This idea is useful even without the graph itself. The actual plotting of the graph, however, is an aid to thought. By plotting one variable versus another, we may see a correlation between the two variables that was not *a priori* obvious and that would not have been included in the analysis without this aid. *Curve fitting* involves the effort to determine the specific form of the correlation, for example, $y = f(x)$. If the data points lie approximately on the hypothesized curve, we have a "good fit."

Graphs also serve as an aid in computation. This kind of graphical technique frequently goes under the name of "calculating charts" or *nomography*, with the graphs being called *nomograms.* The simplest example of the construction of scales explicitly for calculation is the slide rule. Graphical devices, once carefully drawn, often facilitate solution of complicated problems with speed and slight labor. Nomographic techniques are especially useful when many numerical problems of a similar sort are to be solved and when high accuracy is not required.

Graphical techniques have their limitations. A graph should not be used if it does not make the results of the investigation clearer to the reader than if the result had been presented merely in words or in a table. Also, if a graphical technique is less accurate as a means of analyzing data or making a computation than is necessary or is less rapid than another method, some other mode of computation should be used. One of the difficulties with graphical computation is the existence in most charts of regions in which the result has a larger error than is acceptable.

11

DISPLAY GRAPHS

The first function is simply for the *presentation of data.* This aspect of the utility of graphs is widely used in all fields of human activity, not only the purely scientific. Some examples of graphs serving this function are the following.

Horizontal Bar Chart

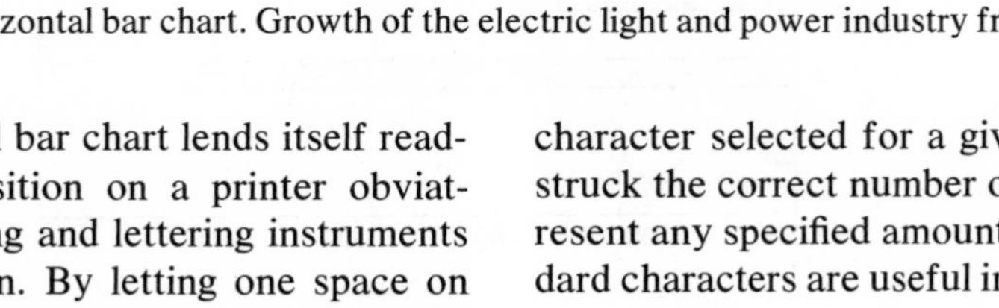

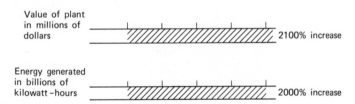

Figure 11.1
Example of horizontal bar chart. Growth of the electric light and power industry from 1902 to 1932.

The horizontal bar chart lends itself readily to composition on a printer obviating the drawing and lettering instruments of a draftsman. By letting one space on the machine represent a unit quantity, the character selected for a given bar can be struck the correct number of times to represent any specified amount. Several standard characters are useful in this context.

Pie Chart or Area Diagram

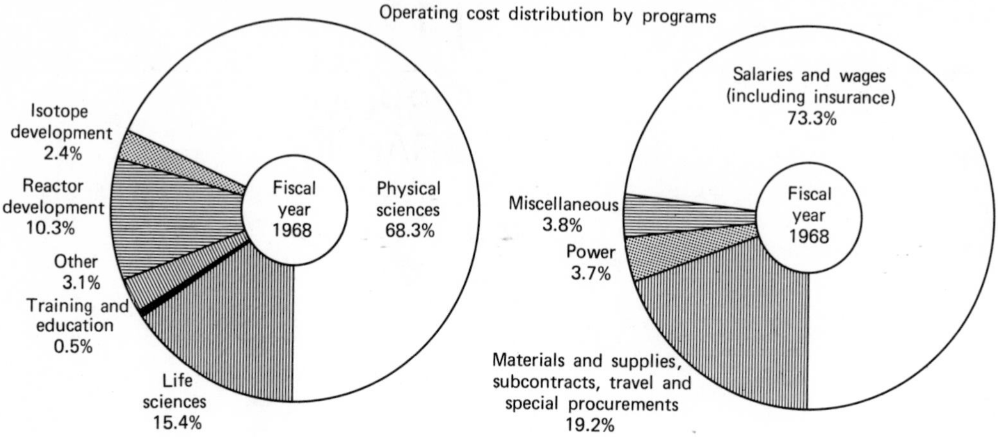

Figure 11.2
Example of pie chart. Area is proportional to percentage.

Volumetric or Solid Diagram

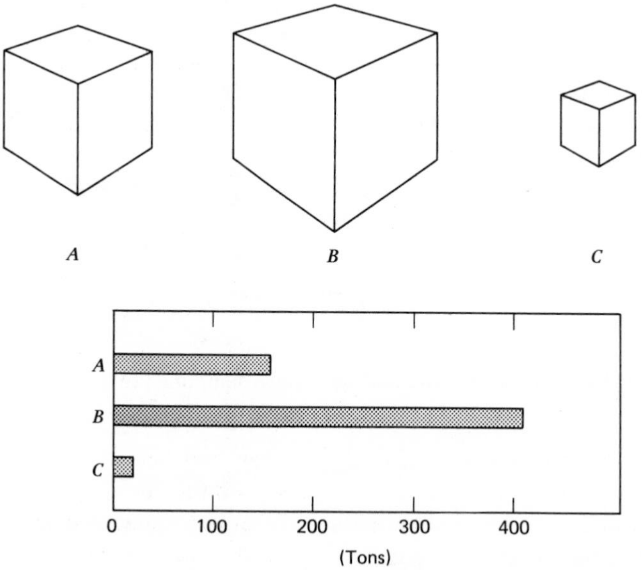

Figure 11.3
The daily output of three different mines shown as a solid diagram (linear dimension proportional to cube root of output) and as a horizontal bar chart (length proportional to output).

The examples in Figures 11.2 and 11.3 convey a relatively small amount of information in a dramatic fashion. The point to be emphasized is made strongly, but not very much analysis can be done on these data. This aspect of graphs is more aesthetic than informative, and these graphs are relatively infrequent in scientific usage.

12

CORRELATION GRAPHS

More information is conveyed when a variate is correlated with time as in the following examples of data presentation.

SILHOUETTE CHART

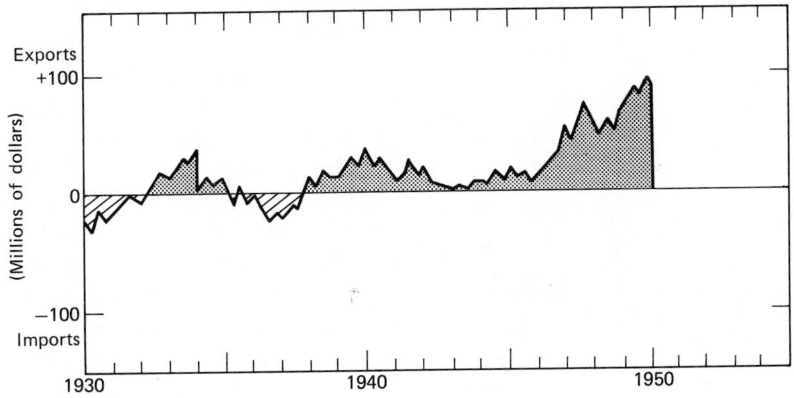

Figure 12.1
Comparison of net exports and imports correlated with time. This may be made more dramatic by indicating negative values in red.

HORIZONTAL BAR CHART

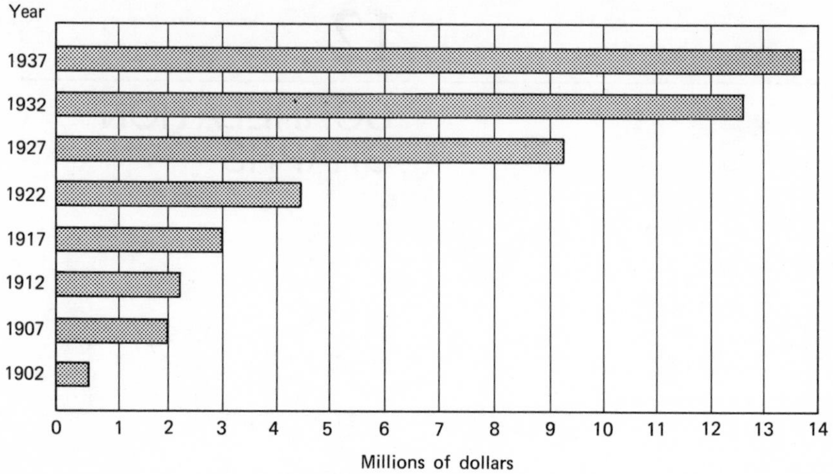

Figure 12.2
Example of horizontal bar chart correlated with time on the vertical axis. We see the growth of the value of plant and equipment in the electric light and power industry from 1902 to 1937.

VERTICAL BAR CHART

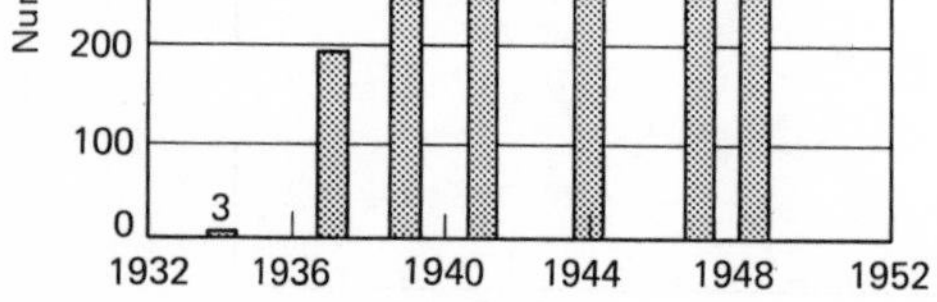

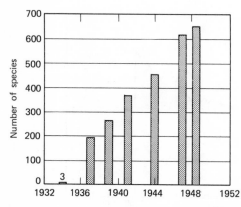

Figure 12.3
Example of a vertical bar chart correlated with time on the horizontal axis. We see the number of radioactive isotopes made in the laboratory as a function of time.

Note that the widths of the bars in Figures 12.2 and 12.3 have no significance. They are taken to be equal and may be any convenient size so long as they do not overlap.

LINE CHART

This graph in Figure 12.4 is really a multiple chart.

1. **Price action.** Each vertical line shows the weekly range of price movement for the particular stock. The horizontal line crossing the vertical indicates the closing price (Friday's) for the week.
2. **Volume.** Each vertical line represents the number of shares traded during the week. Since the volume scale is set for shares in thousands, a vertical line extended up to the 160 line would indicate that 160 thousand shares were traded that week.
3. **19 P-E ratio line.** Shows where Standard Oil would have sold, at any given time, if it traded at 19 times its then-current earnings (price-to-earnings ratio). This arbitrary line has been placed along the chart's approximate high range of market price action.
4. **12 P-E ratio line.** Shows where Standard Oil would have sold, at any given time, if it traded at 12 times its then-current earnings. This line has been

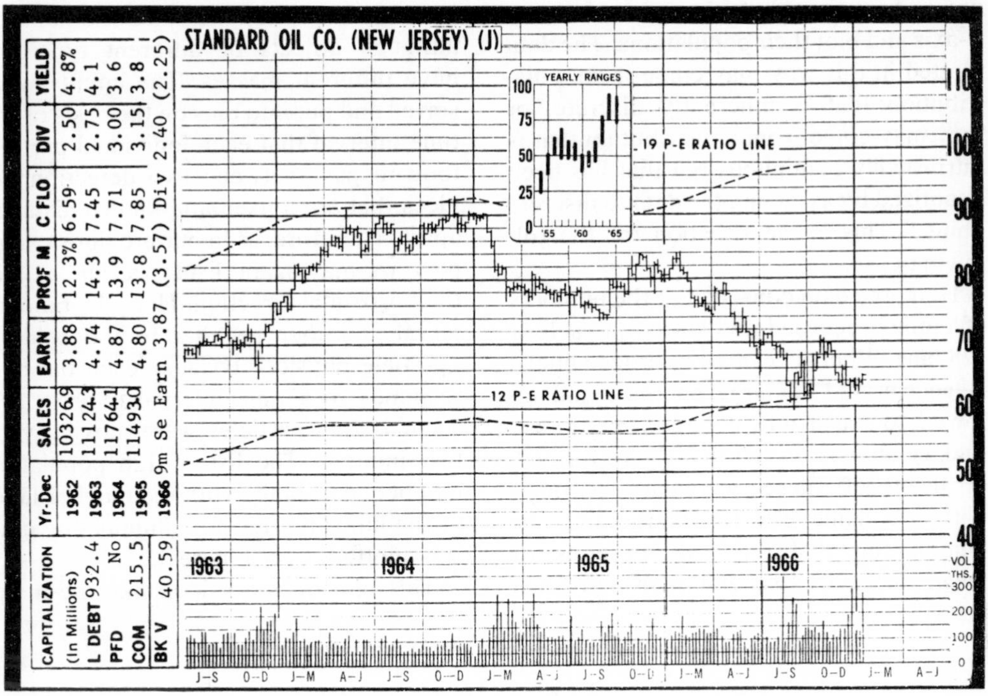

Figure 12.4
A multiple line chart (courtesy of Trendline, a division of Standard & Poor's Corp.) showing various features of the stock market history of a particular security (see text).

placed along the chart's approximate low range of actual price action.

5. **Yearly range chart.** Each vertical line represents the yearly range of price movement adjusted for stock splits and stock dividends.

In Figure 12.4 the information content is high. The price history of the stock is summarized for the period covered. The volume of shares traded is also correlated with the date and the price range of the stock for a given week. The earnings per share of the company are also correlated with the other quantities. The earnings are not explicitly displayed, but two multiples of the earnings are plotted. These latter curves show a new possible use of graphs: to test (intellectual) *speculations.* Here the conjecture involves a possible correlation between the price range of the stock and the two P-E ratio lines.

Search for Correlations

The *search for correlations* between two quantities or among more than two quantities is one of the most important functions of graphical presentations. The last examples are quite typical. Two coordinate axes are used, position along the horizontal axis showing the value of the *abscissa* and distance along the vertical axis showing the value of the *ordinate.* We have so far considered only time as the abscissa.

We have already encountered the specialized plot known as a *histogram* in Figures 7.1 and 7.2. A histogram is a representation of a discrete as opposed to a continuous distribution. We select cells of width Δx, place them *in order* along an x-axis and plot their populations. The data plotted in this way may come from a distribution that is inherently discrete, or they may come from a distribution that is intrinsically continuous but that has been quantized by assigning the data to discrete "bins." This is the situation for the following case.

Here, the data consisted of the numbers of counts recorded and stored in each of 128 bins. When a beam of hydrogen nuclei is accelerated to large energies and allowed to crash into a target, many secondary objects or "particles" are formed. A beam of these secondary particles with the same electric charge may be "momentum-analyzed" by a series of magnetic devices so that each individual has approximately the same momentum, although objects of different type will have different masses. The counts were sorted and put into bins on the basis of the time interval that each individual particle took to pass between two detection devices that are a fixed and known distance apart. Therefore, as in our previous examples, we may again display a time axis in the horizontal direction. Here the time interval is rather peculiar to the field of "particle physics" because each bin corresponds to a time interval of 0.5×10^{-9} seconds (the particles were moving very fast indeed). The list of 128 population numbers correlated with the number label of the bin is not terribly illuminating until it is plotted as in Figure 12.5.

This plot does not look like the histo-

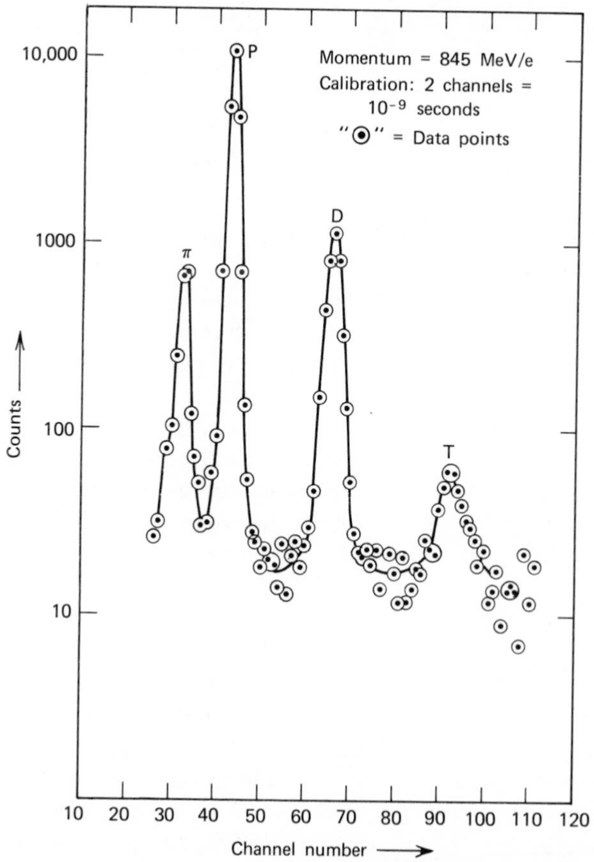

Figure 12.5
Data points indicating number of counts in each channel with smooth curve drawn through the points.

gram of Figure 7.2 because the cell size is so narrow that it is easier to plot the discrete distribution as a smooth, continuous curve. *The plot discloses interesting features of the data*: namely, the existence of four well-defined peaks. This means that there are four transit times that appear to be especially favored. Since the distance is fixed between the two detectors of the experiment, and v = distance/transit time, we can conclude from this graph that four velocities are especially popular. We know that the momentum of a body (even a submicroscopic one, ignoring for the moment relativistic effects), is related to its mass and velocity by the simple formula $p = mv$. Since all particles in the beam had approximately the same momentum, we see that particles of different mass will have different velocities. This histogram of the transit times, then, reveals to us that particles of four different masses were present in the beam. Furthermore, by measuring the positions of the various peaks we may determine the actual transit times and, hence, the values of the four masses. Having identified the four components of the beam (as pi mesons, protons, deuterium nuclei, and tritium nuclei) we may measure their relative abundance by summing the populations of all the cells that we can ascribe to each of the individual peaks.

Note that we have used *data smoothing* in Figure 12.5. The "smoothness of nature" is one of the features that make graphical techniques most useful. We know that experimental errors can cause plotted experimental points to deviate from simple curves. Despite this, even when we do not know the explicit form of the curve, we usually draw a simple, smooth curve through the points. The curve does not have to go through all of the points but should follow them. The region about channels 50 to 60 on Figure 12.5 is shown again in Figure 12.6. Figure 12.6*a* is not correct because the points have been connected with straight lines. The straight line is not unknown in nature but the sharp angle between the straight lines is unphysical. Figure 12.6*b* avoids that difficulty but is much too complicated. Figure 12.6*c* is correct and represents what was actually drawn in Figure 12.5 is shown again in Figure 12.6. Fig- and smooth and *is consistent* with all the points. To be explicit, the experimental

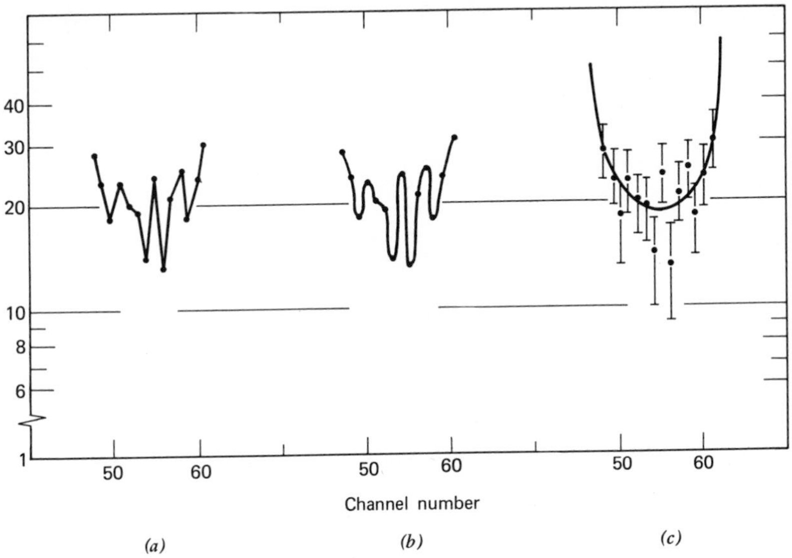

Figure 12.6
Two incorrect curves a and b drawn through the data points and correct smooth curve c (see text).

errors associated with the points have been shown in Figure 12.6*c*.

Figure 12.5 serves several functions. First of all it demonstrates that the experiment could measure time sufficiently well to distinguish particles of different mass. A plot is often used to be *convincing*. Second, the plot *permits further analysis* of the data by enabling the experimenter to determine the masses present and their relative abundances. Third, the plot enables the experimenter to determine features of the "background" in his experiment. This is the part of the plot on which the peaks appear to be sitting. Is it flat, is it small, etc., are all questions one needs to answer to analyze an experiment. The relative abundances can be determined only after subtracting out the background. If the background is not small compared with the peak, then the relative error of the residuum that one wishes to determine may be large. If the background has a pronounced slope with the time axis, the relative abundances after subtraction may have a systematic error in their determination. Another function of this kind of graph, therefore, is that of *checking for systematic errors.*

Fine details appearing on this plot would have escaped the notice of most experimenters if faced only with the table of numbers. The two central peaks appear to be symmetric while the left-hand peak labeled "π" appears to have a "shoulder" on its left-hand side while the peak labeled "T" appears to have an asymmetric tail to higher channel numbers. Both of these observations are real and have significance.

Note the use of *semilogarithmic paper* in Figure12.5. Here, the reason for its use is the *large range of values* that one wished to display. The number of counts in the largest peak was in excess of 10,000 while the number in the smallest channel was of the order of 10 counts. A linear scale is clearly inadequate for this purpose, and so the vertical scale used is proportional to the logarithm of the number of counts. Each *decade* or range of 10 needed in the plot requires another *cycle* of the logarithmic paper. Figure 12.5 uses 5 cycles although 4 cycles would have sufficed had we suppressed the least count (which on a logarithmic scale is not zero but 1). A linear scale was sufficient in the horizontal direction and so only *semi*logarithmic paper was needed. We shall discuss fully logarithmic paper as well as other uses of semilogarithmic paper later. For the moment we need only note that it is also possible to use

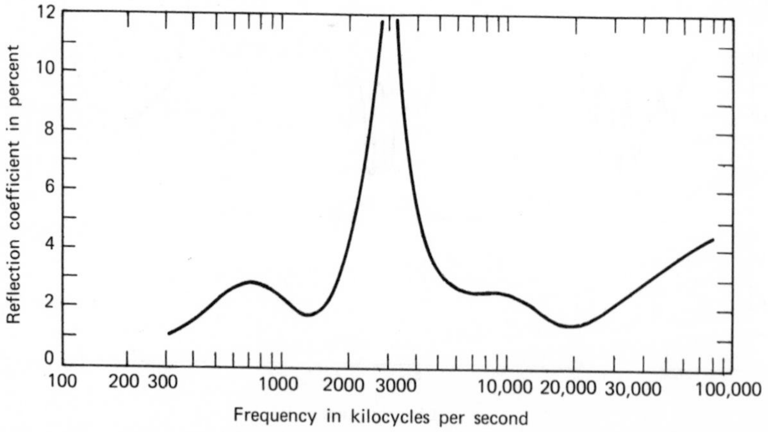

Figure 12.7
Example of logarithmic horizontal scale.

semilogarithmic paper where the linear axis is vertical and the horizontal axis is logarithmic as in Figure 12.7.

Figure 12.7 represents a presentation of data only if the curve summarizes measured values of reflection coefficients for various frequencies. If this chart instead is a summary that is to be utilized in later calculations, then it is a form of *calculating chart* that we shall consider later.

The possibility of suppressing zero should be considered when one has difficulty with the range of values to be plotted. However, to avoid confusion or careless nonobservation on the part of the reader (see Figure 12.8), the zero suppression is sometimes shown by the use of a zigzag line on one of the axes as in Figure 12.6. One may wish to focus attention on two regions that are narrow in range in themselves but that are separated by a large range. Here a break in the scale is occasionally useful as in Figure 12.9, where we call attention to this change in scale.

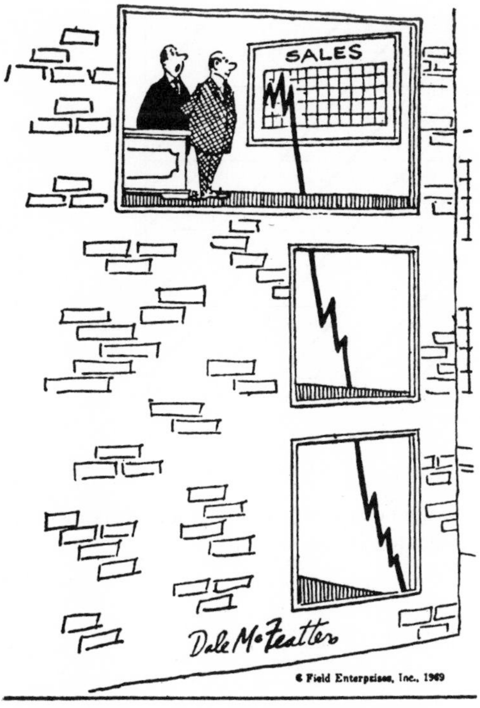

"Unfortunately, that's not all the bad news."

Figure 12.8
Example of zero suppression in graphs.

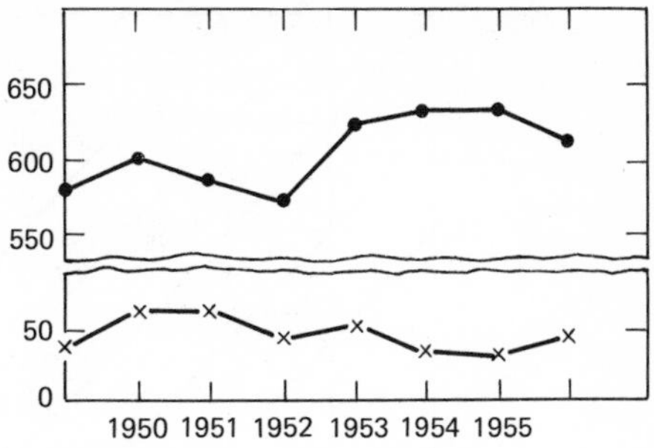

Figure 12.9
Example of change of vertical scale.

SCATTER PLOTS

Another way of plotting experimental data so as to reveal correlations between two quantities is to plot the points one by one as a function of the two quantities as coordinates. Such a *dot* or *scatter diagram* is illustrated in Figure 12.10.

In Figure 12.10, a correlation clearly exists between the quantities N and Z. One can even detect the feature that the relation starts out as an equality given by the straight line $N = Z$. If a scatter plot is made of points characterized by two coordinates X and Y and no correlation between X and Y existed, then the allowed regions of the diagram would be populated uniformly by dots or at least populated so that no simple curve could be drawn through the dots. In Figure 12.11, the melting points of various substances are plotted against their specific gravities. No simple correlation appears to exist. Also, there is no obvious correlation between the specific gravities and the boiling points of various substances as plotted in Figure 12.12. However, there does appear to be a correlation between the boiling points and melting points of various substances as shown in Figure 12.13.

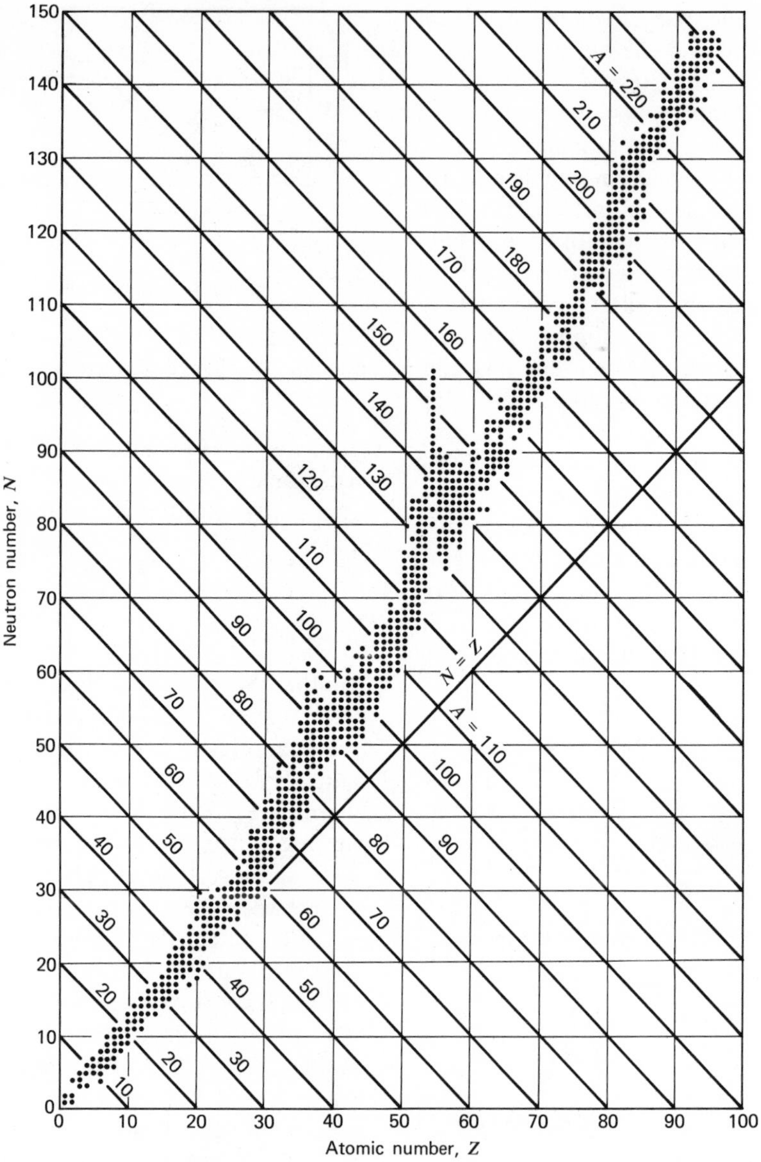

Figure 12.10
A scatter diagram indicating the neutron number, N, of nuclides as a function of atomic number, Z. The sloping lines represent constant values of A. Indicated is the line $N = Z$.

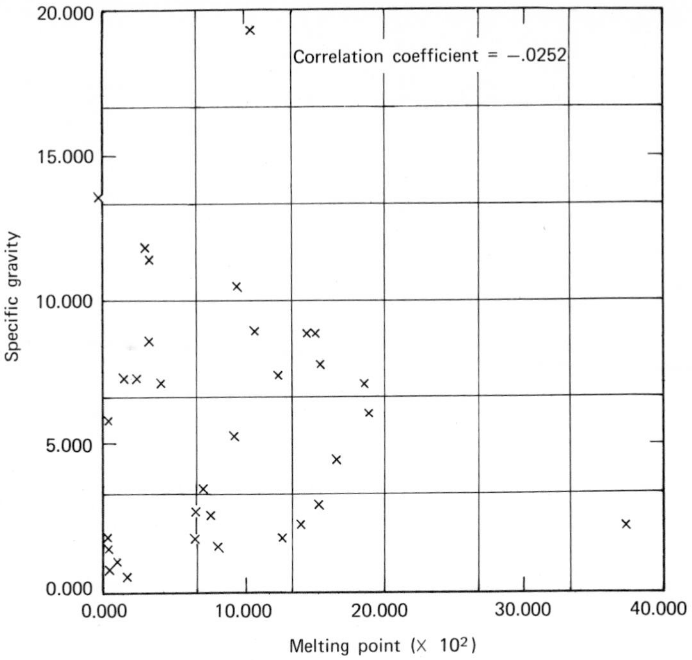

Figure 12.11
A scatter diagram of 33 substances correlating specific gravity with melting point in degrees Celsius. The calculated correlation coefficient [defined in (12.1)] is shown. No strong correlation appears to exist.

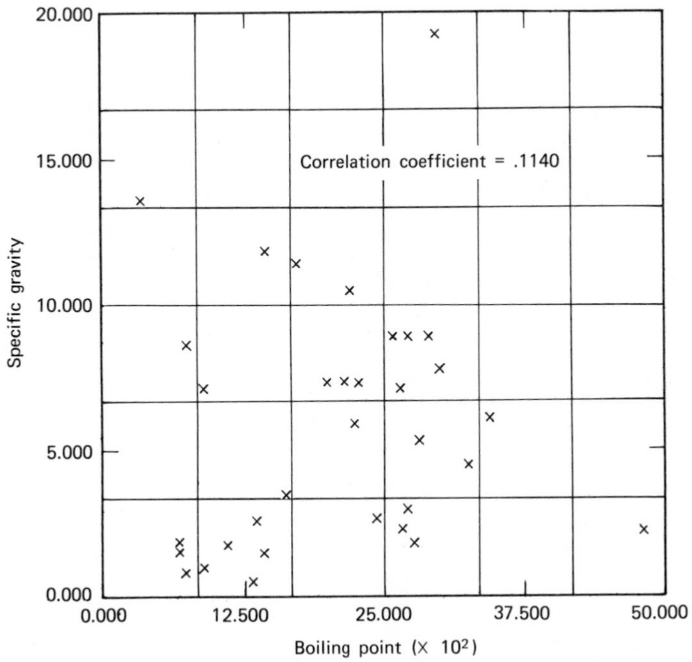

Figure 12.12
A scatter diagram correlating specific gravity with melting point (°C). The calculated correlation coefficient (12.1) is shown. No strong correlation appears to exist.

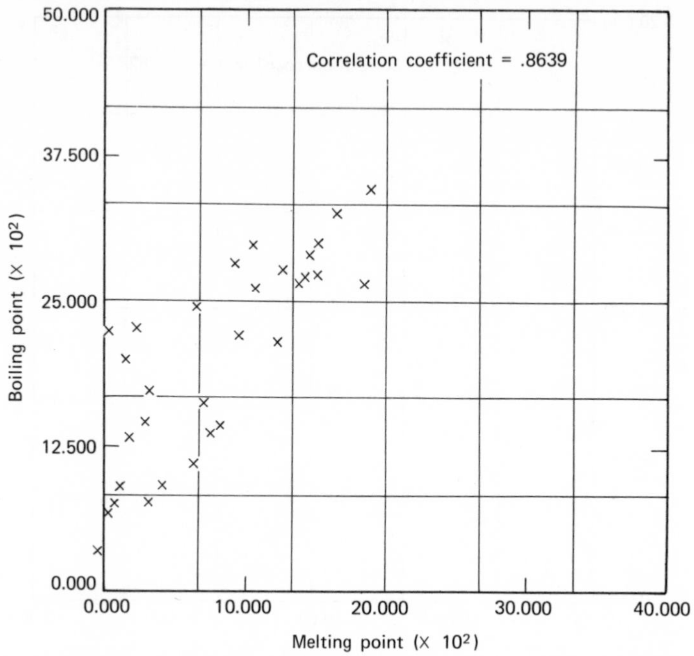

Figure 12.13
A scatter diagram correlating boiling point with melting point (°C). The calculated correlation coefficient (12.1) is shown. There appears to be a correlation between boiling point and melting point.

Correlation Coefficient

For the purpose of assigning a numerical measure to the degree of (linear) correlation between two quantities x and y, we may define the observed (linear) correlation coefficient, r:

$$r(x, y) = \mathrm{Covar}(x, y)/\{\mathrm{Var}(x)\,\mathrm{Var}(y)\}^{1/2} = S_{xy}/(S_x S_y)$$

$$r = \frac{(1/N)\sum_{i=1}^{N}(x_i - \bar{x})(y_i - \bar{y})}{\left[(1/N)\sum(x_i - \bar{x})^2\right]^{1/2}\left[(1/N)\sum(y_i - \bar{y})^2\right]^{1/2}}$$

$$r = \frac{N\sum x_i y_i - \sum x_i \sum y_i}{\left[N\sum x_i^2 - \left(\sum x_i\right)^2\right]^{1/2}\left[N\sum y_i^2 - \left(\sum y_i\right)^2\right]^{1/2}} \qquad (12.1)$$

where N is the number of points (x_i, y_i). Values of the calculated r are shown on Figures 12.11 to 12.13.

If x and y are uncorrelated, we expect r to be close to zero. For perfect correlation (anticorrelation) we expect r to be close to $+1(-1)$.

Even if x and y are uncorrelated, statistical fluctuations might cause the observed r to deviate from zero, especially for small numbers of points N. It is sometimes useful to employ "tables of the correlation coefficient" for various values of N to determine what values of r are *significantly* different from zero.

In this book we shall be more concerned with determining the specific form of a correlation (curve fitting) and shall most often employ the chi-square test of significance to be discussed later.

13

GRAPHS RELATING MORE THAN TWO VARIABLES

Most graphs are necessarily restricted to two dimensions. It often happens, however, that we wish to show graphically the relationship or correlations among more than two variables.

Use of Perspective

Often one is able to represent three dimensions through the use of artistic perspective. An early example is shown in Figure 13.1, which at the time represented a tour de force in the field of scientific illustration. This was actually a representation of a three-dimensional model.

A more recent version of this technique, showing a representation of a model in sections, is shown in Figure 13.2.

The advent of the modern digital computer with analogue output devices has greatly facilitated the use of perspective displays. An example is shown in Figure 13.3, which was made by a standard computer output device.

Perspective drawings do not, however, tell us all that we wish to know. A plot of an important function is shown in Figure 13.4.

Projections

This plot is not very useful for calculations, however, in its "three-dimensional" form. Of greater utility are the *projections* onto the three two-dimensional planes (Figures 13.5*a* and *b*).

The use of projections on two-dimensional planes is a technique that may be extended to an arbitrary number of dimensions. If n variables are involved, it is clearly difficult to indicate information by means of a relief diagram that is restricted to three dimensions One may always, however, graph the projections on planes of the variables taken two at a time.

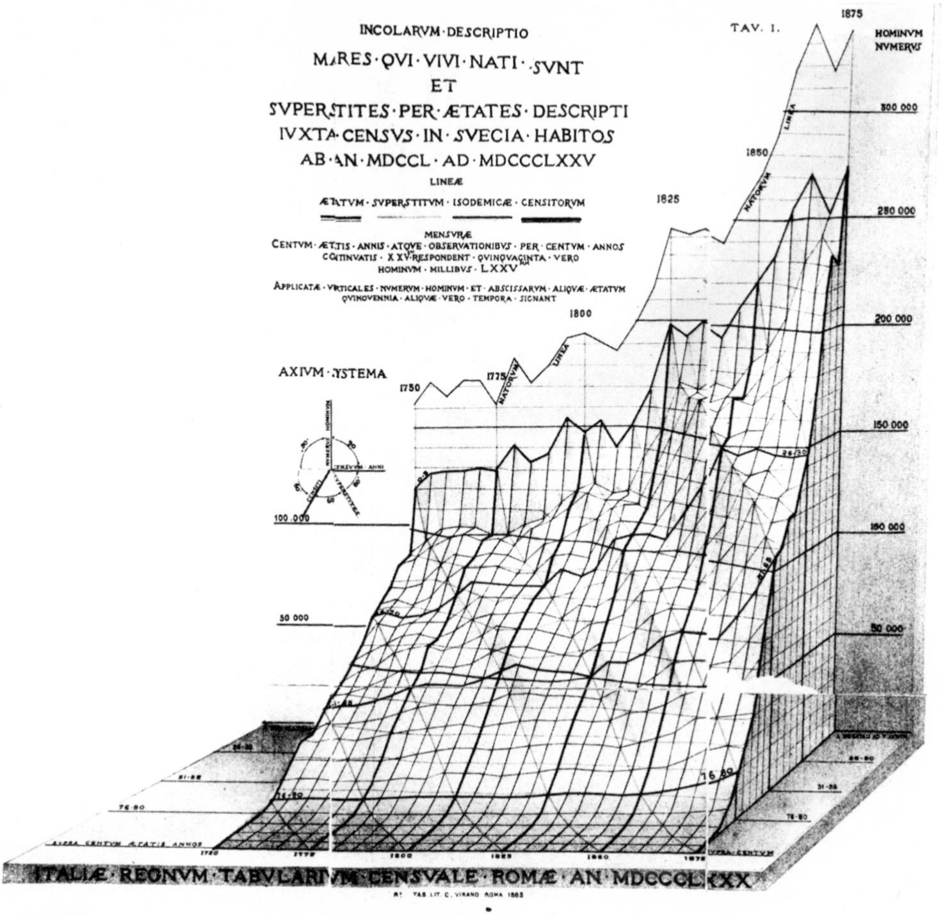

Figure 13.1
Three-dimensional model showing the growth of the population of Sweden from 1750 to 1875. The chart of this model that appeared in the *Journal of the Royal Statistical Society of London* in 1879 was in a brown halftone with black, red, blue and green lines. The three dimensions are the years from 1750 to 1875, the number of persons, and the age of the persons.

Contour Plots

If we have a function of three variables, we may assign a series of values to one of them and then plot on the same sheet the graphs of the resulting curves connecting the other two. These curves are called *contours.* In Figure 13.6 is shown a perspective or relief picture of a complicated mathematical function. We shall not go into a detailed discussion of this function except to note that the altitude is a function of two variables λ_1 and λ_2.

Plotted in Figure 13.7 is a contour chart of the altitudes, that is, the curves in the λ_1-λ_2 plane corresponding to various (constant) values of the altitude.

A familiar example is the case of the perfect gas law $PV = nRT$. In Figure 13.8 is plotted a series of contours i for fixed $T = T_i$.

The general case of plotting contours for the function $F(x, y, z) = 0$ is to hold z fixed at various z_i and graph $F(x, y, z_i)$ for

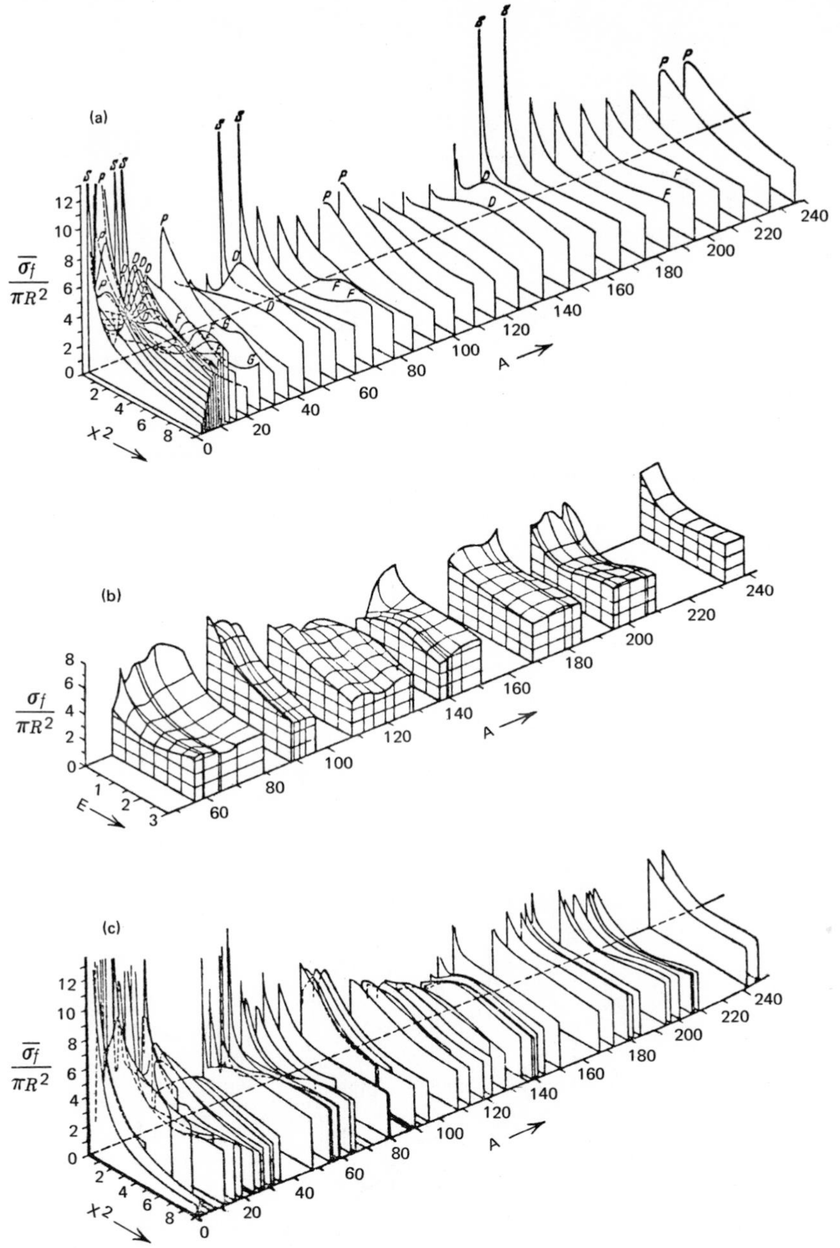

Figure 13.2
(a) Total neutron cross sections plotted versus $x^2 = (R/\gamma)^2$ and A for different choices of a parameter (b) Measured total neutron cross-sections versus E and A. (c) A similar plot using $x^2 = (R/\gamma)^2$ in place of E. [Feshbach, Porter, and Weiskopf, *Phys. Rev.*, **96**, 448 (1954)].

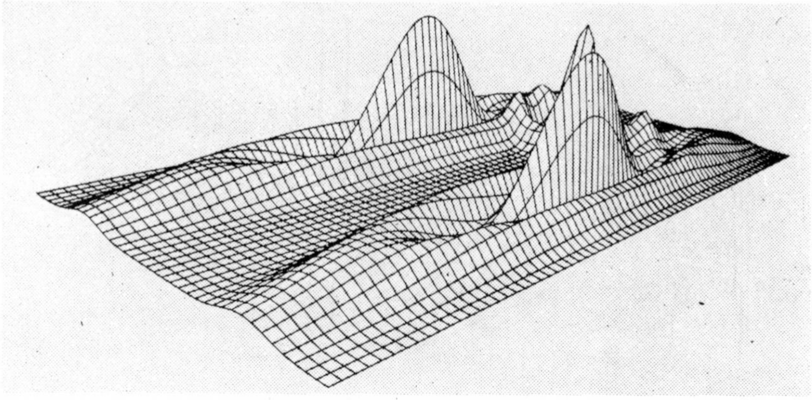

Figure 13.3
A perspective display obtained from a computer plotter.

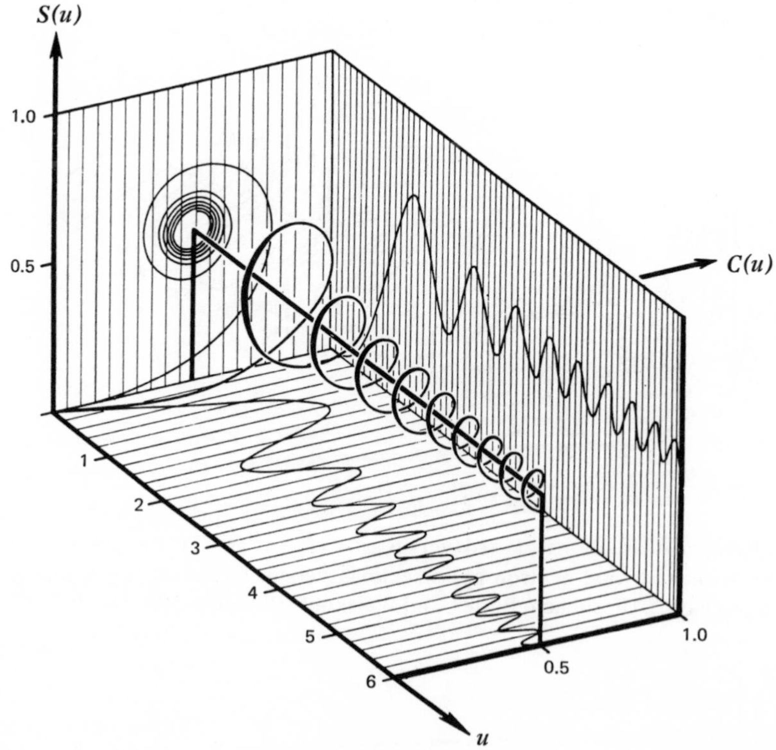

Figure 13.4
A three-dimensional drawing of a curve in space.

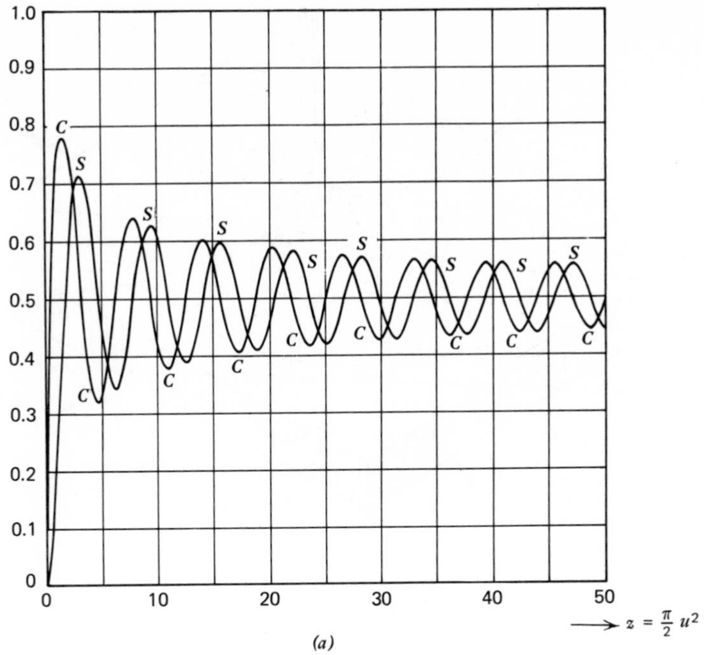

(a)

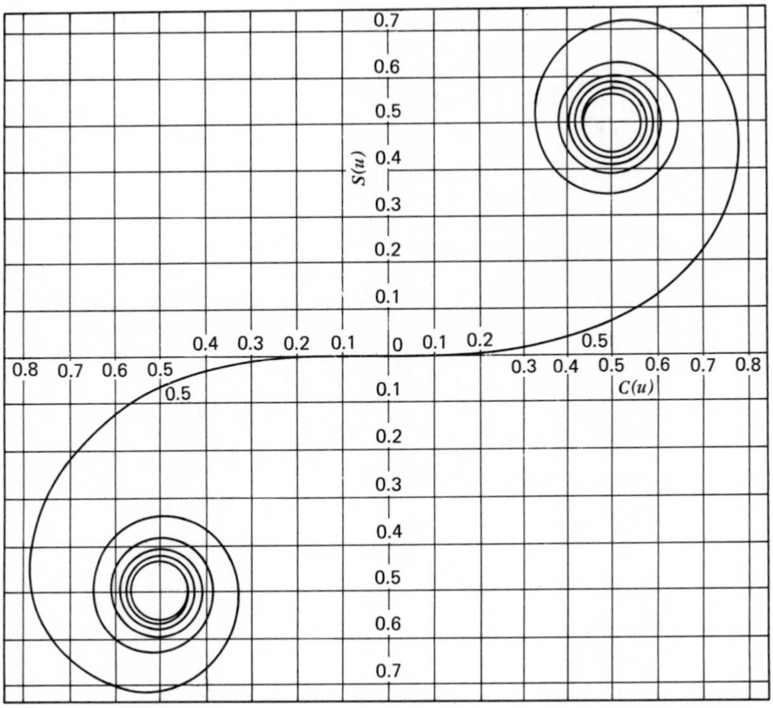

(b)

Figure 13.5
Projections on three planes of the curve in Figure 13.4. (*a*) Fresnel's integrals. (*b*) Cornu's spiral.

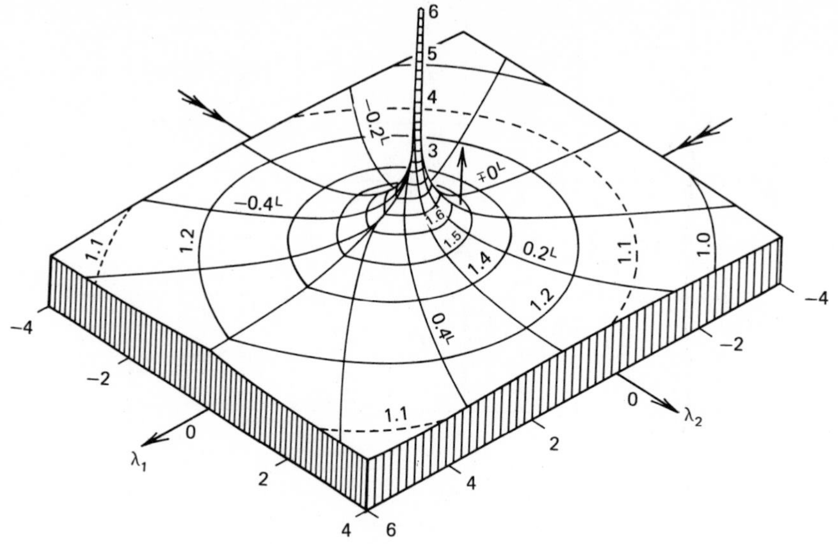

Figure 13.6
A three-dimensional view of a function showing contours.

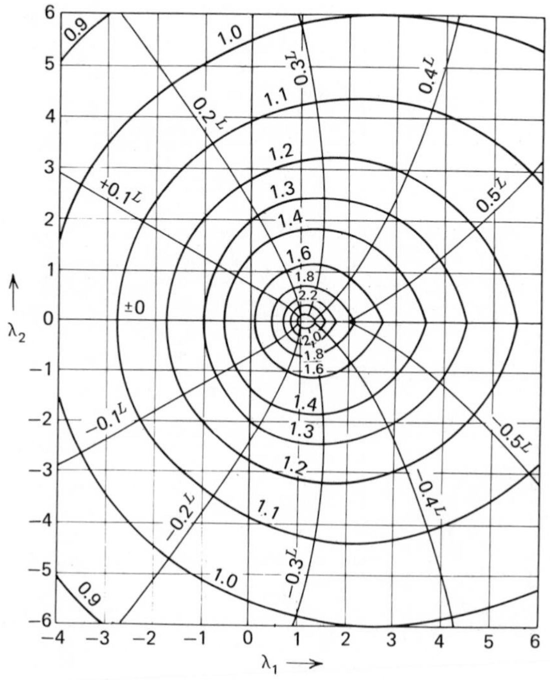

Figure 13.7
Planar plot of the contours of Figure 13.6.

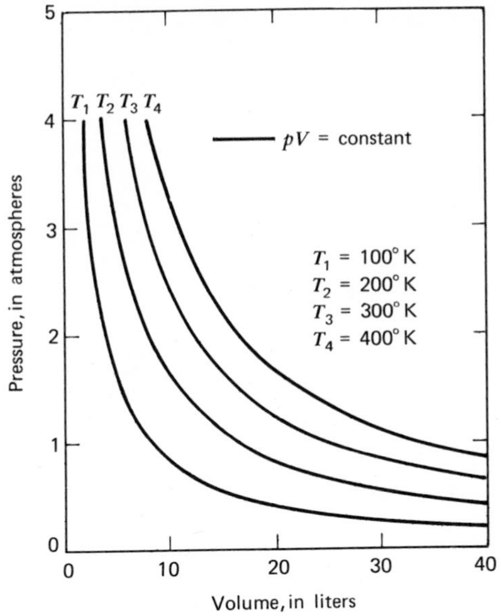

Figure 13.8
Contour curves for the perfect gas law for T = constant. The curves show how the pressure of one mole of an ideal gas changes as its volume is changed, the temperature being held constant (isothermal process).

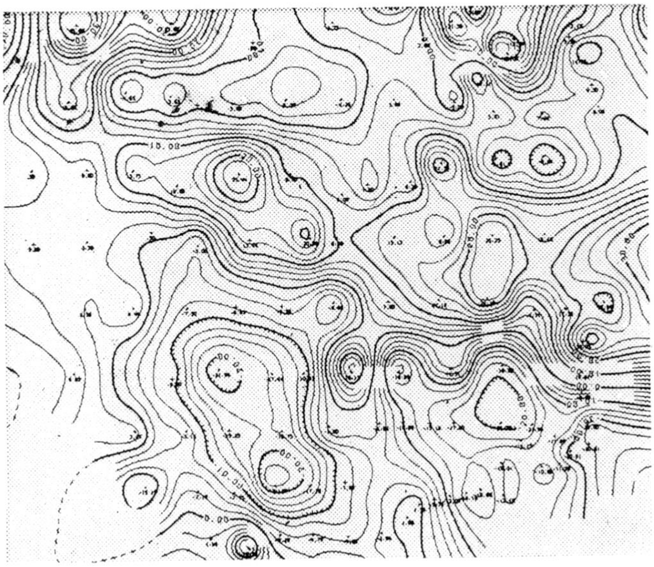

Figure 13.9
A contour map.

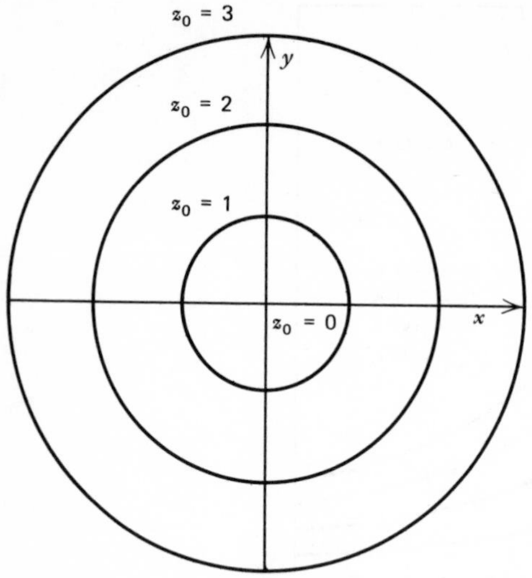

Figure 13.10
A contour plot of the equation $x^2 + y^2 = z^2$.

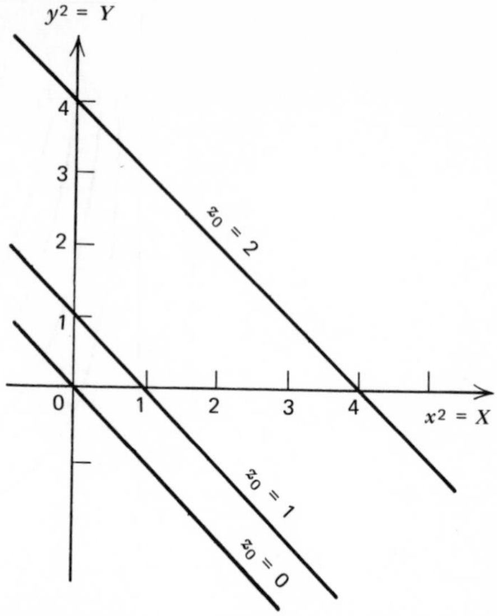

Figure 13.11
A linearized contour plot of the equation $x^2 + y^2 = z^2$.

the different i. Many geographical maps are of this type, showing contours of equal altitude and equal density of rock, for example. The daily newspaper usually contains similar weather maps that show contours of equal barometric pressure (isobars), contours of equal temperature (isotherms), contours of equal precipitation, etc. (Figure 13.9).

Suppose we have the equation

$$P(z) \cdot F(x) + Q(z) \cdot G(y) + R(z) = 0$$

This yields a straight line for each value of $z_i = z_0$ if we use the variables $X = F(x)$, $Y = G(y)$. Since the equation is now linear in the variables, we have the equation $P(z_0)X + Q(z_0)Y + R(z_0) = 0$. For example, the equation

$$x^2 + y^2 = z^2$$

may be represented as a contour plot in z as shown in Figure 13.10. This equation may also be made linear by plotting it, using coordinates such that $X = x^2$ and $Y = y^2$ (Figure 13.11).

14

STRAIGHT-LINE GRAPHS AND FITTING

We have seen that it is usually possible to draw a smooth curve (of unknown form) through experimental points plotted so that a correlation exists between the abscissa and ordinate. It is usually not possible to tell from visual examination of a given curve what is the equation describing it, *unless the curve is a straight line.* We discuss elsewhere analytical techniques of "curve-fitting" by which any functional form may be fitted to experimental data. The straight line, however, is unique since it alone can be identified visually. The straight line is therefore the basis of visual graphical analysis.

Straight Line

For a general description we shall take the rectangular or Cartesian coordinate system of Figure 14.1. The equation relating the coordinates of the points on a line is

$$y = mx + b$$

where $m = \tan\theta$ is the slope of the line and b is the intercept on the y-axis. If x and y have the same dimensions (i.e., are expressed in the same units), then m will be dimensionless. Otherwise, m has the dimensions of y/x.

The slope $m = (y_2 - y_1)/(x_2 - x_1)$ where (x_1, y_1) and (x_2, y_2) are *any* two points on the line. When a plot of empirical data is expected to lie on a straight line, a ruler should be used to draw the "curve" through the points irrespective of how much the points may scatter.

An obvious problem is how to draw a straight line through a set of data points that may not fall exactly on a straight line. One must settle for the "best" straight line that one can get. The considerations of Figure 12.6*c* hold true as much for a straight line as for the drawing of any simple curve. One tries to make as many of the points as possible consistent with the straight line within the errors assignable to the individual points.

If all the points have approximately the same weight (same error), then a simple rule of thumb is to try to arrange the line so that as many points fall above the line as below. In Figure 14.2 we have plotted the speed of a falling object (as measured in a student laboratory) against time. The points scatter somewhat, but a straight

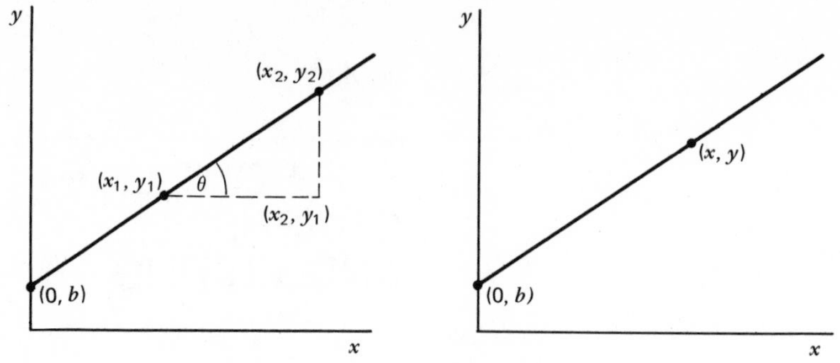

Figure 14.1
A straight line passing through points (x_1, y_1) and (x_2, y_2) and intercepting the y-axis at the point $(0, b)$.

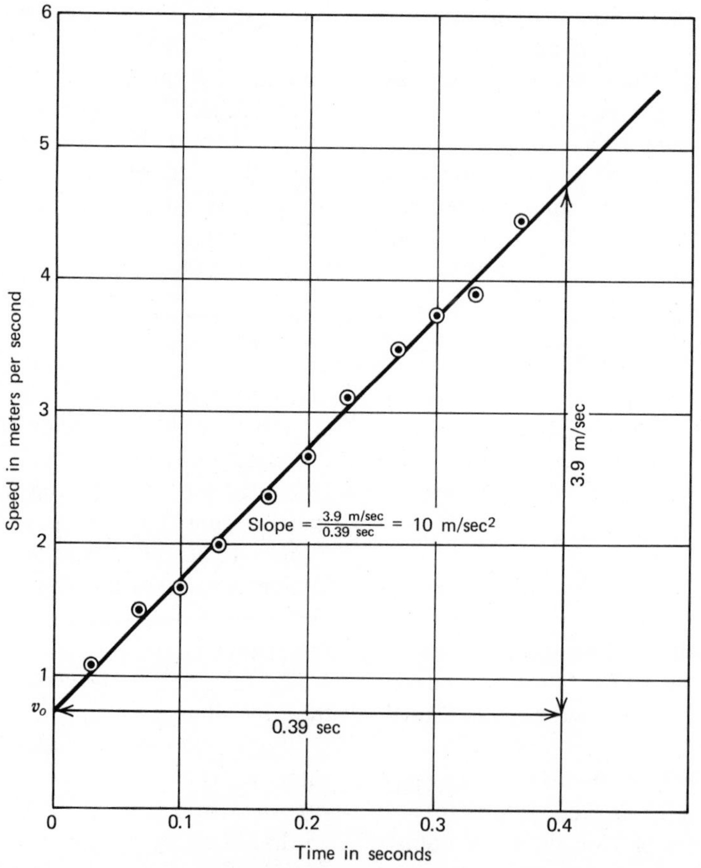

Figure 14.2
Plot of data taken from a student laboratory that measured the speed of a freely falling body as a function of time.

line is certainly suggested by the data. The straight line drawn is symmetrically disposed with respect to the individual points with approximately as many points above the line as below it. We know that a falling object should have its velocity obey the relation $v = v_0 + at$. The graph enables us to read off the slope very quickly, yielding $a = 10\ m/\text{sec}^2$.

It is useful to use a transparent ruler so that points may be seen on both sides of the line. It is important that undue weight not be given to the points at the extrema. *All* the points should be considered when drawing the straight line through them.

More detailed considerations are required when the points have different errors. This matter is gone into in considerable detail in Chapters 26, 32, and 33, in which calculating the best straight *line* is a special case of calculating the best *curve* of a specified type through a set of experimental points. The general rule is to consider the straight line to represent the best value of y for each value of x. If each point (x_i, y_i) is considered to have all of its experimental error associated with the ordinate, we have a set of data points $y_i \pm \Delta y_i$ for the various x_i together with the values $\bar{y}_i = mx_i + b$ from the straight line at x_i. If one forms the sum

$$\sum_i \frac{(\bar{y}_i - y_i)^2}{(\Delta y_i)^2} = \chi^2$$

the straight line that causes this sum to be minimum is the "least-squares" best-fit straight line. We are here concerned not so much with the analytical determination of the best straight line, but with hints for use in a best visual guess of the best straight line.

We may anticipate one helpful result from a later analysis: the "least-squares" straight line passes through the weighted centroid of the points, $(\bar{x}, \bar{y})$ where

$$\bar{x} = \left[\sum_i \frac{x_i}{(\Delta y_i)^2}\right]\left[\sum_i \left(\frac{1}{\Delta y_i}\right)^2\right]^{-1} \tag{32-30}$$

$$\bar{y} = \left[\sum_i \frac{y_i}{(\Delta y_i)^2}\right]\left[\sum_i \left(\frac{1}{\Delta y_i}\right)^2\right]^{-1} \tag{32-31}$$

Figure 14.3 shows data points of unequal statistical weight. These points have *error flags* whose lengths indicate the uncertainties Δy_i. These usually correspond to one standard deviation. Drawn on the plot is the best-fit straight line. Can you find a straight line that yields a smaller least-squares sum?

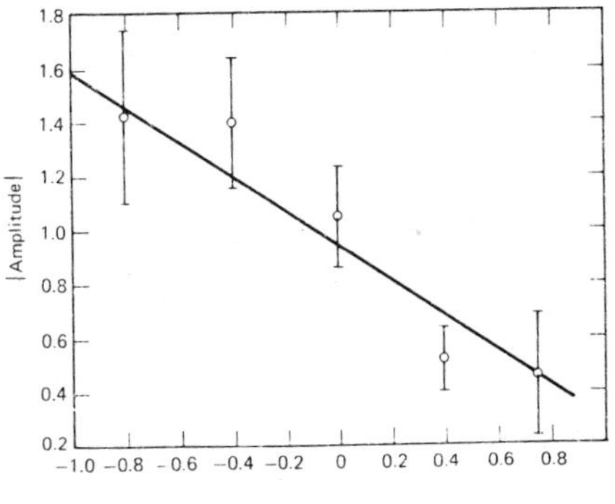

Figure 14.3
A plot of data points of unequal weight. The best-fit straight line is displayed.

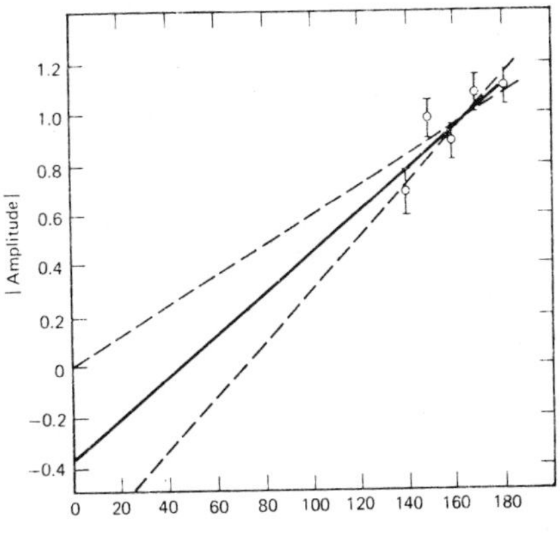

Figure 14.4
A plot of data points taken in a region $E > 140$. The fitted straight line is *extrapolated* to the point $E = 0$. The dotted lines indicate the uncertainties in the determination of the intercept.

It is usually the case that one can draw more than one straight line through the points even if a certain line is believed to be "best." The possible straight lines that are good (if not best) express the uncertainty in the determination of the slope and the intercept. This is shown in Figure 14.4. Another function of graphs is shown in this plot, that of *extrapolation.* It was only possible to make measurements for $E > 135$, yet the vertical axis intercept was needed. The best straight line was merely extended or extrapolated into the region of interest. It is obvious, however, that a large error results. A straight line is clearly better determined by points that are separated as much as possible. In Figure 14.4 the straight line is determined by points clustered at one end of the graph, and the "lever arm" to the vertical axis multiplies the uncertainty in slope to yield a large error in the y-intercept.

Calculation of the Least-Squares Straight Line

If we assume that all the uncertainty in a measurement (x_i, y_i) is in the ordinate y_i with the abscissa x_i being "known," we may *calculate* the best-fit straight line if we have more than two points.

The assumption is that

$$\bar{y}_i = mx_i + b$$

is the best value of each ordinate with each measurement actually yielding y_i. It is required to find the "best" values, m^* and b^*, of m and b for the best-fit straight line.

We minimize the sum (called χ^2), for N points (x_i, y_i),

$$\chi^2 = \sum_{i=1}^{N} \frac{(\bar{y}_i - y_i)^2}{(\Delta y_i)^2} = \sum_{i=1}^{N} \frac{(mx_i + b - y_i)^2}{(\Delta y_i)^2}$$

with respect to m and to b, that is, we get the equations

$$\left.\frac{\partial \chi^2}{\partial m}\right|_{m^*,b^*} = 0$$

and

$$\left.\frac{\partial \chi^2}{\partial b}\right|_{m^*,b^*} = 0$$

When we work out the algebra (which is done more explicitly in Chapter 32) we solve the two equations for the two unknowns m^* and b^* with the results:

$$m^* = \frac{EB - CA}{DB - A^2} \tag{14.1a}$$

$$b^* = \frac{DC - EA}{DB - A^2} \tag{14.1b}$$

where

$$A = \sum_{i=1}^{N} \frac{x_i}{(\Delta y_i)^2}; \qquad B = \sum_{i=1}^{N} \frac{1}{(\Delta y_i)^2}$$

$$C = \sum_{i=1}^{N} \frac{y_i}{(\Delta y_i)^2} \qquad D = \sum_{i=1}^{N} \frac{x_i^2}{(\Delta y_i)^2} \tag{14.2}$$

$$E = \sum_{i=1}^{N} \frac{x_i y_i}{(\Delta y_i)^2}$$

When all the points have the same error

$$\Delta y_i = \text{constant}$$

then relations (14.1) reduce to the form frequently seen:

$$m^* = \frac{N \sum_{i=1}^{N} x_i y_i - \left(\sum_{i=1}^{N} x_i\right)\left(\sum_{i=1}^{N} y_i\right)}{N \sum_{i=1}^{N} x_i^2 - \left(\sum_{i=1}^{N} x_i\right)^2} \tag{14.3a}$$

and

$$b^* = \frac{\left(\sum_{i=1}^{N} x_i^2\right)\left(\sum_{i=1}^{N} y_i\right) - \left(\sum_{i=1}^{N} x_i y_i\right)\left(\sum_{i=1}^{N} x_i\right)}{N \sum_{i=1}^{N} x_i^2 - \left(\sum_{i=1}^{N} x_i\right)^2} \tag{14.3b}$$

for points of equal weight.

We may anticipate another result from Chapter 32 and write the relations for the variances and covariances of the best-fit quantities m^* and b^*:

$$\text{Var}(m^*) = \langle(\Delta m^*)^2\rangle = \frac{B}{DB - A^2} \tag{14.4a}$$

$$\mathrm{Var}(b^*) = \langle(\Delta b^*)^2\rangle = \frac{D}{BD - A^2} \qquad (14.4b)$$

and

$$\begin{aligned}\mathrm{Covar}(m^*, b^*) &= \langle(\Delta b^*)(\Delta m^*)\rangle \\ &= \langle(\Delta m^*)(\Delta b^*)\rangle \\ &= \frac{-A}{BD - A^2} \qquad (14.4c)\end{aligned}$$

where we again use the quantities defined in (14.2).

The quality of the fit of the straight line

$$y_i = m^* x_i + b^*$$

to the data (x_i, y_i) may be seen visually. A numerical measure of the *quality* of the fit may usually be obtained from the number

$$\chi^2(m^*, b^*) = \sum_{i=1}^{N} \frac{(y_i - m^* x_i - b^*)^2}{(\Delta y_i)^2}$$

where we refer to a table of χ^2 probabilities for the case of $N - 2$ "degrees of freedom." Usually, the smaller the χ^2, the better the fit, with the tables giving the "probability" that a *true* curve would nevertheless produce a value $\chi^2(m^*, b^*)$ as large as or larger than that observed.

These matters are discussed at greater length and with examples in Chapters 26, 32, and 33.

Fitting Straight Line Data when Both Variables Have Uncertainties

We should mention the case where there is uncertainty in **both** x **and** y. That is, we are given $(x_i, y_i; \Delta x_i, \Delta y_i)$. Instead of weighting the points by $(y_i)^{-2}$, we define the **effective variance**, σ_i^2:

$$\sigma_i^2 = (\Delta y_i)^2 + m^2(\Delta x_i)^2$$

We consider the sum

$$\chi^2 = \sum_{i=1}^{N} \frac{(mx_i + b - y_i)^2}{\sigma_i^2}$$

and minimize with respect to m and to b.

15

REDUCTION TO STRAIGHT-LINE GRAPHS

Considering the obvious utility of straight-line plots, it is not surprising that much of the use of graphical techniques involves the reduction of data to a form suitable for display as a straight line. Although it is most common to use squared coordinate paper, the horizontal and vertical scales need not be directly proportional to the variables of interest, say, x and y. More or less arbitrary scales may be used. We shall call the generic horizontal variable X and the corresponding vertical variable Y.

Log Plots

The coordinates $X = \log x$ and $Y = \log y$ are convenient for plotting curves of the form $y^r = ax^n$. Taking the logarithms of both sides, we obtain

$$r \log y = n \log x + \log a$$

The graph on paper using the coordinate axes X, Y is the straight line $rY = nX + \log a$. Casting this equation into our usual form $Y = mX + b$, we see that the slope of this straight line is n/r and the intercept on the vertical axis is $(\log a)/r$.

It is more useful to consider a curve of the form $y = Ax^N$. Here the resulting straight line $Y = NX + \log A$ permits a direct measurement of A, N.

Take two points on the *straight line* (which are not necessarily data points): (X_1, Y_1), (X_2, Y_2). The resulting slope is

$$N = \frac{Y_1 - Y_2}{X_1 - X_2} = \frac{\log y_1 - \log y_2}{\log x_1 - \log x_2}$$

and the intercept on the vertical axis is

$$Y_I = \log A = \log y_I$$

Thus

$$A = y_I$$

Here it does not matter to what base the logarithms are taken. There are only two bases that are important for our purposes: logarithms to the base 10 we shall denote "log x"; we shall denote *Napierian* or *natural* logarithms to the base $e = 2.71828183 \cdots$ by "ln x."

To go from natural logarithms (base e) to common logarithms (base 10) we need only make use of the relations

$$10^{\log x} = x \qquad \text{and} \qquad e^{\ln x} = x$$

Taking natural logarithms and common logarithms respectively, we get

$$\log x \ln 10 = \ln x$$

and

$$\ln x \log e = \log x$$

Therefore,

$$\log_{10} x = \frac{\ln x}{\ln 10} = \ln x \log_{10} e$$

$$\ln 10 = 2.30258509 \cdots$$

$$\log_{10} e = 0.434294482 \cdots$$

It is important to note that the logarithms to different bases are *proportional.* Hence, *a straight line on a common logarithm plot is still a straight line on a natural logarithm plot, and conversely.*

► Example 15.1

The following data were taken in a student laboratory from measurements of stroboscopic photographs of a falling body. The positions were measured at equal time intervals of 1/30 sec.

The points ($\log t_i = X$, $\log x_i = Y_i$) are plotted on a linear coordinate paper in Figure 15.1. We have taken t in units of 1/30 sec.

The data points clearly suggest a straight line. The straight line in Figure 15.1 has a slope that is consistent with 2.0. Note that the equation relevant here is:

$$Y = NX + N \log\left(\frac{1}{30}\right) + \log A$$

Table **15.1** Data Taken in a Student Laboratory from Measurements of Stroboscopic Photographs of a Freely Falling Body

t	$\log t$	x Distance in centimeters	$\log x$
3	0.477	4.40	0.643
4	0.602	7.87	0.896
5	0.699	12.42	1.09
6	0.778	17.96	1.25
7	0.845	24.60	1.39
8	0.903	32.18	1.51
9	0.954	40.76	1.61

The last two terms are constant; thus the slope is the same if one factors out the 1/30 or if one leaves it in.

To make the plot of Figure 15.1 it was necessary to take logarithms for plotting on linear coordinate paper. The slope of the line in Figure 15.1 is $N = 1.98$, which is consistent with 2. This shows that the distance varies as the square of the time. The intercept $-0.29 = \log(0.51)$ or $X = 0.51t^2$ where t is measured in units of 1/30 sec.

We expect $x = \frac{1}{2}gt^2$. Therefore $\frac{1}{2}g = 0.51$ or $g = 1.02$, which represents the

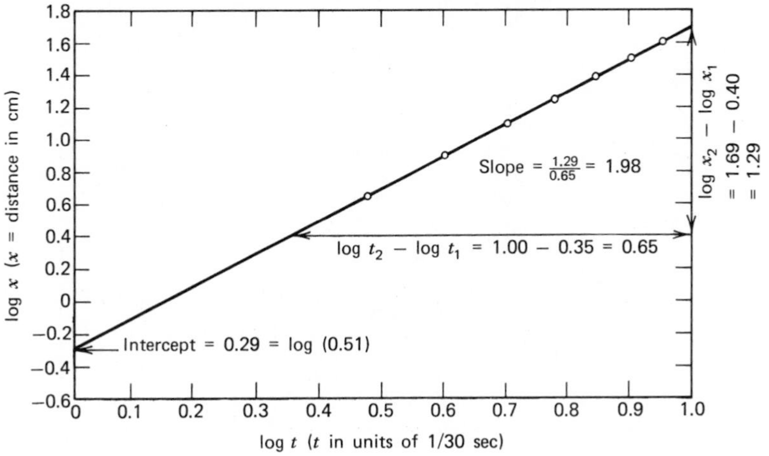

Figure 15.1
The data points are plotted on linear coordinate paper with log t as abscissa and log x as ordinate.

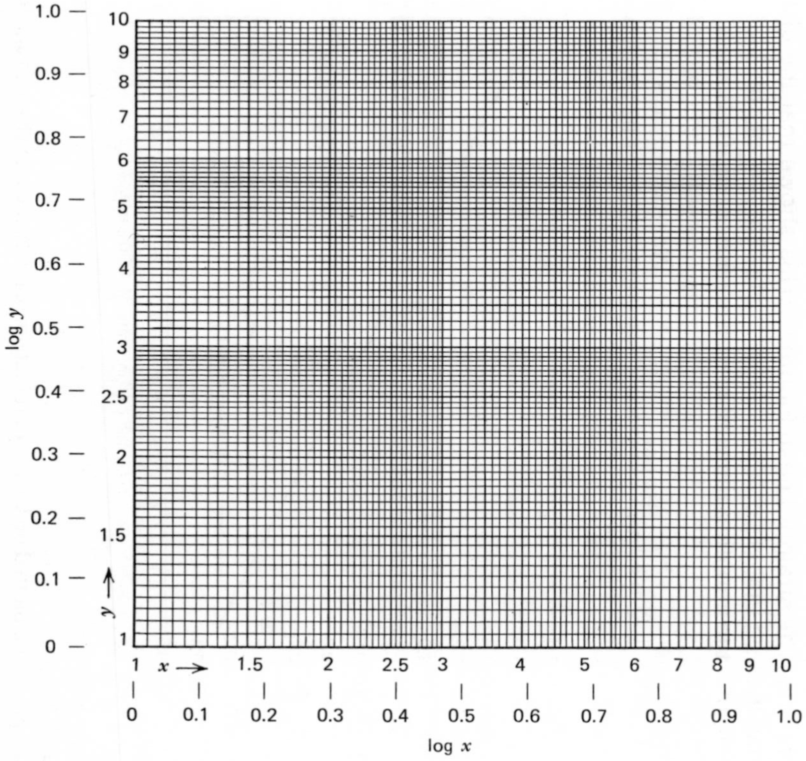

Figure 15.2
Logarithmic ("log-log") graph paper.

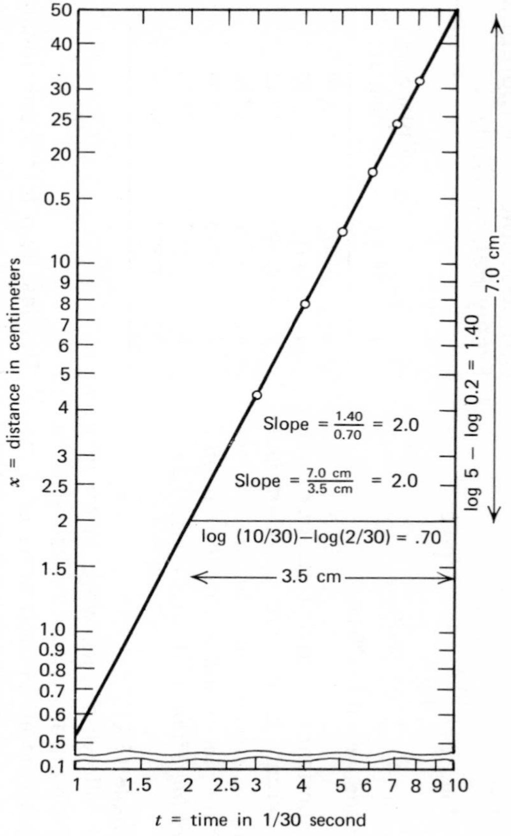

Figure 15.3
The data from Table 15.1 are plotted on log-log graph paper.

acceleration in cm per $(1/30 \text{ sec})^2$.

$$g = \frac{1.02}{(1/30)^2}\frac{\text{cm}}{\text{sec}^2} = 920\frac{\text{cm}}{\text{sec}^2}$$

The slope is well determined. The intercept has a large uncertainty because of the "lever arm" effect: the points determining the straight line are far from the vertical axis. ◄

Another way of plotting logarithms of the data points is to use *logarithmic graph paper.* Here the points are plotted as in Figure 15.1 on scales proportional to log x and log y, but axes show the values x and y. This is shown explicitly in Figure 15.2 where both scales are shown.

The advantage of logarithmic paper is that one need not calculate the logarithms of the data points. The data from Table 15.1 are plotted on log-log paper in Figure 15.3. ("Log-log" refers to the fact that both vertical and horizontal scales are logarithmic.)

Again, as in Figure 15.1 the plot is a straight line. To calculate the slope of the best straight line requires a knowledge of the logarithms of the points *on the straight line used.* In Figure 15.3, the slope is determined from the points (t = 10/30 sec, x = 50 cm) and (t = 2/30 sec, x = 2 cm).

$$\text{The slope} = \frac{\log 50 - \log 2}{\log(10/30) - \log(2/30)}$$
$$= \frac{1.69(9) - 0.30(1)}{-0.47(7) - [-1.17(6)]}$$
$$= \frac{1.39(8)}{0.69(9)} = \frac{1.40}{0.70} = 2.0$$

This, of course, verifies that $x \,\alpha\, t^2$. It is also possible to measure the actual distances on the graph paper and determine the slopes with a ruler. This is also done in Figure 15.3 yielding a slope = 7.0 cm/3.5 cm = 2.0.

The intercept on the vertical axis yields directly the value 0.52 without the necessity for taking antilogarithms as in Figure 15.1.

Of course, one may verify the relation $x \,\alpha\, t^2$ by plotting x versus t^2 and observing the result. This general procedure is discussed at the end of Chapter 13.

► **PROBLEM**

The student should plot the data of Table 15.1 in this way and verify the linear relation between x and t^2. ◄

In the history of science it often happens that a relationship of this form is adduced by the investigator and then verified by an appropriate plot of experimental data. An example of great historical importance is shown in Figure 15.4.

Semilog Plots

Semilogarithmic graph paper is characterized by having one axis linear in a variable to be plotted and the other axis proportional to the logarithm (to any base) of the other quantity to be plotted. We have already seen examples of this in Figures 12.5 and 12.7 where either axis may be vertical and the prime motivation for the use of semilogarithmic plots was the need to display data requiring a large range of values in one variable.

Semilogarithmic paper is especially useful for the graphical analysis of data that are theoretically related by an equation involving the appearance of one of the variates in the exponent, of the general form

$$y = Aa^{Bx}$$

If we take logarithms to any base b, we get the equation

$$\log_b y = \log_b A + (B \log_b a)x$$

In coordinates $Y = \log_b y$ and $X = x$, this plot is a straight line with slope $B \log_b a$. The situation is simplified when $a = b$ and, hence, we are particularly interested in the equations involving $a = 10$ or e.

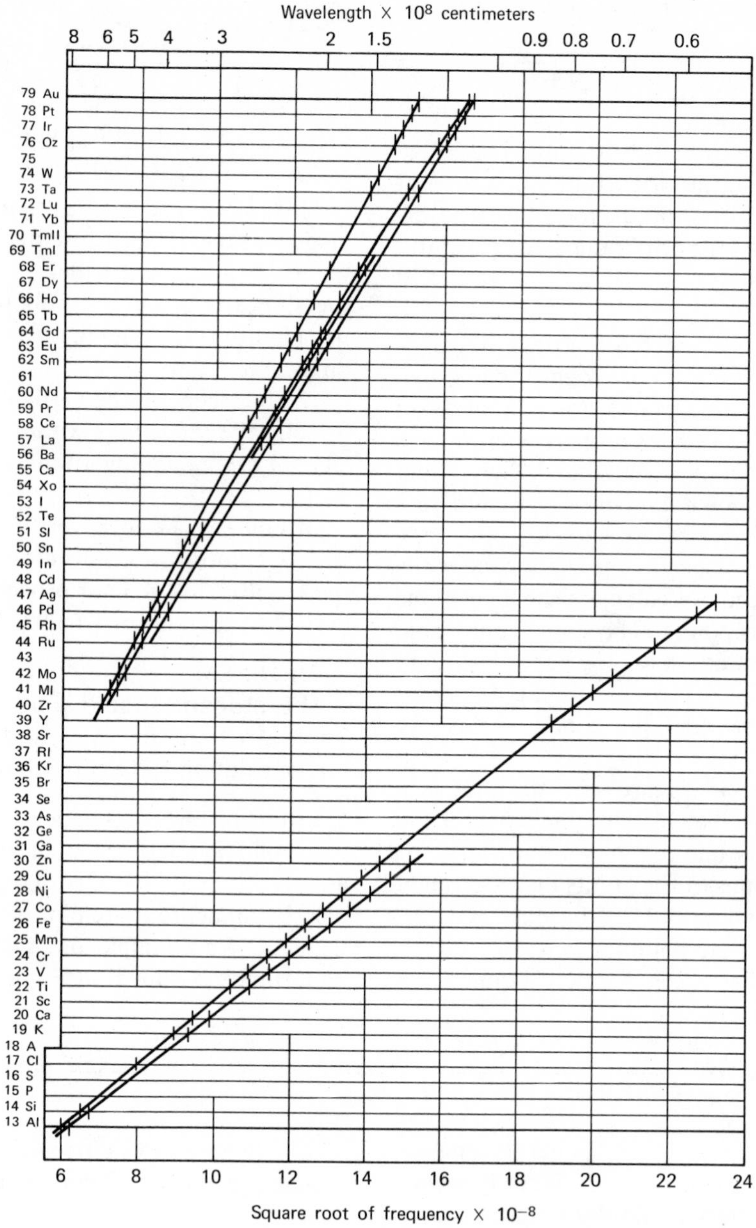

Figure 15.4
Original figure, from a paper by Moseley [*Phil. Mag.*, **26**, 1024 (1913)] showing the relation between atomic number, Z, and the square root of the frequency of X rays.

It is especially useful to consider the class of equations of the form

$$y = Ae^{Bx}$$

We may take natural logarithms of both sides and obtain

$$\ln y = \ln A + Bx$$

Choosing the variables to be plotted as $Y = \ln y$ and $X = x$, we see that a plot of Y versus X should yield a straight line with slope B and vertical axis intercept $\ln A$.

For example, let us consider a simple electrical circuit in which a charged

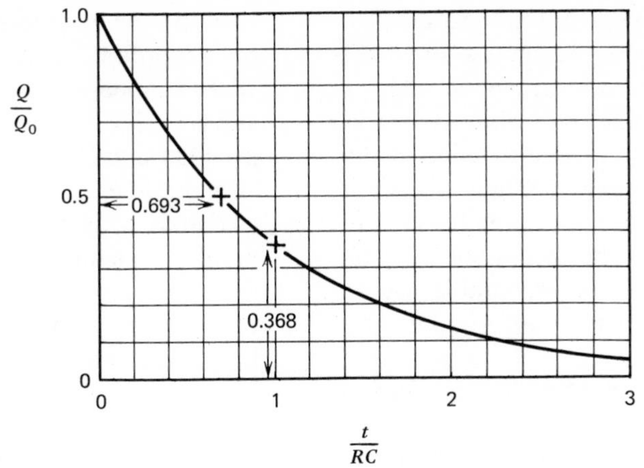

Figure 15.5
A plot of the relation $Q/Q_0 = e^{-t/RC}$ on linear paper.

capacitor C is connected across a resistor R. The charge Q on the capacitor at time t is related to the initial charge Q_0 at time $t = 0$ by the equation

$$Q = Q_0 e^{-t/RC}$$

We say that the charge is decaying exponentially with a *time constant* $\tau = RC$. It is convenient to consider the variables $y = Q/Q_0$ and $x = t/RC$ and we plot y versus x in Figure 15.5.

The corresponding plot on semilogarithmic paper is shown in Figure 15.6 and is a straight line as expected. We first note that the vertical axis intercept is

$$\ln Q/Q_0 = \ln A = \ln(1.0)$$

which tells us that $A = 1.0$. The slope may be obtained from any two points on the straight line (as always, these are not necessarily data points). We arbitrarily choose the points marked A and B in Figure 15.6 and determine the slope:

$$B = \frac{\Delta(\ln y)}{\Delta x} = \frac{\ln y_A - \ln y_B}{x_A - x_B}$$

$$= \frac{\ln(0.67) - \ln(0.165)}{(0.40) - (1.80)}$$

$$= \frac{-0.40 - (-1.80)}{-1.40} = -1.00$$

This analysis of course yields us the original equation

$$y = Q/Q_0 = e^{-x} = e^{-t/RC}$$

There are several points to be noted about Figure 15.6 and the analysis of it.

1. The variables have been *normalized* so that the maximum value plotted corresponds to the maximum value of the

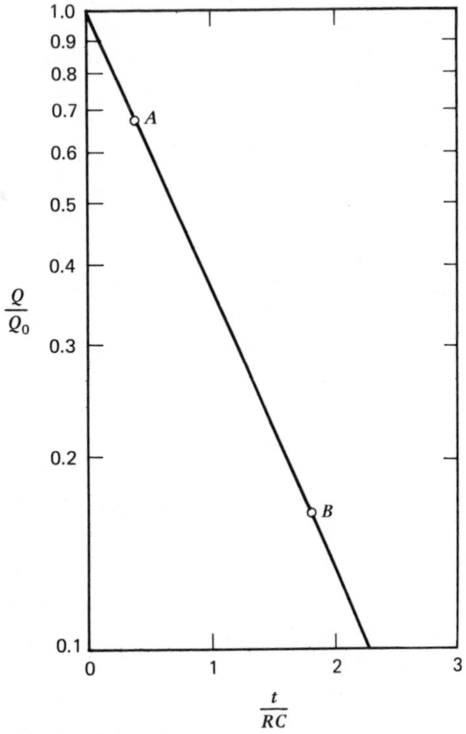

Figure 15.6
A plot of the relation $Q/Q_0 = e^{-t/RC}$ on semilogarithmic paper.

graph. It is always possible to do this, which is the most efficient way to get information from a graphical analysis and to minimize the plotting error, since the maximum use is made of the available graph paper. When this is not done, it is only because it is frequently faster to avoid the arithmetic of normalization. This is acceptable where the plotting error is not significantly worsened.

2. Figures 15.5 and 15.6 do not correspond exactly since the first figure shows a variation in Q/Q_0 of 20 times, while the logarithmic plot shows only one *decade* or factor of 10 (equivalently, order of magnitude) variation. If the first plot represented data, then some data were ignored in the second plot. On the other hand, the plotting error for the data included was less than it would have been if we had used semilogarithmic paper with two *cycles* or *decades* because of the less efficient use of the available space in the latter case. It is frequently judgment and experience that yields the best results. Of course, one is never entitled to ignore data that might systematically affect the results.

3. Note that the points on a semilogarithmic plot do not have equal weight as is usually true for a linear plot of x versus y. We see that the plot in Figure 15.6 itself introduces a difference in that the logarithm of the ordinate of point A may be read to two significant figures while that of point B may be read to three figures. Actually, the danger is that one may give undue weight to points of low ordinate value. The safe procedure is always to assign an error to each point and propagate that error through the computation. The difficulty lies in drawing a straight line on the semilogarithmic plot. The natural inclination is to use points spread as widely as possible to obtain the best possible straight line. It frequently happens that points at the extrema have large errors and should be accorded small statistical weight when compared to other points with smaller "lever arms." Again, the safe procedure to follow in drawing the best straight line is to plot the error associated with the ordinate of each point. Remember that a symmetrical error in y propagates to an asymmetrical error in $\log y$ or $\ln y$, and it is the latter *asymmetrical* error that should be used.

4. The analysis of slope required the knowledge of the natural logarithms of the ordinates of points A and B. We might also have used the extreme points of the plot that yield

$$B = \frac{\ln(1.0) - \ln(0.1)}{0.0 - 2.30} = \frac{0.0 - (-2.30)}{-2.30}$$
$$= -1.00$$

This required only that we know the natural logarithm of 10 (hence, any power of 10) and the x-intercept of the *one-decade down* point. This is the basis of a fast method for estimating the slope of a semilogarithmic plot. For most efficient use of the plot and most accurate reading of the slope, natural logarithms must be used. However, it is possible to estimate the slope quickly without natural logarithms in the following way.

QUICK ANALYSIS 1

Choose a point A on the plot so that another point B is displayed such that y_B and y_A differ by exactly a factor of 10 (B is the *one-decade down* point of A). Take the difference of their abscissas and divide by 2.303 = the natural logarithm of 10. This yields the *time constant* (it is negative if decaying)

$$\tau = \frac{x_A - x_B}{2.303}$$

QUICK ANALYSIS 2

The nature of the exponential is also used here. Choose a point P_{10} such that the ordinate is a multiple of 10. Find the point $P_{3.68}$ such that the ordinate of $P_{3.68}$ is 0.368 of the ordinate of P_{10}. The difference in their abscissas is now directly the *time constant* (negative if decaying):

$$\tau = x_{10} - x_{3.68}$$

The point $P_{3.68}$ corresponds to the e^{-1} point.

We repeat again that these two methods are quick and convenient only. The most accurate results are obtained by using as much as possible of the available plot and calculating the natural logarithms as before.

The equation of exponential decay applies in many different natural phenomena. For an unstable atom or nucleus whose "decay" into another state is governed by random probability (as contrasted with a reaction that "goes" when you push some sort of button) the *mean life* is defined as τ, where

$$N = N_0 e^{-t/\tau}$$

and N is the number of the atoms left at time t if there were N_0 at time $t = 0$. It is conventional to define the *decay rate* $\lambda = 1/\tau$. Another concept closely associated with the decay of radioactive states is that of *half-life* $\tau_{1/2}$, which is defined to be the time after which the number of atoms has been reduced to half of its original value. This may be related to the mean life, since

$$e^{-\tau_{1/2}/\tau} = \frac{1}{2}$$

Taking the natural logarithms of both sides, we obtain

$$\tau_{1/2} = \tau \ln 2 = 0.69315\tau$$
$$\tau = 1.4425\tau_{1/2}$$

QUICK ANALYSIS 3

We may determine the half-life from the plot and multiply by 1.44 to get the mean life.

► Example 15.2

Figure 15.7 shows a plot taken from a research paper. A beam of unstable particles called negative muons was stopped in a target of liquid hydrogen and the time interval between their stopping and subsequent decay was measured and quantized, that is, the times were sorted into bins according to whether they fell within the interval 0 to Δt, Δt to $2\Delta t$, and so on. Figure 15.7, therefore, shows a histogram of the times at which different particles have decayed. The ordinate indicates the number of particle decays in each channel and the abscissa corresponds to the channel number with each channel representing 3.0×10^{-8} sec.

We show elsewhere that the number of counts per channel should be proportional to a term e^{-t/τ_μ} where τ_μ is the mean life of the muon. *The slope of the best straight line on this plot is the relevant quantity*.

There are several remarks to be made about this figure.

1. The peak of the distribution indicates time $t = 0$. There are, nevertheless, counts at $t < 0$ that constitute a background. This is discussed elsewhere, but the background is small and may be neglected for all practical purposes, certainly for what we are interested in here.
2. Note in Figure 15.7 that two points in the vicinity of channel 100 appear well off the line. One need not invoke Chauvenet's criterion or similarly sophisticated rules to judge these to be systematic errors that should be excluded from the analysis. These errors were traced to digital scaling errors in the storage registers associated with these channels. When the error "leaps to the eye," one is on fairly safe ground. *Checking for such systematic errors is one of the most important reasons for plotting data.*

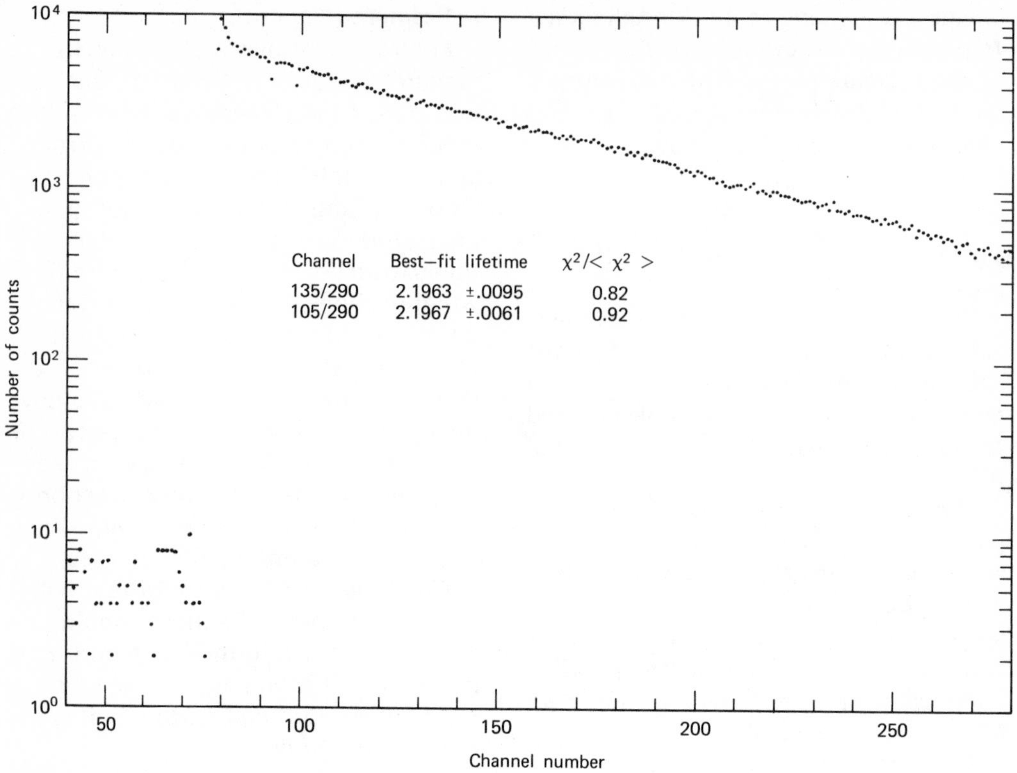

Figure 15.7
A plot on semilog paper of data taken from a "lifetime" experiment.

3. The slope of the line appears to start out much larger around channels 90 to 95 and then to become constant. This is more clearly indicated in Figure 15.8 where only the "bestimated" straight lines are shown. This is a real effect. Some of the negative muons stop in the walls of the target rather than in the target material. Negative muons stopping in the walls are "captured" from the muon population with a rate λ_{capture}. This capture rate depopulates the muons stopping in the wall just as does the decay rate. The number of muons per unit time interval being withdrawn from the muon population by capture and decay is

$$\frac{dN}{dt} = dN_{\text{wall}}/dt + dN_{LH}/dt$$
$$= +\lambda_{\text{wall}} N_{\text{wall}} + \lambda_{LH} N_{LH}$$

(If we had considered the *increase* in the muons we should have used the customary minus sign.)

$$N_{\text{wall}} = N_{0\,\text{wall}} e^{-\lambda_{\text{wall}} t} = f N_0 e^{-\lambda_{\text{wall}} t}$$
$$= \text{number of } \mu^- \text{ in the wall at time } t$$

and

$$N_{LH} = (1-f) N_0 e^{-\lambda_{\text{decay}} t}$$
$$= \text{number of } \mu^- \text{ in the liquid hydrogen at time } t$$

where

f = fraction of stopping μ^- that stop in the wall at $t = 0$
$1-f$ = fraction of μ^- that stop in the liquid hydrogen (LH) at $t = 0$
N_0 = total number of stopping μ^-

$$\lambda_{\text{wall}} = \lambda_{\text{capture}} + \lambda_{\text{decay}} = \frac{1}{\tau_{\text{wall}}}$$

$$\frac{1}{\tau_{\text{wall}}} = \frac{1}{\tau_{\text{capture}}} + \frac{1}{\tau_{\text{decay}}}$$

Suppose $\tau_{\text{capture}} \ll \tau_{\text{decay}}$. In the experiment, the walls were chosen to produce this situation exactly. Then

$$\tau_{\text{wall}} \approx \tau_{\text{capture}}$$

Thus

$$\frac{dN}{dt} = \lambda_{\text{wall}} f N_0 e^{-t/\tau_{\text{wall}}} + \lambda_{LH}(1-f)N_0 e^{-t/\tau_{LH}}$$

The capture rate for muons stopped in the liquid hydrogen is very small and so we have

$$\lambda_{LH} = \lambda_{LH,\text{capture}} + \lambda_{\text{decay}} \approx \lambda_{\text{decay}}$$

and

$$\tau_{LH} \approx \tau_{\text{decay}}$$

Finally

$$\frac{dN}{dt} = A_{\text{wall}} e^{-t/\tau_{\text{wall}}} + A_{\text{decay}} e^{-t/\tau_{\text{decay}}}$$

where

$$A_{\text{wall}} = \lambda_{\text{wall}} f N_0 = \text{constant}$$
$$A_{\text{decay}} = \lambda_{\text{decay}}(1-f)N_0 = \text{constant}$$

At early times the first term is dominant, and the slope is determined by τ_{wall}. This is shown by the steeper straight line in Figure 15.8. Several wall lifetimes after $t = 0$ ($t \approx$ several τ_{wall}) the first term has damped out ($e^{t/\tau_{\text{wall}}} \approx 0$) and the slope is determined by τ_{decay}, which is what we are interested in. This is represented by the second (less steep) straight line in Figure 15.8.

We have indicated on Figure 15.8 various arbitrary points from which we may quickly calculate a rough value for the lifetime, τ_{decay}.

Using Quick Analysis 1 we select points A and B such that their ordinates

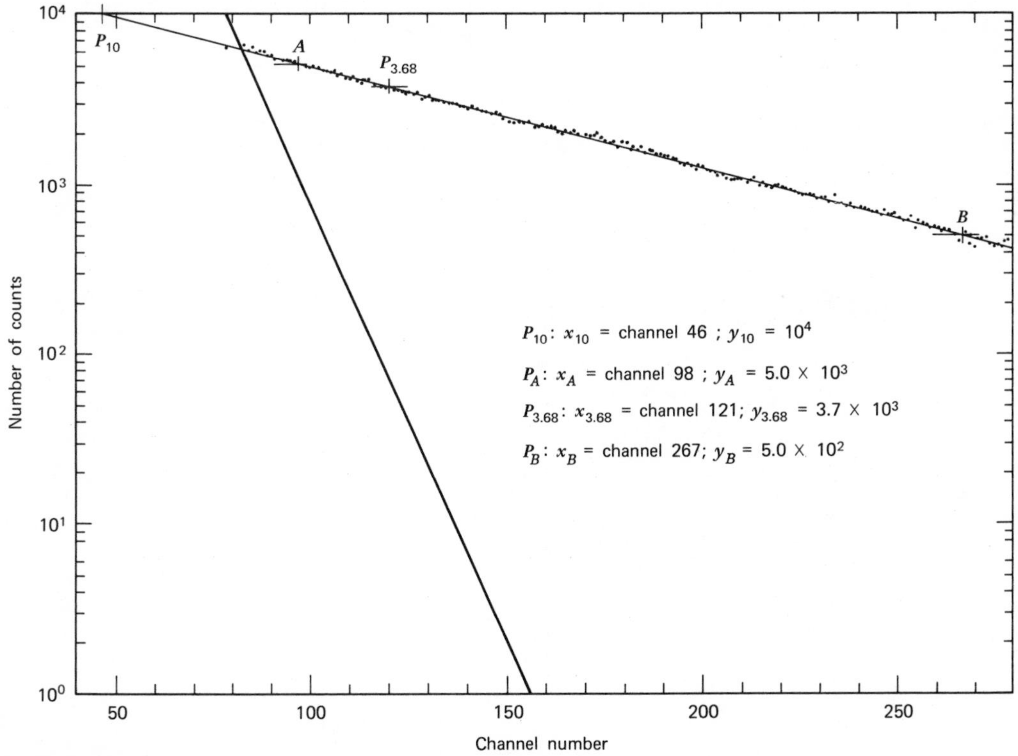

Figure 15.8
Superimposed on the previous figure are straight lines fitted to two exponential lifetimes.

differ by a factor of 10:

$$x_B = \text{channel } 267$$
$$x_A = \text{channel } 98$$

and we obtain

$$\tau_{\text{decay}} = \frac{x_B - x_A}{2.303} = \frac{(267-98)}{2.303} \times 0.0300\ \mu\text{sec}$$
$$= 2.20\ \mu\text{sec}$$

Using Quick Analysis 2 we select points P_{10} and $P_{3.68}$ and note that

$$x_{10} = \text{channel } 46$$
$$x_{3.68} = \text{channel } 121$$

We obtain directly

$$\tau_{\text{decay}} = x_{3.68} - x_{10}$$
$$= (121-46) \times 0.0300\ \mu\text{sec}$$
$$= 2.25\ \mu\text{sec}$$

Using Quick Analysis 3 we choose points P_{10} and P_A whose ordinates are in the ratio 2:1. We get the half-life

$$\tau_{1/2} = x_A - x_{10} = (98-46) \times 0.0300\ \mu\text{sec}$$
$$= 1.5(6)\ \mu\text{sec}$$

from which we obtain

$$\tau_{\text{decay}} = 1.4425\tau_{1/2} = 2.25\ \mu\text{sec}$$

If we are willing to use natural logarithms, we may use the full extent of the graph and choose the extrema P_A and (x = channel 280, $y = 4.15 \times 10^2$):

$$\tau_{\text{decay}} = \frac{\Delta x}{\Delta \ln(y)} = \frac{(280-46.0) \times 0.0300\ \mu\text{sec}}{\ln 10^4 - \ln 415}$$
$$= \frac{234 \times 0.0300\ \mu\text{sec}}{9.210 - 6.028} = \frac{7.02}{3.18(2)}$$
$$= 2.20(6)\ \mu\text{sec}$$

As is indicated on Figure 15.7, a more exact numerical analysis of a type that we discuss elsewhere yields a result that is $2.20 \pm 0.01\ \mu$ sec.

What error would you assign to each graphical determination? Which is most accurate? How precisely should the channel calibration be known? Is the number 3.000×10^{-8} sec per channel precise enough or more precise than is needed? ◄

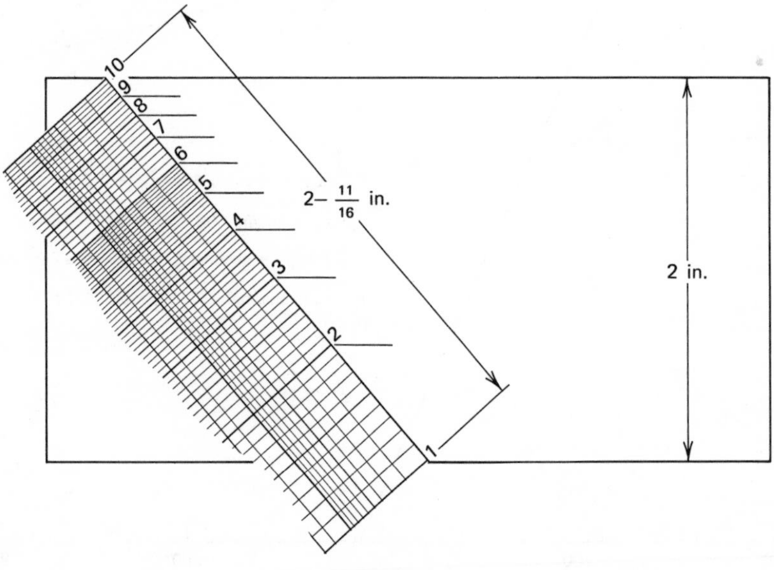

Figure 15.9
Procedure for ruling logarithmic paper to any scale given one cycle of log paper.

► **PROBLEM**
Estimate the lifetime τ_{wall} from the graph. How well determined is this slope? Estimate the error in your result and check the validity of the approximations used in our determination of τ_{decay}. If τ_{wall} was our primary interest, how would you change the measurement to determine it better? ◄

General Use of Logarithmic Scales

The graph paper should be chosen just to contain the data points within the paper. This affords the most efficient use of space and enables the reader to read the graph most readily. When logarithmic paper (either log-log or semilog) with cycles of the right height is not available, paper that is appropriate may be ruled

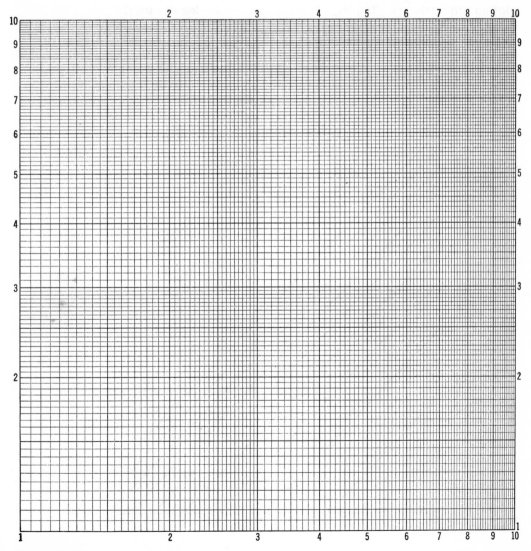

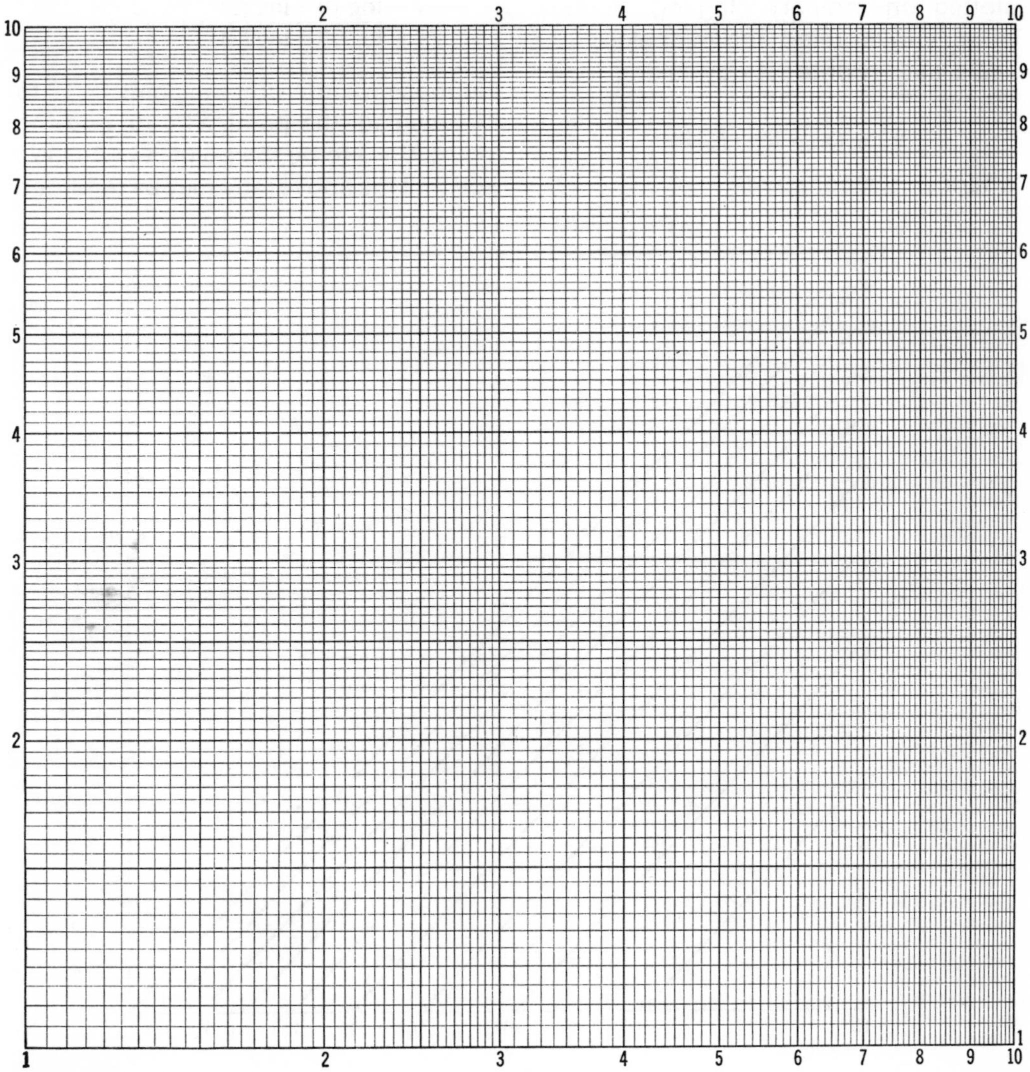

Figure 15.10
One cycle of log-log paper.

with cycles either larger or smaller than the space allotted, provided that a piece of logarithmic graph paper is available such that one cycle takes up more space than is allotted. The procedure for ruling the logarithmic scale desired is shown in Figure 15.9. For convenience, a large single cycle is provided in Figure 15.10 for this purpose.

► **Example 15.3**

Equations of the form

$$x^y = K \tag{A}$$

reduce to straight-line form when plotted on ordinary "log-log" paper; suitably interpreted and relabeled.

We may rewrite (A) by taking logarithms

$$y \log(x) = \log(K) = c$$

where c is a constant.

Taking logarithms again, we obtain

$$\log(y) + \log(\log(x)) = \log c = C$$

where C is a constant.

Clearly, graph paper ruled logarithmically in y and as the logarithm of the logarithm of x will suffice to reduce (A) to straight-line form.

For ordinary log-log paper, the ruling is $X = \log(x)$ so that, for example, the distance between $X = 2$ and $X = 10$ is to the distance between $X = 3$ and $X = 10$ as the ratio

$$\frac{\log 10 - \log(2)}{\log 10 - \log(3)} = 1.337$$

Suppose we interpret the point $X = 2$ to be, instead, the point $X' = 10^2$ and the point $X = 3$ to be the point $X' = 10^3$. We see that the above ratio may be regarded

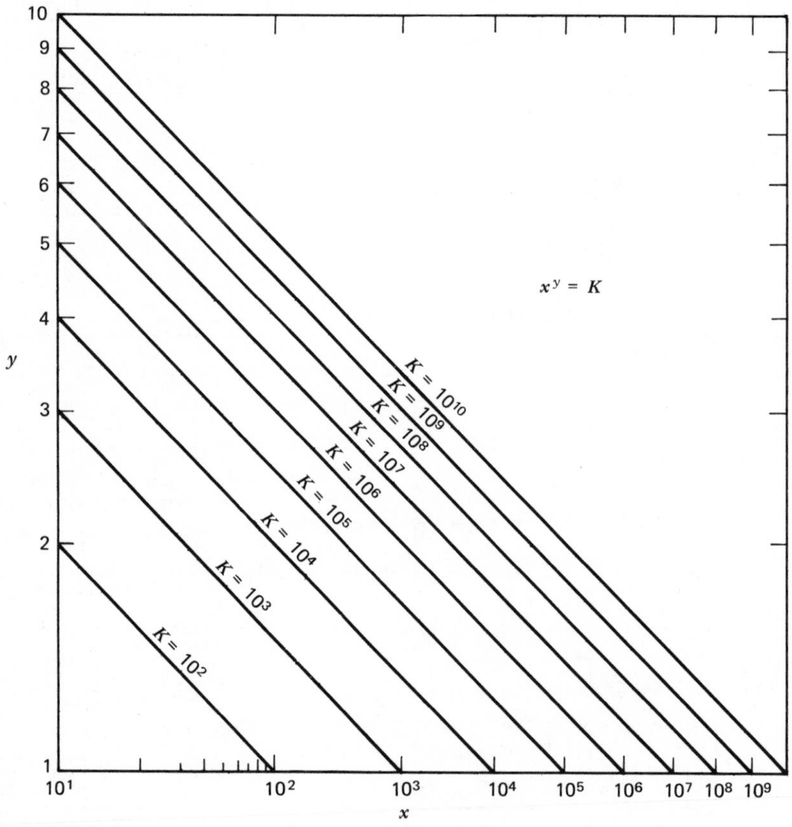

Figure 15.11
Linearized plot of the relation $x^y = K$.

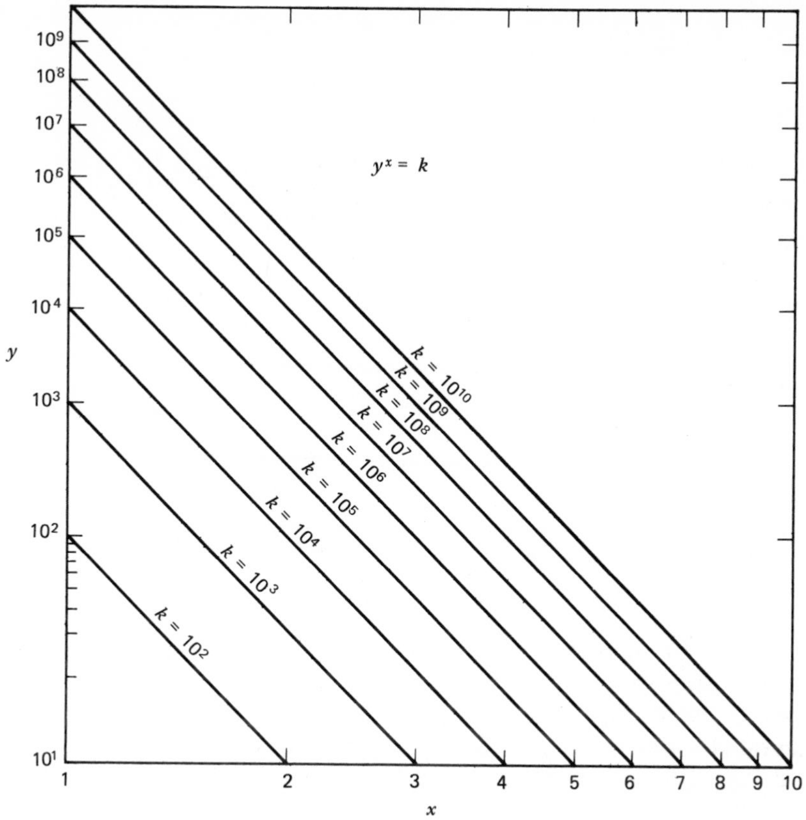

Figure 15.12
Linearized plot of the relation $y^x = k$.

as

$$\frac{\log\{\log 10^{10}\} - \log\{\log(10^2)\}}{\log\{\log 10^{10}\} - \log\{\log(10^3)\}} = 1.337$$

We need merely relabel every point x on a log scale to represent 10^x and we have "log-log by log" paper.

Figure 15.11 shows a plot of equation (A).

Equations of the form

$$y^x = k \tag{B}$$

are also reduced to straight lines by reinterpretation of the logarithmic y-axis. This is illustrated in Figure 15.12 for equation (B). ◄

16

CALCULATING CHARTS

The various kinds of charts (graphs) used in calculations can perhaps be epitomized by analogy to the juxtaposition of two scales calibrated in different units, for example, one scale is linear in inches while another is linear in centimeters, or one is linear in some units while the second is proportional to the logarithm of the scale markings on the first.

Fixed Scales

If we assume that the scales are fixed with respect to each other and that the origins coincide by design, we have an analog "lookup" device. In the examples cited, the first example permits us immediately to read off the equivalent in centimeters of a length in inches (or vice versa) while the second enables us to read off the logarithm of any number on the scale ruled logarithmically (as in Figure 15.2).

Sliding Scales

If we conceptually permit the scales to slide with respect to each other but maintaining parallelism, then we have the possibility of having a computational device. Thus, in Figure 16.1 we have a device for adding two numbers A and B, the total distance from the left-hand origin on the fixed top scale to the desired number A on the bottom scale (with the origin of the bottom scale placed on the desired B on the top scale). For example, Figure 16.1 shows the relation $5+2=7$, which is not very profound but straightforward.

In Figure 16.2 two scales are ruled logarithmically; they enable multiplication to be performed; the example shown is $2\times 3=6$, $3\times 3=9$, etc. This, of course, is the principle of the slide rule. We may also choose to rule one or both of our scales in some more exotic fashion depending on the need. In Figure 16.3, an addition of the squares of two numbers is shown; the example is $4^2+3^2=5^2$.

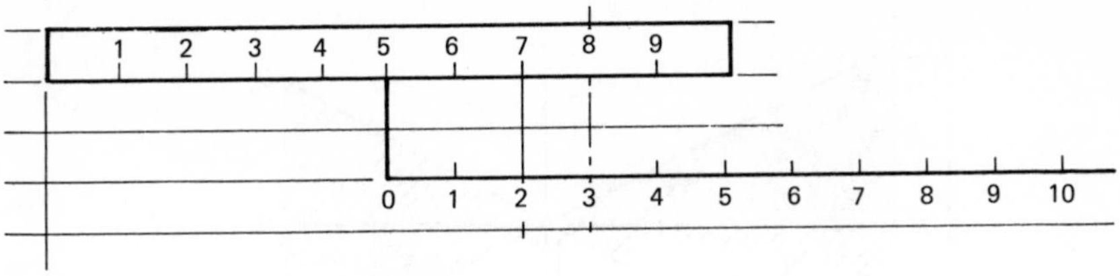

Figure 16.1
The principle of sliding scales. Addition.

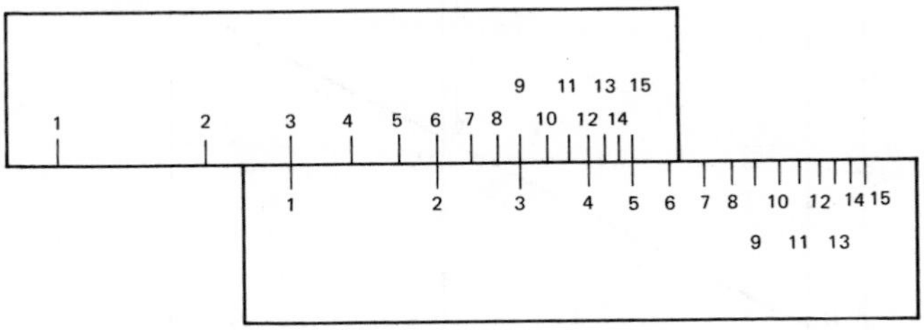

Figure 16.2
The principle of sliding scales. Multiplication.

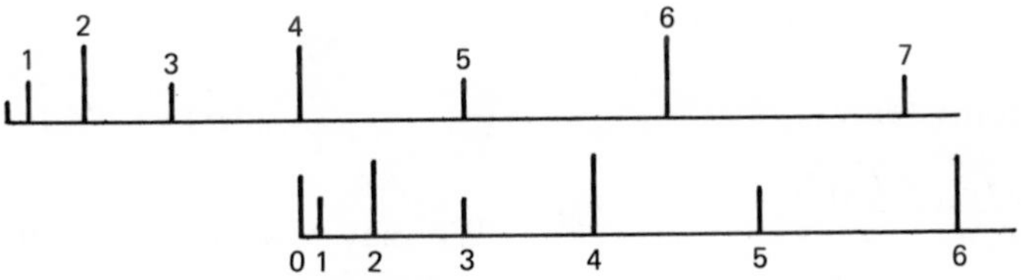

$$X^2 + Y^2 = Z^2$$

Figure 16.3
The principle of sliding scales. Adding in quadrature.

Nomograms

Instead of sliding scales with respect to each other, it is possible to separate two fixed scales in space and connect a point on one with a point on the other by means of a line drawn obliquely to the directionality of the scales. This relationship is equivalent to sliding scales with respect to each other and is the concept underlying the calculating chart known as the *nomogram*. In general, if we wish to operate one number on a second and read the result as a third number, three scales are required. We must investigate the geometrical relationships of the three scales of interest in the construction of a nomogram, using Figure 16.4.

Theorem:

$$Y_s = \frac{aY_2 + bY_1}{a + b}$$

Proof:

$$\Delta H_1P_sQ_1 \sim \Delta Q_2P_sH_2$$

also

$$\Delta P_1H_1P_s \sim \Delta Q_2P_sH_2$$

$$\frac{\overline{Q_1H_1}}{\overline{H_1P_s}} = \frac{\overline{Q_2H_2}}{\overline{H_2P_s}}$$

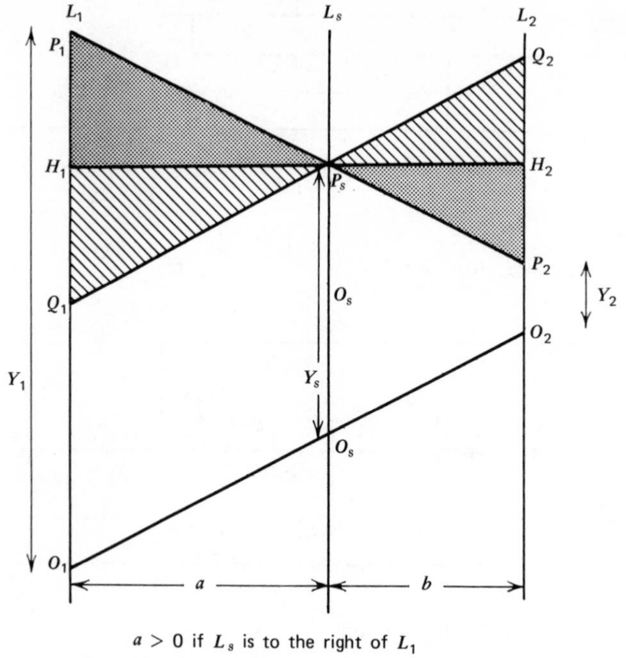

Figure 16.4
The geometry of the nomogram principle.

Keeping signs consistent,

$$\frac{\overline{Q_1H_1}}{a} = \frac{-\overline{Q_2H_2}}{b} \quad (1)$$

$$\frac{\overline{H_1P_1}}{a} = \frac{-\overline{H_2P_2}}{b} \quad (2)$$

Add (1) to (2) to get

$$\frac{\overline{Q_1H_1} + \overline{H_1P_1}}{a} = \frac{-(\overline{Q_2H_2} + \overline{H_2P_2})}{b}$$

which is

$$\frac{\overline{Q_1P_1}}{a} = \frac{-\overline{Q_2P_2}}{b}$$

We rewrite

$$\overline{Q_1P_1} = \overline{O_1P_1} - \overline{O_1Q_1}$$
$$= Y_1 - \overline{O_sP_s} = Y_1 - Y_s$$

and

$$\overline{Q_2P_2} = \overline{Q_2O_2} - \overline{P_2O_2}$$
$$= -Y_s + Y_2 = Y_2 - Y_s$$

Therefore,

$$\frac{Y_1 - Y_s}{a} = \frac{-(Y_2 - Y_s)}{b}$$
$$\Rightarrow bY_1 - bY_s = -aY_2 + aY_s$$

$$Y_s = \frac{(bY_1 + aY_2)}{a + b} \qquad \text{Q.E.D.}$$

We may rewrite the result

$$Y_s = \frac{(aY_2 + bY_1)}{(b + a)}$$

as

$$Y_s = KY_1 + (1 - K)Y_2$$

$$K = \frac{b}{a + b}$$

$$(1 - K) = \frac{a}{a + b}$$

Consider $b = a \Rightarrow K = 1/2$

$$\therefore \; Y_s = \tfrac{1}{2}(Y_1 + Y_2)$$

If all scales are uniformly graduated with the same unit, then Y_s is the *average* of Y_1 and Y_2.

If L_s is graduated with a scale unit 1/2 that of L_1, L_2, then

$$\tfrac{1}{2}y = Y_s = \tfrac{1}{2}(Y_1 + Y_2) \Rightarrow y_s = Y_1 + Y_2$$

where y_s is the reading on the L_s scale. In general, if the unit of graduation on scale L_1 has actual length U_1, then the mark y_1

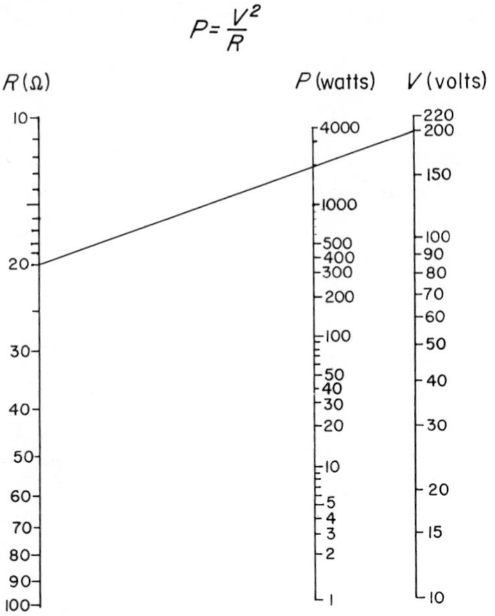

Figure 16.5
A nomogram for the relation $P = V^2/R$.

on the scale will be at an actual distance

$$\overline{O_1P_1} = Y_1 = U_1 y_1$$

from O_1. Also, if U_2 is the actual length of the unit on the L_2 scale and U_s that on the L_s scale, then

$$Y_2 = y_2 U_2; \qquad Y_s = y_s U_s$$

$$U_s y_s = K U_1 y_1 + (1 - K) U_2 y_2$$

$$y_s = \left(\frac{K U_1}{U_s}\right) y_1 + (1 - K)\frac{U_2}{U_s} y_2$$

We can vary K, U_1, U_2, and U_s to construct a nomogram to calculate the formula

$$y_2 = B_1 y_1 + B_2 y_2$$

where

$$B_1 = \frac{K U_1}{U_s} \qquad \text{and} \qquad B_2 = \frac{(1 - K) U_2}{U_s}$$

We may choose U_1 and U_2 to be any value, and then use B_1 and B_2 to determine K and U_s.

We may construct nomograms for

$$y_s = A + B_1 y_1 + B_2 y_2$$

by adding A to all values of y_s.

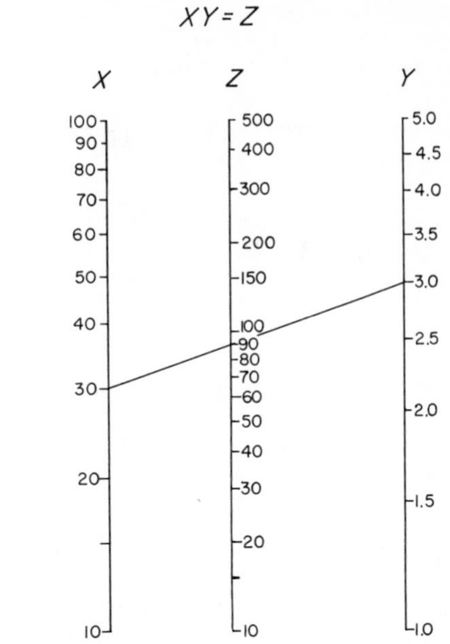

Figure 16.6
A nomogram for the relation $XY = Z$.

LOGARITHMIC SCALES

If $Y_s = \log y_s$, $Y_1 = \log y_1$, and $Y_2 = \log y_2$, then

$$\log y_s = A + B_1 \log y_1 + B_2 \log y_2$$

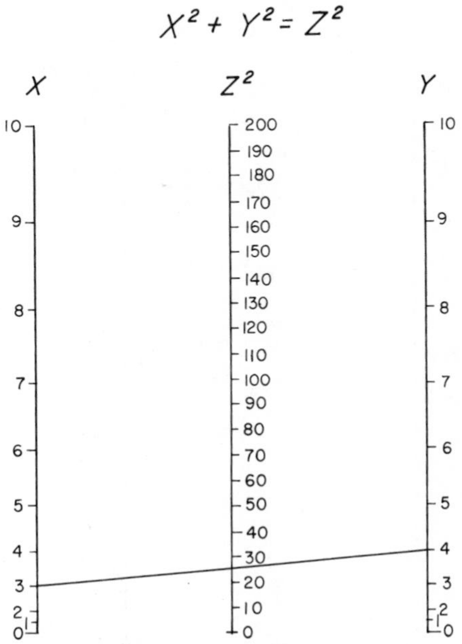

Figure 16.7
A nomogram for the relation $X^2 + Y^2 = Z^2$.

expresses the relation

$$y_s = k y_1^{B_1} y_2^{B_2}; \qquad \log k = A$$

NOMOGRAMS WITH MORE THAN THREE SCALES

$$y_t = y_1 + y_2 + y_3$$
$$y_s = A + B_1 y_1 + B_2 y_2$$

Step by step:

(1) Join y_1 and y_2, get y_s

(2) Join y_3 and y_s, get y_t

Some examples of nomograms are shown in Figures 16.5 to 16.10.

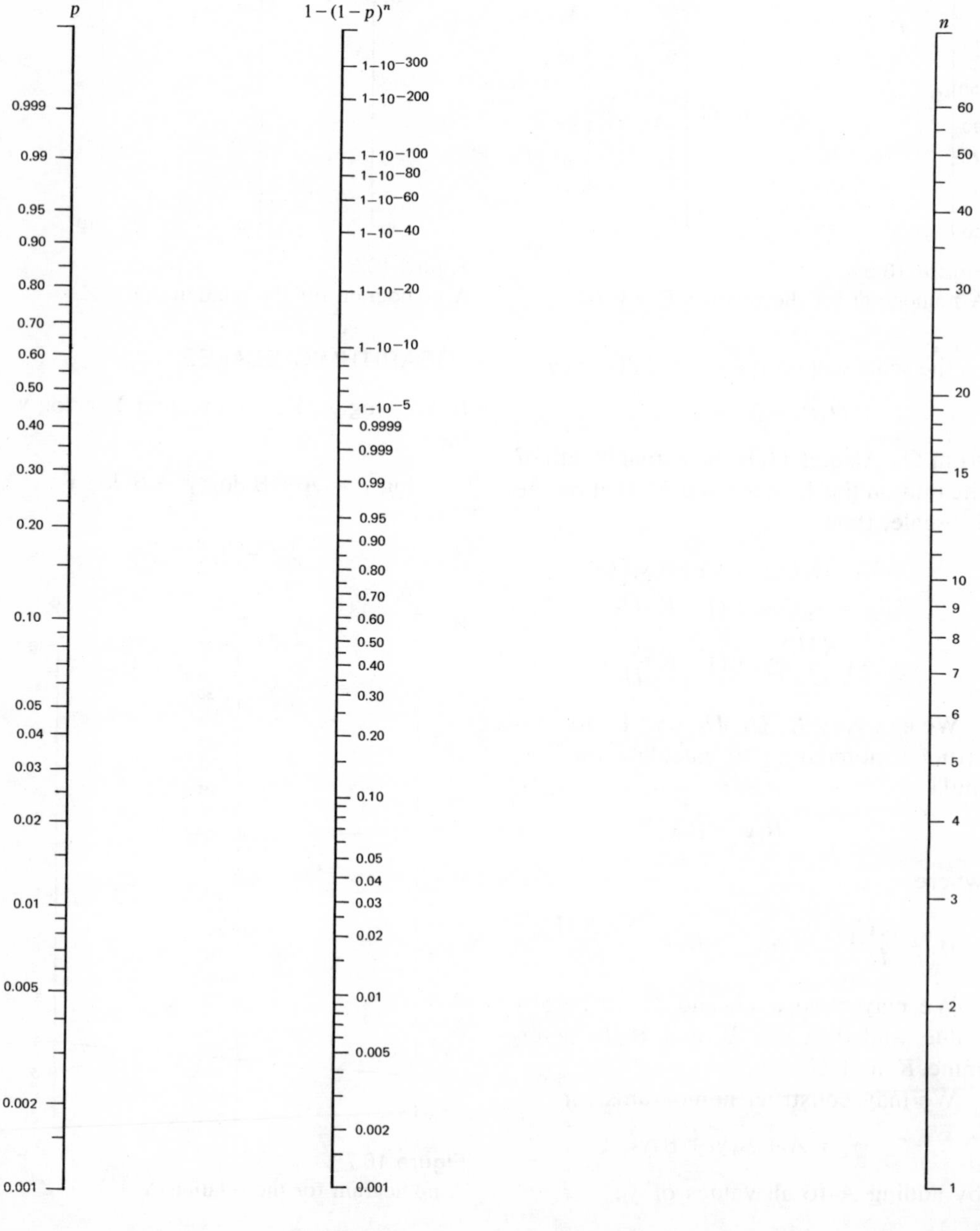

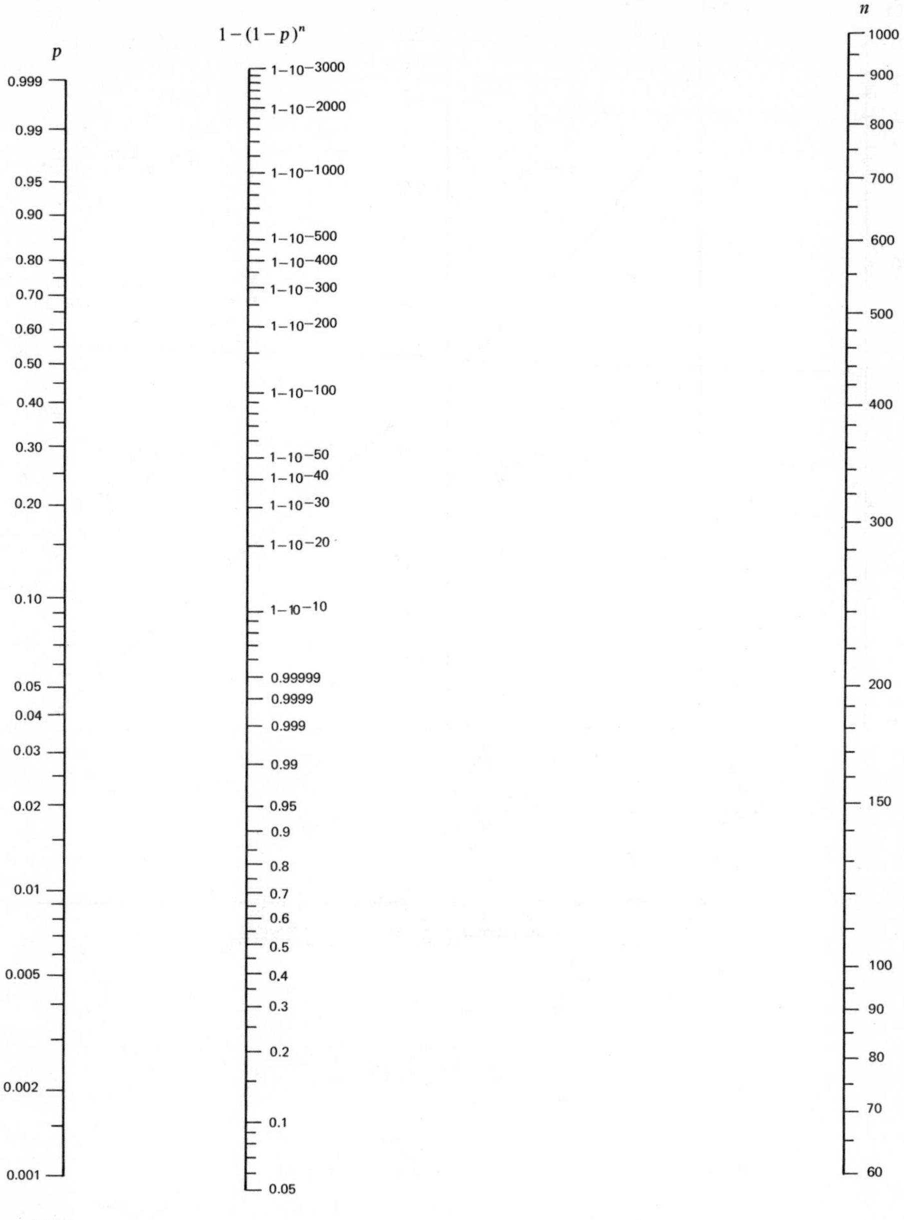

Figure 16.8
Nomogram for calculating $1-(1-p)^n$.

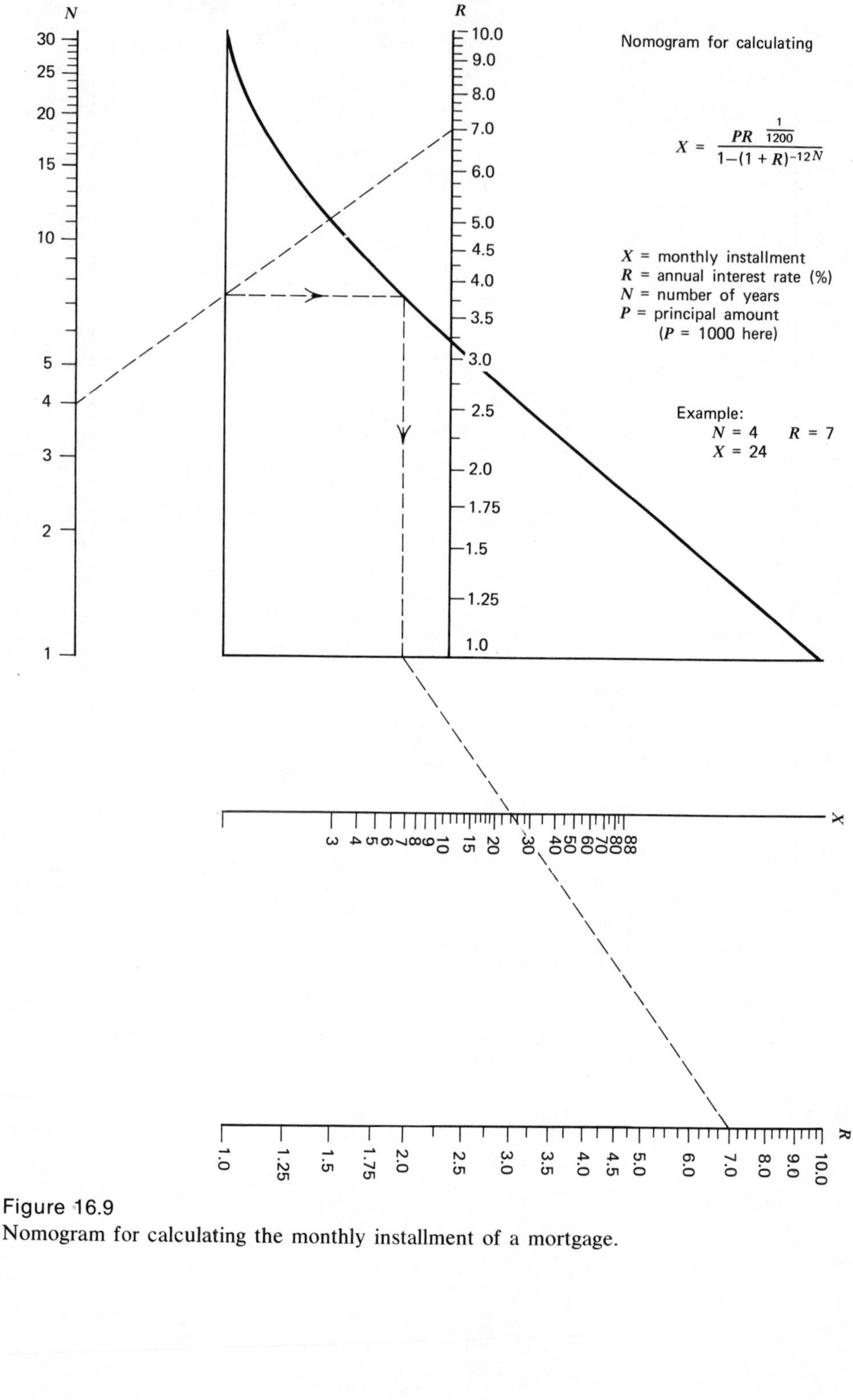

Figure 16.9
Nomogram for calculating the monthly installment of a mortgage.

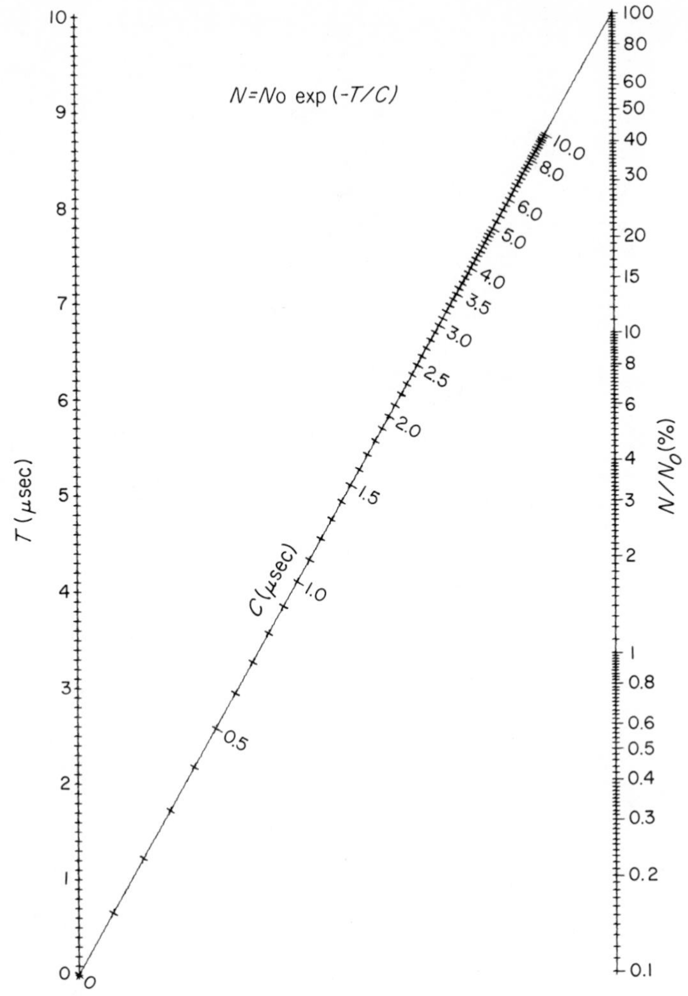

Figure 16.10
Nomogram for calculating lifetimes from exponential decay data.

PART III

Probability

17

THE MEANING OF PROBABILITY

The concept of probability is actually very subtle and controversial. However, it so pervades all of science and scientific reasoning that it is important to make one's first acquaintance with it as early as possible. Therefore, we shall rely on a very intuitive version of the underlying concepts and leave the more sophisticated points of philosophy and rigorous mathematics for another time and place.

Random Phenomena and Random Variables

First, we shall define a random or chance phenomenon to be characterized by the property that repeated occurrences of the phenomenon do not always lead to the same observed outcome. We shall primarily use the "classical" definition of probability (Laplace, 1812): the probability of a random event is the ratio of the number of cases that favor it to the total number of all possible cases when nothing leads us to believe that one of these cases ought to occur more readily than the others. We say that all of these cases are *equally likely.*

A quantity that assumes different numerical values as the outcome of an observation or experiment changes is called a *random variable* or *chance variable.* (The word "random" is sometimes reserved for quantities that are equally probable, for example, "a choice is made *at random*"; on this ground the term "chance variable" is to be preferred. However, in common usage random variable is applied to all such cases, not only those that are equally probable, and we shall defer to the general convention.) A *random* or *chance function* is a function of a random variable and is itself a random variable, assuming different values according to the outcome of an experiment or observation.

FREQUENCY DEFINITION

If n results out of N equally likely possibilities are favorable to some event, then the probability that the event will occur is n/N. This does, of course, not mean that out of n trials we will always have n successes for an event with probability n/N. However, our intuition tells us that the frequency of success expressed as a fraction approaches n/N as N becomes very large. This is an empirical assumption: namely, if an event occurs in more than one way and it is repeated a large number of times under identical conditions, the ratio of the number of times it happens in one way, say "A," to the total number of trials approaches a definite limit as the number of trials increases. This is the basis for the "frequency" definition of probability: the probability is the limiting frequency expressed as a fraction, Prob(A), also called Pr(A).

The connection between frequency of occurrence and probability is thus a natural and intuitively appealing one. We have discussed the idea of discrete and continuous frequency distributions in Chapter 9, and that whole discussion carries over directly into an equivalent discussion of probability.

SOME PROPERTIES AND DEFINITIONS

We first note the obvious features of a probability, Pr (E_i), of any event E_i.

$$0 \le \Pr(E_i) \le 1 \qquad (17.1)$$

The probability is never negative, and when normalized properly, is always less than or equal to unity. The normalization condition is

$$\sum_{i=1}^{k} \Pr(E_i) = 1 \qquad (17.2)$$

when the summation is over *all* the *conceivable* and *exclusive* events $i = 1$ to k. We sometimes refer to these exclusive (or disjoint) events by the name, *primary* or *elementary events* or outcomes.

► **Example 17.1**

If a coin is flipped in a random fashion, the probability of it coming up heads (H) is equal to the probability of it coming up tails (T):

$$\Pr(H) = \Pr(T)$$

and

$$\Pr(H) + \Pr(T) = 1$$

We have not conceded the third possibility that the coin would wind up on its edge. This is one of the dangers of discussions of *a priori* probability, we are tacitly speaking of an ideal coin with zero thickness. As the thickness of the real coin becomes larger, however, we must take note of the possibility that the coin stands on edge:

$$\Pr(T) = \Pr(H);$$

$$\Pr(H) + \Pr(T) + \Pr(\text{Edge}) = 1$$ ◄

The normalization condition is just an expression of the intuitive observation that *something* must happen.

If one event, E_i, itself has a probability of unity, then we call that a *certain* event, one which must happen whenever a trial is attempted. For an ideal but two-headed coin, $\Pr(H) = 1$ and $\Pr(T) = 0$. H is a certain event. If one event itself has a probability of 0, it is called an *impossible* event. In this example, T is an impossible event.

If we consider nonexclusive events (i.e., an occurrence can belong to more than one primary event designation), we can define three outcomes for the tossing of an honest coin: $A = H$, $B = T$, $C = (H \text{ or } T)$. This is nonexclusive, since all outcomes belong to two primary event designations. Also, the event C is a certain event: $\Pr(C) = 1$.

We may consider the *failure* of the event E as itself an event, denoted $\bar{E}$ and called the *complementary event* to E. The probability q of $\bar{E}$ follows immediately from the classical definition. If $\Pr(E) = p = n/N$, then $\Pr(\bar{E}) = q = (N - n)/N = 1 - p$.

Probability Distributions and Their Description

A set of numbers [$\Pr(E_1), \Pr(E_2), \ldots, \Pr(E_n)$] that represent the probabilities of a complete set of primary events associated with an experiment is called the *probability distribution* of the experiment. If a random variable X is associated with the outcomes of the experiment, the random variable X has this set of numbers as its probability distribution. This set of numbers also may be said to define the probability function of X. Any variable that has a probability function is called a *variate.*

Following the discussion in Chapter 9, we can consider plotting the probability of an event E_i versus the event E_i as in Figure 9.1, where we merely relabel the ordinate scale: "relative frequency" becomes "probability." The probability function need not be a discrete function, but may be continuous as shown in Figure 17.1*a* where we denote a continuous *probability density function* (also called *p.d.f.*, *density function, frequency function,* or *differential probability distribution function*) as $f(X)$ to distinguish it from the discrete case. It is not meaningful in the continuous case to speak of the probability of the variate assuming a specific value; it is proper to speak only of the probability of the variate being in a specified interval of

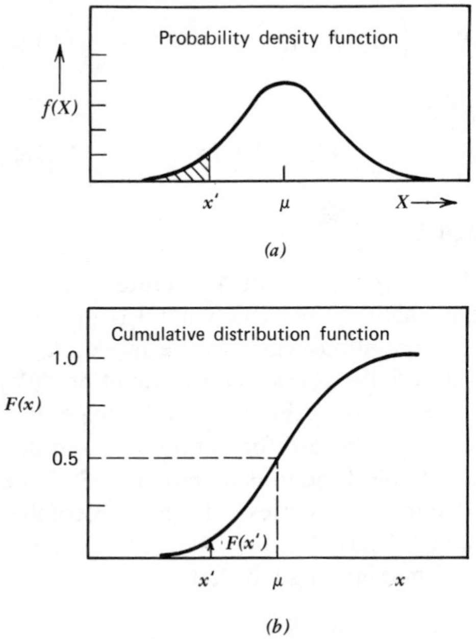

Figure 17.1
(*a*) Plot of the probability density function, $f(X)$ versus X. (*b*) Plot of the cumulative distribution function, $F(x)$ versus x.

values and, hence, a *density* function is used. In this case we use the normalization condition

$$\int_{X_{\min}}^{X_{\max}} f(X)\, dX = 1$$

where the limits define the *range* of the variate appropriate to the situation of interest. This may, of course, be infinite, in which case the limits extend from $-\infty$ to $+\infty$.

In Figure 17.1*b* we have plotted the *cumulative distribution function* (also called *c.d.f.* or *distribution function*).

$$F(x) = \int_{X_{\min}}^{x} f(X)\, dX \qquad (17.3)$$

The cumulative distribution function $F(x)$ represents the (shaded) area under the curve of the probability density function $f(X)$ to the *left* of a specified value, x. It must perforce eventually approach unity as X goes to its maximum value because of the normalization of the density function.

Formally, we can start with the c.d.f. and say that a real-valued function $F(x)$ is called a (univariate) cumulative distribution function if $F(x)$ is nondecreasing. That is,

$$F(x_1) \le F(x_2) \qquad \text{for} \qquad x_1 \le x_2$$

and

$$F(-\infty) = 0 \qquad F(\infty) = 1$$

[Note that $F(x)$ is everywhere continuous from the right, that is, $F(x) = \lim_{\epsilon \to 0^+} F(x + \epsilon)$.]

In terms of probability, the c.d.f. is defined as

$$\Pr\{X \le x\} = F(x) \qquad (17.4)$$

Also,

$$P(x) = \Pr\{X \le x\} = \sum_{x_i \le x} p(x_i) \qquad (17.5)$$

if the random variable X has a discrete probability distribution.

Furthermore,

$$F(x) = \Pr\{X \le x\} = \int_{-\infty}^{x} f(t)\,dt \qquad (17.6)$$

if the random variable X has a continuous probability distribution. We now may define the *probability density function* or *frequency function*

$$f(x) = F'(x) = \text{p.d.f.} \qquad (17.7)$$

We now have the relation in terms of probability:

$$\Pr\{a \le X \le b\} = \int_{a}^{b} f(t)\,dt \qquad (17.8)$$

It is sometimes convenient to define the "upper tail area function" (u.t.a.f.)

$$Q(x) = 1 - F(x) \qquad (17.9)$$

This represents the area under the curve of $f(X)$ plotted against X to the *right* of a specified value x.

Furthermore,

$$F(x) - F(-x) = \int_{-x}^{x} f(t)\,dt \qquad (17.10)$$

This is frequently called $A(x)$, the c.d.f. of $|X|$:

$$A(x) \equiv \int_{-x}^{x} f(t)\,dt \qquad (17.11)$$
$$= \text{c.d.f. of } |X|$$

For $f(t) = f(-t)$ (i.e., an even function of the argument),

$$A(x) = F(x) - Q(x) = 2F(x) - 1 \qquad (17.12)$$

We can extend the formal definition to *multivariate cumulative distribution functions* such that

(a) $F(x_1, x_2, \ldots, x_n) = \Pr\{X_1 \le x_1, X_2 \le x_2, \ldots, X_n \le x_n\}$

(b) $F(x_1, x_2, \ldots, x_n)$ is a nondecreasing function for each x_i

(c) $F(x_1, x_2, \ldots, x_n) = 0$ when any $x_i = -\infty$

(d) $F(\infty, \infty, \ldots, \infty) = 1$

(e) $F(x_1, x_2, \ldots, x_n)$
$= \lim_{\epsilon \to 0^+} F(x_1, \ldots, x_i + \epsilon, \ldots, x_n)$
for any x_i (17.13)

For completeness, we may express the various statistical quantities of interest as we have done in Chapter 9.

MEDIAN

The median of the distribution is readily defined:

$$F(x_{\text{med}}) = \frac{1}{2} \qquad (17.14)$$

MEAN

The mean of the distribution is defined as usual for discrete variates:

$$\langle X \rangle = \sum_i x_i\, p(x_i) \qquad (17.15a)$$

as is the mean value of any arbitrary function of the variate X:

$$\langle \xi(X) \rangle = \sum_i \xi(x_i)\, p(x_i) \qquad (17.16a)$$

These are sometimes called the "expectation values" $E(X)$ or EX and $E(\xi(X))$, respectively. For continuous variates, the definitions are

$$\langle X \rangle = \int X f(X)\, dX \qquad (17.15b)$$

and,

$$\langle \xi(X) \rangle = \int \xi(X) f(X)\, dX \qquad (17.16b)$$

MODE

The mode, X_{mode}, for a discrete variate is that value x_i for which $p(x_i)$ is greatest. For continuous variates the mode, X_{mode}, is that value for which the function $f(X)$ is greatest over the acceptable range. It is defined as before for continuous and differentiable functions as either one of the extreme boundaries of the acceptable range ($X_{\min}$ or $X_{\max}$) or as the abscissa of a local maximum such that

$$\left.\frac{df(X)}{dX}\right|_{X = X_{\text{mode}}} = 0$$

and

$$\left.\frac{d^2 f(X)}{dX^2}\right|_{X = X_{\text{mode}}} < 0 \qquad (17.17)$$

► **Example 17.2**

Consider the exponential probability density function

$$f(X) = e^{-X}$$

over the interval 0 to $+\infty$. The mode corresponds to $X = 0$. ◄

Occasionally a probability distribution is observed as in Figure 17.2 where there are several local maxima. This is an example of a *multimodal* distribution.

MOMENTS

It is convenient to describe a density function $f(x)$ by certain parameters associated with it.

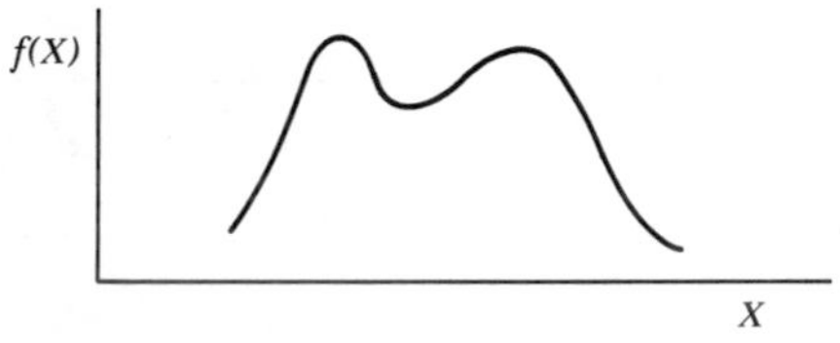

Figure 17.2
Plot of a *bimodal* probability density function, $f(X)$ versus X.

The *moments* are based on powers of the variate and are defined by

$$\mu_r' = \sum_i x_i^r p(x_i) \quad \text{(discrete)} \qquad (17.18a)$$

or

$$\mu_r' = \int x^r f(x)\, dx \quad \text{(continuous)} \qquad (17.18b)$$

For a discrete variate that assumes values spaced at unit intervals, the *factorial moment* is defined by

$$\mu_{(r)}' = \sum_i x_i^{(r)} p(x_i) \qquad (17.19)$$

where

$$x_i^{(r)} = x_i(x_i - 1)(x_i - 2) \cdots (x_i - r + 1) \qquad (17.20)$$

The first moment or *mean* $\langle X \rangle$ is so fundamental that it is usual to consider *moments about the mean* $\mu_1' = \langle X \rangle \equiv \mu$:

$$\mu_r = \sum_i (x_i - \mu)^r p(x_i) \qquad (17.21a)$$

or

$$\mu_r = \int (x - \mu)^r f(x) dx \qquad (17.21b)$$

Chebychev's Inequality

The second moment about the mean, called the variance,

$$\mu^2 = \sigma_x^2 = \langle (X - \langle X \rangle)^2 \rangle \equiv \text{Var}(X) \qquad (17.22)$$

has a special significance for all distributions because of Chebychev's inequality. This is simply derived below for both discrete and continuous probability distributions and may be stated

$$\Pr\{|X - \langle X \rangle| \geq k\sigma_x\} \leq \frac{1}{k^2} \qquad (17.23)$$

This makes a statement about the probability that a random variable will be found at a greater distance from the mean than a given number of standard deviations.

$$\sigma_x = \text{standard deviation} = \sqrt{\mu_2} \qquad (17.24)$$

For example, for any distribution

$$\Pr\{|X - \langle X \rangle| > 2\sigma_x\} \leq \frac{1}{4}$$

This asserts that *any* random variable X will assume a value further from the mean than two standard deviations with a probability less than 1/4. On the average, it will be more than two standard deviations away from the mean less than 1/4 of the time. The Chebychev inequality puts an upper limit on the c.d.f. of $|X - \langle X \rangle|$. This limit is shown as the $1/k^2$ curve in Figure 17.3. Also indicated are the actual c.d.f.'s of two specific distributions to be considered later.

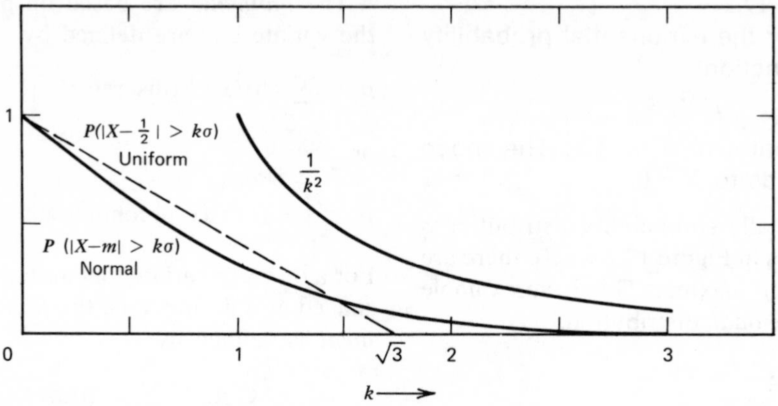

Figure 17.3
Plot of the probability of a deviation from the mean (as ordinate) versus the size of the deviation in units of σ. Shown are the probabilities for the normal and for the uniform distributions and for the Chebychev inequality.

Derivation of Chebychev's Inequality

$$\text{Th}^m: \quad \Pr\{|x - \langle x\rangle| \geq k\sigma_x\} \leq \frac{1}{k^2} \qquad (17.23)$$

DISCRETE CASE

$$\sigma_x^2 = \langle (x - \langle x\rangle)^2\rangle = \sum_i (x_i - \langle x\rangle)^2\, p(x_i)$$

Each term $(x_i - \langle x\rangle)^2$ makes a positive contribution to the sum. Remove those terms for which $|x_i - \langle x\rangle| < k\sigma_x$. Since these terms made a positive contribution

$$\sigma_x^2 \geq \sum_{|x_i - \langle x\rangle| \geq k\sigma_x} \{x_i - \langle x\rangle\}^2\, p(x_i)$$

where each term in the bracket is such that

$$\{x_i - \langle x\rangle\}^2 \geq k^2\sigma_x^2$$

Thus, the right-hand side is further diminished if each bracketed term is replaced by $k^2\sigma_x^2$.

$$\sigma_x^2 \geq \sum_{|x_i - \langle x\rangle| \geq k\sigma_x} k^2\sigma_x^2\, p(x_i)$$

or

$$\sigma_x^2 \geq k^2\sigma_x^2 \sum_{|x_i - \langle x\rangle| \geq k\sigma_x} p(x_i)$$

The summation on the right-hand side is just

$$\Pr\{|x - \langle x_i\rangle| \geq k\sigma_x\}$$

Therefore,

$$\sigma_x^2 \geq k^2\sigma_x^2 \Pr\{|x - \langle x_i\rangle| \geq k\sigma_x\}$$

from which (17.23) follows. Q.E.D.

CONTINUOUS CASE

$$\sigma_x^2 = \int_{-\infty}^{\infty} (x - \langle x\rangle)^2 \varphi(x)\, dx$$

Remove from the integral $|x - \langle x\rangle| < k\sigma_x$. This made a positive contribution and so

$$\sigma_x^2 \geq \int_{|x - \langle x\rangle| \geq k\sigma_x} (x - \langle x\rangle)^2 \varphi(x) dx$$

Since $(x - \langle x\rangle)^2 \geq k^2\sigma_x^2$

$$\sigma_x^2 \geq k^2\sigma_x^2 \int_{|x - \langle x\rangle| \geq k\sigma_x} \varphi(x)\, dx$$

$$= k^2\sigma_x^2 \Pr\{|x - \langle x\rangle| \geq k\sigma_x\}$$

(17.23) follows. Q.E.D.

Symmetrical and Asymmetrical Distributions

Many distributions are symmetrical, that is, the frequency function has the feature

$$f(\mu + x) = f(\mu - x) \qquad (17.25)$$

for all x

For symmetrical distributions, the odd central moments vanish (i.e., $\mu_1 = 0 = \mu_3$, etc.). It is reasonable, therefore, to take the odd central moments or a linear transformation of them as a measure of the asymmetry of a distribution.

The "skewness" is defined as

$$\gamma_1 = \frac{\mu_3}{(\mu_2)^{3/2}} = \frac{\mu_3}{\sigma^3} \qquad (17.26)$$

If a distribution is symmetrical, $\mu_3 = 0 = \gamma_1$. If the long tail of the distribution is on the side of the negative values of the variate, the cubes of the negative values outweigh the cubes of the positive values so that $\mu_3 < 0$, $\gamma_1 < 0$, and we say the distribution has "negative skewness." Also, if the long tail is on the side of positive values, $\mu_3 > 0$, $\gamma_1 > 0$ and the distribution has "positive skewness." This is qualitatively in accord with (9.17) and is illustrated in Figure 17.4.

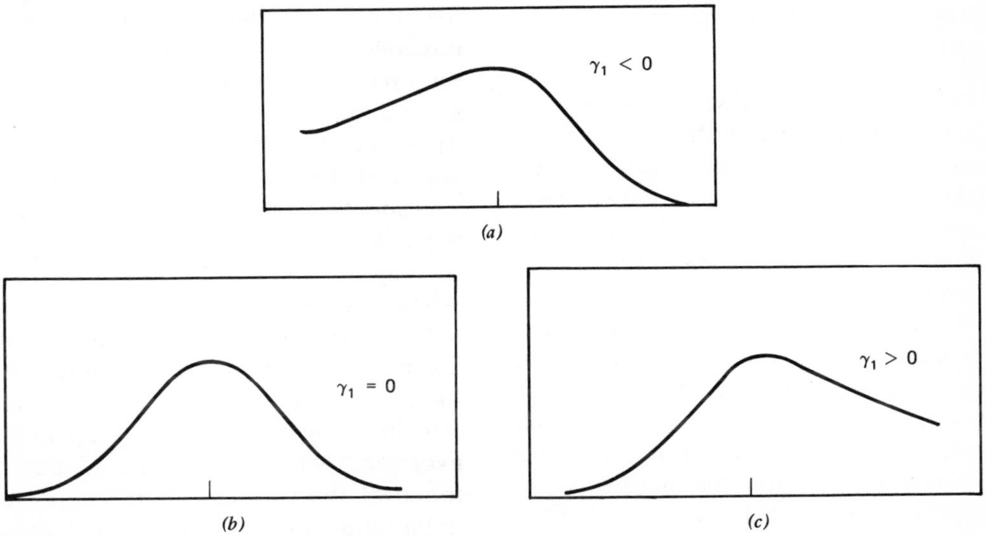

Figure 17.4
Plot of distributions for different values of the skewness parameter, γ_1. (*a*) Negative skewness. (*b*) Zero skewness (symmetry). (*c*) Positive skewness.

Different Kinds of Probability

We are concerned with probability in various ways in experimental science.

A PRIORI PROBABILITY

We are here interested in predicting the results of an experiment by calculating in advance the probabilities for the occurrence of various events. Inevitably we input some prejudgments of the situation (e.g., our expectation that we can neglect a coin landing on its edge) and the ability of the experimenter lies in his ability to make valid approximations and correct assumptions. The important feature of this area is that the probabilities can be *calculated* in advance. Examples of this are legion in the various forms of gambling, which itself gave greatest impetus to the study of probability theory.

A POSTERIORI PROBABILITY—THE ESTIMATION OF PROBABILITIES

To make predictions, we somehow must know the probabilities before the experiment is performed. It often happens, however, that we must perform an experiment to determine the probabilities themselves because the *a priori* calculation either cannot be done or is too difficult to do. Here the probabilities that one obtains for use in later predictions are empirically determined and, hence, have uncertainties that must be incorporated into the predictions. The basic problem here is that of sampling: we cannot get to the limiting frequency in a finite experiment; therefore, we must estimate what that limiting frequency (or probability) is from our observations of a sample of trials.

SUBJECTIVE PROBABILITY OR CREDIBILITY

We have previously discussed hypothesis testing. Here one is testing assumptions that are used to calculate *a priori* probabilities that are then compared with experiment. It is in this regime that *subjective probability* is important. We use the terms "probable" and "improbable" to discuss the various degrees of belief we have in hypotheses or statements. We make statements such as "The sun will rise tomorrow"; "It is probable that it will rain today"; "The hypothesis becomes increasingly probable the more it is successfully tested." This kind of probability is both central to experimental science and difficult to quantify. We shall develop an operational method for determining our "degree of belief" during the course of our discussions without dwelling deeply on the philosophical issues.

Both of the last two kinds of probability, sometimes called **inverse probability**, can be classified under the heading of *statistical inference.* Although we know how nature works when we use *a priori* probabilities to calculate what will happen, the chance mechanism is not completely known in statistical inference. Instead, we observe what has happened and then try to deduce some of the features of the mechanism that made it happen.

In discussing the question of statistical inference and the result of experiment here are some possible questions. For example, in the case of a coin toss one could legitimately ask: "What value of p, the probability for heads, characterizes the coin?" This is sometimes delineated further and called a problem of *point estimation* since we are after a specific value. A variation of this question is to ask if the value of interest is in a specified range or interval. This is the problem of *interval estimation.* Another question that requires a less precise statement about the probability is the question about the underlying hypothesis: "Is the coin fair?" A third possibility is the more practical but still relevant query: "Is the coin fair, biased in favor of heads or biased in favor of tails?" These more practical questions are involved in what is sometimes called *statistical decision theory* because the answers to these questions serve as guides to action.

► **Example 17.3**

An example of *a posteriori* probability estimation is that of determining the frequency of occurrence of various letters in a given language, English for example. The probabilities of occurrence of the various letters cannot be determined *a priori* since English arose as a result of myriad influences (not the least of which concerned the propagation of printers' errors!). We may, however, take a finite sample of an English text and analyze it to determine a best estimate of the individual probabilities. This is a typical problem of statistical inference; we wish to determine an estimate of a value. If we are seeking a definite number (e.g., one of a series of possible discrete values) the problem is one of *point estimation.* In this case, we have no reason but to believe that the probabilities may take on any of a continuum of values and we are more concerned with determining an interval; that is, that the probability, p, for the occurrence of the letter e lies in the interval $0.107 \le p \le 0.127$. This is called the problem of *interval estimation.*

Three detailed studies of this particular subject are summarized in Table 17.1.

Table ***17.1*** Relative Frequencies of Letters in Samples from English Writing

	Percent			
Letters	After Dewey[a]	As in Morse's[b] Day	After Nasvytis[c]	Northwestern University Results Spring 1969 89,602 Letters
e	12.68	13.11	10.70	12.6
t	9.78	10.47	8.20	8.6
a	7.88	8.15	7.78	
o	7.76	8.00	7.18	
i	7.07	6.35	7.50	
n	7.06	7.10	7.15	
s	6.31	6.10	7.24	
r	5.94	6.83	5.61	
h	5.73	5.26	5.76	
l	3.94	3.39	3.84	
d	3.89	3.79	4.17	
u	2.80	2.46	3.17	
c	2.68	2.76	2.98	3.1
f	2.56	2.92	2.51	
m	2.44	2.54	2.90	
w	2.14	1.54	2.02	
y	2.02	1.98	1.96	
g	1.87	1.99	1.79	
p	1.86	1.98	1.79	
b	1.56	1.44	1.28	
v	1.02	0.92	1.69	
k	0.60	0.42	0.94	
x	0.16	0.17	0.49	
j	0.10	0.13	0.59	
q	0.09	0.12	0.53	0.15
z	0.06	0.08	0.23	

[a] G. Dewey, *Relative Frequencies of English Speech Sounds,* Harvard University Press, 1923.
[b] Samuel F. B. Morse.
[c] A. Nasvytis, *Die Gesetzmässigkeiten kombinatorischer Technik,* Berlin-Göttingen Heidelberg, 1954.

We also note that a group of students at Northwestern University studied some 89,602 letters and came up with the results shown in the last column of the table. We shall consider later in more detail the problem of assigning "errors" to our interval estimation. Suffice it to say for now that one way to do this is to divide the data up into independent groups and treat each group so as to provide an individual determination. These groups may then be combined according to the procedures discussed in Part I and an error assignment in terms of standard deviation made. This, in fact, was done for the determination in the last column of Table 17.1. The other groups are quoted to as many significant figures as a similar technique would justify.

Several practical problems arise in this regard. Knowledge of the frequency of occurrence of the letters in English permits one to optimize the placement of keys on a typewriter. Clearly, it helps if the letters that are used most frequently are more conveniently placed.* The transmission of

*The typewriter keyboard in common usage does *not* do this in optimal fashion.

words also involves such considerations. Samuel F. B. Morse, developer of the telegraph, performed one of the first analyses of this type to optimize the transmission of information over telegraph wires. The modern International Morse Code is based largely on his analysis and transmits an English message of 100 letters with an average of about 940 dot units: the duration of a dot equals one dot unit; the dash equals three dot units; the space between dots and dashes of a letter equals one dot unit; and the space between letters equals three dot units. If the symbols had been assigned at random, the same message would run about 1160 dot units, or about 23 percent longer. (The student should see if this statement is at least plausible.) The modern Morse code is shown in Table 17.2.

Another problem of statistical inference that bears on this example is *hypothesis testing*. We can ask such questions as the following. "Is the list of probabilities in the first column consistent with that of the second column and so on?" The reader should ask himself this question and further inquire if any possible systematic differences between studies might have been operative. Another question is, "Is the Northwestern-student-compiled column of probabilities a fair estimate of English as written in modern times?" We may pose the hypothesis to be tested as the positive statement, "The Northwestern column of probabilities should describe a randomly selected sample of English prose (within statistical sampling errors)."

Table 17.2 The Assignment of Dots and Dashes in the International Morse Code

a	·	= e	abab	·–·–	= ā
b	–	= t	baab	–··–	= x
aa	··	= i	abba	·––·	= p
ab	·–	= a	baba	–·–·	= c
ba	–·	= n	bbaa	––··	= z
bb	––	= m	abbb	·–––	= j
aaa	···	= s	babb	–·––	= y
aab	··–	= u	bbab	––·–	= q
aba	·–·	= r	bbba	–––·	= ō
baa	–·	= d	bbbb	––––	= ch
abb	·––	= w			
bab	–·–	= k	aaaaa	·····	= 5
bba	––·	= g	aaaab	····–	= 4
bbb	–––	= o	baaaa	–····	= 6
aaaa	···	= h	aaabb	···––	= 3
aaab	···–	= v	bbaaa	––···	= 7
aaba	··–·	= f	aabbb	··–––	= 2
abaa	·–··	= l	bbbaa	–––··	= 8
baaa	–···	= b	abbbb	·––––	= 1
aabb	··––	= u	bbbba	––––·	= 9
			bbbbb	–––––	= 0

Suppose we are presented with the following selection from *Gadsby*, by Ernest Vincent Wright and published in 1939 by the Wetzel Publishing Company of Los Angeles:

"Upon this basis I am going to show you how a bunch of bright young folks did find a champion; a man with boys and girls of his own; a man of so dominating and happy individuality that Youth is drawn to him as is a fly to a sugar bowl. It is a story about a small town. It is not a gossipy yarn; nor is it a dry, monotonous account, full of such customary 'fill-ins' as 'romantic moonlight casting murky shadows down a long, winding country road.' Nor will it say anything about twinklings lulling distant folds; robins carolling at twilight, nor any 'warm glow of lamplight' from a cabin window. No. It is an account of up-and-doing activity; a vivid portrayal of Youth as it is today; and a practical discarding of that worn-out notion that a "child don't know anything."

This is a sample of some 580 letters without an "e" appearing. What can we say about this test of the stated hypothesis? We shall return to this example in Chapter 18. ◄

18

SOME ARITHMETIC ON COMBINATIONS AND PERMUTATIONS

For our discussion, it is helpful to review some elementary considerations of *a priori* probabilities.

Arrangements: Permutations and Variations

Suppose we have N distinguishable objects, and we wish to count all possible ways of arranging or ordering them. The counting of the possibilities can be likened to the counting of the tiniest branches on a tree: each time a choice is to be made, another branching junction occurs. Starting from the beginning, we have N choices for the first position, $N-1$ for the second, and so on until only one object is left. The total number of possibilities or "permutations" is

$$N(N-1)(N-2)\cdot\ldots\cdot 3\cdot 2\cdot 1 \equiv N! \tag{18.1}$$

called "N-factorial." It is convenient to extend the *definition* of factorial to $N=0$ by noting that

$$(N-1)! = \frac{N!}{N}$$

and so it is consistent to take

$$0! = \frac{1!}{1} = 1 \tag{18.2}$$

as our definition. A further generalization of the factorial function is discussed in Appendix III.

► **Example 18.1**
Given four different letters, how many different "words" (arrangements of four different letters) can be found?

$4! = 4\cdot 3\cdot 2\cdot 1 = 24$ ◄

If we wish to pick and order r objects from N distinguishable objects, we can do this in $N(N-1)\ \cdots\ (N-r+1)$ ways. This is sometimes called the "number of permutations of N things taken r at a time" and denoted

$$P_r^N = N(N-1)\cdots(N-r+1) = \frac{N!}{(N-r)!} \tag{18.3}$$

This is also sometimes called the "number of variations of the rth class of

N elements" denoted

$$^{N}V_r = \frac{N!}{(N-r)!} \qquad (18.3a)$$

► **Example 18.2**

How many trigrams (three letter words) can be formed from 8 letters?

$$^{8}V_3 = \frac{8!}{5!} = 8 \cdot 7 \cdot 6 = 336$$ ◄

This is the number of variations of r elements if none is to occur more than once. To form variations of r elements out of N with replacement (or repetition) means an element may occur up to r times, that is, the number of such variations is N^r.

► **Example 18.3**

The number of trigrams that can be formed from 8 different letters where the trigrams may contain letter repetitions is $8^3 = 512$. ◄

► **Query**

How many possible four letter words are there in English (including obscenities)? ◄

Combinations

We may also ask in how many ways we can choose r objects from a total of N without regard to order. We call this the "number of combinations of N objects taken r at a time" and denote it by $^{N}C_r$ or $\binom{N}{r}$. These are also called the "binomial coefficients" for reasons that are made obvious below. If we multiply this number by $r!$, the number of ways of ordering the r objects, we obtain P_r^N again, that is,

$$^{N}C_r \cdot r! = P_r^N = N(N-1)\cdots(N-r+1)$$

Thus

$$^{N}C_r = \frac{N(N-1)\cdots(N-r+1)}{r!}$$

or

$$^{N}C_r = \binom{N}{r} = \frac{N!}{(N-r)!r!} = {}^{N}C_{N-r} = \binom{N}{N-r} \qquad (18.4)$$

We note that r and $(N-r)$ appear symmetrically here; that is, the number of r combinations of N objects is equal to the number of $(N-r)$ combinations of the N objects.

The Binomial Theorem

At this point it is useful to recall the binomial theorem:

$$(a+b)^N = \sum_{r=0}^{N} \binom{N}{r} a^{N-r} b^r$$
$$= \sum_{r=0}^{N} \frac{N!}{(N-r)!r!} a^{N-r} b^r \qquad (18.5)$$

This may be written as

$$(a+b)^N = \frac{N!}{N!0!} a^N b^0 + \sum_{r=1}^{N} \frac{N!}{(N-r)!r!} a^{N-r} b^r$$
$$= a^N + \sum_{r=1}^{N} \frac{N!}{(N-r)!r!} a^{N-r} b^r$$

For $a = b = 1$

$$(1+1)^N = 2^N = 1 + \sum_{r=1}^{N} \frac{N!}{(N-r)!r!} \qquad (18.6)$$

We may thus inquire about the number of ways in which *something* can be chosen from N distinguishable objects, which is

$$\sum_{r=1}^{N} \binom{N}{r} = \sum_{r=1}^{N} \frac{N!}{r!(N-r)!} = 2^N - 1 \qquad (18.7)$$

This could also have been obtained by noting that two choices exist for each of the objects: each can be either chosen or left. However, we must exclude the one where all are either chosen or all left.

The Laplace Triangle

Note the following relationship for the binomial coefficients:

$$\binom{N}{r}+\binom{N}{r+1} = \frac{N!}{r!(N-r)!} + \frac{N!}{(r+1)!(N-r-1)!} = \frac{N!}{r!(N-r-1)!}\left\{\frac{1}{N-r}+\frac{1}{(r+1)}\right\} = \frac{N!}{r!(N-r-1)!}\left\{\frac{N+1}{(N-r)(r+1)}\right\} = \frac{(N+1)!}{(r+1)!(N-r)!}$$

or

$$\binom{N}{r}+\binom{N}{r+1}=\binom{N+1}{r+1} \quad (18.8)$$

Relationship (18.8) underlies a way of displaying and calculating the binomial coefficients that was first pointed out by Pierre Simon de Laplace.* This "Laplace triangle" is shown in Figure 18.1 where each row represents N, starting with 0, and each column represents r [starting at the left of each row with $\binom{N}{0}$]. For example,

$$\binom{6}{0}=1, \binom{6}{1}=6, \binom{6}{2}=15, \binom{6}{3}=20,$$
$$\binom{6}{4}=\binom{6}{2}=15, \binom{6}{5}=6, \text{ and } \binom{6}{6}=1$$

We note that we may verify (18.8). For example,

$$\binom{6}{3}+\binom{6}{4}=20+15=35=\binom{7}{4}$$

The Laplace triangle enables one to calculate the binomial coefficients by addition and is a convenient aid for calculations on a desert island.

► *Exercise*

Poker is a game played with a deck of *52* cards consisting of four suits (spades, clubs, hearts and diamonds), each of which contains *13* cards, 2, 3, 4, 5, 6, 7, 8, 9, 10, J, Q, K, and A. When considered sequentially the A may be taken to be 1 or A but not both, that is, 10, J, Q, K, A is a *5*-card sequence called a "straight," as is A, 2, 3, 4, 5, but Q, K, A, 2, 3 is not sequential, that is, not a "straight."

A poker hand consists of *5* cards chosen at random. A winning poker hand is the one with a higher "rank" than all other hands that were played through to the end of a particular game.

A "flush" is a *5*-card hand all of the same suit. A "pair" consists of *2* and only *2* cards of the same kind, for example, (J^S, J^C) or (Q^S, Q^D), etc. "Three-of-a-kind" consists of *3* and only *3* cards of the same kind, for example, ($9^S, 9^H, 9^D$). "Four-of-a-kind" is defined similarly. A "full house" is a *5*-card hand consisting of a "pair and three-of-a-kind."

The ranks of the various poker hands are as follows with the highest rank first:

(a) Royal flush (10, J, Q, K, A of one suit)
(b) Straight flush (consecutive sequence of one suit that is not a royal flush)
(c) Four-of-a-kind
(d) Full house
(e) Flush (not a straight flush)
(f) Straight
(g) Three-of-a-kind (not a full house)
(h) Two pairs (not four-of-a-kind)
(i) One pair
(j) No pair

(1) Show the total number of possible poker hands is 2,598,960.
(2) Show the number of possible ways to deal the various hands are:

4 for **a**	10,200 for **f**
36 for **b**	54,912 for **g**
624 for **c**	123,552 for **h**
3,744 for **d**	1,098,240 for **i**
5,108 for **e**	1,302,540 for **j** ◄

**Theorie analytique des probabilities*, 1812.

N	r = 0	r = 1	r = 2	r = 3	r = 4	r = 5	r = 6	r = 7	r = 8	r = 9	r = 10	r = 11	r = 12	r = 13	r = 14	r = 15	r = 16	r = 17	r = 18	r = 19	r = 20
0	1																				
1	1	1																			
2	1	2	1																		
3	1	3	3	1																	
4	1	4	6	4	1																
5	1	5	10	10	5	1															
6	1	6	15	20	15	6	1														
7	1	7	21	35	35	21	7	1													
8	1	8	28	56	70	56	28	8	1												
9	1	9	36	84	126	126	84	36	9	1											
10	1	10	45	120	210	252	210	120	45	10	1										
11	1	11	55	165	330	462	462	330	165	55	11	1									
12	1	12	66	220	495	792	924	792	495	220	66	12	1								
13	1	13	78	286	715	1287	1716	1716	1287	715	286	78	13	1							
14	1	14	91	364	1001	2002	3003	3432	3003	2002	1001	364	91	14	1						
15	1	15	105	455	1365	3003	5005	6435	6435	5005	3003	1365	455	105	15	1					
16	1	16	120	560	1820	4368	8008	11440	12870	11440	8008	4368	1820	560	120	16	1				
17	1	17	136	680	2380	6188	12376	19448	24310	24310	19448	12376	6188	2380	680	136	17	1			
18	1	18	153	816	3060	8568	18564	31824	43758	48620	43758	31824	18564	8568	3060	816	153	18	1		
19	1	19	171	969	3876	11628	27132	50388	75582	92378	92378	75582	50388	27132	11628	3876	969	171	19	1	
20	1	20	190	1140	4845	15504	38760	77520	125970	167960	184756	167960	125970	77520	38760	15504	4845	1140	190	20	1

Figure 18.1
Table of binomial coefficients $\binom{n}{r}$ in Laplace triangle format.

► **Example 18.4**

A group of students was asked to toss 6 coins simultaneously and to record the number of heads observed each time. What do we expect for the occurrences (expressed as a fraction) of 0, 1, 2, 3, 4, 5, and 6 heads respectively, assuming ideal coins and fair tosses?

To calculate the probability of r heads, we must count the number of ways favorable to the event; that is, we count the number of ways a toss of n coins will result in r heads. This is ${}^nC_r = n!/r!(n-r)!$ by (18.4). We must also count the total possible number of outcomes. Since each coin has two possibilities, heads or tails, the total is 2^n.

The probability of obtaining r heads with n coins is thus ${}^nC_r/2^n$. For this case $n = 6$ and $r = 0, 1, \ldots, 6$.

In an experiment 624 students were divided into seven groups with each student making 20 trials. The results are summarized in Table 18.1 where the theoretical probabilities are calculated as above.

We may also use these data to investigate the empirical assumption that the probability represents the observable frequency in the limit of a large number of trials. In Figure 18.2 we graph the measured frequency of one type of event plotted against the accumulated number of trials. ◄

Table 18.1 Results of an Experiment in which a Class of 624 Students was Asked to Toss Six Coins Simultaneously[a]

	Number of Heads							
	0	1	2	3	4	5	6	Total Trials
Group	Percent							
A	28 = 2.1	136 = 10.3	300 = 22.7	378 = 28.6	302 = 22.9	141 = 10.7	35 = 2.7	1,320
B	29 = 2.3	131 = 10.2	282 = 22.0	389 = 30.4	285 = 22.3	143 = 11.2	21 = 1.6	1,280
C	45 = 2.2	185 = 8.7	485 = 22.9	668 = 31.5	482 = 22.7	214 = 10.1	41 = 1.9	2,120
D	30 = 1.4	189 = 9.0	492 = 23.4	632 = 30.1	504 = 24.0	220 = 10.5	33 = 1.6	2,100
E	23 = 1.3	145 = 8.3	430 = 24.7	545 = 31.3	394 = 22.7	168 = 9.7	35 = 2.0	1,740
F	30 = 1.9	142 = 9.2	374 = 24.3	482 = 31.4	329 = 21.4	159 = 10.3	24 = 1.6	1,540
G	25 = 1.1	239 = 10.0	567 = 23.8	715 = 30.0	556 = 23.4	241 = 10.1	37 = 1.6	2,380
Total	210 = 1.7	1167 = 9.4	2930 = 23.5	3809 = 30.5	2852 = 22.8	1286 = 10.3	226 = 1.8	12,480
Theoretical probability	$\frac{1}{64} = 1.6$	$\frac{6}{64} = 9.4$	$\frac{15}{64} = 23.4$	$\frac{20}{64} = 31.2$	$\frac{15}{64} = 23.4$	$\frac{6}{64} = 9.4$	$\frac{1}{64} = 1.6$	

[a] The number of heads observed each time was recorded. The students were divided into seven unequal groups with each student making 20 trials.

Remarks on the Factorial Function

Having introduced the factorial function, (18.1), we note that Table A.III.1 of Appendix III lists values of $n!$ to four places of decimals for $n = 1$ to 100 and the common logarithms of $n!$ to five places of decimals for $n = 1$ to 200.

A very good estimate of $n!$ for n large is obtained from *Stirling's approximation*:

$$n! \sim (2\pi n)^{1/2}\, n^n e^{-n} \qquad (18.9)$$

which is good for n as small as 10. This

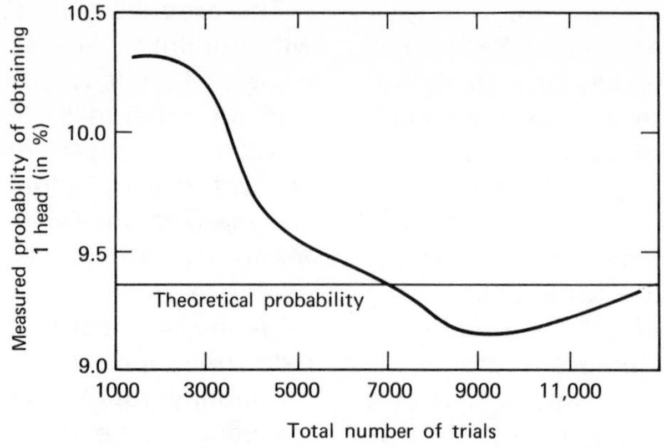

Figure 18.2
Plot of the measured frequency (in %) of obtaining one head versus the accumulated number of trials.

was first derived by Stirling in 1730.* It is discussed in Appendix III, but it is useful to quote Stirling's own statement of the accuracy of approximation (circa 1764): "If the expression $n!$ be replaced by the formula [(18.9)], the true value will have been divided by a number lying between 1 and $1/(10n)$."

Table 18.2 indicates how the percentage error of this approximation decreases as n increases.

Table 18.2[a] Comparison of the Exact Value of $n!$ with the Value Obtained from Stirling's Approximation for Values of n between 1 and 100

n	$n!$	$(2\pi n)^{1/2} n^n e^{-n}$	Difference	$\frac{\text{Difference}}{n!}$
1	1	0.922	0.078	0.08
2	2	1.919	0.081	0.04
5	120	118.019	1.981	0.02
10	$(3.6288)10^6$	$(3.5986)10^6$	$(0.0302)10^6$	0.008
100	$(9.3326)10^{157}$	$(9.3249)10^{157}$	$(0.0077)10^{157}$	0.0008

[a] Taken from W. Feller, *Probability Theory and Its Applications*, John Wiley, 1st Edition, New York, 1950.

**Methodus Differentialus*, 1730.

Some Necessary Specifications in Combinatorial Analysis

No discussion of combinatorial analysis is complete without a use of the example of an urn with N distinguishable balls, numbered from 1 to N. We ask the question: In how many ways can one draw a sample of n balls from such an urn?

1. We must specify whether the sampling is done *without replacement* or *with replacement.* This is to ask whether we physically remove n balls at a time from the urn or do we draw one, *note* its nature, and replace it, winding up with a sample of *n notations.* The total number of ways of sampling with replacement is N^n.
2. A second important consideration is to give a rule for deciding whether two

possible samples are distinguishable. This is to reply to the initial query with the rejoinder: What's a "way"?

Whether two possible samples are distinguishable or not depends on whether the samples are regarded as being *ordered*, *unordered*, or *partitioned*.

A sample is said to be *ordered* if attention is paid not only to the numbers that appear on the balls in the sample, but also to the order in which the numbers appear (i.e., the order in which the balls are drawn).

A sample is said to be *unordered* if attention is paid only to the numbers and not to the order.

A sample is said to be *partitioned* if one divides the n positions (first, second, $\ldots$, nth) in which a ball may be drawn into k groups of sizes $n_1, n_2, \ldots, n_k$, (so that $n_1 + n_2 + \cdots + n_k = n$) which are considered ordered, but within each group of positions the numbers that are drawn on these positions are considered unordered. *The number of partitioned samples* is

$$\frac{N!}{n_1!n_2!\cdots n_k!} \qquad (18.10)$$

►Example 18.5

Suppose we have an urn with *4* balls numbered 1, 2, 3, 4. A sample of *3* balls is to be drawn.

(a) Sampling without replacement with the samples ordered:
The possible samples that could be drawn are

123	231	341	412
124	234	342	413
132	241	312	423
134	243	314	421
142	213	321	431
143	214	324	432

This total number is given by (18.3)

$$P_3^4 = 4 \cdot 3 \cdot 2 = 24$$

(b) Sampling without replacement, samples unordered:
Possibilities are

123, 124, 134, and 234

This number is given by (18.4)

$$^4C_3 = \frac{4 \cdot 3 \cdot 2}{1 \cdot 2 \cdot 3} = 4$$

(c) Sampling with replacement, samples ordered:

$$4^3 = 64 \text{ ways}$$

(d) Sampling with replacement, samples unordered:

$$^{N+r-1}C_r = {}^{4+3-1}C_r = \frac{6 \cdot 5 \cdot 4}{1 \cdot 2 \cdot 3} = 20 \text{ ways}$$

(e) Sampling without replacement, samples are partitioned: (a, b, c) positions are 1st, 2nd, 3rd.
A partition specifies that only certain of the partitions are ordered; for example, suppose 2, 3 are unordered. This means that 123 = 132. The sample is partitioned so that there are two groups of partitions, the first group consisting of the first position and the second group consisting of the second and third positions.
The possible samples are

123	213	312	412
124	214	314	413
134	234	324	423

This corresponds to

$$\frac{N!}{n_1!n_2!\cdots n_k!(N-n_k)!} = 12$$

◄

►Example 18.6
Sock problem

There are *10* individual socks lying loose in the drawer of a bachelor living alone: two identical pairs of green, one pair of yellow, one pair of pink, and one pair of black. Stumbling about in the dark, the bachelor draws three socks at random from the drawer. What is the probability of getting

(a) A pair of yellow socks?
(b) A pair of pink socks?
(c) A pair of black socks?
(d) A pair of green socks?
(e) A pair of any color?

The number of possible draws is

$$^{10}C_3 = \binom{10}{3} = \frac{10 \cdot 9 \cdot 8}{1 \cdot 2 \cdot 3} = 120$$

(a) The number of ways of drawing *2 yellow socks + any* is

$$^2C_2 \times {}^8C_1 = 1 \times 8 = 8$$

$$\therefore \text{Pr}(2 \text{ yellow}) = \frac{8}{120} = 0.0666$$

(b) and **(c)** are equivalent to **(a)** so

$$\text{Pr}(2 \text{ pink}) = \text{Pr}(2 \text{ black}) = 8/120 = 0.0666$$

(d) The number of ways of drawing *2 green socks + any* (including a third green sock) is

$$^4C_2 \times {}^8C_1 = \frac{4 \cdot 3}{1 \cdot 2} \times 8 = 48$$

$$\therefore \text{Pr}(2 \text{ or more green}) = \frac{48}{120} = 0.4000$$

(e) If any pair is drawn a pair of another color is precluded, since only three socks are drawn. The various probabilities in **(a)**, **(b)**, **(c)** and **(d)** are, therefore, independent.

$$\begin{aligned}\text{Pr(any pair)} &= \text{Pr}(2 \text{ yellow}) + \text{Pr}(2 \text{ pink}) + \text{Pr}(2 \text{ black}) + \text{Pr}(2 \text{ or more green}) \\ &= 72/120 = 0.6000\end{aligned}$$

◄

The Multinomial Theorem

The binomial theorem may be "turned around" and proved as follows:

$$(x+y)^n = \sum_{i=0}^{n} C_i x^{n-i} y^i$$

where it is required to find C_i.

This may be seen to be equivalent to determining the number of ways of dividing n factors into two groups.

The first term is immediately seen to be x^n since there is one power of x from each factor.

The general term, multiplying $x^{n-i}y^i$, arises from selecting a power of x from $(n-i)$ of the factors and a power of y from each of the remaining i factors. To get C_i we must count the ways of dividing n factors into two groups so that one group contains i factors (each providing a power of y) and the other $(n-i)$ factors (each providing a power of x). This is just $\binom{n}{i}$ so that

$$(x+y)^n = \sum_{i=0}^{n} \binom{n}{i} x^{n-i} y^i$$

We may now use this mode of reasoning to prove the "multinomial theorem."

Here we have the problem of expanding

$$(x_1 + x_2 + \cdots + x_k)^n = \sum_{n_1, n_2, \ldots, n_k} C x_1^{n_1} x_2^{n_2} \cdots x_k^{n_k}$$

where

$$\sum_{i=1}^{k} n_i = n.$$

Here n_1 of the n factors provides a power of x_1, n_2 of the n factors provides a power of x_2, etc.

The coefficient is the number of ways of dividing n factors into k groups containing $n_1, n_2, \ldots, n_k$ factors, respectively, which is what we have called the "number of partitioned samples" (18.10).

Therefore, each C is

$$\frac{n!}{n_1! n_2! \cdots n_k!}$$

and we may write each term as

$$\frac{n!}{n_1!n_2!\cdots n_k!}x_1^{n_1}x_2^{n_2}\cdots x_k^{n_k} = n!\prod_{i=1}^{k}\frac{x_i^{n_i}}{n_i!}$$

The result is the multinomial theorem:

$$(x_1+x_2+\cdots+x_n)^n = \sum_{n_1,n_2,\ldots,n_k} n!\prod_{i=1}^{k}\frac{x_i^{n_i}}{n_i!} \qquad (18.11)$$

where each index goes between 1 and n but subject to the constraint that

$$\sum_{i=1}^{k} n_i = n \qquad (18.12)$$

Before leaving this section let us use our combinatorial techniques on an earlier example.

► **Example 18.7**

We may return (for the first of several visits) to the scene of the problem mentioned at the end of Chapter 17, the selection from *Gadsby.*

At this time we shall apply a combinatorial argument to the problem. Let us assume that the 580-letter passage is equivalent to sampling an equivalent size selection from the 89,602 letters used in the Northwestern sample wherein were found 11,290 occurrences of the letter e.

Given the Northwestern sample, the obvious question has an obvious answer: the 580-letter selection is "expected" to have 73 occurrences of the letter e. The important question, however, relates to the deviation from 73 that one might expect to find in random samples of this size.

Let us consider only the two possibilities that a given letter is e, called event E, or not-e, called event $\bar{E}$. The original sample of N (e.g., 89,602) letters had M (e.g., 11,290) E's and $(N-M)$ occurrences of $\bar{E}$. Let us consider the drawing of a sample of n (e.g., 580) letters from the total. There are $\binom{N}{n}$ such samples of size n. A particular sample of n elements contains m events E and $(n-m)$ events $\bar{E}$. In order that we should have a sample with n elements and m events E, the m events must be chosen at random from the M events in the original sample. This can happen in $\binom{M}{m}$ ways. Also, the $(n-m)$ elements are chosen from the $(N-M)$ events $\bar{E}$ in the original sample, which can occur in $\binom{N-M}{n-m}$ different ways.

Among the $\binom{N}{n}$ samples of size n there will, therefore, be $\binom{M}{m}\binom{N-M}{n-m}$ samples, which contain m events of type E. The quotient of this to the total number of ways of drawing a sample of n yields the relative frequency (or probability) of samples of size n containing m E-type events:

$$\Pr(m) = \frac{\binom{M}{m}\binom{N-M}{n-m}}{\binom{N}{n}} = \frac{({}^{M}C_m)({}^{N-M}C_{n-m})}{{}^{N}C_n} = \frac{M!(N-M)!(N-n)!n!}{(M-m)!m!(N-M-n+m)!(n-m)!N!} \qquad (18.13)$$

Let us recall Stirling's formula (18.9):

$$n! \approx n^n e^{-n}\sqrt{2\pi n}$$

We may take natural logarithms and obtain

$$\ln n! \approx n\ln n - n + \frac{1}{2}\ln n + \frac{1}{2}\ln 2\pi$$
$$\approx \left(n+\frac{1}{2}\right)\ln n - n \qquad (18.14)$$
$$\ln n! \approx n(\ln n - 1) \qquad \text{for } n \text{ large}$$

For our problem, $m = 0$. Thus (18.13) reduces to

$$\Pr(0) = \frac{(N-M)!(N-n)!}{(N-M-n)!N!}$$

$$\ln \Pr(0) = \ln(N-M)! + \ln(N-n)! - \ln(N-M-n)! - \ln N!$$

Here $N = 89{,}602$, $M = 11{,}290$, $n = 580$

$$\therefore \ln P \approx \ln(78{,}312!) + \ln(89{,}022!) - \ln(77{,}732!) - \ln(89{,}602!)$$
$$\approx 804{,}143 + 925{,}530 - 797{,}610 - 932{,}142$$
$$\approx -78.38$$

This yields a probability $P \approx 9.1 \times 10^{-35}$, which is very small indeed! It is thus hard to imagine that this selection corresponds to a random sample of English prose, let alone one that comes from the original sample of 89,602. Actually, the full title of the book from which the selection was taken is *Gadsby, a story of over 50,000 words without using the letter e.* ◄

19

EVENT CALCULUS —THE LOGIC OF PROBABILITY

Although we are stressing the intuitive nature of probability, the introduction of some mathematical concepts now taught in primary school can be justified as providing a vocabulary for our intuitive discussion.

Definitions

A *description space*,* S, is a *list* or *set* of all possible descriptions of the outcome of a random experiment or observation one of which must occur; for example, for the toss of a coin $S_1 = \{H, T\}$. More complete is $S_2 = \{H, T, E\}$ and even more so is $S_3 = \{H, T, E, L\}$ where we include the possibility that the coin is lost. The elements of S must be *exclusive*; that is, not more than one can occur. Before we can ask any questions about the outcome of the experiment, we must *choose* a description space. Suppose we choose S_2. Will the outcome be: (1) H; (2) T; (3) E; (4) either H or T; (5) either H or E; (6) either T or E; (7) either H or T or E; (8) neither H nor T nor E?

The *null set* $\emptyset$ is the set containing no elements.

Given a set S, another set A is said to be a *subset* of S, denoted $A \subset S$ if every member of A is a member of S. For future reference we note that for all sets $S, \emptyset \subset S$ and $S \subset S$.

*This is usually called a *sample space*. We prefer to reserve "sample" for sampling.

► **Example**

For the set $S_2 = \{H, T, E\}$, the possible subsets are (1) $\{H\}$; (2) $\{T\}$; (3) $\{E\}$; (4) $\{H, T\}$; (5) $\{H, E\}$; (6) $\{T, E\}$; (7) S_2; (8) $\emptyset$. We state the general theorem:

If a set S has n elements, there are 2^n possible subsets of S. (19.1)

The reader should prove this using the notions relevant to (18.7) and (18.8). ◄

We have introduced these ideas for the following reason. Any question that can be asked about the outcome of an experiment S can be posed in the form: Will the outcome be a member of the set E where E is a subset of S?

Note that nothing in this mathematical formalism tells you how to choose S. This is where the experience and judgment of the observer is needed. For example, for a given coin toss, which S is most relevant, $\{H, T\}, \{H, T, E\}$, or $\{H, T, E, L\}$? When in doubt the description space should be made larger instead of smaller.

UNIFORM PROBABILITY DISTRIBUTION

We may complete our formal description of the classical definition of probability by denoting the "size" or number of members of a set S by Nu[S]; for example, if $S = \{A_1, \ldots, A_n\}$ then $\text{Nu}[S] = n$.

Clearly, $\text{Nu}[\emptyset] = 0$.

We say that a description space is *discrete* if Nu[S] is *finite.* If we say that a description space has a *uniform probability distribution,* we mean that each description is *equally likely* in the classical sense; that is, $\Pr[\{E_i\}] = 1/\text{Nu}[S]$ for $i = 1, \ldots, \text{Nu}[S]$.

It then follows that the probability of an event E defined on a discrete description space is written

$$\Pr[E] = \frac{\text{Nu}[E]}{\text{Nu}[S]} \tag{19.1a}$$

If we speak of a continuous variate having a *uniform distribution* on an interval $a \le X \le b$, written X is $U(a, b)$, we mean that

$$f(X) = \begin{cases} \dfrac{1}{b-a} & \text{for} \quad a \le X \le b \\ 0 & \text{otherwise} \end{cases} \tag{19.1b}$$

We emphasize that this is normalized since

$$\int_{-\infty}^{\infty} f(X)dX = \int_a^b \frac{1}{b-a} dX = 1$$

The probability that $\alpha \le X \le \beta$ is

$$\Pr[\alpha \le X \le \beta] = \frac{\beta - \alpha}{b - a}$$

The expectation value or mean of this distribution is

$$E[X] = \langle X \rangle = \int Xf(X)dX$$
$$= \frac{1}{b-a}\left[\frac{X^2}{2}\right]_a^b = \frac{b+a}{2} \tag{19.1c}$$

The variance, $\text{Var}(X) = \sigma^2(X)$, is

$$\sigma^2(X) = \langle X^2 \rangle - [\langle X \rangle]^2$$
$$\langle X^2 \rangle = \int X^2 f(X)dX = \frac{1}{b-a}\left.\frac{X^3}{3}\right|_a^b$$
$$= \frac{(b^3 - a^3)}{(b-a)3}$$
$$\therefore \sigma^2(X) = \frac{(b^3 - a^3)}{3(b-a)} - \left(\frac{b+a}{2}\right)^2 = \frac{(b-a)^2}{12} \tag{19.1d}$$

The corresponding cumulative distribution function is

$$F(x) = \begin{cases} 0 & \text{for} \quad x \le a \\ \dfrac{x-a}{b-a} & \text{for} \quad a \le x \le b \\ 1 & \text{for} \quad x \ge b \end{cases} \tag{19.1e}$$

ORDERED AND UNORDERED SETS

In describing events we must often specify whether the *order* of a set of objects is important. Formally we define a description $(z_1, \ldots, z_n)$ as an "n-tuple" where the ith component is z_i. An n-tuple is said to be *ordered* or *unordered,* according to whether attention is paid to the order in which the components are written. The ordered n-tuple $(z_1, \ldots, z_n)$ may also be written as a *vector,* (z).

Two ordered n-tuples $(z_1, \ldots, z_n)$ and $(z'_1, \ldots, z'_n)$, are equal if and only if $z_1 = z'_1$, $z_2 = z'_2$, etc.

Two unordered n-tuples are equal if they have the same components; for example, $(2,3,2) = (2,2,3)$ if both are unordered.

► **Example 19.1**

What is the probability that each hand in a bridge game contains exactly one ace? (There are 4 players, each with 13 cards.)

The description space S = set of 52-tuples where no two components of the 52-tuples are alike.

Partition: 1st player's hand = 13
2nd player's hand = 13
3rd player's hand = 13
4th player's hand = 13

$$\text{The number of bridge hands} = \text{Nu}[S] = \frac{52!}{13!13!13!13!}$$

Let A be the event that each hand contains exactly one ace.

$$\text{Nu}[A] = \frac{4!(48)!}{(12!)(12!)(12!)(12!)}$$

$$\Pr[A] = \frac{\text{Nu}[A]}{\text{Nu}[S]} = \frac{4!(13)^4}{52 \cdot 51 \cdot 50 \cdot 49}$$ ◄

COMPLEMENT

The *complement* of an event E, denoted $\bar{E}$ (also in the literature as $-E$, E^*, E^c, E', and $\tilde{E}$) is the event that consists of the descriptions that do not belong to E; for example, $S = \{1,2,3,4,5,6\}$, $E = \{2,4,6\}$, then $\bar{E} = \{1,3,5\}$.

It is convenient to introduce the *Venn diagram* as in Figure 19.1. The rectangle represents the description space S while the circle represents the event E. The shaded area represents $\bar{E}$.

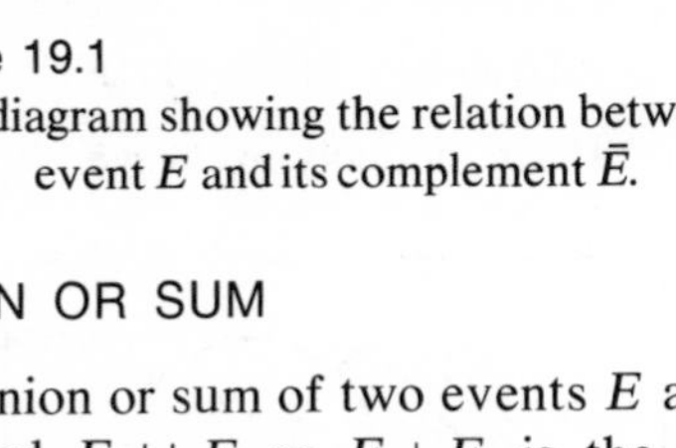

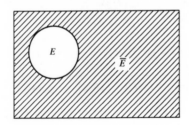

Figure 19.1
Venn diagram showing the relation between an event E and its complement $\bar{E}$.

UNION OR SUM

The union or sum of two events E and F, denoted $E \cup F$ or $E + F$, is the event whose members are descriptions that belong to ***either*** E ***or*** F.

► **Example**

$S = \{1,2,3,4,5,6\}$; $E = \{2,4,6\}$; $F = \{3,5\}$, then $E \cup F = \{2,3,4,5,6\}$. ◄

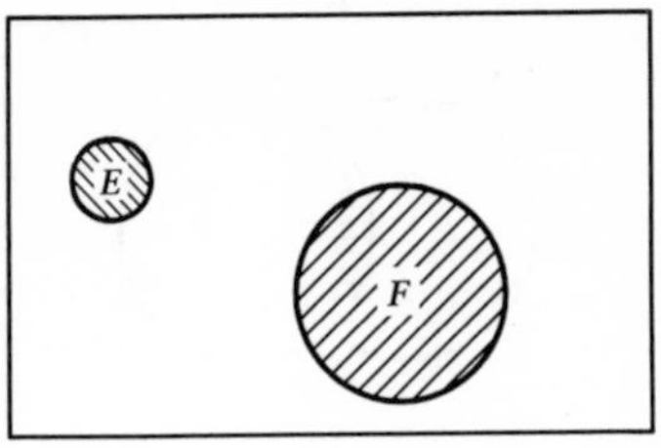

Figure 19.2
Venn diagram showing the union of two events, $E \cup F$, when the events are exclusive.

This is illustrated in Figure 19.2 in which the circles represent the events E and F as being mutually exclusive; that is, they have no elements in common. The union $E \cup F$ is the entire shaded region.

Note that the union is also defined when the events are not exclusive.

► **Example**

$S = \{1,2,3,4,5,6\}$; $E = \{2,4,6\}$; $F = \{3,4,5\}$. Here $E \cup F = \{2,3,4,5,6\}$. ◄

This is illustrated in Figure 19.3 where the shaded region again represents $E \cup F$.

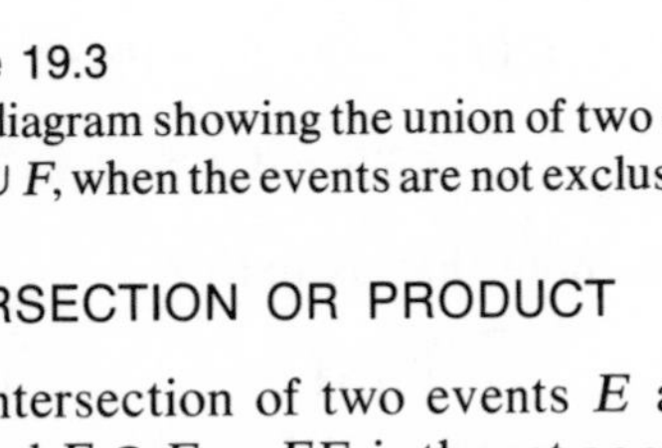

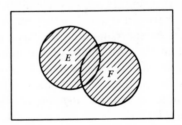

Figure 19.3
Venn diagram showing the union of two events, $E \cup F$, when the events are not exclusive.

INTERSECTION OR PRODUCT

The intersection of two events E and F, denoted $E \cap F$ or EF, is the set consisting of descriptions that are members of ***both*** E ***and*** F. To say that the intersection or product of two events has occurred is to say that they have occurred simultaneously. *Example*: $S = \{1,2,3,4,5,6\}$; $E = \{2,4,6\}$; $F = \{3,4,5\}$; $E \cap F = \{4\}$.

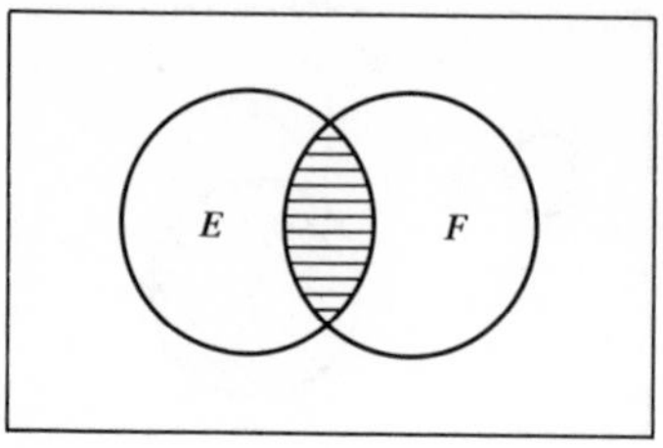

Figure 19.4
Venn diagram showing the intersection of two events, $E \cap F$.

This is illustrated in Figure 19.4 where the shaded area is now the intersection.

We see that events E and F are *mutually exclusive* if and only if $E \cap F = EF = \emptyset$. In Figure 19.2, $E \cap F = EF = \emptyset$.

We note that the concepts of union and intersection generalize readily; for example, we may have $E \cup F \cup G \cup H \cdots$ and $EFGH \cdots$.

PROBABILITY

Let us now introduce the idea of probability. Given a random situation that is described by S, probability is a function that to every event E assigns a real number $\Pr(E)$ between 0 and 1 inclusive. We introduce as an assumption the basic property that the probability of the union of two *mutually exclusive* events E and F is the sum of their probabilities. This is sometimes called the "*defining relation of probability*":

$$\Pr[E \cup F] = \Pr[E] + \Pr[F]$$

$$\text{if} \quad EF = \emptyset \tag{19.2}$$

We also note that as before

$$\Pr[S] = 1 \tag{19.2a}$$

We are not concerned with stressing the axiomatic development and so the proofs of some of our subsequent statements, which follow from (19.2), are indicated after (19.7).

$$\Pr[\emptyset] = 0 \tag{19.3}$$

ADDITIVE RULE FOR PROBABILITIES

$$\Pr[E \cup F] = \Pr[E] + \Pr[F] - \Pr[EF] \tag{19.4}$$

where E, F are not mutually exclusive.

SUBTRACTIVE RULE FOR PROBABILITIES

$$\Pr[E - F] = \Pr[E] - \Pr[EF] \tag{19.5}$$

where $E - F$ refers to elements that are in E but are not in F. This is indicated in Figure 19.5 where the shaded region indicates $E - F$.

We may extend the addition rule to three non-mutually-exclusive events A, B, C:

$$\begin{aligned}\Pr(A + B + C) = \Pr(A) + \Pr(B) + \Pr(C) \\ - \Pr(AB) - \Pr(AC) \\ - \Pr(BC) + \Pr(ABC)\end{aligned} \tag{19.6}$$

In general for n events $A_1, \ldots, A_n$:

$$\begin{aligned}\Pr[A_1 + \cdots + A_n] &= \sum_{i=1}^{n} \Pr[A_i] - \sum_{\substack{i,j=1 \\ i \neq j}}^{n} \Pr[A_i A_j] \\ &+ \sum_{\substack{i,j,k \\ i \neq j \neq k}} \Pr[A_i A_j A_k] + \cdots \\ &+ (-1)^{n+1} \Pr(A_1 A_2 A_3 \cdots A_n)\end{aligned} \tag{19.7}$$

Proofs of (19.3), (19.4), and (19.5)

Starting with the defining relation (19.2), we have

$$\Pr[E \cup F] = \Pr[E] + \Pr[F]$$

if E and F are mutually exclusive.

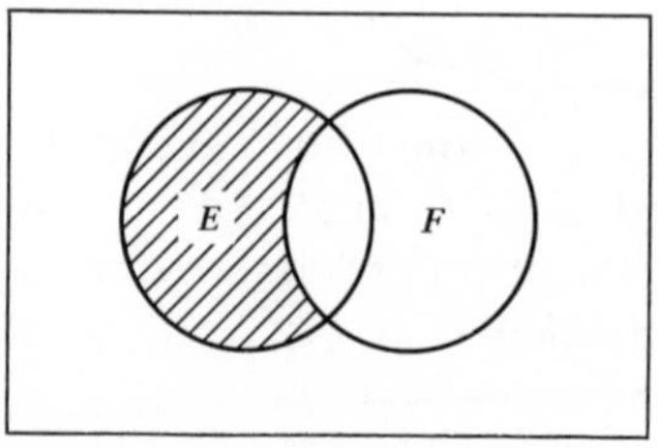

Figure 19.5
Venn diagram showing the event $E - F$.

Theorem: $\Pr[\emptyset] = 0$

Proof: Starting with (19.2), let $E = \emptyset$, $F = \emptyset$. $EF = \emptyset$ so the requirement of (19.2) is met. $E \cup F = \emptyset \; \therefore \Pr[\emptyset] = 2\Pr[\emptyset]$, which can only occur if $\Pr[\emptyset] = 0$ which is (19.3), Q.E.D.

Theorem: $\Pr[E - F] = \Pr[E] - \Pr[EF]$

Proof: $E - F$ and EF are mutually exclusive,

$\therefore$ By (19.2) $\Pr[E - F] + \Pr[EF] = \Pr[E]$

$\therefore \Pr[E - F] = \Pr[E] - \Pr[EF]$

which is (19.5). Q.E.D.

Theorem: $\Pr[E \cup F] = \Pr[E] + \Pr[F] - \Pr[EF]$

Proof: $E - F$ and F are mutually exclusive. Thus, using (19.2) we have

$$\Pr[E \cup F] = \Pr[F] + \Pr[E - F] = \Pr[F] + \Pr[E] - \Pr[EF] \text{ by (19.5).}$$

But this result is (19.4). Q.E.D.

DeMORGAN'S LAWS

I. The complement of the intersection (product) of events equals the union (sum)

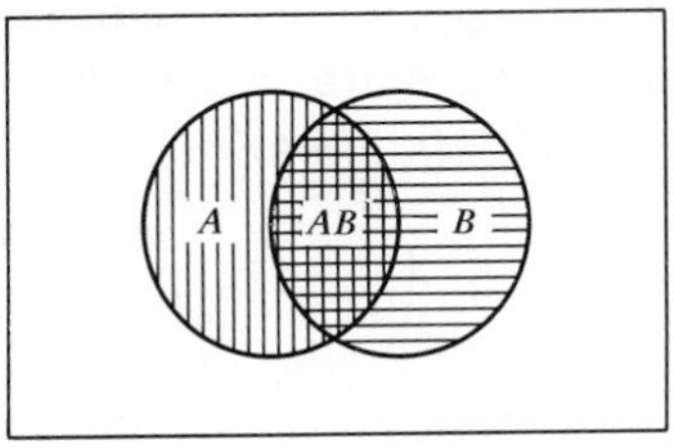

Figure 19.6
Venn diagram showing relation 19.8.

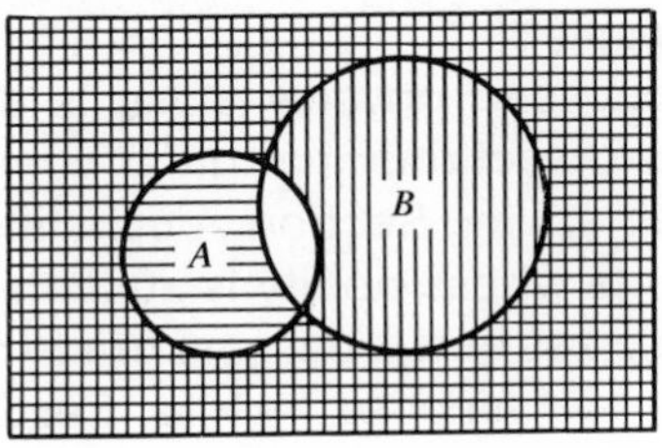

Figure 19.7
Venn diagram showing relation 19.9.

of the complements of the events; that is,

$$(AB)^* = \overline{(AB)} = \bar{A} \cup \bar{B} \quad (19.8a)$$

This is easily seen by reference to Figure 19.6.

II. The complement of the union of events = the intersection of complements; that is,

$$\overline{(A \cup B)} = \bar{A}\bar{B}. \quad (19.9a)$$

This is illustrated in Figure 19.7 where $\bar{A}$ is indicated by vertical hatching and $\bar{B}$ is indicated by horizontal hatching and $\overline{(A \cup B)} = \bar{A}\bar{B}$ is cross-hatched.

Note that these may be expressed as logical statements:

I. NOT (A and B) = (NOT A) or (NOT B)
II. NOT (A or B) = (NOT A) and (NOT B)

Expressed in terms of probabilities we write

$$\Pr(\overline{AB}) = \Pr(\bar{A} \cup \bar{B}) \quad (19.8b)$$

and

$$\Pr(\overline{A \cup B}) = \Pr(\bar{A}\bar{B}) \quad (19.9b)$$

Conditional Probabilities: Dependence and Independence

We have thus far considered the probabilities of events alone, in vacuo as it were. When we consider probabilities in situ, we find that we must investigate the important situation where one probability can depend on another. Suppose we consider the probability of the simultaneous occurrence of two events A and B. We may say that the probability of B may be *unconditional*, written $\Pr(B)$, or *conditional*, denoted $\Pr_A(B)$, which means that this is the probability of B *assuming A has occurred.**

*This is usually written $P(B|A)$. We prefer to reserve the argument for other designations and to maintain a distinction between *probability* (Pr) and the event or *function P*. We also use $P_A(x_i)$ to denote the probability of a *sequence* of x_i.

► **Example 19.2**

Suppose a physics class is divided into two sections that differ somewhat in their ability. Section A consists of 70 students whose ability is such that 80% of the class gets a passing grade on any given examination while Section B consists of 30 students whose ability is such that only 60% get a passing grade on a typical examination.

Let us consider that, after a typical examination, the exam books are mixed together in a random fashion. When the professor sits down to grade the books, what is the probability that a given book will receive a passing grade?

Define the event, P, that the book receives a passing grade while $\bar{P} = F$ is the event describing a failing grade. We can calculate the probability of P by counting the total number of passing grades in the total class including both sections and dividing by the total number of students:

$$\Pr(P) = \frac{(0.80)(70) + (0.60)(30)}{100} = \frac{(56+18)}{100} = 0.74$$

Let us now investigate the case that the professor knows from which section the exam book came, that is, he is given the occurrence of event A (the book came from section A) or of event B (the book came from section B) where $\bar{A} = B$ and $\bar{B} = A$. If the exam book is known to come from A, $\Pr_A(P) = 0.80$ where $\Pr_A(P)$ is the *conditional* probability of event P assuming that A has occurred. Also, $\Pr_B(P) = 0.60 = \Pr_{\bar{A}}(P)$.

We thus see that the question of the probability of the exam grade being P depends on the state of our knowledge. The *unconditional* probability of P is $\Pr(P) = 0.74$. The two conditional probabilities are $\Pr_A(P) = 0.80$ and $\Pr_B(P) = 0.60$. Also, the unconditional probability of F is $\Pr(F) = \Pr(\bar{P}) = 1 - \Pr(P) = 26/100$ with the conditional probabilities being $\Pr_A(F) = 0.20$ and $\Pr_B(F) = 0.40$.

We may also have the professor ask another question of a given exam book: Was it written by a student in section A or in B? The unconditional probabilities of A and of B are easily seen to be $\Pr(A) = 70/100 = 0.70$ and $\Pr(B) = \Pr(\bar{A}) = 0.30$. Suppose, however, that the book has been graded by the professor before he undertakes to answer the question of the book's origin. If the book has received a P, we calculate the probability that it came from A as follows, since there are 74 total P grades and 56 of them come from A: $\Pr_P(A) = 56/74 = 0.76$. If the grade is F, we get $\Pr_F(A) = \Pr_{\bar{P}}(A) = 14/26 = 0.54$.

For completeness,

$$\Pr_P(B) = \frac{18}{74} = 0.24$$

and

$$\Pr_{\bar{P}}(B) = \frac{12}{26} = 0.46$$

What can we say about the probability that the exam book came from A *and* is a pass? There are 56 such books out of a total of 100. Therefore,

$$\Pr(A \text{ and } P) = \Pr(AP) = \frac{56}{100} = \frac{74}{100}\frac{56}{74} = \Pr(P) \cdot \Pr_P(A)$$ ◄

This is a general statement, true for any events A and B and may serve to define the conditional probability.

DEFINITION OF $\Pr_B(A)$, $\Pr_A(B)$

$$\Pr(AB) = \Pr(B) \cdot \Pr_B(A) = \Pr(A) \cdot \Pr_A(B) \tag{19.10}$$

where we have used the obvious symmetry.

We may state this in words: the (compound) probability that both of two events occur is equal to the probability of one event multiplied by the conditional probability of the second calculated on the assumption that the first has occurred.

Going back to our example, we verify that

$$\Pr(P)\cdot\Pr_P(A)=\frac{74}{100}\frac{56}{74}=\frac{70}{100}\frac{80}{100}$$
$$=\Pr(A)\cdot\Pr_A(P)=0.56$$

which checks all parts of (19.10).

We note that (19.10) generalizes to the *chain rule of compound probabilities*:

$$\Pr(ABCD\cdots)=\Pr(A)\cdot\Pr_A(B)\cdot\Pr_{AB}(C)\cdot\Pr_{ABC}(D)\cdots \qquad (19.11)$$

We must recognize some other relations from the example we have been considering, for example

$$\Pr(A)=\frac{70}{100}=\Pr(P)\cdot\Pr_P(A)+\Pr(\bar{P})\cdot\Pr_{\bar{P}}(A)$$
$$=\frac{74}{100}\frac{56}{74}+\frac{26}{100}\frac{14}{26}=\frac{56+14}{100}$$

This is an instance of the following general theorem.

THEOREM OF TOTAL PROBABILITY

$$\Pr(A)=\Pr(B)\Pr_B(A)+\Pr(\bar{B})\Pr_{\bar{B}}(A) \quad \text{for any } A, B \qquad (19.12)$$

In general, if $S=E_1\cup E_2\cup E_3\cup\cdots\cup E_n$ where the subsets E_i are mutually exclusive (i.e., $E_iE_j=\emptyset$), then

$$\Pr(A)=\sum_{i=1}^{n}\Pr(E_i)\Pr_{E_i}(A) \qquad (19.13)$$

In our example about examination books we see that (19.12) is also verified by

$$\Pr(B)=\Pr(P)\Pr_P(B)+\Pr(\bar{P})\Pr_{\bar{P}}(B)$$
$$=\frac{74}{100}\frac{18}{74}+\frac{26}{100}\frac{12}{26}=\frac{18+12}{100}=0.30$$

which is our known answer. Also, we may calculate the probabilities of P and F

$$\Pr(P)=\Pr(A)\Pr_A(P)+\Pr(\bar{A})\Pr_{\bar{A}}(P)$$
$$=\frac{70}{100}\frac{80}{100}+\frac{30}{100}\frac{60}{100}$$
$$=0.56+0.18=0.74$$

which is our known answer. It is left as an exercise to verify the relation for $\Pr(F)$.

INDEPENDENCE

Now, it may very well happen that $\Pr_B(A)=\Pr(A)$ that is, the event A is not at all affected by the event B. In this case, the events A and B are *independent*, and we have

$$\Pr(AB)=\Pr(A)\cdot\Pr(B) \qquad (19.14)$$

for mutually independent events. This is extendable indefinitely for mutually independent events A, B, C, . . .

$$\Pr(ABC\cdots)=\Pr(A)\Pr(B)\Pr(C)\cdots \qquad (19.15)$$

PROBLEMS OF THE ". . . AT LEAST ONE . . ." TYPE

Another problem we should investigate at this point is the ". . . at least one . . ." type.

► **Example 19.3**
Tossing three coins

Suppose we consider as a trial the tossing of three coins. We wish to find the probability that there will be *at least one* head in this trial. We might think of writing this as $\Pr(H_1$ or H_2 or $H_3)$ and using (19.6) where H_i refers to getting a head on the ith toss. This procedure does in fact work, but the full number of terms in (19.6) must be used since the events whose union we are considering are not mutually exclusive. The student should solve this problem using (19.6) as an exercise. Far better, however, is to recognize that the event "*at-least-one-head*" is the complement of the event "*all-tails*" and that

$$\Pr(\textit{at-least-one-head})=1-\Pr(\textit{three tails})=1-\Pr(T_1T_2T_3)$$

where $\Pr(T_1T_2T_3)=\Pr(T_1)\Pr(T_2)\Pr(T_3)$, since the tosses are independent. We get for our answer

$$\Pr(\textit{at-least-one-head})=1-\left(\frac{1}{2}\right)\left(\frac{1}{2}\right)\left(\frac{1}{2}\right)=\frac{7}{8}$$

◄

In general, the event (A_1 or A_2 or $\cdots$ or A_n), where the A_i are mutually independent, is complementary to the event ($\bar{A}_1$ and $\bar{A}_2$ and $\cdots$ and $\bar{A}_n$). Another way of stating this is that either the event (*at least one of* A_k *occurs*) or the event (*all of* $\bar{A}_k$ *occur*) must occur:

$$\begin{aligned}\Pr(A_1 \text{ or } A_2 \text{ or } \cdots \text{ or } A_n) &= 1 - \Pr(\bar{A}_1 \text{ and } \bar{A}_2 \text{ and } \cdots \text{ and } \bar{A}_n) \\ &= 1 - [1 - \Pr(A_1)][1 - \Pr(A_2)] \times \cdots \\ &\quad \times \cdots [1 - \Pr(A_n)] \qquad (19.16)\end{aligned}$$

When all the events A_k have the same probability, p, the formula becomes

$$\Pr(A_1 \text{ or } A_2 \text{ or } \cdots \text{ or } A_n) = 1 - (1-p)^n \qquad (19.17)$$

Note that Figures 16.8a and 16.8b are nomograms for (19.17).

► **Example 19.4**
Salvo probability

Suppose that an infantryman has a probability of $\frac{1}{2}\%$ of downing a helicopter with a single rifle shot. How many rifles should be fired simultaneously (in salvo) to have even odds of downing a helicopter?

For each shot, the probability is $1 - 0.005 = 0.995$ that the helicopter will not be downed. The probability that none of n shots downs the helicopter is $(0.995)^n$. We want the probability to be greater than or equal to 1/2 by choosing n large enough. We, therefore, wish to solve the following equation for n:

$$1 - (1 - 0.005)^n \geq 0.5$$

Using natural logarithms, we get the equation

$$n \ln (0.995) \leq \ln (0.5)$$

and

$$n \geq 138$$

We may check this result using Figure 16.8b. ◄

► **Example 19.5a**
Birthday problem

A problem illustrating the more general relation (19.16) is the one popularized by Gamow:* What is the probability that at least two people out of a randomly chosen group of 24 will have the same birthdate?

This is $1 - \Pr$(the birthdates are all different). If we number the possible birthdates $A_1, \ldots, A_{24}$ and note that we want the probability that they are chosen all to be different, we get

$$\frac{365}{365}\frac{364}{365}\frac{363}{365}\cdots\left(\frac{365-24+1}{365}\right) = \frac{365!}{365^{24}341!}$$

since the first birthdate may be chosen at will, but the second birthdate may be chosen from all dates but the first, the third may be chosen from all dates but the first two, and so on.

We may take logarithms and use the approximation (18.14):

$$\begin{aligned}\ln \Pr(\text{all 24 birthdates are different}) &= \ln(365!) - 24\ln(365) - \ln(341!) \\ &= 365(\ln 365 - 1) - 24\ln 365 - 341(\ln 341 - 1) \\ &= -0.8095\end{aligned}$$

or

$$\Pr(\text{all 24 birthdates are different}) = 0.445$$

whereas direct calculation yields 0.462.

Therefore, the probability of interest is

$$\Pr(\text{at least two birthdates are the same}) = 1 - 0.46 = 0.54$$

The odds are better than even! ◄

OTHER RELATIONS IN PROBABILITY

For completeness, we note several other logical relations of interest:

$$\begin{aligned}\Pr(\bar{A}_1\bar{A}_2 \cdots \bar{A}_n) &= \Pr(\bar{A}_1)\Pr(\bar{A}_2)\cdots\Pr(\bar{A}_n) \\ &= [1 - \Pr(A_1)][1 - \Pr(A_2)] \times \cdots \\ &\quad \times \cdots [1 - \Pr(A_n)] \qquad (19.18)\end{aligned}$$

where the $\bar{A}_k$ are mutually independent. We also note that, for A and B mutually

*George Gamow, *One Two Three . . . Infinity*, Viking, New York, 1947.

independent,

$$\begin{aligned}\Pr(\bar{A}\bar{B}) &= [1-\Pr(A)][1-\Pr(B)] \\ &= 1-[\Pr(A)+\Pr(B)-\Pr(A)\Pr(B)] \\ &= 1-\Pr(A\cup B) \text{ by (19.4) where } A \text{ and } B \text{ are independent.} \\ &= \Pr(\overline{A\cup B}), \text{ which is just relation (19.8a)}\end{aligned}$$

Furthermore,

$$\begin{aligned}\Pr(A \text{ or } B\text{, but not both}) &= \\ &= \Pr(A)[1-\Pr(B)] + \Pr(B)[1-\Pr(A)] \\ &= \Pr(A)+\Pr(B)-2\Pr(A)\Pr(B)\end{aligned} \quad (19.19)$$

and

$$\begin{aligned}\Pr(A \text{ or } B \text{ or both}) &= \Pr(A)\Pr(B) + \Pr(A \text{ or } B\text{, but not both}) \\ &= \Pr(A)+\Pr(B)-\Pr(A)\Pr(B)\end{aligned}$$

which is just (19.4) for A and B independent.

INDEPENDENT AND DEPENDENT EVENTS

►**Example 19.5b**
An urn problem illustrating independent and dependent events

An urn contains N balls of which N_1 are white (W) and the remainder black (B). A sample of two balls is drawn, first one and then the second. Let A be the event that the first ball drawn is white. Let B be the event that the second ball drawn is white. As usual, $\Pr(AB)$ is the probability that both balls drawn are white.

Case 1: Sampling with replacement (the first ball drawn is replaced before the second draw).

The description space is $S = \{z_1, z_2\}$ where $z_i = 1, 2, \ldots, N_1, N_1+1, \ldots, N$ with the first N_1 elements representing white balls and the last $(N-N_1)$ black balls. Note that in this case a label may be repeated, that is, a given label may appear in both z_1 and z_2.

We may count the elements of the various subsets of interest:

$$\begin{aligned}\text{Nu}[S] &= N^2 \\ \text{Nu}[AB] &= N_1^2 \\ \text{Nu}[A] &= N_1 N \\ \text{Nu}[B] &= N N_1\end{aligned}$$

from which we may calculate the probabilities of interest:

$$\Pr(AB) = \left(\frac{N_1}{N}\right)^2$$

$$\Pr(A) = \frac{N_1}{N}$$

and

$$\Pr(B) = \frac{N_1}{N}$$

We note that

$$\Pr_A(B) = \frac{\Pr(AB)}{\Pr(A)} = \frac{N_1}{N} = \Pr(B)$$

Therefore, B is independent of A as expected for this case.

Case 2: Sampling without replacement (the first ball drawn is not replaced before the second draw).

As in Case 1, the description space is $S = \{z_1, z_2\}$ with z_i as above, but here no repetition of a label in the description is allowed. We count as usual

$$\begin{aligned}\text{Nu}[S] &= N(N-1) \\ \text{Nu}[AB] &= N_1(N_1-1) \\ \text{Nu}[A] &= N_1(N-1) \\ \text{Nu}[B] &= (N-1)N_1\end{aligned}$$

which yield the probabilities of interest

$$\Pr(AB) = \left(\frac{N_1}{N}\right)\left(\frac{N_1-1}{N-1}\right)$$

$$\Pr(A) = \frac{N_1}{N}$$

$$\Pr(B) = \frac{N_1}{N}$$

We can calculate the conditional probability

$$\Pr_A(B) = \frac{\Pr(AB)}{\Pr(A)} = \left(\frac{N_1}{N}\right)\left(\frac{N_1-1}{N-1}\right)\frac{1}{(N_1/N)} = \frac{N_1-1}{N-1}$$

Here, $\Pr_A(B) \neq \Pr(B)$ since $\Pr(B)$ depends on the outcome of the first draw. In fact,

$$\Pr_{\bar{A}}(B) = \frac{\Pr(\bar{A}B)}{\Pr(\bar{A})} = \frac{N_1}{N-1}$$

Also,

$$\Pr_B(A) = \frac{\Pr(AB)}{\Pr(B)} = \frac{N_1-1}{N-1}$$

which is not the same as $\Pr(A)$.
Thus, in Case 1 the two draws are independent events, whereas in Case 2 they are mutually dependent. ◄

► **Example 19.6**
Successive tosses of a coin

Let us consider successive flips of a coin. These events are independent and thus the probability of getting two heads in two tosses is

$$\Pr(HH) = \Pr(H)\cdot\Pr(H) = \left(\frac{1}{2}\right)^2 = \frac{1}{4}^*$$

Consider the situation that a run of 10 consecutive heads is observed. What is the probability that the 11th toss will be a head? Obviously the answer is still $\Pr(H) = 1/2$ since the 11th toss is independent of all the others. There are usually naive gamblers around, however, who will bet on tails because of the "law of averages." In actual fact, in such a practical gambling situation, you might actually prefer to bet on an 11th head on the following grounds.

Let us expand our model of the "chance mechanism" that describes the situation. Let us define the following event probabilities (which are really hypothesis probabilities) in this case:

$\Pr(A)$ = probability that the coin is fair $= 1-\epsilon$

$\Pr(B)$ = probability that the coin is biased (in favor of heads) $= \epsilon$

$\Pr_A(H)$ = probability that the next toss will be a head on the assumption of a fair coin $= 1/2$

$\Pr_B(H)$ = probability that the next toss will be a head on the assumption that the coin is biased in favor of heads $> 1/2$

Using (19.12) and recognizing that A and B are complementary, we get the total probability for heads to be

$$\Pr(H) = \Pr(A)\Pr_A(H) + \Pr(B)\Pr_B(H) = (1-\epsilon)\frac{1}{2} + \epsilon\Pr_B(H) = \frac{1}{2} + \epsilon\left[\Pr_B(H) - \frac{1}{2}\right]$$

Therefore, $\Pr(H) \geq 1/2$ and is exactly equal to 1/2 only if $\epsilon = 0$.

Note that we have tacitly assumed that the possible bias was in favor of heads given that 10 consecutive heads have occurred. The reader should ask himself if this is logically required or merely a good guess [otherwise, $\Pr_B(H) < 1/2$].

This kind of argument actually operates in the procreation of children. Even ignoring the fact $\Pr(\text{Boy}) \approx 0.51$ rather than 1/2, there is a systematic bias in some couples in favor of one or

*There is an amusing paradox in the historical archives of probabilistic discussions. We calculate the probability of the (HH) sequence of two-coin tosses by counting the number of favorable cases and dividing this by the total number of cases. There is clearly only one favorable case, (HH). An unfavorable case is (HT), in which the head appears first and is then followed by a tail. Now consider when a tail appears first. Clearly no point is served by a second toss since it must inevitably fail the hypothesis because of the first toss. We denote this trial by (T-). We now calculate

$$\Pr(HH) = \frac{HH}{(HH)+(HT)+(T\text{-})} = \frac{1}{3}$$

What happened? Where lies the fallacy in this argument? (Let the timid reader be assured that there *is* a fallacy!)

the other sex. The longer a run of, say, boys lasts the more probable (or credible in the sense of subjective probability) is the hypothesis that a systematic bias exists and, thus, it becomes (slightly) more probable that the run of boys will continue than that it will change. We shall see below in more detail how a measurement or a trial affects our *a priori* estimate of the probability. ◄

BAYES' THEOREM

Suppose a sequence of events $E_1, \ldots, E_i, \ldots, E_n$ forms a complete system of events, that is, one and only one of the events E_i must occur. For the purpose of the following discussion we may consider these to be n hypotheses, *one* of which *must* obtain. Let us assume that we know the individual unconditional probabilities $\Pr(E_i)$. We also know that each event or hypothesis, E_i, implies a certain probability for some event A: the conditional probability, $\Pr_{E_i}(A)$, which is the probability of A calculated on the assumption that E_i is a true hypothesis (or that the event E_i has occurred). If we perform an experiment and the event A has occurred, then this requires a reevaluation of the hypotheses E_i. In other words, we wish to calculate the quantity $\Pr_A(E_i)$, the (new) probability of E_i on the assumption that A has occurred.

This certainly accords with our commonsense intuition for the case in which only one of the hypotheses, E_K, yields an appreciable probability for the occurrence of A. If A actually occurs, we tend to believe more in the hypothesis E_K than in the others. If repetition of the experiment indicates the continued occurrence of A, then we tend to believe more in the hypothesis E_K at the expense of believing less in the other hypotheses.

Equation (19.10) enables us to write

$$\Pr(AE_i) = \Pr(A)\Pr_A(E_i) = \Pr(E_i)\Pr_{E_i}(A)$$

which yields the quantity of interest

$$\Pr_A(E_i) = \frac{\Pr(E_i)\Pr_{E_i}(A)}{\Pr(A)} \quad (1 \le i \le n) \tag{19.20}$$

We may substitute from (19.13) and obtain the usual form of *Bayes' theorem*:

$$\Pr_A(E_i) = \frac{\Pr(E_i)\Pr_{E_i}(A)}{\sum_{j=1}^{n} \Pr(E_j)\Pr_{E_j}(A)} \quad (1 \le i \le n) \tag{19.21}$$

This formula purports to show how to modify *a priori* probabilities in the light of experiment. The philosophical import of Bayes' theorem is quite great and still somewhat controversial in some of its possible ramifications (which, nevertheless, does not detract from the validity of the theorem itself). We shall, accordingly, elucidate its significance through some noncontroversial examples.

► **Example 19.7**
Three urns each containing two balls

Three urns, indistinguishable from the outside, each contains two balls. In one urn both balls are black; in a second both are white, while the third urn contains one of each.

An urn is chosen at random and a ball is drawn which proves to be white. What is the probability that the other ball in the urn is also white?

(a) We can solve this problem by very simple considerations. There are three white balls in toto and each is as likely to be chosen as the others. Two of these are in an urn such that the other in the urn is also white. The probability is thus $\Pr(WW) = 2/3$.
(b) We may also solve this in another fashion. Call E_{WW} the event that the urn with two white balls is chosen, and E_{WB} and E_{BB} the other events in an obvious notation. Let W be the event that a white ball is drawn, and let $B = \bar{W}$ be the event that a black

ball is drawn. We note the obvious

$$\Pr(E_{WW}) = \Pr(E_{BB}) = \Pr(E_{BW}) = \frac{1}{3}$$

and

$$\Pr(W) = \frac{\mathrm{Nu}[W]}{\mathrm{Nu}[W] + \mathrm{Nu}[B]} = \frac{1}{2} = \Pr(B)$$

in the absence of other information. But if the (WW) urn has been chosen, then $\Pr_{E_{WW}}(W) = 1$. Therefore, (19.20) yields

$$\Pr_W(E_{WW}) = \frac{\Pr(E_{WW})\Pr_{E_{WW}}(W)}{\Pr(W)} = \frac{1/3 \cdot 1}{1/2} = \frac{2}{3}$$

that is, the conditional probability that the second draw will yield a white ball *given* that the first ball drawn was white is 2/3.

(c) We may also use (19.21) directly to obtain

$$\Pr_W(E_{WW}) = \frac{\Pr(E_{WW})\Pr_{E_{WW}}(W)}{\Pr(E_{WW}) \cdot \Pr_{E_{WW}}(W) + \Pr(E_{BW}) \cdot \Pr_{E_{BW}}(W) + \Pr(E_{BB}) \cdot \Pr_{E_{BB}}(W)} = \frac{1/3 \cdot 1}{(1/3)1 + (1/3)(1/2) + (1/3)0} = \frac{2}{3}$$

This illustrates the use of Bayes' formula but we have seen that the problem could be solved without it. ◄

Let us return to our example of a coin-tossing that yields 10 consecutive heads. Common sense tells us that the longer the run of heads lasts, the more likely it is that the coin is not fair. Bayes' theorem enables us to say *how much* more likely.

To be specific, let us consider only two hypotheses: E_1 is the hypothesis that the coin is fair; E_2 is the hypothesis that the coin has two heads. We define A to be the event that 10 consecutive heads in a row have been observed. We start out (before the experiment) with the *a priori* or antecedent probability estimates that $\epsilon = \Pr(E_2)$ and $1 - \epsilon = \Pr(E_1)$. Using (19.21), we obtain

$$\Pr_A(E_1) = \frac{\Pr(E_1)\Pr_{E_1}(A)}{\Pr(E_1)\Pr_{E_1}(A) + \Pr(E_2)\Pr_{E_2}(A)} = \frac{(1-\epsilon)(1/2)^{10}}{(1-\epsilon)(1/2)^{10} + \epsilon} < (1-\epsilon)$$

and $$\Pr_A(E_2) = \frac{\epsilon \cdot 1}{(1-\epsilon)(1/2)^{10} + \epsilon} > \epsilon$$

We observe that the experimental result has caused the probability of E_1 to decrease while the probability of E_2 has increased. We note that Bayes' formula does not enable us to *calculate* the probability of hypothesis E_1 or E_2, but it does permit us to *modify* or reassess our *antecedent* estimates in the light of experience.

For example, if we adjudge the antecedent probability to be 0.01 that the coin in question is two-headed, we have as our *a posteriori* estimate (after 10 consecutive heads) that

$$\Pr(E_2) = \frac{0.01}{(0.99)(1/2)^{10} + 0.01} \approx 0.91$$

If, however, we adjudged the antecedent probability of a two-headed coin to be only 0.001, then we get $\Pr_A(E_2) \approx 1/2$ after the run of 10 consecutive heads. What do we mean that $\Pr_A(E_2) \approx 1/2$? If we had 1000 coins of which 999 are fair while one is two-headed, we would simulate our problem with the antecedent probability 0.001. On the average, if we made 1000 trials of 10 tosses each, we would observe 10 consecutive heads approximately twice. One of these trials would involve a fair coin, and the other would involve the two-headed coin. Having observed such a sequence of 10 heads, therefore, we have one chance in two of having the two-headed coin.

► Example 19.8
Diagnosis of disease

To illustrate our theme that these ideas permeate all areas of empirical science, let us consider the case of a patient in a hospital who is believed to suffer from one of three possible diseases: D_1, D_2, or D_3. Based on the patient's history and symptoms, the probabilities of the diseases are believed to be:

$$\Pr(D_1) = 0.50$$
$$\Pr(D_2) = 0.30$$
$$\Pr(D_3) = 0.20$$

The diagnostician in charge of the case subjects the patient to a test that is known to yield a positive result 10% of the time in instances of disease D_1, 30% of the time in cases of disease D_2, and 80% of the time in cases of disease D_3. The analysis of the test proves the result to be positive. What can be said about the likelihood of the three diseases after the test?

Let A be the event that the test yielded a positive result. We require

$$\Pr_A(D_1) = \frac{\Pr(D_1)\Pr_{D_1}(A)}{\Pr(D_1)\Pr_{D_1}(A) + \Pr(D_2)\Pr_{D_2}(A) + \Pr(D_3)\Pr_{D_3}(A)}$$
$$= \frac{(0.5)(0.1)}{(0.5)(0.1) + (0.3)(0.3) + (0.2)(0.8)} = \frac{0.05}{0.30} \approx 0.17$$

$$\Pr_A(D_2) = \frac{(0.3)(0.3)}{0.30} = \frac{0.09}{0.30} = 0.30$$

$$\Pr_A(D_3) = \frac{(0.20)(0.80)}{0.30} = \frac{0.16}{0.30} \approx 0.53$$

We note that $\quad \Pr_A(D_1) + \Pr_A(D_2) + \Pr_A(D_3) = 1,$

and the probability of D_3 has been enhanced at the expense of the probability of D_1 as a result of the test.

Suppose the test had been performed twice rather than once with B representing the event that both results are positive. Here $\Pr_{D_1}(B) = (0.1)^2 = 0.01$, $\Pr_{D_2}(B) = (0.3)^2 = 0.09$, and $\Pr_{D_3}(B) = (0.8)^2 = 0.64$.

We now obtain

$$\Pr_B(D_1) = \frac{\Pr(D_1)\Pr_{D_1}(B)}{\Pr(D_1)\Pr_{D_1}(B) + \Pr(D_2)\Pr_{D_2}(B) + \Pr(D_3)\Pr_{D_3}(B)}$$
$$= \frac{(0.5)(0.01)}{(0.5)(0.01) + (0.3)(0.09) + (0.2)(0.64)} = \frac{0.005}{0.160} \approx 0.03$$

$$\Pr_B(D_2) = \frac{(0.30)(0.09)}{0.160} = \frac{0.027}{0.160} = 0.17$$

$$\Pr_B(D_3) = \frac{(0.2)(0.64)}{0.160} = \frac{0.128}{0.160} = 0.80$$

where we have increased the probability of D_3 further at the expense of the probabilities of both D_1 and D_2. We are obviously on the right track to obtain a diagnosis.

We could also have treated the second test as one trial subsequent to the first, but using as antecedent probabilities the first results obtained ($\mathrm{Pr}_A(D_i)$), that is, $\mathrm{Pr}(D_1) = (0.50/0.30)$; $\mathrm{Pr}(D_2) = (0.09/0.30)$, and $\mathrm{Pr}(D_3) = (0.16/0.30)$. Thus

$$\mathrm{Pr}_B(D_1) = \frac{(0.05/0.30)(0.1)}{(0.05/0.30)(0.1) + (0.09/0.30)(0.30) + (0.16/0.30)(0.80)}$$

and $\mathrm{Pr}_B(D_1) = (0.005/0.160)$ as above. Also, $\mathrm{Pr}_B(D_2)$ and $\mathrm{Pr}_B(D_3)$ have their values as before. Thus we see that another way of viewing the situation is that the probabilities of the various hypotheses actually change during the course of testing. ◄

"I figure there's a 40% chance of showers, and a 10% chance we know what we're talking about."

BAYESIAN INFERENCE

We must caution the unwary, however, that (19.21) does not yield more than is warranted by the judgment of the user. The circumstance that we can obtain a quantitative estimate of what is called the inverse probability should not be allowed to obscure the fact that judgment enters importantly in determining what are the possible hypotheses and in inputting *a priori* probabilities. Philosophers of science are quick to point out that it is difficult in most problems to assign values to the antecedent probabilities and that the usefulness of Bayes' theorem is usually limited.

Many of the objections raised to "bayesian inference" are actually attacks on *Bayes' Postulate*,* which is quite a different idea. This says that, if we have no *a*

*This is sometimes called the Principle of Indifference or the Principle of Insufficient Reason.

priori knowledge of the probabilities of the alternative hypotheses, we may assume them to be equally probable. This assumption, of course, is often not justified and leads to logical inconsistencies. Bayes was personally doubtful about this postulate and it was published only posthumously. Despite its shaky nature, Laplace appears to have believed in Bayes' Postulate for some time and used it to derive the erroneous Laplace law of succession which, for example, purported to enable one to *calculate* the odds that the sun would rise tomorrow. We must carefully distinguish between Bayes' Theorem, which is a consequence of the idea of conditional probability, and Bayes' Postulate, which is an *ad hoc* assumption usually not valid.

Despite this caveat, we can make a few further remarks concerning what actually happens to the credibility of a hypothesis that has been subjected to experimental test. We can carry along the idea that the probabilities are constantly changing as a result of the tests that have been performed, that is, as a result of all our current knowledge. We may regard this as an iterative procedure.

Formally, let us consider that only two hypotheses are possible and that the *a priori* probabilities for hypotheses A and B respectively are denoted by a superscript 0 to indicate that they exist before the experiment: $\Pr^0(A) + \Pr^0(B) = 1$. Suppose we perform an experiment and obtain a *series* of empirical values x_i. We note that the probability of this *sequence*, under the assumption that hypothesis A is correct, is $P_A(x_i)$ where we have changed notation slightly to indicate that it is the probability of a whole *sequence* of values that we are concerned with here. The next step in the iteration is to obtain $\Pr^1(A)$ as a result of the experiment which we can do using Bayes' theorem (19.21).

$$\Pr^1(A) = \frac{\Pr^0(A)P_A(x_i)}{\Pr^0(A)P_A(x_i) + \Pr^0(B)P_B(x_i)} \tag{19.22}$$

and

$$\Pr^1(B) = \frac{\Pr^0(B)P_B(x_i)}{\Pr^0(A)P_A(x_i) + \Pr^0(B)P_B(x_i)}$$

As a result of an experiment we increase our confidence in A at the expense of B or conversely. The goal is to perform experiments that distinguish between A and B, that is, experiments for which

$$P_A(x_i) \neq P_B(x_i)$$

and the more different these probabilities are, the better. This is an important goal of experimental design.

LIKELIHOOD FUNCTIONS AND THE PRINCIPLE OF MAXIMUM LIKELIHOOD

A most important procedure arises where the predictions of the hypotheses A and B can be written in functional form. Here we get

$$P_A(x_i) = f_A(x_1)f_A(x_2)\cdots f_A(x_n) = \prod_{i=1}^{n} f_A(x_i) \tag{19.23a}$$

for the probability of a sequence of n values x_i where $f_A(x_k)$ represents the probability for the individual value x_k on the assumption of hypothesis A.

This case arises particularly when it is possible to parameterize a problem so that when we say that we are distinguishing between different hypotheses, we really mean that we are distinguishing between different values of a parameter that appears in a similar form in each of the hypotheses. In such a case, we are back to our original problem of Part I, since we wish to find from the data the best value of a parameter; this now corresponds to the hypothesis in which we have the most confidence. For generality, therefore, let us append the parametric label, "c," to our various probabilities to include this most important case.

In these terms, we can write (19.22) as follows:

$$\Pr^1(A\,;c_A)=\frac{\Pr^0(A\,;c_A)\prod_{i=1}^{n} f_A(x_i\,;c_A)}{\Pr^0(A\,;c_A)\prod_{i=1}^{n} f_A(x_i\,;c_A)+\Pr^0(B\,;c_B)\prod_{i=1}^{n} f_B(x_i\,;c_B)}$$

and (19.23b)

$$\Pr^1(B\,;c_B)=\frac{\Pr^0(B\,;c_A)\prod_{i=1}^{n} f_B(x_i\,;c_B)}{\Pr^0(A\,;c_A)\prod_{i=1}^{n} f_A(x_i\,;c_A)+\Pr^0(B\,;c_B)\prod_{i=1}^{n} f_B(x_i\,;c_B)}$$

where $f_A(x_k\,;c_A)$ represents the individual probability of each value of x_k on the assumption of hypothesis A (including the specified value of the parameter, c_A).

We tend to believe in hypothesis A over hypothesis B according to whether the ratio $\Pr^1(A\,;c_A)/\Pr^1(B\,;c_B)$ gets large. This ratio,

$$\frac{\Pr^1(A\,;c_A)}{\Pr^1(B\,;c_B)}=\frac{\Pr^0(A\,;c_A)\prod_{i=1}^{n} f_A(x_i\,;c_A)}{\Pr^0(B\,;c_B)\prod_{i=1}^{n} f_B(x_i\,;c_B)} \tag{19.24}$$

is called the *likelihood ratio* and, perhaps misleadingly, the *betting odds*.

We also define in this regard the *likelihood function*,

$$\mathcal{L}_A(x_i\,;c_A)=\prod_{i=1}^{n} f_A(x_i\,;c_A) \tag{19.25}$$

defined for the particular experimentally observed sequence of numbers x_i and the parameter c_A, that is, one value of the parameter under the assumption of hypothesis A. The procedure based on this function is called that of *maximum likelihood* and was developed primarily by R. A. Fisher. This utilizes the intuitively appealing statement that the hypothesis is *most likely* that yields the maximum likelihood function of all the functions considered. If we seek to determine the "best value" of a parameter, it suffices to calculate the likelihood function for each of the possible values of the parameter and choose the value for which the likelihood function is maximum. Another way of stating this *principle of maximum likelihood* is that the sequence of observations that actually occurs in an experiment is the one with the maximum probability. The corollary of this principle relevant for statistical inference is that the best value of a parameter as determined by a series of experimental measurements will cause the likelihood function for that series of measurements to be a maximum.

As usual, it is possible to find a local maximum (for a likelihood function that has derivatives) at a value or set of values of the parameter c for which the derivative with respect to c is zero and the second derivative is negative:

$$\left.\frac{\partial \mathcal{L}_A(x_i\,;c_{Aj})}{\partial c_{Aj}}\right|_{c_j^*}=0$$

and (19.26)

$$\left.\frac{\partial^2 \mathcal{L}_A(x_i\,;c_{Aj})}{\partial c_{Aj}^2}\right|_{c_j^*}<0$$

We should perhaps emphasize that the quantities $f_A(x_i\,;c_A)$ are *a posteriori* probabilities, that is, *after* a result x_i has been observed it states how probable it was that this result is found. An important distinction between ordinary (*a priori*) probabilities and *a posteriori* probabilities is that the former are well-defined numbers (even if unknown) whereas the latter have the character of being random variables. The term "*likelihood*" distinguishes, and is reserved for, *a posteriori* probabilities.

Expectation Values—Recursion Relations

If we know the frequency function $p(x_i)$ for a discrete variate, or the probability density function $f(x)$ for a continuous variate, we may calculate the expected value of a function of the variate, $\xi(x)$, from equations (17.16a) or (17.16b). We shall use these relations in considering particular probability distributions.

It may happen that we do not know the underlying probability distribution. It is still possible in some cases to calculate expected values using a recursion or recurrence argument. This argument is valid when trials are independent and unvarying.

► **Example 19.9**
Average number of trials to get r successes

Suppose we consider an event that can take on two values: success or failure. The probability for success on a given trial is p, that for failure on a given trial is $q = 1 - p$. What is the average number of trials required to get r successes?

We first consider the number of trials to the first success. Call this n_1. We may consider that only two cases, A and B, are possible:

In event A, the first trial yields a success. Here $n_1 = 1$.

In event B, the first trial yields a failure, and we must wait a further number of trials, n_B, before a success. Here $n_1 = n_B + 1$.

We may find the average by weighting the probabilities:

$$\langle n_1 \rangle = \Pr(A)1 + \Pr(B)(n_B + 1)$$

But

$$\Pr(A) = p, \Pr(B) = q = 1 - p$$

and

$$n_B = \langle n_1 \rangle$$

since we are starting over again and the trials are independent. This is the central point of this technique and constitutes the recursion relation.

$$\therefore \langle n_1 \rangle = p + (1-p)[\langle n_1 \rangle + 1]$$

and

$$\langle n_1 \rangle = \frac{1}{p}$$

We now see that the average number of trials to the first success is $\langle n_1 \rangle = 1/p$. The average number of additional trials needed to get the second success is $\langle n_2 \rangle = 1/p$; therefore the average number of trials needed to get r successes is

$$\langle n \rangle = \langle n_1 \rangle + \langle n_2 \rangle + \cdots + \langle n_r \rangle = r/p$$ ◄

► **Example 19.10**
Fair payoff for a game played with one die

A game is to be played with one honest die. A player pays \$1 for the privilege of making one toss and gets a payoff only when he has achieved a fixed number of fours, say 3. What should the payoff be for the game to be fair?

Each toss has a probability, $p = 1/6$, of getting a four. The average number of tosses required to get 3 fours is $n = r/p = 18$. $\therefore$ The player pays an average of \$18 to get to the payoff that should, therefore, be \$18. ◄

► **Example 19.11**
Average number of trials to get r consecutive successes

Consider the problem of getting r consecutive successes. How many trials are required, on the average, to achieve this?

We may set up a recursion relation similar to that in **Example 19.9** by weighting the various n_i's to get $\langle n \rangle$:

$$\langle n \rangle = \sum_{i=0}^{r} \Pr(i) n_i$$

where i is the number of successes before failure $i = r$, of course, denotes r consecutive successes.

We may enumerate these by what is known as a "tree diagram" shown in Figure 19.8 on page 138.

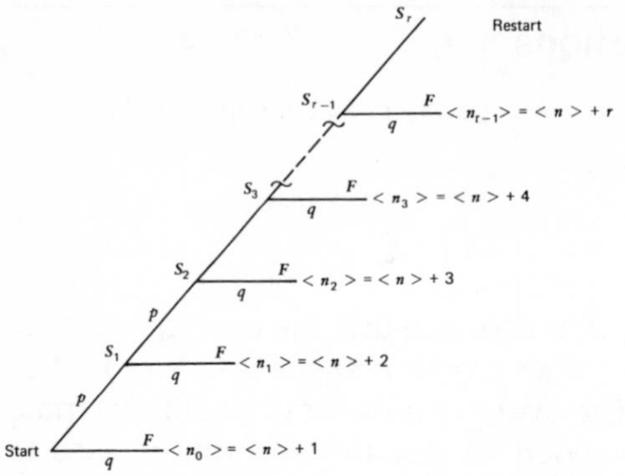

Figure 19.8 Tree diagram for binomial trials with probability, p, of success S and probability, $q = 1 - p$, of failure F.

We write each trial as a branching point with the lines denoting the probability leading to either success or failure.

The probability of getting r straight successes is p^r. The number of trials corresponding to this probability is r. The last term in the summation, $i = r$, is rp^r.

The first term represents a success followed by a failure. Each time we get a failure we start counting over again. The average number of trials required from this point is $\langle n\rangle+$(the number of trials to get to this point). For example,

$$n_0 = \langle n\rangle + 1 \text{ with } \Pr(0) = q$$

Also

$$n_1 = \langle n\rangle + 2 \text{ with } \Pr(1) = pq$$
$$n_2 = \langle n\rangle + 3 \text{ with } \Pr(2) = p^2q, \text{ etc.}$$
$$\therefore \langle n\rangle = rp^r + q[\langle n\rangle + 1] + pq[\langle n\rangle + 2] + \cdots + \cdots + p^{r-1}q[\langle n\rangle + r]$$ ◄

► **Example 19.12**
Coin tossing

How many tosses of a fair coin are needed on the average to get three consecutive heads?

Here

$$\langle n\rangle = 3(\tfrac{1}{2})^3 + \tfrac{1}{2}[\langle n\rangle + 1] + \tfrac{1}{2}\tfrac{1}{2}[\langle n\rangle + 2] + (\tfrac{1}{2})^2\tfrac{1}{2}[\langle n\rangle + 3] = 14$$ ◄

► **Example 19.13**
Bubble gum problem

As an inducement to buy bubble gum, the manufacturer encloses in each package a picture of a baseball player or manager and encourages the purchaser to accumulate a complete set of cards. Assume that there is a total of 10 cards (nine baseball players plus a manager for a specific local team) and that the manufacturer distributes equal numbers of cards throughout his packages at random. What is the average number of purchases required to make a complete collection of the 10 cards?

We may construct a summation

$$\langle n\rangle = \langle n_1\rangle + \langle n_2\rangle + \langle n_3\rangle + \cdots + \langle n_{10}\rangle$$

where each $\langle n_i\rangle$ represents the average number of purchases required to get a new card after having $i-1$ already. From a previous example (**19.9**) we see that each

$$\langle n_i\rangle = \frac{1}{p_i}$$

where p_i is the probability of getting a new card.

For the first purchase, the probability is unity that a new card is obtained. For the second card, the probability of getting a new card on each purchase is 9/10. For the third card, the probability of getting a new card is 8/10; for the fourth card, the probability is 7/10; and so on. The summation, therefore, is

$$\langle n\rangle = 1 + \frac{10}{9} + \frac{10}{8} + \cdots + \frac{10}{1}$$

This is the summation of an harmonic progression that may be computed to be

$$n = 10\left(\sum_{i=1}^{10} 1/i\right) = 29.29$$

or, slightly more than 29 purchases on the average are required to complete the collection. ◄

20

JOINT PROBABILITY DISTRIBUTIONS AND FUNCTIONS OF RANDOM VARIABLES

We have defined a *random variable* to be a quantity with no definite value but with an *ensemble* of number values which it assumes in accord with a known probability distribution. A random variable is defined on some description space S on which the probability function has been defined as in Chapter 19. Thus far we have concentrated on a probability distribution whose argument, called a random variable, has been restricted to one quantity. In general, we may be interested in the joint probability distribution of two or more random variables. Much of what is interesting can be elucidated by considering two random variables and the generalization to more than two is straightforward.

Joint and Marginal Probability Distributions

Suppose we have two random variables, X and Y, which can assume the values $x_1, \ldots, x_i, \ldots, x_n$ and $y_1, \ldots, y_j, \ldots, y_m$, respectively, if the individual probability distributions are discrete or are defined on the intervals $(x_{\min}, x_{\max})$ and $(y_{\min}, y_{\max})$ if continuous. The individual probability distributions are $X(x_i)$ and $Y(y_j)$, respectively. The 2-tuple (in the sense of Chapter 19) denoted by (x_k, y_ℓ) defines an event or *random vector* whose probability may be denoted

$$\Pr(X = x_k, Y = y_\ell) = p(x_k, y_\ell) \qquad (20.1)$$

where $p(x, y)$ is called the joint probability distribution of X and Y.

This may be represented as in Table 20.1

We shall consider the various probability distributions to be discrete for the purpose of description, but the reader should always remember that these definitions can equally well be made for continuous distributions.

The individual probability distributions $X(x)$ and $Y(y)$ satisfy the usual conditions (17.1) and (17.2). The joint probability distribution satisfies similar conditions:

$$p(x_k, y_\ell) \geq 0 \qquad \text{and} \qquad \sum_{i,j} p(x_i, y_j) = 1 \qquad (20.2)$$

Table 20.1 Joint Probabilities, $p(x, y)$ for Different Values of (x, y)

$y \backslash x$	x_1	x_2	$\cdots$	x_n	$Y(y_j)$
y_1	$p(x_1, y_1)$	$p(x_2, y_1)$	$\cdots$	$p(x_n, y_1)$	$\sum_i p(x_i, y_1) = Y(y_1)$
y_2	$p(x_1, y_2)$	$p(x_2, y_2)$	$\cdots$	$p(x_n, y_2)$	$\sum_i p(x_i, y_2) = Y(y_2)$
$\vdots$	$\vdots$	$\vdots$		$\vdots$	$\vdots$
y_n	$p(x_1, y_n)$	$p(x_2, y_n)$	$\cdots$	$p(x_n, y_n)$	$\sum_i p(x_i, y_n) = Y(y_n)$
$X(x_i)$	$\sum_j p(x_1, y_j) = X(x_1)$	$\sum_j p(x_2, y_j) = X(x_2)$		$\sum_j p(x_n, y_j) = X(x_n)$	

Furthermore,

$$\sum_j p(x_k, y_j) = \Pr\{x = x_k\} = X(x_k) \quad (20.3a)$$

and

$$\sum_i p(x_i, y_\ell) = \Pr\{y = y_\ell\} = Y(y_\ell) \quad (20.3b)$$

The frequency function $X(x_k)$ in (20.3a) is the probability that $X = x_k$ *no matter what the value of Y is.* $X(x)$ is a one-dimensional distribution called the *marginal* distribution of X attached to the two-dimensional joint distribution of X and Y.

► **Example 20.1**

Consider the events $(x, y) = (-1, -1)$, $(+1, -1)$, $(0, 0)$, $(-1, +1)$, $(+1, +1)$ each with probability 1/5. We summarize this distribution in Table 20.2.

Table 20.2 Joint Probabilities for Example 20.1

$y \backslash x$	-1	0	$+1$	$Y(y)$
-1	1/5	0	1/5	2/5
0	0	1/5	0	1/5
$+1$	1/5	0	1/5	2/5
$X(x)$	2/5	1/5	2/5	

This is indicated graphically in Figure 20.1 ◄

The cumulative distribution function is frequently useful:

$$F_X(x_k) = \Pr(X \le x_k) = \sum_{x_i \le x_k} X(x_i)$$
$$= \sum_{x_i \le x_k} \sum_{\text{all } y_j} p(x_i, y_j) \quad (20.3c)$$

Expectation Values

The expectation value of $z = \phi(x, y)$ is defined in direct analogy to the single random variable case of (17.16):

$$\langle \phi(x, y) \rangle = \langle z \rangle = \sum_{i,j} \phi(x_i, y_j) p(x_i, y_j) \quad (20.4)$$

We shall assume that all sums and integrals used are absolutely convergent.

We may consider some commonly used expectation values.

ax + b

$$\langle ax + b \rangle = \sum_i p(x_i)(ax_i + b)$$
$$= a \sum p(x_i) x_i + b \sum p(x_i)$$
$$= a\langle x \rangle + b \quad (20.5)$$

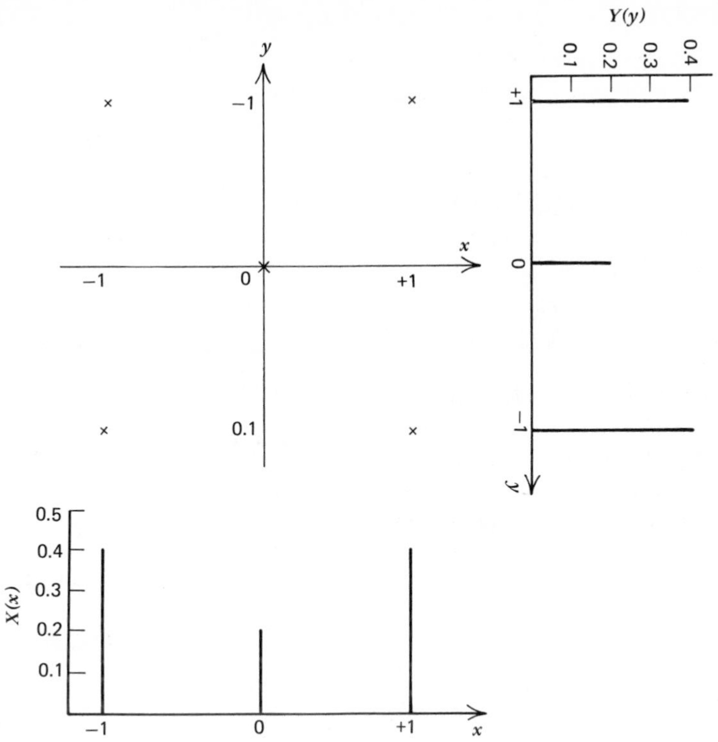

Figure 20.1
Graph of Table 20.2.

x* + *y

$$\langle x + y\rangle = \sum_{i,j} (x_i + y_j)p(x_i, y_j) = \sum_{i,j} x_i p(x_i, y_j) + \sum_{i,j} y_j p(x_i, y_j) = \sum_i x_i X(x_i) + \sum_j y_j Y(y_j)$$
$$= \langle x\rangle + \langle y\rangle \tag{20.6}$$

This generalizes readily to

$$\langle x_1 + x_2 + \cdots + x_n\rangle = \langle x_1\rangle + \langle x_2\rangle + \cdots + \langle x_n\rangle \tag{20.7}$$

Note the relations (20.5) to (20.7) have not required that the random variables be independent. However, for a product of x and y we must consider them mutually independent.

Independence

We now consider the idea of independence of random variables. A necessary and sufficient condition that x and y be independent is the relation

$$p(x_i, y_j) = X(x_i)Y(y_j) \tag{20.8}$$

This is just a revision of relation (19.14). We might remark that an equivalent condition that is also both necessary and sufficient for independence is the relation of the cumulative distribution functions

$$F(x, y) = F_X(x)F_Y(y) \tag{20.8a}$$

xy

$$\langle xy\rangle = \sum_{i,j} p(x_i, y_j)(x_i y_j)$$
$$= \Big(\sum_i X(x_i)x_i\Big)\Big(\sum_j Y(y_j)y_j\Big)$$
$$= \langle x\rangle\langle y\rangle \tag{20.9}$$

for x, y mutually independent.

This generalizes to

$$\langle (x_1 \cdots x_n) \rangle = \langle x_1 \rangle \cdots \langle x_n \rangle$$
$$\text{for } x_1, \ldots, x_n \text{ mutually independent} \tag{20.9b}$$

We remark what the meaning is in this context of the conditional probability. We can refer to the conditional probability that $y = y_\ell$ given that $x = x_k$ as follows:

$$\Pr_{x=x_k}(y = y_\ell) = \frac{p(x_k, y_\ell)}{X(x_k)} \tag{20.10}$$

This, of course, is just a restatement of (19.10).

The converse of (20.9) is *not* true. If $\langle xy \rangle = \langle x \rangle \langle y \rangle$, x and y are *uncorrelated* but they are not necessarily independent. We may demonstrate this by citing a specific case in Example 20.2.

► **Example 20.2**

Consider the events (x, y), each with probability 1/5, of Example 20.1. We calculate the various probabilities

$$X(-1) = p(-1,-1) + p(-1,0) + p(-1,+1) = \frac{1}{5} + 0 + \frac{1}{5} = \frac{2}{5}$$

$$X(0) = \frac{1}{5}$$

$$X(+1) = \frac{2}{5}$$

$$Y(-1) = \frac{2}{5}$$

$$Y(0) = \frac{1}{5}$$

$$Y(+1) = \frac{2}{5}$$

x and y are uncorrelated, since

$$\langle x \rangle = \frac{-1 \cdot 2}{5} + \frac{0 \cdot 1}{5} + \frac{1 \cdot 2}{5} = 0$$

$$\langle y \rangle = \frac{-1 \cdot 2}{5} + \frac{0 \cdot 1}{5} + \frac{1 \cdot 2}{5} = 0$$

$$\langle xy \rangle = \sum x_i y_j p(x_i, y_j) = (-1)(-1)\frac{1}{5} + (-1)(+1)\frac{1}{5} + (1)(-1)\frac{1}{5} + (1)(1)\frac{1}{5} = 0$$

That is,

$$\langle x \rangle \langle y \rangle = \langle xy \rangle$$

However, $p(0,-1) = 0$ whereas

$$X(0)Y(-1) = (1/5)(2/5) = 2/25$$

Therefore, x and y are not independent! ◄

Covariance

We return to the definition of covariance as in (10.1)

$$\text{Covar}(x, y) = \sigma_{xy} = \langle (x - \langle x \rangle)(y - \langle y \rangle) \rangle = \langle xy \rangle - \langle x \rangle \langle y \rangle \tag{20.11}$$

If x and y are independent then (20.9) yields the relation

$$\sigma_{xy} = 0 \qquad \text{if } x \text{ and } y \text{ are independent} \tag{20.12}$$

The converse is not necessarily true, that is, the vanishing of the covariance does not guarantee independence.

$$\text{Covar}(x, y) = 0$$

if and only if x, y are uncorrelated

The covariance does not vanish if knowledge of the value of x (or y) gives us some information about y (or x). From the definition (20.11) we see that Covar $(x, y) > 0$ if values of x such that $x > \langle x \rangle$ and values of y such that $y > \langle y \rangle$ appear together. (Also, values of $x < \langle x \rangle$ tend to appear with values of $y < \langle y \rangle$.) This is shown schematically in Figure 20.2*a* when $x > \langle x \rangle$ on the average implies that $y > \langle y \rangle$, and $x < \langle x \rangle$ on the average implies that $y < \langle y \rangle$.

Also, Covar $(x, y) < 0$ if values $(x < \langle x \rangle)$ are correlated with $(y > \langle y \rangle)$ and $(x > \langle x \rangle)$ correlated with $(y < \langle y \rangle)$. This is indicated schematically in Figure 20.2*b*.

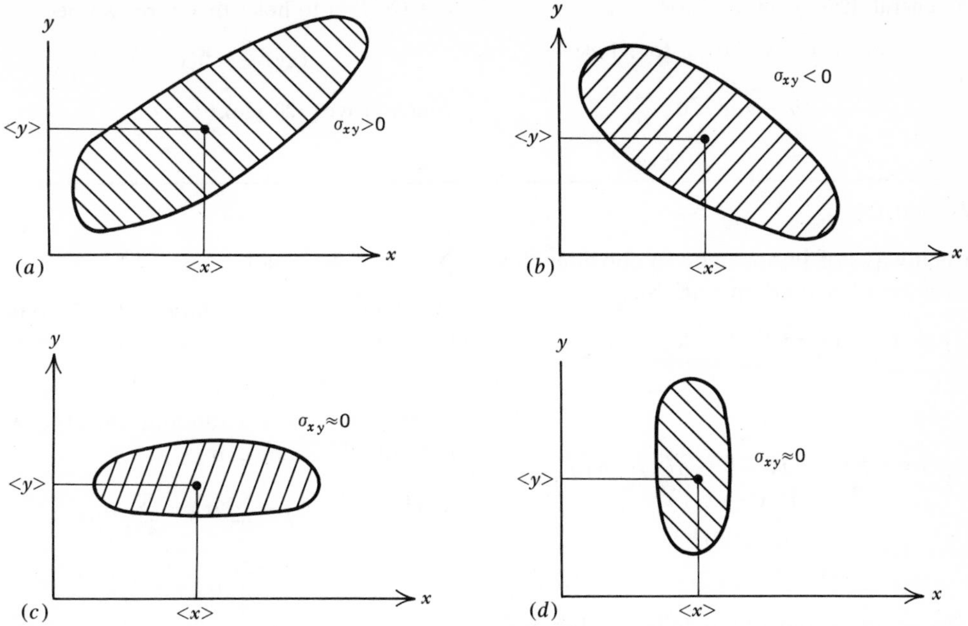

Figure 20.2
(*a*) Distribution of (*x*, *y*) for positive covariance. (*b*) Distribution of (*x*, *y*) for negative covariance. (*c*) Distribution of (*x*, *y*) for zero covariance. (*d*) Distribution of (*x*, *y*) for zero covariance.

The covariance vanishes if knowledge of one variate tells us nothing about the other. This is shown schematically in Figures 20.2*c* and *d*. Note that the distribution of *x*, *y* is roughly symmetric about $x = \langle x \rangle$ and $y = \langle y \rangle$ (that is, we may fold the distribution about these lines) when $\sigma_{xy} \approx 0$. If we know $x > \langle x \rangle$, it tells us nothing about *y* etc.

CORRELATION COEFFICIENT

To investigate whether and the extent to which quantities are correlated, it is useful to define the correlation coefficient:

$$\rho(x, y) = \frac{\text{Covar}(x, y)}{\sqrt{\text{Var}(x)\,\text{Var}(y)}} \tag{20.13}$$

This is convenient to use since, as we show below,

$$-1 \le \rho(x, y) \le +1 \tag{20.13a}$$

and

$$\rho(x, y) = 0 \tag{20.13b}$$

when *x*, *y* are uncorrelated.

We may write $\rho(x, y)$ alternatively:

$$\rho(x, y) = \frac{\langle (x - \langle x \rangle)(y - \langle y \rangle) \rangle}{\{\langle (x - \langle x \rangle)^2 \rangle \langle (y - \langle y \rangle)^2 \rangle\}^{1/2}} \tag{20.13c}$$

$$\rho(x, y) = \frac{\langle xy \rangle - \langle x \rangle \langle y \rangle}{\sigma_x \sigma_y} \tag{20.13d}$$

We may show (20.13a) by proving that

$$[\langle xy \rangle]^2 \le \langle x^2 \rangle \langle y^2 \rangle \tag{20.13e}$$

for all random variables *x* and *y* such that $\langle x^2 \rangle$ and $\langle y^2 \rangle$ exist and are finite. Equation 20.13e is a form of the Cauchy Schwartz inequality. We prove (20.13e) by considering

$$0 \le \langle (ax + y)^2 \rangle = a^2 \langle x^2 \rangle + 2a \langle xy \rangle + \langle y^2 \rangle \tag{20.13f}$$

where the nonnegativity holds for all *a* and where we have used (20.5) and (20.7). When the equality holds, we may solve for *a*:

$$a = \frac{-\langle xy \rangle}{\langle x^2 \rangle} \pm \frac{1}{2\langle x^2 \rangle} \{4\langle xy \rangle^2 - 4\langle x^2 \rangle \langle y^2 \rangle\}^{1/2}$$
$$= \frac{-\langle xy \rangle}{\langle x^2 \rangle} \pm \frac{1}{\langle x^2 \rangle} \{\langle xy \rangle^2 - \langle x^2 \rangle \langle y^2 \rangle\}^{1/2}$$

In general, if a quadratic form

$$au^2 + bu + c > 0 \qquad \text{for all } u$$

then

$$b^2 - 4ac \leq 0$$

For (20.13f) to hold, therefore, we need

$$\langle xy\rangle^2 - \langle x^2\rangle\langle y^2\rangle \leq 0$$

which proves (20.13e).

Variance

It is always of importance to consider the variance of a random variable:

$$\text{Var}(x) = \langle (x - \langle x\rangle)^2\rangle = \langle x^2\rangle - \langle x\rangle^2 = \sigma^2$$

$ax + b$

$$\begin{aligned}\text{Var}(ax + b) &= \langle (ax + b - a\langle x\rangle - b)^2\rangle \\ &= \langle a^2(x - \langle x\rangle)^2\rangle \\ &= a^2 \sum_i p(x_i)(x_i - \langle x\rangle)^2 \\ &= a^2\,\text{Var}(x) \qquad (20.14)\end{aligned}$$

This relation and (20.6) tells us that the quantity

$$x^* = \frac{x - \langle x\rangle}{\sigma} \qquad (20.15)$$

has mean zero and variance unity. We also note that x^* is a dimensionless quantity and is called the *normalized variable corresponding to x.*

$x + y$

Let $x' = x - \langle x\rangle$, $y' = y - \langle y\rangle$.

$\text{Var}(x') = \text{Var}(x)$ and

$\text{Var}(y') = \text{Var}(y)$ by (20.14)

$$\begin{aligned}\text{Var}(x + y) &= \text{Var}(x' + y' + \langle x\rangle + \langle y\rangle) \\ &= \text{Var}(x' + y') \text{ by (20.14)} \\ &= \langle [(x' + y') - \langle x' + y'\rangle]^2\rangle\end{aligned}$$

But $\langle x' + y'\rangle = \langle x'\rangle + \langle y'\rangle = 0$

$$\begin{aligned}\therefore \text{Var}(x + y) &= \langle [(x' + y')^2]\rangle \\ &= \langle [(x')^2 + (y')^2 + 2x'y']\rangle \\ &= \langle [(x')^2]\rangle + \langle [(y')^2]\rangle + 2\langle [x'y']\rangle\end{aligned}$$

and

$$\begin{aligned}\text{Var}(x + y) = \text{Var}(x) &+ \text{Var}(y) \\ &+ 2\,\text{Covar}(x,y)\end{aligned} \qquad (20.16)$$

If x, y are uncorrelated

$$\text{Var}(x + y) = \text{Var}(x) + \text{Var}(y)$$

$\sum x_i = x_1 + \cdots + x_n$

We may generalize relation (20.16) to note that for

$$S = x_1 + x_2 + \cdots + x_n$$

where each x_i is a random variable with individual variance $\sigma_{x_i}^2$

$$\begin{aligned}\text{Var}\left(S = \sum_i^n x_i\right) &= \sum_{i=1}^n \sigma_{x_i}^2 + \sum_{\substack{j,k=1 \\ j\neq k}}^n \sigma_{x_j x_k} \\ &= \sum_{i=1}^n \sigma_{x_i}^2 + 2\sum_{\substack{j,k \\ j<k}} \sigma_{x_j x_k}\end{aligned} \qquad (20.17)$$

where we have made use of the symmetry of the covariance with respect to its arguments and counted each pair in the sum only once. In general, there are $\binom{n}{2}$ terms in the sum

$$\sum_{\substack{j,k \\ j<k}} \sigma_{x_j x_k}$$

$\bar{x} = \frac{1}{n}\sum_{i=1}^n x_i$

Combining results (20.14) and (20.17), we calculate the variance of an arithmetic average:

$$\text{Var}(\bar{x}) = \frac{1}{n^2}\sum_{i=1}^n \sigma_{x_i}^2 + \frac{2}{n^2}\sum_{\substack{j,k=1 \\ j<k}} \sigma_{x_j x_k} \qquad (20.17a)$$

If the x_i are uncorrelated, the second term vanishes. If, in addition, each x_i has the same variance σ_x^2, then we have the usual result

$$\sigma_{\bar{x}}^2 = \frac{\sigma_x^2}{n} \qquad (20.17b)$$

$\sum_{i=1}^n a_i x_i = a_1 x_1 + \cdots + a_n x_n$

We generalize further to the case

$$S = \sum_{i=1}^n a_i x_i = a_1 x_1 + \cdots + a_n x_n$$

where a_i may be positive or negative. We find

$$\mathrm{Var}\left(S = \sum a_i x_i\right) = \sum_{i=1}^{n} a_i^2 \,\mathrm{Var}(x_i) + \sum_{\substack{j,k=1 \\ j\neq k}}^{n} a_j a_k \,\mathrm{Covar}(x_j, x_k) = \sum_{i=1}^{n} a_i^2 \sigma_{x_i}^2 + 2\sum_{\substack{j,k \\ j<k}} a_j a_k \sigma_{x_j x_k} \tag{20.17c}$$

► **Example 20.3**

Suppose $A_1 = x_1 + x_2$ and $A_2 = x_1 - x_2$ with x_1, x_2 being independent with individual variances $\sigma_{x_1}^2$, $\sigma_{x_2}^2$.

We, first of all, note that A_1 and A_2 are correlated:

$$\begin{aligned}\mathrm{Covar}(A_1, A_2) &= \langle (A_1 A_2 - \langle A_1\rangle \cdot \langle A_2\rangle)\rangle \\ &= \langle (x_1 + x_2)(x_1 - x_2) - (\bar{x}_1 + \bar{x}_2)(\bar{x}_1 - \bar{x}_2)\rangle \\ &= \langle [x_1{}^2 - (\bar{x}_1)^2] - [x_2{}^2 - (\bar{x}_2)^2]\rangle \\ &= \langle x_1{}^2\rangle - \langle x_1\rangle^2 - \{\langle x_2{}^2\rangle - \langle x_2\rangle^2\} \\ &= \sigma_{x_1}^2 - \sigma_{x_2}^2\end{aligned} \tag{20.17d}$$

What can we say about the average

$$\bar{A} = \tfrac{1}{2}(A_1 + A_2)?$$

Using (20.17c) we obtain

$$\begin{aligned}\mathrm{Var}(\bar{A}) &= \tfrac{1}{4}\{\mathrm{Var}(A_1) + \mathrm{Var}(A_2) + 2\,\mathrm{Covar}(A_1, A_2)\} \\ &= \tfrac{1}{4}\{[\sigma_{x_1}^2 + \sigma_{x_2}^2] + [\sigma_{x_1}^2 + \sigma_{x_2}^2] + 2[\sigma_{x_1}^2 - \sigma_{x_2}^2]\} \\ &= \sigma_{x_1}^2\end{aligned}$$

For this simple case, of course

$$\bar{A} = \tfrac{1}{2}[x_1 + x_2 + x_1 - x_2] = x_1$$

and we check that

$$\mathrm{Var}(\bar{A}) = \mathrm{Var}(x_1)$$ ◄

► **Example 20.4**

Suppose it is required to measure the weights of two objects x_1 and x_2 with a balance and set of accurate standard weights that are known to arbitrary precision. Let us assume that the error in any weighing is a constant σ. (We mean that σ represents the standard deviation of a distribution of repeated weighings of the same object.) The precision of a balance may be characterized as $1/\sigma$ so that a smaller error implies a higher precision. Call the actual weights of the two objects w_1 and w_2.

If we weigh the objects separately, we obtain the measurements w_1' and w_2', which may be regarded as independent random variables. We thus get the results $w_1' \pm \sigma$ and $w_2' \pm \sigma$. The average $\frac{1}{2}(w_1' + w_2') = \bar{w}'$ is seen by (20.17a) to have a variance

$$\sigma_{\bar{w}'}{}^2 = \frac{1}{2^2}(\sigma^2 + \sigma^2) = \frac{\sigma^2}{2}$$

that is, the average has a standard deviation (of the mean) of $\sigma/\sqrt{2}$.

Suppose now we consider two independent weighings of the sum and difference of the objects

$$y_1' = w_1 + w_2 + \text{random error}$$
$$y_2' = w_1 - w_2 + \text{random error}$$

The measurements y_1', y_2' may here be considered as independent random variables with $\mathrm{Var}(y_1') = \mathrm{Var}(y_2') = \sigma^2$. From these two independent measurements, we may obtain estimates of the two weights:

$$\hat{w}_1 = \tfrac{1}{2}(y_1' + y_2')$$
$$\hat{w}_2 = \tfrac{1}{2}(y_1' - y_2')$$
$$\mathrm{Var}(\hat{w}_1) = \mathrm{Var}(\hat{w}_2) = \frac{1}{2^2}\sum_{i=1}^{2} \mathrm{Var}(y_i') = \frac{\sigma^2}{2}$$

since $\mathrm{Covar}(y_1', y_2') = 0$.

We see at once that the error on the individual weights is smaller by the factor $1/\sqrt{2}$ than the error obtained by separate measurement.

From relation 20.17d we note that

$$\mathrm{Covar}(\hat{w}_1, \hat{w}_2) = \sigma_{y_1'}^2 - \sigma_{y_2'}^2 = 0$$

that is, the estimates $\hat{w}_1$ and $\hat{w}_2$ are *uncorrelated*. This enables us to obtain the variance of the mean,

$$\bar{\hat{w}} = \tfrac{1}{2}(\hat{w}_1 + \hat{w}_2)$$

$$\mathrm{Var}(\bar{\hat{w}}) = \frac{1}{2^2}\sum_{i=1}^{2}\mathrm{Var}(\hat{w}_i) + 0$$
$$= \frac{1}{2^2}2\,\mathrm{Var}(\hat{w}_1)$$
$$= \frac{1}{2}\left(\frac{\sigma^2}{2}\right) = \frac{\sigma^2}{4}$$

We see that

$$\sigma_{\bar{\hat{w}}}{}^2 = \frac{\sigma^2}{4} = \frac{1}{2}\sigma_{\hat{w}'}{}^2$$

that is, the error on the mean is also improved by the factor, $1/\sqrt{2}$, over that obtained by separate weighing.

We say that the second method is statistically *more efficient* than the first in that it results in a smaller variance with the same number of measurements. ◄

WEIGHTED AVERAGE

$$\bar{x} = \sum_{i=1}^{n} a_i x_i, \quad a_i = \frac{1/\sigma_i^2}{\sum_{i=1}^{n} 1/\sigma_i^2}$$

Suppose we have n independent measurements x_i where the variances are, in general, not equal, that is, the variances are σ_i^2, $i = 1, \ldots, n$. We wish to construct a weighted average, $\bar{x}$, which has minimum variance, where $\Sigma_{i=1}^{n} a_i = 1$.

Since the x_i are independent, we get

$$\mathrm{Var}(\bar{x}) = \sum_{i=1}^{n} a_i^2 \sigma_i^2 \qquad (20.17\text{e})$$

We seek to prove that the weight

$$a_i^0 = \frac{1/\sigma_i^2}{\sum_{i=1}^{n} 1/\sigma_i^2} \qquad (20.17\text{f})$$

provides a minimum variance.

We may minimize (20.17e) subject to the constraint, $\Sigma_{i=1}^{n} a_i = 1$, using the method of Lagrange multipliers discussed in Appendix A. However, we may demonstrate that (20.17f) provides a minimum variance in a more elementary fashion.

In general, the variance of a random variable Z satisfies the inequality

$$E(Z^2) - [E(Z)]^2 = \mathrm{Var}(Z) \geq 0$$

or,

$$[E(Z)]^2 \leq E(Z^2)$$

Consider the sum

$$S = \sum_{i=1}^{n} A_i$$

and assume that the random variable Z takes on the values a_i/A_i with probabilities A_i/S,

$$\Pr\left(Z = \frac{a_i}{A_i}\right) = \frac{A_i}{S}$$

The probabilities are clearly normalized since

$$\sum_{i=1}^{n}\frac{A_i}{S} = \frac{1}{S}\sum_{i=1}^{n} A_i = 1$$

The expectation value of Z is

$$E(Z) = \sum_{i=1}^{n}\frac{a_i}{A_i}\frac{A_i}{S} = \frac{\sum_{i=1}^{n} a_i}{S}$$

The expectation value of Z^2 is

$$E(Z^2) = \sum_{i=1}^{n}\left(\frac{a_i}{A_i}\right)^2\frac{A_i}{S} = \frac{1}{S}\sum_{i=1}^{n}\frac{a_i^2}{A_i}$$

Therefore we have the general relation

$$\left(\frac{\sum_{i=1}^{n} a_i}{S}\right)^2 \leq \frac{1}{S}\sum_{i=1}^{n}\frac{a_i^2}{A_i}$$

Let

$$A_i = 1/\sigma_i^2, \; S = \sum_{i=1}^{n} 1/\sigma_i^2$$

and

$$\mathrm{Var}\left(\sum_{i=1}^{n} a_i x_i\right) = \sum_{i=1}^{n} a_i^2\sigma_i^2$$
$$= \sum_{i=1}^{n} a_i^2/A_i \geq \frac{(\Sigma a_i)^2}{S} = \frac{1}{S}$$

for all choices of a_i.

Now choose

$$a_i^0 = A_i/S = \frac{1/\sigma_i^2}{S}$$

so

$$\mathrm{Var}\left(\sum_{i=1}^{n} a_i^0 x_i\right) = \sum_{i=1}^{n}(a_i^0)^2/A_i = \sum_{i=1}^{n}\frac{A_i^2}{S^2 A_i}$$
$$= \sum_{i=1}^{n}\frac{A_i}{S^2} = \frac{1}{S^2}\sum_{i=1}^{n} A_i = \frac{1}{S}$$

$$\therefore \mathrm{Var}\left(\sum_{i=1}^{n} a_i x_i\right) \geq \frac{1}{S} = \mathrm{Var}\left(\sum_{i=1}^{n} a_i^0 x_i\right)$$

for all a_i and the choice a_i^0 of (20.17f) provides a minimum. Q.E.D.

Recall that we proved this weighting by induction in our discussion of the "principle of least-squared error" (minimum variance) in Chapter 10.

DIFFERENCE OF TWO MEASURED QUANTITIES, $\bar{x} - \bar{y}$

Although this is included in the general case where a_i may be positive or negative, we wish to emphasize the importance of this case.

Suppose that we make n_X independent measurements of a quantity X with $\text{Var}(X) = \sigma_X^2$ and n_Y independent measurements of a quantity Y with $\text{Var}(Y) = \sigma_Y^2$.

We wish to consider $\bar{x} - \bar{y}$ where

$$\bar{x} = \frac{1}{n_X}\sum_i^{n_X} x_i$$

$$\bar{y} = \frac{1}{n_Y}\sum_j^{n_Y} y_j$$

$\text{Var}(\bar{x} - \bar{y}) = \text{Var}(\bar{x}) + \text{Var}(\bar{y}) - 2\,\text{Covar}(\bar{x}, \bar{y})$ where the last term vanishes since $\bar{x}$, $\bar{y}$ are independent.

$$\therefore \text{Var}(\bar{x} - \bar{y}) = \frac{\sigma_X^2}{n_X} + \frac{\sigma_Y^2}{n_Y} \qquad (20.17g)$$

xy

As always with products, it is convenient to assume the random variables are *independent.*

$$\begin{aligned}\text{Var}(xy) &= \langle (xy)^2\rangle - [\langle xy\rangle]^2 \\ &= \langle x^2\rangle\langle y^2\rangle - \langle x\rangle^2\langle y\rangle^2\end{aligned}$$

since x, y are independent.

Since

$$\text{Var}(x) = \langle x^2\rangle - \langle x\rangle^2$$

and

$$\text{Var}(y) = \langle y^2\rangle - \langle y\rangle^2$$

we may substitute

$$\begin{aligned}\text{Var}(xy) = \langle x^2\rangle\langle y^2\rangle & \\ -[-\text{Var}(x) + \langle x^2\rangle][-\text{Var}(y) + \langle y^2\rangle] & \\ = \langle x^2\rangle\langle y^2\rangle - \langle x^2\rangle\langle y^2\rangle & \\ -[\text{Var}(x)\,\text{Var}(y) - \text{Var}(y)\langle x^2\rangle & \\ - \text{Var}(x)\langle y^2\rangle] & \\ = -\text{Var}(x)\,\text{Var}(y) & \\ + \text{Var}(x)\{\text{Var}(y) + \langle y\rangle^2\} & \\ + \text{Var}(y)\{\text{Var}(x) + \langle x\rangle^2\} &\end{aligned}$$

$$\begin{aligned}\therefore \text{Var}(xy) = \text{Var}(x)\langle y\rangle^2 + \text{Var}(y)\langle x\rangle^2 & \\ + \text{Var}(x)\,\text{Var}(y) & \qquad (20.18)\end{aligned}$$

This is just (10.11a).

Calculus of Probability Density Functions (Univariate)

We may note some important functions of one variable, X, where $f_X(x)$ is the p.d.f. for X. Also, $F_X(x)$ is the c.d.f. for X.

$Z = X^2$

We take X to be real and note that $\Pr(Z = X^2 < 0) = 0$.

$$\begin{aligned}F_Z(z) &= \Pr(Z = X^2 < z) = \Pr(|X| < \sqrt{z}) \\ &= \Pr(-\sqrt{z} < X < +\sqrt{z}) \\ &= \Pr(X < \sqrt{z}) - \Pr(X < -\sqrt{z})\end{aligned}$$

$$\begin{aligned}F_{Z=X^2}(z) &= F_X(\sqrt{z}) - F_X(-\sqrt{z}) && z \geq 0 \\ &= 0 && z < 0\end{aligned} \qquad (20.19)$$

The p.d.f. may be obtained as usual:

$$\begin{aligned}f_Z(z) &= \frac{d}{dz}F_Z = \frac{d}{dz}\{F_X\sqrt{z}) - F_X(-\sqrt{z})\} \\ &= f_X(\sqrt{z})\left[\frac{1}{2\sqrt{z}}\right] - f_X(-\sqrt{z})\left[\frac{-1}{2\sqrt{z}}\right]\end{aligned}$$

We may write

$$\begin{aligned}f_{Z=X^2}(z) &= \frac{1}{\sqrt{z}}\{f_X(\sqrt{z}) + f_X(-\sqrt{z})\} && z > 0 \\ &= 0 && z \leq 0\end{aligned} \qquad (20.20)$$

$Z = \sqrt{X}$

$$\begin{aligned}F_{Z=\sqrt{X}}(z) &= \Pr(Z = \sqrt{X} < z) \\ &= \Pr(X < z^2) \\ &= F_X(z^2)\end{aligned} \qquad (20.21)$$

Therefore,

$$f_Z(z) = \frac{d}{dz}F_Z(z) = f_X(z^2)[2z]$$

and

$$f_{Z=\sqrt{X}}(z) = 2zf_X(z^2) \qquad (20.22)$$

$Z = aX + b$

$$F_{Z=aX+b}(z) = F_X\left(\frac{z-b}{a}\right) \quad (20.23)$$

Again,

$$f_Z(z) = \frac{d}{dx} F_Z(z) = \frac{d}{dx} F_X\left(\frac{z-b}{a}\right)\left[\frac{1}{a}\right]$$

and

$$f_{Z=aX+b}(z) = \frac{1}{a} f_X\left(\frac{z-b}{a}\right) \quad (20.24)$$

If $X^* = (X - \langle X \rangle)/\sigma(X)$, the normalized variable defined in (20.15), then

$$F_{X^*}(x^*) = F_X\left(\frac{x^* - \langle x \rangle}{\sigma(x)}\right) \quad (20.25)$$

and

$$f_{X^*}(x^*) = \frac{1}{\sigma(x)} f_X\left(\frac{x^* - \langle x \rangle}{\sigma(x)}\right) \quad (20.26)$$

Calculus of Probability Density Functions (Multivariate)

Consider X, Y two random variables with joint p.d.f. $f_{XY}(x, y)$. If $Z = g(X, Y)$, then

$$F_Z(z) = \iint_{\substack{\text{Region in} \\ X-Y \text{ plane} \\ \text{where } g(x,y)<z}} f_{XY}(x, y)\, dx\, dy \quad (20.27)$$

$Z = X + Y$

For $Z = X + Y$,

$$F_Z(z) = \int_{-\infty}^{\infty} dx \int_{-\infty}^{z-x} dy\, f_{XY}(x, y)$$

We change variables.
Let

$$u = x + y \qquad du = dy$$
$$y = -\infty \to u = -\infty$$
$$y = z - x \to u = z$$

and get

$$F_Z(z) = \int_{-\infty}^{\infty} dx \int_{-\infty}^{z} f_{XY}(x, u - x)\, du \quad (20.28)$$

We assume that we can interchange the order of integration and obtain:

$$F_Z(z) = \int_{-\infty}^{\infty} dx \int_{-\infty}^{z} dx\, f_{XY}(x, u - x) \quad (20.29a)$$

We want

$$f_Z(z) = \frac{d}{dz} F_Z(z)$$

and recall that

$$\frac{d}{dx} \int_a^x h(u)\, du = h(x)$$

Thus

$$f_Z(z) = \frac{d}{dz}\left[\int_{-\infty}^{z} du \int_{-\infty}^{\infty} dx\, f_{XY}(x, u - x)\right]$$
$$= \int_{-\infty}^{\infty} dx\, f_{XY}(x, z - x) \quad (20.30a)$$

Note that we could have started with the relation

$$F_Z(z) = \int_{-\infty}^{\infty} dy \int_{-\infty}^{z-y} dx\, f_{XY}(x, y)$$

and, by a similar route, obtained the relations

$$F_Z(z) = \int_{-\infty}^{z} dv \int_{-\infty}^{\infty} dy\, f_{XY}(v - y, y) \quad (20.29b)$$

and

$$f_Z(z) = \int_{-\infty}^{\infty} dy\, f_{XY}(z - y, y) \quad (20.30b)$$

Note that the marginal distributions come out as before

$$f_X(x) = \int_{-\infty}^{\infty} dy\, f_{XY}(x, y) \quad (20.30c)$$

and

$$f_Y(y) = \int_{-\infty}^{\infty} dx\, f_{XY}(x, y) \quad (20.30d)$$

which agrees with (20.3a) and (20.3b).

The necessary and sufficient condition that X, Y are *independent random variables* is

$$f_{XY}(x, y) = f_X(x) f_Y(y) \quad (20.8b)$$

which is just (20.8). It is left as an exercise to show the equivalence of the statements (20.8) and (20.8a).

$Z = X - Y$

For $Z = X - Y$, we may obtain

$$F_{Z=X-Y}(z) = \int_{-\infty}^{z} du \int_{-\infty}^{\infty} f_{XY}(x, x-u)\, dx \quad (20.31a)$$

$$= \int_{-\infty}^{z} dv \int_{-\infty}^{\infty} f_{XY}(x+y, y)\, dy \quad (20.31b)$$

and

$$f_{Z=X-Y}(z) = \int_{-\infty}^{\infty} dx f_{XY}(x, x-z) \quad (20.32a)$$

$$= \int_{-\infty}^{\infty} dy f_{XY}(z+y, y) \quad (20.32b)$$

$Z = XY$

For $Z = XY$

$$F_Z(z) = \int_{-\infty}^{\infty} dx \int_{-\infty}^{z/x} f_{XY}(x, y)\, dy$$

$$= \int_{-\infty}^{\infty} \int_{-\infty}^{z} f_{XY}\left(x, \frac{u}{x}\right) \frac{du}{x}$$

with the substitution $u = xy$.

Assuming as usual that we can interchange the order of integration, we have

$$F_{Z=XY}(z) = \int_{-\infty}^{z} du \int_{-\infty}^{\infty} \frac{dx}{x} f_{XY}\left(x, \frac{u}{x}\right) \quad (20.33)$$

which yields

$$f_{Z=XY}(z) = \int_{-\infty}^{\infty} \frac{dx}{x} f_{XY}\left(x, \frac{z}{x}\right) \quad (20.34a)$$

We may also obtain

$$f_{Z=XY}(z) = \int_{-\infty}^{\infty} \frac{dy}{y} f_{XY}\left(\frac{z}{y}, y\right) \quad (20.34b)$$

We have glossed over a small subtlety since we need the region where $x/y < z$ and to insure this we must take absolute values. This results in the formulas:

$$f_{Z=XY}(z) = \int_{-\infty}^{\infty} f_{XY}\left(x, \frac{z}{x}\right) \frac{1}{|x|}\, dx \quad (20.35a)$$

$$= \int_{-\infty}^{\infty} f_{XY}\left(\frac{z}{y}, y\right) \frac{1}{|y|}\, dy \quad (20.35b)$$

$Z = X/Y$

For $Z = X/Y$

$$F_{Z=X/Y}(z) = \int_{-\infty}^{\infty} dy \int_{-\infty}^{yz} dx f_{XY}(x, y)$$

$$= \int_{-\infty}^{z} du \int_{-\infty}^{\infty} |y| f_{XY}(uy, y)\, du$$

from which it follows by interchanging the order of integration that

$$f_{Z=X/Y}(z) = \int_{-\infty}^{\infty} |y| f_{XY}(yz, y)\, dy \quad (20.36a)$$

Alternatively,

$$F_{Z=X/Y}(z) = \int_{-\infty}^{\infty} dx \int_{x/z}^{\infty} f_{XY}(x, y)\, dy$$

$$= \int_{-\infty}^{\infty} dx \int_{0}^{z} \frac{|x|}{u^2} f_{XY}\left(x, \frac{x}{u}\right) du$$

Again, interchanging the order of integration, we get

$$f_{Z=X/Y}(z) = \int_{-\infty}^{\infty} \frac{|x|}{z^2} f_{XY}\left(x, \frac{x}{z}\right) dx \quad (20.36b)$$

FUNCTIONS OF RANDOM VARIABLES: PROBABILITY DENSITY FUNCTIONS

We may summarize our results:

$Z = X^2$

$$f_{Z=X^2}(z) = \frac{1}{2\sqrt{z}} \{f_X(\sqrt{z}) + f_X(-\sqrt{z})\} \qquad z > 0$$

$$= 0 \qquad z \le 0 \quad (20.20)$$

$Z = \sqrt{X}$

$$f_{Z=\sqrt{X}}(z) = 2z f_X(z^2) \quad (20.22)$$

$Z = aX + b$

$$f_{Z=aX+b}(z) = \frac{1}{a} f_X\left(\frac{z-b}{a}\right) \quad (20.24)$$

$X^* = \dfrac{X - \langle X \rangle}{\sigma(X)}$

$$f_{X^*}(x^*) = \frac{1}{\sigma(x)} f_X\left(\frac{x^* - \langle x \rangle}{\sigma(x)}\right) \quad (20.26)$$

$Z = X + Y$

$$f_{Z=X+Y}(z) = \int_{-\infty}^{\infty} f_{XY}(x, z-x)\, dx$$

$$= \int_{-\infty}^{\infty} f_{XY}(z-y, y)\, dy \quad (20.30)$$

$Z = X - Y$

$$f_{Z=X-Y}(z) = \int_{-\infty}^{\infty} f_{XY}(x, x-z)\,dx$$
$$= \int_{-\infty}^{\infty} f_{XY}(z+y, y)\,dy \qquad (20.32)$$

$Z = XY$

$$f_{Z=XY}(z) = \int_{-\infty}^{\infty} f_{XY}\left(x, \frac{z}{x}\right)\frac{1}{|x|}\,dx$$
$$= \int_{-\infty}^{\infty} f_{XY}\left(\frac{z}{y}, y\right)\frac{1}{|y|}\,dy \qquad (20.35)$$

$Z = X/Y$

$$f_{Z=X/Y}(z) = \int_{-\infty}^{\infty} \frac{|x|}{z^2} f_{XY}\left(x, \frac{x}{z}\right)\,dx$$
$$= \int_{-\infty}^{\infty} |y| f_{XY}(yz, y)\,dy \qquad (20.36)$$

CONVOLUTION

When X, Y are *independent* random variables we use the relation (20.8b). For $Z = X + Y$, (20.30) becomes

$$f_{Z=X+Y}(z) = \int_{-\infty}^{\infty} f_X(x) f_Y(z-x)\,dx \qquad (20.37)$$

This relationship is of more general interest. $f_Z(z)$ is called the *convolution of the functions f_X and f_Y.*

Definition of uniform distribution. $U(a, b)$

We have previously considered in relations (19.1) the *uniform probability distribution.* We shall say that a random variable X is uniformly distributed on the interval (a, b), written $U(a, b)$, if it obeys relations (19.1).

► **Example 20.5**

Consider X, Y independently and identically distributed $U(-a/2, a/2)$. Find $f_{Z=X+Y}(z)$.

Here $f_X(x) = \dfrac{1}{a}$ if $|x| \le \dfrac{a}{2}$
$= 0$ otherwise

$f_Y(y) = \dfrac{1}{a}$ if $|y| \le \dfrac{a}{2}$
$= 0$ otherwise

which is schematically indicated in Figure 20.3

$$f_{Z=X+Y}(z) = \int_{-\infty}^{\infty} f_X(x) f_Y(z-x)\,dx$$
$$= \frac{1}{a}\int_{-a/2}^{a/2} f_Y(z-x)\,dx$$

The function $f_Y(y)$ is defined to be nonzero when the argument is such that $|y| \le a/2$. This yields the nonzero interval to be

$$-\frac{a}{2} \le z - x \le +\frac{a}{2}$$

or

$$z - \frac{a}{2} \le x \le z + \frac{a}{2}$$

We have the additional requirement that the integrand is nonzero when

$$-\frac{a}{2} \le x \le +\frac{a}{2}$$

The integrand is nonzero only when

$$-a \le z \le a$$

Case 1: $-a \le z \le 0.$

Here

$$x \le z + \frac{a}{2}$$

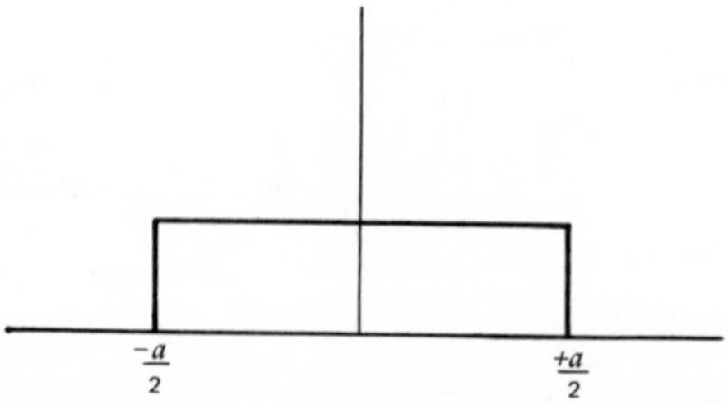

Figure 20.3
Plot of the probability density function for variate uniformly distributed between $-a/2$ and $+a/2$.

and

$$f_Z(z) = \frac{1}{a} \int_{-a/2}^{z+a/2} \frac{1}{a}\, dx$$
$$= \frac{1}{a^2}(z + a) \qquad -a \leq z \leq 0$$

Case 2: $0 \leq z \leq a$.

Here

$$z - \frac{a}{2} \leq x \leq \frac{a}{2}$$

Therefore,

$$f_Z(z) = \frac{1}{a} \int_{z-a/2}^{a/2} \frac{1}{a}\, dx$$
$$= \frac{1}{a^2}(a - z) \qquad 0 \leq z \leq a$$

Overall,

$$f_{Z=X+Y}(z) = \frac{1}{a^2}(a - |z|) \qquad |z| \leq a$$
$$= 0 \qquad |z| > a \tag{20.38}$$

This is indicated schematically in Figure 20.4. ◄

GENERAL TRANSFORMATION OF VARIABLES: UNIVARIATE

In general, if we have a p.d.f. $f_X(x)$ and we wish to get the p.d.f. $g(y)$ for a variate Y where

$$Y = Y(x)$$

we find the inverse function

$$X = X(y)$$

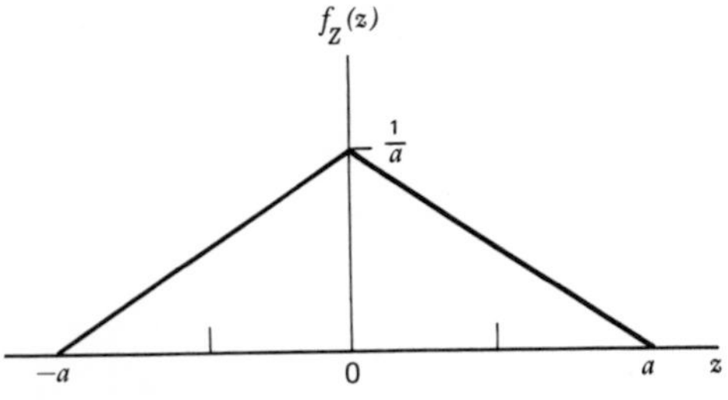

Figure 20.4
Plot of the probability density function, $f(z)$, for $Z = X + Y$ where X and Y are uniformly distributed as in Figure 20.3.

and write

$$g_Y(y)\, dy = f_X(x)\, dx$$
$$= f_X[X(y)] \left| \frac{dX(y)}{dy} \right| dy \tag{20.39}$$

► *Exercise*
Use (20.39) to obtain (20.20), (20.22), and (20.24). ◄

GENERAL TRANSFORMATION OF VARIABLES: MULTIVARIATE

In general, if we have a joint p.d.f. $f(x_1, x_2, \ldots, x_n)$ and we wish to find the p.d.f. for $y = (y_1, y_2, \ldots, y_n) = (\mathbf{y})$ with the relations $y_i = y_i(x_1, x_2, \ldots, x_n)$ for each i, we find the inverse relations $x_i = x_i(y_1, y_2, \ldots, y_n)$ and write

$$g(y_1, y_2, \ldots, y_n)$$
$$= f[x_1(\mathbf{y}), x_2(\mathbf{y}), \ldots, x_n(\mathbf{y})]$$
$$\times \left| \frac{\partial(x_1, x_2, x_3, \ldots, x_n)}{\partial(y_1, y_2, y_3, \ldots, y_n)} \right| \tag{20.40}$$

where the last term is the absolute value of the Jacobian of the transformation

$$(x_1, x_2, \ldots, x_n) \to (y_1, y_2, \ldots, y_n)$$

where the Jacobian is defined as the determinant

$$\begin{vmatrix} \frac{\partial x_1}{\partial y_1} & \frac{\partial x_1}{\partial y_2} & \cdots & \frac{\partial x_1}{\partial y_n} \\ \frac{\partial x_2}{\partial y_1} & \frac{\partial x_2}{\partial y_2} & \cdots & \frac{\partial x_2}{\partial y_n} \\ \vdots & \vdots & & \vdots \\ \frac{\partial x_n}{\partial y_1} & \frac{\partial x_n}{\partial y_2} & \cdots & \frac{\partial x_n}{\partial y_n} \end{vmatrix} = \frac{\partial(x_1, x_2, \ldots, x_n)}{\partial(y_1, y_2, y_3, \ldots, y_n)} \tag{20.41}$$

► *Exercise*
Use (20.40) to obtain (20.30), (20.32), (20.35) and (20.36). ◄

LINEAR ORTHOGONAL TRANSFORMATION OF VARIABLES

It is sometimes true that the variates $(x_1, \ldots, x_n)$ and $(y_1, \ldots, y_n)$ are related linearly.

$$y_i = \sum_{j=1}^{n} C_{ij} x_j \qquad i = 1, \ldots, n \tag{20.42}$$

We may write the coefficients C_{ij} as a matrix

$$(C_{ij}) = \begin{pmatrix} C_{11} & C_{12} \cdots & C_{1n} \\ C_{21} & C_{22} \cdots & C_{2n} \\ \vdots & \vdots & \vdots \\ C_{n1} & C_{n2} \cdots & C_{nn} \end{pmatrix}$$

and consider that x_i and y_i are vectors $\mathbf{x}, \mathbf{y}$ such that $\mathbf{x}$ is transformed into $\mathbf{y}$ by the matrix $\underline{\mathbf{C}}$. A brief review of matrix algebra is given in Appendix A and we shall use the ideas summarized there.

We may write the matrix equation

$$\mathbf{y} = \underline{\mathbf{C}}\mathbf{x} \tag{20.43}$$

as a shorthand expression for (20.42). It often is useful to consider situations in which the C_{ij} are constants such that

$$\sum_{i=1}^{n} y_i^2 = \sum_{i=1}^{n} x_i^2 \tag{20.44}$$

When this is true, we say that $\mathbf{x}, \mathbf{y}$ are related by a *linear, orthogonal transformation.*

It may be shown that the Jacobian of a linear orthogonal transformation is unity.

$$|\underline{\mathbf{C}}| = 1 \tag{20.45}$$

Relation (20.44) shows that linear orthogonal transformations preserve distances.

► Example 20.6

A simple example of a linear orthogonal transformation is a rigid rotation of a coordinate system as in Figure 20.5.

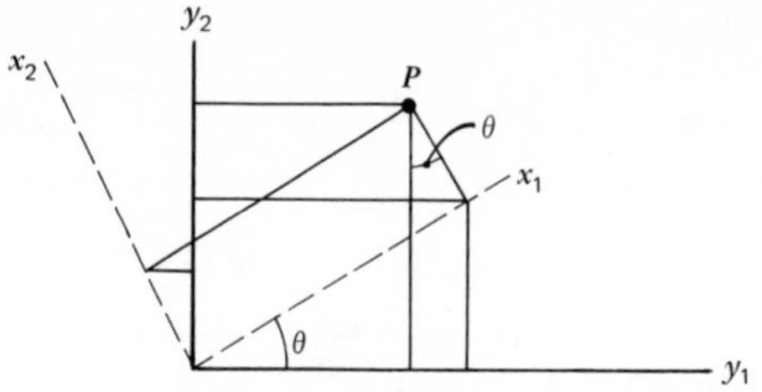

Figure 20.5
Rectangular coordinate system (y_1, y_2) together with rectangular coordinate system (x_1, x_2) rotated through angle θ.

The point P has coordinates (x_1, x_2) in one system and coordinates (y_1, y_2) in the rotated system. We note that

$$y_1 = x_1 \cos\theta - x_2 \sin\theta$$

$$y_2 = x_1 \sin\theta + x_2 \cos\theta$$

which may be written

$$\begin{pmatrix} y_1 \\ y_2 \end{pmatrix} = \begin{pmatrix} \cos\theta & -\sin\theta \\ \sin\theta & +\cos\theta \end{pmatrix} \begin{pmatrix} x_1 \\ x_2 \end{pmatrix}.$$

We may verify that

$$y_1^2 + y_2^2 = x_1^2 + x_2^2$$

The Jacobian of this transformation is

$$\begin{vmatrix} \cos\theta & -\sin\theta \\ \sin\theta & +\cos\theta \end{vmatrix} = \cos^2\theta - (-\sin^2\theta) = 1$$

◄

21

GEOMETRICAL PROBABILITY, RANDOM NUMBERS, AND MONTE CARLO EXPERIMENTS

We have encountered before the concept of *equally likely* cases in computing probabilities. In the history of probability theory this was quite early applied to problems of geometry. We have already considered the *uniform distribution* (19.1b, c, d, e), and this underlies most problems of geometrical probability.

A simple case is to consider a line segment of length l between the left end called 0, and the right end called l. Suppose we choose a point x, on the line *at random*. Few of us would quibble with the assertion that the probability is 0.5 that the point is to be found in the right half of the line segment.

Again, suppose a square target is set up with side l and an inscribed circle of radius $l/2$. Suppose shots are fired *at random*, and we ask for the probability that a shot that hits the target occurs within the circle. It is quite straightforward to take the ratio of the areas and obtain a probability

$$\Pr(\text{circle}) = \frac{\pi(l/2)^2}{l^2} = \frac{\pi}{4}$$

If we introduce one oxygen molecule into an evacuated cubical box of side L (and we assume that the walls will not absorb the molecule and that enough time has elapsed so that it has "forgotten" at which point it was introduced) then the probability of being in a volume V is just V/L^3.

In each of these problems of geometrical probability it is assumed that there is a constant probability density (probability of being in a unit volume) and that the probability is obtained by multiplying this density by the volume of interest. We are considering "volume" in its general sense, not restricting it to three dimensions. Thus, the volume may be a linear element, $\int dx$, an area $\iint dA = \iint dx\,dy$ or a three-dimensional volume $\iiint dV = \iiint dx\,dy\,dz$. We may, of course, generalize further to the multidimensional case: $\int \cdots \int dV_n = \int_1 \cdots \int_n dx_1 \cdots dx_n$.

Buffon's Needle

A famous problem concerns an experiment in which needles (or toothpicks in an economical form suitable for student use) are dropped on a ruled grid. The number

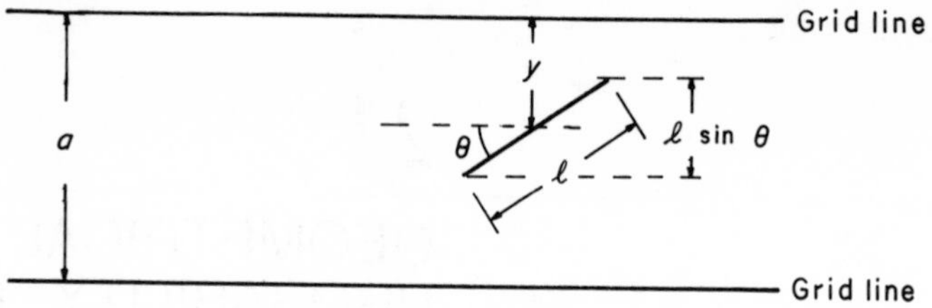

Figure 21.1
Buffon's needle of length ℓ dropped on grid with spacing a.

of times a needle intersects a line is noted and compared with the total number of needles dropped.

If we have a grid of parallel lines with separation a onto which we drop needles of length ℓ, we can classify the position of any needle according to the distance y from its midpoint to the nearest line, and signify its orientation by the angle θ $(0 \leq \theta \leq \pi/2)$, which it makes with the direction of the grid lines (Figure 21.1).

If the needle were perpendicular to the grid lines in position 1 of Figure 21.2, it could be moved a distance ℓ in a direction perpendicular to the grid lines to position 2 and still intersect a line. It could then be moved a distance $(a - \ell)$ in the same direction to position 3 before intersecting the next grid line (Figure 21.2). The intersections and nonintersections are in a ratio $\ell/(a - \ell)$. The proportion of intersections to *total* possibilities would then be $\ell/(\ell + (a - \ell)) = \ell/a$.

If the needle were not perpendicular to the grid lines, the effective length in this direction would be $\ell \sin\theta$ (Fig. 21.1). For all needles of this same characteristic angle, the probability of an intersection would be $\ell \sin\theta/a$. Since the angle is arbitrary when the needles are dropped randomly, with each angle being equally likely, we are interested in the average value of $\ell \sin\theta$, $(0 < \theta < \pi/2)$. The average value of $\sin\theta$ between 0 and $\pi/2$ is just $2/\pi$,

$$\frac{\int_0^{\pi/2} \sin\theta\, d\theta}{\int_0^{\pi/2} d\theta} = \frac{-\cos\theta}{\pi/2}\Bigg]_0^{\pi/2} = \frac{-(-1)}{\pi/2} = \frac{2}{\pi} \tag{21.1}$$

so that our probability of intersection is $\ell(2/\pi)/a$. If M is the number of needles that cross a line and N is the total number of needles dropped, then the ratio of in-

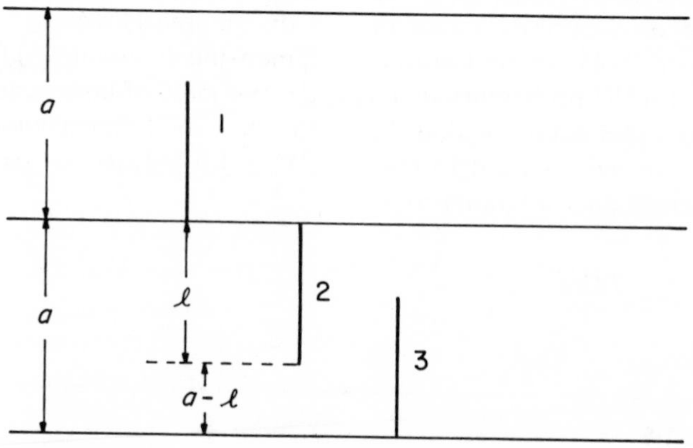

Figure 21.2
Buffon's needle when the needle is perpendicular to the grid lines.

tersections to total trials is

$$\frac{M}{N} = \frac{2\ell}{\pi a} \tag{21.2}$$

or, solving for π:

$$\pi = \frac{2\ell}{a}\frac{N}{M} \tag{21.3}$$

We can present an alternate derivation. Consider again the position of the needle designated as in Figure 21.1 by distance y and angle θ. For an intersection to occur, we must have the condition

$$y \leq \frac{\ell}{2}\sin\theta$$

From this relationship, you can convince yourself that those needles that intersect a line lie in the shaded area under the sine curve in Figure 21.3. The ratio of intersections to total possibilities is just the ratio of the area under the sine curve to the total area of the box

$$0 \leq \theta \leq \frac{\pi}{2}$$

$$0 \leq y \leq \frac{a}{2}$$

The area under the sine curve is given by

$$\int_0^{\pi/2} \left(\frac{\ell}{2}\sin\theta\right) d\theta = \frac{\ell}{2}\int_0^{\pi/2} \sin\theta \, d\theta$$

Then

$$\frac{M}{N} = \frac{\frac{\ell}{2}\int_0^{\pi/2} \sin\theta\, d\theta}{\left(\frac{a}{2}\right)\left(\frac{\pi}{2}\right)} = \frac{2\ell}{\pi a} \tag{21.4}$$

Or, again

$$\pi = \frac{2\ell}{a}\frac{N}{M}$$

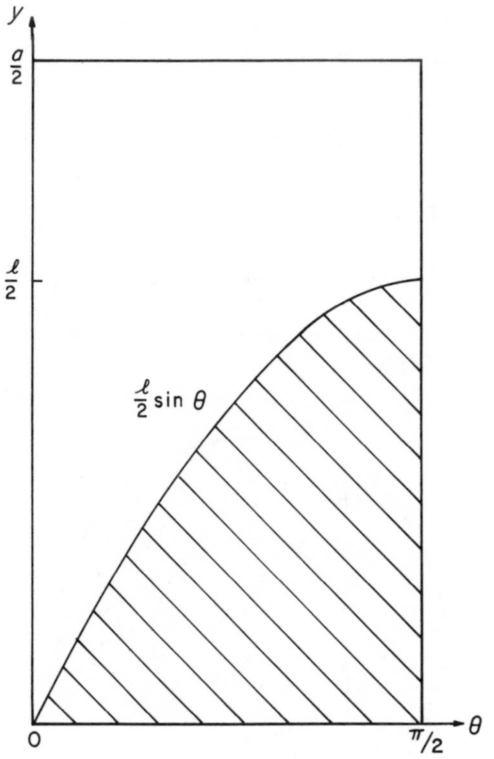

Figure 21.3
The rectangular area represents all possible combinations of (y, θ). The shaded area beneath the curve represents those combinations (y, θ) for which the needle intersects a grid line.

A famous story* connected with the needle experiment concerns an Italian mathematician named Lazzarini who made 3408 tosses of needles 5.0 cm long on a ruled grid, with a spacing of 6.0 cm, and counted 1808 needles that intersected lines. The statement is then usually made that for this experiment,

$$\pi = \frac{3408}{1808} \times \frac{2 \times 5}{6} = 3.1415929$$

which is wrong only in the seventh decimal place.

Can you comment on this statement? In particular, what must have been the situation if he had tossed one more needle? Do you believe Lazzarini? How well should he have measured the length of the needles and the spacing of the grid?

Perhaps more dramatic in this regard is the (apochryphal) experiment of the Ruthenian idiot savant Kerenyi who performed the experiment with the spacing equal to the needle length. He made only 355 tries and observed 226 intersections.

*This is well-described in T. H. O'Beirne, *Puzzles and Paradoxes*, Oxford University Press, 1965.

His value for $\pi = 2 \times 355/266 = 3.1415929$. This, like Lazzarini's result, is wrong only in the seventh decimal place. Clearly Kerenyi was a better calculator than Lazzarini, since he achieved the same accuracy with only 1/10 the labor!

We shall return to the question of determining how close to the right answer the result of a finite number of trials would be. For the moment let us remark that we would expect an ideal experiment to yield M/N arbitrarily close to $2\ell/\pi a$ as $N \to \infty$.

THE INVERSE PROBLEM

Before we leave Buffon's needle we note that the problem may be turned around. Obviously, we know π to as many places as we care to compute (as of 1972, 10^6 digits!). We may use the experimental result and relation (21.3) to compute ℓ given a, or a given ℓ, or the ratio ℓ/a.

When we seek to measure a length with a meter stick, we assume that the error in the measurement depends on the size of the smallest division. We now see that we need only have a standard length with no scale markings at all! All that is required is that the overall length of the measuring stick be known. If the stick is then thrown "randomly" at the unknown length and the intersections noted, an arbitrarily large number of trials will produce an arbitrarily good measurement of a.

We may see also that this experiment permits us to evaluate integrals as shown in (21.1) and (21.4). If a student knew no calculus but could measure he could get a very good value for the integral

$$\int_0^{\pi/2} \sin\theta \, d\theta$$

Very often integrals are not simply soluble and this kind of "statistical" integration may be employed.

Bertrand's Paradox

To illustrate the point that problems in geometrical probability have their difficulties, F. L. Bertrand in 1899 identified the following problem:

What is the probability that a "random chord" in a circle will be greater than a side of an inscribed equilateral triangle?

The problem is illustrated in Figure 21.4.

METHOD 1: CHOICE OF AN ANGLE AT RANDOM

We may fix one end of the chord at point P and choose "at random" the other end, Q.

(a) This is equivalent to choosing at random the angle θ that can vary between 0 and 2π. If we assume that all values of θ are equally probable (tantamount to assuming a uniform distribution for θ between 0 and 2π) we have a simple way of deciding whether $\overline{QP} > s$, where s is the side of the inscribed equilateral triangle:

$$\overline{QP} > s \qquad \text{when} \qquad \frac{4\pi}{3} > \theta > \frac{2\pi}{3}$$

We have only to calculate $\Pr\{4\pi/3 > \theta > 2\pi/3\}$.

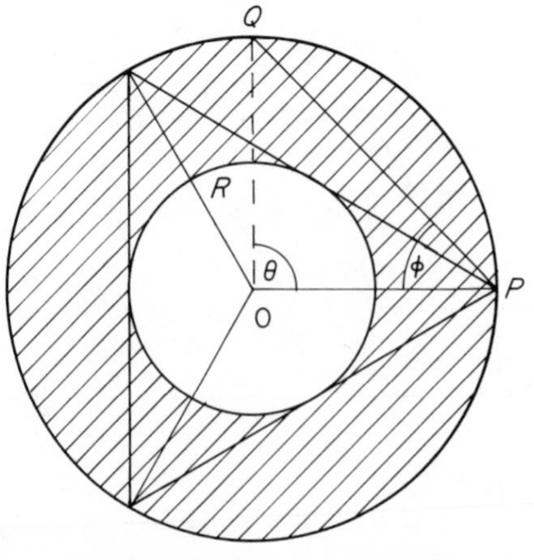

Figure 21.4
Geometry for Bertrand's problem.

Clearly,

$$\Pr\left\{\frac{4\pi}{3} > \theta > \frac{2\pi}{3}\right\} = \frac{4\pi/3 - 2\pi/3}{2\pi} = \frac{1}{3}$$

(b) We may also choose at random the angle $\phi = \measuredangle OPQ$. Here $\overline{QP} > s$ when $\pi/6 > \phi > -\pi/6$ where the entire range of ϕ is between $\pi/2$ and $-\pi/2$. Here

$$\Pr\{\overline{QP} > s\} = \Pr\{\pi/6 > \phi > -\pi/6\}$$

If we assume ϕ uniformly distributed over the interval $\pi/2$ to $-\pi/2$, we get

$$\Pr\{\pi/6 > \phi > -\pi/6\} = \frac{\pi/6 + \pi/6}{\pi/2 - (-\pi/2)} = \frac{1}{3}$$

We see that choosing an angle at random yields a probability of 1/3.

METHOD II: CHOICE OF A MIDPOINT AT RANDOM

Alternatively, we say that the midpoint of the chord is chosen "at random." Thus, if the midpoint lies in the unshaded region within the circle of radius $R/2$, the chord $\overline{QP}$ will be greater than s. If we assume that the midpoint is uniformly distributed over the area, we get

$$\Pr\{\overline{QP} > s\} = \Pr\{\text{midpoint is in unshaded area}\} = \frac{\pi(R/2)^2}{\pi R^2} = \frac{1}{4}$$

The answers obtained from methods I and II are different!

Randomness and Random Numbers Drawn from the Uniform Distribution

Although the existence of such paradoxes as Bertrand's gave geometrical probability a bad image, all that is required is to specify more precisely what constitutes an operation made "at random."

If we consider a random variable x with a frequency function $p(x_i)$ in the discrete case, or a p.d.f. $f(x)$ in the continuous case, we may think of generating random numbers from the distribution of x. The starting point is usually to generate a random number from the uniform distribution defined in (19.1b, c, d, and e). When we speak of a random number without further specification, we shall mean one drawn from the interval $0 < x < 1$ with the p.d.f. $f(x) = 1$.

We shall not dwell on how to generate random numbers. Conceptually we may think of a spinner as in Figure 21.5, which has equal probability of choosing single digits 0, 1, 2, 3, 4, 5, 6, 7, 8, 9. We emphasize for this simple example that the numbers need not be sequential nor the circumference divided into only 10 pieces. It is only necessary that the total circumference represented by each digit be 1/10th of the total.

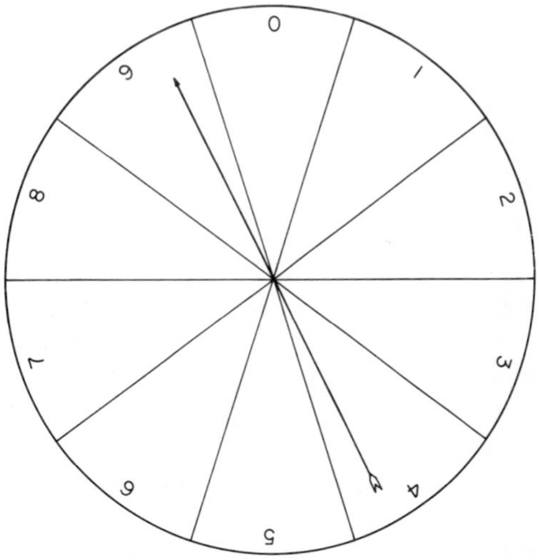

Figure 21.5
Spinner to generate random digits.

We may also think of using a roulette wheel numbered from 0 to 9. One of the largest tables of random numbers published* used an "electronic roulette wheel." The similarity of these random

A Million Random Digits, the Rand Corporation, The Free Press, Glencoe, Illinois, 1955.

number generating techniques to gambling led to the appellation, "Monte Carlo" methods, after the famous gambling establishment.

All digital computer installations today have random number generating routines. Since a computer program is basically deterministic, however, the mathematician John Von Neumann correctly pointed out that the generation of random numbers by computation is a contradiction in terms. For many purposes, however, such computer-generated "pseudorandom" numbers are random enough. We might remark that one reason for performing the *tour de force* of extending the number of digits computed for a transcendental number such as π is to provide a source of computable random numbers. For transcendental numbers there is no deterministic sequence of digits to be expected, and π computed to 10^6 digits should provide a sequence of 10^6 digits that are truly random.

We are glossing over some points in this discussion. We wish to simulate the uniform distribution which is continuous. We can approximate a continuous distribution by choosing groups of successive digits using our conceptual random number generator to produce one digit at a time. For example, if we choose 5 digit random numbers, there are 10^5 numbers possible to span the unit interval. If we choose n digits, there are 10^n numbers available to span the unit interval. We can, therefore, approach the continuous case as closely as we need.

There is, of course, nothing sacred about *decimal* digits, and we may generate random *binary* digits (*bits*) by successively drawing balls from an urn containing equal numbers of white (0) and black (1) balls or by utilizing a spinner with the circumference divided into two equal regions. We may also proceed to generate binary digits from random decimal digits and vice versa.

Simulation of Probability Problems: Monte Carlo Experiments

Using an appropriate source of random numbers we may perform experiments to investigate probabilities. We start by assigning to numbers their appropriate weights as the *a priori* probabilities. We may then proceed to count the results and use the observed frequencies to inform us about the probabilities of interest.

► **Example 21.1**
Sock problem

Consider the "sock problem" of Chapter 18 (Example 18.6). There are 10 individual socks of which 2 are black, 2 are yellow, 2 are pink, and 4 are green. We may assign numbers to represent the various socks such as: black (1) = 0, black (2) = 1, yellow (1) = 2, yellow (2) = 3, pink (1) = 4, pink (2) = 5, green (1) = 6, green (2) = 7, green (3) = 8, and green (4) = 9. The experiment was to draw three socks at random. This corresponds to choosing triplets of numbers. We must note that the real-world situation constitutes sampling without replacement whereas the random numbers constitute a case of sampling with replacement; that is, with the random numbers it is possible to select the same sock more than once although this is not possible in the real experiment. One way to handle this problem is to throw away any triplet of random numbers in which there is duplication.

We may ask how many triplets we might expect to discard this way, and we calculate the fraction of triplets without duplication to be

$$\frac{(10)(9)(8)}{10^3} = 0.72$$

The 200 triplets of random decimal digits listed in Table 21.1 were examined with these results.

Table **21.1** 200 Triplets of Random Decimal Digits

820	854	655	346	985	016	505	685
209	271	493	522	362	220	642	795
141	950	633	101	748	373	405	289
429	509	699	100	318	991	684	726
824	931	587	967	421	209	094	188
028	326	002	953	926	162	230	469
992	078	146	773	519	262	421	994
741	352	214	311	409	100	269	478
961	494	395	846	055	278	611	976
584	434	305	230	674	875	704	146
820	766	723	468	185	397	747	404
002	089	836	219	216	412	704	320
380	919	801	048	672	574	842	332
784	498	609	035	685	228	150	884
703	415	764	466	119	234	443	605
718	884	524	131	469	037	072	134
636	623	011	411	320	870	004	719
679	011	949	106	961	289	193	172
366	486	088	056	859	991	967	814
663	414	695	411	466	626	563	733
786	050	868	488	854	775	818	373
607	085	046	091	965	192	146	512
603	071	529	593	678	431	375	183
519	676	742	501	882	815	328	267
405	928	469	771	072	910	533	504

140 triplets without duplicates
65 triplets with no pairs
46 triplets contained 2 green + anything
8 triplets contained 2 black + anything
10 triplets contained 2 yellow + anything
11 triplets contained 2 pink + anything
6 triplets contained 3 green socks

In Table 21.2 we compare these results with the theoretical probabilities calculated in Example 18.6.

The last probability in the table is obtained from $(4)(3)(2)/(10)(9)(8) = 0.033$.

We might also note that the average of $\Pr(2B)$, $\Pr(2Y)$, and $\Pr(2P)$ is 0.069, whereas the expected value is 0.067. ◄

We should perhaps remark that the difference between the columns marked "Observed" and "Theoretical" are not

Table **21.2**

	Observed	Theoretical
Pr(no dupl)	$\frac{140}{200} = 0.70$	0.72
Pr(no pair)	$\frac{65}{140} = 0.464$	0.400
$\Pr(2G)$	$\frac{46}{140} = 0.328$	0.400
$\Pr(2B)$	$\frac{8}{140} = 0.057$	0.067
$\Pr(2Y)$	$\frac{10}{140} = 0.071$	0.067
$\Pr(2P)$	$\frac{11}{140} = 0.078$	0.067
$\Pr(3G)$	$\frac{6}{140} = 0.043$	0.033

due to any mistakes or errors. The differences only reflect the inherent variability of random experiments. If we used a very much larger number of trials, the differences would likely be smaller (see the discussions of the "Frequency Definition of Probability" and the "Law of Large Numbers"). If we do not know the theoretical values, of course, then the frequencies observed in a random experiment are all that we have.

It is frequently true that we are interested in the average or expected value of a quantity. We may perform a random experiment and determine the quantity of interest. If the experiment is repeated n times, the n quantities determined may be averaged to estimate the expectation value desired. It is important to note that this procedure can be employed when the expectation value may not be calculable analytically or when it is difficult or inconvenient to do so. Even when an analytical solution may be obtained, it is sometimes helpful to get a "feeling" for what is involved by performing a Monte Carlo experiment.

► **Example 21.2**
Random walk

Consider that we are able to move only along a straight line, the x-axis, a step of one unit at a time, but with the direction chosen randomly. We ask: How far from the origin might we expect to be after n steps? We may use Table 21.1 by interpreting even numbers to be + (a unit step in the positive direction) and odd numbers, − (a unit step in the negative direction). The first 25 numbers are:

8208546553469850165056852.

We interpret these as:

++++−++−−−++−+−+−+−+−++−+.

Twenty-five steps leave us at $x = +5$.

How can we understand this result? We are interested in the distance from the origin, r. We may write each step as x_i where $|x_i| = 1$ and where x_i may be positive or negative with equal probability.

$$r^2 = (x_1 + x_2 + x_3 + \cdots + x_n)^2$$

This may be written as

$$r^2 = (x_1^2 + x_2^2 + \cdots + x_n^2) + (\text{cross terms})$$

where the cross terms are of the form

$$\sum_{i,j=1}^{n} x_i x_j \qquad i \neq j$$

Because the cross terms always have the magnitude $|x_i x_j| = 1$, but appear, on the average, with equal numbers of + and − signs, the cross terms average out to 0.

$$\therefore \langle r^2 \rangle = \sum_{i=1}^{n} x_i^2 = n|x_i^2| = n$$

$$\therefore \langle r \rangle = \sqrt{n}$$ ◄

We expect that the average distance from the origin will be proportional to the square root of the number of steps.

A two-dimensional version of this random walk problem is well-known.

► **Example 21.3**
Drunkard's walk

As charmingly discussed by George Gamow,* we consider a drunkard who has somehow arrived at a lamp post in the middle of a large city square. He sets off in an unpredictable way making one step at a time of length l in a randomly chosen direction. What is his expected distance from the lamp post after n steps?

Here we let x_i, y_i be the projections of the ith step on the respective axes. The final distance after n steps is obtained from

$$r^2 = \left(\sum x_i\right)^2 + \left(\sum y_i\right)^2$$
$$= \sum x_i^2 + \sum y_i^2 + (\text{cross terms})$$

The cross terms average out to zero as

***One Two Three . . . Infinity*, Viking Press, 1947.

before. We are left with

$$\langle r^2 \rangle = \sum x_i^2 + \sum y_i^2$$
$$= n\overline{x^2} + n\overline{y^2}$$

where $\overline{x^2}$, $\overline{y^2}$ are the averages of the squared projections on the axes

$$\overline{x^2} = \frac{\int_0^{2\pi} l^2 \cos^2 \theta \, d\theta}{\int_0^{2\pi} d\theta}$$
$$= l^2 \frac{(2\pi/2)}{2\pi} = \frac{l^2}{2}$$

and

$$\overline{y^2} = \frac{\int_0^{2\pi} l^2 \sin^2 \theta \, d\theta}{\int_0^{2\pi} d\theta}$$
$$= \frac{l^2}{2}$$

where l is the step size.

$$\therefore \langle r^2 \rangle = n\left(\frac{l^2}{2} + \frac{l^2}{2}\right)$$
$$= nl^2$$

or

$$\langle r \rangle = l\sqrt{n}$$ ◄

This is characteristic of random walk problems: the average distance after n steps is proportional to the square root of the number of steps.

► *Exercise*

Simulate the Drunkard's Walk using random numbers. ◄

We return to other problems previously considered.

► Example 21.4

Coin tossing (Cf. Example 19.12)

We may consider the problem discussed in Chapter 19. For a fair coin, what is the average number of tosses required to get three consecutive heads? Since the coin is fair, we take the probability of heads and tails to be equal. We may use Table 21.1 by assigning even numbers to heads and odd numbers to tails. We count the number of trials (digits) until three consecutive heads appear (three consecutive even digits) and then start again. From Table 21.1 we get the results: $n = 3, 38, 3, 49, 4, 24, 6, 12, 4, 19, 25, 11, 6, 16, 21, 11, 12, 3, 33, 8, 8, 4, 10, 5, 11, 10, 9, 24, 11, 4, 32, 3, 3, 16, 13, 3, 10, 5, 3, 21, 17, 37, 7$, and 9. These average to $\langle n \rangle = 13.25$, which should be compared with the expectation value of 14. ◄

► Example 21.5

Birthday problem (Cf. Example 19.5)

We may simulate the birthdate problem of Chapter 19 using Table 21.1. We may number the days in the year from 1 to 365 and consider only triplets in the range [001, 365]. If we look at the first 24 acceptable triplets, we find a match at 209, that is, July 28. If we look at the next 24 acceptable triplets, we see no match. ◄

► Example 21.6

Bubble gum problem (Cf. Example 19.13)

We may simulate the bubble gum problem of Chapter 19 (Example 19.13) again using Table 21.1. Here the problem is to determine the average number of purchases necessary to get a complete set of 10 baseball cards if they are distributed in equal numbers and at random throughout all the packages of gum. We may label each card by a specific digit and use the random numbers of Table 21.1 to observe how many digits must be looked at to get a complete set of 10. The results for the numbers of Table 21.1 are: 29, 34, 19, 21, 28, 26, 25, 19, 20, 36, 47, 27, 21, 23, 68, 41, 38, 21, 20, and 17. These observations average to 29.0. This compares well with the expectation value of 29.3 computed in Chapter 19. ◄

► *Exercise*

Continue the previous experiments using more trials. Consider how to modify the assignments in the case of 11 socks, 12 baseball cards, etc. ◄

Operations with Random Numbers

We shall denote a random number* drawn from the interval (0, 1) as

$$X = \text{RAND}(0, 1) \quad (21.5)$$

This should be considered as related to the notation

$$X \text{ is } U(0, 1)$$

of Chapter 20 (Example 20.5) and relations 19.1b.

*When we say "random number" without further description, we mean "*uniformly distributed* random number."

We expect that the mean and variance of $X = \text{RAND}(A, B)$ are given by

$$\bar{X} = \frac{A + B}{2} \quad (19.1c)$$

$$\sigma^2(X) = \frac{(B - A)^2}{12} \quad (19.1d)$$

A table of random numbers is contained in Appendix IV.

The sum of two random numbers, X, Y is *not* a random number. Rather, it is distributed as indicated in Figure 20.4 and relation (20.38).

Sums of Random Numbers

Let us consider generally

$$Z_n = X_1 + X_2 + \cdots + X_n \quad (21.6)$$

where $X_i = \text{RAND}(0, 1)$ for each i.

We note that

$$\bar{Z}_n = \sum_{i=1}^{n} \bar{X}_i = \frac{n}{2} \quad (21.7)$$

by (20.7) and

$$\text{Var}(Z_n) = \sum_{i=1}^{n} \text{Var}(X_i)$$

$$= \frac{n}{12} = \sigma_{Z_n}^{\ 2} \quad (21.8)$$

by (20.17) and (19.1d) since the X_i are independent.

We plot the frequency histogram for a sample of 2500 values of Z_n defined by (21.6) in Figures 21.6 to 21.19 for n = 1–10, 15, 20, 25, and 30.

Superimposed on the histograms are gaussian curves of the form

$$f(Z) = K \exp\left[\frac{-(Z - \bar{Z})^2}{2\,\text{Var}(Z)}\right]$$

where $\bar{Z}$ and Var(Z) are given by relations (21.7) and (21.8). We shall return to the matter of the gaussian curves in Chapter 25 on the normal distribution, but we remark even now that the curves represent an increasingly good fit to the histograms as n gets larger.

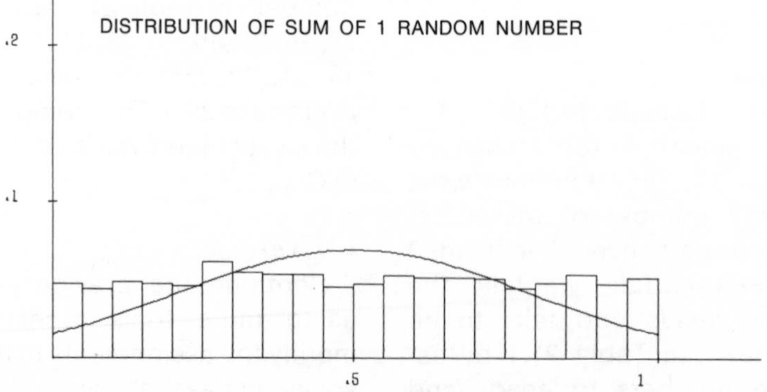

Figure 21.6

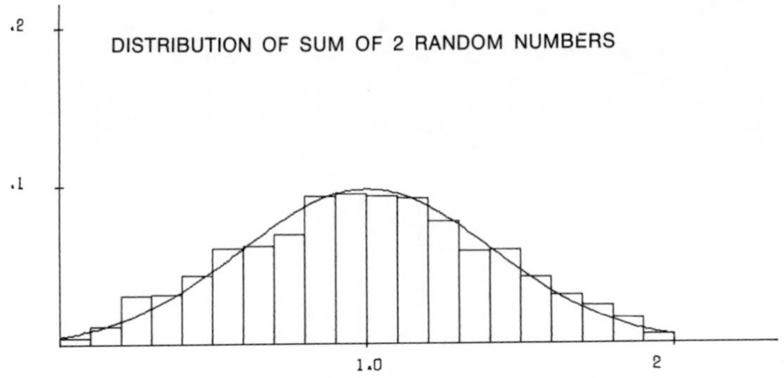

Figure 21.7

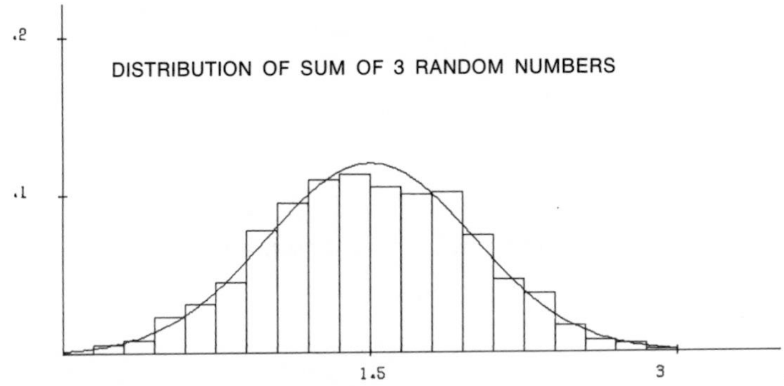

Figure 21.8

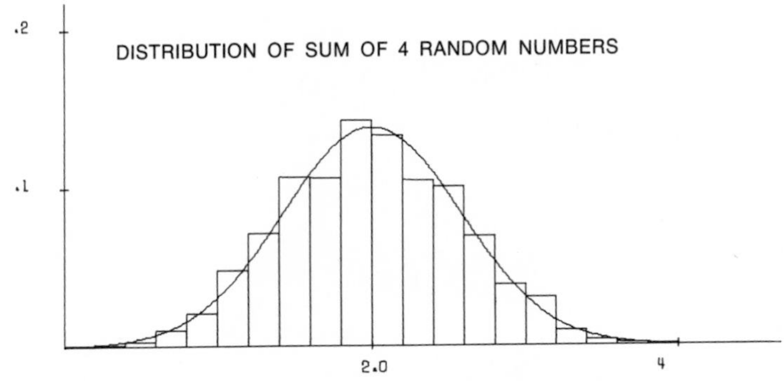

Figure 21.9

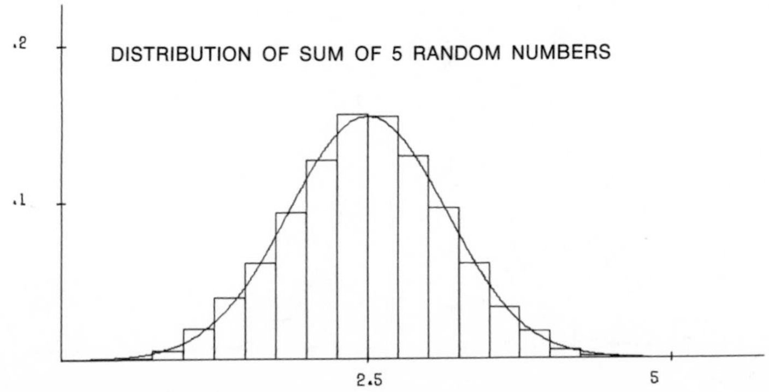

Figure 21.10

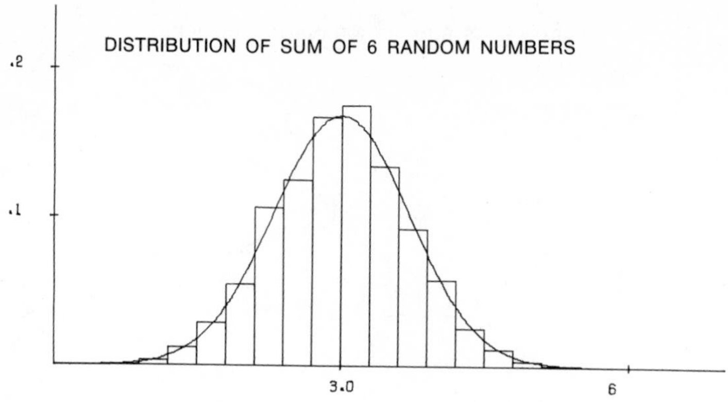

Figure 21.11

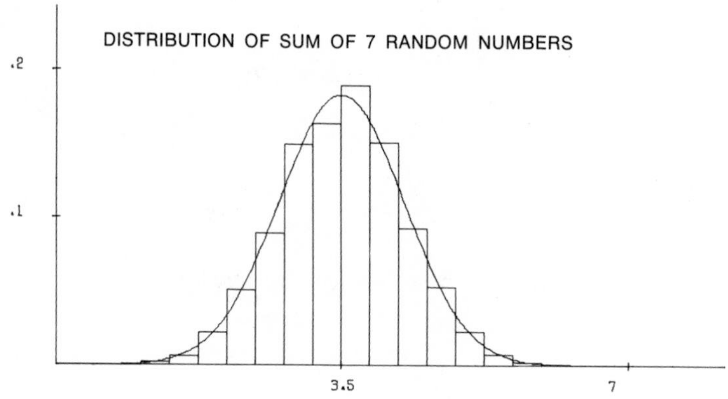

Figure 21.12

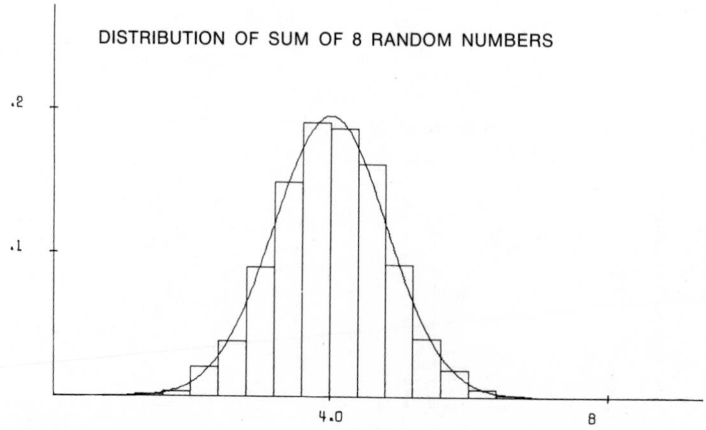

Figure 21.13

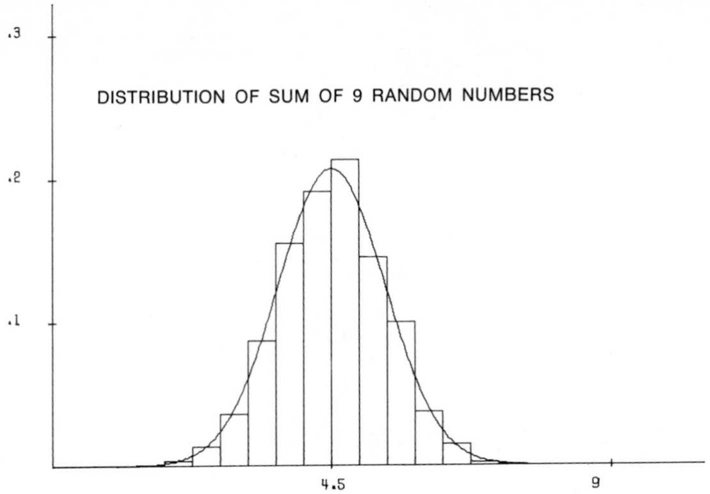

Figure 21.14

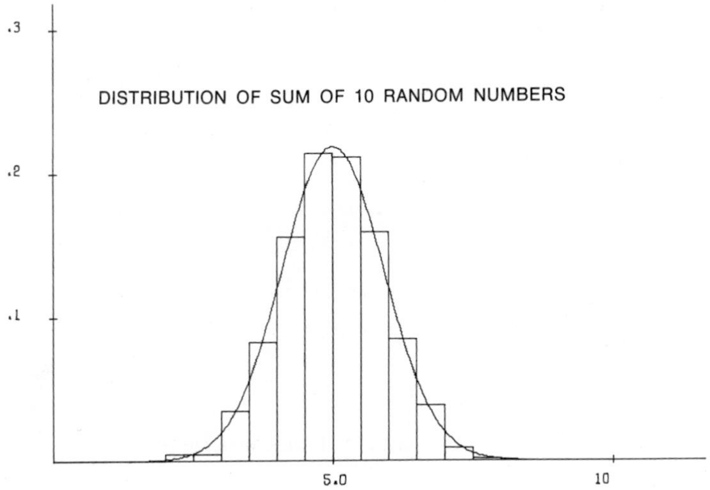

Figure 21.15

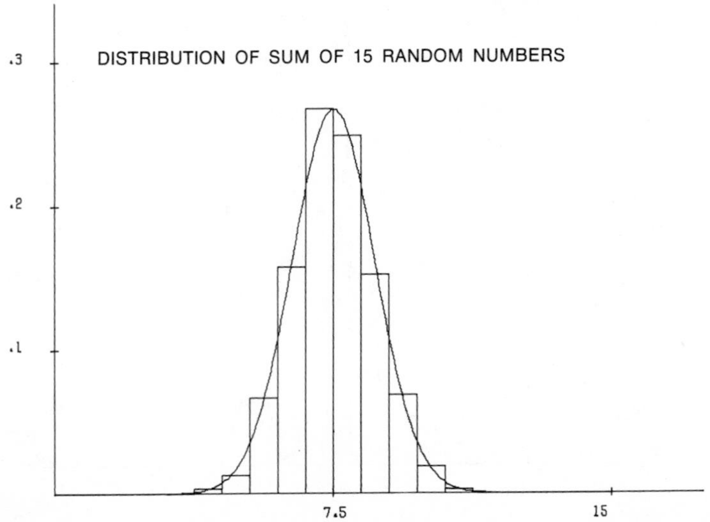

Figure 21.16

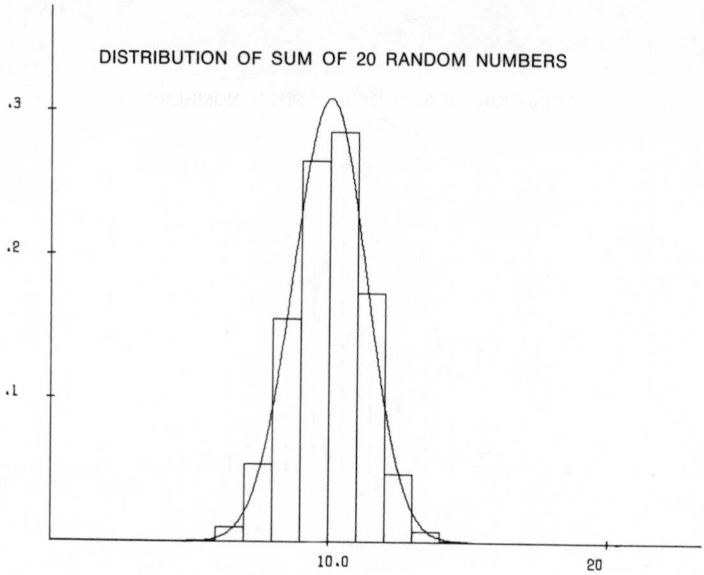

Figure 21.17

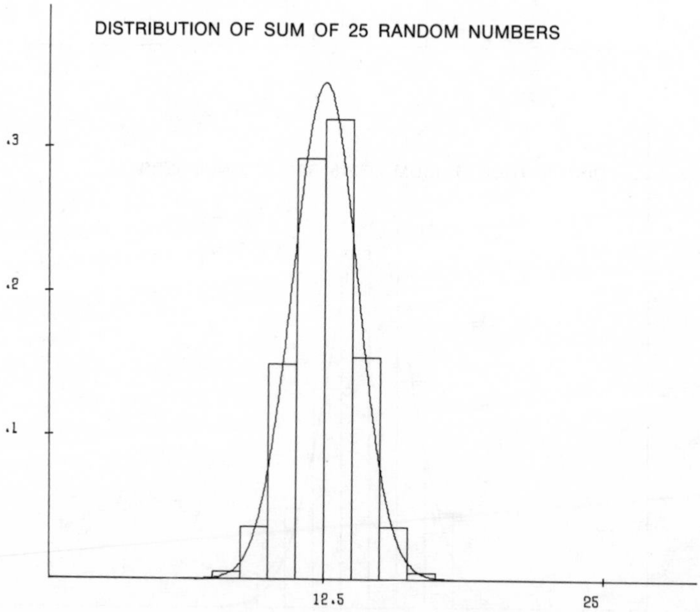

Figure 21.18

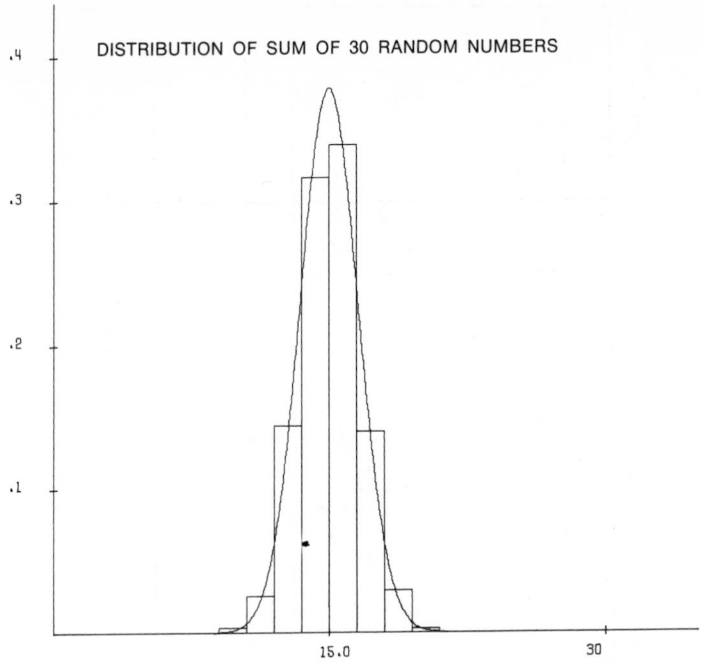

Figure 21.19

Figures 21.6 to 21.19 contain histograms of observed distributions of sums of random numbers. Superimposed are gaussian curves with mean and variance given by relations 21.7 and 21.8, respectively.

If we considered the X_i to be truly $U(0, 1)$ instead of RAND(0, 1) it can be shown* that the p.d.f. of Z_n is rigorously

$$f_{Z_n}(Z) = \left\{\binom{n}{0}[c(Z)]^{n-1} - \binom{n}{1}[c(Z-1)]^{n-1} + \cdots + (-1)^n\binom{n}{n}[c(Z-n)]^{n-1}\right\} \Big/ [(n-1)!]$$

where

$$c(x) = \begin{cases} 0 & \text{if} \quad x \leq 0 \\ x & \text{if} \quad x \geq 0 \end{cases}$$

*See Dwass, *Probability Theory and Applications*, W. A. Benjamin, 1970, p. 279.

Random Numbers Drawn from an Arbitrary Distribution

Once we have a source of random numbers drawn from the uniform distribution, it is straightforward to generate random numbers drawn from any arbitrary distribution.† This is the basis of any simulation or "Monte Carlo" experiment.

DISCRETE CASE

Consider a random variable Z with a frequency function $p(Z_i)$ in the discrete case such as shown in Figure 21.20*a*. We wish to generate random numbers from the distribution of Z. We start with x = RAND(0, 1) and choose x_i.

Consider the cumulative distribution function of Z as shown in Figure 21.20*b*. Locate x_i on the vertical axis and project a horizontal line until it intersects the cumulative distribution function. We assert that **the value of $z = Z_i$ corresponding to this intersection is a random number drawn from the frequency distribution $p(Z_i)$.**

To verify that the numbers Z_i generated in this way are drawn from the distribu-

†Note. We are now considering random numbers that are *not* uniformly distributed.

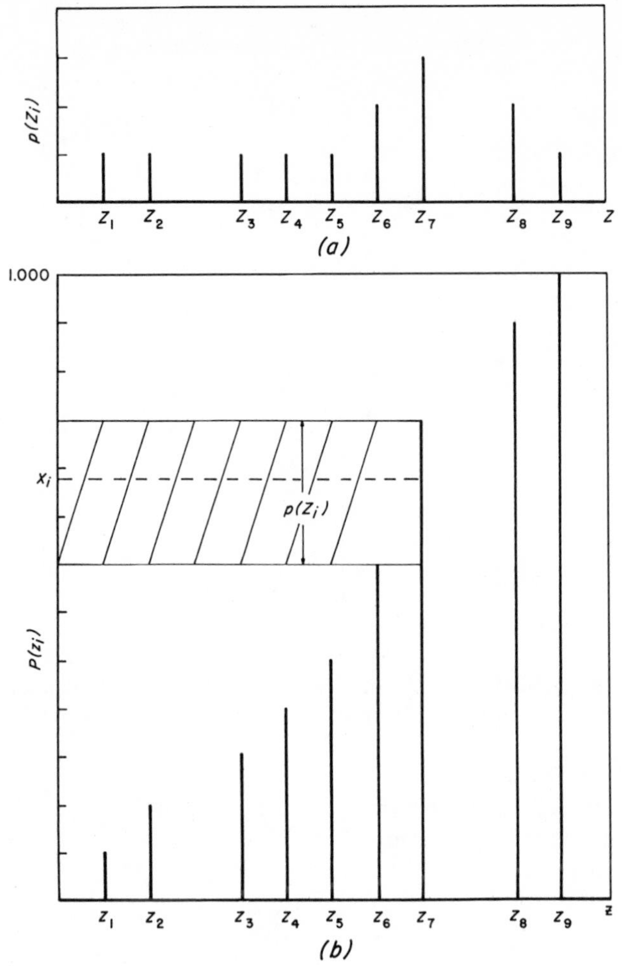

Figure 21.20
(*a*) Frequency function for arbitrary discrete distribution of *Z*. (*b*) Cumulative distribution function corresponding to (*a*). It illustrates choice of a random number drawn from the arbitrary distribution of *Z*.

tion $p(Z_i)$, let us consider the probability that a value Z_i will be chosen. This will occur when x_i is such that

$$P(z_{i-1}) < x_i \leq P(z_i)$$

But

$$P(z_i) - P(z_{i-1}) = p(Z_i)$$

so the length of the interval containing values of x_i that yield Z_i is $p(Z_i)$. The probability that x_i is in any interval when x is $U(0, 1)$ is proportional to the length of the interval, $p(Z_i)$. The probability of selecting Z_i is thus $p(Z_i)$, and each Z_i selected is independent of the others.

CONTINUOUS CASE

Consider a random variable Z with p.d.f. $f(Z)$ in the continuous case as shown in Figure 21.21*a*. We wish to generate random numbers from the distribution $f(Z)$. Again we start with x_i where $x =$ RAND(0, 1).

We consider the cumulative distribution function $F(z)$ as shown in Figure 21.21*b* and locate x_i on the vertical scale. As before, we assert that Z_i, obtained graphically by drawing a horizontal line from x_i until it intersects $F(z)$ and determining the value $z = Z_i$ that corresponds to the

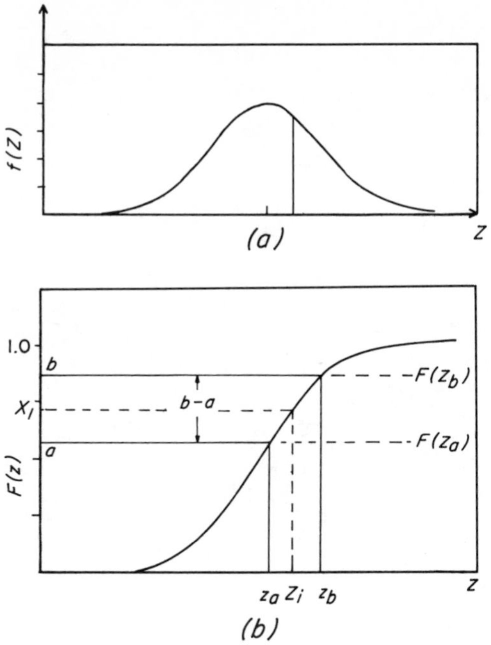

Figure 21.21
(a) Probability density function for arbitrary continuous distribution in Z. (b) Cumulative distribution function corresponding to the p.d.f. of (a). Illustrates choice of a random number drawn from the arbitrary (continuous) distribution of Z.

intersection, is distributed in the desired way. Analytically we are solving

$$x_i = F(x) \qquad (21.9)$$

for $z = Z_i$, where $x_i = \text{RAND}(0, 1)$.

Consider that x_i is in an interval (a, b). The probability for this is just the length of the interval since x is drawn from the uniform distribution:

$$\Pr(a < x_i < b) = b - a$$

Therefore, Z_i is in the interval (z_a, z_b) with probability $b - a = F(z_b) - F(z_a)$.

Therefore, $\Pr(z_a < Z_i < z_b) = F(z_b) - F(z_a)$ and the numbers Z_i are drawn from the distribution with p.d.f. $f(Z)$.

► **Example 21.7**
Random Normal Deviates and Gaussian-distributed Random Numbers

For example, the standardized normal distribution is defined by

$$f(t) = \frac{1}{\sqrt{2\pi}} e^{-t^2/2} \qquad \text{(p.d.f.)} \quad (21.10)$$

and

$$F(Z) = \frac{1}{\sqrt{2\pi}} \int_{-\infty}^{Z} e^{-t^2/2}\, dt \qquad \text{(c.d.f.)} \quad (21.10a)$$

If random numbers x are generated and the equation

$$x_i = F(z) \qquad (21.10b)$$

solved for $z = Z_i$, the Z_i will be random numbers drawn from the standard normal distribution. We shall have occasion to use these numbers called *"random normal deviates"*:

$$Z = \text{GAUSS}(0, 1). \qquad (21.11)$$

A table of these is contained in Appendix IV. ◄

Anticipating Chapter 25, we can generalize and generate numbers called *"gaussian-distributed random numbers"*

$$Z' = \text{GAUSS}(\text{mean}, \sigma^2) \qquad (21.12)$$

where the notation indicates the mean and variance of the distribution from which they are drawn:

$$Z' = (Z\sigma + \text{mean}) \qquad (21.13)$$

Correlations

Consider the quantities X, Y where X, Y are each independently RAND(0, 1). Plotted as a scatter diagram in Figure 21.22 are 1000 points (X, Y).

We may compute the means of X, Y and find (for these 1000 pairs of random numbers):

$$\bar{X} = 0.508$$

$$\bar{Y} = 0.505$$

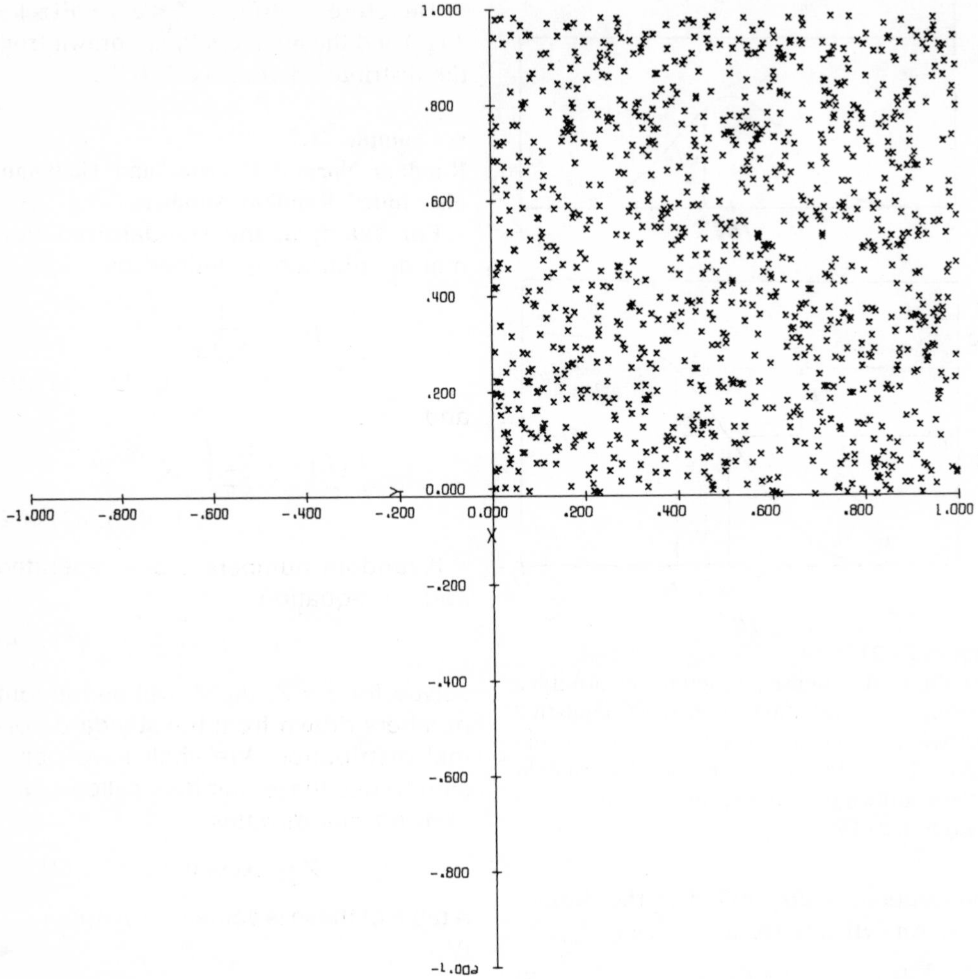

Figure 21.22
Scatter plot of points (X, Y) where X and Y are each uniformly distributed random variables on the interval $(0, 1)$. $\bar{X} = 0.50845$; $\bar{Y} = 0.50513$; $\rho(x, y) = 0.02507$.

which is consistent with the expected values of 0.500 [for X, Y both $U(0, 1)$].

We may check the correlation coefficient (20.13) with

$$\frac{\sum (X - \bar{X})(Y - \bar{Y})}{\left[\sum (X - \bar{X})^2\right]^{1/2} \left[\sum (Y - \bar{Y})^2\right]^{1/2}}$$

and get

$$\rho(X, Y) = +0.025$$

which is consistent with the value of 0 expected since X, Y are independent. We also note that the scatter plot in Figure 21.22 reveals no obvious structure (correlation) over the allowed region.

We may generate random quantities that are *not* independent of each other.

► Example 21.8
Sum and difference of uniformly-distributed random numbers

Consider the quantities

$$A = X + Y$$

and $$\tag{21.14}$$

$$B = X - Y$$

which are clearly not independent since they are both derived from X and Y.

For the case

$$X = \text{RAND}(0, 1) \qquad Y = \text{RAND}(0, 1)$$

see the scatter diagram of 1000 points (A, B) plotted in Figure 21.23. The projections on the A and B axes are histogrammed.

We compute

$$\bar{A} = 1.014$$
$$\bar{B} = 0.003$$

and

$$\rho(A, B) = -0.024$$

using (20.13). The expected values are $\langle A \rangle = 1.000$ and $\langle B \rangle = 0.000$.

For $\rho(A, B)$, we use relations (20.17d), (20.17c), and (20.13) to get the expected value

$$\langle \rho(A, B) \rangle = \frac{\text{Covar}(A, B)}{\sqrt{\text{Var}(A)\,\text{Var}(B)}} = \frac{\sigma_x^2 - \sigma_y^2}{\sqrt{\sigma_x^2 + \sigma_y^2}\sqrt{\sigma_x^2 + \sigma_y^2}} = \frac{(\sigma_x + \sigma_y)(\sigma_x - \sigma_y)}{(\sigma_x^2 + \sigma_y^2)} \tag{21.15}$$

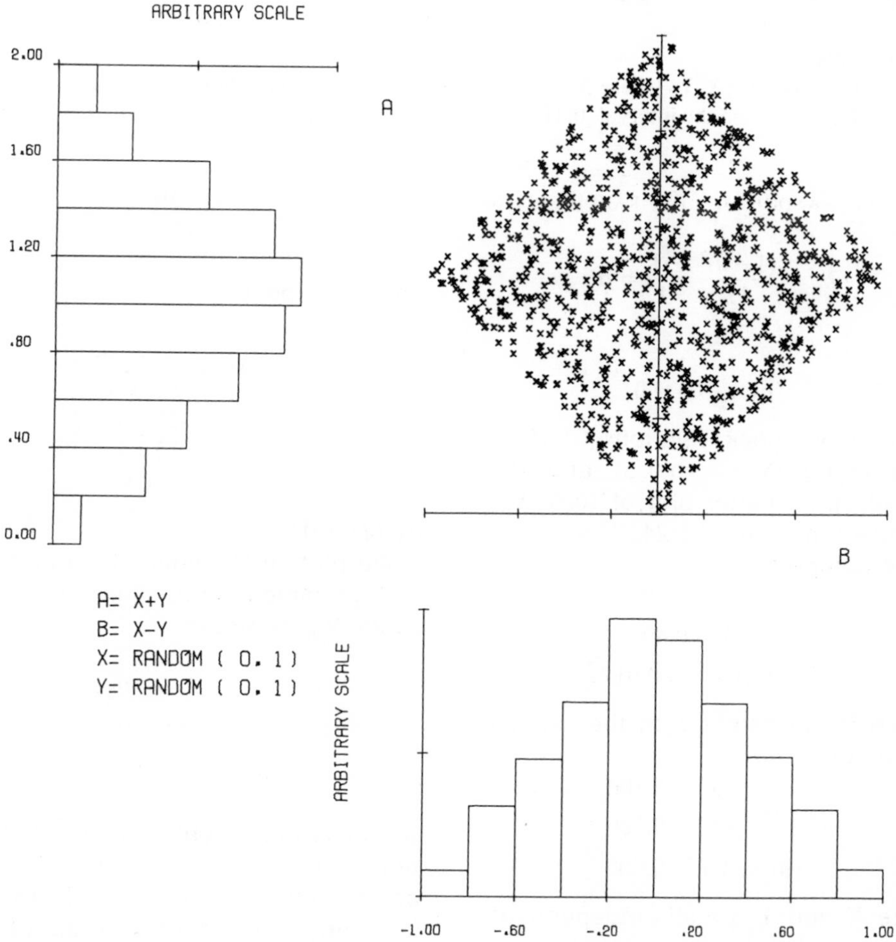

Figure 21.23
Scatter plot of points (A, B) where $A = X + Y$, $B = X - Y$ and X and Y are uniformly distributed random numbers. The histograms show the projections on the A and B axes. A and B appear to be uncorrelated. $\bar{A} = 1.01358$; $\bar{B} = 0.00332$; $\rho(A, B) = -0.0238$.

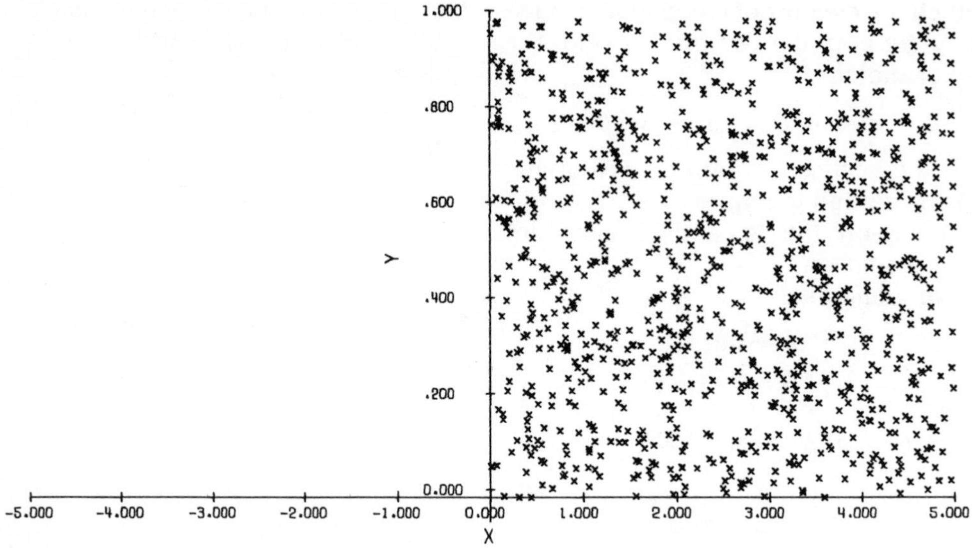

Figure 21.24
Scatter plot of points (X, Y) where X, Y are each uniformly distributed random numbers, X on the interval (0,5) and Y on the interval (0,1). $\bar{X} = 2.4923$; $Y = 0.49174$; $\rho(\bar{X}, Y) = -0.01673$.

For $X = \text{RAND}(0, 1)$, $Y = \text{RAND}(0, 1)$ we have

$$\sigma_x^2 = \frac{(1-0)^2}{12} = \frac{1}{12} = \sigma_y^2$$

and

$$\langle \rho(A, B) \rangle = 0$$

We may check relation (21.15) by considering $X = \text{RAND}(0, 5)$ and $Y = \text{RAND}(0, 1)$. A scatter plot of 1000 (X, Y) is shown in Figure 21.24.

We compute

$$\bar{X} = 2.492$$
$$\bar{Y} = 0.492$$
$$\rho(X, Y) = -0.017$$

which is consistent with the expected values of

$$\langle x \rangle = 2.500$$
$$\langle y \rangle = 0.500$$
$$\langle \rho(X, Y) \rangle = 0.000$$

since X and Y are still independent.

The situation is different for the quantities A and B since now

$$\sigma_x^2 = \frac{(5-0)^2}{12} = \frac{25}{12};$$

$$\sigma_y^2 = \frac{(1-0)^2}{12} = \frac{1}{12}$$

and we expect

$$\langle \rho(A, B) \rangle = \frac{\sigma_x^2 - \sigma_y^2}{\sigma_x^2 + \sigma_y^2} = \frac{25-1}{25+1} = \frac{24}{26} = +0.923$$

by (21.15).

We plot 1000 points (A, B) and their histogrammed projections in Figure 21.25. We compute

$$\bar{A} = 2.984$$
$$\bar{B} = 2.000$$
$$\rho(A, B) = 0.925$$

which is consistent with our expectations. Figure 21.25 shows an obvious correlation between A and B. The plot is clearly not symmetrical about axes parallel to the A and B axes. ◄

We may redo this example using gaussian-distributed random numbers.

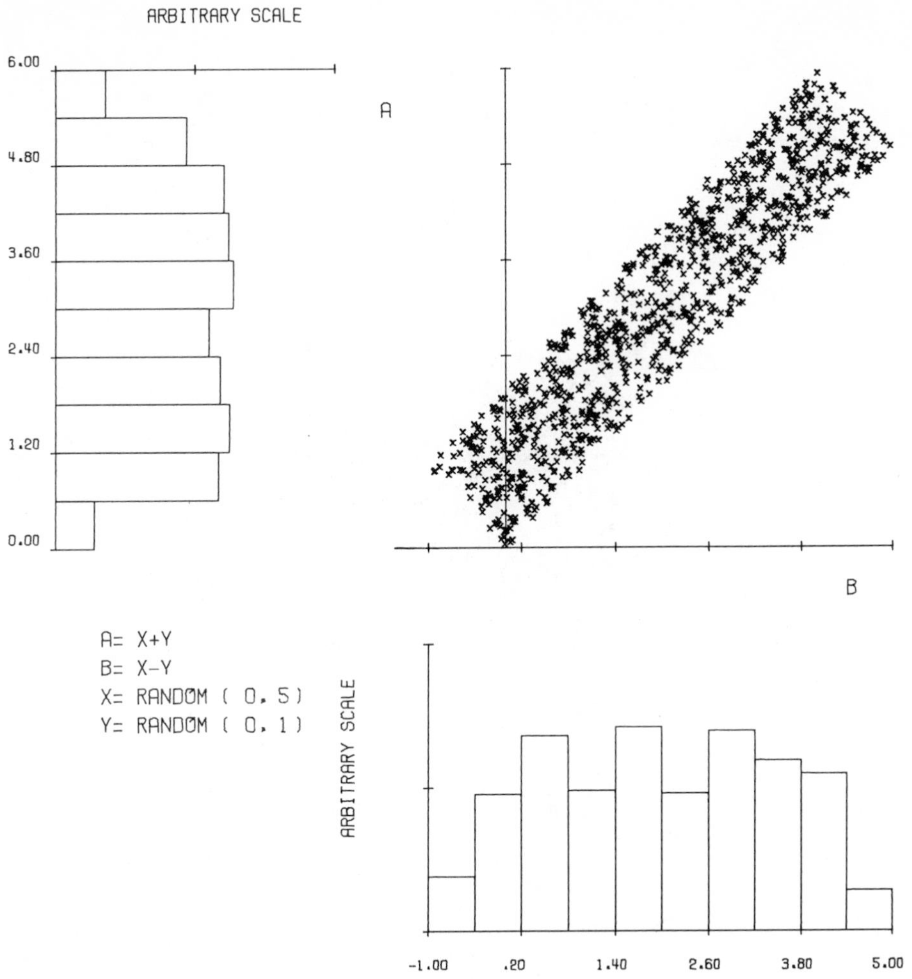

Figure 21.25
Scatter plot of points (A, B) where $A = X + Y$ and $B = X - Y$ with X uniformly distributed on the interval (0,5) and Y uniformly distributed on the interval (0,1). The histograms show the projections on the A, B axes. A and B appear to be positively correlated. $\bar{A} = 2.98404$; $\bar{B} = 2.00056$; $\rho(A, B) = 0.92465$.

► **Example 21.8**
Sum and difference of gaussian-distributed random numbers

Consider the quantities $X =$ GAUSS(0, 1) and $Y =$ GAUSS(0, 1). We plot 1000 points (X, Y) as a scatter diagram in Figure 21.26. We compute the means and the correlation coefficient to be:

$$\bar{X} = -0.036$$
$$\bar{Y} = -0.032$$

and

$$\rho(X, Y) = +0.028$$

These results are consistent with the values of 0 expected (in the latter case because the quantities are independent). We note that the scatter plot is symmetrical indicating no obvious correlation between X and Y.

We again define the sum and difference quantities:

$$A = X + Y$$

and

$$B = X - Y.$$

We plot 1000 points (A, B) as a scatter

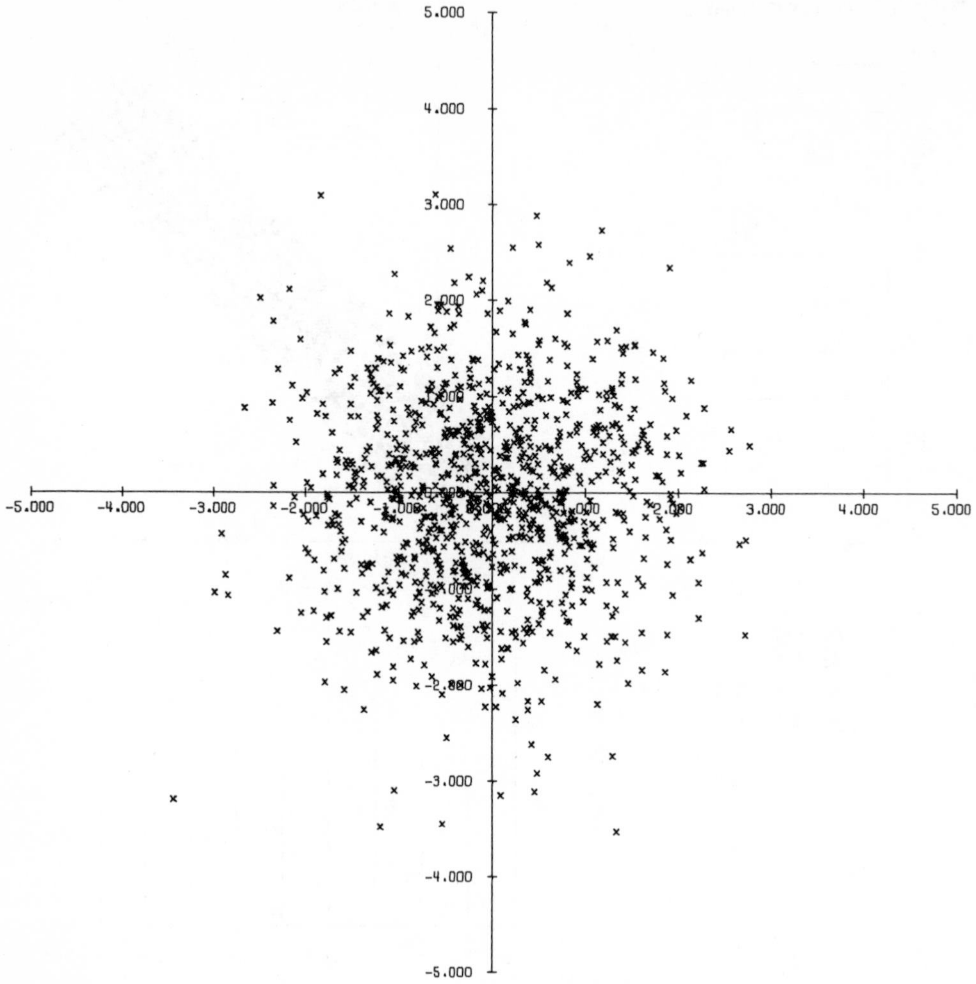

Figure 21.26
Scatter plot of points (X, Y) where X, Y are random numbers drawn from the gaussian distribution: $X = \text{GAUSS}(0,1)$, $Y = \text{GAUSS}(0,1)$. $\bar{X} = 0.03550$; $\bar{Y} = 0.03183$; $\rho(X, Y) = 0.02781$.

diagram in Figure 21.27 with the projections on the A and B axes histogrammed. We compute the quantities

$$\bar{A} = -0.068$$
$$\bar{B} = -0.004$$

and

$$\rho(A, B) = +0.001$$

These are consistent with 0 which is expected since $\text{Var}(X) = \text{Var}(Y)$ in relation (21.15). Once more we note that the scatter plot is symmetrical indicating no correlation between A and B. We stress again that A and B are not independent, merely uncorrelated.

We may investigate relation (21.15) again by choosing the X, Y such that the correlation coefficient does not vanish. Choose $X = \text{GAUSS}(0, 1)$ and $Y = \text{GAUSS}(0, 4)$. This choice insures that the variances of X and Y are not equal and that the correlation coefficient of (A, B) need not vanish by (21.15).

We plot 1000 points (X, Y) in Figure 21.28 and note that the plot appears to be symmetrical about both the X and Y axes. We compute the results

$$\bar{X} = +0.004$$
$$\bar{Y} = -0.026$$

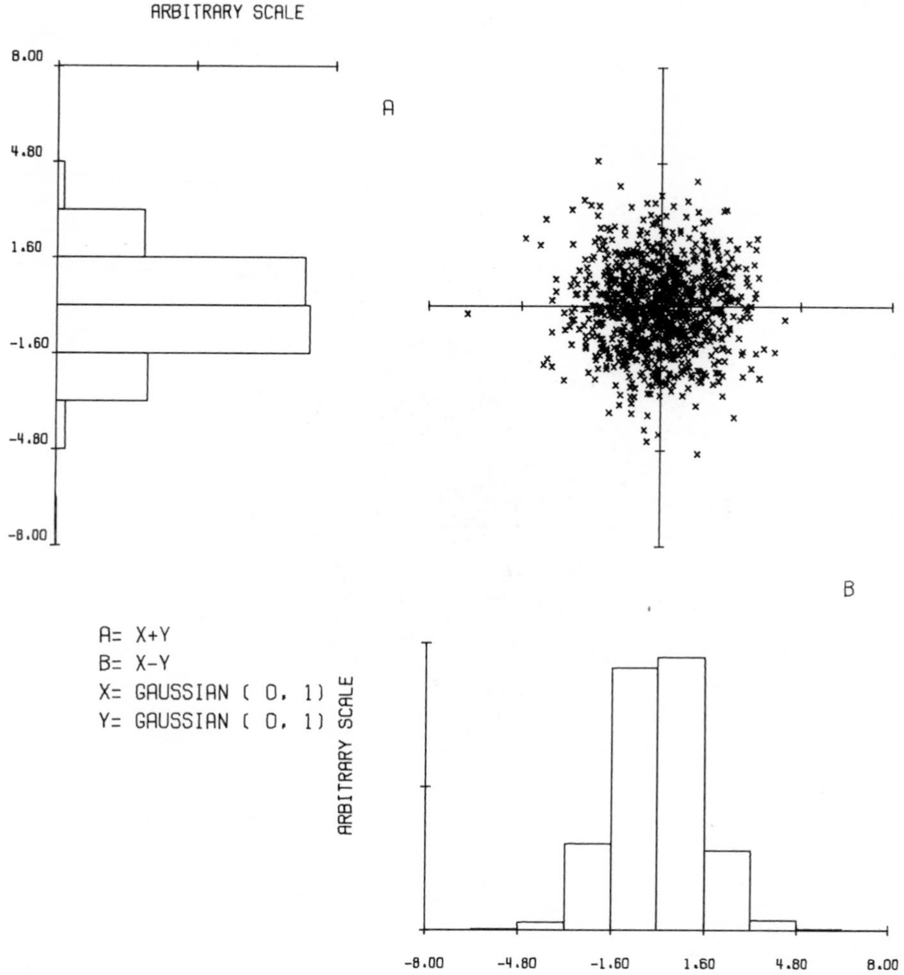

Figure 21.27
Scatter plot of points (A,B) where $A = X+Y$ and $B = X-Y$ with $X = \text{GAUSS}(0,1)$ and $Y = \text{GAUSS}(0,1)$. The histograms show the projections on the A, B axes. A and B appear to be uncorrelated. $\bar{A} = -0.06732$; $\bar{B} = -0.00367$; $\rho(A, B) = 0.00122$.

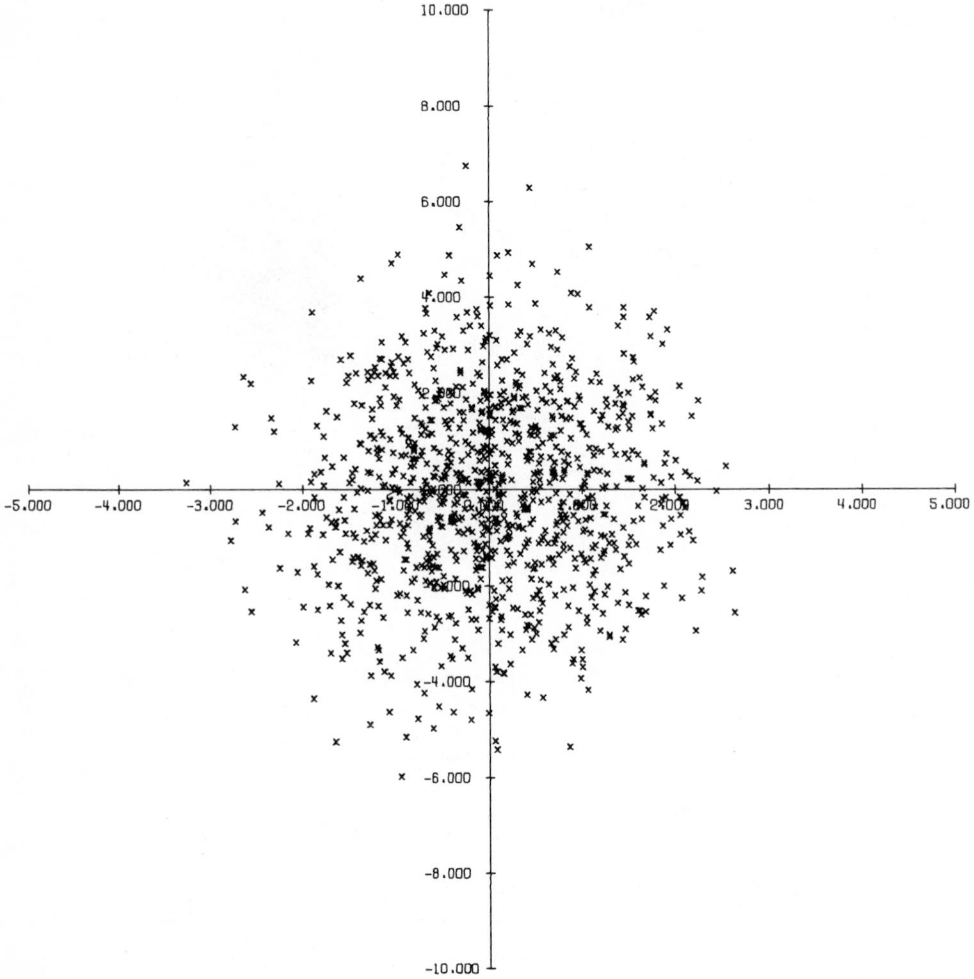

Figure 21.28
Scatter plot of points (X, Y) where $X = \text{GAUSS}(0, 1)$ and $Y = \text{GAUSS}(0, 4)$. $\bar{X} = 0.00413$; $\bar{Y} = 0.02566$; $\rho(X, Y) = 0.05164$.

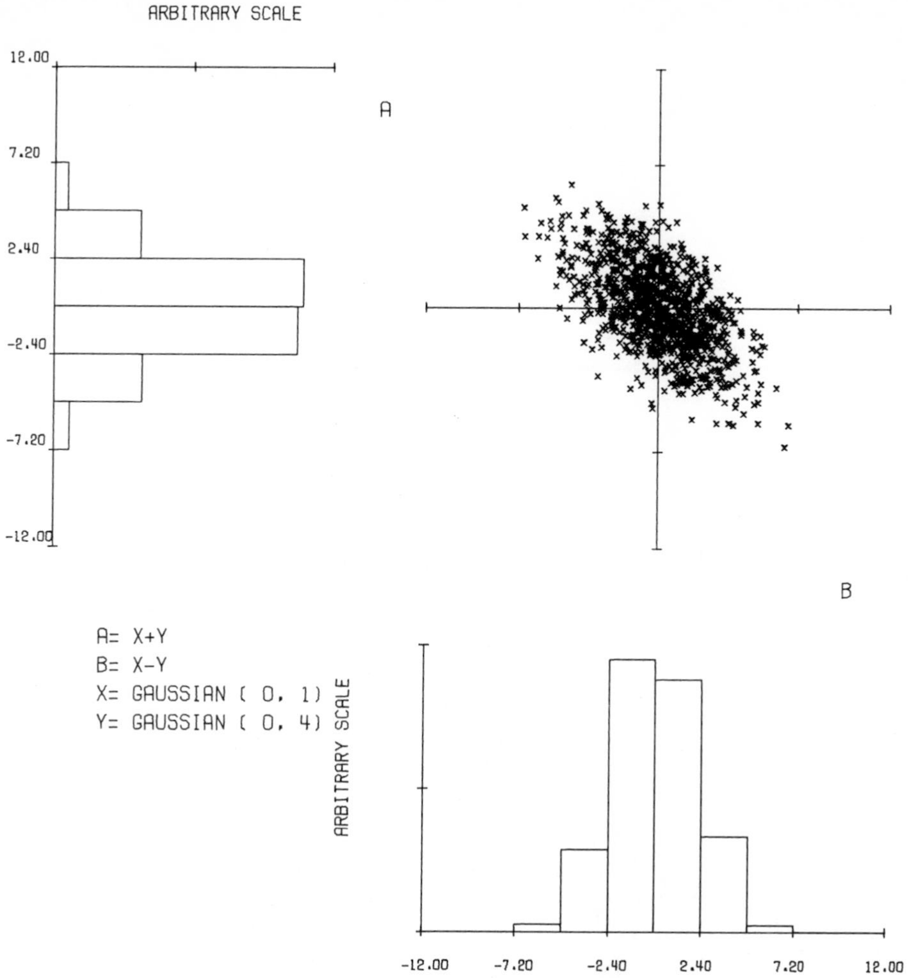

Figure 21.29
Scatter plot of points (A,B) where $A = X+Y$ and $B = X-Y$ with $X = \text{GAUSS}(0,1)$ and $Y = \text{GAUSS}(0,4)$. The histograms show the projections on the A,B axes. A and B appear to be negatively correlated. $\bar{A} = 0.02153$; $\bar{B} = 0.02979$; $\rho(A,B) = -0.58523$.

and

$$\rho(X, Y) = +0.052$$

all of which are consistent with zero.

The test of (21.15) involves the quantities $A = X+Y$ and $B = X-Y$. We plot 1000 points (A, B) as a scatter diagram in Figure 21.29 with the projections on the A, B axes histogrammed. The scatter plot is not symmetrical about the A and B axes, indicating that there is probably a correlation. We compute the results

$$\bar{A} = -0.022$$
$$\bar{B} = +0.030$$

and

$$\rho(A, B) = -0.585$$

By (21.15) we expect to get

$$\langle \rho(A, B) \rangle = \frac{\sigma_x^2 - \sigma_y^2}{\sigma_x^2 + \sigma_y^2} = \frac{1-4}{1+4} = \frac{-3}{5} = -0.600$$

◄

PART IV

Some Probability Distributions and Applications

In this part we shall consider some important probability distributions that arise in many situations of interest. We shall consider in some detail three discrete distributions (binomial, hypergeometric, and Poisson) and three continuous distributions (normal, chi-square, and Student's t). We shall also treat other useful distributions (binomial distribution of Poisson, Cauchy, negative binomial, multinomial, exponential, interval, Weibull, lognormal, F, bivariate normal, and multivariate normal) and special forms of distributions (folded and truncated). Each probability distribution represents a *model* for the situation of concern. The model is good or bad depending on the validity of the assumptions on which the model is predicated. In this chapter we shall endeavor to make specific what the various assumptions of each probability model are so that we may ascertain when the model is an exact description of the situation in the real world and when it is a good approximation. It is also of interest to determine what the properties of the various distributions are and how calculations may be performed with them.

22

THE BINOMIAL DISTRIBUTION

Let us consider a trial such that only two outcomes are possible, one that we shall call a success and the other that we shall call a failure. We suppose further that the probability for success in a given trial is p and the probability for failure is $q = 1 - p$.

Definitions

We consider a series of n mutually independent trials each made under these conditions and ask for the probability of exactly r successes and $n - r$ failures. Each of these independent two-possibility trials is called a *binomial* trial, a *dichotomous* trial, or a *Bernoulli* trial after Jacob Bernoulli who first published a study of the problem.*

Each trial is independent and so the probability of a *specific* sequence [e.g., starting off with r successes followed by $(n - r)$ failures] is $p^r q^{n-r}$. However, the order of the sequence is irrelevant. Any order will do, and each possible order has the same probability of occurring, $p^r q^{n-r}$. We must, therefore, multiply this probability by the number of ways that n trials can be divided into r successes and $(n - r)$ failures. By (18.4) this number is just $\binom{n}{r}$, and the overall probability required is

$$P_{n,p}(r) = \binom{n}{r} p^r q^{n-r} = \frac{n!}{r!(n-r)!} p^r q^{n-r} \qquad (22.1)$$

This is the probability that n Bernoulli trials will yield exactly r successes.

Relation (22.1) defines the *binomial* or *Bernoulli* distribution. We may plot $P_{n,p}(r)$ against r to see what this distribution looks like for a given n and p.

Since the probability depends on the parameters n,p this is sometimes written

$$\mathbf{b}(r; n, p) = \binom{n}{r} p^r q^{n-r} \qquad (22.1a)$$

We note that the binomial distribution is normalized as it stands. By the binomial

**Ars Conjectandi,* posthumously published in 1713.

theorem (18.5) we have

$$(p+q)^n = \sum_{r=0}^{n} \binom{n}{r} p^r q^{n-r}$$
$$= \sum_{r=0}^{n} P_{n,p}(r)$$
$$= [p+(1-p)]^n = 1^n = 1 \qquad (22.2)$$

Consider the probability of getting r heads in n tosses of a fair coin. Here $p = q = 1/2$ and

$$P_{n,1/2}(r) = \binom{n}{r}\left(\frac{1}{2}\right)^r\left(\frac{1}{2}\right)^{n-r}$$
$$= \binom{n}{r}\Big/2^n.$$

which we note was obtained earlier as a special case (Example 18.4).

We plot the binomial distribution $P_{n,1/2}(r)$ versus r in Figure 22.1 for $n = 2, 6$ and 10. Recall that Table 18.1 lists the probabilities for $n = 6$. The probabilities for $n = 2$, and $n = 10$ are left as an exercise.

Although Figure 22.1 *looks* like a histogram, it is *not*. This is a ***discrete distribution*** such that r occurs only at the *integer* values shown.

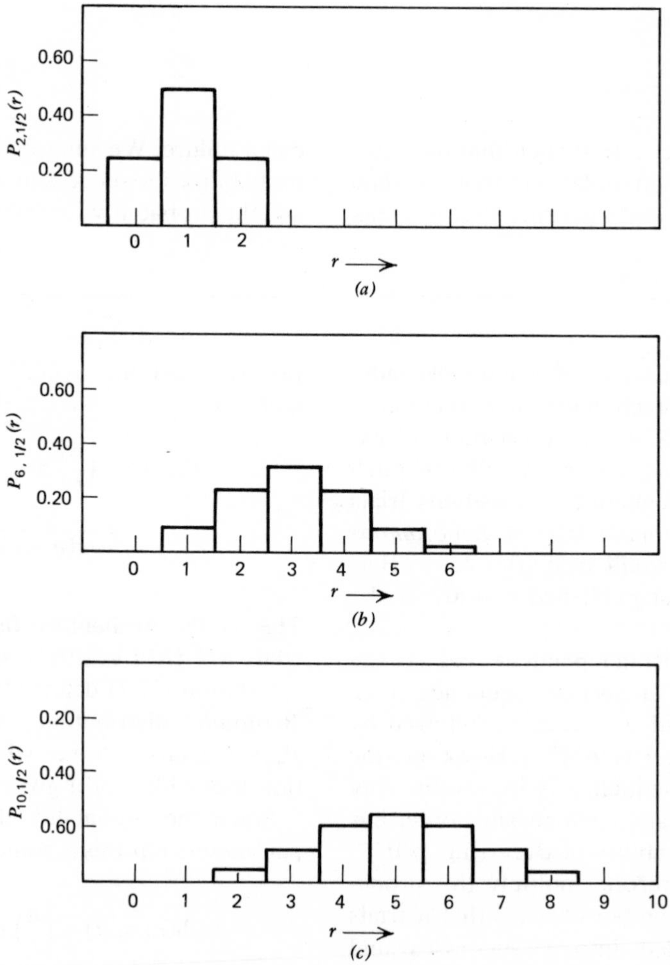

Figure 22.1
Discrete frequency distribution for binomial distribution with $p = q = 1/2$. (*a*) $n = 2$. (*b*) $n = 6$. (*c*) $n = 10$.

► **Example 22.1**
Selection from Gadsby

Let us consider again the selection from *Gadsby* discussed in Chapters 17 and 18. Recall that this is a passage of 580 letters without the letter "e". We may consider this a binomial problem in which the possibilities are either "e" or "not *e*." Taking the estimate of the probability of occurrence of "e" to be *0.126*, we may calculate the probability that this selection is a random occurrence:

$$P_{580,0.126}(0) = \binom{580}{0}(0.126)^0(0.874)^{580}$$

or

$$\begin{aligned}\ln P_{580,0.126}(0) &= 580 \ln(0.874)\\ &= 580(-0.1346)\\ &= -78.1\end{aligned}$$

implying

$$P_{580,0.126}(0) \approx 12.1 \times 10^{-35}.$$

This agrees quite well with the estimate obtained in Chapter 18 where the probability was calculated to be 9.1×10^{-35}. ◄

Reproductive Property of the Binomial Distribution

If

$$\Pr(X = x) = \binom{n_x}{x} p^x q^{n_x - x}$$

and

$$\Pr(Y = y) = \binom{n_y}{y} p^y q^{n_y - y}$$

and X, Y are independent, then $Z = X + Y$ is binomially distributed with

$$\Pr(Z = z) = \binom{n}{z} p^z q^{n-z}$$

where

$$z = x + y$$

and

$$n = n_x + n_y$$

In general, if $x_1, x_2, \ldots, x_n$ are each mutually independent binomially distributed random variables with probability p, that is, $\mathbf{b}(x_i; n_i, p) = \Pr\{X_i = x_i\}$ for each i then

$$X = \sum_i X_i$$

is binomially distributed with probability p and

$$\Pr(X = x) = \mathbf{b}(x; n, p)$$

where $n = \sum n_i$.

Probability of a Range of Values

It is frequently of interest to calculate the probability of r occurring in a range of values, $[a, b]$, that is,

$$\Pr[a \le r \le b] = \sum_{r=a}^{b} P_{n,p}(r)$$

► **Example 22.2**
Probability of between 15 and 21 heads in 36 tosses

Consider the probability of getting a number of heads between *15* and *21* in *36* tosses of a fair coin.

$$\begin{aligned}\Pr[15 \le r \le 21] &= \sum_{r=15}^{r=21} \binom{36}{r}\left(\frac{1}{2}\right)^r \left(\frac{1}{2}\right)^{36-r}\\ &= \left(\frac{1}{2}\right)^{36} (36!) \sum_{r=15}^{21} \frac{1}{r!(36-r)!}\\ &= \left(\frac{1}{2}\right)^{36} (36!) \left\{2\left[\frac{1}{15!21!} + \frac{1}{16!20!} + \frac{1}{17!19!}\right] + \frac{1}{18!18!}\right\}\\ &= \left(\frac{1}{2}\right)^{36} \frac{(36!)}{(15!)(18!)}\\ &\quad \times \left\{2\left[\frac{1}{(21)(20)(19)} + \frac{1}{(16)(20)(19)} + \frac{1}{(17)(16)(19)}\right] + \frac{1}{(18)(17)(16)}\right\}\end{aligned}$$

$$= \left(\frac{1}{2}\right)^{36} \frac{(36!)}{(15!18!)}$$
$$\times \{2[0.0001253 + 0.0001645 + 0.0001935] + 0.0002042\}$$
$$= \left(\frac{1}{2}\right)^{36} \frac{(36!)}{(15!(18!)} \{0.001171\}$$
$$= 646.577(0.001171) \text{ using logarithms and Table A.III.1}$$
$$= 0.7571$$ ◀

It is possible to use tables of the cumulative binomial distribution function

$$\mathbf{B}(r; n, p) = \Pr(R \le r) = \sum_{R \le r} \mathbf{b}(R; n, p)$$

or tables of partial sums

$$1 - \mathbf{B}(r; n, p) = \sum_{S=r}^{n} P_{n,p}(S)$$
$$= \sum_{S=r}^{n} \binom{n}{S} p^S q^{n-S}$$

The U.S. Government Printing Office has a publication (National Bureau of Standards, Applied Mathematics Series; 6) that tabulates such "partial sums" of the binomial distribution for $n = 2$ to 49 and $p = 0.01$ to 0.50. From these tables

$$\sum_{S=r}^{n} \binom{n}{S} p^S q^{n-S} = 0.8785075$$

for

$$n = 36,\ p = 0.5,\ r = 15$$

and

$$= 0.1214925$$

for

$$n = 36,\ p = 0{,}5,\ r = 22$$

$$\Pr\{15 \le r \le 21\} = \sum_{S=15}^{36} \binom{36}{S} \left(\frac{1}{2}\right)^S \left(\frac{1}{2}\right)^{36-S} - \sum_{S=22}^{36} \binom{36}{S} \left(\frac{1}{2}\right)^S \left(\frac{1}{2}\right)^{36-S}$$
$$= 0.7570$$

which checks Example 22.2.

Symmetry and Asymmetry

We note that the distributions in Figure 22.1 are symmetric about the midpoint, which shifts to the right as n increases and is itself equal to $n/2$. The areas of all the distributions are, of course, the same because of the normalization (22.2). As n gets larger, the range of horizontal width at the base of the distribution gets larger (and, in fact, is $n + 1$) while the distribution gets lower as it gets wider.

We remark that the symmetry of the distribution is a consequence of $p = q = 1/2$. Here is an example in which the probabilities p and q are not the same. Let us consider that a five-card poker hand consists of four aces and a deuce. A trial consists of shuffling the five cards and drawing one, considering it a success if the deuce is drawn and a failure if an ace is drawn. The drawn card is replaced before the next draw so that each draw is independent of the others. In a sequence of n draws, we wish to find the probabilities of r deuces. Here $p = 1/5 = 0.20$. We plot the values of $P_{n,0.2}(r)$ versus r for $n = 2$, 6, and 10 in Figure 22.2 where the values are summarized in Table 22.1.

In Figure 22.2 the plots are not symmetric although the symmetry appears to be least for the lowest value of n and seems to improve as n gets larger. We note the same feature which appeared in Figure 22.1: namely, the distribution gets lower and wider as n increases.

Table ***22.1***

r	0	1	2	3	4	5	6	7	8	9	10
$P_{2,1/5}(r)$	0.64	0.32	0.04								
$P_{6,1/5}(r)$	0.262	0.393	0.246	0.082	0.015	0.0015	0.0001				
$P_{10,1/5}(r)$	0.1074	0.2684	0.3020	0.2015	0.0881	0.0264	0.0055	0.0008	0.0001	$<10^{-4}$	$<10^{-4}$

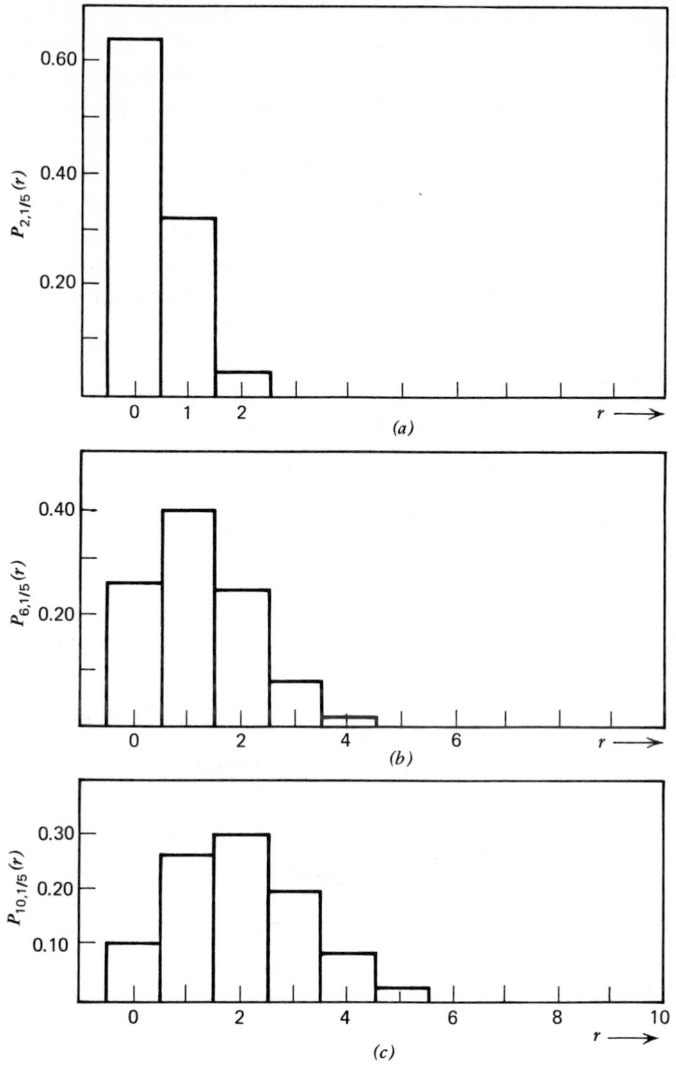

Figure 22.2
Plot of the discrete values of $P_{n,0.2}(r)$ versus r. (a) $n = 2$. (b) $n = 6$. (c) $n = 10$.

Expectation Value of the Binomial Distribution

We may ask what our *expectation* is for the number of successes in n Bernoulli trials with individual probability p. This is given by (17.6) where we have a discrete distribution function as in Chapter 9:

$$\langle r \rangle = \sum_{r=0}^{n} r P_{n,p}(r) \tag{22.3}$$

We have encountered a similar expression in (22.2) when we considered the normalization. Here we have an extra factor of r. We shall try to convert (22.3) into one of the forms of (22.2) in order to be able to use the normalization condition. We first note that the first term ($r = 0$) makes no contribution. We then proceed to separate out a factor of $r(n/r)p$:

$$\begin{aligned}\langle r \rangle &= \sum_{r=1}^{n} r P_{n,p}(r) = \sum_{r=1}^{n} r \frac{n!}{r!(n-r)!} p^r q^{n-r} \\ &= \sum_{r=1}^{n} r(n/r)p \frac{(n-1)!}{(r-1)!(n-r)!} p^{r-1} q^{n-r} \\ &= np \sum_{r=1}^{n} \frac{(n-1)!}{(r-1)!(n-r)!} p^{r-1} q^{n-r} \end{aligned} \tag{22.4}$$

Now introduce new variables m and s such that $m = n - 1$ and $s = r - 1$. The range of the running variable s in the sum is $s = 0$ (for $r = 1$) to $s = n - 1 = m$ (for $r = n$). Note that $m - s = (n - 1) - (r - 1) = n - r$.

We may now rewrite (22.4) in terms of the new variables:

$$\langle r\rangle = np \sum_{s=0}^{m} \frac{m!}{s!(m-s)!} p^s q^{m-s} \qquad (22.5)$$

But the sum in (21.5) is just $\sum_{s=0}^{m} P_{m,p}(s) = 1$ by (22.2).

Therefore, the expectation value for r is

$$\langle r\rangle = \sum_{r=0}^{n} rP_{n,p}(r) = np \qquad (22.6)$$

This result is intuitively obvious, since the value of r that we expect on the average is just the number of trials multiplied by the probability per trial.

Note that the distributions in Figure 22.1 have $\langle r\rangle$ as the mean, mode, and median because of the symmetry. The distributions of Figure 22.2, however, are *not* symmetric and the expectation value or mean of r is not even a possible value for $n = 6$ where $\langle r\rangle = 6 \cdot (1/5) = 1.2$. For both sets of distributions, (22.6) explains how the midpoint shifts to the right as n increases.

Variance of the Binomial Distribution

Having obtained the mean or expectation value of the binomial distribution, we must address ourselves to the problem of determining the *variance* of this distribution.

We may calculate the expectation value of r^2:

$$\langle r^2\rangle = \sum_{r=0}^{n} r^2 P_{n,p}(r) = \sum_{r=1}^{n} r^2 P_{n,p}(r)$$

since the $r = 0$ term makes no contribution as before. Again we try to use the same trick we used in obtaining (22.6): to cast the formula in terms of the normalization sum that appeared in (22.2).

$$\langle r^2\rangle = np \sum_{r=1}^{n} r \frac{(n-1)!}{(r-1)!(n-r)!} p^{r-1} q^{n-r} \qquad (22.7)$$

As before, we make the substitution $m = n - 1$ and $s = r - 1$:

$$\langle r^2\rangle = np \sum_{s=0}^{m} \frac{(s+1)m!}{s!(m-s)!} p^s q^{m-s}$$
$$= np \sum_{s=0}^{m} (s+1) P_{m,p}(s)$$

Therefore,

$$\langle r^2\rangle = np \left\{ \sum_{s=0}^{m} sP_{m,p}(s) + \sum_{s=0}^{m} P_{m,p}(s) \right\} \qquad (22.8)$$

The first term in the bracket is mp by (22.6) and the second term is unity by (22.2). Thus we have

$$\langle r^2\rangle = np\{mp + 1\} = np\{(n-1)p + 1\} \qquad (22.9)$$

However, the variance of r

$$\sigma^2(r) = \sum_{r=0}^{n} (r - \langle r\rangle)^2 P_{n,p}(r) = \langle r^2\rangle - \langle r\rangle^2$$

by (7.8) or (9.16). Thus

$$\sigma^2(r) = np(np - p + 1) - (np)^2 = np(1-p)$$

using (22.6) and (22.9).

The final result for the ***variance*** is

$$\sigma^2(r) = np(1-p) = npq \qquad (22.10)$$

The simple result (22.10) tells us how the width of the distribution changes with n. For $n = 10$ and $p = 1/2$ we get $\sigma = \sqrt{npq} = \sqrt{10 \cdot 1/2 \cdot 1/2} = \sqrt{2.5}$, which characterizes the width of the distribution shown in Figure 22.1*c*.

We may ask about other features of the binomial distribution.

The Expectation Value for the Number of Trials Required for a Specified Number of Successes

Suppose that it is desired to achieve exactly r successes and that the trials will proceed until this occurs. What is the expectation for n, the average number of trials necessary to produce r successes?

We make invoke the theorem of total probability to answer this question:

$$\langle n\rangle = \sum_{n=r}^{\infty} n \Pr(n) \qquad (22.11)$$

If r successes occur in r trials, then we get a contribution to this sum of r multiplied by the Prob (r successes in r trials) which is rp^r. Thus

$$\begin{aligned}\langle n\rangle &= rp^r + (r+1)p^r qr \\ &\quad + (r+2)p^r q^2 \frac{(r+1)r}{2} + \cdots \\ &= rp^r \sum_{j=0}^{\infty} \frac{(r+j)!}{r!j!} q^j\end{aligned}$$

which we note is

$$\langle n\rangle = rp^r(1-q)^{-(r+1)} = \frac{r}{p} \qquad (22.12)$$

This is intuitively consistent with (22.6) and checks Example 19.9 of Chapter 19.

The Mode of the Binomial Distribution

An important problem that we should consider is how to determine the value of r for which (22.1) is maximum; that is, we wish to maximize $P_{n,p}(r)$ with respect to r. Equivalently, we wish to find the *mode*, or most probable value of r, for the binomial distribution.

This is a discrete distribution. The mode, r_M, has the properties

$$P_{n,p}(r_M) \geq P_{n,p}(r_M - 1) \qquad (22.13a)$$

and

$$P_{n,p}(r_M) \geq P_{n,p}(r_M + 1) \qquad (22.13b)$$

We may use (22.13) to establish relations for the ratios

$$\frac{P_{n,p}(r_M)}{P_{n,p}(r_M - 1)} \geq 1 \qquad (22.14a)$$

and

$$\frac{P_{n,p}(r_M)}{P_{n,p}(r_M + 1)} \geq 1 \qquad (22.14b)$$

Equation (22.14a) may be rewritten after explicit substitution by (22.1)

$$\frac{p}{q}\frac{n - r_M + 1}{r_M} \geq 1$$

from which we obtain the inequality

$$pn - r_M p + p \geq r_M q = r_M - r_M p$$

or

$$np + p \geq r_M \qquad (22.15a)$$

The second ratio, (22.14b), may be rewritten

$$\frac{q}{p}\frac{r_M + 1}{(n - r_M)} \geq 1$$

from which we obtain the inequality

$$qr_M + q \geq np - r_M p$$

which yields $r_M(p+q) \geq np - q$

or

$$r_M \geq np - q \qquad (22.15b)$$

We may now write the total relation for the mode, r_M

$$np + p \geq r_M \geq np - q \qquad (22.16)$$

This analysis, by the way, shows that $P_{n,p}(r)$ increases as r increases so long as r does not attain the value $np - q$. As soon as r exceeds the limit $np - q$, the probability $P_{n,p}(r)$ starts to decrease as r increases until it reaches its least value, $P_{n,p}(n)$. This shows the general behavior of the binomial distribution and proves that a maximum exists and satisfies (22.16).

The interval $(np - q)$ to $(np + p)$ has unit length. Since $p \leq 1$, it is usually the case that the endpoints of this interval are not integers and that one and only one integer lies between them. The mode, r_M, is thus usually a unique point. In certain exceptional instances the limit $np - q$ is an integer, and in these cases (22.16) yields two values for r_M that differ by unity. These two values are the most probable with equal probabilities. For example, for $n = 15$, $p = 1/2$, (22.16) yields the two modal values $r_M = 7$ and $r_M = 8$ with equal and maximal probabilities of approximately 0.196. For each of the examples in Figures 22.1 and 22.2, the modes are uniquely determined by (22.16).

► **Example 22.3**
Pepys' problem

Samuel Pepys, famous for his *Diary*, carried on an extensive correspondence with the intellectual leaders of his day. F. N. David, in the book *Games, Gods and Gambling*, cites the following query sent by Pepys in London to Isaac Newton in Cambridge on November 22, 1693:

> "The Question:
> A has 6 dice in a box with which he is to fling a six;
> B has in another box 12 dice with which he is to fling 2 sixes;
> C has in another box 18 dice with which he is to fling 3 sixes.
> Q.—whether B and C have not as easy a chance as A at even luck?"

Assuming that what was meant was that A is to throw *exactly* 1 six, B *exactly* 2 sixes, and C *exactly* 3 sixes, the answer to Pepys' question is given by the binomial probabilities:

$$A:\ P_{6,1/6}(1) = \binom{6}{1}\left(\frac{1}{6}\right)^1\left(\frac{5}{6}\right)^5 = 6\cdot 1\left(\frac{1}{6}\right)(0.4019) = 0.4019$$

$$B:\ P_{12,1/6}(2) = \binom{12}{2}\left(\frac{1}{6}\right)^2\left(\frac{5}{6}\right)^{10} = \left(\frac{12\cdot 11}{1\cdot 2}\right)\left(\frac{1}{6}\right)^2(0.1615) = 0.2961$$

$$C = \ P_{18,1/6}(3) = \binom{18}{3}\left(\frac{1}{6}\right)^3\left(\frac{5}{6}\right)^{15} = \left(\frac{18\cdot 17\cdot 16}{1\cdot 2\cdot 3}\right)\times\left(\frac{1}{6}\right)^3(0.064905) = 0.2452$$

According to this interpretation, A has the easiest time, B the next, and C the hardest time.

Newton pointed out the ambiguity between *exactly one* and *at least one*, and Pepys verified that he meant the latter interpretation in a letter written on December 9, 1693 in which he restated the question.

> "I should avoid some of the ambiguities that commonly hang about our discoursings of it, by changing the characters of the dice from numbers to letters and supposing them instead of 1, 2, 3, . . . etc to be branded with the 6 initial letters of the alphabet, A, B, C, D, E, F. And the case should then be this:
>
> "Peter, a criminal convict doomed to die, Paul his friend prevails for his having the benefit of one throw only for his life, upon dice so prepared; with a choice of any one of these three chances for it, viz.:
>
> "One F at least on six such dice,
> Two F's at least on twelve such dice, or
> Three F's at least on eighteen such dice.
> Question: Which one of these chances should Peter in this case choose?"

The answer to this question involves the calculation of complementary probabilities:

In the case of 6 dice, the probability of at least one is

$$(1 - \Pr(0)).$$

$$\Pr(0) = P_{6,1/6}(0) = \binom{6}{0}\left(\frac{1}{6}\right)^0\left(\frac{5}{6}\right)^6 = 0.3349.$$

$$1 - \Pr(0) = \underline{0.6651}.$$

For 12 dice, the probability of at least two is

$$[1 - \Pr(0) - \Pr(1)].$$

$$\Pr(0) = \left(\frac{5}{6}\right)^{12} = 0.11216.$$

$$\Pr(1) = \binom{12}{1}\left(\frac{1}{6}\right)\left(\frac{5}{6}\right)^{11} = 0.26918.$$

$$\Pr(\text{at least two}) = 1 - (0.11216 + 0.26918) = \underline{0.6187}$$

For 18 dice, the probability of at least three is

$$[1 - \Pr(0) - \Pr(1) - \Pr(2)].$$

$$\Pr(0) = \left(\frac{5}{6}\right)^{18} = 0.03756.$$

$$\Pr(1) = \binom{18}{1}\left(\frac{1}{6}\right)\left(\frac{5}{6}\right)^{17} = 3(0.04507).$$

$$\Pr(2) = \binom{18}{2}\left(\frac{1}{6}\right)^{2}\left(\frac{5}{6}\right)^{16}.$$

$$\Pr(2) = \left(\frac{18 \cdot 17}{1 \cdot 2}\right)\left(\frac{1}{36}\right)(0.05409) = 0.2299.$$

$$[1 - \Pr(0) - \Pr(1) - \Pr(2)] = \underline{0.5973}.$$ ◄

Rare Events

If we consider a rare event E with a probability of $1/n$ where n is a "large" number, we do not expect the event to occur in a single trial, that is, we are surprised if it does. However, in a sequence of trials, its chance of occurring becomes more likely.

In particular, what is the probability of E occurring at least once in n trials?

$(1 - 1/n)^n$ is the probability that E does not occur at all.

$$\therefore \Pr\{E \text{ once or more}\} = 1 - \left(1 - \frac{1}{n}\right)^n$$

But if n is large,

$$\left(1 - \frac{1}{n}\right)^n \sim e^{-1}$$

$$\therefore Pr\{E \text{ once or more}\} \cong 1 - e^{-1} = 0.63$$

Let us consider one version of Chauvenet's criterion, which is to throw away an event whose probability of occurrence is $\leq 1/2n$. The probability of a legitimate occurrence in n trials of an event whose probability is $1/2n$ is

$$\begin{aligned} 1 - \left(1 - \frac{1}{2n}\right)^n &= 1 - \left\{\left(1 - \frac{1}{2n}\right)^{2n}\right\}^{1/2} \\ &\cong 1 - e^{-1/2} \qquad \text{for } n \text{ large} \\ &= 1 - 0.606 = 0.39 \end{aligned}$$

The reader should judge for himself if the straightforward application of this criterion is wise.

Inverse Probability

In the sense of Chapter 17, we distinguish between problems in *probability* and problems in *statistics*. The former involve deductive reasoning only, while the latter also entail inductive reasoning. The former case presupposes that a chance mechanism is known and calculates from it predicted experimental results. The latter starts with the experimental results and attempts to determine something about the chance mechanism.

For example, if we are given that a coin has probability p of landing heads and $q = (1 - p)$ of landing tails, we may ask what is the probability of getting r heads in n trials. The answer is given by (22.1). This is a problem in probability.

A problem in *statistics* would start with the observation of r heads in n trials and ask what is the probability p of getting a head on a single throw. Unlike the probability problem that has a definite answer given by the formula, the answer to the statistical question is not so sharp. We cannot say that, on the basis of the observed result, p has some definite value. *A priori*, we can only state generally that

$$0 \leq p \leq 1$$

If $r \neq 0$, we can, furthermore, eliminate $p = 0$, and if $r \neq n$ we can also eliminate $p = 1$. We shall see immediately that the most likely or expected value of p is just $r/n = p$, but the most that we can hope to do is to define an interval $[p_1, p_2]$ in which p is likely to be. We can hope to find a p.d.f. $f_P(p)$ such that $f_P(p)dp$ measures the probability that p is in an interval dp at p.

To be somewhat more correct, we define a *statistic* to be an **estimate** of p, which we may denote p_{est}, or $\hat{p}$.

We write this statistic as

$$p_{est} = \frac{r}{n} \tag{22.17}$$

Note that

$$E\left[\frac{r}{n}\right] = \frac{\langle r \rangle}{n} = \frac{np}{n} = p$$

or

$$E[p_{est}] = p \tag{22.18}$$

The statistic, p_{est}, is a random variable. Furthermore, $p_{est} = r/n$ is an ***unbiased*** *estimator* of p since its ***expectation value is equal to p.***

Moreover,

$$\text{Var}(p_{est}) = \text{Var}\left(\frac{r}{n}\right) = \frac{1}{n^2}\text{Var}(r)$$

$$= \frac{npq}{n^2} = \frac{pq}{n} \tag{22.19}$$

using (22.10).

By (22.19),

$$\lim_{n\to\infty} \text{Var}(p_{est}) = 0 \tag{22.20}$$

Because $\text{Var}(p_{est}) \to 0$ as $n \to \infty$, we say that p_{est} is a ***consistent*** *statistic.* **The variance of a *consistent* statistic vanishes as the number of trials, *n*, becomes arbitrarily large.**

Law of Large Numbers

The "**Law of Large Numbers**" insures that the estimate r/n *converges to the probability* p as the number of trials increases. The quantity $r/n = f_E$ may be considered to be, in general, the relative frequency of the event E, where $\Pr(E) = p$.

As we have seen, f_E is a random variable such that $E(f_E) = \langle f_E \rangle = p$. By Chebychev's inequality, (17.23), we have

$$\Pr\{|f_E - \langle f_E \rangle| \geq k\sigma\} \leq \frac{1}{k^2}$$

Let $\epsilon = k\sigma$, where $\epsilon > 0$ and $\sigma^2 = \text{Var}(f_E) = p(1-p)/n$ by (22.19).

$$\therefore \Pr\{|f_E - p| \geq \epsilon\} \leq \frac{\sigma^2}{\epsilon} = \frac{p(1-p)}{n\epsilon^2} \tag{22.21}$$

Consider the probability of the complementary event

$$\Pr\{|f_E - p| < \epsilon\} \geq 1 - \frac{p(1-p)}{n\epsilon^2} \tag{22.22}$$

We may take the limit

$$\lim_{n\to\infty} \Pr\{|f_E - p| < \epsilon\} = 1 \qquad \text{for any } \epsilon > 0 \tag{22.23}$$

The convergence here is somewhat different from that in, for example,

$$\left(1 - \frac{1}{n}\right)^n \xrightarrow[n\to\infty]{} e^{-1}$$

In (22.23) we have that f_E **converges "*in probability*"** to p. This is sometimes written

$$f_E \xrightarrow[n\to\infty]{\text{in } p} p \tag{22.24}$$

to distinguish it from the more familiar usage of convergence. This "**Law of Large Numbers**" is sometimes called *Bernoulli's theorem* and may be stated in two ways. As we have derived it, (22.23) says this:

The probability of f_E being closer to p than any $\epsilon > 0$ becomes arbitrarily close to unity as n increases indefinitely.

We may also use (22.21) to say that

The probability of f_E deviating from p by more than any ϵ, where $\epsilon > 0$, becomes arbitrarily small as n increases indefinitely.

The Frequency Definition of Probability

We see that the relative frequency converges to the probability p for n very large. A practical problem involves estimating how close f_E is to p for a finite number of trials. We may set ourselves equally the problem of determining the number of trials, n, needed to have some assurance that f_E is close to p. Equation (22.22) permits us to do this.

Suppose we ask how many trials we need to have a probability of at least P_0 that f_E differs from p by less than ϵ: $\Pr\{|f_E - p| < \epsilon\} \geq P_0$.

$$\therefore P_0 \leq 1 - \frac{p(1-p)}{n\epsilon^2}$$

or

$$1 - P_0 \geq \frac{p(1-p)}{n\epsilon^2}$$

from (21.22).

Thus,

$$n\epsilon^2 \geq \frac{p(1-p)}{1-P_0}$$

and

$$n \geq \frac{p(1-p)}{\epsilon^2(1-P_0)} \tag{22.25}$$

When we know p we may calculate the n needed for a given ϵ and P_0. If we have **no** knowledge of p at all, we may take

$$n \geq \frac{\max\{p(1-p)\}}{\epsilon^2(1-P_0)} = \frac{1/4}{\epsilon^2(1-P_0)} \tag{22.26}$$

as a **safe** limit for n for **any** p.

We should emphasize that relations (22.25) and (22.26) set very *conservative* estimates on n. It may happen that the n we calculate produces a probability

$$\Pr\{|f_E - p| < \epsilon\}$$

that is noticeably *greater* than P_0. The relations (22.25) and (22.26), like Chebychev's inequality itself, hold for **any** distribution. If we have specific knowledge of the distribution we may be able to estimate an n smaller than indicated by (22.26), which produces the specified probability P_0.

► **Example 22.4**
Coin tossing

How many times should a fair coin be tossed to insure that the probability is $\geq 68\%$ that the relative frequency of heads is within 10% of the theoretical probability?

From (22.26)

$$n \geq \frac{1}{4(0.05)^2(0.32)} \cong 310$$

We may check this for the binomial distribution

$$\Pr\left\{0.45 \leq \frac{r}{n} \leq 0.55\right\}$$
$$\cong \Pr\{140 \leq r \leq 170\} \equiv P_{140,170}$$
$$= \sum_{r=140}^{170} \binom{310}{r} \left(\frac{1}{2}\right)^r \left(\frac{1}{2}\right)^{310-r}$$
$$= \left(\frac{1}{2}\right)^{310} (310)! \sum_{140}^{170} \frac{1}{r!(310-r)!}$$

Since this is tedious to calculate, the reader is referred to the normal approximation in Chapter 25 for a convenient treatment with less labor. For the moment, however, let us attack this problem directly.

$$P_{140,170} = \left(\frac{1}{2}\right)^{310} (310)! \left[2\left\{\sum_{r=140}^{154} \frac{1}{r!(310-r)!}\right\} + \frac{1}{155!155!}\right]$$

by symmetry

$$= \left(\frac{1}{2}\right)^{310} (310)! \frac{1}{140!170!} \times \left\{2\left[1 + \frac{170}{141} + \cdots + \frac{170 \cdot 169 \cdots 157}{141 \cdot 142 \cdots 154}\right] + \frac{170 \cdot 169 \cdots 156}{141 \cdot 142 \cdots 155}\right\}$$

$$P_{140,170} = \left(\frac{1}{2}\right)^{310} (310!)(140!170!)^{-1}(86.7245)$$

We use Stirling's approximation (18.9) and natural logarithms to estimate

$$\ln x! \approx -x + x \ln x + \frac{1}{2}\ln(2\pi x)$$

yielding

$$\ln 310! = 1472.1246$$
$$\ln 170! = 706.5726$$
$$\ln 140! = 555.2197$$

We use

$$\ln\left(\frac{1}{2}^{310}\right) = -214.8756 \quad \text{and}$$
$$\ln(86.7245) = 4.4627$$

to get

$$\ln P_{140,170} = -0.0806$$

or

$$P_{140,170} = 0.92$$

We note two things about this calculation:

(1) it is very tedious, and
(2) the probability, 0.92, is greater than the 68% asked for. However, the inequality is satisfied. ◄

When a specific distribution may be designated, it is often sufficient to relate the *variance* to the number of trials.

► **Example 22.5**
Radioactive decay

A radioactive atom may decay into more than one final state, that is, into more than one set of decay products. For example, $A \rightarrow a$ and $A \rightarrow b$. Suppose it is known that radioactive atom A decays into a certain channel b with a probability of about 1/200. This is called the *branching fraction*, f. How many decays of the radioactive atom A must be observed to determine f to 1%?

When we say "determine f to 1%," we mean have $\sigma_f = (0.01)f$. By (22.19),

$$\operatorname{Var}\left(\frac{r}{n}\right) = \sigma^2 = \frac{p(1-p)}{n}$$

This yields

$$n = p(1-p)/\sigma^2$$

or

$$n \approx \frac{(1/200)(1-1/200)}{(1/100)^2(1/200)^2}$$
$$= 2 \times 10^6 \text{ decays}$$ ◄

The Law of Large Numbers and the Sample Mean

We have previously considered the use of the sample mean

$$\bar{x} = \frac{1}{n}\sum_{i=1}^{n} x_i$$

with its variance given by (20.17b):

$$\sigma_{\bar{x}} = \frac{\sigma^2}{n}$$

where x is such that $E(x) = \mu$ and $\operatorname{Var}(x) = \sigma^2$. Using Chebychev's inequality, (17.23) for a sample size of n, we get

$$\Pr\left\{|\bar{x} - \mu| \geq \frac{k\sigma}{\sqrt{n}}\right\} \leq \frac{1}{k^2} \qquad (22.27)$$

Similar to before, we define

$$\varepsilon = \frac{k\sigma}{\sqrt{n}} \qquad (22.28)$$

and write

$$\Pr\{|\bar{x} - \mu| \geq \varepsilon\} \leq \frac{\sigma^2}{n\varepsilon^2}$$

The probability of the complementary event is

$$1 - \frac{\sigma^2}{n\varepsilon^2} \leq \Pr\{|\bar{x} - \mu| < \varepsilon\}$$
$$= \Pr\{-\varepsilon < \bar{x} - \mu < +\varepsilon\} \qquad (22.29)$$

For any ε, we may choose n large enough so that $\bar{x}$ is within ε of μ. That is, as $n \rightarrow \infty$, the probability tends to unity that $\bar{x} \rightarrow \mu$. *The sample mean as an estimate of the population mean is an **unbiased** statistic.*

Since $\sigma^2/\sqrt{n} = \sigma_{\bar{x}}^2$, of course, the sample mean is a ***consistent** statistic* since $\operatorname{Var}(\bar{x}) \rightarrow 0$ as $n \rightarrow \infty$.

23

THE HYPERGEOMETRIC DISTRIBUTION

We may express the generic situation of interest as an urn problem. Suppose an urn contains a *population* of n balls of which n_1 are white and $n_2 = n - n_1$ are black. We draw a random *sample* of r balls without replacement. We wish to determine the probability that the group so chosen will contain exactly k white balls. Many sampling situations are equivalent to this problem.

Definition of the Probability

The sample with the desired feature contains k white balls and $r - k$ black ones. The white balls may be chosen in $\binom{n_1}{k}$ ways and the black balls in $\binom{n-n_1}{r-k}$ ways. The total number of ways in which a sample with the desired property may be chosen is the product of these. The total number of ways in which the sample of r balls may be drawn is just $\binom{n}{r}$, and so the probability required is

$$\Pr(k) = \frac{\binom{n_1}{k}\binom{n-n_1}{r-k}}{\binom{n}{r}} \tag{23.1}$$

We note that we have encountered this distribution before in treating the problem of the "e-less" prose selection from *Gadsby*, equation (18.13).

Implicit in (23.1) is the idea that k is less than r and less than n_1. The definition must be extended to make the probability zero for values of k that do not satisfy these conditions.

The variate k is said to have the *hypergeometric* distribution, and the probability is sometimes written in the following form to specify the parameters:

$$\mathbf{h}(k; r, p, n) = \frac{\binom{pn}{k}\binom{qn}{r-k}}{\binom{n}{r}} \tag{23.1a}$$

where

$$p = n_1/n \qquad \text{and} \qquad q = 1 - p$$

and k can assume the values $0, 1, \ldots, r$ if $r \le pn$ and $0, 1, \ldots, pn$ if $r > pn$. Here we speak of a population of size n having a certain property that occurs with probability p. We draw a sample of size r and ask what is the probability that k items in the sample have the required property.

We note that (23.1) and (23.1a) are normalized:

$$\sum_{k=0}^{m} \Pr(k) = \sum_{k=0}^{m} \mathbf{h}(k; r, p, n) = 1 \tag{23.2}$$

where m is the maximum value that k can assume ($m = r$ if $r \leq n_1$ and $m = n_1$ otherwise).

► Example 23.1
Fittings in a bin

Pipe fittings with right-handed and left-handed threads are kept in the same bin. Suppose six right-handed and four left-handed fittings are in the bin and two fittings are randomly drawn. What is the probability that there will be one of each?

$$\begin{aligned}\Pr(1) &= \binom{6}{1}\binom{10-6}{2-1}\Big/\binom{10}{2} \\ &= \frac{6!}{1!5!}\frac{4!}{1!3!}\Big/\frac{10!}{2!8!} \\ &= \frac{6 \cdot 4}{10 \cdot 9/2} = 0.533\end{aligned}$$

Suppose there are 60 right-handed and 40 left-handed fittings in the bin and two are again randomly drawn. What is now the probability that there will be one of each?

$$\begin{aligned}\Pr(1) &= \binom{60}{1}\binom{100-60}{2-1}\Big/\binom{100}{2} \\ &= \frac{60!}{1!59!}\frac{40!}{1!39!}\Big/\frac{100!}{2!98!} \\ &= \frac{(60)(40)}{100 \cdot 99/2} = 0.485\end{aligned}$$

◄

► Example 23.2
Salesman's strategy

A salesman for an optical supply company is mailing out sample kits to opthalmologists. He has two kits left, and there are two opthalmologists in a particular town. Unfortunately, while he has the names and addresses of all the 100 M.Ds in town, he does not have them listed by specialty and, in desperation, chooses two at random and mails out the kits. What is the probability that the kits go to neither of the two opthalmologists?

$$\begin{aligned}\Pr(0) &= \binom{2}{0}\binom{100-2}{2-0}\Big/\binom{100}{2} \\ &= \frac{2!}{0!2!}\frac{98!}{2!96!}\Big/\frac{100!}{2!98!} \\ &= \frac{(98)(97)}{(100)(99)} = 0.960\end{aligned}$$

What is the probability of the kits going to both opthalmologists?

$$\begin{aligned}\Pr(2) &= \binom{2}{2}\binom{100-2}{2-2}\Big/\binom{100}{2} \\ &= \frac{2!}{2!0!}\frac{98!}{0!98!}\Big/\frac{100!}{2!98!} \\ &= 1/(50)(99) = 0.0002\end{aligned}$$

◄

RECURSION RELATION

Formula (23.1) is sometimes tedious to compute. Because the probability depends on three parameters, it is not very convenient to tabulate many values of the hypergeometric distribution. It is frequently useful to note and utilize the recursion or recurrence relation between $\Pr(k)$ and $\Pr(k-1)$ for the same values of n, n_1, and r:

$$\Pr(k) = \frac{(n_1 - k + 1)(r - k + 1)}{k[(n - n_1) - (r - k)]} \Pr(k-1) \tag{23.3}$$

These recursion relations are convenient to use on a digital computer.

► Example 23.3
Use of the recursion relation

In Example 23.2 we calculated $\Pr(0) = 0.960$. Here we have the parameters ($n_1 = 2$, $n = 100$, $r = 2$). We use (23.3) to get

$$\begin{aligned}\Pr(1) &= \frac{(2-1+1)(2-1+1)}{1[(100-2)-(2-1)]}\Pr(0) \\ &= \left(\frac{4}{97}\right)(0.960) = 0.0396\end{aligned}$$

and

$$\Pr(2) = \frac{(2-2+1)(2-2+1)}{2[(100-2)-(2-2)]}\Pr(1)$$

$$= \frac{0.0396}{(2)(98)} = 0.0002$$

which agrees with our previous calculation. ◄

Expectation Value and Variance

We shall show in what follows that the mean or expected value of k is

$$E(K) = \frac{rn_1}{n} = rp \qquad (23.4)$$

and that the variance of k is

$$\text{Var}(K) = rpq\left(\frac{n-r}{n-1}\right) \qquad (23.5)$$

We may calculate the mean directly as follows.

$$E(K) = \sum_{k=0}^{r} k \Pr(k) \qquad \text{assuming } r \leq n_1$$

$$= \frac{1}{\binom{n}{r}} \sum_{k=0}^{r} k\binom{n_1}{k}\binom{n-n_1}{r-k} \qquad (23.6)$$

The first term is 0 and so

$$E(K) = \frac{1}{\binom{n}{r}} \sum_{k=1}^{r} k \frac{n_1!}{k!(n_1-k)!}\binom{n-n_1}{r-k}$$

$$= \frac{1}{\binom{n}{r}} \sum_{k=1}^{r} \frac{n_1!}{(k-1)!(n_1-k)!}\binom{n-n_1}{r-k}$$

$$= \frac{n_1}{\binom{n}{r}} \sum_{k=1}^{r} \frac{(n_1-1)!}{(k-1)!(n_1-k)!}\binom{n-n_1}{r-k}$$

$$= \frac{n_1}{\binom{n}{r}} \sum_{k=1}^{r} \binom{n_1-1}{k-1}\binom{n-n_1}{r-k}$$

Let $s = k-1$, $Y = n_1 - 1$, $m = n-1$, and $t = r-1$. The sum over s goes from $s = 0$ to $s = r-1 = t$. We get

$$E(K) = \frac{n_1}{\binom{n}{r}} \sum_{s=0}^{t} \binom{Y}{s}\binom{m-Y}{t-s} \qquad (23.7)$$

From (23.2) we have

$$\sum_{k=0}^{r} \binom{n_1}{k}\binom{n-n_1}{r-k} = \binom{n}{r} \qquad (23.8)$$

and so we rewrite (23.7) to get

$$E(K) = \frac{n_1}{\binom{n}{r}}\binom{m}{t} = \frac{n_1}{\binom{n}{r}}\binom{n-1}{r-1}$$

$$= n_1 \frac{r!(n-r)!}{n!} \frac{(n-1)!}{(r-1)!(n-r)!}$$

which yields the result (23.4):

$$E(K) = \frac{n_1 r}{n} = rp \qquad \textbf{Q.E.D.}$$

We may proceed to obtain the variance directly from the familiar relation

$$\text{Var}(K) = E(K^2) - [E(K)]^2 \qquad (23.9)$$

We calculate the first term:

$$E(K^2) = \sum_{k=0}^{r} k^2 \Pr(k)$$

which we rewrite

$$= \sum_{k=0}^{r} k(k-1+1)\Pr(k)$$

$$= \sum_{k=0}^{r} k(k-1)\Pr(k) + \sum_{k=0}^{r} k\Pr(k)$$

$$= \frac{1}{\binom{n}{r}} \sum_{k=0}^{r} k(k-1)\binom{n_1}{k}\binom{n-n_1}{r-k} + rp$$

using (23.4) and (23.6).

The terms $k = 0$ and $k = 1$ do not contribute to the sum and so

$$E(K^2) = \frac{1}{\binom{n}{r}} \sum_{k=2}^{r} k(k-1)\frac{n_1!}{k!(n_1-k)!}\binom{n-n_1}{r-k}$$
$$+ rp$$

$$= \frac{n_1(n_1-1)}{\binom{n}{r}} \sum_{k=2}^{r} \frac{(n_1-2)!}{(k-2)!(n_1-k)!}\binom{n-n_1}{r-k}$$
$$+ rp$$

$$= \frac{n_1(n_1-1)}{\binom{n}{r}} \sum_{k-2=0}^{r-2} \binom{n_1-2}{k-2}\binom{n-n_1}{r-k} + rp$$

By (23.8) the sum is

$$\sum_{k-2=0}^{r-2} \binom{n_1-2}{k-2}\binom{n-n_1}{r-k} = \binom{n-2}{r-2}$$

$$\therefore E(K^2) = \frac{n_1(n_1-1)}{\binom{n}{r}}\binom{n-2}{r-2} + rp$$

$$= \frac{n_1(n_1-1)}{n!} r! \frac{(n-r)!(n-2)!}{(r-2)!(n-r)!} + rp$$

$$= \frac{n_1(n_1-1)r(r-1)}{n(n-1)} + rp \qquad (23.10)$$

Substitute (23.10) and (23.4) into (23.9) and get

$$\text{Var}(K) = \frac{n_1(n_1-1)r(r-1)}{n(n-1)} - \frac{n_1^2 r^2}{n^2} + \frac{n_1 r}{n}$$

$$= \frac{nn_1(n_1-1)r(r-1) + n_1 rn(n-1) - n_1^2 r^2(n-1)}{n^2(n-1)}$$

$$= \frac{n_1 r(n-n_1)(n-r)}{n^2(n-1)}$$

$$= r\left(\frac{n_1}{n}\right)\left(1-\frac{n_1}{n}\right)\left(\frac{n-r}{n-1}\right)$$

$$= rpq\left(\frac{n-r}{n-1}\right)$$

which is (23.5). **Q.E.D.**

Instead of obtaining the mean and variance of k by direct computation using (23.1), we may utilize some of our previous results.

Define $S = x_1 + x_2 + \cdots + x_r$ to be a random variable consisting of the sum of random variables x_i that are each either 0 or 1 according to whether the ith ball drawn is black or white, respectively. Thus, $S = k$ refers to the event that exactly k white balls have been drawn in a sample of r.

Each x_i represents the result of a single draw and so

$$\Pr(x_i = 1) = \frac{n_1}{n} \qquad (23.11)$$

The mean and variance of the probability distribution of x_i are

$$\langle x_i \rangle = \frac{n_1}{n} = p = 1 - q \qquad (23.12)$$

by (22.6) with $n = 1$ and

$$\sigma^2(x_i) = p(1-p) = pq = \left(\frac{n_1}{n}\right)\left(1-\frac{n_1}{n}\right) \qquad (23.13)$$

by (22.10) for $n = 1$.

Consider $i \neq j$. Then $x_i x_j = 1$ only if the ith and jth balls drawn in the sample are both white and is $= 0$ otherwise.

$$\Pr(x_i x_j = 1) = \Pr(x_i = 1)\Pr_{x_i=1}(x_j = 1)$$

$$= \left(\frac{n_1}{n}\right)\left(\frac{n_1-1}{n-1}\right). \qquad (23.14)$$

This yields

$$\langle x_i x_j \rangle = \left(\frac{n_1}{n}\right)\left(\frac{n_1-1}{n-1}\right) \qquad (23.15)$$

by (22.6) since $(x_i x_j)$ is binomially distributed with p given by (23.14). We further note the covariance of (x_i, x_j) from (20.11):

$$\text{Covar}(x_i, x_j) = \langle x_i x_j \rangle - \langle x_i \rangle\langle x_j \rangle$$

$$= \frac{n_1}{n}\left(\frac{n_1-1}{n-1}\right) - \left(\frac{n_1}{n}\right)^2$$

$$= \frac{-n_1(n-n_1)}{n^2(n-1)} \qquad (23.16)$$

Now

$$\langle S \rangle = \sum_{i=1}^{r} \langle x_i \rangle = \frac{rn_1}{n} \qquad (23.17)$$

by (20.7) which is the expectation value desired.

By (20.17) we write

$$\text{Var}(S) = \sum_{i=1}^{r} \sigma_i^2 + 2\sum_{i<j}^{r} \text{Covar}(x_i x_j)$$

$$= r\left\{\left(\frac{n_1}{n}\right)\left(1-\frac{n_1}{n}\right)\right\}$$

$$+ 2\frac{r(r-1)}{2}\left\{\frac{-n_1}{n^2}\left(\frac{n-n_1}{n-1}\right)\right\}$$

$$= r\frac{n_1}{n}\left(1-\frac{n_1}{n}\right)\left[1-\frac{(r-1)}{n-1}\right]$$

$$= rpq\left(\frac{n-r}{n-1}\right) \qquad (23.18)$$

which is the required *variance of the hypergeometric distribution.* **Q.E.D.**

► Example 23.4
Gadsby

Returning to our analysis of *Gadsby* in Chapter 18, we note that, pursuant to the assumption that the 580-letter passage came from among the 89,602 letters in the Northwestern sample, the expectation value for the number of letters-e in the passage is

$$\langle k \rangle = (580)\left(\frac{11290}{89602}\right) = 73.0$$

as before.

$$\sigma^2(k) = 580\left(\frac{11290}{89602}\right)\left(1-\left(\frac{11290}{89602}\right)\right) \times \left(\left(\frac{89602-580}{89602-1}\right)\right)$$

$$\approx (73.0)(0.874)\left(\frac{89022}{89601}\right) = 63.5$$

Therefore,

$$\sigma \approx 8$$

The observed deviation from the mean divided by the standard deviation is

$$(73.0 - 0)/7.97 = 9.16$$

That is, the observation of no "e's" is 9.16 standard deviations away from the expected number. ◄

Some examples of the hypergeometric distribution are shown in Figures 23.1*a* to 23.1*e* for $p = 0.4$.

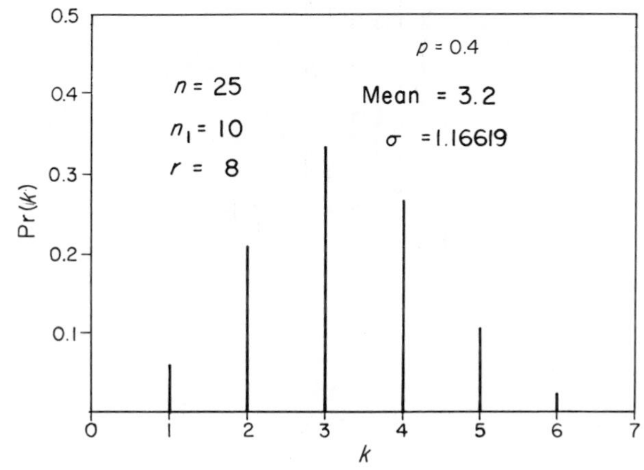

(*a*)

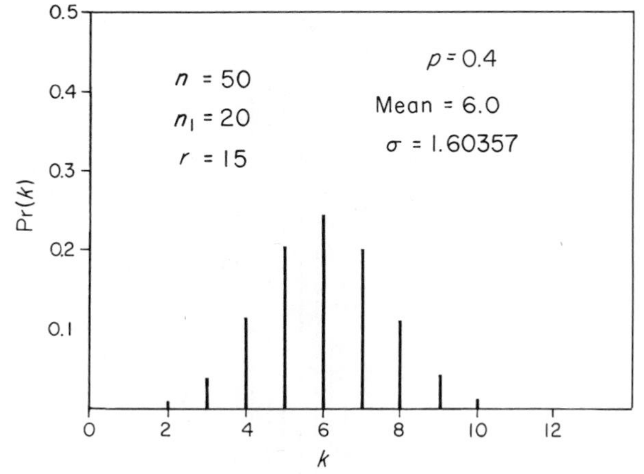

(*b*)

Figure 23.1
Plot of the discrete frequency distribution $\Pr(k) = \mathbf{h}(k; r, p, n)$ versus k, where $p = n_1/n = 0.4$. (*a*) $n = 25$, $n_1 = 10$, $r = 8$. (*b*) $n = 50$, $n_1 = 20$, $r = 15$. (*c*) $n = 100$, $n_1 = 40$, $r = 30$. (*d*) $n = 200$, $n_1 = 80$, $r = 60$. (*e*) $n = 400$, $n_1 = 160$, $r = 120$.

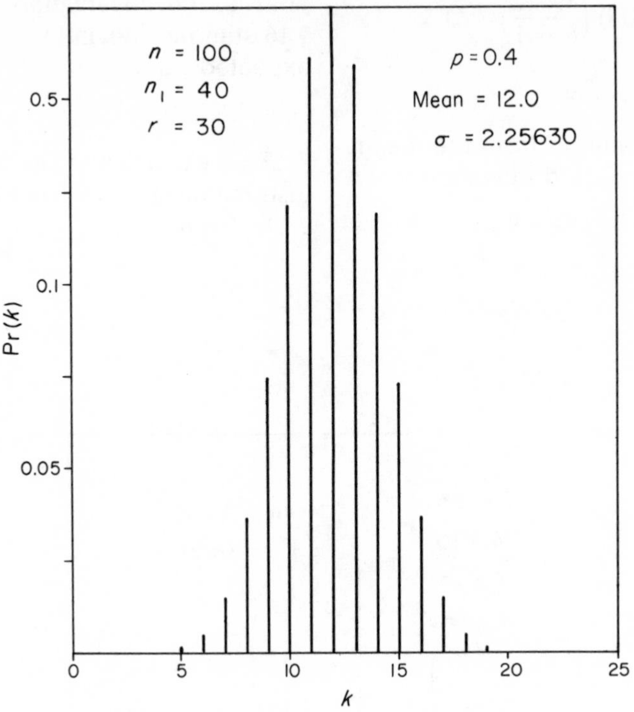

(c)

Figure 23.1(c) (d) and (e)

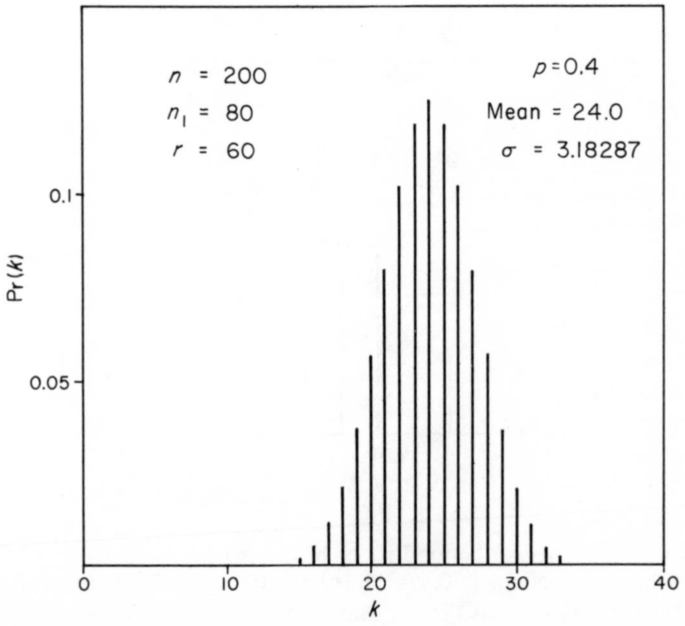

(d)

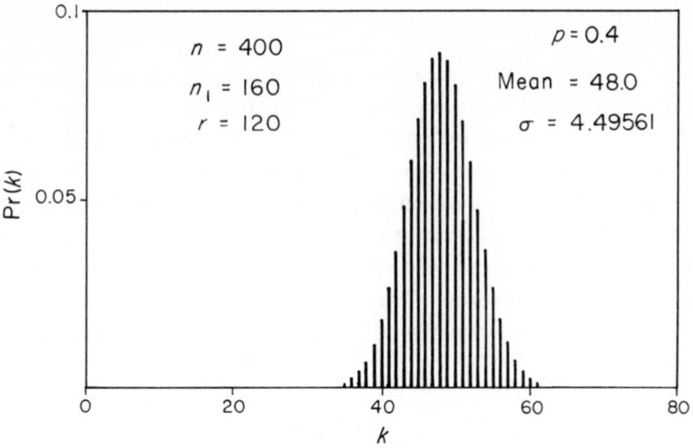

(e)

Binomial Approximation to the Hypergeometric Distribution

When the population n is large with respect to the sample size r, then the hypergeometric distribution is well-approximated by the binomial*:

$$\mathbf{h}(k\,;r,p,n) \approx \mathbf{b}(k\,;r,p) \qquad \frac{r}{n} \ll 1 \quad (23.19)$$

We may see this for the following reasons. The hypergeometric distribution describes drawing a sample of size r *without replacement* from the population. The binomial distribution may be thought of as describing a sample of 1 being drawn, *replacing it*, and repeating the operation r times. The difference is that the probability changes in the case of the drawing without replacement and does not change in the case of drawing with replacement. If the sample to be drawn is small with respect to the population size, the probability does not change very much and, as r/n becomes small, the probability is essentially constant.

*See definitions in (22.1a) and (23.1a).

► **Example 23.5**
Fittings in a bin

In Example 23.1, we considered bins containing 60% right-handed fittings and 40% left-handed fittings and drew samples of two out of each. We asked for the probability of getting one right-handed fitting. For a bin containing 10 fittings the probability is

$$\mathbf{h}(1;2,0.6,10) = 0.533$$

For a bin containing 100 fittings the probability is

$$\mathbf{h}(1;2,0.6,100) = 0.485$$

For a bin containing 1000 fittings the probability is

$$\mathbf{h}(1;2,0.6,1000) = 0.4804$$

If we consider that the probability of getting a right-handed fitting does not change after the first fitting is drawn, we get the binomial probability

$$\mathbf{b}(1;2,0.6) = \binom{2}{1}(0.6)^1(0.4)^1 = 0.4800$$

◄

Inverse Probability

Suppose a population contains n members, and a fraction p of the members has a certain property. We wish to estimate the fraction p without looking at every member of the population, that is, we wish to draw a random sample of size r and count the number, k, of members of the sample with the property of interest. The fraction k/r will represent an estimate of the population **proportion** p.

We may define the statistic

$$w = \frac{k}{r} \equiv \hat{p} \qquad (23.20)$$

where k is a random variable whose frequency function is $\mathbf{h}(k; r, p, n)$. We note that

$$\langle k \rangle \equiv \mu_k = rp$$

and

$$\sigma_k^2 = \text{Var}(k) = rpq\left(\frac{n-r}{n-1}\right)$$

We see that

$$\langle w \rangle = \langle \hat{p} \rangle = \mu_w = \frac{\langle k \rangle}{r} = p \qquad (23.21)$$

And so w is an unbiased statistic with variance

$$\sigma_w^2 = \frac{pq}{r}\left(\frac{n-r}{n-1}\right) \qquad (23.22)$$

The *statistics* (23.11), (23.12), and (23.13) are useful for such problems as sampling to estimate the percentage of people who will buy a certain toothpaste, vote for a certain candidate, or watch a certain television program. However, the *normal approximation to the hypergeometric distribution* is usually employed, and we defer consideration of an example until after the introduction of the normal approximation.

Because the hypergeometric distribution is not as convenient as other distributions, we usually employ an approximation to consider specific examples. Generally, we shall find that the Poisson distribution is a good approximation to the hypergeometric when p is small. The normal approximation will usually be found to be good otherwise, especially when p is close to 0.5.

For the moment let us employ only Chebychev's inequality (17.23) to decide how large a sample to take to get an estimate of p. As always we cannot make absolute statements about how close the statistic, $w = \hat{p}$, is to p; we can set our criterion in the form that we wish the probability to be small (less than α) that w differs from p by more than an amount, δ. Formally, we want

$$\Pr\{|w - p| \geq \delta\} \leq \alpha$$

Equation (17.23) assures us that

$$\Pr\{|w - p| \geq k\sigma_w\} \leq \frac{1}{k^2}$$

For example, we are sure that the probability is less than 1/4 that w and p will differ by more than $2\sigma_w$. If this is our criterion, we may proceed to ask how large the sample size, r, should be.

► **Example 23.6**
A polling problem

Suppose the town of Evanston has 40,000 registered voters and an upcoming referendum is going to be close ($p \approx 0.5$). The local newspaper wishes to take a random sample to determine p and wishes to have a probability of 75% or more that the estimate will differ from the true p by less than 0.01. How large a sample is needed?

Here $\delta = 0.01$. We may define $\delta/\sigma_w = \beta = 1/p \approx 1/0.5$ so

$$\sigma_w^2 = \frac{\delta^2}{\beta^2} = \frac{(0.01)^2}{2^2} = (0.005)^2$$

since $\beta = 2$. Also $pq \approx 0.25$.

By (23.22)

$$\sigma_w^2 = \frac{pq}{r}\left(\frac{n-r}{n-1}\right) = (0.005)^2$$

$$= \frac{0.25}{r}\left(\frac{n}{n-1} - \frac{r}{n-1}\right)$$

or

$$(0.005)^2 \approx \frac{0.25}{r} - \frac{0.25}{40000}$$

since $n/(n-1) \approx 1$.

$$\therefore 3.1 \times 10^{-5} \cong \frac{0.25}{r}$$

$$r \geq \frac{0.25}{0.31} \times 10^{+4} = 8060$$

The newspaper must poll more than 8000 of the voters. ◄

24

THE POISSON DISTRIBUTION

The Poisson distribution is a good model for describing random phenomena where the probability of occurrence is **small** and **constant**. It arises as

1. The exact model underlying various physical phenomena involving events in time and space.

2. An approximation to the binomial distribution where the number of trials, n, is large and the probability for success on a single trial, p, is small. The Poisson distribution is important for the description of rare events.

The important assumptions of the model are related to the adjectives above, *constant* and *small*. *Constant* refers to the probability of an event in a particular region of time or space being **independent** of what the probability is for a different region of time or space. *Small* refers to the ability to choose an interval so small that we may **neglect** the probability that **two** events occur in it. That is, we may choose an interval so that only zero or one occurrence is possible. Furthermore, the probability is **linear** in the interval, that is, the probability of one occurrence is proportional to the size of the interval.

Exact Model

Consider that the occurrence or nonoccurrence of an event at a point x is independent of what happens away from x. x may be a point in time or in space. (If x is time, this says that the probability is independent of the history before time x; the probability is independent of what happens after time x by considerations of causality, that is, it cannot depend on future happenings.) Consider an interval dx so small that the probability of observing two events in dx is negligible. Furthermore, the probability of a single event in dx is $\lambda\, dx$ where λ is a constant. The probability of getting 0 events in dx is, of course, $1-\lambda\, dx$. $\boldsymbol{\lambda}$ is a (constant) **probability *density***. It may have the nature of a probability *per* unit time, a probability *per* unit length, a probability *per* unit area, etc.

We wish to calculate the probability, $P_r(x)$, of observing r events in the interval $(0, x)$.

Consider first the probability of getting no events in $(0, x)$, $P_0(x)$. We can get no events in the interval $(0, x + dx)$ if and only if there are no events in the interval $(0, x)$ **and** none in the interval $(x, x + dx)$. Since the probabilities are independent, the desired probability is the *product*

$$P_0(x + dx) = P_0(x)(1 - \lambda\, dx)$$

We get the relation

$$\frac{P_0(x + dx) - P_0(x)}{dx} = -\lambda P_0(x)$$

If we let dx become arbitrarily small,

$$\lim_{dx \to 0} \frac{P_0(x + dx) - P_0(x)}{dx} = \frac{dP_0(x)}{dx}$$

or

$$\frac{dP_0(x)}{dx} = -\lambda P_0(x)$$

This may be integrated simply, since

$$\frac{dP_0(x)}{P_0(x)} = -\lambda\, dx$$

to get

$$P_0(x) = Ae^{-\lambda x}$$

where A is an arbitrary constant.
It is *certain* that no events occur in the interval (0, 0) so

$$P_0(0) = 1 \qquad \text{at } x = 0$$

which fixes the constant, $A = 1$.

$$\therefore P_0(x) = e^{-\lambda x}.$$

Consider next the probability, $P_1(x)$, of getting one and only one event in the interval $(0, x)$. There are *two* ways of getting one event in the interval $(0, x + dx)$: one in $(0, x)$ and none in $(x, x + dx)$ ***or*** none in $(0, x)$ and one in $(x, x + dx)$. Therefore,

$$P_1(x + dx) = P_1(x)(1 - \lambda\, dx) + P_0(x)\lambda dx$$

We may again form the derivative by letting $dx \to 0$:

$$\lim_{dx \to 0} \frac{P_1(x + dx) - P_1(x)}{dx} = \frac{dP_1(x)}{dx}$$

which yields

$$\frac{dP_1(x)}{dx} = -\lambda P_1(x) + \lambda e^{-\lambda x}$$

This has the solution

$$P_1(x) = \lambda x e^{-\lambda x}$$

as may be verified by substitution.

We may extend this argument to the general case. There are only *two* ways of getting r events in the interval $(0, x + dx)$. One way is to have $(r - 1)$ events in the interval $(0, x)$ and one event in the interval $(x, x + dx)$. The second way is to have r events in the interval $(0, x)$ and none in the interval $(x, x + dx)$. The probability may be written

$$P_r(x + dx) = P_{r-1}(x)\lambda\, dx + P_r(x)(1 - \lambda\, dx)$$

which yields the general relation

$$\frac{dP_r(x)}{dx} = -\lambda P_r(x) + \lambda P_{r-1}(x)$$

We may verify by substitution that this equation has the solution

$$P_r(x) = \frac{(\lambda x)^r}{r!} e^{-\lambda x} \qquad (24.1)$$

We note that this is normalized, since

$$\sum_{r=0}^{\infty} P_r(x) = e^{-\lambda x}\left\{\sum_{r=0}^{\infty} \frac{(\lambda x)^r}{r!}\right\}$$
$$= e^{-\lambda x}\{e^{+\lambda x}\}$$

since the bracketed quantity is the series expansion of the exponential.

$$\therefore \sum_{r=0}^{\infty} P_r(x) = 1 \qquad (24.2)$$

We might expect that the *average* number of occurrences in the interval $(0, x)$ is just λx. This is true as we shall later prove from the relation

$$E(r) = \sum_{r=0}^{\infty} r P_r(x)$$

This will be shown to be

$$E(r) = \lambda x \equiv \mu \qquad (24.3)$$

We may, therefore, write

$$P_r = \frac{\mu^r e^{-\mu}}{r!} \equiv \mathbf{p}(r; \mu) \qquad (24.4)$$

where **p(r; μ) stands for the Poisson probability of r given the parameter μ.** This is plotted in Figures 24.1*a* to 24.1*l* for μ = 0.5 through 10.0. Note that the Poisson distribution is quite asymmetrical for small values of μ, but becomes increasingly symmetrical as μ increases.

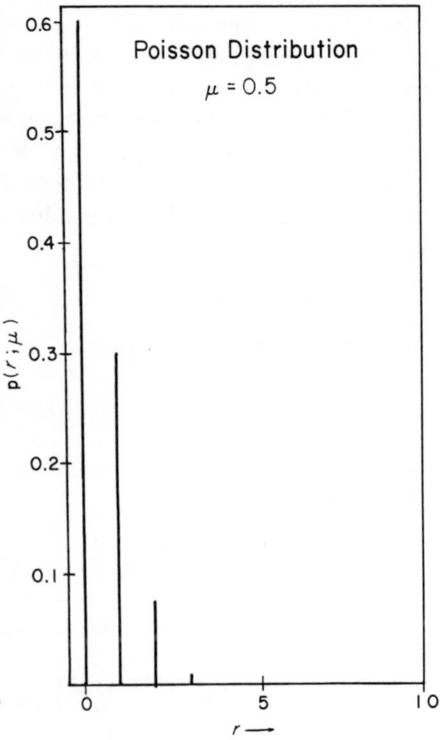

(*a*)

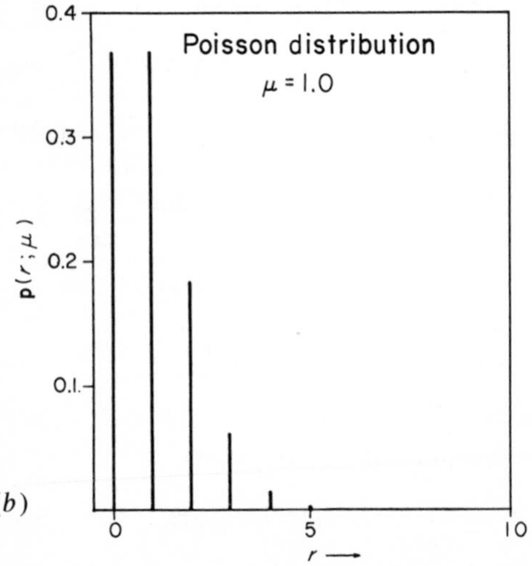

(*b*)

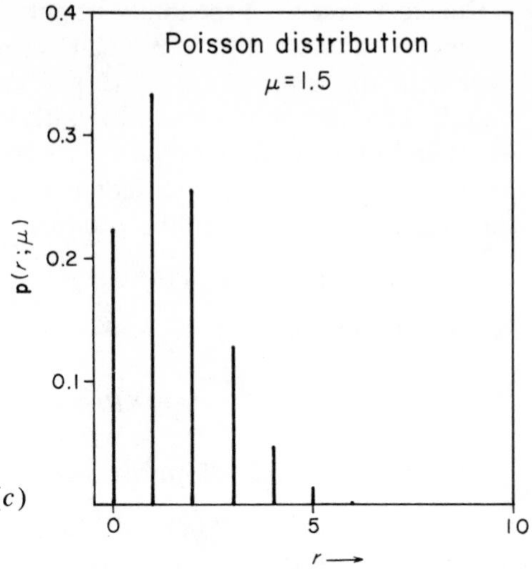

(*c*)

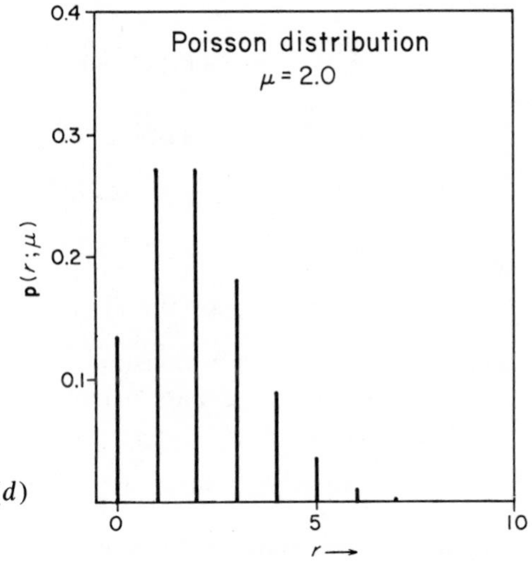

(*d*)

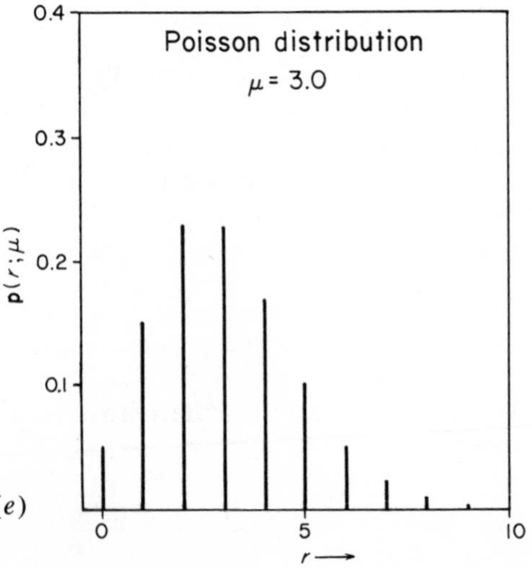

(*e*)

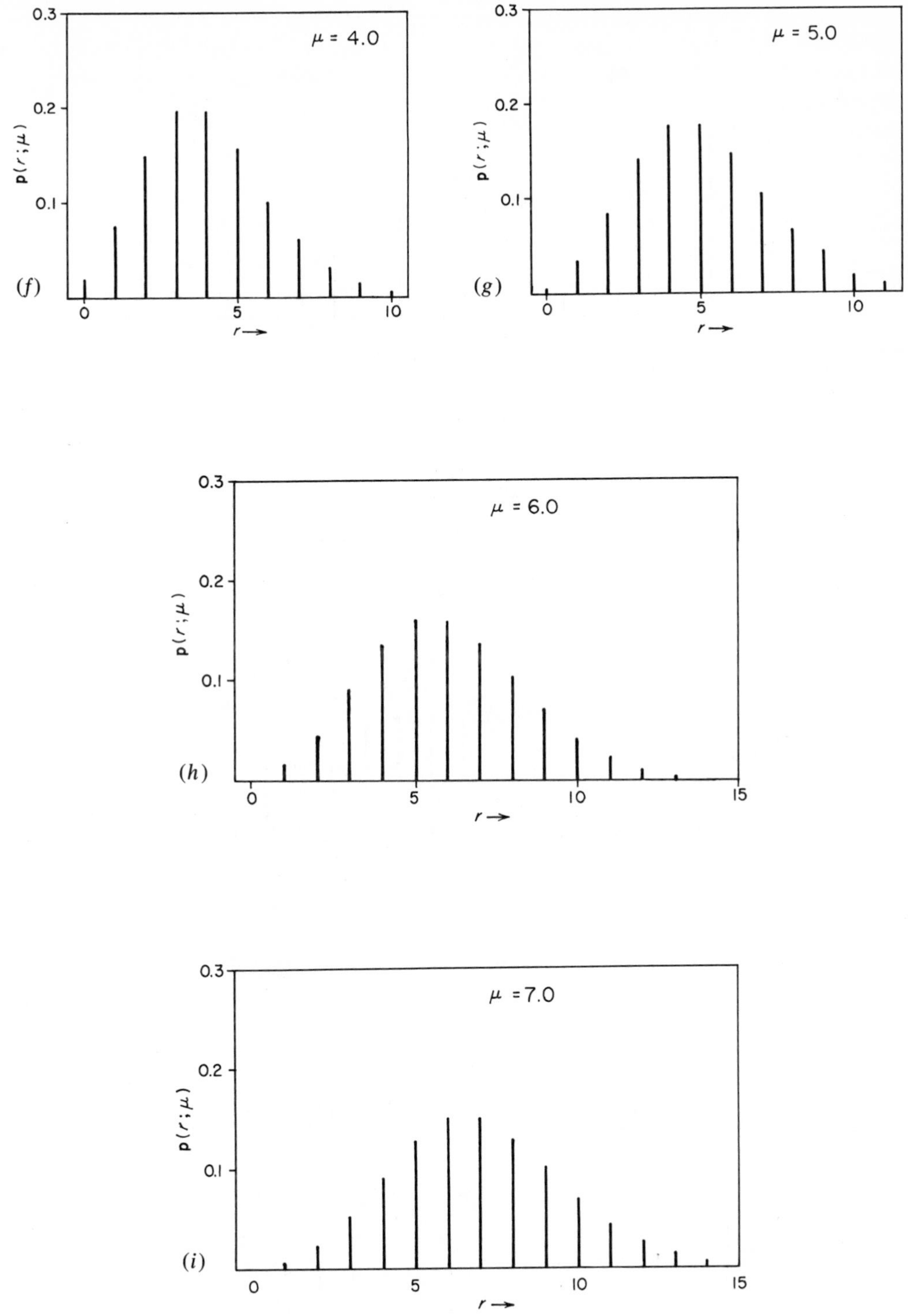

Figure 24.1
Plot of the discrete frequency function of the Poisson distribution $p(r;\mu)$ versus r for different values of μ. (*a*) $\mu = 0.5$. (*b*) $\mu = 1.0$. (*c*) $\mu = 1.5$. (*d*) $\mu = 2.0$. (*e*) $\mu = 3.0$. (*f*) $\mu = 4.0$. (*g*) $\mu = 5.0$. (*h*) $\mu = 6.0$. (*i*) $\mu = 7.0$. (*j*) $\mu = 8.0$. (*k*) $\mu = 9.0$. (*l*) $\mu = 10.0$.

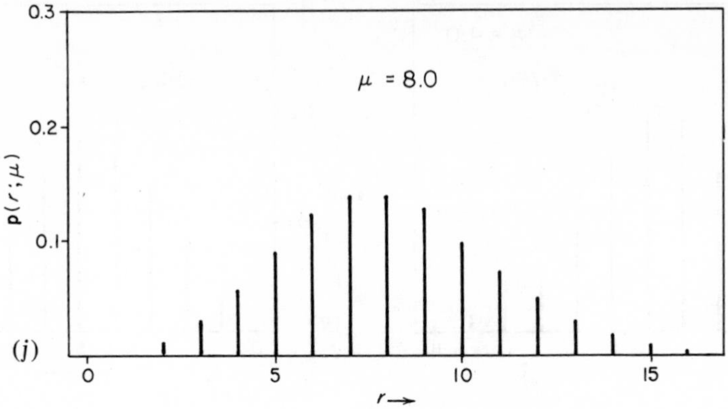

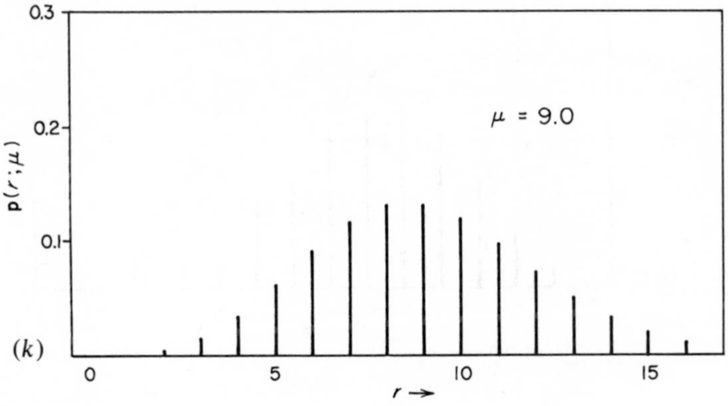

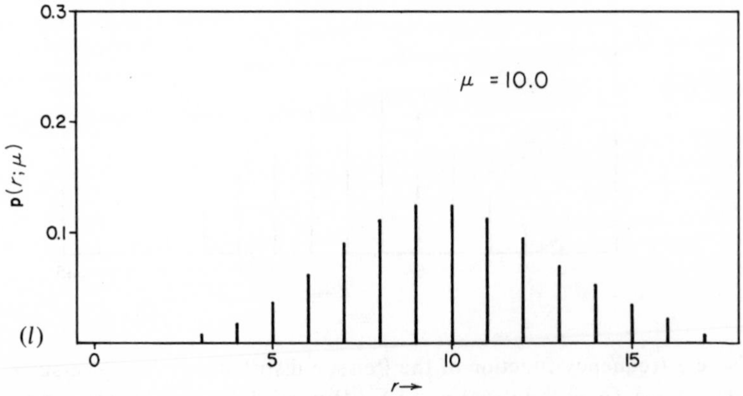

Figure 24.1(*j*) (*k*) and (*l*)

► **Example 24.1**
Number of raisins in a piece of cake

What is the probability of finding two raisins in a piece of cake of volume 5 cm^3, cut from a large cake in which the density of raisins is $2/\text{cm}^3$? Assume the raisins are tiny and randomly distributed.

Let S be the number of raisins in a 5 cm^3 volume. $\lambda = 2\text{ cm}^{-3}$ is the density.
The expected value per 5 cm^3 volume is $\mu = \lambda x = 10$ raisins
This is the Poisson parameter and

$$\Pr(S = k) = \frac{e^{-\mu}\mu^k}{k!}$$

$$\therefore \Pr(S = 2) = \frac{e^{-10}(10)^2}{2!} = 0.0023$$

Why was it necessary to assume the raisins are tiny? ◄

► **Example 24.2**
A cosmic ray measurement

A long-term cosmic ray experiment is set up so that the detector counts an average of 20 cosmic rays per day. The detector system is read out and reset daily for one year. How often would you expect it to happen that only 10 cosmic ray events in a day have been observed?

Here the average value expected per day is $\mu = 20$.

$$\Pr(10) = \frac{e^{-20}(20)^{10}}{10!} = \frac{e^{-20}10^{10}2^{10}}{(3628800)} = 0.00582$$

$$\langle N \rangle = n \Pr(10) = 365\ (0.00582) = 2.12$$

$\therefore$ we expect to observe 10 events twice in a year. ◄

► **Example 24.3**
A radioactivity experiment

In a radioactivity experiment on a source of constant average intensity, a total of 19,500 α-rays were observed in 50 hours of continuous observation. The time of arrival of each α-ray was recorded on a tape so that the number of α-rays recorded in each successive 1-minute interval can be determined.

(a) What is the average number of α-rays per 1-minute interval?
(b) In how many 1-minute intervals would you expect to observe no α-rays?
(c) In how many intervals would you expect to observe one α-ray?
(d) In how many would you expect to observe six α-rays?

(a) The average number for a 1-minute interval is

$$\mu = \frac{19500}{(50)(60)} = 6.5\alpha\text{-rays/min}$$

(b) $\Pr(0) = \mathbf{p}(0; 6.5) = \dfrac{e^{-6.5}(6.5)^0}{0!} = 0.00150$

There are $50 \times 60 = 3000$ intervals so

$$\langle n_0 \rangle = \text{expected number with none} = (3000)(0.0015) = 4.5$$

(c) $\Pr(1) = \mathbf{p}(1; 6.5) = \dfrac{e^{-6.5}(6.5)^1}{1!} = .00977$

$$\langle n_1 \rangle = \text{expected number with 1} = 3000 \Pr(1) = 29.3$$

(d) $\Pr(6) = \mathbf{p}(6; 6.5) = \dfrac{e^{-6.5}(6.5)^6}{6!} = 0.157$

$\langle n_6 \rangle = 3000 \Pr(6) = 472.4$ ◄

Poisson Approximation to the Binomial Distribution

For the physical realization of a situation in which this approximation is applicable, we can hearken back to the binomial distribution defined in (22.1):

$$P_{n,p}(r) = \frac{n!}{r!(n-r)!} p^r q^{n-r}$$

which is the probability of r successes in

n trials where the probability of success is p and that of failure is $q = 1 - p$. Each trial was taken to be independent. We may consider the limiting case of the binomial distribution where

$$\begin{aligned} &p \text{ is small } (p \ll 1) \\ &n \text{ is large } (n \gg 1) \text{ and} \\ &n \gg r \end{aligned} \qquad (24.5)$$

With these relations

$$n - r + 1 \approx n$$

and

$$\frac{n!}{(n-r)!} = n(n-1)(n-2)\cdots(n-r+1) \approx n^r$$

since there are r terms, each approximately equal to n.

We note that the binomial theorem gives

$$(1-p)^{n-r} = 1 - p(n-r) + \frac{p^2}{2!}(n-r)(n-r-1) + \cdots$$

while the expansion of e^x yields

$$e^{-p(n-r)} = 1 - p(n-r) + \frac{p^2}{2!}(n-r)^2 + \cdots$$

Therefore, relations (24.5) imply

$$(1-p)^{n-r} \approx e^{-p(n-r)} \approx e^{-pn}$$

But $pn = \langle r \rangle$ by (22.6).

$$\therefore (1-p)^{n-r} \approx e^{-\langle r \rangle}$$

We may now see how (22.1) is modified by (24.5):

$$P_{n,p}(n) = \frac{n!}{r!(n-r)!} p^r (1-p)^{n-r} \approx \frac{(np)^r}{r!} e^{-\langle r \rangle}$$

This is defined as the Poisson distribution

$$P_r = \Pr\{r \text{ successes}\} = \frac{\langle r \rangle^r e^{-\langle r \rangle}}{r!} \qquad (24.5\text{a})$$

where $\langle r \rangle$ is the expected number.

Formally, we may say that a random variable X is Poisson distributed with parameter $\langle r \rangle > 0$ if

$$\Pr(X = r) = \frac{\langle r \rangle^r e^{-\langle r \rangle}}{r!} \qquad r = 0, 1, \ldots \qquad (24.6)$$

We may check the normalization of (24.6):

$$\sum_{r=0}^{\infty} P_r = \left\{ \sum_{r=0}^{\infty} \frac{\langle r \rangle^r}{r!} \right\} e^{-\langle r \rangle}$$

and we note the bracketed term is the series for $e^{+\langle r \rangle}$. Therefore,

$$\sum_{r=0}^{\infty} P_r = \{e^{+\langle r \rangle}\} e^{-\langle r \rangle} = 1 \qquad (24.7)$$

Figures 24.2a to 24.2d show comparisons of the Poisson and Binomial distributions. We make the approximation

$$\mathbf{b}(r; n, p) \approx \mathbf{p}(r; \langle r \rangle) \qquad \text{for } np = \langle r \rangle \qquad (24.8)$$

The approximation is quite good for $p < 0.1$ and $np > 5$. It is very good for $p = 0.05$ even when $np < 5$.

► **Example 24.4**
Failure of detectors

A typical nuclear physics experiment utilizes 50 independent counters (instruments sensitive to ionizing radiation). The counters are checked between data runs and are observed to have a one percent chance of failure between checks (i.e., $p = 0.01$ for failure).

What is the probability that all the counters are operative throughout a data run?

Here we may use the binomial distribution with $n = 50$, $p = 0.01$

$$\mathbf{b}(0; 50, 0.01) = \binom{50}{0} (0.01)^0 (0.99)^{50} = 0.6050$$

The Poisson approximation may be invoked here with $np = 0.5$:

$$\mathbf{p}(0; 0.5) = \frac{(0.5)^0 e^{-0.5}}{0!} = 0.6065$$

Suppose that a data run is salvageable if one counter fails but is not useful if two or more counters fail. What is the probability that the data run is spoiled by counter failure?

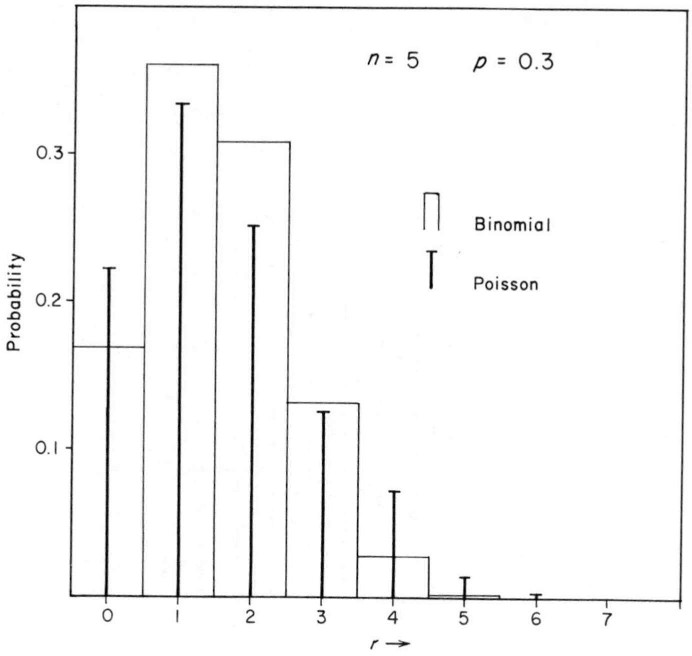

(a)

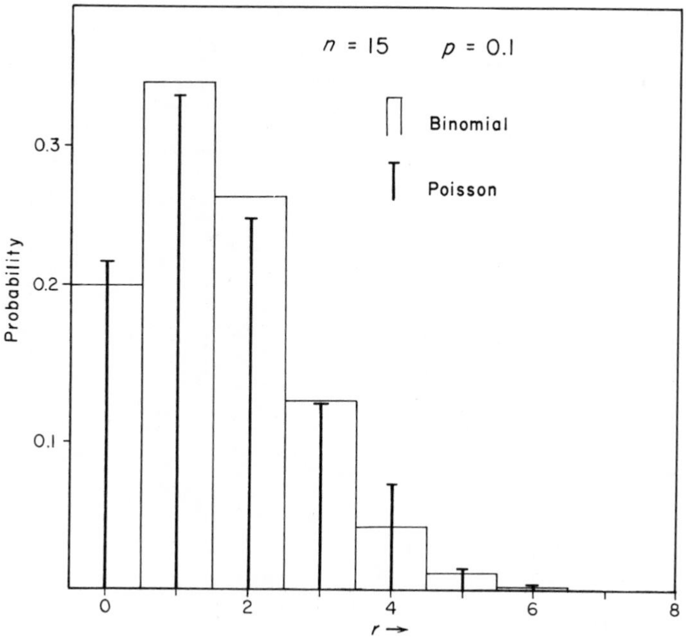

(b)

Figure 24.2
Comparison of the discrete frequency functions of the Poisson and binomial distributions. (*a*) $n = 5$, $p = 0.3$. (*b*) $n = 15$, $p = 0.1$. (*c*) $n = 30$, $p = 0.3$. (*d*) $n = 30$, $p = 0.05$.

(c)

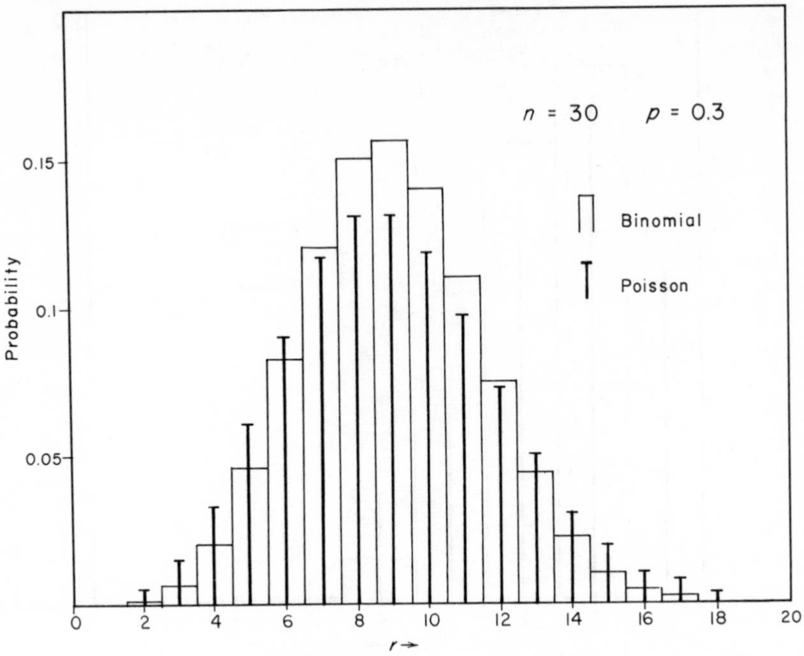

(d)

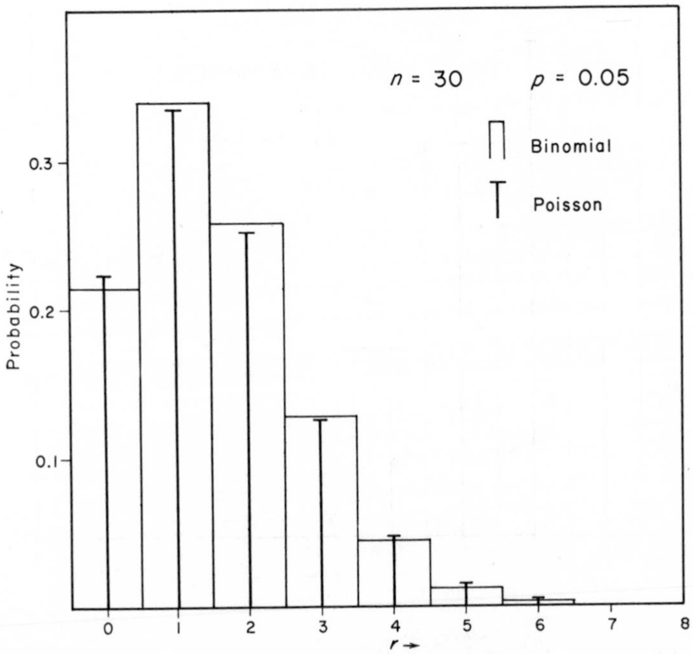

Figure 24.2(*c*) and (*d*).

We require the probability that two or more counters fail:

$$\Pr(r \geq 2) = 1 - \Pr(0) - \Pr(1)$$

By using the binomial distribution, we get

$$\begin{aligned}\Pr(r \geq 2) &= 1 - \mathbf{b}(0; 50, 0.01) \\ &\quad - \mathbf{b}(1; 50, 0.01) \\ &= 1 - 0.6050 \\ &\quad - \frac{50!}{1!49!} \\ &\quad \times (0.01)^1(0.99)^{49} \\ &= 1 - 0.6050 - 0.3056 = 0.0894\end{aligned}$$

By using the Poisson approximation, we get

$$\begin{aligned}\Pr(r \geq 2) &= 1 - 0.6065 - \frac{(0.5)^1 e^{-0.5}}{1!} \\ &= 1 - 0.6065 - 0.3032 \\ &= 0.0903\end{aligned}$$

Thus, about 9% of the time a data run will be spoiled by counter failure. ◄

► **Example 24.5**
Incidence of colorblindness

Suppose colorblindness appears in 1% of the people in a certain population.

(**a**) What is the probability that a sample of 100 will contain
  (**i**) no colorblind people
  (**ii**) two or more colorblind people
(**b**) How large must a random sample be if the probability of its containing a colorblind person is to be ≥ 0.95?

We may consider this a binomial problem with $p = 0.01$. If r = number of color-blind people in a sample of size 100, then

(**a**) $\Pr(r = 0) = \mathbf{b}(0; 100, .01)$

$$= \binom{100}{0}(0.01)^0(0.99)^{100}$$

$$= (0.99)^{100} = 0.3660$$

$$\Pr(r = 1) = \mathbf{b}(1; 100, 0.01)$$

$$= \binom{100}{1}(0.01)^1(0.99)^{99}$$

$$= (0.99)^{99} = .3697$$

$$\Pr(r \geq 2) = 1 - \Pr(r = 0) - \Pr(r = 1)$$

$$\therefore \Pr(r \geq 2) = 0.2643$$

As a Poisson problem, we expect to find on the average

$$\mu = np = 100(0.01) = 1$$

color-blind person in a sample of 100.

$$\Pr(r = 0) = \mathbf{p}(0; 1) = \frac{e^{-1}(1)^0}{0!} = 0.3679$$

This is to be compared with 0.3660.

$$\Pr(r = 1) = \mathbf{p}(1; 1) = \frac{e^{-1}(1)^1}{1!} = 0.3679$$

so

$$\begin{aligned}\Pr(r \geq 2) &= 1 - \mathbf{p}(0; 1) - \mathbf{p}(1; 1) \\ &= 1 - 0.7358 = 0.2642\end{aligned}$$

The binomial problem yielded 0.2643.

(**b**) Let S_n be the number of color-blind people in a sample of size n. It is required to find n such that

$$\Pr(S_n \geq 1) \geq 0.95$$

We note that

$$\Pr(S_n \geq 1) = 1 - \Pr(S_n = 0)$$

Binomial

$$\Pr(S_n = 0) = \mathbf{b}(0; n, 0.01) = (0.99)^n$$

$\therefore$ we require n such that

$$1 - (0.99)^n \geq 0.95.$$

We may write this

$$0.05 \geq (0.99)^n$$

or

$$\log(0.05) \geq n \log(0.99).$$

We may solve this inequality for n recalling that $\log(0.99) < 0$.

$$\therefore n \geq \frac{\log(0.05)}{\log(0.99)} = 298.07$$

$\therefore$ n must be larger than 298.

Poisson

$$\Pr(S_n = 0) = \mathbf{p}(0; 0.01n) = \frac{e^{-.01n}(0.01)^0}{0!}$$

$$\therefore 1 - e^{-0.01n} \geq 0.95$$

and

$$0.05 \geq e^{-0.01n}.$$

This yields

$$\ln(0.05) \geq -0.01n$$

or

$$n \geq -100 \ln(0.05) = 299.57.$$

This means that n must be larger than 299. ◄

Expectation Value for the Poisson Distribution

Thus far we have taken the definition $\langle r \rangle$ from the binomial distribution. We should really calculate the expectation value of $X = r$ for (24.6):

$$\begin{aligned} E(X) &= \sum_{r=0}^{\infty} r \Pr(X = r) \\ &= \sum_{r=0}^{\infty} \frac{r\langle r \rangle^r e^{-\langle r \rangle}}{r!} \\ &= \sum_{r=1}^{\infty} r \frac{\langle r \rangle^r e^{-\langle r \rangle}}{r!} \end{aligned}$$

$$= e^{-\langle r \rangle} \langle r \rangle \sum_{r=1}^{\infty} \frac{\langle r \rangle^{r-1}}{(r-1)!} \tag{24.9}$$

If we set $S = r - 1$, we can write

$$\sum_{r=1}^{\infty} \frac{\langle r \rangle^{r-1}}{(r-1)!} = \sum_{S=0}^{\infty} \frac{\langle r \rangle^S}{S!} = e^{\langle r \rangle}$$

Therefore,

$$E(X) = e^{-\langle r \rangle} \langle r \rangle e^{\langle r \rangle} = \langle r \rangle \tag{24.10}$$

Variance for the Poisson Distribution

We note that (24.9) and (24.10) yield the relation

$$\langle r \rangle = \sum_{r=0}^{\infty} \frac{r\langle r \rangle^r}{r!} e^{-\langle r \rangle}$$

or

$$e^{\langle r \rangle} \langle r \rangle = \sum_{r=0}^{\infty} \frac{r\langle r \rangle^r}{r!} \tag{24.10a}$$

We may calculate the variance of the Poisson distribution

$$\begin{aligned} \langle r^2 \rangle &= \sum_{r=0}^{\infty} r^2 \frac{\langle r \rangle^r}{r!} e^{-\langle r \rangle} = e^{-\langle r \rangle} \sum_{r=0}^{\infty} \frac{r}{(r-1)!} \langle r \rangle^r \\ &= e^{-\langle r \rangle} \sum_{r=1}^{\infty} \frac{r\langle r \rangle^r}{(r-1)!} \end{aligned}$$

If we set $S = r - 1$, we get

$$\sum_{r=1}^{\infty} \frac{r\langle r \rangle^r}{(r-1)!} = \langle r \rangle \sum_{S=0}^{\infty} \frac{(S+1)\langle r \rangle^S}{S!}$$

$$= \langle r \rangle \left\{ \sum_{S=0}^{\infty} \frac{S\langle r \rangle^S}{S!} + \sum_{S=0}^{\infty} \frac{\langle r \rangle^S}{S!} \right\}$$

The first term in the bracket may be evaluated by (24.10a) to be $e^{\langle r \rangle}\langle r \rangle$ while the second term is just $e^{\langle r \rangle}$.

Therefore

$$\begin{aligned} \langle r^2 \rangle &= e^{-\langle r \rangle} \langle r \rangle \{ e^{\langle r \rangle} \langle r \rangle + e^{\langle r \rangle} \} \\ &= \langle r \rangle^2 + \langle r \rangle \end{aligned}$$

But, as always,

$$\sigma^2 = \langle r^2 \rangle - \langle r \rangle^2$$

so

$$\sigma^2 = \langle r \rangle \tag{24.11}$$

the variance for the Poisson distribution.

Poisson Approximation to the Hypergeometric Distribution

The hypergeometric distribution (23.1*a*) is often quite inconvenient to use directly since its computation involves several factorials. Since it depends on *three* parameters, it is also inconvenient to tabulate. The Poisson distribution can be used to *approximate* the hypergeometric

$$\mathbf{h}(k; r, p, n) \approx \mathbf{p}(k; rp) \tag{24.12}$$

where $p = n_1/n$ is small and np is moderately large.

The approximation should be very good for $p \leq 0.1$, $np > 5$.

► **Example 24.6**
Reprise of Example 23.1

Consider Example 23.1 in Chapter 23. There we calculated that

$$\mathbf{h}(1; 2, 0.6, 10) = 0.5333$$

The Poisson approximation in this case is $\mathbf{p}(1; 1.2) = (1.2)^1 e^{-1.2}/1! = 0.3614$. We note that this is not terribly good.

The second part of this example considered the case where $n = 100$:

$$\mathbf{h}(1; 2, 0.6, 100) = 0.5858$$

The Poisson approximation is still not very good but is better than for $n = 10$.

Let us consider an extension of this example in which the bin contains 600 right-handed fittings and 400 left-handed fittings and, again, two are drawn. Here

$$\mathbf{h}(1; 2, 0.6, 1000) = \binom{600}{1}\binom{1000-600}{2-1}\bigg/\binom{1000}{2}$$

$$= \frac{(600)(400)}{(1000)(999)/2} = 0.4804$$

Still, the approximation is not very good since $p = 0.6$. ◄

► **Example 24.7**
Reprise of Example 23.2

Consider Example 23.2 in Chapter 23. Here we have $r = 2$, $p = 0.02$, and $n = 100$. We calculated that $\mathbf{h}(0; 2, 0.02, 100) = 0.9602$.

The Poisson approximation yields

$$\mathbf{p}(0; 0.04) = \frac{(0.04)^0 e^{-0.04}}{0!} = 0.9608$$

This approximation is quite good.

In general, however, approximating small probabilities does not usually yield good (percentage) results. The second part of Example 23.2 yielded $\mathbf{h}(2; 2, 0.02, 100) = 0.0002$. Here the Poisson approximation gives

$$\mathbf{p}(2; 0.04) = \frac{(0.04)^2 e^{-0.04}}{2!} = 0.0008$$

◄

Reproductive Property of the Poisson Distribution

If

$$\Pr(X = x) = \frac{\langle x\rangle^x e^{-\langle x\rangle}}{(x!)}$$

and

$$\Pr(Y = y) = \frac{\langle y\rangle^y e^{-\langle y\rangle}}{(y!)}$$

and X, Y are independent then $Z = X + Y$ is Poisson-distributed with

$$\langle z\rangle = \langle x\rangle + \langle y\rangle$$

that is,

$$\Pr(Z = z) = \frac{\langle z\rangle^z e^{-\langle z\rangle}}{(z!)}$$

where

$$\langle z\rangle = \langle x\rangle + \langle y\rangle.$$

In general, if $x_1, x_2, \ldots, x_n$ are each mutually independent Poisson-distributed random variables with parameter $\langle x_i\rangle$, then

$$X = \sum_i X_i$$

is Poisson-distributed with parameter

$$\langle x\rangle = \sum_i \langle x_i\rangle$$

and

$$\Pr(X = x) = \mathbf{p}(x; \langle x\rangle)$$

where

$$\langle x\rangle = \sum_i \langle x_i\rangle.$$

Radioactive Decay and the Exponential Decay Distribution

A classical application of the Poisson distribution is the decay of radioactive atoms. In this case we have a large number of atoms each of which has the same probability, λ, of decaying in unit time. The probability of $x = 0, 1, 2$, etc. decays in a given time interval ΔT is given by (24.6) where

$$\Pr(X = x) = \frac{(\lambda \Delta T)^x e^{-\lambda \Delta T}}{x!}$$

The decay constant, λ, is the probability that any particular atom will decay in unit time. If there are N radioactive atoms at any given time, we may define the *activity* to be the total number of disintegrations per unit time, which is λN. The activity is also the rate at which the atoms are depleted so

$$\frac{dN}{dt} = -\lambda N \qquad (24.13)$$

where the negative sign expresses depletion.

We should emphasize certain nontrivial assumptions.

1. The probability of decay is independent of age, that is, $\lambda \neq \lambda(t)$. This is equivalent to our assumption of *independent* trials. Note that this is contrary to all experience in **biological** systems.
2. λ is the same for all atoms of the species.

These assumptions enable us to integrate (24.13) immediately:

$$N = N_0 e^{-\lambda t} \qquad (24.14)$$

where $N = N_0$ at time $t = 0$.

If $N_0 = N$ at any time t_0 at which we start counting:

$$N = N_0 e^{-\lambda(t - t_0)} \qquad (24.14a)$$

If we take ΔT to be a small time interval such that

$$\Delta T \ll \frac{1}{\lambda}$$

then $(\lambda \Delta T)$ is the probability that a particular atom will decay in ΔT.

The probability of survival for $\Delta T = (1 - \lambda \Delta T)$

The probability of survival for $2\Delta T = (1 - \lambda \Delta T)^2$

The probability of survival for $t = n\Delta T$ is

$$(1 - \lambda \Delta T)^n = (1 - \lambda \Delta T)^{t/\Delta T}$$

As we take ΔT smaller

$$\lim_{\Delta T/t \to 0} (1 - \lambda \Delta T)^{t/\Delta T} = e^{-\lambda t}$$

which shows that $e^{-\lambda t}$ is the probability of *survival* for each atom.

The Binomial Distribution of Poisson (BDOP)

It happens that we may *relax* the requirement that p is the same for all atoms (trials). In 1837 Poisson considered the problem of n Bernoulli trials (two possible outcomes) but with the probability *varying* at each trial. In this case let p_i be the probability that the ith atom decays in a particular time interval Δt at t, (or that the ith trial is a success). **As usual we take $p_i \ll 1$.** Although, previously, we took $\langle r \rangle = np$, here

$$\langle r \rangle = \sum_{i=1}^{n} p_i \qquad \text{where} \qquad n \gg 1 \quad (24.15)$$

Again, we ask for the probability P_r, that r decays are noted in Δt at t (or that r successes are noted).

The probability of no decays is just

$$P_0 = \prod_{i=1}^{n} (1 - p_i) \approx \prod_{i=1}^{n} e^{-p_i} \qquad (24.16)$$

since $p_i \ll 1$ for each p_i.

$$\therefore P_0 = e^{-\sum_i p_i} = e^{-\langle r \rangle}$$

The probability of the ith atom *and only* the ith atom decaying is the **product** of the probability that the ith atom decays and the probability that all the others do not.

$$\text{Prob (only } i\text{th)} = \frac{p_i \prod_{j=1}^{n} (1 - p_j)}{(1 - p_i)}$$

The probability of one and only one decay is:

$$P_1 = \sum_{i=1}^{n} \frac{p_i}{(1-p_i)} \prod_{j=1}^{n} (1-p_j)$$
$$= \sum_{i=1}^{n} \frac{p_i}{(1-p_i)} P_0 \qquad \text{by (24.16)}$$
$$P_1 \approx \left\{ \sum_{i=1}^{n} \frac{p_i}{1-p_i} \right\} e^{-\langle r \rangle}$$
$$\approx \left[\sum_{i=1}^{n} p_i \right] e^{-\langle r \rangle} = \langle r \rangle \, e^{-\langle r \rangle}$$

We calculate the probability of getting r decays (or r successes) by a similar method:

$$P_r = \frac{1}{r!} \sum_{\alpha_1=1}^{n} \cdots \sum_{\alpha_r=1}^{n} \left[\left(\frac{p_{\alpha_1}}{1-p_{\alpha_1}} \right) \times \left(\frac{p_{\alpha_2}}{1-p_{\alpha_2}} \right) \cdots \left(\frac{p_{\alpha_r}}{1-p_{\alpha_r}} \right) \right]$$

where we have divided by $r!$ since the **order doesn't matter**. Setting $(1-p_{\alpha_i}) \approx 1$, we use (24.10) and get

$$P_r \approx \frac{e^{-\langle r \rangle}}{r!} \left(\sum_{\alpha_1=1}^{n} p_{\alpha_1} \right) \left(\sum_{\alpha_2=1}^{n} p_{\alpha_2} \right) \cdots \left(\sum_{\alpha_r=1}^{n} p_{\alpha_r} \right)$$
$$= \frac{e^{-\langle r \rangle} \langle r \rangle^r}{r!}$$

This is just (24.6).

In general, the **binomial distribution of Poisson** (BDOP) describes trials that can have only *two* outcomes (e.g. success and failure), but where the probability of success can *vary* from trial to trial. We may compare the binomial distribution of Poisson with the Bernoulli (or simple binomial) distribution with the same mean probability. That is, there are n trials with constant probability, p, for the Bernoulli case, whereas there are n trials with probability p_i, $i = 1, \ldots, n$ for the binomial distribution of Poisson. For this comparison the relation holds that

$$np = \sum_{i=1}^{n} p_i \qquad (24.17)$$

The variance of the *Bernoulli* distribution is

$$npq = np(1-p)$$

as usual, whereas the variance for the **binomial distribution of Poisson** is

$$\sum_{i=1}^{n} p_i q_i = \sum_{i=1}^{n} p_i(1-p_i) \qquad (24.18)$$

DISTINGUISHING BDOP FROM BERNOULLI

We are sometimes faced with a dichotomous situation in which we wish to check that the probability is, indeed, constant from trial to trial, that is, we wish to determine if we are dealing with a situation for which the simple binomial (*Bernoulli*) distribution is a good model or one that requires the **binomial distribution of Poisson.** We may test this by noting that the variance for the binomial distribution of Poisson is *always less* than that of the Bernoulli distribution of relation (24.17).

We may define the "**variance of probability**" in the n trials to be

$$\sigma_p^2 = \frac{1}{n} \sum_{i=1}^{n} (p_i - p)^2$$
$$= \sum_{i=1}^{n} \frac{p_i^2}{n} - \frac{2p}{n} \sum_{i=1}^{n} p_i + \frac{np^2}{n}$$
$$= (1/n) \sum_{i=1}^{n} p_i^2 - p^2 \qquad (24.19)$$

This quantity can never be negative although it can be zero if p_i = constant for all i.

$$\text{Variance (BDOP)} = \sum_{i=1}^{n} p_i q_i$$
$$= \sum_{i=1}^{n} p_i(1-p_i)$$
$$= np - \sum_{i=1}^{n} p_i^2$$
$$= np - np^2 - n\sigma_p^2$$
$$= np(1-p) - \sum_{i=1}^{n} (p_i - p)^2$$
$$= npq - n\sigma_p^2$$
$$= \text{Variance (BERN)} - n\sigma_p^2 \qquad (24.20)$$

The observed variance can be checked against the (presumed) variance, npq, where p is the mean probability.

Interval Distribution

Suppose events are occurring at random with a mean rate that is small and that has the constant value R per unit time. We wish to describe the distribution in size of the time intervals between successive events. The probability that there will be no events during time t, that is, none in $(0, t)$ is

$$P_0 = \frac{e^{-Rt}(Rt)^0}{0!} = e^{-Rt}$$

The probability that there will be one event in the time interval $(t, t+dt)$ is $R\,dt$. We require the combined probability that there will be no events during t and one event between t and $t + dt$. This is

$$P(t)\,dt = Re^{-Rt}\,dt$$

The p.d.f. is

$$P(t) = Re^{-Rt} \qquad (24.21)$$

which is sometimes called the "interval distribution." This means that small time intervals are more probable than long ones.

Inverse Probability

We have thus far considered the problem from the point of view of probability, that is, we start with the known distribution and calculate the expectation value and variance, (24.10) and (24.11), which happen to be the same, $\langle r\rangle$. We may examine the **inverse**, or *statistical*, problem, which is more interesting from the viewpoint of the experimental scientist. We may state the inverse problem as follows.

DISTRIBUTION OF THE ESTIMATOR

If we observe a certain number of successes, x, what is the probability that the Poisson parameter, $\langle r\rangle$, lies in a certain interval, say, u to $u + du$? Note that we have assumed that the Poisson probability model is applicable:

$$P_x(u)\,du = \frac{u^x e^{-u}}{x!}\,du \qquad (24.22)$$

Here x is a fixed integer number and (24.22) is a *continuous* probability density function *in u*. Note that $\langle r\rangle$ is *not* truly a random variable, since it is a fixed number that just happens to be unknown to us. However, our *estimate* of $\langle r\rangle$ is a random variable, and we should regard U as our estimate of $\langle r\rangle$, so that

$$f_U(u; x) = \frac{u^x e^{-u}}{x!} \qquad (24.23)$$

is the p.d.f. for U given the observation x.

We must check that $f_U(u; x)$ is properly normalized:

$$\begin{aligned}\int_0^\infty f_U(u; x)\,du &= \int_0^\infty \frac{u^x e^{-u}}{x!}\,du \\ &= \frac{1}{x!}\int_0^\infty u^x e^{-u}\,du \\ &= \frac{x!}{x!} \qquad \text{by (A.III.5)} \\ &= 1 \qquad (24.24)\end{aligned}$$

MEAN AND MODE OF THE ESTIMATOR

We may calculate the "*best estimator*" of $\langle r\rangle$, which is $E(U) = \langle u\rangle$:

$$\langle u\rangle = \int_0^\infty u f_U(u; x)\,du = \frac{1}{x!}\int_0^\infty u^{x+1} e^{-u}\,du$$

$$= \frac{(x+1)!}{x!} \qquad \text{by (A.III.5)}$$

$$\langle u\rangle = x + 1 \qquad (24.25)$$

This is, perhaps, a surprising result. **The best estimate for $\langle r\rangle$ is *not* the observed number of counts, x, but $x + 1$.**

This is because the distribution (24.23) is *skewed*, not symmetrical. Therefore, the expectation value is *not* the same as the *most probable* value.

We may calculate the modal (or, most probable) value of u:

$$0 = \frac{d}{du} f_U(u; x) = \frac{-e^{-u}u^x}{x!} + x\frac{u^{x-1}e^{-u}}{x!}$$

or

$$\frac{u^x}{x!}(x - u) = 0$$

Since $u \neq 0$,

$$u_{\text{mode}} = x \tag{24.26}$$

The most probable value for u is x, the observed number of successes, but is not equal to the expectation value.

VARIANCE OF THE ESTIMATOR

We also need the variance of the distribution defined in (24.23).

$$\langle u^2 \rangle = \int_0^\infty u^2 f_U(u; x)\, du = \frac{1}{x!}\int_0^\infty u^{x+2}e^{-u}\, du$$

$$= \frac{(x+2)!}{x!} \quad \text{by (A.III.5)}$$

$$= (x+2)(x+1) \tag{24.27}$$

$$\text{Var}(U) = \sigma^2(u) = \langle u^2 \rangle - \langle u \rangle^2$$

$$= (x+2)(x+1) - (x+1)^2$$

by (24.25) and (24.27).

This yields

$$\sigma^2 = (x+1) \tag{24.28}$$

The case of x large

The value of (24.23) at the maximum (24.26) is

$$f_U(u = x) = \frac{e^{-u}u^u}{u!}$$

If $u = x \gg 1$, we may use Stirling's approximation (18.9):

$$u! \approx (2\pi u)^{1/2} u^u e^{-u}$$

and get

$$f_U(u = x) \approx \frac{e^{-u}u^u}{(2\pi u)^{1/2}u^u e^{-u}} = \frac{1}{(2\pi u)^{1/2}}$$

This is the maximum value of the p.d.f. (24.23):

$$f_U(u = x) \approx \frac{1}{(2\pi\sigma^2)^{1/2}} \quad \text{for } u = x \text{ large} \tag{24.29}$$

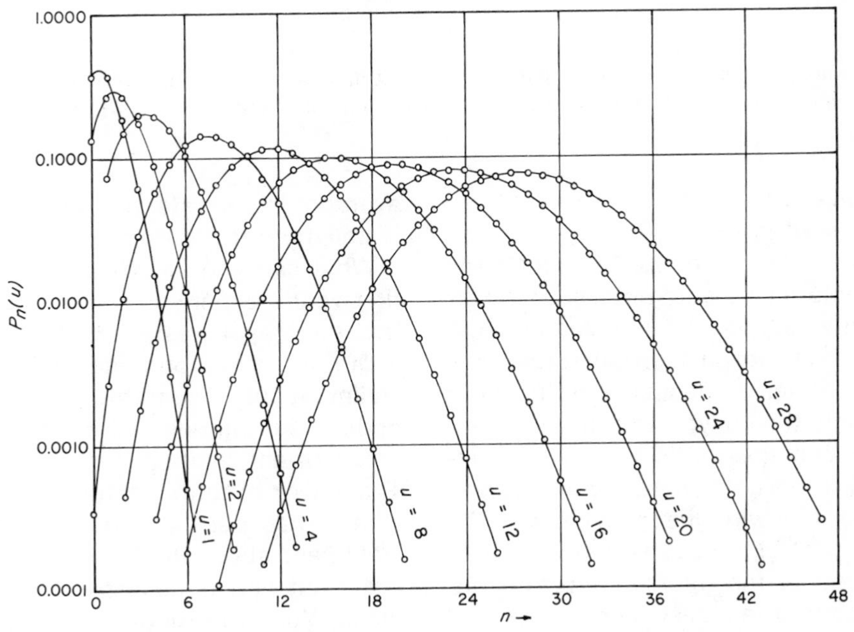

Figure 24.3
Discrete frequency distribution $P_n(u)$ plotted versus n. If u is the mean expected result, then $P_n(u)$ is the probability that exactly n counts will be obtained where n is an integer.

We emphasize that we have considered ***two*** distributions, (24.6) and (24.23). The first is a *discrete* distribution plotted as $P_n(u)$ against n in Figure 24.3: **if u is the mean expected result, then $P_n(u)$ is the probability that exactly n counts will be obtained where n is an integer number.** The second is a *continuous* distribution plotted as $P_n(u)$ **against u** in Figure 24.4: **if n counts are actually obtained, then $P_n(u)\,du$ is the probability that the true mean is between u and $(u + du)$.**

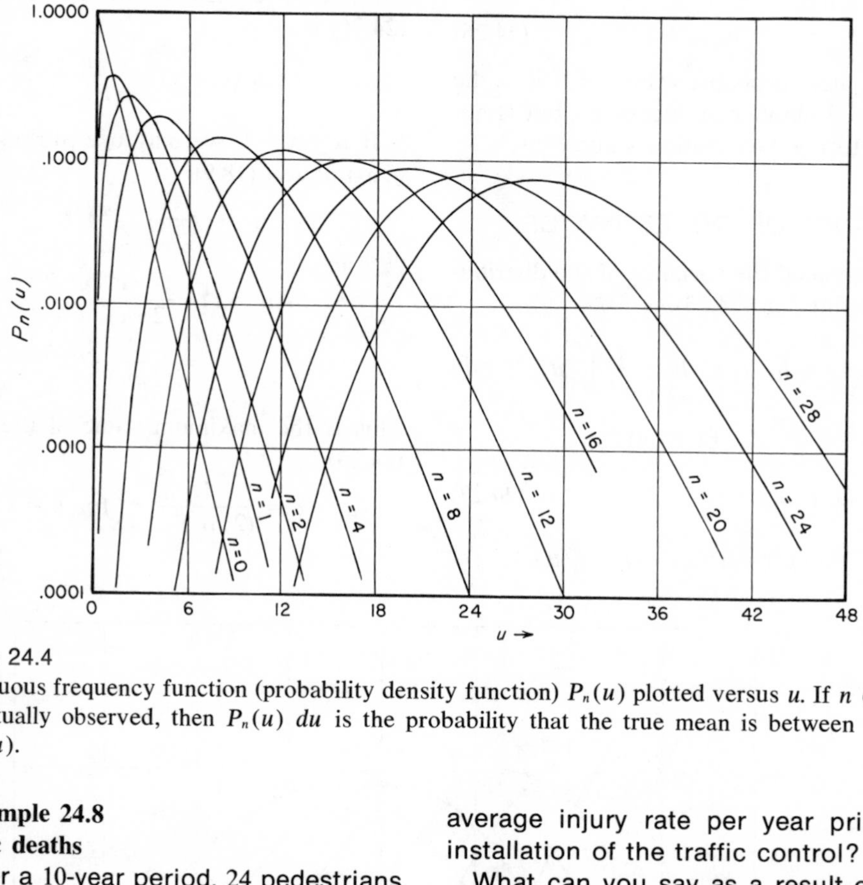

Figure 24.4
Continuous frequency function (probability density function) $P_n(u)$ plotted versus u. If n counts are actually observed, then $P_n(u)\ du$ is the probability that the true mean is between u and $(u + du)$.

► **Example 24.8**
Traffic deaths

Over a 10-year period, 24 pedestrians were injured by automobiles at a certain intersection in town. A campaign by a local citizen's group forced the town council to install a traffic light although the mayor and his staff insisted that the traffic light was completely irrelevant to the problem. During the year following the installation of the traffic control, no pedestrian accidents occurred at the intersection. You are brought in as a statistical consultant to mediate the dispute between the mayor and his opponents.

What is your best estimate as to the average injury rate per year prior to installation of the traffic control?

What can you say as a result of the first year's experience? Can it be said that the mayor was wrong?

During the second year after the installation of the traffic control two pedestrian injuries were noted as a result of accidents at the intersection. The mayor announced that the rate was what it had been and that his original statement about the traffic light being irrelevant and unnecessary was vindicated. You are brought back as a consultant. Comment on the mayor's remarks. Supposing that the mayor is right, how probable is it that the observations are the result of random chance? Assuming that the situation is new as of the installation of the traffic control, what is your best present estimate for the expected number of accidents per year based on the two-year data?

Suppose that, 10 years later, it is noted that a total of 12 injuries have occurred since the installation of the traffic control. Should the mayor be criticized in retirement?

The Poisson distribution is applicable to this problem since the number of pedestrians, n, is presumably large while the probability, p, of a single pedestrian being injured is small.

Using (24.25) we estimate $24 + 1 = 25$ injuries per 10-year period (or 2.5 per year). This estimate has a standard deviation of 5.0 by (24.28).

We calculate the probability of getting 0 if the expected value is 2.5:

$$\mathbf{p}(0; 2.5) = \frac{e^{-2.5}}{0!}(2.5)^0 = 0.082$$

We usually reject at the level of 0.05 and so we must say that the mayor has not been disproved by the results of the first year. We also note from Figure 24.4 that an observation of 0 still implies a substantial probability (0.05) that the true number is 3.

After two years, two injuries have occurred and we test the mayor's hypothesis of no change by calculating

$$\mathbf{p}(2; 5) = \frac{e^{-5}(5)^2}{2!} = 0.084$$

Figure 24.4 also indicates a substantial probability for 5 based on 2 events observed. The mayor's reputation is still safe.

After two years, however, your estimate of the average rate of injury after installation should be $2 + 1 = 3$ by (24.25) or an average yearly rate of 1.5.

After a 10-year span, however, the observation of 12 injuries whereas the mayor would have expected 25 is another matter.

$$\mathbf{p}(12; 25) = \frac{e^{-25}(25)^{12}}{12!} = 0.0017$$

This is too small a probability to accept the hypothesis that the average rate is 2.5 per year.

The new estimate is 1.3 injuries per year with $\sigma = 0.36$.

The lesson is that it is very hard to draw conclusions about rare events.

However, the mayor's statement was that *nothing would change*, not necessarily that the rate would be 2.5 per year.

We shall see in Chapter 25 that a measurement of 2.5 with a $\sigma = 0.5$ is not very inconsistent with a measurement of 1.3 with a $\sigma = 0.36$. ◄

Cumulative Poisson-Distribution Function

It is frequently useful to consider the cumulative distribution function for the Poisson distribution.

$$\mathbf{P}(r; \mu) \equiv \Pr(R \leq r) = \sum_{x=0}^{x=r} p(x; \mu) \qquad (24.30)$$

This is sometimes tabulated in the form of the partial sum

$$\sum_{x=r}^{x=\infty} \mathbf{p}(x; \mu) = 1 - \mathbf{P}(r; \mu) = \sum_{x=r}^{\infty} \frac{e^{-\mu}\mu^x}{x!} \qquad (24.31)$$

Tables of these partial sums may be found in *Poisson's Exponential Binomial*

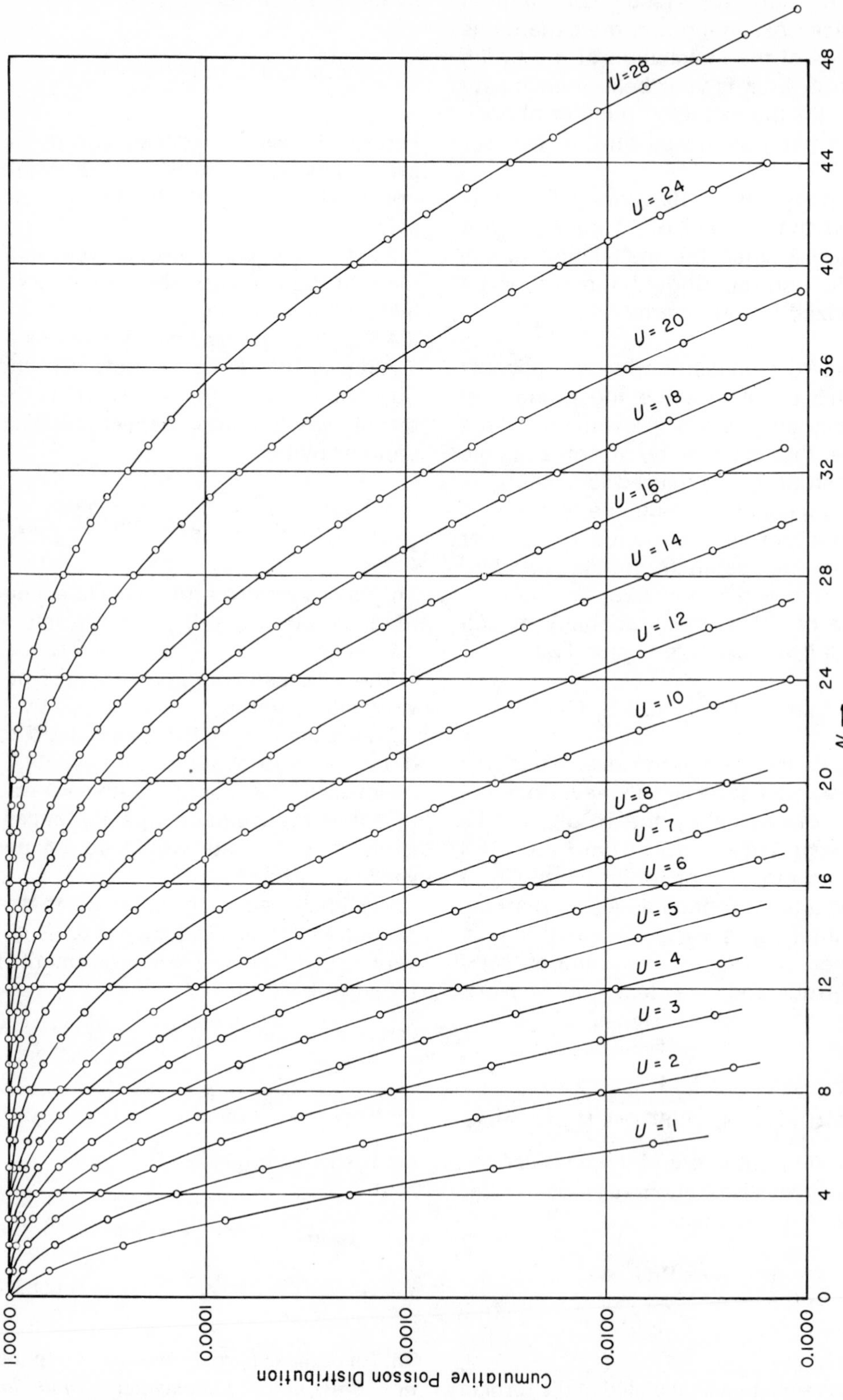

Figure 24.5
Plot of the (discrete) cumulative Poisson-distribution function $\sum_{x=N}^{\infty} \frac{e^{-U} U^x}{x!}$ versus N.

*Limit.** We plot some of these partial sums

$$\sum_{x=N}^{\infty} \frac{e^{-U}U^x}{x!} \qquad \text{versus} \qquad N$$

in Figure 24.5.

► **Example 24.9**
Telephone installations

A company has 400 employees each of whom spends an average of 4% of his time on his private telephone. Assume that these calls are randomly spread throughout the working day (i.e., ignore the effects of lunch hours, coffee breaks, overlap with other time zones, etc.). It is proposed to move these people to a new building and to economize by removing the 400 private telephone lines and to replace them with a smaller number of lines with electronic switching that would automatically make available a free line when a telephone is picked up. A person would notice the change only if he wanted to make a call when *all* available lines were in use.

It has been ascertained that people are not unduly annoyed if they can get a line when they want it more than 96% of the time.

How many lines should be installed to fulfill the 96% requirement?

The average number of calls in progress at any given time under the old system is $m = np = (400)(0.04) = 16$.

If X represents the number of calls potentially in progress at any given time and x_0 is the number of lines installed, we want to find x_0 such that the following relation is satisfied:

$$\Pr(X > x_0) \leq 0.04$$

That is, we wish to determine x_0 where $m = 16$ in the relation

$$\sum_{x=x_0}^{400} \frac{e^{-m}m^x}{x!} \leq 0.04$$

We may take the difference between

$$\sum_{x=x_0}^{\infty} \quad \text{and} \quad \sum_{x=x_0}^{400}$$

to be negligible and note from Figure 24.5 that

$$1 - P(N; U) = 0.0582$$

for

$$U = 16 \text{ and } N = 23$$

and

$$= 0.0367$$

for $U = 16$ and $N = 24$.

We see, therefore, that 24 lines will suffice to meet our requirement. ◄

We shall see in Chapter 25 that it is frequently possible to *approximate* the cumulative Poisson distribution by the normal distribution (e.g. see Example 25.6).

► **Example 24.10**
A cosmic ray measurement

Cosmic ray particles pass through a detector "at random," but with an average rate of 5 every 6 hours. The number of particles is recorded in a "binary scaler" that has n stages and can record up to 2^n particles after which it resets to zero and starts over again.

(a) What is the probability of observing no cosmic rays in a given 1-hour period?
(b) What is the probability of observing no cosmic rays in a 2-hour period?
(c) An experiment is to be set up in which the scaler is to be read every 24 hours. For reasons of economy it is desired to minimize the number of stages in each scaler. How many stages should be supplied if the scaler is to be found to have "overflowed" less than 3% of the time?

(a) $$\mathbf{p}\left(0; \frac{5}{6}\right) = \mathbf{p}(0; 0.833) = \frac{e^{-0.833}(0.833)^0}{0!}$$
$$= 0.435.$$

*E. C. Molina, *Poisson's Exponential Binomial Limit*, Van Nostrand, 1942.

(b) $\mathbf{p}\left(0; \frac{10}{6}\right) = \mathbf{p}(0; 1.666) = \frac{e^{-1.666}(1.666)^0}{0!}$

$= 0.189.$

(c) The expected number of cosmic rays is 20. We may use Figure 24.5 and draw a line corresponding to a probability of 0.03. We note that this line intersects the $U = 20$ curve above $N = 30$. That is, for an expected rate of 20 we expect fluctuation ≥ 30 less than 3% of the time. For ≥ 32 the probability is 0.008. Therefore, 5 stages will suffice since $2^5 = 32$. ◄

25

THE GAUSSIAN OR NORMAL DISTRIBUTION

We now turn to a continuous probability distribution that has perhaps the most universal applicability. Either as the exact probability model or as a good approximation to the exact probability model, the normal distribution provides a good description for many situations.

We shall first present a heuristic derivation of the normal distribution starting from certain assumptions. Because these assumptions are reasonable for many situations, it follows that the normal distribution is a reasonable probability model for those situations. We shall see also that the normal distribution may be obtained as a limiting form of distributions that we have previously considered.

It happens that the mathematics of the normal distribution is especially convenient. Furthermore, procedures derived on the assumption of normality exhibit "*robustness*", that is, their applicability is *not very sensitive to deviations from normality.* We shall describe some of the important properties of the normal distribution and introduce certain useful tables and graphs. We shall then discuss various situations in which it is appropriate to employ the normal distribution and consider some applications of its usage.

Derivation of the Gaussian Distribution from Certain Assumptions

We shall make several assumptions and observe the distribution that emerges as a consequence of these assumptions.

We should keep before us as a guide our intuitive ideas on the distribution of errors when we make a series of N measurements of a quantity. Note carefully that our intuition leads us to make several (reasonable) *assumptions.* Once the assumptions are stated, however, what follows is a consequence of mathematics that is quite rigorous.

Consider a set of observations x'_i that have an average value $\bar{x}$ and consider the p.d.f. for $Z = x'_i - \bar{x}$, $f_Z(z)$, where Z is a random variable. We do not yet know the form of $f_Z(z)$, but we may say

$$\Pr\left(z - \frac{\Delta z}{2} < Z < z + \frac{\Delta z}{2}\right) = f_Z(z)\Delta z$$

where Δz is a small interval centered at z. We shall take the interval Δz so small that not more than one value occurs in the interval and we shall pass, in this limit, between the discrete and continuous descriptions. Thus, the probability that we have the value z_i is $f_Z(z_i)\Delta z = p(z_i)$. The *likelihood function* defined as in (19.23) and (19.25) to be the probability that N values have the distribution $z_1, z_2, \ldots, z_N$ is

$$\mathscr{L} = \prod_{i=1}^{N} f_Z(z_i)\Delta z = (\Delta z)^N \prod_{i=1}^{N} f_Z(z_i) \tag{25.1}$$

From this we obtain

$$\ln \mathscr{L} = \sum_{i=1}^{N} \ln f_Z(z_i) + N \ln \Delta z \tag{25.2}$$

We assume that the distribution is *symmetrical*, which tells us that the most probable value is identical with the average or expectation value. (This is consistent with our guiding model since positive deviations from the mean are *a priori* as likely as negative deviations.) When $\mathscr{L}$ is at a maximum, so is $\ln \mathscr{L}$, and we may find the maximum of either.

$$\frac{d(\ln \mathscr{L})}{d\bar{x}} = 0 \tag{25.3}$$

which yields

$$\sum_{i=1}^{N} \frac{\partial(\ln \mathscr{L})}{\partial z_i} \frac{dz_i}{d\bar{x}} = 0 \tag{25.4}$$

where the last term in (25.2) is constant. Since $Z = x'_i - \bar{x}$,

$$\frac{dz_i}{d\bar{x}} = -1$$

We rewrite (25.4),

$$\sum_{i=1}^{N} \left\{ \frac{1}{f_Z(z_i)} \frac{df_Z(z_i)}{dz_i} \right\} = \sum_{i=1}^{N} \phi(z_i) = 0 \tag{25.5}$$

where

$$\phi(z_i) = \frac{1}{f_Z(z_i)} \frac{df_Z(z_i)}{dz_i} \tag{25.6}$$

To find the form of $\phi(z_i)$, we may expand $\phi(z_i)$ in a power series in z_i:

$$\phi(z_i) = \mathscr{A}_0 + \mathscr{A}_1 z_i + \mathscr{A}_2 z_i^2 + \cdots + \mathscr{A}_n z_i^n \tag{25.7}$$

where

$$\sum_{i=1}^{N} \phi(z_i) = 0 = N\mathscr{A}_0 + \mathscr{A}_1 \sum_{i=1}^{N} z_i + \mathscr{A}_2 \sum_{i=1}^{N} z_i^2 + \cdots \tag{25.8}$$

To *insure* that the sum in (25.8) is zero, it is *sufficient* to set each term equal to zero separately. To insure this we may set each $\mathscr{A}_i = 0$ except for $\mathscr{A}_1$. $\mathscr{A}_1$ need not be zero because

$$\sum_{i=1}^{N} z_i = \sum_{i=1}^{N} (x'_i - \bar{x}) = 0 \tag{25.9}$$

This requirement yields the form

$$\phi(z_i) = \mathscr{A}_1 z_i = \frac{1}{f_Z(z_i)} \frac{df_Z(z_i)}{dz_i} \tag{25.10}$$

and

$$\ln f_Z(z) = \tfrac{1}{2}\mathscr{A}_1 z^2 + \ln K$$

or, finally,

$$f_Z(z) = K e^{\mathscr{A}_1 z^2/2} \tag{25.11}$$

where K is, so far, an arbitrary constant. We must make more assumptions in order to proceed further. We must *assume that a maximum exists.* Our assumption of symmetry then informs us that the maximum occurs at $z = 0$.

Furthermore, we assume that $f(z) \to 0$ as $z \to \pm\infty$.

These assumptions lead us to conclude that

$$\mathscr{A}_1 < 0$$

which we insure by setting

$$\mathscr{A}_1 = -2h^2 < 0 \tag{25.12}$$

We specify that the p.d.f. in (25.11) be normalized to fix the constant, K:

$$K \int_{-\infty}^{\infty} e^{-h^2 z^2} dz = 1 \tag{25.13}$$

By the change of variable $u = hz\sqrt{2}$, we may rewrite this to

$$\frac{K}{h\sqrt{2}} \left\{ \int_{-\infty}^{\infty} e^{-u^2/2} du \right\} = 1 = \frac{K}{h\sqrt{2}} \{\sqrt{2\pi}\}$$

where we have evaluated the bracketed integral using (A.III.9) of Appendix III.

This yields $K = h/\sqrt{\pi}$ and we have

$$f(z) = \frac{h}{\sqrt{\pi}} e^{-h^2 z^2} \tag{25.14}$$

This is one form of the **Gaussian or normal p.d.f.**. We note that the mean of this distribution is at $z = 0$ and that another parameter, h, called the *modulus of precision*, specifies it further. The Gaussian is therefore, **specified by two parameters** that we may exhibit explicitly.

VARIANCE

Consider the variance, σ^2, of this distribution:

$$\sigma^2 = \frac{h}{\sqrt{\pi}} \int_{-\infty}^{\infty} z^2 e^{-h^2 z^2} \, dz \quad (25.15)$$

and we make the substitution, $u = h^2 z^2$, which yields

$$\sigma^2 = \frac{2h}{\sqrt{\pi}} \int_0^{\infty} \frac{du}{2h^2 \sqrt{\frac{u}{h^2}}} e^{-u} \left(\frac{u}{h^2}\right)$$
$$= \frac{1}{h^2 \sqrt{\pi}} \int_0^{\infty} u^{1/2} e^{-u} \, du$$

The integral is defined in (A.III.18b) to be the gamma function, $\Gamma(3/2)$, which is evaluated in (A.III.23) to be $\frac{1}{2}\sqrt{\pi}$.

Therefore,

$$\sigma^2 = \frac{1}{h^2 \sqrt{\pi}} \frac{\sqrt{\pi}}{2} = \frac{1}{2h^2} \quad (25.16)$$

THE GAUSSIAN PROBABILITY DENSITY FUNCTION

It is frequently convenient to parametrize the Gaussian by its mean value, m, and its variance, σ^2.

Gaussian p.d.f.:

$$f(x) = \frac{1}{\sqrt{2\pi\sigma^2}} e^{-(x-m)^2/2\sigma^2} \quad (25.17)$$

Equation (25.17) is the ubiquitous p.d.f. that we first alluded to in Chapter 7 and that is plotted as a curve superposed on histograms in Figures 7.2 to 7.4, where the total area is normalized to NA, the number of measurements multiplied by the width of a histogram interval, and the mean and standard deviation are calculated from (7.1) and (7.5), respectively.

The Gaussian p.d.f. is also called the normal p.d.f. and is sometimes written

$$\mathbf{n}(x; m, \sigma) = \frac{1}{\sqrt{2\pi\sigma^2}} e^{-(x-m)^2/2\sigma^2} \quad (25.17a)$$

Figure 25.1 shows $\mathbf{n}(x; m, \sigma)$ plotted against x for $m = 0$ and 7 and $\sigma = 1, 2, 3$, and 10.

When we take $m = 0$ and $\sigma = 1$, we get the standardized normal p.d.f.,

$$\mathbf{n}(x; 0, 1)$$

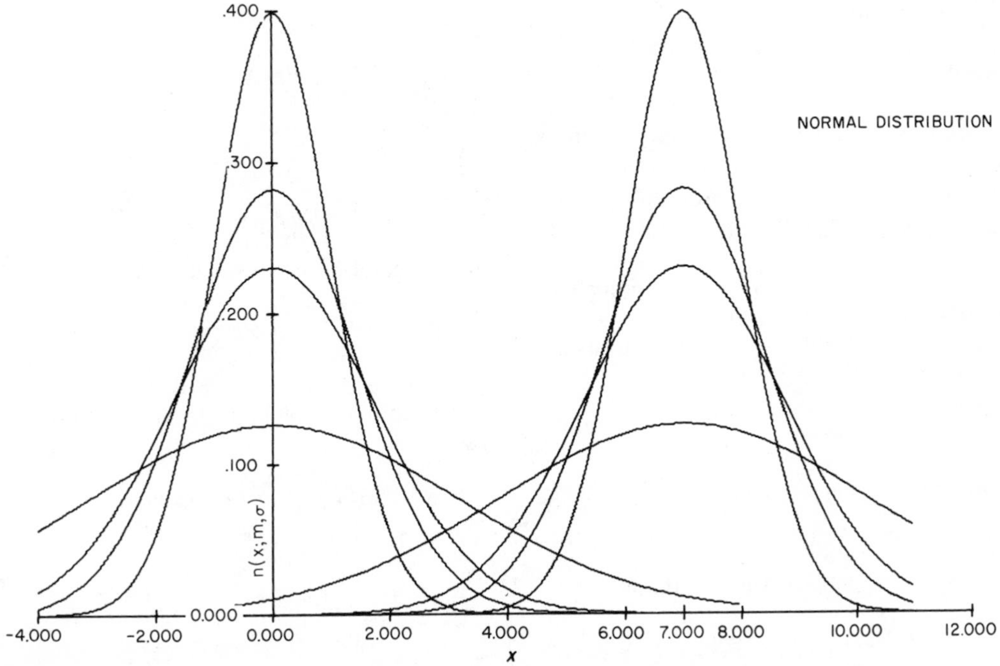

Figure 25.1
Plot of $\mathbf{n}(x;m,\sigma)$ versus x for $m = 0$ and 7 and for $\sigma = 1,2,3$, and 10.

which is often written

$$\varphi(x) = \frac{1}{\sqrt{2\pi}} e^{-x^2/2} \qquad (25.17b)$$

We note that this form has *no* parameters and is hence conveniently tabulated or plotted as a function of x.

We remark that

$$\mathbf{n}(x; m, \sigma) = \frac{1}{\sigma}\varphi\left(\frac{x-m}{\sigma}\right) \qquad (25.17c)$$

Relation of the Mean Deviation to the Standard Deviation

We are now in a position to verify relation (7.6) that related the mean deviation, a, to the standard deviation, σ.

$$a = \langle |z-m| \rangle = \frac{h}{\sqrt{\pi}} \int_{-\infty}^{\infty} e^{-h^2z^2} |z-m|\, dz$$

$$= \frac{2h}{\sqrt{\pi}} \int_0^{\infty} z\, e^{-h^2z^2}\, dz \qquad \text{since } m = 0$$

$$= \frac{2h}{\sqrt{\pi}} \left[-\frac{1}{2}\right]\left[\frac{e^{-u^2}}{h^2}\right]_0^{\infty} = \frac{1}{h\sqrt{\pi}}$$

By using (25.16), we get

$$\frac{\sigma}{a} = \frac{1}{h\sqrt{2}} h\sqrt{\pi} = \sqrt{\pi/2} = (1.57080)^{1/2}$$

or

$$\frac{\sigma}{a} = 1.2533 \qquad (25.18)$$

which agrees with (7.6).

Derivation of Gaussian Distribution from the Binomial Distribution

As is often true for other discoveries in science, the Gaussian distribution is not named after the mathematician who first discovered it. Abraham De Moivre* gave the first example of a normal frequency distribution by considering the following problem.

Suppose a fair coin is tossed N times, where N is a very large and even ($N = 2n$) number. What is the probability of exactly $(\frac{1}{2}N - s)$ heads and $(\frac{1}{2}N + s)$ tails!

In Example 18.4 we saw that the probability of obtaining r heads with $2n$ fair coins is $\binom{2n}{r}/2^{2n}$. The probability of exactly n heads and n tails in $N = 2n$ tosses is

$$P(n,n) = \frac{1}{2^{2n}} \frac{(2n)!}{n!n!}$$

Using Stirling's approximation,

$$n! \sim (2\pi n)^{1/2} n^n e^{-n} \qquad (18.9)$$

which was first published in 1730, this becomes

$$P(n,n) \approx \left(\frac{2}{\pi n}\right)^{1/2} = \left(\frac{2}{\pi N}\right)^{1/2}$$

**Doctrine of Chances*, 1756.

The probability of $(n - s)$ heads and $(n + s)$ tails in $2n$ tosses is

$$P(n-s, n+s) = \frac{1}{2^{2n}} \frac{(2n)!}{(n-s)!(n+s)!}$$

Let

$$R = \frac{P(n-s, n+s)}{P(n,n)}$$

$$= \left\{\frac{(2n)!}{2^{2n}(n-s)!(n+s)!}\right\} \Big/ \left\{\frac{(2n)!}{2^{2n}n!n!}\right\}$$

$$= \frac{n!n!}{(n-s)!(n+s)!}$$

Take natural logarithms and obtain

$$\ln R = 2\ln(n!) - \ln[(n-s)!] - \ln[(n+s)!]$$

We replace the logarithmic factorials by Stirling's approximation in the form of (18.14) and get

$$\begin{aligned} \ln R &\approx 2n(\ln n - 1) - (n-s)\{\ln(n-s) - 1\} \\ &\quad - (n+s)\{\ln(n+s) - 1\} \\ &= 2n \ln n - 2n - (n-s) \\ &\quad \times [\ln n + \ln(1 - s/n) - 1] \\ &\quad - (n+s)[\ln n + \ln(1 + s/n) - 1] \\ &= -\{(n-s)[\ln(1 - s/n)] \\ &\quad + (n+s)[\ln(1 + s/n)]\} \end{aligned}$$

We can expand the log term using the series expansion

$$\ln(1+x) = x - \frac{x^2}{2} + \frac{x^3}{3} - \frac{x^4}{4} + \cdots$$

for

$$-1 < x \leq +1$$

This yields

$$\ln R \approx -\left\{(n-s)\left[\frac{-s}{n} - \frac{s^2}{2n^2} + \cdots\right] + (n+s)\left[\frac{s}{n} - \frac{s^2}{2n^2} + \cdots\right]\right\}$$

$$= -\frac{s^2}{n}$$

$$= -\frac{2s^2}{N}$$

$$\therefore R \approx e^{-2s^2/N}$$

or

$$P(n-s, n+s) \approx \left(\frac{2}{\pi N}\right)^{1/2} e^{-2s^2/N}$$

The conclusion of De Moivre's analysis is that the probability of exactly $(\frac{1}{2}N - s)$ heads and $(\frac{1}{2}N + s)$ tails is approximately

$$\left(\frac{2}{\pi N}\right)^{1/2} e^{-2s^2/N}$$

If we observe that the mean number of heads expected is

$$m = N/2$$

then we may cast this result in the form of (25.17) by taking

$$\sigma^2 = \frac{N}{4}$$

Expressing De Moivre's result in continuous form we note that the probability that the number of heads will be between

$$\tfrac{1}{2}N + x\sqrt{N} \quad \text{and} \quad \tfrac{1}{2}N + (x+dx)\sqrt{N}$$

is approximately $(2/\pi)^{1/2} e^{-2x^2}\, dx$.

We shall see later that De Moivre's example is a special case of the normal approximation to the binomial distribution where, in general,

$$m = Np \quad \text{and} \quad \sigma^2 = Npq$$

in (25.17).

By similar use of Stirling's approximation, the general case of a binomially distributed variate X

$$\Pr(X = r) = \mathbf{b}(r; n, p) = \binom{n}{r} p^r (1-p)^{n-r}$$

becomes, for large n

$$\cong [1/(2\pi np(1-p))^{1/2}]\, e^{-(r-np)^2/2np(1-p)} \tag{25.19}$$

Derivation of Gaussian Distribution from the Poisson Distribution

In the relation

$$P_r = \frac{e^{-\langle r\rangle}}{r!} \langle r\rangle^r \tag{24.5a}$$

let $x \equiv r - \langle r\rangle$:

$$P_r = P_{\langle r\rangle + x} = \frac{e^{-\langle r\rangle}\langle r\rangle^{\langle r\rangle + x}}{(x + \langle r\rangle)!}$$

$$= \frac{e^{-\langle r\rangle}\langle r\rangle^{\langle r\rangle}}{\langle r\rangle!}\left\{\frac{\langle r\rangle}{\langle r\rangle+1} \cdot \frac{\langle r\rangle}{\langle r\rangle+2} \cdot \cdots \cdot \frac{\langle r\rangle}{\langle r\rangle+x}\right\}$$

$$= P_{\langle r\rangle}\left\{\frac{\langle r\rangle}{\langle r\rangle+1} \cdots \frac{\langle r\rangle}{\langle r\rangle+x}\right\}$$

But

$$P_{r=\langle r\rangle} = \frac{e^{-\langle r\rangle}\langle r\rangle^{\langle r\rangle}}{\langle r\rangle!} = \frac{1}{(2\pi\langle r\rangle)^{1/2}}$$

if

$$\langle r\rangle \gg 1$$

since we may invoke (18.9) to approximate $\langle r\rangle!$.

We also note

$$\frac{\langle r\rangle}{\langle r\rangle+1} \cdot \frac{\langle r\rangle}{\langle r\rangle+2} \cdots \frac{\langle r\rangle}{\langle r\rangle+x} = \frac{1}{1+1/\langle r\rangle} \cdot \frac{1}{1+2/\langle r\rangle} \cdots \frac{1}{1+x/\langle r\rangle}$$

For $x/\langle r\rangle \ll 1$,

$$1 + \frac{x}{\langle r\rangle} \approx e^{x/\langle r\rangle}$$

$$\therefore P_{\langle r\rangle + x} \approx \frac{1}{(2\pi\langle r\rangle)^{1/2}} \times \exp\left\{-\left[\frac{1}{\langle r\rangle} + \frac{2}{\langle r\rangle} + \cdots + \frac{x}{\langle r\rangle}\right]\right\}$$

$$= \frac{1}{(2\pi\langle r\rangle)^{1/2}} \exp\left(\frac{-1}{\langle r\rangle}\sum_{i=1}^{x} i\right)$$

But

$$\sum_{i=1}^{x} i = \frac{x}{2}(1+x) \approx \frac{x^2}{2} \qquad \text{for } x \gg 1.$$

and we get

$$P_{\langle r\rangle+x} \approx \frac{1}{(2\pi\langle r\rangle)^{1/2}} e^{-x^2/2\langle r\rangle} \qquad (25.20)$$

which gives us the Gaussian distribution especially since $\langle r\rangle = \sigma^2$ by (24.11). We needed $\langle r\rangle \gg x \gg 1$. That is, *in the limit of a large number of observed successes, Poisson → Gaussian.* x is the deviation from the mean number of counts. The distribution is maximum at $x = 0$ and is symmetric about $\langle r\rangle$.

Inverse Probability

We may also consider the limit of the inverse Poisson distribution (24.23):

$$f_U(u; N) = \frac{u^N e^{-u}}{N!}$$

where the fixed number, N, has been observed.

Let $y \equiv u - N$; thus $dy = du$. We take the limit

$$\frac{y}{N} \ll 1 \qquad \text{and} \qquad N \gg 1$$

$$f_U(u; N)\,du = f_Y(y; N)\,dy = \frac{u^N e^{-u}}{N!}\,dy$$

Again using Stirling's approximation (18.9)

$$N! \approx (2\pi N)^{1/2} N^N e^{-N}$$

$$f_Y(y; N)\,dy = \frac{e^{-u} u^N e^N}{(2\pi N)^{1/2} N^N}\,dy$$

$$= \frac{u^N e^{-(u-N)}}{N^N (2\pi N)^{1/2}}\,dy$$

We rewrite this:

$$f_Y(y; N)\,dy = \frac{(y+N)^N e^{-y}}{N^N (2\pi N)^{1/2}}\,dy$$

Note that

$$\frac{(y+N)^N}{N^N} = \left(1 + \frac{y}{N}\right)^N$$

and

$$N \ln\left(1 + \frac{y}{N}\right) = N\left[\frac{y}{N} - \left(\frac{y}{N}\right)^2 \frac{1}{2} + \cdots\right]$$

$$\text{for } \frac{y}{N} < 1$$

$$\therefore \frac{(y+N)^N}{N^N} \approx e^{y - y^2/2N}$$

and

$$f_Y(y; N)\,dy = \frac{e^{y-y^2/2N} e^{-y}}{(2\pi N)^{1/2}}\,dy$$

$$= \frac{e^{-y^2/2N}}{(2\pi N)^{1/2}}\,dy$$

Again we ask the question about inverse probabilities:

If N is the measured number, how far will u be from N where U is the estimate of $\langle r\rangle$? Equivalently, how far will $y = u - N$ be from zero?

$$\sigma^2 = N + 1 \approx N \qquad \text{for } N \gg 1 \text{ by } (24.28)$$

$$f_Y(y; \sigma)\,dy = \frac{e^{-y^2/2\sigma^2}}{\sqrt{2\pi\sigma^2}}\,dy \qquad (25.21)$$

We, therefore, answer the question and say that **the probability that U is in the range $(N - a)$ to $(N + b)$ is given by**

$$\Pr(N - a < U < N + b)$$

$$= \int_{-a}^{b} \frac{1}{\sqrt{2\pi\sigma^2}} e^{-y^2/2\sigma^2}\,dy \qquad (25.21\text{a})$$

The range $(-a$ to $+b)$ is sometimes called the *confidence interval* corresponding to the probability on the right-hand side.

It is left as an exercise to show

$$\langle y^2\rangle = \sigma^2$$

and

$$\langle y\rangle = 0 = \langle y^{2j+1}\rangle \qquad \text{for } j = 0, 1, 2, \ldots$$

NORMALIZED FORM OF THE INVERSE DISTRIBUTION

Note that we can measure y in units of standard deviation.

$$t \equiv \frac{y}{\sigma}$$

which enables us to obtain the important form for the Gaussian,

$$f_T(t)\,dt = \frac{e^{-t^2/2}}{\sqrt{2\pi}}\,dt \equiv \varphi(t)\,dt. \qquad (25.22)$$

We note

$$\langle t^2 \rangle = 1 \quad (25.23)$$

and

$$\langle t \rangle = \langle t^{2j+1} \rangle = 0 \quad (25.24)$$

Recalling (25.17b) we see that *the direct and inverse distributions have the* ***same*** *form.*

Some Properties of the Normal Distribution

Formal definition:
A random variable X is said to be normally distributed with mean m and variance σ^2 if the probability that X is less than or equal to x is

$$\begin{aligned}\Pr\{X \leq x\} &= \frac{1}{\sigma\sqrt{2\pi}} \int_{-\infty}^{x} e^{-(t-m)^2/2\sigma^2}\, dt \\ &= F_x(x) \\ &= \frac{1}{\sqrt{2\pi}} \int_{-\infty}^{(x-m)/\sigma} e^{-t^2/2}\, dt \\ &\equiv \Phi\left(\frac{x-m}{\sigma}\right) \quad (25.25)\end{aligned}$$

This is the *cumulative distribution function.*

If X is normally distributed with mean m and variance σ^2, we write

$$X \text{ is } N(m, \sigma^2) \quad (25.26)$$

From the defined cumulative distribution function we get the p.d.f. (25.17):

$$\begin{aligned}\frac{\partial}{\partial x}\Phi\left(\frac{x-m}{\sigma}\right) &= \frac{1}{\sigma}\varphi\left(\frac{x-m}{\sigma}\right) \\ &= \frac{1}{\sigma\sqrt{2\pi}} e^{-(x-m)^2/2\sigma^2} = f(x) \\ &= \mathbf{n}(x; m, \sigma)\end{aligned}$$

The p.d.f. is symmetric about m

$$\varphi\left(\frac{m+x}{\sigma}\right) = \varphi\left(\frac{m-x}{\sigma}\right) \quad (25.27)$$

Since

$$f'(x) = \frac{-1}{\sqrt{2\pi}\sigma^3} e^{-(x-m)^2/2\sigma^2}[x-m]$$

and

$$f''(x) = \frac{1}{\sqrt{2\pi}\sigma^3} e^{-(x-m)^2/2\sigma^2}\left\{\frac{(x-m)^2}{\sigma^2} - 1\right\}$$

the p.d.f. has one and only one maximum $(f'(x) = 0)$ at

$$x_{\max} = m \quad (25.28)$$

and two inflection points $[f''(x) = 0]$ at

$$x_{\text{inflection}} = m \pm \sigma \quad (25.29)$$

Graphs of (25.25) and (25.17) are shown in Figures 25.2. The function (25.17) is tabulated in Table A.VI.1 and is plotted on log scales in Figures 25.3*a* and *b*.

We recall the definitions (17.3), (17.9), and (17.11) and apply them to the Gaussian **p.d.f.**:

$$\varphi(x) = \frac{1}{\sqrt{2\pi}} e^{-x^2/2}$$

P is the **c.d.f. of X**:

$$P(x) = \int_{-\infty}^{x} \varphi(t)\, dt \equiv \Phi(x) \quad (25.30)$$

$Q = 1 - P$ is the **upper-tail area function**:

$$Q(x) = \int_{x}^{\infty} \varphi(t)\, dt \quad (25.31)$$

$A = P - Q$ is the **c.d.f. of |X|**:

$$A(x) = \int_{-x}^{x} \varphi(t)\, dt \quad (25.32)$$

We note the relations:

$$\begin{aligned}P(x) + Q(x) &= 1 \\ P(-x) = Q(x) &= 1 - P(x) \quad (25.33) \\ A(x) &= 2P(x) - 1\end{aligned}$$

We also define the **error function**

$$\operatorname{erf} x = 2P(x\sqrt{2}) - 1 \equiv \theta(x) \quad (25.34)$$

The functions defined in equations (25.30), (25.31), (25.32), and (25.34) must be evaluated numerically.

For example,

$$\begin{aligned}A(x) &= \frac{1}{\sqrt{2\pi}} \int_{-x}^{x} e^{-t^2/2}\, dt \\ &= \frac{2}{\sqrt{2\pi}} \int_{0}^{x} e^{-t^2/2}\, dt\end{aligned}$$

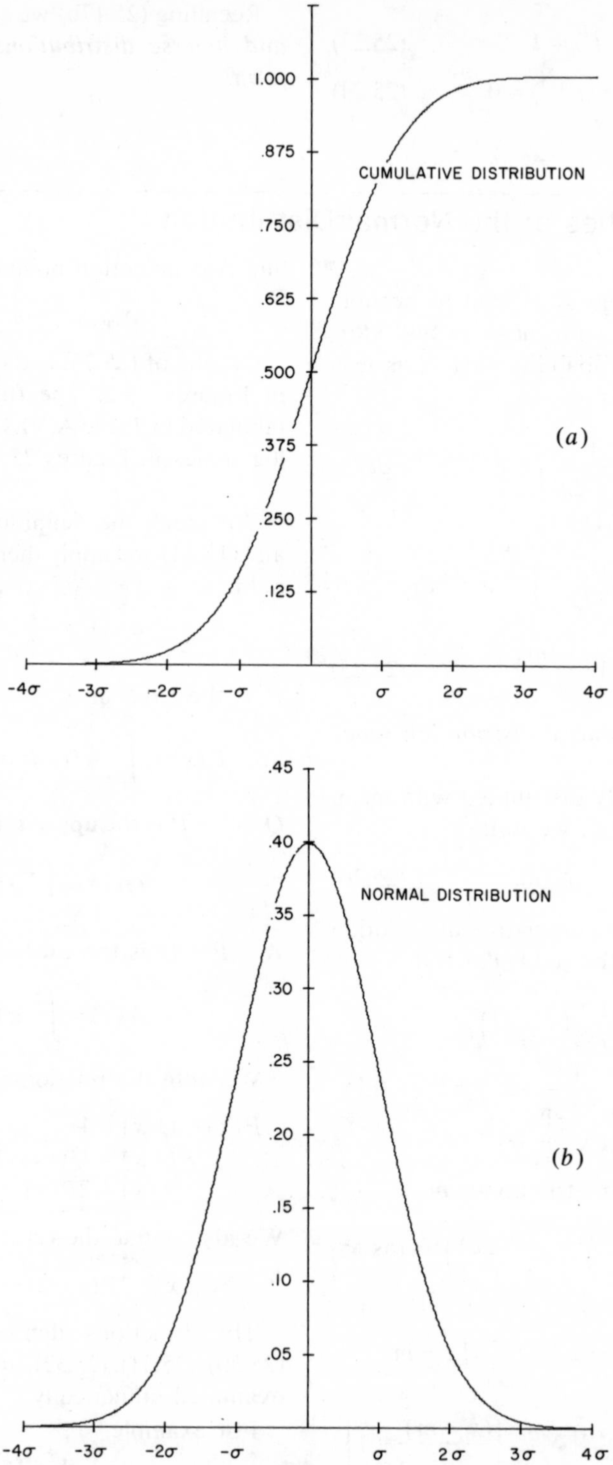

Figure 25.2
Plots of the *cumulative distribution function* (a) and the *probability density function* (b) of the normal distribution. (a) $\Phi(z) = \Phi((x - m)/\sigma)$ versus z. The normal cumulative distribution function. (b) $\varphi(z) = \varphi((x - m)/\sigma)$ versus z. The normal probability density function.

Let $z\sqrt{2} = t$,

$$\therefore A(x) = \frac{2}{\sqrt{\pi}} \int_0^{x/\sqrt{2}} e^{-z^2}\, dz \qquad (25.35)$$

The integrand may be expanded in a series, and the terms integrated individually and summed:

$$A(x) = \frac{2}{\sqrt{\pi}} \times \int_0^{x/\sqrt{2}} \left(1 - z^2 + \frac{z^4}{2!} - \frac{z^6}{3!} + \cdots\right) dz$$

$$A(x) = \frac{2}{\sqrt{\pi}} \sum_{j=0}^{\infty} \frac{(-1)^j (x/\sqrt{2})^{2j+1}}{j!(2j+1)} \qquad (25.36)$$

This may also be written

$$A(x) = \frac{2}{\sqrt{\pi}} e^{-x^2/2} \sum_{j=0}^{\infty} \frac{(x/\sqrt{2})^{2j+1} 2^j}{(2j+1)!!} \qquad (25.37)$$

where

$(2j+1)!! = (2j+1)(2j-1)(2j-3)\cdots(3)(1)$.

(25.37) converges somewhat faster than (25.36). The function $A(z)$, where z is the standardized variable with mean 0 and

(a)

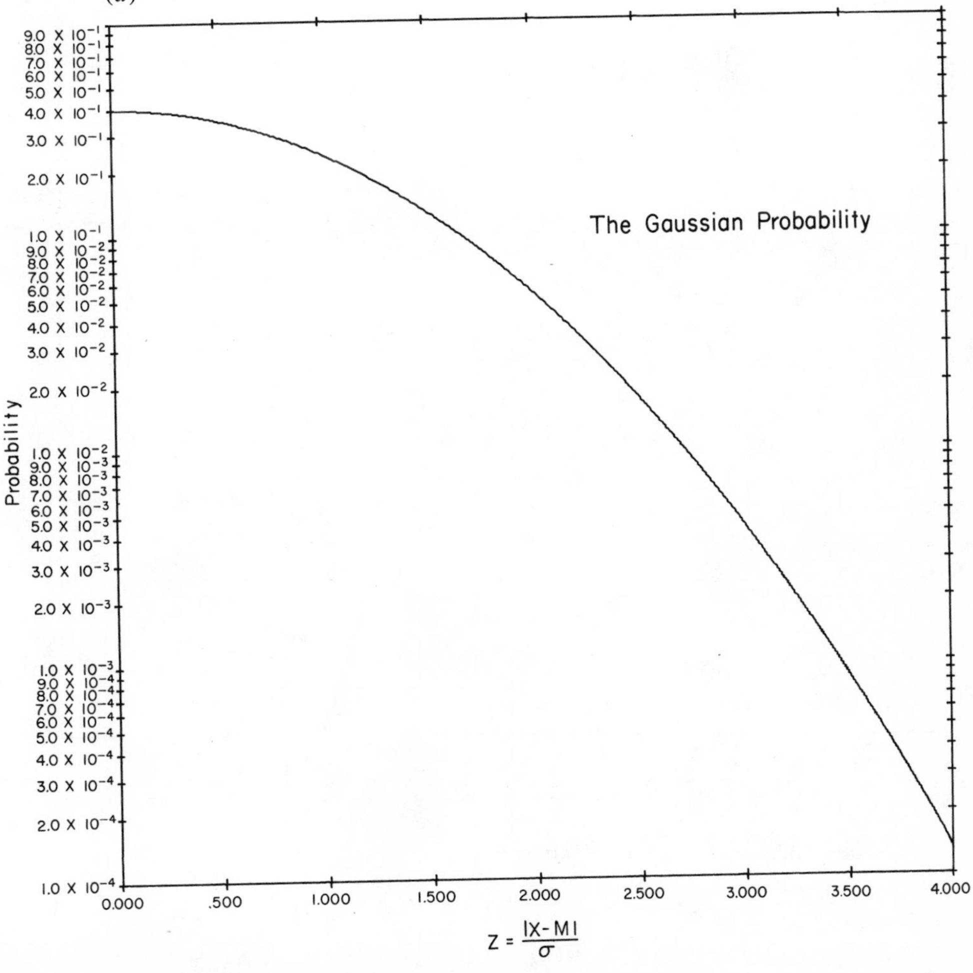

Figure 25.3(a) and (b)
Plot of the normal probability density function $\varphi(z)$ versus z on vertical log scales.

(*b*)

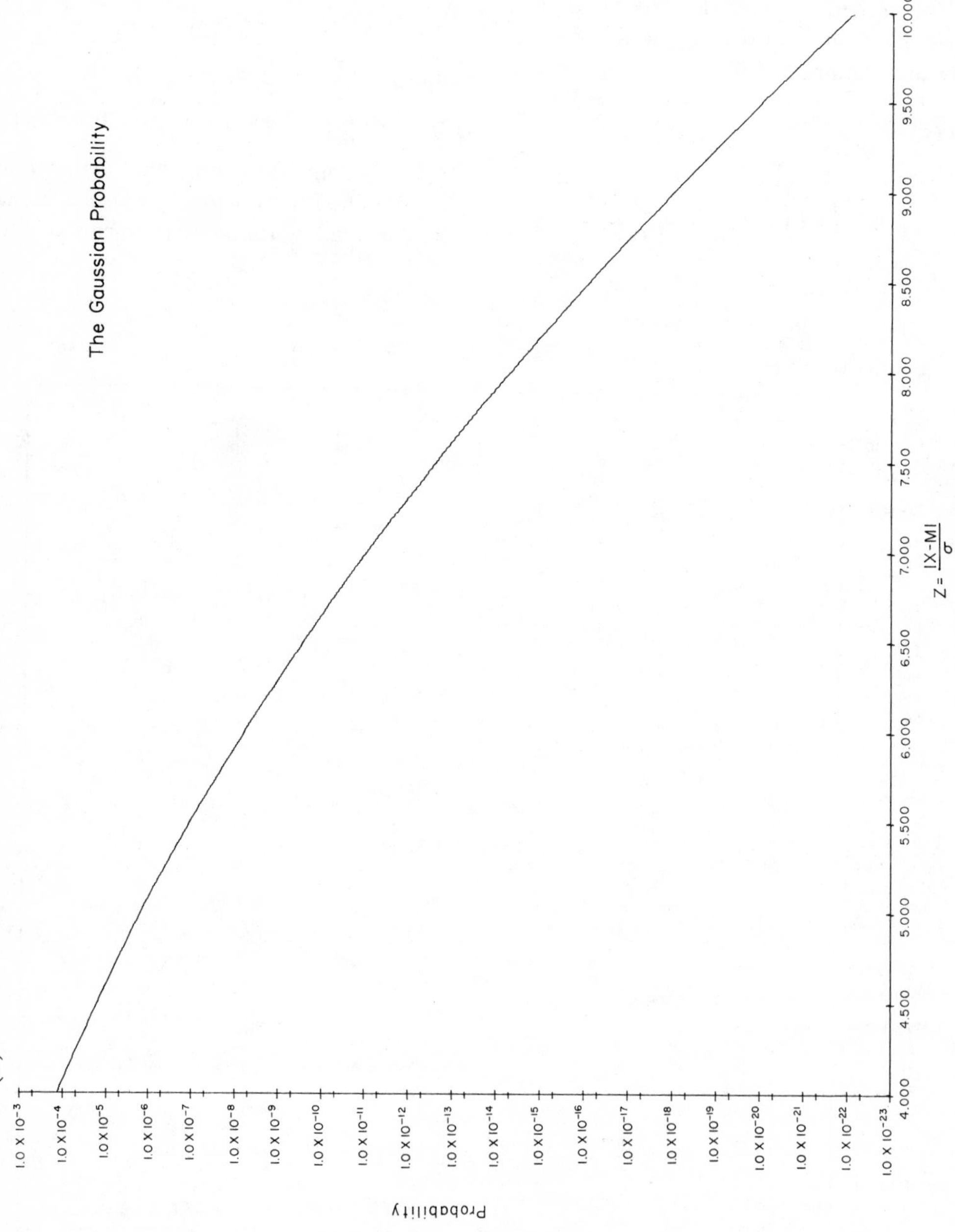

Figure 25.3 *(continued)*

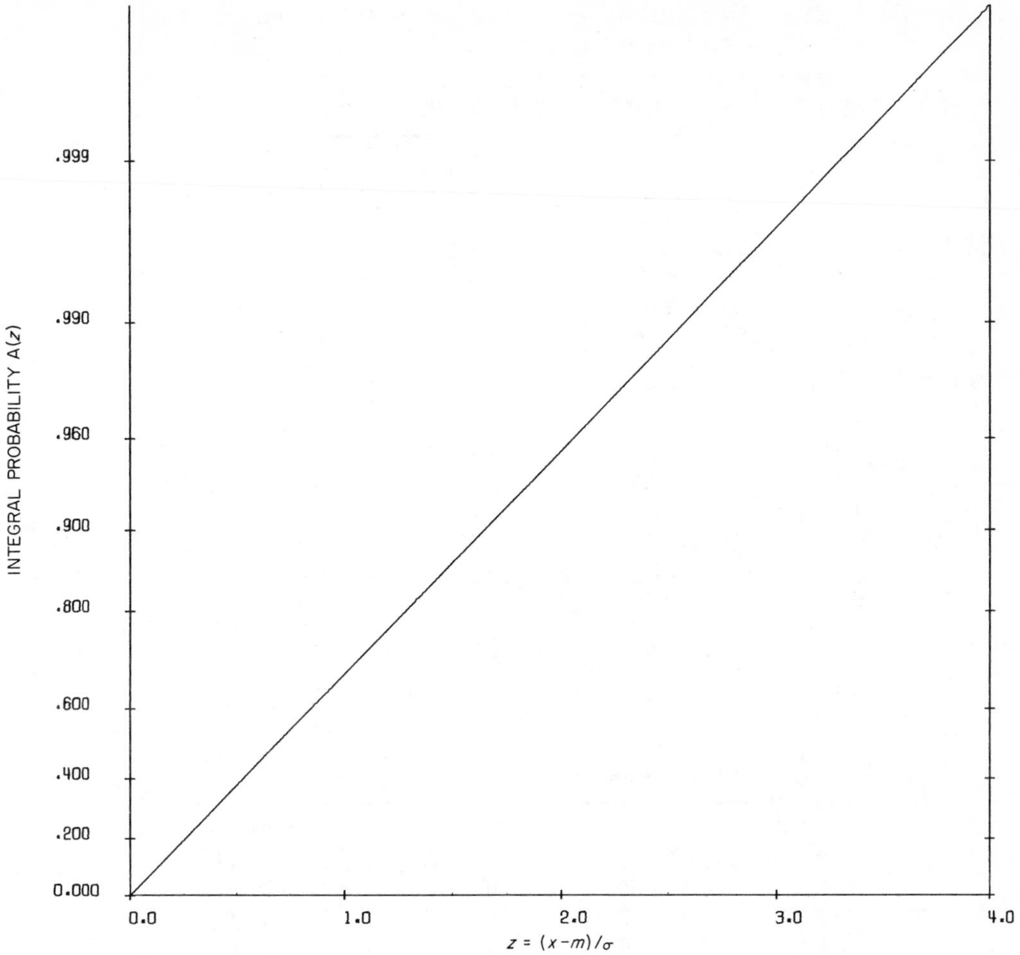

Figure 25.4
Linearized plot of $A(z)$ (vertical) versus $z = (x - \mu)/\sigma$ (horizontal).

variance 1, is tabulated in Table A.VI.2. $A(z)$ is plotted against $z = |x - m|/\sigma$ in Figure 25.4. z is sometimes called the "**normal deviate**."

An interesting quantity is $1 - A(x)$. We have defined this in equation (A.III.7) to be

$$E(x) = \sqrt{\frac{2}{\pi}} \int_x^{\infty} e^{-u^2/2}\, du$$

$$= \sqrt{\frac{2}{\pi}} \left\{ \int_0^{\infty} e^{-u^2/2}\, du - \int_0^{x} e^{-u^2/2}\, du \right\}$$

or

$$E(x) = 1 - A(x) \qquad (25.38)$$

since

$$A(\infty) = \sqrt{\frac{2}{\pi}} \int_0^{\infty} e^{-u^2/2} = 1$$

For a normally distributed variable X, $E(x)$ expresses the probability that $|X| > x$, where x is some positive number.

$$E(x) = \Pr\{|X| > x\}$$
$$\text{where } X \text{ is } N(0, 1) \text{ and } x > 0 \qquad (25.39)$$

A short summary of these probabilities is listed in Table 25.1. This table, for example, tells us that a normally distributed variate will deviate from its mean value by more than one standard deviation only 31.73% of the time. This, by the way, is the justification for the statement made back in Chapter 1 that a standard deviation error implies that 68 out of 100

Table 25.1 Probability of occurrence of statistical deviations relative to the standard deviation.

$t = \lvert\text{deviation}\rvert/\sigma$	$\Pr(T \geq t)$	$t = \lvert\text{deviation}\rvert/\sigma$	$\Pr(T \geq t)$
0.67448	.5000	2.5	.0124
0.7	.4839	2.6	.00932
0.8	.4237	2.7	.00693
0.9	.3681	2.8	.00511
1.0	.3173	2.9	.00373
1.1	.2713	3.0	.00270
1.2	.2301	3.1	.00194
1.3	.1936	3.2	.00137
1.4	.1615	3.3	.000967
1.5	.1336	3.4	.000674
1.6	.1096	3.5	.000465
1.7	.0891	3.6	.000318
1.8	.0719	3.7	.000216
1.9	.0574	3.8	.000145
2.0	.0455	3.9	.0000962
2.1	.0357	4.0	.0000634
2.2	.0278	5.0	.000000573
2.3	.0214	6.0	.0000000020
2.4	.0164	7.0	.0000000000026

$\Pr(T \geq t)$	$t = \lvert\text{deviation}\rvert/\sigma$	$\Pr(T \geq t)$	$t = \lvert\text{deviation}\rvert/\sigma$
1.0	0	0.01	2.57583
0.9	0.12566	10^{-3}	3.29053
0.8	0.25335	10^{-4}	3.89059
0.7	0.38532	10^{-5}	4.41717
0.6	0.52440	10^{-6}	4.89164
0.5	0.67448	10^{-7}	5.32672
0.4	0.84162	10^{-8}	5.73073
0.3	1.03643	10^{-9}	6.10941
0.2	1.28155	10^{-10}	6.46695
0.1	1.64485	10^{-11}	6.80650
0.05	1.95996	10^{-12}	7.13051

repetitions of a measurement will yield results within one standard deviation of the quoted value.

Similarly, we may show that

$$P(x) = \tfrac{1}{2} + \frac{1}{\sqrt{2\pi}} \sum_{n=0}^{\infty} \frac{(-1)^n x^{2n+1}}{n!2^n(2n+1)} \tag{25.40}$$

or,

$$P(x) = \tfrac{1}{2} + (x) \sum_{n=0}^{\infty} \frac{x^{2n+1}}{1 \cdot 3 \cdot 5 \cdots (2n+1)}$$

$$x \geq 0 \tag{25.41}$$

where (25.41) converges faster than (25.40). $P(z) = \Phi(z)$ is tabulated in Table A.VI.3 against values of the normal deviate, $z = (x - \mu)/\sigma$.

REPRODUCTIVE PROPERTY OF THE NORMAL DISTRIBUTION

We have the general rule that the *sum* of **two independent, normally distributed random variables is normally distributed.** Specifically, if X is $N(m_1, \sigma_1^2)$ and Y is

$N(m_2, \sigma_2^2)$, then

$$Z = X + Y \quad \text{is} \quad N(m_1 + m_2, \sigma_1^2 + \sigma_2^2) \tag{25.42}$$

This may be proved straightforwardly (but tediously) as follows:

$$f_X(x) = \frac{1}{\sqrt{2\pi}\sigma_1} e^{-(1/2)[(x-m_1)/\sigma_1]^2}$$

$$f_Y(y) = \frac{1}{\sqrt{2\pi}\sigma_2} e^{-(1/2)[(y-m_2)/\sigma_2]^2}$$

$$f_{Z=X+Y}(z)$$
$$= \int_{-\infty}^{\infty} dx f_X(x) f_Y(z-x) \qquad \text{by (20.37)}$$
$$= \int_{-\infty}^{\infty} dx \frac{1}{\sqrt{2\pi}\sigma_1} e^{-(1/2)[(x-m_1)/\sigma_1]^2} \times \frac{1}{\sqrt{2\pi}\sigma_2} e^{-(1/2)[(z-x-m_2)/\sigma_2]^2}$$

We note as a lemma the algebraic identity

$$e^{-(1/2)[(x-M_1)/\sigma_1]^2} e^{-(1/2)[(x-M_2)/\sigma_2]^2} = e^{-(1/2)(M_1-M_2)^2/(\sigma_1^2+\sigma_2^2)} e^{-(1/2)[(x-M^*)/\sigma^*]^2}$$

where

$$M^* = \frac{M_1/\sigma_1^2 + M_2/\sigma_2^2}{1/\sigma_1^2 + 1/\sigma_2^2} = \frac{M_1\sigma_2^2 + M_2\sigma_1^2}{\sigma_1^2 + \sigma_2^2}$$

and

$$\sigma^{*2} = \left[\frac{1}{\sigma_1^2} + \frac{1}{\sigma_2^2}\right]^{-1} = \frac{\sigma_1^2\sigma_2^2}{\sigma_1^2 + \sigma_2^2}$$

$$\therefore f_{Z=X+Y}(z) = \frac{1}{\sqrt{2\pi}} \frac{1}{\sigma_1} \frac{1}{\sqrt{2\pi}} \frac{1}{\sigma_2} \int_{-\infty}^{\infty} dx \times \exp\left\{\frac{-(1/2)[z-(m_1+m_2)]^2}{\sigma_1^2+\sigma_2^2}\right\} \times e^{-(1/2)[(x-M^*)/\sigma^*]^2}$$
$$= \frac{1}{\sqrt{2\pi}} \frac{1}{\sqrt{\sigma_1^2+\sigma_2^2}} \times \exp\left\{\frac{-(1/2)[z-(m_1+m_2)]^2}{\sigma_1^2+\sigma_2^2}\right\} \times \frac{1}{\sqrt{2\pi}} \times \frac{\sqrt{\sigma_1^2+\sigma_2^2}}{\sigma_1\sigma_2} \int_{-\infty}^{\infty} dx\, e^{-(1/2)[(x-M^*)/\sigma^*]^2}$$
$$= \frac{1}{\sqrt{2\pi}} \frac{1}{\sqrt{\sigma_1^2+\sigma_2^2}} \times \exp\left\{\frac{-(1/2)[z-(m_1+m_2)]^2}{\sigma_1^2+\sigma_2^2}\right\} \times \left\{\frac{1}{\sqrt{2\pi}\sigma^*} \int_{-\infty}^{\infty} dx\, e^{-(1/2)[(x-M^*)/\sigma^*]^2}\right\}$$

$$f_{Z=X+Y}(z) = \frac{1}{\sqrt{2\pi}} \frac{1}{\sqrt{\sigma_1^2+\sigma_2^2}} \times \exp\left\{\frac{-[z-(m_1+m_2)]^2}{2(\sigma_1^2+\sigma_2^2)}\right\}$$

since the bracketed integral is unity as shown in Appendix III.

Therefore, Z is $N(m_1 + m_2, \sigma_1^2 + \sigma_2^2)$ **Q.E.D.**

Also, we may show that the ***difference* of two independent, normally distributed random variables is normally distributed.** Specifically, if X is $N(m_1, \sigma_1^2)$ and Y is $N(m_2, \sigma_2^2)$, then

$$Z = X - Y \quad \text{is} \quad N(m_1 - m_2, \sigma_1^2 + \sigma_2^2) \tag{25.43}$$

We may also generalize (25.42) to the statement that the ***sum* of *any* number of independent normally distributed random variables is itself normally distributed.**

If X_i is $N(m_i, \sigma_i^2)$ and the X_i are all independent, then

$$Z = \sum_{i=1}^{n} X_i \quad \text{is} \quad N\left(\sum_{i=1}^{n} m_i, \sum_{i=1}^{n} \sigma_i^2\right) \tag{25.44a}$$

We further see that the X_i may appear with $\pm$ in the sum and the mean of Z will accordingly have the terms appear $\pm m_i$.

A corollary of these results is that the ***mean of a distribution of variates x_i, each of which is normally distributed, is itself normally distributed.*** In particular, if x_i is $N(\bar{x}, \sigma^2)$ then

$$\bar{x} \quad \text{is} \quad N(\bar{x}, \sigma^2/n) \tag{25.44b}$$

Normal Deviate Test for the Difference of Two Sample Means

In what follows we shall use the example of the weighing of two coins and attempt to determine if the two weights are equal. The discussion, however, applies to any property of two objects or two populations where the property may be quantified that is, where we may assign a number to the property.

Suppose we have made n_1 normally distributed measurements of the weight of a coin, C_1, each measurement x_i being characterized by a standard deviation σ_1. We have also n_2 normally distributed measurements of the weight of a second coin, C_2, each measurement y_i being characterized by a standard deviation σ_2. *Are the data (the two observed sample means,* $\bar{x}$ and $\bar{y}$) *consistent with the hypothesis that the weights of the two coins are equal?*

We shall consider the testing of this hypothesis ("*The weights of the coins are equal.*") more carefully in Part V, but for now we are interested in considering the observed difference of the sample means

$$\bar{x} - \bar{y}$$

and asking if this difference is **significant.** That is, how often would a difference this large or larger arise *by random chance* if the weights are actually equal?

We see that

$$\bar{x} \quad \text{is} \quad N(\bar{x}, \sigma_1^2/n_1)$$

and

$$\bar{y} \quad \text{is} \quad N(\bar{y}, \sigma_2^2/n_2)$$

By (25.43) the difference

$$d = \bar{x} - \bar{y}$$

is normally distributed

$$d \text{ is } N([\bar{x} - \bar{y}], [\sigma_1^2/n_1 + \sigma_2^2/n_2])$$

We similarly ask, assuming d is

$$N(0, \sigma_1^2/n_1 + \sigma_2^2/n_2):$$

How probable is it that $(\bar{x} - \bar{y})$ is as large as or larger than d_{observed}? We shall not distinguish the sign (we shall later call this a "two-tailed test") and so require the probability

$$\Pr(|\bar{x} - \bar{y}| \geq |d_{\text{observed}}|).$$

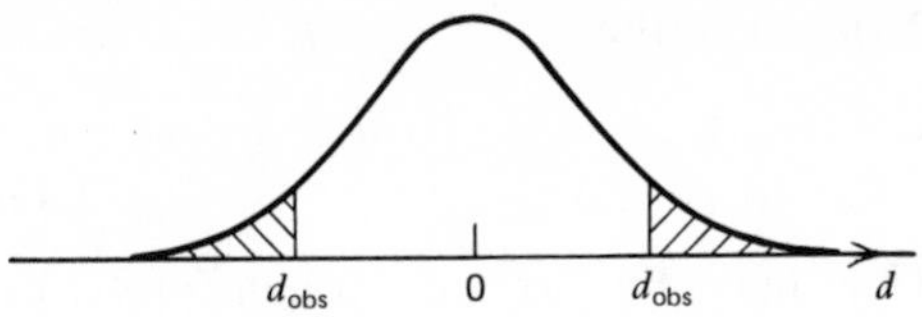

Figure 25.5
Plot of the probability density function of the normal distribution $\varphi(z)$ versus z. The area of the shaded region represents the probability of deviations as large as or larger than $|d_{\text{obs}}|$.

This is the area shaded in Figure 25.5 which is

$$1 - A\left(\frac{d_{\text{observed}}}{\sqrt{\sigma_1^2/n_1 + \sigma_2^2/n_2}}\right) = 1 - A(z_{\text{obs}})$$

from (25.32).

We may obtain $A(z_{\text{obs}})$ from Table A.VI. 2. If $1 - A(z_{\text{obs}})$ is large enough, for example, $1 - A(z_{\text{obs}}) \geq 0.05$, then we say that d_{observed} is *consistent with zero,* and the **observed weights are consistent with the assumption of equality.**

► Example 25.1
Coin weighing

A class of freshman students was asked to weigh a sample of U.S. pennies of two different dates, 1964 and 1970, in an effort to determine if there was a significant difference between the 1964 pennies and the 1970 pennies. The test was performed twice using a beam balance (BB) and then an analytical balance (AB) of higher precision.

The various measurements, w_i, were tabulated and the quantities $\bar{w}$, s, and $\Delta\bar{w}$ were computed from the usual relations

$$\bar{w} = \frac{1}{N}\sum_{i=1}^{N} w_i$$

$$s = \left\{\sum_{i=1}^{N} \frac{(w_i - \bar{w})^2}{N-1}\right\}^{1/2}$$

and

$$\Delta\bar{w} = \left(\frac{s^2}{N}\right)^{1/2}$$

These are summarized in Table 25.2.

Table 25.2 Summary of data on coin weighing by a freshman class.

Identification	Number of Measurements	Mean $\bar{w}$	Standard Deviation s	Standard Error $\Delta\bar{w}$
1964 Beam Balance (*BB*)	21	3.0733	0.0939	0.0205
1970 Beam Balance (*BB*)	33	3.0968	0.1066	0.0183
1964 Analytical Balance (*AB*)	22	3.0806	0.0377	0.0080
1970 Analytical Balance (*AB*)	34	3.1045	0.0305	0.0059

BEAM BALANCE DATA

For the beam balance data we make the substitution

$$\sigma_1 = s_{1970} \qquad \sigma_2 = s_{1964}$$

together with $\bar{x} = \bar{w}_{1970}$, $\bar{y} = \bar{w}_{1964}$.

We form the statistic

$$t = \frac{(\bar{w}_{1970} - \bar{w}_{1964})}{\sqrt{\sigma_1^2/n_1 + \sigma_2^2/n_2}} = \frac{(3.0968 - 3.0733)}{\sqrt{(0.018)^2 + (0.020)^2}}$$
$$= 0.87$$

From Table A.VI.2 we see that

$$A(0.87) = 0.616$$

or

$$\Pr(|t| \geq 0.87) = 1 - A(0.87) = 0.384$$

that is, a Gaussian has 38.4% of its area at deviations $|t| \geq 0.87$ on the assumption that the mean is zero ($\bar{w}_{1964} = \bar{w}_{1970}$).

The *BB* data are, therefore, *consistent* with the hypothesis of equal means.

ANALYTICAL BALANCE DATA

For the *AB* data we form

$$t = \frac{\bar{w}_{1970} - \bar{w}_{1964}}{[(\Delta\bar{w}_{1970})^2 + (\Delta\bar{w}_{1964})^2]^{1/2}}$$
$$= \frac{(3.1045 - 3.0806)}{[(0.0059)^2 + (0.0080)^2]^{1/2}} = 2.39$$

From the table we see that $A(2.39) = 0.983$ or

$$\Pr(|t| \geq 2.39) = 0.017$$

that is, the assumption of equal means would produce a deviation as large as or larger than that observed only 1.7% of the time. **We would *tend to reject* the hypothesis of equal means on the basis of the *AB* data.** ◄

Normal Approximation to the Binomial Distribution

Following De Moivre's derivation, if x is a binomially distributed random variable with probability p, mean np, and variance $np(1-p)$, then as $n \to \infty$

$$Z = \frac{x - np}{\sqrt{np(1-p)}} \tag{25.45}$$

becomes normally distributed

$$Z \text{ is } N(0, 1)$$

We may also say that

$$x \text{ becomes } N[np, np(1-p)]$$

that is,

$$\mathbf{b}(x; n, p)$$
$$= \binom{n}{x} p^x (1-p)^{n-x} \qquad \text{For } n \text{ large}$$
$$\approx \mathbf{n}[x; np, \sqrt{np(1-p)}] \tag{25.46}$$
$$= [2\pi np(1-p)]^{-1/2} e^{-(x-np)^2/2np(1-p)}$$
$$= [np(1-p)]^{-1/2} \varphi\left(\frac{x - np}{np(1-p)}\right)$$

by (24.17c).

The approximation improves as n increases and is quite good for p that is not too close to 0 or 1. Comparison of the binomial and normal distributions is shown in Figures 25.6.

► **Example 25.2**

What is the probability of getting exactly 24 heads in 48 tosses of a fair coin? By direct calculation, $\mathbf{b}(24; 48, 1/2) = 0.1146$.

The argument, $(x - np)/[np(1-p)]$, is $(24-24)/\sqrt{12} = 0$.

$$(12)^{-1/2}\varphi(0) = 0.3989/3.464$$
$$= 0.1152 \quad \text{from Table A.VI.1}$$

See Example 25.5. ◄

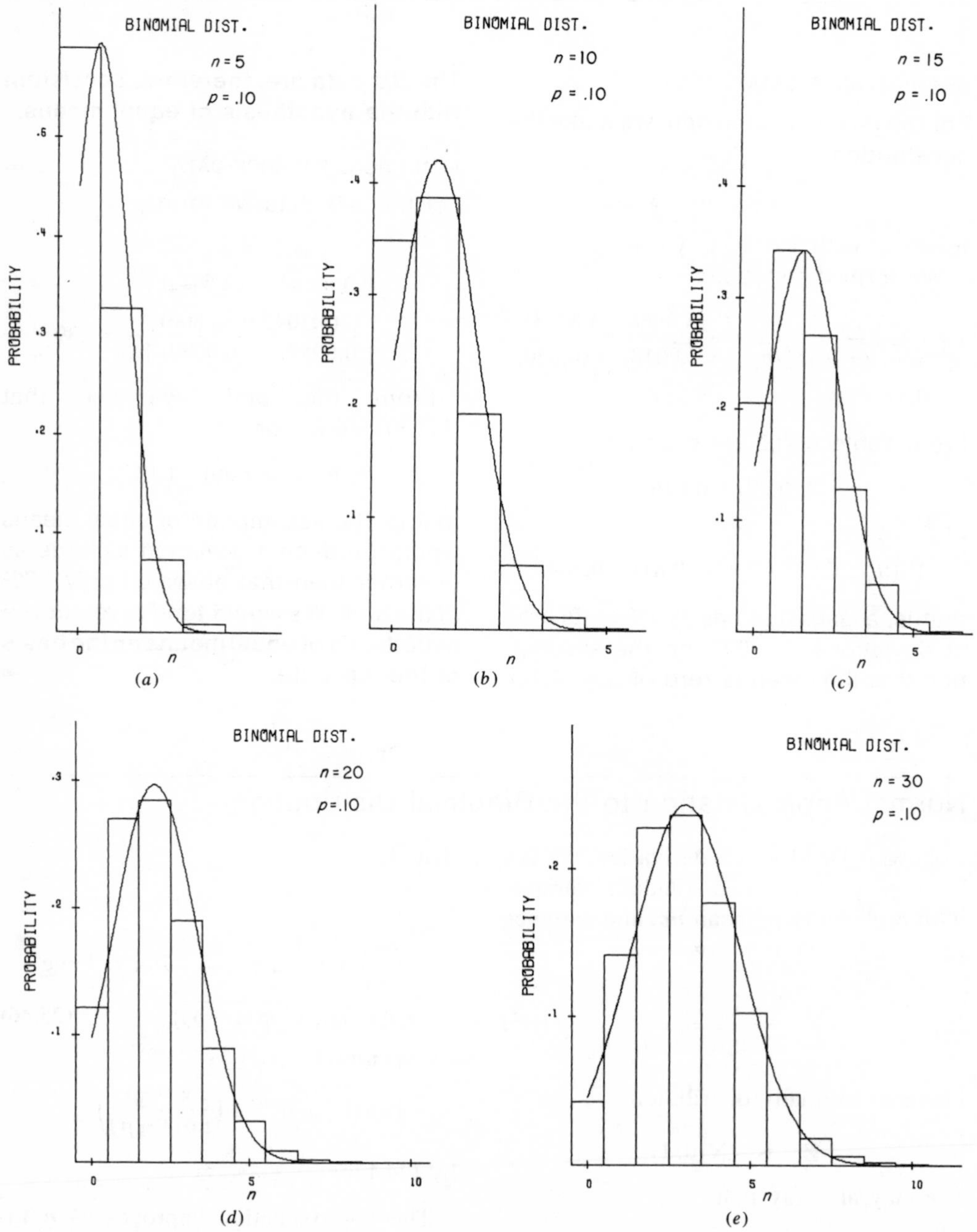

Figure 25.6(a) to (e)

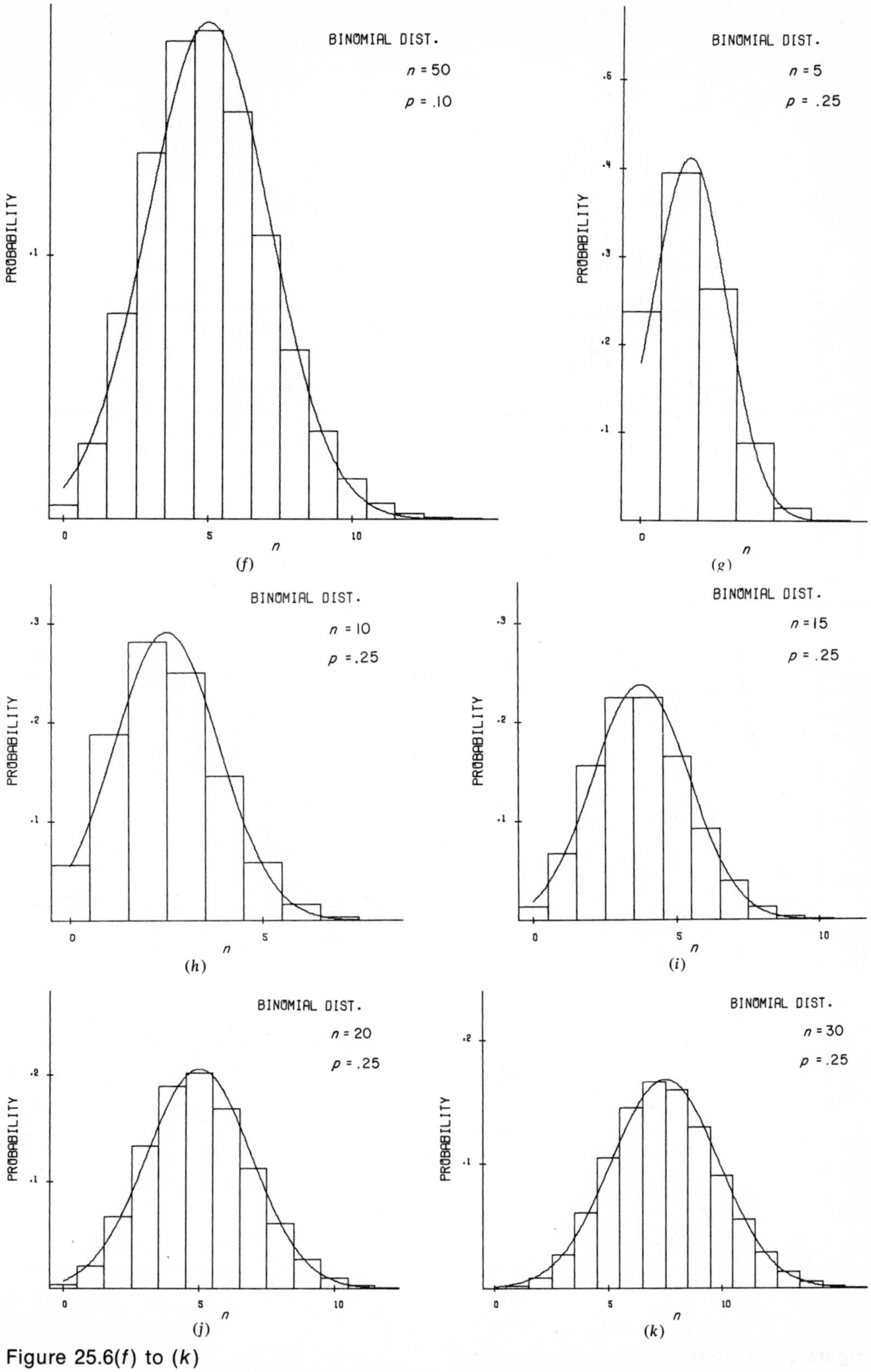

Figure 25.6(f) to (k)

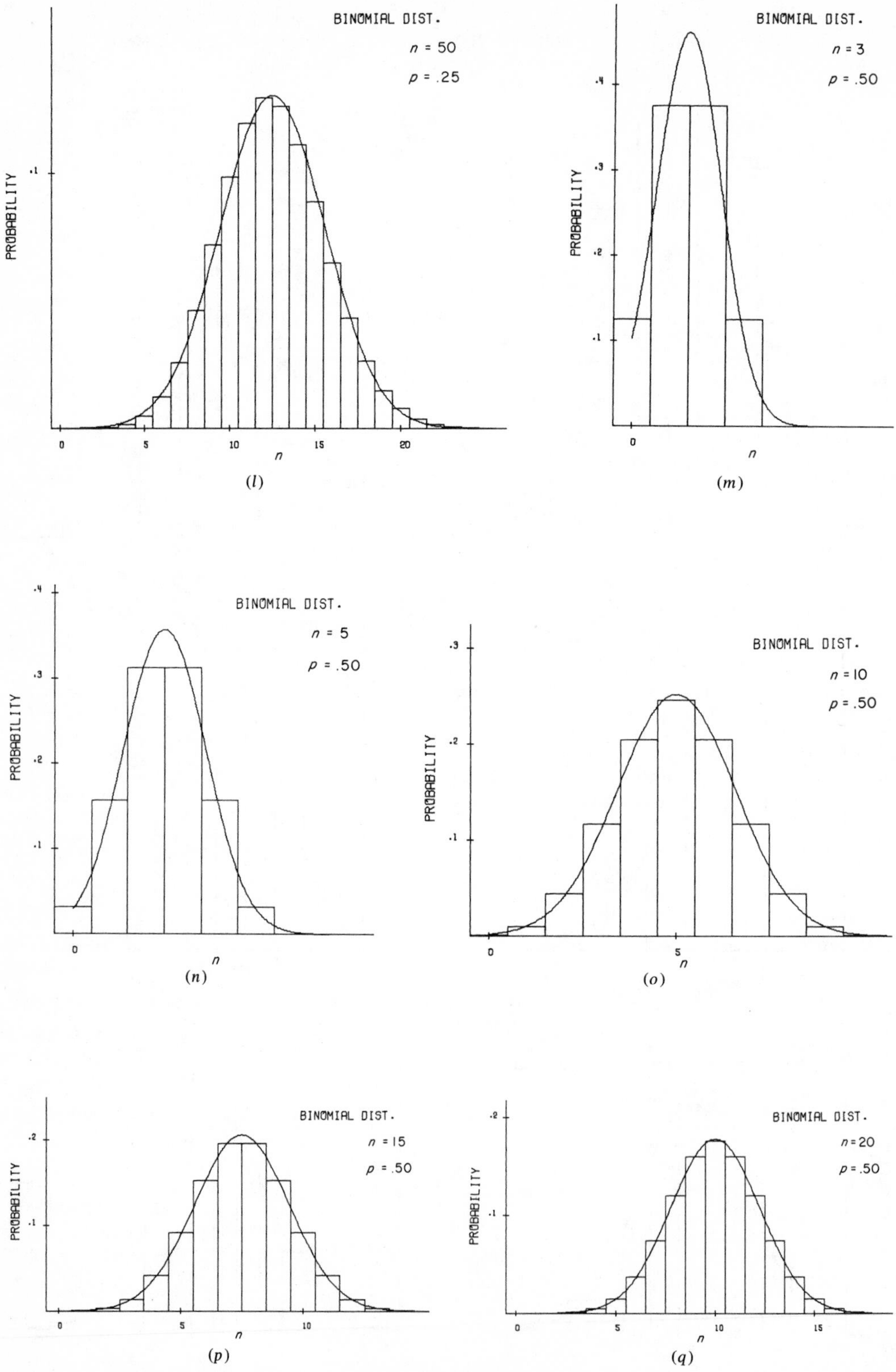

Figure 25.6(*l*) to (*q*)

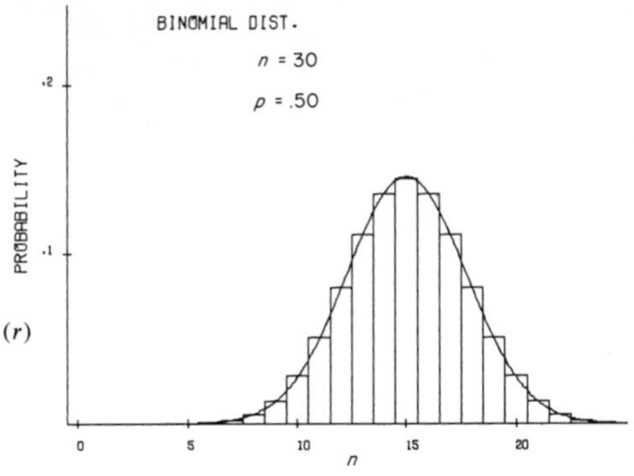

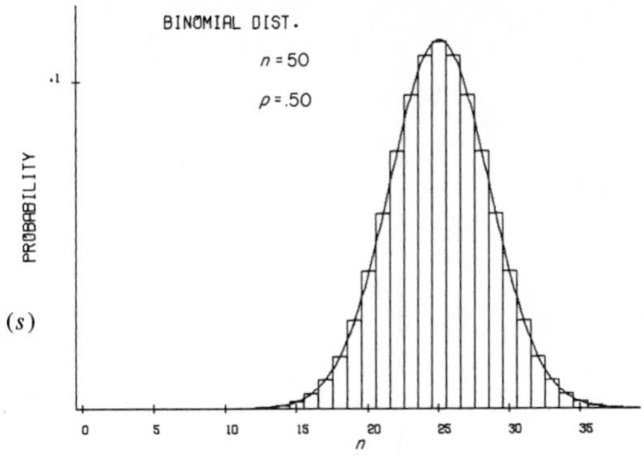

Figures 25.6
Comparison of the (discrete) binomial and (continuous) normal frequency functions. (*a*) $n = 5$, $p = 0.10$. (*b*) $n = 10$, $p = 0.10$. (*c*) $n = 15$, $p = 0.10$. (*d*) $n = 20$, $p = 0.10$. (*e*) $n = 30$, $p = 0.10$. (*f*) $n = 50$, $p = 0.10$. (*g*) $n = 5$, $p = 0.25$. (*h*) $n = 10$, $p = 0.25$. (*i*) $n = 15$, $p = 0.25$. (*j*) $n = 20$, $p = 0.25$. (*k*) $n = 30$, $p = 0.25$. (*l*) $n = 50$, $p = 0.25$. (*m*) $n = 3$, $p = 0.5$. (*n*) $n = 5$, $p = 0.50$. (*o*) $n = 10$, $p = 0.50$. (*p*) $n = 15$, $p = 0.50$. (*q*) $n = 20$, $p = 0.50$. (*r*) $n = 30$, $p = 0.50$. (*s*) $n = 50$, $p = 0.50$.

HALF-INTEGER CORRECTIONS IN THE NORMAL APPROXIMATION

To use the various tables of the standardized normal distribution it is always convenient to normalize the variable as in (20.15), for example, $x \to x^*$ where

$$x^* = \frac{x - \langle x \rangle}{\sigma} = \frac{x - np}{\sqrt{np(1-p)}}$$

Because the binomial distribution is **discrete** while the normal distribution is **continuous**, when calculating ranges of probabilities we must make the "*half-integer corrections for continuity.*"

Suppose we wish to compute

$$\Pr(a \le X \le b) = P_B(a \le X \le b)$$

using the normal approximation where X is binomially distributed. We note that

$$\Pr(a \le X \le b) = \sum_{r=a}^{b} \mathbf{b}(r; np) \qquad (25.47)$$

This is indicated in Figure 25.7 where (25.47) is represented by the area under the histogram between $(a - 1/2)$ on the left and $(b + 1/2)$ on the right. If we were simply to use the normal approximation, we might write

$$\begin{aligned}
\sum_{r=a}^{b} \mathbf{b}(r;n;p) &= P_B(a \le X \le b) \\
&= P_B\left(\frac{a - \langle x \rangle}{\sigma} \le X^* \le \frac{b - \langle x \rangle}{\sigma}\right) \\
&\cong P_N\left(\frac{a - \langle x \rangle}{\sigma} \le X^* \le \frac{b - \langle x \rangle}{\sigma}\right) \\
&= \int_{(a-\langle x \rangle)/\sigma}^{(b-\langle x \rangle)/\sigma} \varphi(t)\,dt \\
&= \Phi\left(\frac{b - \langle x \rangle}{\sigma}\right) - \Phi\left(\frac{a - \langle x \rangle}{\sigma}\right)
\end{aligned}$$

This, however, represents the area under the curve between a and b. A *better* approximation is to include the areas in the intervals $[a - 1/2, a]$ and $[b, b + 1/2]$. Therefore, *a better approximation includes the half-integer corrections*

$$P_B(a \le X \le b) \approx P_N\left(\frac{a - 1/2 - \langle x \rangle}{\sigma} \le x^* \le \frac{b + 1/2 - \langle x \rangle}{\sigma}\right) \qquad (25.48)$$

If a and b are large, then the half-integer correction is small.

► **Example 25.3**

Consider Example 22.2 of Chapter 22 where it was required to find the probability of getting a number of heads, X,

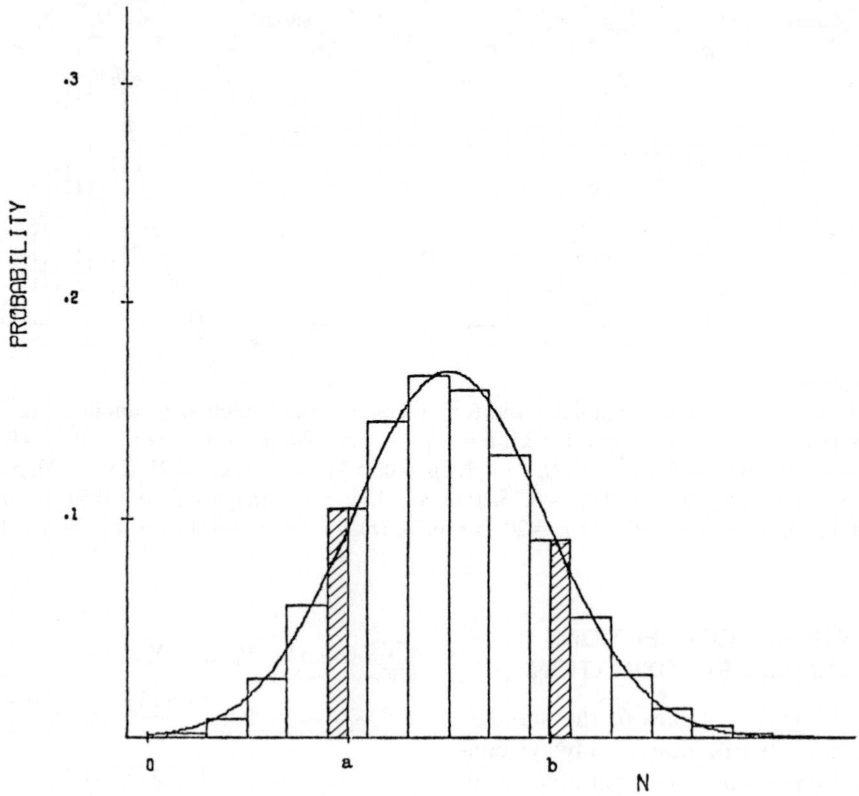

Figure 25.7
Comparison of the binomial and normal distributions. Demonstration of the "half-integer corrections for continuity" (see text).

between 15 and 21 in 36 tosses of a fair coin.

The mean and variance are $np = 36 \cdot 1/2 = 18$ and $npq = 36 \cdot 1/2 \cdot 1/2 = 9$, respectively.

We set

$$P_B(15 \leq X \leq 21)$$
$$= P_B\left(\frac{15-18}{3} \leq X^* \leq \frac{21-18}{3}\right)$$
$$\approx P_N\left(\frac{14.5-18}{3} \leq X^* \leq \frac{21.5-18}{3}\right)$$
$$= P_N(-1.17 \leq Z \leq 1.17)$$
$$= \Phi(+1.17) - \Phi(-1.17)$$
$$= P(+1.17) - [1 - P(+1.17)]$$

using Table A.VI.3

$$= A(1.17)$$

using Table A.VI.2

$$= 0.7580$$

We may compare this with the approximation neglecting the half-integer corrections

$$P_B(15 \leq X \leq 21)$$
$$= P_B\left(\frac{15-18}{3}\right) \leq X^* \leq \frac{21-18}{3}\right)$$
$$\approx P_N\left(\frac{15-18}{3} \leq X^* \leq \frac{21-18}{3}\right)$$
$$= P_N(-1 \leq Z \leq +1)$$
$$= A(1)$$
$$= 0.6827$$

In Chapter 22 we obtained 0.7571 by direct calculation. ◄

► **Example 25.4**

What is the probability that $45 \leq x \leq 55$ in 100 tosses?

$$\begin{aligned}
&\Pr\{45 < x < 55\} \\
&= \sum_{r=45}^{55} \binom{100}{r} \left(\frac{1}{2}\right)^r \left(\frac{1}{2}\right)^{100-r} \\
&= \left(\frac{1}{2}\right)^{100} 100! \sum_{r=45}^{55} \frac{1}{r!(100-r)!} \\
&\equiv P_{\pm 10\%}
\end{aligned}$$

By brute force and a desk calculator, we get

$$\begin{aligned}
P_{\pm 10\%} &= \left(\frac{1}{2}\right)^{100} 100! \left\{\frac{1}{45!55!} + \frac{1}{46!54!} + \cdots \right. \\
&\quad \left. + \frac{1}{49!51!} + \frac{1}{50!50!} + \frac{1}{51!49!} + \cdots + \frac{1}{55!45!}\right\} \\
&= \left(\frac{1}{2}\right)^{100} 100! \left\{2\left[\frac{1}{45!55!} + \cdots + \frac{1}{49!51!}\right] \right. \\
&\quad \left. + \frac{1}{50!50!}\right\} = \left(\frac{1}{2}\right)^{100} \frac{100!}{45!50!} \{2A + B\}
\end{aligned}$$

where

$$\begin{aligned}
A &= \frac{1}{(55)(54)(53)(52)(51)} + \frac{1}{(46)(54)(53)(52)(51)} \\
&\quad + \frac{1}{(47)(46)(53)(52)(51)} + \frac{1}{(48)(47)(46)(52)(51)} \\
&\quad + \frac{1}{(49)(48)(47)(46)(51)} \\
&= \{2.39549 + 2.86417 \\
&\quad + 3.29075 + 3.63354 + 3.85600\}10^{-9}
\end{aligned}$$

$$B = \frac{1}{(50)(49)(48)(47)(46)} = 3.93312 \times 10^{-9}$$

and

$$\{2A + B\} = 3.60130 \times 10^{-8}$$

Taking logarithms and using Table A.III.1,

$$\begin{aligned}
\log P_{\pm 10\%} &= \log(100!) - 100 \log 2 - \log(45!) \\
&\quad - \log(50!) + \log(3.60130) - 8.\dot{0} \\
&= 157.9700 - 30.10300 - 56.07781 \\
&\quad - 64.48307 + 0.55646 - 8.00000 \\
&= -0.13742 \\
&= 9.86258 - 10
\end{aligned}$$

Therefore, $P_{\pm 10\%} = 0.7288$, by direct calculation.

By using the normal approximation with half-integer corrections, we get

$$\begin{aligned}
P_B(45 \le X \le 55) &= P_B\left(\frac{45\text{–}50}{5} \le X^* \le \frac{55-50}{5}\right) \\
&\approx P_N\left(\frac{45 - 1/2 - 50}{5} \le X^* \right. \\
&\quad \left. \le \frac{55 + 1/2 - 50}{5}\right) \\
&= P_N(-1.10 \le Z \le +1.10) \\
&= A(1.10) \\
&= 0.7287
\end{aligned}$$

◄

Also, ***individual*** *binomial probabilities are improved by the half-integer correction.* That is,

$$\begin{aligned}
\mathbf{b}(x; n, p) &= \Phi\left(\frac{x + 1/2 - np}{\sqrt{npq}}\right) \\
&\quad - \Phi\left(\frac{x - 1/2 - np}{\sqrt{npq}}\right) \\
&= A\left(\frac{x - np + 1/2}{\sqrt{npq}}\right), \qquad (25.49)
\end{aligned}$$

using Table A.VI.2.

This is a better approximation than (25.46).

► **Example 25.5**

Consider again $\mathbf{b}(24; 48, 1/2) = 0.1146$ as in Example 25.2. Using the half-integer correction, formula (25.49), we get

$$\begin{aligned}
\mathbf{b}(24; 48, 1/2) &= A((24 - 24 + 1/2)/\sqrt{12}) \\
&= A(0.1443)
\end{aligned}$$

Using linear interpolation between the values $A(0.14) = 0.1113$ and $A(0.15) = 0.1192$, we get $A(0.1443) = 0.1147$. This result is closer to the exact calculation than was obtained with (25.46).

For $p = 0.5$, the normal approximation is good even for $np \ge 5$.

For example,

$$\mathbf{b}(5; 10, 1/2) = 0.2461$$

by direct calculation

$$\begin{aligned}
&= A\left(\frac{5 - 5 + 1/2}{10 \cdot 1/2 \cdot 1/2}\right) \\
&= A(0.316) \qquad \text{by (25.49)}
\end{aligned}$$

By linear interpolation between $A(0.31) = 0.2434$ and $A(0.32) = 0.2510$, we get

$$A(0.316) = 0.2480$$

◄

The Central Limit Theorem (Normal Convergence Theorem)

The general applicability of the normal distribution derives from a remarkable theorem called the *central limit theorem* or the **normal convergence theorem**. We shall not discuss its proof* nor go very deeply into the conditions required for it to hold.† Suffice it to say that it holds under very general conditions and applies to almost all the distributions we might encounter (with one exception noted).

We have seen that the sum of n independent normally distributed random variables

$$x = \sum_{i=1}^{n} x_i$$

is normally distributed (the reproductive property of the normal distribution). If n is large, and given certain assumptions, the sum will be normally distributed even if the individual x_i are not themselves normally distributed.

This is the subject of the *central limit theorem*, which states the following.

Suppose n ***independent*** random variables x_i have arbitrary distributions such that the means, $\langle x_i \rangle = \mu_i$, and variances, $\text{Var}(x_i) = \sigma_i^2$, all exist (but are not infinite). Then

$$z_n = \frac{\sum_{i=1}^{n} x_i - \sum_{i=1}^{n} \mu_i}{\left(\sum_{i=1}^{n} \sigma_i^2\right)^{1/2}}$$

with $\langle z_n \rangle = 0$ and $\sigma(z_n) = 1$, has p.d.f.

$$f(z_n) \to \frac{1}{\sqrt{2\pi}} e^{-z_n^2/2} \qquad \text{as} \qquad n \to \infty \tag{25.50}$$

The theorem says that

$$z_n \text{ becomes } N(0, 1) \qquad \text{as} \qquad n \to \infty$$

That is, the p.d.f.

$$f(z)_n \to \varphi(z_n) \qquad \text{as} \qquad n \to \infty$$

Also, the sum, $x = \sum x_i$, becomes normally distributed.

The ***condition*** for this convergence to the normal distribution to hold is that **no single variance, $\boldsymbol{\sigma_i^2}$, dominates *and*** that

$$\sum_{i=1}^{n} \sigma_i^2 \to \infty \qquad \text{as} \qquad n \to \infty$$

We might remark for completeness that the central limit theorem also holds for certain classes of *dependent* random variables.

We have seen that the normal approximation to the binomial holds very well for n large. This followed from the de Moivre derivation, but we could have also started from the central limit theorem. We could have written the binomial distribution $\mathbf{b}(x; n, p)$ as the distribution for $x = \sum x_i$ because of the reproductive property of the binomial distribution.

NORMAL APPROXIMATION TO THE POISSON DISTRIBUTION

We have also seen that the normal distribution may be directly derived from the Poisson distribution. Because of the reproductive property of the Poisson distribution we may regard a Poisson-distributed random variable X with parameter μ to be the sum of n Poisson variables, x_i, each with parameter μ/n. This indicates that the central limit theorem may be used to invoke the normal approximation to the Poisson distribution as the Poisson parameter, μ, gets large.

$$\begin{aligned}\Pr(X = x) = \mathbf{p}(x; \mu) &= \frac{\mu^x e^{-\mu}}{x!} \\ &\approx \mathbf{n}(x; \mu, \mu^{1/2}) \\ &= \frac{1}{(\mu)^{1/2}} \varphi\left(\frac{x-\mu}{\sqrt{\mu}}\right)\end{aligned} \tag{25.51}$$

as $\mu \to$ large.

*For some proofs see Harald Cramér, *Mathematical Methods of Statistics*, Princeton University Press, Princeton, New Jersey, 1946, pp. 213–220.

†See also B. V. Gnedenko and A. N. Kolmogorov, *Limit Distributions for Sums of Independent Random Variables*, English translation by K. L. Chung, Addison-Wesley, Reading, Mass., 1954.

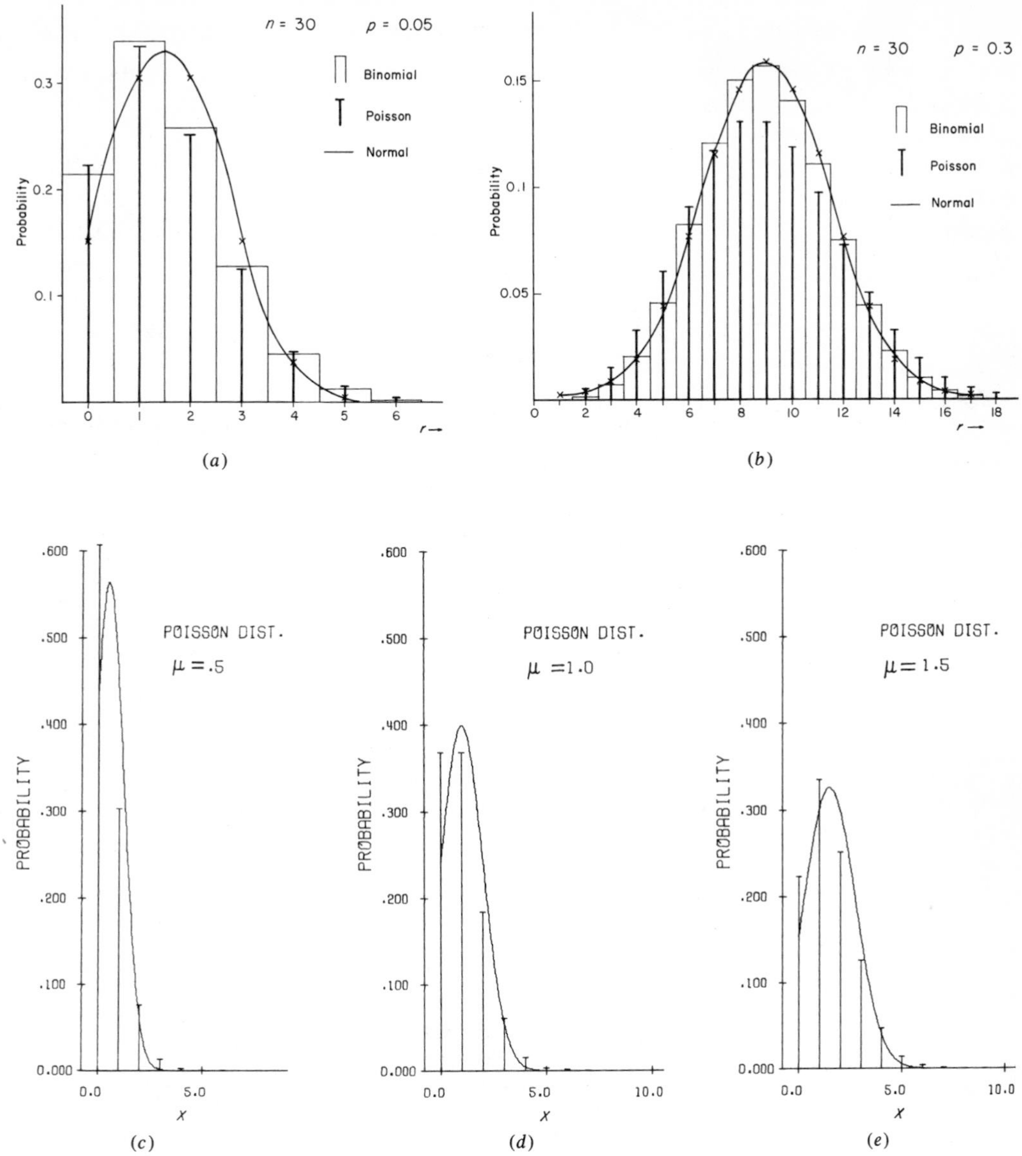

Figures 25.8
Comparison of the (discrete) Poisson distribution and the normal approximation to it. (*a*) Plots of the (discrete) binomial and Poisson-frequency functions with the (continuous) normal p.d.f. $n = 30$, $p = 0.05$. (*b*) Plots of the binomial and Poisson frequency functions and the normal p.d.f. for $n = 30$, $p = 0.3$. (*c*) Comparison of the Poisson and normal frequency functions $\mu = 0.5$. (*d*) Comparison of the Poisson and normal frequency functions, $\mu = 1.0$. (*e*) Comparison of the Poisson and normal frequency functions, $\mu = 1.5$. (*f*) Comparison of the Poisson and normal frequency functions, $\mu = 2.0$. (*g*) Comparison of the Poisson and normal frequency functions, $\mu = 3.0$. (*h*) Comparison of the Poisson and normal frequency functions, $\mu = 4.0$. (*i*) Comparison of the Poisson and normal frequency functions, $\mu = 5.0$. (*j*) Comparison of the Poisson and normal frequency functions, $\mu = 6.0$. (*k*) Comparison of the Poisson and normal frequency functions, $\mu = 7.0$. (*l*) Comparison of the Poisson and normal frequency functions, $\mu = 8.0$. (*m*) Comparison of the Poisson and normal frequency functions, $\mu = 9.0$. (*n*) Comparison of the Poisson and normal frequency functions, $\mu = 10.0$.

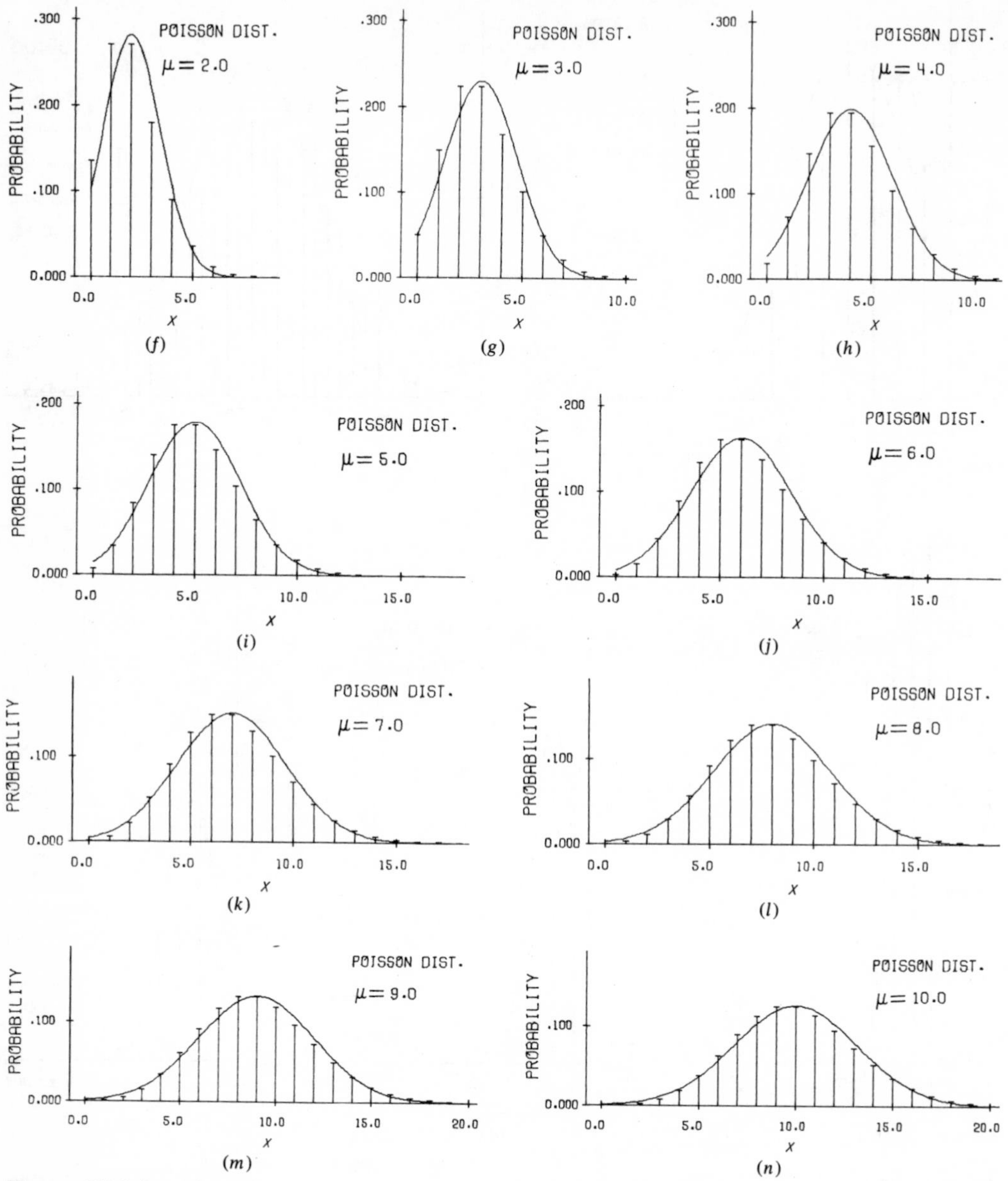

Figure 25.8 (*continued*)

We compare the Poisson distribution and the normal approximation in Figures 25.8*a* to *n*.

The normal approximation to the Poisson distribution is especially useful when it is required that a **range** of probabilities be calculated. For example,

$$P_P(a \le x \le b) = \sum_{x=a}^{b} \mathbf{p}(x;\mu)$$

$$\approx \int_a^b \mathbf{n}(x;\mu,\mu^{1/2})\,dx$$

alternatively,

$$P_P(a \le x \le b) = P_P\left(\frac{a-\mu}{\sqrt{\mu}} \le x^* \le \frac{b-\mu}{\sqrt{\mu}}\right)$$

$$\approx \Phi\left(\frac{b-\mu}{\sqrt{\mu}}\right) - \Phi\left(\frac{a-\mu}{\sqrt{\mu}}\right) \quad (25.52)$$

(25.52) is a convenient form to use with Tables A.VI.2 and A.VI.3.

► **Example 24.6**

Recall Example 23.9 in Chapter 24, involving telephone lines. It was required to find x_0 where

$$\Pr(x_0 \le X \le 400) = \sum_{x=x_0} \mathbf{p}(x; \mu) \le 0.04$$

where $\mu = np = (400)(.04) = 16$.

We may rewrite this as

$$P_P\left(\frac{x_0 - \mu}{\sqrt{\mu}} \le X^* \le \frac{400 - \mu}{\sqrt{\mu}}\right)$$
$$\approx P_N\left(\frac{x_0 - \mu}{\sqrt{\mu}} \le X^* \le \frac{400 - \mu}{\sqrt{\mu}}\right)$$
$$= \Phi\left(\frac{400 - 16}{4}\right) - \Phi\left(\frac{x_0 - 16}{4}\right)$$
$$= \Phi(96) - \Phi\left(\frac{x_0 - 16}{4}\right)$$
$$\approx 1.000 - \Phi\frac{(x_0 - 16)}{4} \le 0.04$$

or

$$0.96 \le \Phi\frac{(x_0 - 16)}{4}$$

From Table A.VI.3

$$\Phi(Z) \ge 0.96 \quad \text{for} \quad Z = \frac{x_0 - 16}{4} \ge 1.76$$

$$\therefore \frac{x_0 - 16}{4} \ge 1.76$$

or

$$x_0 \ge 23$$

which agrees with the previous result. ◄

NORMAL APPROXIMATION TO THE HYPERGEOMETRIC DISTRIBUTION

The justification of the normal approximation to the hypergeometric distribution from the central limit theorem is somewhat more involved. In Chapter 23 we saw that the variate X described by the hypergeometric distribution could be considered as the sum of random variables. [see (23.11)]. These, however, are not independent [cf. (23.16)]. Nevertheless, the central limit theorem applies in a more generalized form, and the normal distribution is a good approximation to the hypergeometric distribution where n is large, that is

$$\mathbf{h}(x; r, p, n) \approx \mathbf{n}(n; \mu, \sigma) = \frac{1}{\sigma}\varphi\left(\frac{x - \mu}{\sigma}\right) \tag{25.53}$$

where

$$\mu = rp \qquad \text{and} \qquad \sigma^2 = rp(1 - p)\left(\frac{n - r}{n - 1}\right) \tag{25.54}$$

by (23.4) and (23.5).

The comparison (25.53) is shown for some examples in Figures 25.9.

The normal distribution is especially good when p is close to 0.5. Here $n \ge 10$ yields an approximation good enough for many purposes.

Because the hypergeometric is a *discrete* distribution very much like the binomial, the **half-integer corrections** may be made; a more accurate approximation is

$$\mathbf{h}(x; r, p, n) = \Phi\left(\frac{x + 1/2 - \mu}{\sigma}\right) - \Phi\left(\frac{x - 1/2 - \mu}{\sigma}\right) \tag{25.55}$$

If we are interested in the *range* of values for a variate described by the hypergeometric distribution, then we use the cumulative hypergeometric distribution

$$\mathbf{H}(x; r, p, n) = \sum_{X=0}^{x} \mathbf{h}(X; r, p, n) = \Pr(X \le x) \tag{25.56}$$

to get

$$\Pr(a \le X \le b) = \sum_{x=a}^{b} \mathbf{h}(X; r, p, n) = \mathbf{H}(b; r, p, n) - \mathbf{H}(a; r, p, n) \tag{25.57}$$

Alternatively,

$$\Pr(a \le X \le b) = \Pr\left(\frac{a - \mu}{\sigma} \le X^* \le \frac{b - \mu}{\sigma}\right)$$
$$\approx \int_{a-1/2}^{b+1/2} \mathbf{n}(x; r, p, n)\, dx$$
$$= \Phi\left(\frac{b + 1/2 - \mu}{\sigma}\right) - \Phi\left(\frac{a + 1/2 - \mu}{\sigma}\right) \tag{25.58}$$

(a)

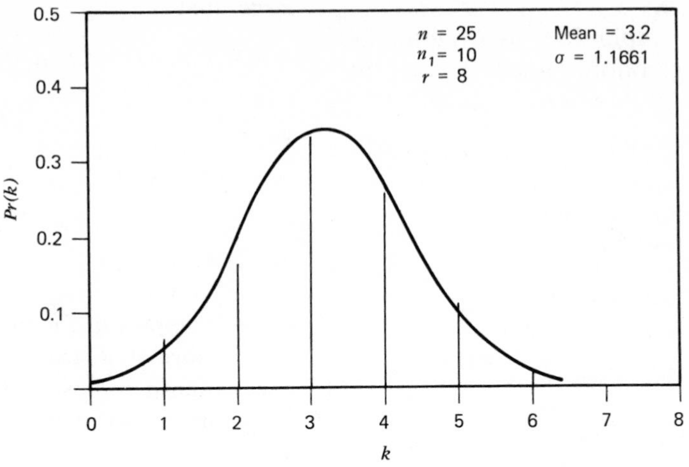

0.5
0.4
0.3
0.2
0.1
Pr(k)
n = 25
n1= 10
r = 8
Mean = 3.2
σ = 1.1661
0
1
2
3
4
5
6
7
8
k

(b)

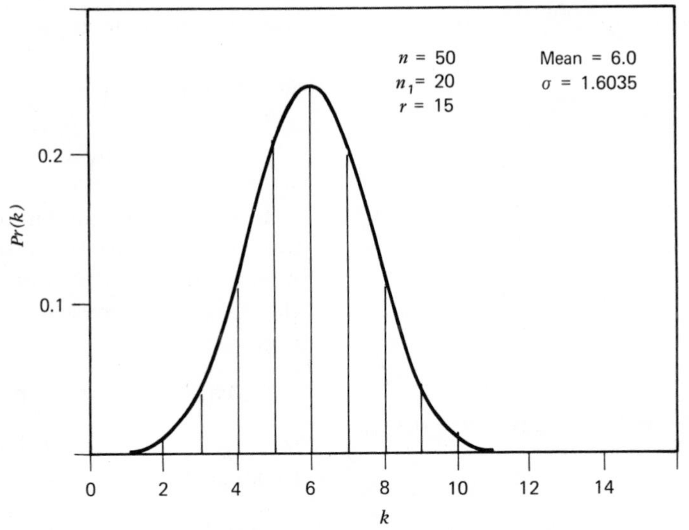

0.2
0.1
Pr(k)
n = 50
n1= 20
r = 15
Mean = 6.0
σ = 1.6035
0
2
4
6
8
10
12
14
k

(c)

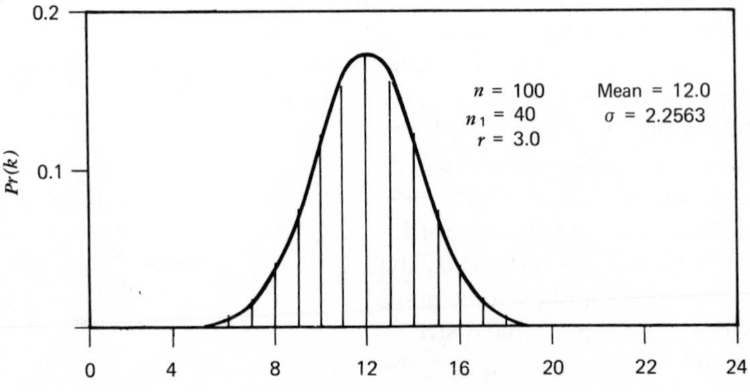

0.2
0.1
Pr(k)
n = 100
n1 = 40
r = 3.0
Mean = 12.0
σ = 2.2563
0
4
8
12
16
20
22
24
k

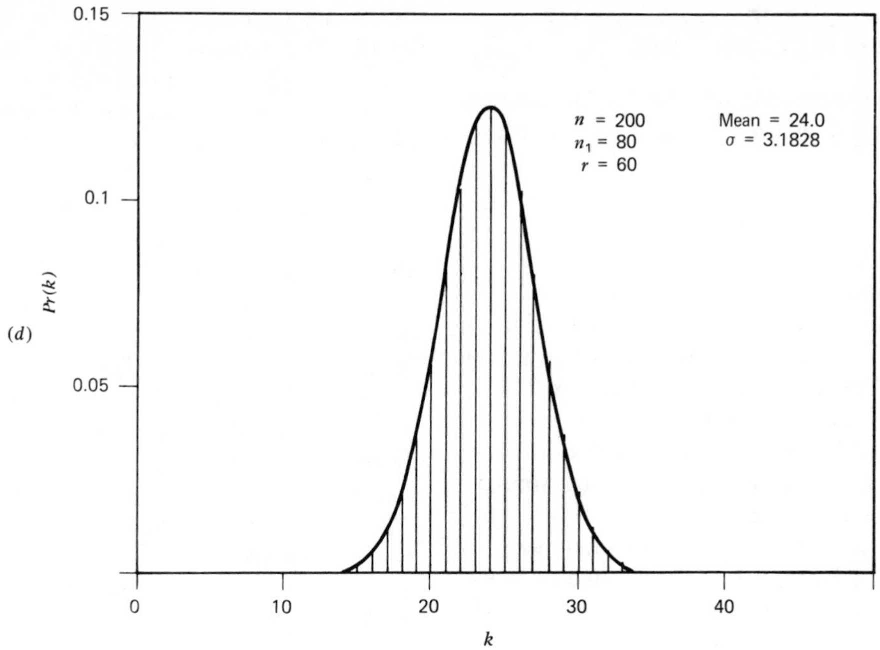

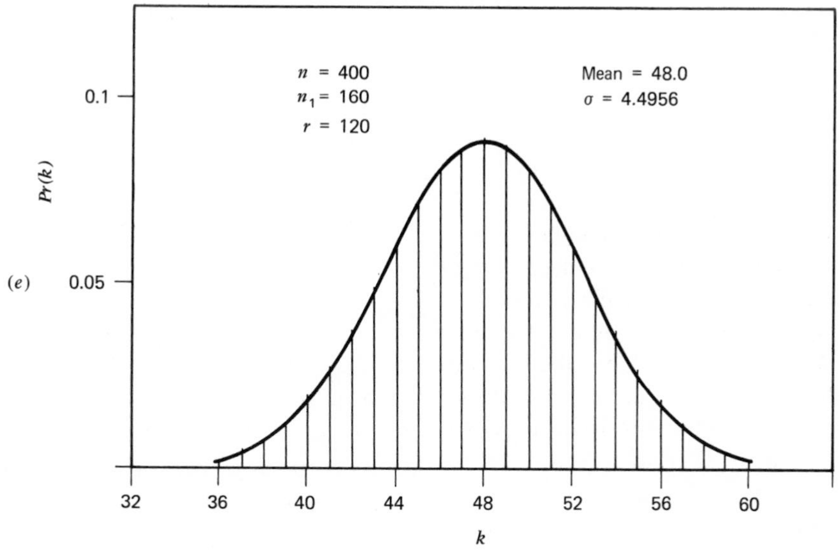

Figures 25.9
Comparison of the (discrete) hypergeometric distribution and the (continuous) normal approximation to it. (*a*) $n = 25$, $n_1 = 10$, $r = 8$. (*b*) $n = 50$, $n_1 = 20$, $r = 15$. (*c*) $n = 100$, $n_1 = 40$, $r = 3.0$. (*d*) $n = 200$, $n_1 = 80$, $r = 60$. (*e*) $n = 400$, $n_1 = 160$, $r = 120$.

where μ and σ are given by (25.54) and

$$\mathbf{H}(x; r, p, n) \approx \Phi\left(\frac{x + 1/2 - \mu}{\sigma}\right) \tag{25.59}$$

NORMAL APPROXIMATION TO OTHER DISTRIBUTIONS

We have considered sums of **independent** *uniformly distributed* random variables

$$x = \sum_{i=1}^{n} x_i$$

in Chapter 21. The underlying distribution of the individual x_i is constant, that is, quite unlike the normal distribution. Nevertheless, we see in Figures 21.6 to 21.19 that the distribution is well described by the normal distribution curve superimposed on the plots. The approximation is quite good for $n \geq 3$. We remark that the curve plotted on the histograms is of the form $\mathbf{n}(x; \mu, \sigma)$ multiplied by a normalization constant, A, to equate the areas under the histogram and the curve. For the uniform distribution

$$\mu = \sum_{i=1}^{n} \mu_i = \frac{n(1.0-0)}{2} = \frac{n}{2} \tag{25.60}$$

and

$$\sigma^2 = \sum_{i=1}^{n} \sigma_i^2 = \frac{n(1.0-0)^2}{12} = \frac{n}{12}$$

In general, if n is **large enough** the individual variables in the sum *need not be independent.* We may usually take any sum of *more than 25* variables to represent a variate that is well-described as normally distributed. For some distributions without peculiar characteristics (e.g., the uniform) the number may safely be set smaller.

We note in particular, that the **mean** *of almost any distribution is approximately normally distributed.*

The central limit theorem underlies the universal applicability of the normal distribution. **Whenever a variate may be assumed to be the result of a large number of small effects, the distribution is approximately normal.**

Galton devised a mechanical demonstration of this. A schematic version of "**Galton's Quicunx**" is shown in Figure 25.10. (The "quicunx" refers to a pattern used for laying out orchards of fruit trees.) A more rigorous model suitable for demonstration is shown in Figure 25.11. Instead of a large but finite grid of obstacles, this model uses the *microscopic* granularity of a "smooth surface" to produce the deflection of the individual balls. The deflections are essentially *infinitely large* in number and, individually, *infinitesimal* in amount. The requirements of the central limit theorem are met almost exactly. By varying the angle to the horizontal, the width of the resulting Gaussian distribution may be changed. The width may also be changed by varying the distance the balls traverse. (This may be regarded as a *random walk* away from the center.)

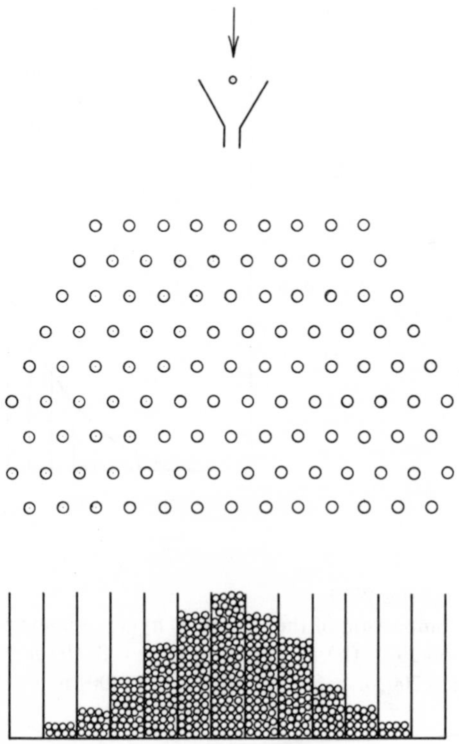

Figure 25.10
Schematic illustration of "Galton's Quicunx" demonstration of the central limit theorem.

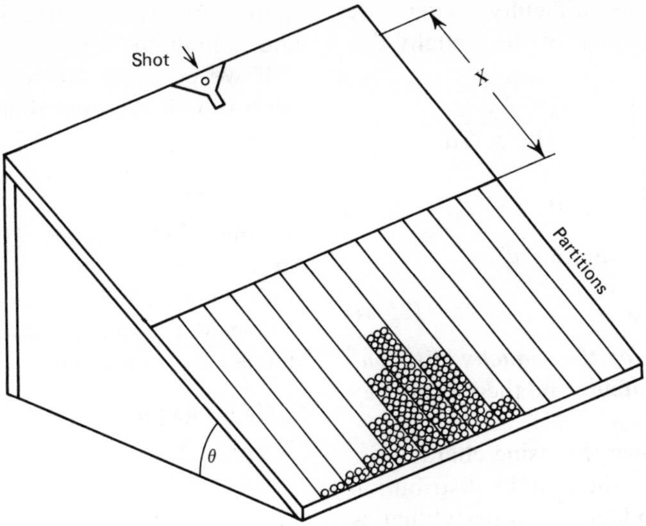

Figure 25.11
Improved demonstration of the central limit theorem (see text).

EXCEPTIONS TO THE CENTRAL LIMIT THEOREM

Consider the **Cauchy distribution**

$$f(x) \equiv \mathbf{c}(x; \mu, \alpha) = \frac{1}{\pi\alpha}\frac{1}{1+(x+\mu)^2/\alpha^2}$$
$$-\infty < x < +\infty \quad (25.61)$$

This distribution* has certain peculiarities that bear on the applicability of the central limit theorem.

We note first that the Cauchy distribution is *normalized*:

$$\int_{-\infty}^{\infty} f(t)\,dt = \frac{1}{\pi\alpha}\int_{-\infty}^{\infty}\frac{1}{1+(t-\mu)^2/\alpha^2}dt$$
$$= \frac{1}{\pi\alpha}\int_{-\infty}^{\infty}\frac{\alpha\,du}{1+u^2}$$

where $u \equiv \dfrac{t-\mu}{\alpha}$

$$= \frac{1}{\pi}\left|\tan^{-1} u\right|_{-\infty}^{\infty}$$
$$= \frac{1}{\pi}\left[\frac{\pi}{2}-\left(\frac{-\pi}{2}\right)\right]$$
$$= 1 \quad (25.62)$$

*See Chapter 27.

If we consider the *mean* of the Cauchy distribution, we get

$$\langle x\rangle = \int_{-\infty}^{\infty} tf(t)\,dt$$
$$= \int_{-\infty}^{\infty}[(t-\mu)+\mu]f(t)\,dt$$
$$= \frac{1}{\pi\alpha}\left\{\int_{-\infty}^{\infty}\frac{(t-\mu)\,dt}{1+(t-\mu)^2/\alpha^2}\right\}+\mu$$
$$= \mu + \frac{1}{\pi\alpha}\left|\frac{1}{2}\ln\left[1+\frac{(t-\mu)^2}{\alpha^2}\right]\right|_{-\infty}^{\infty}$$
$$= \mu + \infty - \infty$$

This is *indeterminate*, that is, the mean does not exist, strictly speaking.

However, since $f(x)$ is symmetrical about μ, it is convenient to *define* the mean

$$\langle x\rangle = \lim_{R\to\infty}\int_{-R}^{+R} tf(t)\,dt$$
$$= \lim_{R\to\infty}\int_{-R}^{R}[(t-\mu)+\mu]f(t)dt$$
$$= \mu + \lim_{R\to\infty}\frac{1}{\pi\alpha}\left|\ln\left[1+\frac{(t-\mu)^2}{\alpha^2}\right]\right|_{-R}^{R}$$
$$= \mu \quad (25.63)$$

Also the mode and the median are both equal to μ.

We also have difficulty when we evaluate the *variance* of the Cauchy distribution.

$$\sigma^2(x) = \frac{1}{\pi\alpha}\int_{-\infty}^{\infty} \frac{(t-\mu)^2 dt}{1+(t-\mu)^2/\alpha^2}$$
$$= \frac{\alpha^2}{\pi}\int_{-\infty}^{\infty} \frac{u^2 du}{1+u^2}$$
$$= \frac{\alpha^2}{\pi}|u - \arctan u|_{-\infty}^{\infty}$$
$$= \infty - \alpha^2 = \infty \qquad (25.64)$$

The *variance of the Cauchy distribution*, is, in fact, **infinite** and there is no way to define this away.

It may be shown (by using characteristic functions) that the Cauchy distribution exhibits a reproductive property that is quite different from that noted for the binomial, Poisson, and normal distributions. In fact, if X_1 and X_2 are two independent random variables, each Cauchy-distributed with p.d.f. given by (25.61), then their mean $\bar{X} = (X_1 + X_2)/2$ is Cauchy-distributed with the p.d.f. (25.61).

Also, if $\bar{X} = (1/n)\Sigma X_i$, with each X_i, having a p.d.f. given by (25.61), then the p.d.f. of $\bar{X}$ is given again by (25.61).

This is quite unusual: *the* ***mean*** *of any number of independent Cauchy random variables has the same distribution as any* **one** *of them.*

In our discussion of the sample mean in Chapter 22 we considered

$$\Pr\left(\left|\frac{\Sigma X_i}{n} - \mu\right| \le \varepsilon\right) \qquad \text{as} \qquad n \to \infty$$

For the Cauchy case,

$$\Pr\left(\left|\frac{\Sigma X_i}{n} - \mu\right| \le \varepsilon\right) = \Pr(|X_i - \mu| \le \varepsilon)$$

and $\bar{X}$ does **not** approach μ with unity probability as $n \to \infty$. We have the remarkable (and inconvenient) result that **the average of *n* Cauchy-distributed observations is no better than a single observation** in estimating the "population mean" or "true value." That is, the limiting distribution of the mean of a large number of Cauchy variates is not normal. The central limit theorem does not apply because the requirements are not met (existence of a mean and noninfinite variance).

If we consider the requirement of the central limit theorem that

$$\sum_{i=1}^{n} \sigma_i^2 \to \infty \qquad \text{as} \qquad n \to \infty$$

we may follow Arley (1950)* and construct another example where the central limit theorem fails to apply.

Consider **Laplace's distribution** that is defined to have a p.d.f.

$$f(x) = \mathbf{l}(x;\mu,\alpha)$$
$$= \frac{1}{2\alpha} e^{-|x-\mu|/\alpha} \qquad -\infty < x < +\infty \qquad (25.65)$$

We may check that *this is normalized,*

$$\int_{-\infty}^{\infty} \mathbf{l}(x;\mu,\alpha)\,dx = 1 \qquad (25.66)$$

the mean is μ,

$$\int_{-\infty}^{\infty} x\,\mathbf{l}(x;\mu,\alpha) = \mu \qquad (25.67)$$

and that *the variance is given by*

$$\sigma^2(x) = \int_{-\infty}^{\infty} (t-\mu)^2 \mathbf{l}(t;\mu,\alpha)\,dt = 2\alpha^2 \qquad (25.68)$$

Consider the special case

$$\mu = 0 \qquad \alpha = 1$$

such that the random variable x is described by the p.d.f.

$$f(x) = \tfrac{1}{2} e^{-|x|} \qquad -\infty < x < +\infty \qquad (25.69)$$

We consider the sum

$$Z = \sum_{i=1}^{n} x_i$$

where

$$x_i = \frac{2}{\pi}\frac{1}{(2i-1)}x$$

Therefore

$$\sum_{i=1}^{n} \langle x_i \rangle = \langle x \rangle = 0$$

*N. Arley and K. R. Buch, *Introduction to the Theory of Probability and Statistics*, John Wiley, 1950.

and

$$\sum_{i=1}^{n} \sigma_i^2 = \left(\frac{2}{\pi}\right)^2 2 \sum_{i=1}^{n} \left(\frac{1}{(2i-1)}\right)^2$$

It may be shown* that

$$\sum_{i=1}^{n} (2i-1)^{-2} \to \pi^2/8 \qquad \text{as} \qquad n \to \infty$$

and

$$\sum_{i=1}^{n} \sigma_i^2 \to 1 \qquad \text{as} \qquad n \to \infty$$

*E. T. Whittaker and G. N. Watson, *Modern Analysis*, Cambridge University Press, 1935.

*This **violates** one of the conditions on the central limit theorem.*

It may be shown† that, as $n \to \infty$

$$f(Z) \to \frac{1}{2\cosh(\pi Z/2)} = \{e^{\pi Z/2} + e^{-\pi Z/2}\}^{-1}$$

which is obviously *different* from

$$f(Z) \to \mathbf{n}(Z; 0, \sqrt{2})$$

†See Arley (1950), p. 110.

26

THE CHI-SQUARE DISTRIBUTION

Consider a set of n variates x_i, $i = 1, \ldots, n$, which are *independent* and each of which is $N(\bar{x}_i, \sigma_i^2)$ (i.e., *each is normally distributed* with mean $\bar{x}_i$ and variance σ_i^2).

We are interested in determining "**best values**" of $\bar{x}_i$ from n individual *measurements* or *data points* x_i, and we shall regard the $\bar{x}_i$ as *estimates* of the true mean in order to consider the $\bar{X}_i$ as *random variables* in the sense of our discussion of inverse probabilities in Chapter 25.

Chi-Square and Minimization

By (25.21), the probability that such $\bar{x}_i$ lies in an interval $d\bar{x}_i$ is given by

$$P(x_i, \bar{x}_i, \sigma_i)\, d\bar{x}_i = \frac{\exp\left\{-\frac{1}{2}\left[\frac{x_i - \bar{x}_i}{\sigma_i}\right]^2\right\} d\bar{x}_i}{\sqrt{2\pi\sigma_i^2}} \tag{26.1}$$

We form a "hyperspace" in the n coordinates $\bar{x}_i$, and calculate the probability, G, that [($\bar{x}_1$ lies in $d\bar{x}_1$) *and* ($\bar{x}_2$ lies in $d\bar{x}_2$) *and* $\cdots$ *and* ($\bar{x}_n$ lies in $d\bar{x}_n$)]:

$$G(x_1, \ldots, x_n; \bar{x}_1, \ldots, \bar{x}_n; \sigma_1, \ldots, \sigma_n)\prod_{i=1}^{n} d\bar{x}_i = \prod_{i=1}^{n} \frac{\exp\left[-\frac{1}{2}\left(\frac{x_i - \bar{x}_i}{\sigma_i}\right)^2\right] d\bar{x}_i}{\sqrt{2\pi\sigma_i^2}}$$

$$= \exp\left[-\sum_{i=1}^{n}\frac{1}{2}\left(\frac{x_i - \bar{x}_i}{\sigma_i}\right)^2\right]\prod_{i=1}^{n}\frac{1}{\sqrt{2\pi\sigma_i^2}}\, d\bar{x}_i \tag{26.2}$$

We define "chi-square":

$$\chi^2 \equiv \sum_{i=1}^{n}\left(\frac{x_i - \bar{x}_i}{\sigma_i}\right)^2 \tag{26.3}$$

We rewrite (26.2)

$$G(x_1, \ldots, x_n; \bar{x}_1, \ldots, \bar{x}_n; \sigma_1, \ldots, \sigma_n) = e^{-\chi^2/2}\prod_{i=1}^{n}\frac{1}{(2\pi\sigma_i^2)^{1/2}} \tag{26.4}$$

Note that, if we wish to obtain the "best values" of $\bar{x}_i$, we must *maximize* G with respect to the estimates $\bar{x}_i$. Since the probability, G, is the likelihood function, we are **maximizing the likelihood**. This is equivalent to ***minimizing*** $\boldsymbol{\chi^2}$ with respect to $\bar{x}_i$.

If the $\bar{x}_i$ are all independent, the condition becomes

$$\frac{\partial\chi^2}{\partial\bar{x}_i} = 0 \qquad i = 1, \ldots, n \tag{26.5}$$

or

$$\frac{-2}{\sigma_i^2}(x_i - \bar{x}_i) = 0$$

which yields the result

$$\bar{x}_i = x_i \qquad \text{and} \qquad \chi^2_{\min} = 0 \qquad (26.5a)$$

This result is reasonable since the best estimate should be the measurement itself, for the case when we have only one measurement.

CHI-SQUARE WITH (HOLONOMIC) CONSTRAINTS

The *measurements* x_i are all independent, but the $\bar{x}_i$ need *not* be. We could have m relations among the n values $\bar{x}_i$, $i = 1, \ldots, n$ of the general form

$$\varphi_j(\bar{x}_1, \ldots, \bar{x}_n) = 0 \qquad j = 1, \ldots, m \quad (26.6)$$

Each of these is called a *constraint equation*. The constraint is called "*holonomic*" because a constraint *equation* (equality) exists. In general, then, we would maximize G with respect to $\bar{x}_i$ *subject to the constraints* (26.6). These would introduce *correlations* among the $\bar{x}_i$.

For example, the $\bar{x}_i$ could all be equal, that is, we would consider n measurements of the *same* quantity. Here we would have $(n-1)$ constraint equations of the form:

$$\bar{x}_1 = \bar{x}_2; \bar{x}_1 = \bar{x}_3; \ldots; \bar{x}_1 = \bar{x}_n \quad (26.7)$$

► **Example 26.1**
Three measurements of the same quantity
Consider the simple case in which we make three measurements, (x_1, x_2, and x_3) of the same quantity which we estimate by $\bar{x}$. We assume that the errors are normally distributed and that the measurements have equal weight, that is, they have the same standard deviation, δ, although we do not know *a priori* what this is.

We may form

$$\chi^2 = \frac{(x_1 - \bar{x})^2}{\delta^2} + \frac{(x_2 - \bar{x})^2}{\delta^2} + \frac{(x_3 - \bar{x})^2}{\delta^2}$$

and minimize this by setting

$$\frac{d\chi^2}{d\bar{x}} = 0 = \frac{-2(x_1 - \bar{x})}{\delta^2} + \frac{-2(x_2 - \bar{x})}{\delta^2} + \frac{-2(x_3 - \bar{x})}{\delta^2}$$

which yields

$$\frac{(x_1 + x_2 + x_3)}{3} = \bar{x}$$

that is, the best estimate is obtained by *averaging* the measurements. ◄

We can treat the general case of constraints (26.6) in two ways. The relation (26.4) could be reexpressed in terms of $(n-m)$ *independent* variables $\bar{y}_i$ using relations (26.6). This would yield

$$\frac{\partial \chi^2}{\partial \bar{y}_i} = 0 \qquad i = 1, \ldots, (n-m) \quad (26.8)$$

► **Example 26.2**
Measurements of the angles of a triangle
Consider that independent measurements have been made of the angles of a triangle (flat space assumed) yielding the results $(\theta_1 \pm \sigma_1)$, $(\theta_2 \pm \sigma_2)$, and $(\theta_3 \pm \sigma_3)$.

What are the best values for the angles, α, β, γ?

$$\chi^2 = \frac{(\theta_1 - \alpha)^2}{\sigma_1^2} + \frac{(\theta_2 - \beta)^2}{\sigma_2^2} + \frac{(\theta_3 - \gamma)^2}{\sigma_3^2}$$

and

$\pi = \alpha + \beta + \gamma$ is the constraint.

$$\therefore \chi^2 = \frac{(\theta_1 - \alpha)^2}{\sigma_1^2} + \frac{(\theta_2 - \beta)^2}{\sigma_2^2} + \frac{(\theta_3 - \pi + \alpha + \beta)^2}{\sigma_3^2}$$

We now have only two *independent* angles: α, β.

$$\frac{\partial \chi^2}{\partial \alpha} = 0 = \frac{2(\theta_1 - \alpha)}{\sigma_1^2}(-1) + \frac{2(\theta_3 - \pi + \alpha + \beta)}{\sigma_3^2}$$

$$\frac{\partial \chi^2}{\partial \beta} = 0 = \frac{2(\theta_2 - \beta)}{\sigma_2^2}(-1) + \frac{2(\theta_3 - \pi + \alpha + \beta)}{\sigma_3^2}$$

$$\frac{(\theta_1 - \alpha)}{\sigma_1^2} = \frac{(\theta_2 - \beta)}{\sigma_2^2}$$

or

$$\beta = \left\{ \frac{\theta_2}{\sigma_2^2} - \frac{\theta_1}{\sigma_1^2} + \frac{\alpha}{\sigma_1^2} \right\} \sigma_2^2$$

$$\therefore \frac{\theta_1 - \alpha}{\sigma_1^2} = \frac{\theta_3 - \pi + \alpha}{\sigma_3^2} + \frac{\sigma_2^2}{\sigma_3^2} \left\{ \frac{\theta_2}{\sigma_2^2} - \frac{\theta_1}{\sigma_1^2} + \frac{\alpha}{\sigma_1^2} \right\}$$

$$\frac{\theta_1}{\sigma_1^2} + \frac{\theta_1 \sigma_2^2}{\sigma_3^2 \sigma_1^2} - \frac{\theta_2}{\sigma_3^2} - \frac{\theta_3}{\sigma_3^2} - \frac{\pi}{\sigma_3^2} = \alpha \left(\frac{1}{\sigma_1^2} + \frac{1}{\sigma_3^2} + \frac{\sigma_2^2}{\sigma_3^2 \sigma_1^2} \right)$$

and

$$\theta_1 \left(\frac{1}{\sigma_1^2} + \frac{\sigma_2^2}{\sigma_3^2 \sigma_1^2} + \frac{1}{\sigma_3^2} \right) - \frac{(\theta_1 + \theta_2 + \theta_3 - \pi)}{\sigma_3^2} = \alpha \left(\frac{\sigma_3^2 + \sigma_1^2 + \sigma_2^2}{\sigma_3^2 \sigma_1^2} \right)$$

We get

$$\alpha = \left(\frac{\sigma_1^2 \sigma_3^2}{\sigma_1^2 + \sigma_2^2 + \sigma_3^2} \right) \left\{ \theta_1 \left[\frac{\sigma_3^2 + \sigma_2^2 + \sigma_1^2}{\sigma_3^2 \sigma_1^2} \right] - \frac{(\theta_1 + \theta_2 + \theta_3 - \pi)}{\sigma_3^2} \right\}$$

$$\alpha = \theta_1 - \frac{\sigma_1^2}{\sigma_1^2 + \sigma_2^2 + \sigma_3^2} (\theta_1 + \theta_2 + \theta_3 - \pi)$$

But the three angles are treated symmetrically; thus we may write

$$\alpha_i = \theta_1 - \frac{\sigma_i^2}{\sigma_1^2 + \sigma_2^2 + \sigma_3^2} (\theta_1 + \theta_2 + \theta_3 - \pi)$$

where

$$\alpha_1 = \alpha$$

$$\alpha_2 = \beta$$

$$\alpha_3 = \gamma$$ ◄

The more general method of minimizing χ^2 subject to constraints is that of **Lagrange multipliers** reviewed in Appendix A. Here each constraint equation, $\varphi_j = 0$, is multiplied by λ_j and added. The resulting minimization relation is

$$\frac{\partial \chi^2}{\partial \bar{x}_i} + \sum_{j=1}^{m} \lambda_j \frac{\partial \varphi_j}{\partial \bar{x}_i} = 0 \qquad i = 1, \ldots, n \qquad (26.9)$$

Here we have $(n + m)$ unknown quantities:

$$\bar{x}_i, i = 1, \ldots, n$$

and

$$\lambda_j, j = 1, \ldots, m$$

We have $(n + m)$ equations:

m relations (26.6)

and n relations of the form (26.9):

$$-2 \frac{(x_i - \bar{x}_i)}{\sigma_i} + \sum_{j=1}^{m} \lambda_j \frac{\partial \varphi_j}{\partial \bar{x}_i} = 0 \qquad (26.10)$$

$$i = 1, \ldots, n$$

and we may solve for the unknowns $\bar{x}_i$ (and λ_j).

► **Example 26.3**

Measurement of the sides of a right triangle

Consider that we measure the three sides of a right triangle and get the values a, b, and c. We assume the measurements are normally distributed with standard deviation σ.

What are the best estimates (x, y, z) for the lengths of the three sides where z is the estimate for the hypotenuse?

These measurements are subject to the constraint

$$\varphi(x, y, z) = x^2 + y^2 - z^2 = 0 \qquad \text{(a)}$$

We form

$$\chi^2 = \left(\frac{x - a}{\sigma} \right)^2 + \left(\frac{y - b}{\sigma} \right)^2 + \left(\frac{z - c}{\sigma} \right)^2$$

The minimum is obtained from (26.9):

$$\frac{\partial \chi^2}{\partial x} + \lambda \frac{\partial \varphi}{\partial x} = 0 = 2 \left(\frac{x - a}{\sigma} \right) \frac{1}{\sigma} + \lambda 2x \qquad \text{(b)}$$

$$\frac{\partial \chi^2}{\partial y} + \lambda \frac{\partial \varphi}{\partial y} = 0 = 2 \left(\frac{y - b}{\sigma} \right) \frac{1}{\sigma} + \lambda 2y \qquad \text{(c)}$$

$$\frac{\partial \chi^2}{\partial z} + \lambda \frac{\partial \varphi}{\partial z} = 0 = 2 \left(\frac{z - c}{\sigma} \right) \frac{1}{\sigma} - \lambda 2z \qquad \text{(d)}$$

The four equations (a), (b), (c), and (d) may be solved for the four unknowns: x, y, z, and λ.

From (b)

$$\frac{x}{\sigma^2} - \frac{a}{\sigma^2} + 2\lambda x = 0$$

$$x \left(\frac{1}{\sigma^2} + 2\lambda \right) = \frac{a}{\sigma^2}$$

and

$$x(1 + 2\lambda\sigma^2) = a \qquad \text{(e)}$$

Similarly,

$$y(1+2\lambda\sigma^2)=b \qquad \text{(f)}$$

and

$$z(1-2\lambda\sigma^2)=c \qquad \text{(g)}$$

Then

$$x=a\left[\frac{1}{1+2\lambda\sigma^2}\right]=a\left[1-\frac{2\lambda\sigma^2}{1+2\lambda\sigma^2}\right]$$

$$y=b\left[\frac{1}{1+2\lambda\sigma^2}\right]=b\left[1-\frac{2\lambda\sigma^2}{1+2\lambda\sigma^2}\right]$$

and

$$z=c\left[\frac{1}{1-2\lambda\sigma^2}\right]=c\left[1+\frac{2\lambda\sigma^2}{1-2\lambda\sigma^2}\right]$$

From (a)

$$\frac{a^2}{(1+2\lambda\sigma^2)^2}+\frac{b^2}{(1+2\lambda\sigma^2)^2}-\frac{c^2}{(1-2\lambda\sigma^2)^2}=0$$

and

$$\frac{a^2+b^2}{(1+2\lambda\sigma^2)^2}=\frac{c^2}{(1-2\lambda\sigma^2)^2}$$

We take the square root of both sides,

$$\frac{\sqrt{a^2+b^2}}{1+2\lambda\sigma^2}=\frac{c}{1-2\lambda\sigma^2}$$

$$\therefore \frac{\sqrt{a^2+b^2}}{c}=\frac{1+2\lambda\sigma^2}{1-2\lambda\sigma^2}=1+\frac{4\lambda\sigma^2}{1-2\lambda\sigma^2}$$

and

$$\frac{2\lambda\sigma^2}{1-2\lambda\sigma^2}=\frac{1}{2}\left(\frac{\sqrt{a^2+b^2}}{c}-1\right)=\frac{\sqrt{a^2+b^2}-c}{2c}$$

Also,

$$\frac{c}{\sqrt{a^2+b^2}}=\frac{1-2\lambda\sigma^2}{1+2\lambda\sigma^2}=1-\frac{4\lambda\sigma^2}{1+2\lambda\sigma^2}$$

$$\therefore \frac{2\lambda\sigma^2}{1+2\lambda\sigma^2}=\frac{1}{2}\left(1-\frac{c}{\sqrt{a^2+b^2}}\right)=\frac{\sqrt{a^2+b^2}-c}{2\sqrt{a^2+b^2}}$$

Therefore,

$$x=a\left[1-\frac{\sqrt{a^2+b^2}-c}{2\sqrt{a^2+b^2}}\right]$$

$$y=b\left[1-\frac{\sqrt{a^2+b^2}-c}{2\sqrt{a^2+b^2}}\right]$$

and

$$z=c\left[1+\frac{\sqrt{a^2+b^2}-c}{2c}\right]$$

Finally,

$$x=a\left[1-\frac{(\sqrt{a^2+b^2}-c)(\sqrt{a^2+b^2}+c)}{2\sqrt{a^2+b^2}(\sqrt{a^2+b^2}+c)}\right]$$

$$=a\left[1-\frac{a^2+b^2-c^2}{2(a^2+b^2+c\sqrt{a^2+b^2})}\right]$$

$$y=b\left[1-\frac{a^2+b^2-c^2}{2(a^2+b^2+c\sqrt{a^2+b^2})}\right]$$

$$z=c\left[1+\frac{a^2+b^2-c^2}{2(c^2+c\sqrt{a^2+b^2})}\right]$$ ◄

n independent measurements of the same quantity

Because of its importance, let us extend Example 26.1. For (26.7),

$$\bar{x}_i=\bar{x} \qquad (26.11)$$

Here

$$\chi^2=\sum_{i=1}^{n}\left(\frac{x_i-\bar{x}}{\sigma_i}\right)^2$$

The condition for a minimum is

$$\frac{\partial\chi^2}{\partial\bar{x}}=0=\sum_{i=1}^{n}\frac{(x_i-\bar{x})}{\sigma_i^2}$$

$$\sum_{i=1}^{n}\left(\frac{x_i}{\sigma_i^2}\right)=\bar{x}\sum_{i=1}^{n}\frac{1}{\sigma_i^2}$$

and

$$\bar{x}=\frac{\sum_{i=1}^{n}(x_i/\sigma_i^2)}{\sum_{i=1}^{n}(1/\sigma_i^2)} \qquad (26.12)$$

for the best estimate.

We may define

$$\frac{1}{\bar{\sigma}^2}\equiv\frac{1}{\sigma_{\bar{x}}^2}=\sum_{i=1}^{n}\frac{1}{\sigma_i^2} \qquad (26.13)$$

where we note that

$$\operatorname{Var}(\bar{x})=\sigma_{\bar{x}}^2=\left[\sum_{i=1}^{n}(1/\sigma_i^2)\right]^{-1}$$

by (20.17c).

If all the σ_i are equal

$$\sigma_i^2=\sigma^2 \qquad \text{for } i=1,\ldots,n \qquad (26.14)$$

then

$$\bar{\sigma}^2=\frac{\sigma^2}{n}=\sigma_{\bar{x}}^2 \qquad (26.15)$$

and

$$\bar{x}=\frac{\Sigma x_i}{n} \qquad (26.16)$$

Note that for n measurements with equal error, the normal distribution yields the arithmetic average as the best estimator.

We have obtained these results earlier on more general grounds *not requiring the normal distribution.* See relations (10.4h), (10.4i), and Chapter 8.

One-standard-deviation limits for one-parameter χ^2

We note what happens if we allow our estimate of the mean, $\hat{\mu} = \bar{x}$, to vary by one standard deviation:

$$\begin{aligned}
\chi^2(\bar{x} \pm \Delta\bar{x}) &= \sum_{i=1}^{n} \frac{(x_i - \bar{x} \pm \Delta\bar{x})^2}{\sigma_i^2} \\
&= \sum_{i=1}^{n} \frac{[(x_i - \bar{x}) \pm \Delta\bar{x}]^2}{\sigma_i^2} \\
&= \sum_{i=1}^{n} \frac{(x_i - \bar{x})^2}{\sigma_i^2} \pm 2(\Delta\bar{x}) \sum_{i=1}^{n} \frac{(x_i - \bar{x})}{\sigma_i^2} \\
&\quad + (\Delta\bar{x})^2 \sum_{i=1}^{n} \left(\frac{1}{\sigma_i^2}\right) \\
&= \sum_{i=1}^{n} \frac{(x_i - \bar{x})^2}{\sigma_i^2} + (\Delta\bar{x})^2 \sum_{i=1}^{n} \left(\frac{1}{\sigma_i^2}\right) \\
&= \chi^2_{\text{min}} + 1 \qquad (26.17)
\end{aligned}$$

since $(\Delta\bar{x})^2 = \left\{\sum_{i=1}^{n} (1/\sigma_i^2)\right\}^{-1}$

Therefore, for a one-parameter chi-square problem the one-standard-deviation "error" on the parameter is determined by where χ^2 increases from its minimum by 1. Also, $\chi^2_{\text{min}} + 4$ determines where the "error" on one parameter corresponds to two standard deviations.

Expected value of χ^2

We may expand χ^2 about its minimum value χ_0^2:

$$\begin{aligned}
\chi^2 = \chi_0^2 &+ \sum_{i=1}^{n} \left.\frac{\partial\chi^2}{\partial\bar{x}_i}\right|_0 (\bar{x}_i - \bar{x}_i^0) \\
&+ \frac{1}{2} \sum_{\substack{i=1 \\ j=1}}^{n} \left.\frac{\partial^2\chi^2}{\partial\bar{x}_i \partial\bar{x}_j}\right|_0 (\bar{x}_i - \bar{x}_i^0)(\bar{x}_j - \bar{x}_j^0) \\
&+ \cdots
\end{aligned}$$

$$\begin{aligned}
\chi^2 - \chi_0^2 &= \frac{1}{2} \sum_{i,j=1}^{n} \left.\frac{\partial^2\chi^2}{\partial\bar{x}_i \partial\bar{x}_j}\right|_0 (\bar{x}_i - \bar{x}_i^0)(\bar{x}_j - \bar{x}_j^0) \\
&\quad + \text{higher order terms} \qquad (26.18)
\end{aligned}$$

since the first derivatives vanish by (26.8).

This relation defines a quadratic surface (a hyperparaboloid) in $(n + 1)$-dimensional space. If we draw a plane

$$\chi^2 = \chi_0^2 + 1 \qquad (26.19)$$

we intersect the hyperparaboloid in a curve that is a hyperellipse. The values of $\bar{x}_i$ corresponding to the boundary of the hyperellipse define the one-standard-deviation limits on $\bar{x}_i^0$.

For n measurements of the same quantity

$$\chi_0^2 = \sum_{i=1}^{n} \frac{(x_i - \bar{x})^2}{\sigma_i^2}$$

and

$$\bar{x} = \bar{\sigma}^2 \sum_{i=1}^{n} \frac{x_i}{\sigma_i^2}$$

by (26.12) and (26.13).

Algebraic manipulation yields the result

$$\chi_0^2 = \sum_{i=1}^{n} \frac{x_i^2}{\sigma_i^2} - \bar{\sigma}^2 \left[\sum_{i=1}^{n} \frac{x_i}{\sigma_i^2}\right]^2$$

We now wish to take averages (or, find the expectation value of both sides of the equation).

$$\langle\chi_0^2\rangle = \left\langle \sum_{i=1}^{n} \frac{x_i^2}{\sigma_i^2} \right\rangle - \left\langle \bar{\sigma}^2 \left[\sum_{i=1}^{n} \frac{x_i}{\sigma_i^2}\right]^2 \right\rangle \qquad (26.20)$$

where we have used the $\langle x \rangle$ form of denoting expectation value.

We note that the expectation value

$$\langle x_i x_j \rangle = \bar{x}^2 + \sigma_i^2 \delta_{ij}$$

where

$$\begin{aligned}
\delta_{ij} &= 1 \quad \text{if} \quad i = j \\
&= 0 \quad \text{if} \quad i \neq j
\end{aligned}$$

since x_i and x_j are independent if $i \neq j$ and

$$\langle x_i x_j \rangle = \langle x_i \rangle \langle x_j \rangle = \bar{x}^2$$

and

$$\langle x_i^2 \rangle = \bar{x}^2 + \sigma_i^2 \qquad \text{by (7.8)}$$

We rewrite (26.20)

$$\begin{aligned}
\langle\chi_0^2\rangle &= \sum_{i=1}^{n} \left(\frac{\bar{x}^2 + \sigma_i^2}{\sigma_i^2}\right) - \bar{\sigma}^2 \sum_{i,j} \frac{\langle x_i x_j \rangle}{\sigma_i^2 \sigma_j^2} \\
&= \frac{\bar{x}^2}{\bar{\sigma}^2} + n - \bar{\sigma}^2 \sum_{i,j} \left(\frac{\bar{x}^2 + \sigma_i^2 \delta_{ij}}{\sigma_i^2 \sigma_j^2}\right)
\end{aligned}$$

$$= \frac{\bar{x}^2}{\bar{\sigma}^2} + n - \bar{\sigma}^2\bar{x}^2 \sum_{i,j} \frac{1}{\sigma_i^2}\frac{1}{\sigma_j^2} - \bar{\sigma}^2 \sum_{i,j} \frac{\delta_{ij}}{\sigma_j^2}$$

$$= \frac{\bar{x}^2}{\bar{\sigma}^2} + n - \bar{\sigma}^2\bar{x}^2 \frac{1}{\bar{\sigma}^2}\frac{1}{\bar{\sigma}^2} - \bar{\sigma}^2 \frac{1}{\bar{\sigma}^2}$$

$$= \frac{\bar{x}^2}{\bar{\sigma}^2} + n - \frac{\bar{x}^2}{\bar{\sigma}^2} - 1$$

Therefore,

$$\langle \chi_0^2 \rangle = n - 1 \qquad (26.21)$$

We remark that n data points provide n degrees of freedom. The expectation value for $\chi^2_{\min}$ is the *number of degrees of freedom − 1.*

We note, moreover, that we used up one degree of freedom by making the specification of the mean value of x, $\bar{x}$ in (26.16).

We have, therefore, shown for this case that

$$E(\chi^2) = N$$

where N is the number of *independent* degrees of freedom.

Probability Density Functions for n Independent Degrees of Freedom

Let us consider again the relation (26.2) where we stipulate that there are $n + m$ variables initially and that all the $n + m$ variables have been reduced by the m constraint equations to yield n *effectively independent* degrees of freedom ($\bar{x}_i$, $i = 1, \ldots, n + m$ have been replaced by $\bar{y}_j$, $j = 1, \ldots, n$). We assume the y_j are *given,* that is, are numbers that are *not* "free parameters" to be determined from the data:

$$G(y_1, \ldots, y_n; \bar{y}, \ldots, \bar{y}_n; \sigma_1, \ldots, \sigma_n) \times d\bar{y}_1 \cdots d\bar{y}_n$$

$$= \prod_{i=1}^{n} \frac{1}{\sqrt{2\pi\sigma_i^2}} e^{-(1/2)\chi^2} \, d\bar{y}_1, \ldots, d\bar{y}_n$$

where

$$\chi^2 = \sum_{i=1}^{n} \left(\frac{y_i - \bar{y}_i}{\sigma_i} \right)^2$$

Define

$$Z_i \equiv \frac{y_i - \bar{y}_i}{\sigma_i} \qquad (26.22)$$

and get

$$G(Z_1, \ldots, Z_n) dZ_1 \cdots dZ_n$$

$$= \left(\frac{1}{\sqrt{2\pi}} \right)^n e^{-(1/2)\chi^2} \, dZ_1 \cdots dZ_n \qquad (26.23)$$

where

$$\chi^2 = \sum_{i=1}^{n} Z_i^2 \qquad (26.24)$$

ONE DEGREE OF FREEDOM

Consider $n = 1$:

Let $z = Z_1 = \chi$

$$f_Z(z)\, dz = \frac{1}{\sqrt{2\pi}} e^{-(1/2)z^2} \, dz$$

which is just the normal p.d.f. of the form (25.22).

We may also write this:

$$f(\chi)\, d\chi = \frac{1}{\sqrt{2\pi}} e^{-(1/2)\chi^2} d\chi \qquad (26.25a)$$

We are interested in

$$f_{X^2=Z^2}(\chi^2) = \frac{1}{2\sqrt{\chi^2}} \{ f_Z(\sqrt{\chi^2}) + f_Z(-\sqrt{\chi^2}) \} \qquad \text{by (20.20)}$$

$$= \frac{1}{2\sqrt{\chi^2}} \left\{ \frac{1}{\sqrt{2\pi}} e^{-(1/2)\chi^2} + \frac{1}{\sqrt{2\pi}} e^{-(1/2)\chi^2} \right\}$$

$$\therefore f_{X^2}(\chi^2) d(\chi^2) = \frac{1}{\sqrt{2\pi}} \frac{e^{-(1/2)\chi^2}}{\sqrt{\chi^2}} d(\chi^2)$$

$$n = 1 \qquad (26.25b)$$

Let us define this to be the *p.d.f. of the chi-square distribution with one degree of freedom* (**D.O.F.**) and denoted by

$$\mathbf{CS}(\chi^2; 1) = \frac{1}{\sqrt{2\pi}} \frac{e^{-(1/2)\chi^2}}{\sqrt{\chi^2}} \qquad (26.25c)$$

TWO DEGREES OF FREEDOM

For $n = 2$, the c.d.f. for Z_1 and Z_2 is

$$F(Z_1, Z_2) = \int_{-\infty}^{Z_1}\int_{-\infty}^{Z_2} G\,dZ_1\,dZ_2$$
$$= \left(\frac{1}{\sqrt{2\pi}}\right)^2 \int_{-\infty}^{Z_1}\int_{-\infty}^{Z_2} e^{-(1/2)Z^2}\,dZ_1\,dZ_2 \qquad (26.26a)$$

where

$$Z^2 = Z_1^2 + Z_2^2 \qquad (26.26b)$$

In this two-dimensional problem we may consider the axes to be Z_1 and Z_2. Equation (26.26b), then, is the equation of a circle where Z is the radius. We can, therefore, transform to polar coordinates Z, θ:

$$F(Z, \theta) = \frac{1}{2\pi}\int_0^Z e^{-(1/2)Z^2}\,Z\,dZ \int_0^\theta d\theta$$

We integrate over θ to obtain the *marginal distribution* of Z alone:

$$F(Z) = \int_0^{2\pi} d\theta \left\{\frac{1}{2\pi}\int_0^Z e^{-(1/2)Z^2}\,Z\,dZ\right\}$$
$$\therefore f_Z(Z)\,dZ = Ze^{-Z^2/2}\,dZ$$

We may write this

$$f(\chi)\,d\chi = \chi e^{-\chi^2}\,d\chi \qquad (26.27a)$$

However, we want the p.d.f. of χ^2; therefore we equate

$$f_Z(Z)\,dZ = \tfrac{1}{2}e^{-Z^2/2}\,d(Z^2)$$
$$= f(\chi^2)\,d(\chi^2)$$
$$\therefore f_{\chi^2}(\chi^2)\,d(\chi^2) = \tfrac{1}{2}e^{-\chi^2/2}\,d(\chi^2) \qquad n = 2 \qquad (26.27b)$$

We follow the convention of (26.25b) and define

$$\mathbf{CS}(\chi^2; 2) = \tfrac{1}{2}e^{-\chi^2/2} \qquad (26.27c)$$

This is the *p.d.f. of the chi-square distribution with two degrees of freedom.*

THREE DEGREES OF FREEDOM

For $n = 3$, the c.d.f. for Z_1, Z_2, Z_3 is

$$F(Z_1, Z_2, Z_3) = \int_{-\infty}^{Z_1}\int_{-\infty}^{Z_2}\int_{-\infty}^{Z_3} G\,dZ_1\,dZ_2\,dZ_3$$
$$= \left(\frac{1}{\sqrt{2\pi}}\right)^3 \int_{-\infty}^{Z_1}\int_{-\infty}^{Z_2}\int_{-\infty}^{Z_3} e^{-(1/2)Z^2}\,dZ_1\,dZ_2\,dZ_3$$

where

$$Z^2 = Z_1^2 + Z_2^2 + Z_3^2$$

We note that Z is the radius in a three-dimensional space where Z_1, Z_2, and Z_3 are the three rectangular coordinates. It is natural to transform to *spherical polar coordinates* using Z, θ, and φ where Z is the radius, θ is the polar angle, and φ is the azimuth. The volume element $dZ_1\,dZ_2\,dZ_3$ becomes

$$Z^2\,dZ\,\sin\theta\,d\theta\,d\varphi = Z^2\,dZ\,d\Omega$$

and the c.d.f.

$$F(Z, \theta, \varphi) = \left[\frac{1}{\sqrt{2\pi}}\right]^3 \int_0^Z Z^2 e^{-(1/2)Z^2}\,dZ$$
$$\times \int_0^\theta \sin\theta\,d\theta \int_0^\varphi d\varphi$$

or

$$F(Z, \Omega) = \left[\frac{1}{\sqrt{2\pi}}\right]^3 \int_0^Z Z^2 e^{-(1/2)Z^2}\,dZ \int_0^\Omega d\Omega$$

We obtain the marginal distribution of Z alone by integrating over Ω leaving

$$F(Z) = \frac{4\pi}{(2\pi)^{3/2}}\int_0^Z Z^2 e^{-(1/2)Z^2}\,dZ$$

From this we get

$$f_Z(Z)\,dZ = \frac{2}{\sqrt{2\pi}}\,Z^2 e^{-(1/2)Z^2}\,dZ$$

or

$$f(\chi)\,d\chi = \frac{2}{\sqrt{2\pi}}\,\chi^2 e^{-\chi^2/2}\,d\chi \qquad (26.28a)$$

We rewrite this to obtain $f(\chi^2)$:

$$f_Z(Z)\,dZ = \frac{2}{\sqrt{2\pi}}\sqrt{Z^2}\,\frac{e^{-(1/2)Z^2}\,d(Z^2)}{2}$$
$$= \frac{\sqrt{Z^2}}{\sqrt{2\pi}}\,e^{-(1/2)Z^2}\,d(Z^2)$$
$$= f_{\chi^2}(\chi^2)\,d(\chi^2)$$

where $\chi^2 = Z^2$.

$$f_{\chi^2}(\chi^2)\,d(\chi^2) = \frac{\sqrt{\chi^2}}{\sqrt{2\pi}}\,e^{-\chi^2/2}\,d(\chi^2) \qquad n = 3 \qquad (26.28b)$$

from which we obtain *the p.d.f. of the chi-square distribution with three degrees of freedom*

$$\mathbf{CS}(\chi^2; 3) = \frac{(\chi^2)^{1/2}}{(2\pi)^{1/2}} e^{-\chi^2/2} \qquad (26.28c)$$

MORE THAN THREE DEGREES OF FREEDOM

When $n > 3$, it is necessary to consider a "hyperspace" of dimensionality greater than 3. Although it is not possible to visualize this in an "intuitive" way, we may yet calculate the volume element mathematically. We shall cite the result without proof for (26.24):

$$\chi^2 = Z_1^2 + Z_2^2 + \cdots + Z_n^2$$

It turns out that

$$f(\chi)\,d\chi = \frac{\chi^{n-1}}{\left[\frac{n-2}{2}\right]!\,2^{(n-2)/2}} e^{-\chi^2/2}\,d\chi \qquad (26.29a)$$

We note that the exponent of χ is $(n-1)$, that is, *one less* than the number of independent variates. Expressed in terms of χ^2 we get

$$f_{\chi^2}(\chi^2)d(\chi^2) = \frac{(\chi^2)^{(1/2)(n-2)}e^{-\chi^2/2}}{[(n/2)-1]!2^{n/2}}\,d(\chi^2) \qquad (26.29b)$$

or

$$\mathbf{CS}(\chi^2; n) = \frac{(\chi^2)^{(1/2)(n-2)}}{\Gamma(n/2)2^{n/2}} e^{-\chi^2/2} \qquad (26.29c)$$

where $\Gamma(n/2) = [(n/2)-1]!$ as in Appendix III.

This is called the **p.d.f. of the "chi-square distribution with *n* degrees of freedom."** We state again what is the significance of (26.29c): if n variates $X_1 \cdots X_n$ are independent and each $N(0,1)$ (i.e., each normally distributed with mean 0 and variance 1) then $\chi^2 = \Sigma X_i^2$ is distributed with (26.29c) as p.d.f. The p.d.f., (26.29c), is plotted in Figure 26.1 as $F(X, N)$ versus X where $X = \chi^2$ and $N = n$ = number of *independent* degrees of freedom.

Mean, Mode and Variance of the Chi-Square Distribution

In general, the *mean value* of χ^2 for n independent degrees of freedom is

$$\langle \chi_n^2 \rangle = E(\chi_n^2) = n \qquad (26.30)$$

This is straightforwardly shown as follows using some results derived in Appendix III:

$$\begin{aligned}
\langle \chi^2 \rangle &= \int_0^\infty (\chi^2)\mathbf{CS}(n)d(\chi^2) \qquad \text{since } \chi^2 > 0 \\
&= \int_0^\infty \frac{(\chi^2)(\chi^2)^{(n/2)-1}}{\Gamma(n/2)2^{n/2}} e^{-\chi^2/2} d(\chi^2) \\
&= \frac{2}{\Gamma(n/2)} \int_0^\infty \left(\frac{\chi^2}{2}\right)^{n/2} e^{-(\chi^2/2)} d\left(\frac{\chi^2}{2}\right) \\
&= \frac{2}{\Gamma(n/2)} \Gamma\left(\frac{n}{2}+1\right) \qquad \text{by (A.III.1)} \\
&= \frac{2(n/2)}{\Gamma(n/2)} \Gamma\left(\frac{n}{2}\right) \qquad \text{by (A.III.2)} \\
&= n. \qquad \textbf{Q.E.D.}
\end{aligned}$$

The *modal* or **most probable** value of χ^2 is easily shown from (26.29c) to be:

$$\chi^2_{\text{mode}} = n - 2 \qquad (26.31)$$

For example,

$$\mathbf{CS}(\chi^2; n) = f(x; n) = x^{(1/2)(n-2)} e^{-x/2} [\Gamma(n/2)2^{n/2}]^{-1}$$

$$\left.\frac{\partial f(x; n)}{\partial x}\right|_{x = x_{\text{mode}}} = 0 \qquad x_{\text{mode}} = n - 2 \qquad \textbf{Q.E.D.}$$

The *variance* turns out to be

$$\sigma^2(\chi^2; n) = 2n \qquad (26.32)$$

This is easily shown as follows:

$$\begin{aligned}
\langle (\chi^2)^2 \rangle &= \int_0^\infty (\chi^2)^2 \mathbf{CS}(n)d(\chi^2) \\
&= \frac{2}{\Gamma(n/2)} \int_0^\infty \frac{(\chi^2)^{(n/2)+1}}{2^{n/2}} e^{-\chi^2/2} d\left(\frac{\chi^2}{2}\right) \\
&= \frac{4\Gamma(n/2+2)}{\Gamma(n/2)} \qquad \text{by (A.III.1)} \\
&= \frac{4}{\Gamma(n/2)} \left(\frac{n}{2}+1\right) \\
&\quad \times \left(\frac{n}{2}\right) \Gamma(n/2) \qquad \text{by (A.III.2)} \\
&= (n+2)(n) \\
\sigma^2 &= \langle (\chi^2)^2 \rangle - (\langle \chi^2 \rangle)^2 \\
&= (n+2)n - n^2 = 2n \qquad \textbf{Q.E.D.}
\end{aligned}$$

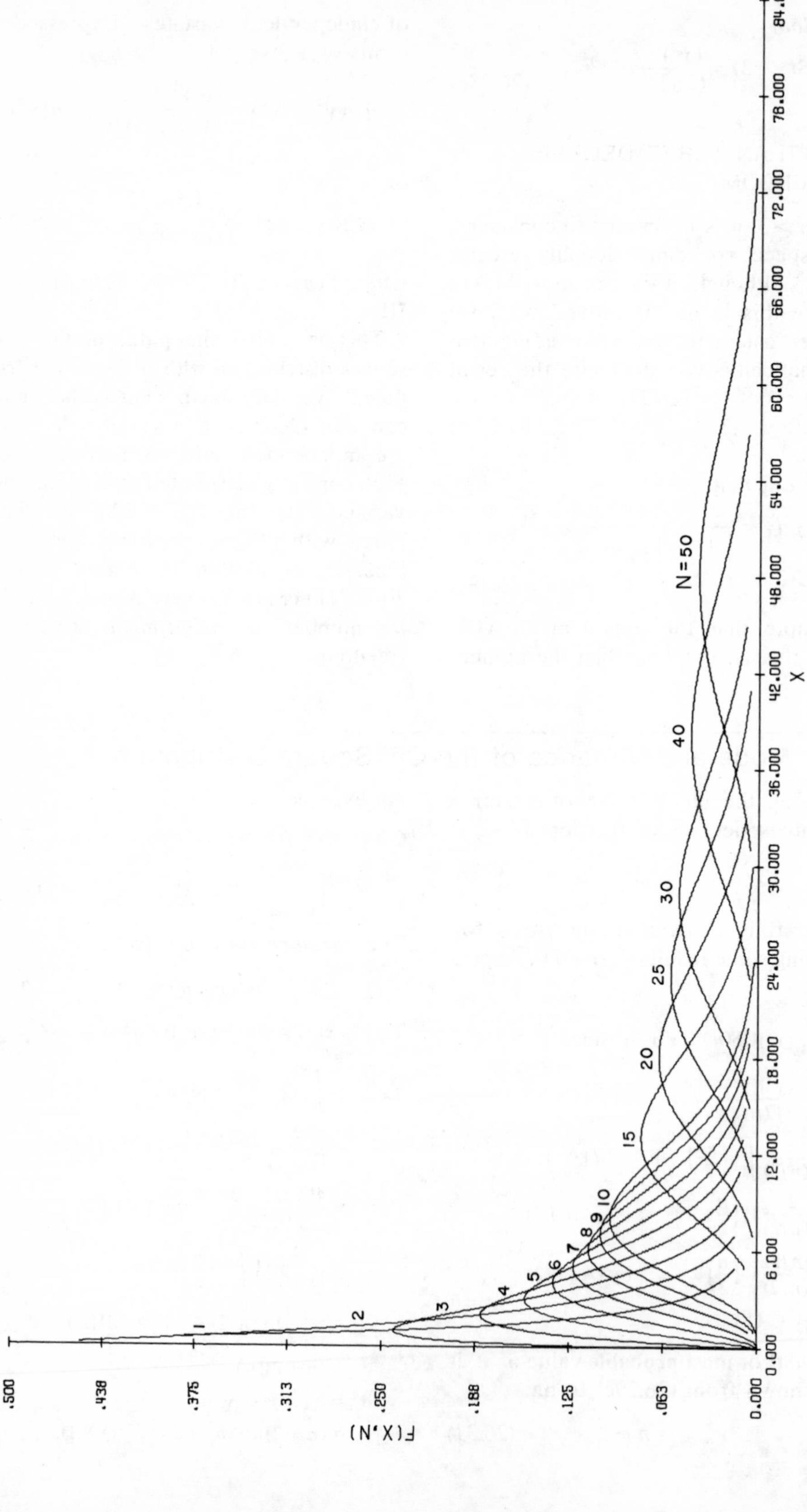

Figure 26.1
The probability density function of the chi-square distribution plotted as $F(X,N)$ versus X for different values of N, the number of independent degrees of freedom. X = value of χ^2.

REPRODUCTIVE PROPERTY OF THE CHI-SQUARE DISTRIBUTION

If the k quantities X_i^2, $i = 1, \ldots, k$, are each independent, chi-square-distributed random variables with n_i degrees of freedom respectively, then

$$X^2 = \sum_{i=1}^{k} X_i^2$$

is chi-square-distributed with $n = \Sigma_{i=1}^{k} n_i$ degrees of freedom.

Computations with the Chi-Square Distribution

We mention an obvious point: the *p.d.f. for the normal distribution* is simply related to the *p.d.f. of the chi-square distribution for n* = 1.

We stress that all our results are referred to *n independent* degrees of freedom. Let us assume that there are $n + m$ terms in the sum (26.3) where we are given each $x_i, \bar{x}_i$:

$$\chi^2 = \sum_{i=1}^{n+m} \left(\frac{x_i - \bar{x}_i}{\sigma_i} \right)^2$$

(a) if there are m independent relations among the $\bar{x}_i$ of the form of (26.6), then χ^2 is distributed according to $\mathbf{CS}(\chi^2; n)$.

(b) if each $\bar{x}_i$ is given in the form $\bar{x}_i = f_i(a_1, \ldots, a_m)$ where the a_j are parameters to be determined from the data, then χ^2 is again distributed with a p.d.f. $\mathbf{CS}(\chi^2; n)$.

We are frequently interested in the "*upper-tail area function*" defined in (17.9):

$$\begin{aligned} Q(\chi^2) &= \Pr(X^2 > \chi^2) \\ &= 1 - F_{X^2}(\chi^2) \\ &= \int_{\chi^2}^{\infty} \mathbf{CS}(\chi^2; n)\, d(\chi^2) \qquad (26.33) \end{aligned}$$

This is the probability that χ^2 will be greater than or equal to a given value. If a given value of χ^2 is observed, we ask what is the probability of getting a value that large or larger by **random chance** alone. We define the probability in terms of a *confidence level.* The larger χ^2 is, the less we believe the probability is that the value arose by chance. We can never "*prove*" that a given set of $\bar{x}_i$ fits the data, but we can *exclude* a set $\bar{x}_i$ on the grounds that it is too improbable that chance fluctuations yielded a value of χ^2 as large as that observed. It is *customary* to set $Q(\chi^2) = 0.05$ as the level at which to consider rejecting a set of $\bar{x}_i$.

The evaluation of the integral in (26.33) is tedious, and we shall merely note some results. The basic idea is to use generalized polar coordinates and integrate successively by parts, considering separately the cases where n is even and n is odd. For n even,

$$Q(\chi^2) = e^{-\chi^2/2} \left[1 + (\tfrac{1}{2}\chi^2) + \frac{1}{2!}(\tfrac{1}{2}\chi^2)^2 + \cdots + \frac{1}{[(n/2) - 1]!}(\tfrac{1}{2}\chi^2)^{(n/2)-1} \right]$$

or

$$Q(\chi^2) = e^{-\chi^2/2} \sum_{i=0}^{N} \frac{(\chi^2/2)^i}{i!}$$

where $N = (n/2) - 1$, n even. (26.34a)

For n odd,

$$Q(\chi^2) = \sqrt{\frac{2}{\pi}} \int_{\chi}^{\infty} e^{-x^2/2}\, dx + \sqrt{\frac{2}{\pi}}\, e^{-\chi^2/2} \times \left[\chi + \frac{\chi^3}{3} + \frac{\chi^5}{3 \cdot 5} + \cdots + \frac{1}{3 \cdot 5 \cdots (n-2)} \chi^{n-2} \right]$$

For n odd, we may also write Q as an infinite series

$$Q(\chi^2) = 1 - \frac{1}{\Gamma(N+1)} \times \sum_{i=0}^{\infty} \left\{ (-1)^i \frac{(\chi^2/2)^{N+i+1}}{i!(N+i+1)} \right\}$$

where

$$N = \frac{n}{2} - 1 \qquad n \text{ odd} \qquad (26.34b)$$

Using (26.34) we may tabulate values of $Q(\chi^2, n)$ given χ^2 and n. This is shown in

Table A.VII.1. A more compact form is to calculate χ^2 given Q and n. This requires an extra interpolation calculation iterating with Newton's formula:

$$x_{k+1} = x_k - \frac{f(x_k)}{f'(x_k)}$$

where f is obtained from (26.34a,b). The results are tabulated in Table A.VII.2.

We have also plotted Q, χ^2, and n as a graph in Figure A.VII.1.

Approximations to the Chi-Square Distribution

We have remarked in (26.30) and (26.32) that the mean and variance of χ_n^2 are n and $2n$ respectively. It may be shown that, in the limit of large n, χ^2 becomes normally distributed with mean n and variance $2n$. That is,

$$Y_2 = \frac{\chi_n^2 - n}{\sqrt{2n}} \qquad (26.35)$$

becomes $N(0, 1)$ in the limit $n \to \infty$.

We may inquire how rapidly Y_2 approaches a normally distributed variable as n gets large. It is quite straightforward to compute both the p.d.f.

$$\mathbf{n}(\chi_n^2; n, 2n)$$

for comparison with $\mathbf{CS}(\chi_n^2; n)$ and the *upper-tail area function* for comparison with (26.33). We shall call this the "**second approximation**" to χ^2.

A better approximation was suggested by R. A. Fisher who first showed that the random variable $\sqrt{2\chi^2}$ becomes normally distributed with mean $\sqrt{2n-1}$ and variance 1 as $n \to \infty$. Another way of expressing this is that

$$Y_1 = \sqrt{2\chi_n^2} - \sqrt{2n-1} \qquad (26.36)$$

becomes $N(0, 1)$ in the limit $n \to \infty$.

The convergence of Y_1 is more rapid than Y_2; thus we shall call this the "*first approximation*" to χ^2.

To compute the p.d.f. of χ^2 according to the *first approximation*, we may use some of the results of Chapter 20.

$$X = \sqrt{2\chi_n^2} \quad \text{is} \quad N(\sqrt{2n-1}, 1)$$

$$\therefore f_X(x) = \frac{1}{\sqrt{2\pi}} \exp\{-\tfrac{1}{2}(x - \sqrt{2n-1})^2\}$$

If $Z_1 = X^2$,

$$f_{Z_1}(z_1) = \frac{1}{2\sqrt{z_1}}[f_X(+\sqrt{z_1}) + f_X(-\sqrt{z_1})] \qquad z > 0$$

by (20.20).

If $Z = \frac{1}{2}Z_1 = \chi_n^2$,

$$f_Z(z) = 2f_{Z_1}(2z)$$

by (20.24).

We obtain, for $z = \chi_n^2 > 0$,

$$f_Z(z; n) = \frac{1}{\sqrt{\pi z}} e^{-z-n+1/2} \cosh\left(2\sqrt{2z(2n-1)}\right)$$

We compare the p.d.f.'s for the rigorous χ^2 distribution and for the two approximations in Figures 26.2*a* to *g* for $n = 10, 20, 30, 40, 50, 70$, and 100. We see that the first approximation is superior to the second and that both improve as n gets larger.

More important in practice is the *upper-tail area function* defined in (17.9) and (26.33), sometimes called the *confidence level* for the observed χ^2.

The upper-tail area function **for the first approximation** is easily obtained as follows:

$$\begin{aligned}
\Pr(X_n^2 > \chi^2) &= \Pr\left(\sqrt{2X_n^2} > \sqrt{2\chi^2}\right) \\
&= \Pr\left(\sqrt{2X_n^2} - \sqrt{2n-1} > \sqrt{2\chi^2} - \sqrt{2n-1}\right) \\
&= \Pr\left(Y_1 > \sqrt{2\chi^2} - \sqrt{2n-1}\right) \\
&\approx 1 - \Phi\left(\sqrt{2\chi^2} - \sqrt{2n-1}\right) \\
&= \frac{1}{\sqrt{2\pi}} \int_{y_1 = \sqrt{2\chi^2} - \sqrt{2n-1}}^{\infty} e^{-t^2/2}\, dt
\end{aligned}$$

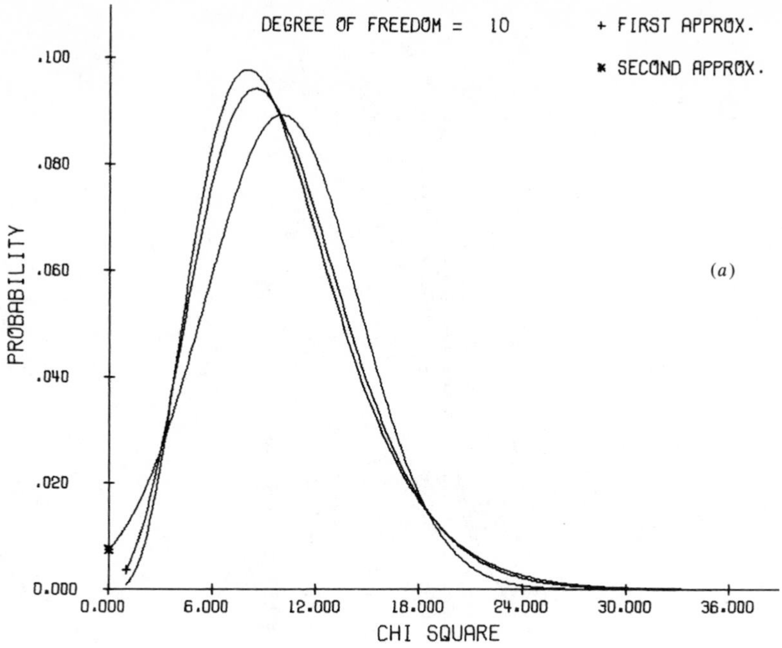

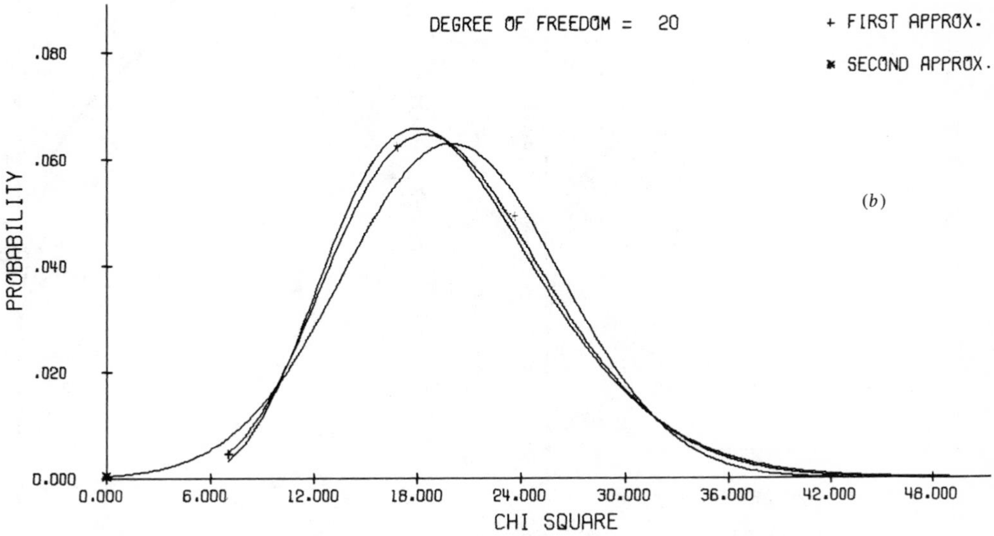

Comparison of the exact chi-square p.d.f. with the p.d.f.'s for two approximations for different values of n, the number of independent degrees of freedom. (*a*) $n = 10$. (*b*) $n = 20$. (*c*) $n = 30$. (*d*) $n = 40$. (*e*) $n = 50$. (*f*) $n = 70$. (*g*) $n = 100$.

(*c*)

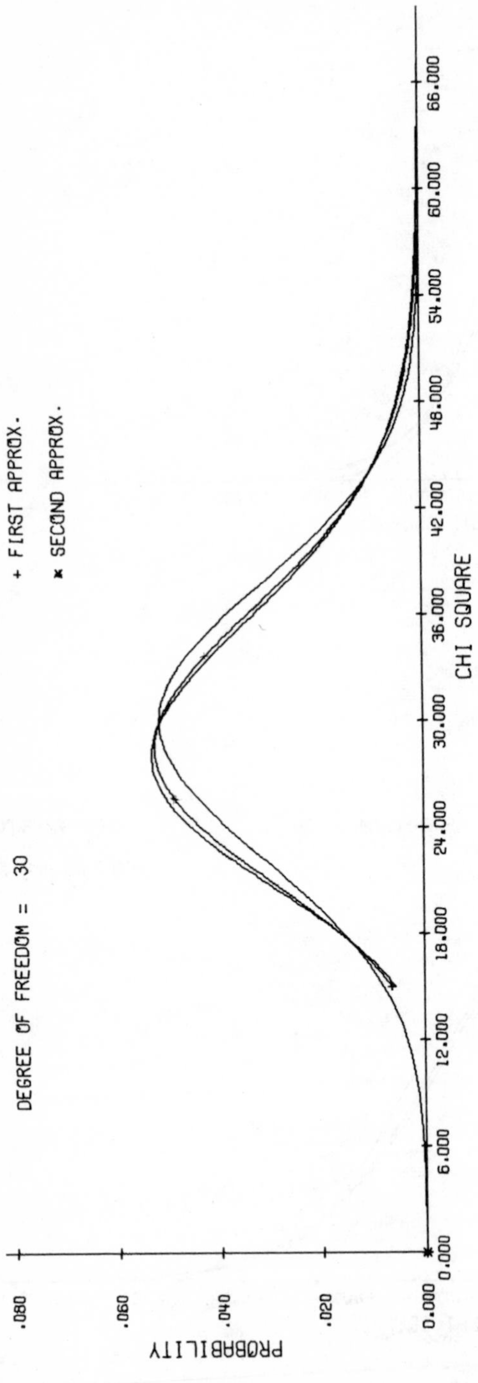
DEGREE OF FREEDOM = 30
+ FIRST APPROX.
✱ SECOND APPROX.
PROBABILITY
CHI SQUARE
.080
.060
.040
.020
0.000
0.000
6.000
12.000
18.000
24.000
30.000
36.000
42.000
48.000
54.000
60.000
66.000

(*d*)

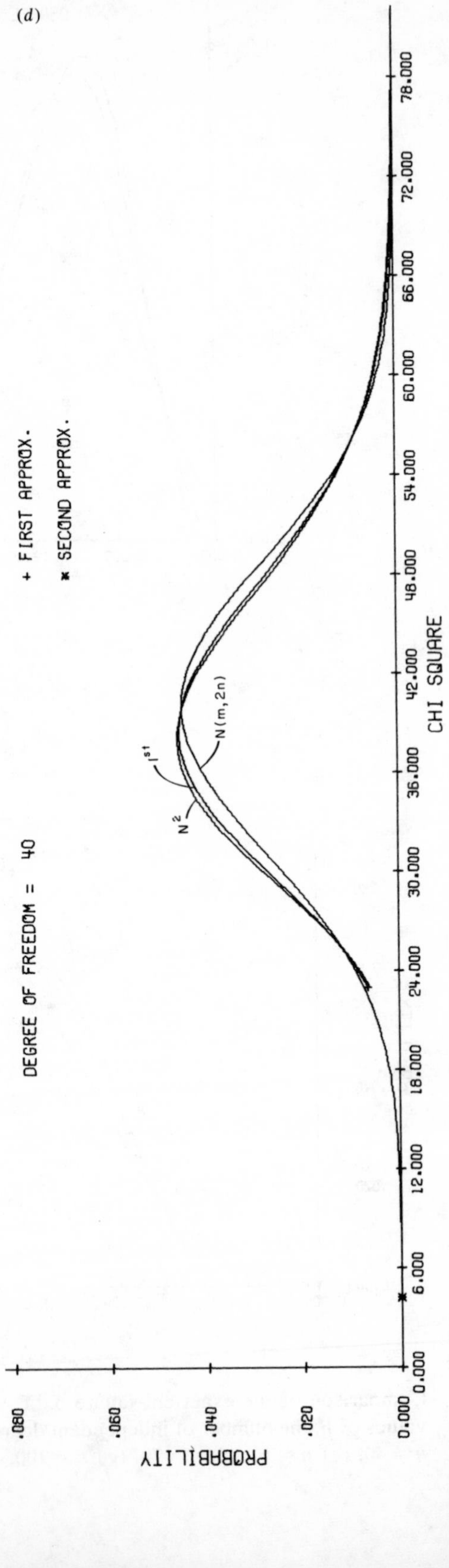
DEGREE OF FREEDOM = 40
+ FIRST APPROX.
✱ SECOND APPROX.
PROBABILITY
CHI SQUARE
N(m,2n)
.080
.060
.040
.020
0.000
0.000
6.000
12.000
18.000
24.000
30.000
36.000
42.000
48.000
54.000
60.000
66.000
72.000
78.000

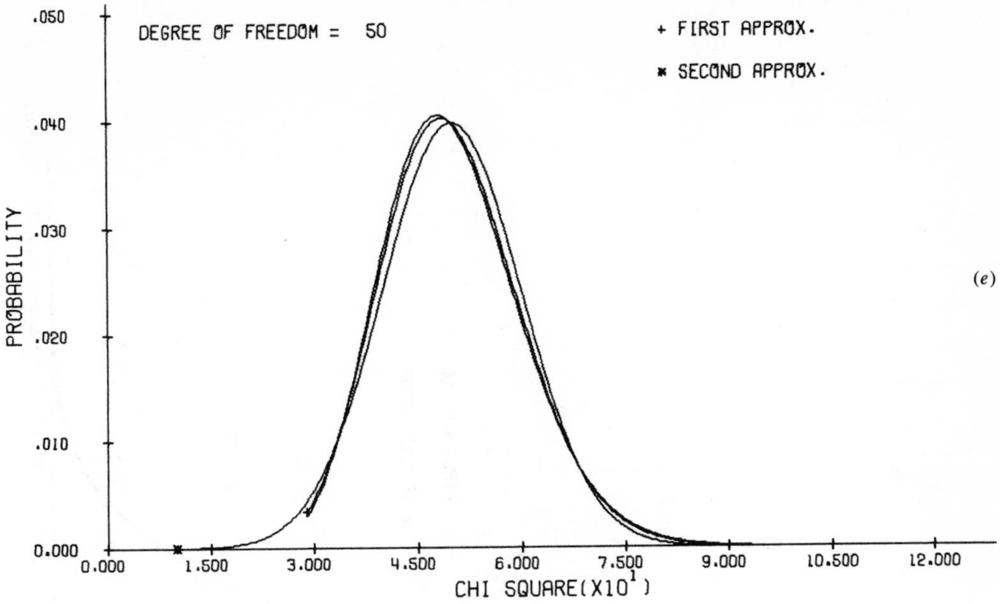

(*e*)

Figure 26.2(*c*) (*d*) (*e*) and (*f*).

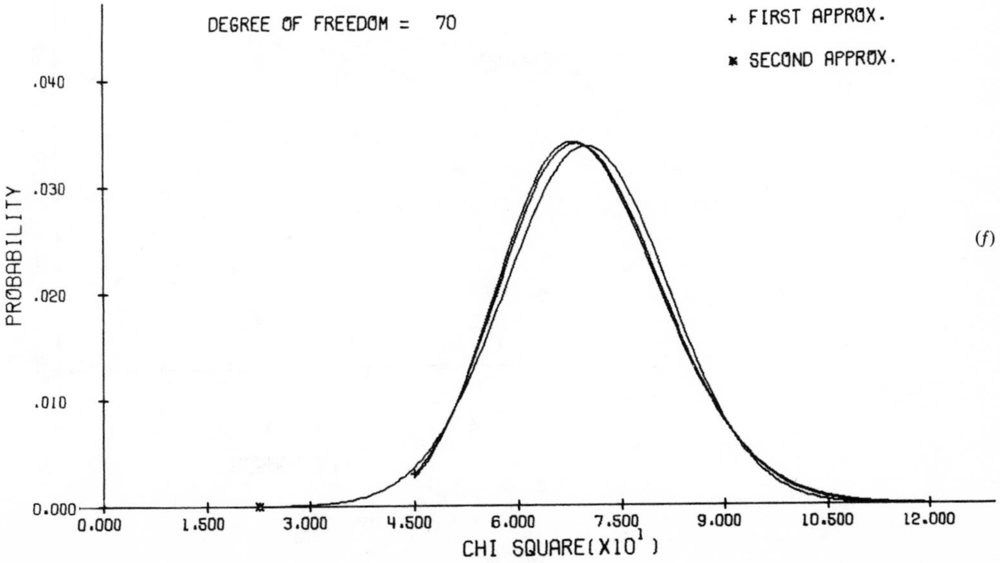

(*f*)

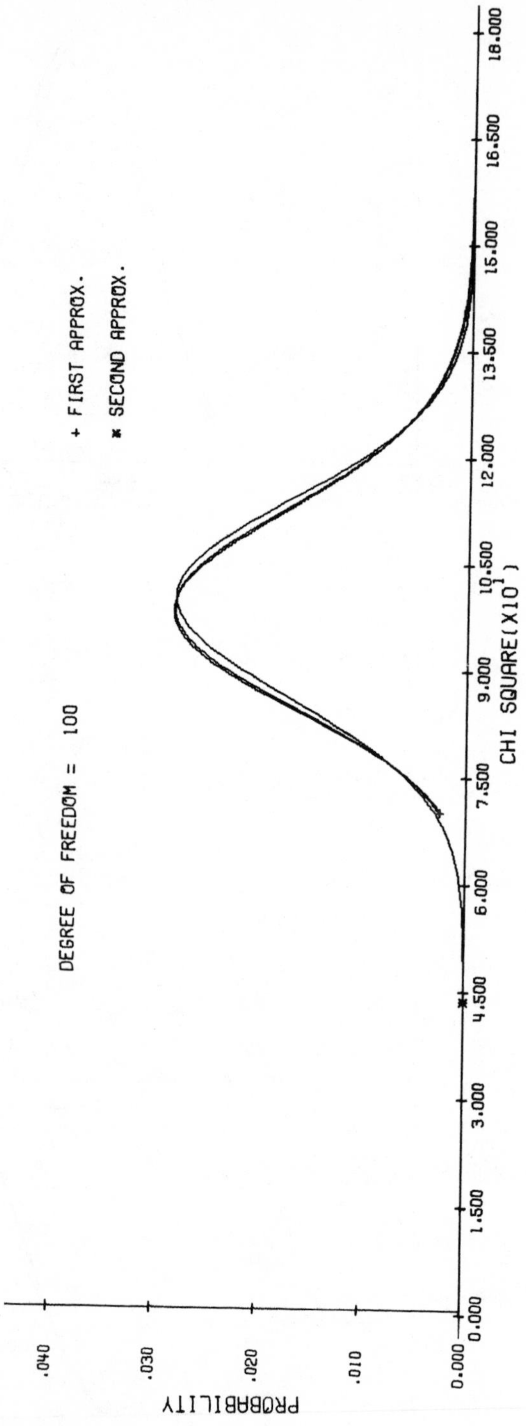
DEGREE OF FREEDOM = 100
+ FIRST APPROX.
* SECOND APPROX.
PROBABILITY
0.000
.010
.020
.030
.040
0.000
1.500
3.000
4.500
6.000
7.500
9.000
10.500
12.000
13.500
15.000
16.500
18.000
CHI SQUARE(X10^1)

(g)

since Y_1 is (approximately) normally distributed.

Also, **for the second approximation,**

$$\Pr(X_n^2 > \chi^2) = \Pr\left(\frac{X_n^2 - n}{\sqrt{2n}} > \frac{\chi^2 - n}{\sqrt{2n}}\right)$$
$$= \Pr\left(Y_2 > \frac{\chi^2 - n}{\sqrt{2n}}\right)$$
$$\approx 1 - \Phi\left(\frac{\chi^2 - n}{\sqrt{2n}}\right)$$
$$= \frac{1}{\sqrt{2\pi}} \int_{y_2 = (\chi^2 - n)/\sqrt{2n}}^{\infty} e^{-t^2/2}\, dt$$

since Y_2 is (approximately) normally distributed.

Therefore, we may write both approximations in the form

$$\Pr(X_n^2 > \chi^2) \approx 1 - \Phi(y) = \frac{1}{\sqrt{2\pi}} \int_y^{\infty} e^{-t^2/2}\, dt \qquad (26.37)$$

where

$$y_1 = \sqrt{2\chi^2} - \sqrt{2n - 1}$$

and

$$y_2 = \frac{\chi^2 - n}{\sqrt{2n}}$$

The two approximations are compared with the rigorous *upper-tail area function* in Figures 25.3*a* and *b* for $n = 30, 50, 70, 100, 150,$ and 200. The approximations are quite good for reasonable confidence levels (values of Q). Both approximations **underestimate** the confidence level for small levels with the first approximation being better.

That this regime is not totally devoid of interest is shown by the following report.*

"The best exchange I have heard at this Conference was in a discussion between Dr. Lovelace and Dr. Moorhouse. Lovelace said that he had computed that in a "best fit" which Moorhouse had obtained to some data the likelihood that this fit be correct was 10^{-166}. To this Moorhouse replied that Lovelace will be happy to learn that a check he made on one of Lovelace's "best fits" came out much better: the likelihood for it to be a correct fit was 10^{-24}. This illustrates an important point. Both these gentlemen have actually made excellent fits to large amounts of data and have obtained very interesting and reliable results. The point is, however, that one has to be very careful in interpreting χ^2 probabilities in a literal sense."

*G. Goldhaber, *Proceedings of the XIIIth International Conference on High Energy Physics*, University of California Press, 1967, p. 103.

The Sample Variance

Suppose we have a sample of n independent values x_i drawn from a population that is $N(\mu, \sigma^2)$. The sample mean is $\bar{x} = (1/n)\Sigma\, x_i$. Consider the *sum of squares about the mean.*

$$w = \sum_{i=1}^{n} (x_i - \bar{x})^2 = \sum_{i=1}^{n} x_i^2 - n(\bar{x})^2 \qquad (26.38)$$

We wish to obtain the distribution that describes w.

Recall our discussion of linear orthogonal transformations in Chapter 20:

$$y_i = \sum_{j=1}^{n} C_{ij} x_j \qquad i = 1 \cdots n \qquad (20.42)$$

with

$$\sum_{i=1}^{n} y_i^2 = \sum_{j=1}^{n} x_j^2 \qquad (20.44)$$

and

$$|\underline{\mathbf{C}}| = 1 \qquad (20.45)$$

In particular,

$$y_1 = \sum_{j=1}^{n} C_{1j} x_j$$

Suppose we choose the first variable, y_1, such that

$$y_1 = K_1 \sum_{j=1}^{n} x_j \qquad (26.39)$$

where K_1 is a constant.

(a)

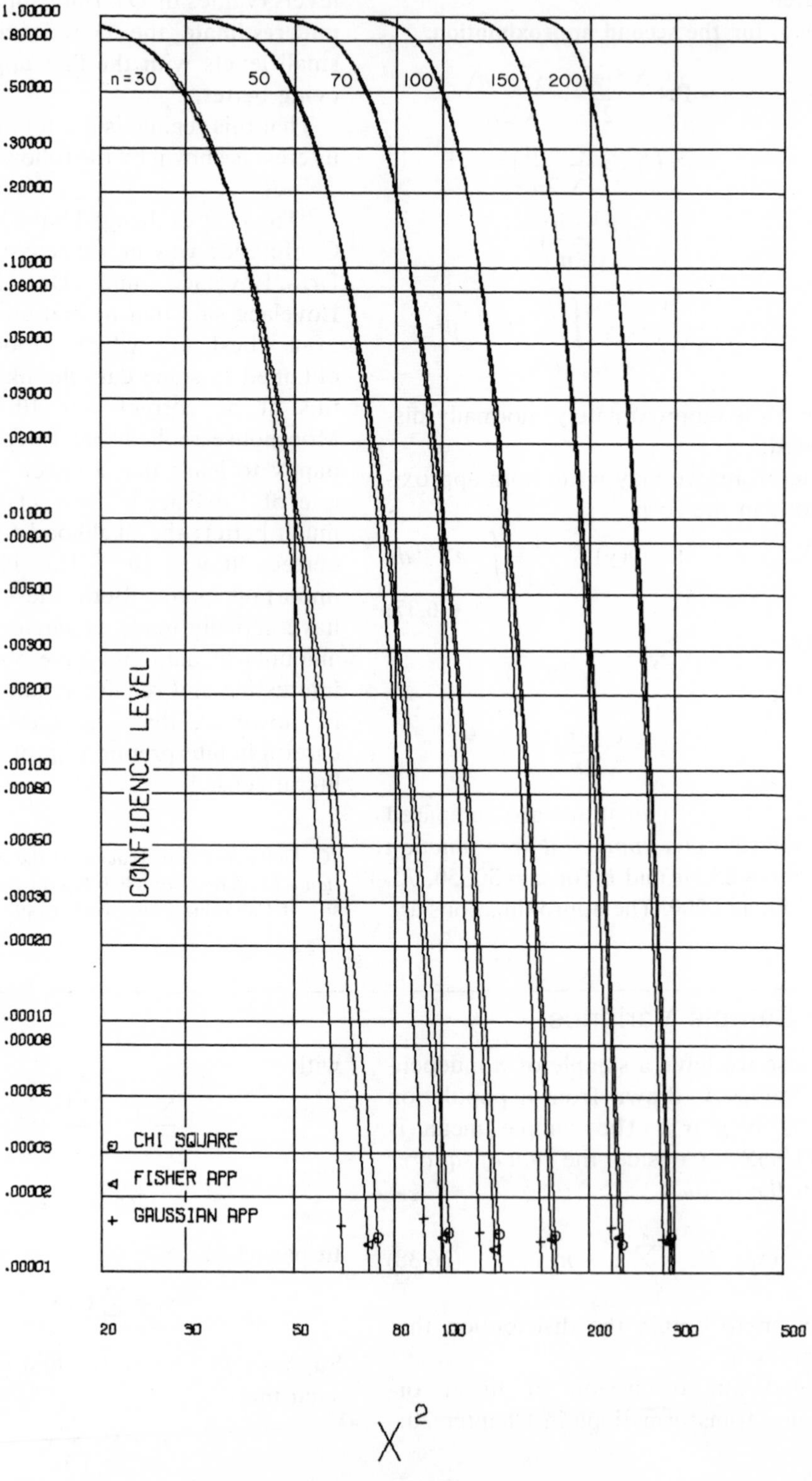
1.00000
.80000
.50000
.30000
.20000
.10000
.08000
.05000
.03000
.02000
.01000
.00800
.00500
.00300
.00200
.00100
.00080
.00050
.00030
.00020
.00010
.00008
.00005
.00003
.00002
.00001
n=30
50
70
100
150
200
CONFIDENCE LEVEL
CHI SQUARE
FISHER APP
GAUSSIAN APP
20
30
50
80
100
200
300
500
X²

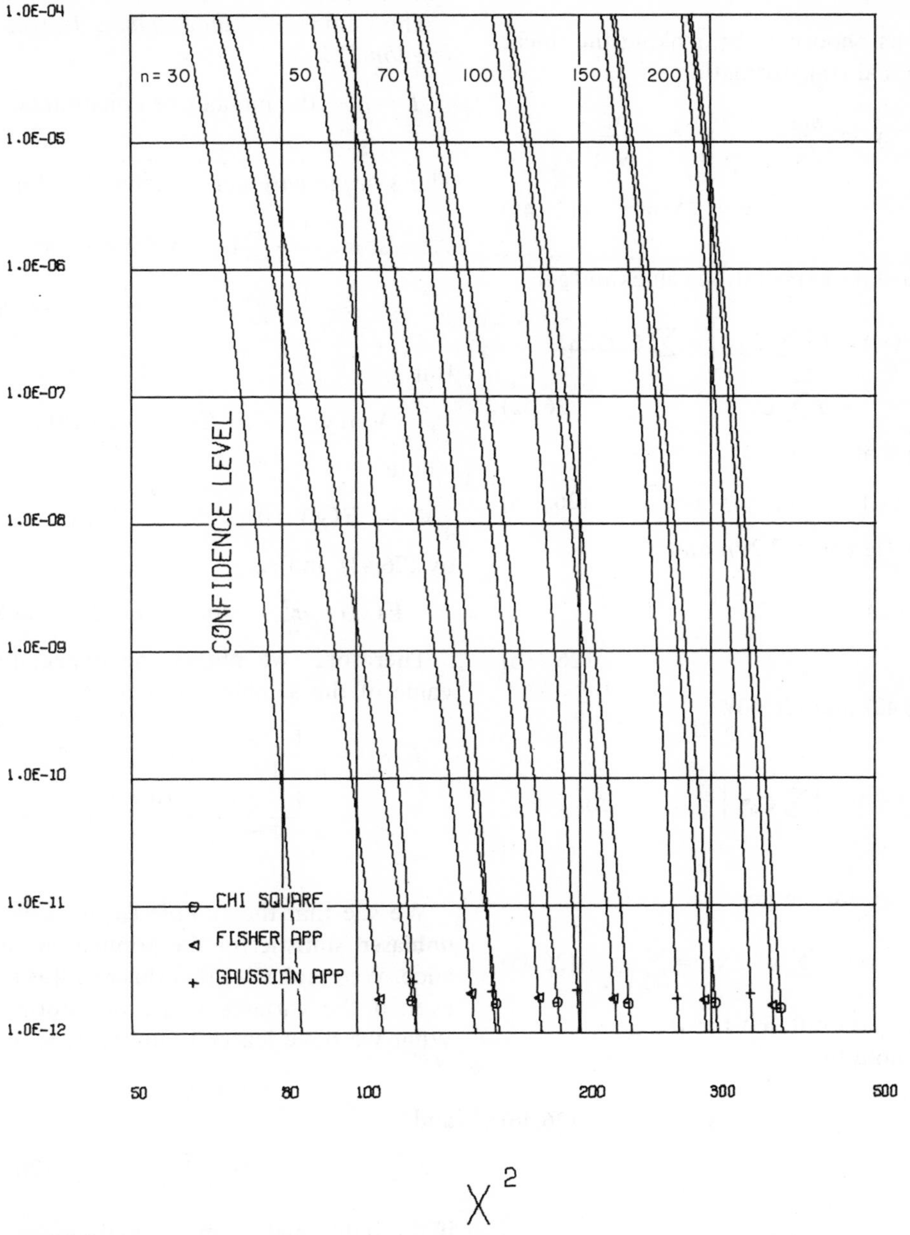

(*b*)

Figure 26.3
Comparison of the *exact* chi-square upper-tail area function with the upper-tail area functions for two approximations for different values of *n*, the number of degrees of freedom. $n = 30, 50, 70, 100, 150, 200$. (*a*) The range of the ordinate (confidence level) is 1.0 to 10^{-5}. (*b*) The range of the ordinate (confidence level) is 10^{-4} to 10^{-12}.

In this case, for $y_2, y_3, \ldots, y_n$ we have

$$\sum_{j=1}^{n} C_{ij} = 0 \qquad \text{for } i \geq 2 \qquad (26.40)$$

Let us choose y_1 by making the linear orthogonal transformation

$$y_1 = \frac{x_1}{\sqrt{n}} + \frac{x_2}{\sqrt{n}} + \cdots + \frac{x_n}{\sqrt{n}}$$
$$= \frac{1}{\sqrt{n}} \sum_{i=1}^{n} x_i = \bar{x}\sqrt{n} \qquad (26.41)$$

We may take expectation values and get

$$E(y_i) = E\left[\sum C_{ij}x_j\right] = \sum C_{ij}E(x_j)$$
$$= \mu \sum_{i=1}^{n} C_{ij} \qquad (26.42)$$

noting that

$$E(y_1) = E(\bar{x})\sqrt{n} = \mu\sqrt{n} \qquad (26.43a)$$

by (26.41) and (22.29), and

$$E(y_i) = \mu \sum_{i=2}^{n} C_{ij} = 0 \qquad \text{for } i \geq 2 \qquad (26.43b)$$

by (26.42) and (26.40).
Also,

$$\text{Var}(y_i) = \text{Var}\left\{\sum_{i=2}^{n} C_{ij}x_i\right\} = \sigma^2 \qquad \text{for } i \geq 2 \qquad (26.44)$$

Using (26.38) we get

$$w = \sum_{i=1}^{n} y_i^2 - y_1{}^2 = \sum_{i=2}^{n} y_i^2 \qquad (26.45)$$

where y_i is $N(0; \sigma^2)$ for $i \geq 2$.
We note that

$$\sum_{i=2}^{n} \frac{y_i^2}{\sigma^2} = \chi^2_{n-1} \qquad (26.46)$$

by (26.3).

$$\therefore w = \sum_{i=2}^{n} y_i^2 = \sigma^2\chi^2_{n-1} \qquad (26.47)$$

We summarize this result as follows. The sum $w = \Sigma_{i=1}^{n}(x_i - \bar{x})^2$ of n nonindependent normal random variables, $(x_i - \bar{x})$, is equivalent to the sum of squares of $(n - 1)$ independent normal variables, y_i. The dependence comes from the relation

$$\sum_{i=1}^{n} (x_i - \bar{x}) = 0 \qquad (26.48)$$

which represents a **constraint.**

The number of "*independent degrees of freedom*," k, is

$$k = n - (\text{the number of constraints}) \qquad (26.49)$$

The *sample variance* is defined to be

$$s^2 = \frac{1}{n-1} \sum_{i=1}^{n} (x_i - \bar{x})^2 = \frac{1}{n-1} w$$
$$= \frac{1}{n-1} \sum_{i=2}^{n} y_i^2 \qquad (26.50)$$

But

$$\text{Var}(y_i) = \sigma^2 = E(y_i^2) - [E(y_i)]^2$$

where

$$E(y_i) = 0 \qquad \text{for} \qquad i \geq 2$$

by (26.43) and so

$$E(y_i^2) = \sigma^2 \qquad \text{for} \qquad i \geq 2 \qquad (26.51)$$

Therefore, we obtain the expectation value of the sample variance

$$E(s^2) = \frac{1}{n-1} \sum_{i=2}^{n} E(y_i)^2$$
$$= \frac{1}{n-1} \sum_{i=2}^{n} \sigma^2 = \frac{(n-1)}{n-1}\sigma^2 = \sigma^2 \qquad (26.52)$$

We see that the *sample variance* is an **unbiased statistic** for the population variance. We emphasize that this result is true even if the variates x_i are *not* normal.* What we have learned, however, is that

$$w = \sigma^2\chi^2_{n-1}$$

and

$$s^2 = \frac{\sigma^2\chi^2_{n-1}}{n-1} \qquad (26.53)$$

if the individual x_i are normal.

When the central limit theorem applies,

$$s^2 \approx \frac{\sigma^2\chi^2_{n-1}}{n-1} \qquad (26.54)$$

in general, for x_i described by most distributions.

*See Chapter 8.

When the x_i are normal, either rigorously or asymptotically, we may get the *variance of the sample variance*:

$$\mathrm{Var}(s^2) = \left(\frac{\sigma^2}{n-1}\right)^2 \mathrm{Var}(\chi^2_{n-1})$$

$$\mathrm{Var}(s^2) = \left(\frac{\sigma^2}{n-1}\right)^2 2(n-1)$$

by (26.32).

$$\mathrm{Var}(s^2) = \frac{2(\sigma^2)^2}{n-1} = \sigma_{s^2}{}^2 \qquad (26.55a)$$

The *standard deviation of the sample variance*, where the x_i are normally distributed is

$$SD(s^2) = \sigma^2 \sqrt{\frac{2}{(n-1)}} \qquad (26.55b)$$

which is (7.9a).

27

STUDENT'S t DISTRIBUTION

In our treatment of the normal distribution we have considered two related examples. One was to verify that a set of n values x_i is *consistent with being drawn from a normal distribution.* The second was to consider two samples, n values x_i and m values y_i, and to see if they were consistent with being drawn from normally distributed populations **with the same mean**. In both cases we tested the hypotheses assuming that the population variance was known and obtainable by calculating the sample variance. In most cases this is quite adequate. We must note, however, that this is rigorously not correct since we usually do not know the population variance but can only *estimate* it by the sample variance. If the sample size is small, this estimate may not be as close as we would like. The general problem, *where the population variance is unknown,* was first treated in 1905 by W. S. Gossett, who published his analysis under the pseudonym, "Student" (His employers, the Guinness Breweries of Ireland, had a policy of keeping all their research as proprietary secrets. The importance of his work argued for its being published, but it was felt that anonymity would protect the company).

Definition of t and its p.d.f.

Suppose there are n values x_i, ($i = 1, \ldots, n$), with sample mean

$$\bar{x} = \frac{1}{n}\sum_{i=1}^{n} x_i \qquad (27.1)$$

We wish to test the "null hypothesis" that this sample is drawn from a normally distributed population, $N(\mu_0, \sigma^2)$.

As in Chapter 25 we could define the normal deviate

$$z = \frac{\bar{x} - \mu}{\sigma/\sqrt{n}} \qquad (27.2)$$

as a standardized variable which is $N(0, 1)$.

If σ were known, we could use tables of the cumulative normal distribution (e.g., Tables A.VI.2 and A.VI.3) to test whether z is significantly different from zero. Here, however, we face up to the fact that we do not know σ^2 but must *estimate* it by

$$s^2 = \frac{1}{n-1}\sum_{i=1}^{n}(x_i - \bar{x})^2 \qquad (27.3)$$

The estimate of the standard deviation of $\bar{x}$ is

$$(SD \text{ of } \bar{x})_{\text{est}} = \left(\frac{\sigma}{\sqrt{n}}\right)_{\text{est}} = \frac{s}{\sqrt{n}} = \sqrt{\frac{s^2}{n}} \qquad (27.4)$$

We may form a test statistic to correspond to (27.2):

$$t = \frac{\bar{x} - \mu}{s/\sqrt{n}} \qquad (27.5)$$

We wish to evaluate the distribution function of t assuming the null hypothesis.

$$t = \left(\frac{\bar{x} - \mu}{\sigma/\sqrt{n}}\right) \Big/ \sqrt{\frac{s^2}{\sigma^2}} = \frac{Z}{\sqrt{U}} \qquad (27.6a)$$

where $Z = \dfrac{\bar{x} - \mu}{\sigma/\sqrt{n}}$ is $N(0, 1)$ and

$$U = \frac{s^2}{\sigma^2} = \frac{\chi^2_{n-1}}{n-1} \qquad (27.6b)$$

by (26.53).

We see that t is a function of two independent random variables whose distributions are known (normal and chi-square). This permits us to use relation (20.8).

The *joint probability density function* of Z and U is

$$f_{ZU}(z, u) = f_Z(z)(n-1) \times \mathbf{CS}[(n-1)U; n-1]$$

where we recall

$$\mathbf{CS}(\chi^2; n) = \frac{(\chi^2)^{(1/2)(n-2)}e^{-\chi^2/2}}{\Gamma(n/2)2^{n/2}}$$

and $f_Z(z) = (1/\sqrt{2\pi})e^{-z^2/2}$

and we specify the ranges

$$-\infty \le Z \le +\infty$$

and

$$0 \le U \le \infty$$

We have the *joint probability distribution* of Z and U. We wish to transform to the distribution of t and U.

$$g_{tU}(t, u) = f_{ZU}[z(t, u), u(t, u)]\left|\frac{\partial(z, u)}{\partial(t, u)}\right|$$

by (20.40).

We then can integrate to get the *marginal distribution*

$$f(t) = \int_0^{\infty} g(t, u)\, du$$

The result of some manipulation is

$$f(t; n-1) = \frac{\left(\frac{n-2}{2}\right)!}{\left(\frac{n-3}{2}\right)!\sqrt{\pi(n-1)}} \times \left[1 + \frac{t^2}{n-1}\right]^{-n/2} \qquad -\infty < t < +\infty \qquad (27.7)$$

where t is said to have $(n-1)$ degrees of freedom since we started with χ^2_{n-1}.

For completeness we write the p.d.f. for the "**t distribution for n degrees of freedom**"

$$f(t; n) \equiv \mathbf{St}(t; n) = \frac{\left(\frac{n-1}{2}\right)!}{\left(\frac{n-2}{2}\right)!\sqrt{\pi n}}\left[1 + \frac{t^2}{n}\right]^{-(n+1)/2} \qquad -\infty < t < +\infty \qquad (27.8)$$

Cauchy Distribution

For $n = 1$, Student's t distribution reduces to

$$\mathbf{St}(t; 1) = \frac{1}{\pi}\frac{1}{1+t^2} \equiv \mathbf{c}(t; 0, 1) \qquad (27.9)$$

which we have earlier encountered as the *Cauchy distribution* (25.61). At that time we were mainly interested in it as a "freak" distribution for which the central limit theorem did not apply.

We recall that the *mean did not exist* although we could *define* the mean to be the center of symmetry of the distribution, in this case [by (25.63)]

$$\langle\langle t; 1\rangle\rangle = 0 \qquad (27.10)$$

The *variance is infinite* by (25.64)

$$\sigma^2(t;1) = \infty \qquad (27.11)$$

Perhaps surprisingly, the Cauchy distribution arises in several situations in the real world: via Student's *t*, in the theory of atomic and nuclear transitions, etc.

► **Example 27.1**
Enfilading fire
The Cauchy distribution may be obtained from the physical situation shown in Figure 27.1.

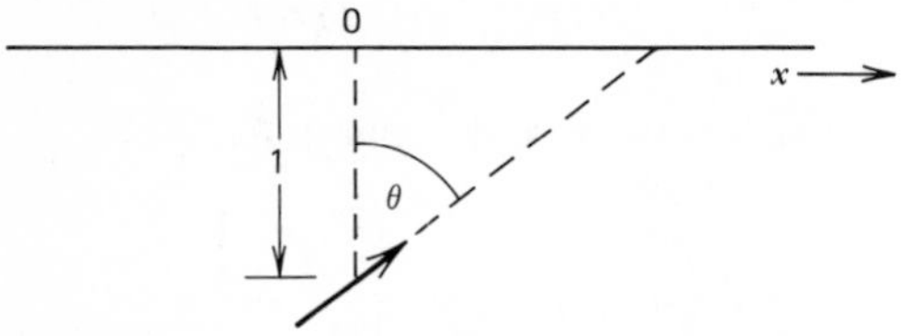

Figure 27.1
Geometry for the example of enfilading fire (see text).

We consider a rotating machine gun mounted at a unit distance from a long wall with the arrow indicating the direction of fire. The gun is rotated at a constant angular velocity, $d\theta/dt = \omega$, and is fired at a constant rate. It is required to find the distribution of hits along the wall (x-axis).
A hit will occur for

$$-\frac{\pi}{2} \leq \theta \leq +\frac{\pi}{2}$$

where θ is $U(-\pi/2, +\pi/2)$, that is, *uniformly distributed* over this range so that

$$f(\theta) = \frac{1}{\pi} \qquad -\frac{\pi}{2} \leq \theta \leq +\frac{\pi}{2}$$

Since the distance from the wall is unity, $\tan x = \theta$.

$$\theta = \arctan x = \theta(x)$$

and $d\theta/dx = 1/(1+x^2)$

We obtain the p.d.f. for x by (20.39):

$$g(x)dx = f[\theta(x)]\left|\frac{d\theta}{dx}\right|dx = \frac{1}{\pi}\frac{1}{1+x^2}$$
$$-\infty \leq x \leq +\infty$$

This is just the *Cauchy distribution.* ◄

COMPARISON OF THE CAUCHY AND STUDENT'S t DISTRIBUTIONS WITH THE NORMAL

We have shown in (25.83) that the Cauchy distribution is normalized

$$\frac{1}{\pi}\int_{-\infty}^{\infty}\frac{dt}{1+t^2} = 1 \qquad (27.12)$$

We plot the Cauchy p.d.f. (27.7) in Figure 27.2. Because the Cauchy distribution has an infinite variance its width is characterized by its "**full width at half-maximum**," also called the "*half-width,*" Γ. When $t = \Gamma/2 = 1$

$$\mathbf{St}\left(\frac{\Gamma}{2}; 1\right) = \mathbf{c}\left(\frac{\Gamma}{2}; 0, 1\right) = \frac{1}{2}\mathbf{c}(0; 0, 1) = \frac{1}{2\pi} \qquad (27.13)$$

We may look at this same parameter for the *normal* p.d.f.:

$$\mathbf{n}\left(t = \frac{\Gamma}{2}; \mu, \sigma\right) = \frac{1}{2}\mathbf{n}(t = \mu; \mu, \sigma) = \frac{1}{2}\frac{1}{\sigma}\varphi(0)$$
$$= \frac{1}{\sqrt{2\pi\sigma^2}} \times \exp\left[-\left(\frac{\Gamma}{2} - \mu\right)^2 \Big/ 2\sigma^2\right]$$

This yields

$$\frac{1}{2\sigma}\frac{1}{\sqrt{2\pi}} = \frac{1}{\sqrt{2\pi\sigma^2}}\exp\left[-\left(\frac{\Gamma}{2} - \mu\right)^2 \Big/ 2\sigma^2\right]$$

or

$$\frac{\left(\frac{\Gamma}{2} - \mu\right)^2}{2\sigma^2} = \ln 2$$

$$\therefore \frac{\Gamma}{2} - \mu = \sigma\sqrt{2\ln 2} = 1.1774\sigma \qquad (27.14)$$

For $\mu = 0$

$$\Gamma_{\text{normal}} = 2.3548\sigma \qquad (27.15)$$

A normal p.d.f. with the same "*half-*

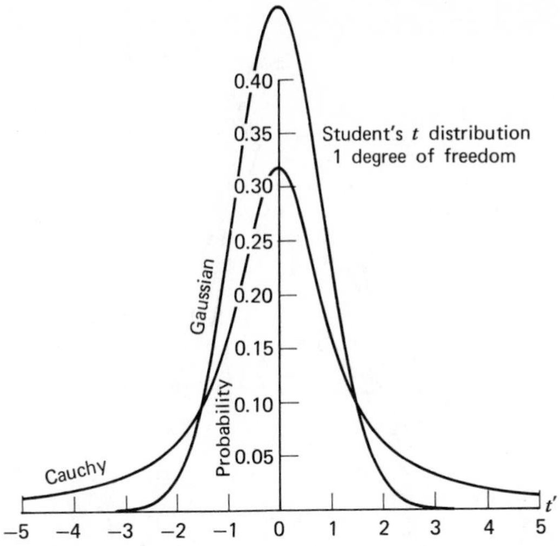

Figure 27.2
Plot of the p.d.f. for the Cauchy distribution $\mathbf{St}(t;1)$ versus t. Also plotted in this figure is a normal (Gaussian) p.d.f. with the same "half-width" as the Cauchy curve, $\mathbf{n}(t;0,\Gamma/2.3548)$ versus t.

width" as the Cauchy curve is plotted on Figure 27.2:

$$\mathbf{n}\left(t;0,\frac{\Gamma}{2.3548}\right) \qquad \text{versus} \qquad t$$

Note that the Gaussian curve is more peaked than is the Cauchy.

In Figure 27.3 we plot the usual Gaussian

$$\mathbf{n}(x;0,1) \qquad \text{versus} \qquad x$$

together with a Cauchy curve chosen so that

$$\Gamma = 2.3548$$

We plot the general Student's t distribution

$$\mathbf{St}(t;n) \qquad \text{versus} \qquad t$$

in Figure 27.4 for $n = 1, 2, 3, 4, 5, 6, 8, 10, 12, 16, 20, 30, 50$. It should be noted that the p.d.f. *approaches the normal* p.d.f. as $n \to \infty$

$$\mathbf{St}(t;n) \to \varphi(t) \qquad n \to \infty \quad (27.16)$$

We emphasize that the Student's t distribution involves *greater probabilities of large deviations from the mean* than does

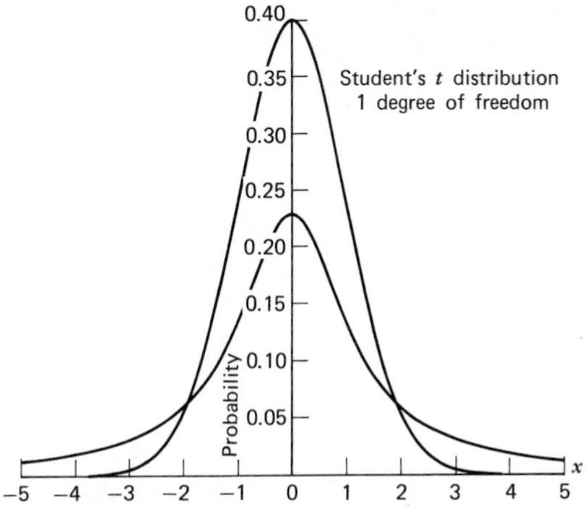

Figure 27.3
Plot of the more usually normalized Gaussian p.d.f. $\mathbf{n}(x;0,1)$ versus x together with a Cauchy curve chosen so that $\Gamma = 2.3548$.

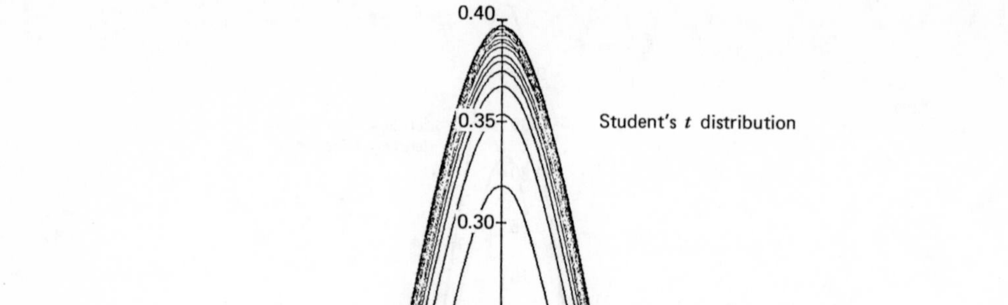

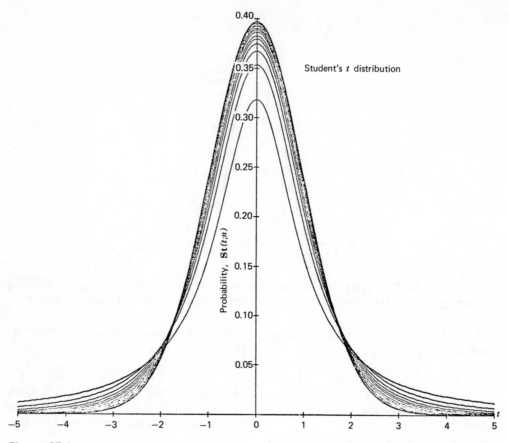

Figure 27.4
Plot of the p.d.f. for the general Student's t distribution $\mathbf{St}(t\,;n)$ versus t for various values of the number of degrees of freedom, $n = 1, 2, 3, 4, 5, 6, 8, 10, 12, 16, 20, 30$ and 50.

the normal distribution for small n, that is, for small samples **the normal distribution may underestimate the probabilities of large deviations.** We shall return to this important point.

MOMENTS OF STUDENT'S t DISTRIBUTION

Because of the symmetry of $\mathbf{St}(t\,;n)$ we may always take

$$\langle t \rangle = 0 \tag{27.17}$$

The variance may be obtained for $n \neq 1, 2$:

$$\mathrm{Var}(t\,;n>2) = \frac{n}{n-2} \qquad n>2 \tag{27.18}$$

CUMULATIVE DISTRIBUTION FUNCTION

The probability that $t < x$ is defined as usual:

$$P(x\,;n) = \int_{-\infty}^{x} \mathbf{St}(t\,;n)\,dt \tag{27.19}$$

We have integrated this numerically and tabulated the results in Table A.VIII.1. This represents the probability that t is *less than* a number, x.

$$P(x\,;n) = \Pr(t<x) \tag{27.20}$$

To get the probability that t is *greater than* a number we use

$$\begin{aligned}\Pr(t>x) &= 1-\Pr(t<x) = Q(x\,;n)\\ &= 1-P(x\,;n)\end{aligned} \tag{27.21}$$

Only values $x > 0$ are tabulated, but we note that **St**$(t; n)$ is **symmetrical**. Hence

$$\begin{aligned} \Pr(t < -x) &= \Pr(t > +x) \\ &= 1 - \Pr(t < +x) \\ &= Q(x; n) = 1 - P(x; n) \end{aligned} \qquad (27.22)$$

Furthermore, we may get the probability that $|t| < x$ from (17.11) and (17.12):

$$\begin{aligned} \Pr(|t| < x) &= \Pr(-x < t < +x) \\ &= P(x; n) - [1 - P(x; n)] \\ &= 2P(x; n) - 1 \end{aligned} \qquad (27.23)$$

Therefore,

$$\Pr(|t| > x) = 2[1 - P(x; n)] \qquad (27.24)$$

We may invert the tables and use an iterative technique to tabulate the "*percentage points*" of the cumulative Student's t distribution. Here we tabulate the values of x versus F such that $F = \Pr(t < x)$ for various n. This is done in Table A.VIII.2.

Applications of Student's t Distribution

The usual application is to make a "**Student's t test**" which may take on several forms.

(A) TEST THAT A SAMPLE OF n VALUES x_i COMES FROM A NORMALLY DISTRIBUTED POPULATION WITH MEAN μ [x is $N(\mu, \sigma^2)$]

Another way of expressing this is to test the hypothesis that the mean of the sample is consistent with being equal to the mean, μ, of the assumed population.

We form the *sample mean* and the *sample variance*

$$\bar{x} = \frac{1}{n} \sum x_i$$

and

$$s^2 = \frac{1}{n-1} \sum (x_i - \bar{x})^2$$

If we knew the variance of the assumed parent population, we could form the normal deviate statistic (27.2) and use the tables of the normal distribution. If we do not know the population variance, we might approximate it by the sample variance when n is large enough and still use the normal tables. However, when n is small (the "*small sample case*"), we instead form the statistic t as in (27.5). For this case we use the tables appropriate to **St**$(t; n-1)$, that is, a sample of size n means that we are considering a student's t distribution with $(n-1)$ degrees of freedom.

We are always interested in determining if the observed difference between $\bar{x}$ and μ is large enough to be significant or if there is a substantial probability (usually taken arbitrarily to be greater than or equal to 0.05) that the observed difference, or a larger one, may happen through random fluctuation.

► **Example 27.2**

We wish to test the (null) hypothesis that the length of a rod in a student laboratory is 10.5 cm. A sample of 10 measurements by students yields the values (in cm): 10.1, 10.1, 10.5, 10.7, 10.9, 11.1, 11.2, 11.4, and 11.4.

We are testing the hypothesis that the lengths are normally distributed with mean 10.5 cm and variance σ^2. The average

$$\bar{x} = \frac{1}{10} \sum_{i=1}^{10} x_i = 10.86 \text{ cm}$$

The alternative hypothesis is that the length of the table is > 10.5 cm. We wish to determine if $\bar{x}$ is sufficiently greater than $\mu = 10.5$ to reject the null hypothesis or is it simply a random fluctuation that $\bar{x} > \mu$? We ask for the probability of a random fluctuation of this size assuming that $\mu = 10.5$.

We find the sample variance

$$s^2 = \frac{1}{n-1}\sum (x_i - \bar{x})^2$$
$$= \frac{1}{n-1}\left\{\sum x_i^2 - n(\bar{x})^2\right\}$$
$$= \frac{1}{9}\{1181.58 - 1179.40\} = \frac{2.18}{9} = 0.242$$

or $s = 0.492$.

If we assume that $s = \sigma$, we could construct the normal deviate

$$z = \frac{\bar{x} - \mu}{\sigma/\sqrt{n}} = \frac{(10.86 - 10.5)\sqrt{10}}{(0.492)} = 2.31$$

The table of the cumulative normal distribution function yields

$$\Pr(z < 2.31) = P(2.31) = 0.9896$$

or

$$\Pr(z > 2.31) = 0.0104$$

We emphasize that this assumption is *not rigorous.*

Instead, we construct the test statistic

$$t = \frac{\bar{x} - \mu}{s/\sqrt{n}} = 2.31$$

and use the tables of the cumulative student t distribution for $n - 1 = 9$ degrees of freedom.

$$\Pr(t < 2.30; 9) = 0.9765$$

or

$$\Pr(t > 2.30; 9) = 0.0235$$

We see that the probability is more than twice as great as calculated from the normal approximation. Since we usually set 0.05 as the (arbitrary) limit for belief, we tend to reject the hypothesis since there is only a $2\frac{1}{2}\%$ chance that the hypothesis is true and that the result arose by random error. ◄

The differences between Student's t test and the normal deviate test become more pronounced for n smaller; that is, smaller samples.

► **Example 27.3**

An executive needs a typist who can type 50 words per minute (wpm). The typing pool sends over an applicant who claims to average 50 wpm. The executive's secretary tests the applicant and three trials yield the results (in wpm): 47, 48, 49. Query: Is the applicant's claim inconsistent?

The sample average is $\bar{x} = 48$. The sample variance is

$$s^2 = \tfrac{1}{2}[(-1)^2 + 0 + 1^2] = 1$$

We form the statistic

$$t = \frac{(\bar{x} - \mu)\sqrt{n}}{s} = \frac{2\sqrt{3}}{1} = 3.46$$

The executive secretary knows a smattering of statistics and assumes that $\sigma = s$ and $t = z$. She recalls that $\Pr(z > 3.46)$ for a normal deviate is less than 3.4×10^{-4}. She is just about to dismiss the applicant when her boss shows up and points out that the t distribution should be used. For this, $\Pr(t > 3.46)$ is greater than (0.04). This is considerably different. ◄

(B) TEST THE DIFFERENCE BETWEEN TWO SAMPLE MEANS

Suppose we are given a sample of n_1 observations x_i and n_2 observations y_i. We assume that the x_i and y_i are normally distributed.

$$x \text{ is } N(\mu_1, \sigma_1^2); \qquad y \text{ is } N(\mu_2, \sigma_2^2)$$

We wish to determine if the observations are consistent with the hypothesis that the population means are equal, that is, $\mu_1 = \mu_2$.

As usual, we form the sample means

$$\bar{x} = \frac{1}{n_1}\sum^{n_1} x_i \qquad \text{and} \qquad \bar{y} = \frac{1}{n_2}\sum^{n_2} y_i$$

If we knew the population variances, σ_1^2 and σ_2^2, we could form the statistic

$$z = \frac{\bar{x} - \bar{y}}{[(\sigma_1^2/n_1) + (\sigma_2^2/n_2)]^{1/2}} \qquad (27.25)$$

as in the normal deviate test of Chapter 25. We could then use the tables of the cumulative normal function to calculate

$$\Pr(|Z| > z) = 1 - A(z) \quad (27.26)$$

If we do not know the σ_1 and σ_2 as given, then we proceed differently. We form the sample variances

$$s_1^2 = \frac{1}{n_1 - 1} \sum (x_i - \bar{x})^2$$

and

$$s_2^2 = \frac{1}{n_2 - 1} \sum (y_i - \bar{y})^2$$

If n_1 and n_2 are each large, we can use the normal approximation (27.25) and (27.26) with the relations $\sigma_1^2 \approx s_1^2$ and $\sigma_2^2 \approx s_2^2$. We are interested in the cases where we may not invoke the normal approximation, that is, the *small sample* cases with n_1, n_2 small.

Case 1: Assume $\sigma_1^2 = \sigma_2^2 = \sigma^2$, but unknown

Since σ^2 is not known we make a "*pooled estimate*" of the variance, σ^2:

$$s^2 = \frac{(n_1 - 1)s_1^2 + (n_2 - 1)s_2^2}{n_1 + n_2 - 2} \quad (27.27)$$

This is an unbiased estimate of the *common* population variance σ^2.

We then form the **test statistic**

$$(t; n_1 + n_2 - 2) = \frac{\bar{x} - \bar{y}}{\sqrt{s^2/n_1 + s^2/n_2}} \quad (27.28)$$

corrresponding to $(n_1 + n_2 - 2)$ *degrees of freedom.*

We may justify the use of Student's t distribution to describe the statistic in (27.28) as follows.

The numerator is the difference of two means and is normally distributed. (This difference is often *asymptotically* normal even if the population x and y are not assumed rigorously normal). The denominator is $s\sqrt{1/n_1 + 1/n_2}$ which is independent of the numerator. We have seen in Chapter 26 that

$$s_1^2 \approx \frac{1}{n_1 - 1} \sigma^2 \chi^2_{n_1 - 1}$$

and

$$s_2^2 \approx \frac{1}{n_2 - 1} \sigma^2 \chi^2_{n_2 - 1}$$

That is, s_1^2 and s_2^2 are *chi-square distributed* with $(n_1 - 1)$ and $(n_2 - 1)$ degrees of freedom, respectively.

$$s^2 \approx \frac{\sigma^2}{(n_1 + n_2 - 2)} \{\chi^2_{n_1 - 1} + \chi^2_{n_2 - 1}\}$$

and we may note the reproductive property of the χ^2 distribution:

If X_1 is $\chi^2_{n_1}$ and X_2 is $\chi^2_{n_2}$, then

$$Z = X_1 + X_2 \qquad \text{is} \qquad \chi^2_{n_1 + n_2}$$

Here s^2 is chi-square distributed with $(n_1 - 1) + (n_2 - 1) = (n_1 + n_2 - 2)$ degrees of freedom. This shows that (27.28) is **Student's t distributed with $(n_1 + n_2 - 2)$ degrees of freedom.**

For a common variance, therefore, we use (27.28) and the tables of the Student's t distribution to calculate

$$\Pr(|T| > t) = 2P(t; n_1 + n_2 - 2) - 1 \quad (27.29)$$

Case 2: $\sigma_1^2 \neq \sigma_2^2$

This case is more complicated. We may state, however, that the statistic

$$t' = \frac{\bar{x} - \bar{y}}{\sqrt{s_1^2/n_1 + s_2^2/n_2}} \quad (27.30)$$

may be used where the number of degrees of freedom is given by the **integer** *nearest to*

$$n' = \left\{ \frac{(s_1^2/n_1 + s_2^2/n_2)}{\dfrac{[s_1^2/n_1]^2}{n_1 - 1} + \dfrac{[s_2^2/n_2]^2}{n_2 - 1}} \right\} - 2 \quad (27.31)$$

The statistic t' is *approximately* **Student's t distributed with n' degrees of freedom** and

$$\Pr(|T| > t') = 2P(t'; n') - 1 \quad (27.32)$$

►Example 27.4

A machine on an assembly line stamps out gaskets with an opening, x which is approximately normally distributed [x is $N(\mu, \sigma^2)$]. As the plant manager walked through the plant he noticed that the machine operator appeared to brush against the machine although he continued on as usual. The manager accused the operator of inadvertently changing the setting on the machine, which the operator vehemently denied. To forestall an unpleasant labor-management dispute, the shop steward offered to measure six gaskets that were still in the "out" box, and made prior to the alleged incident, and to compare them with seven gaskets made after the fact. Query: Are the mean gasket lengths consistent with being the same?

Here the variance of the two distributions may safely be **assumed** to be the same ($\sigma_1 = \sigma_2 = \sigma$) although not known *a priori*.

The steward measured the 6 "before" gaskets to be (in cm) $x_i = 11.23$, 11.37, 11.04, 11.14, 11.49, and 11.26 and the 7 "after" gaskets to be $y_i = 11.57$, 11.03, 11.26, 11.45, 11.45, 11.38 and 11.19.

We get $\bar{x} = 11.255$ cm and $\bar{y} = 11.306$ cm.

The *sample variances* are

$$S_1^2 = \frac{1}{6-1}\sum_1^6 (x_i - \bar{x})^2 = \frac{1}{5}\left\{\sum_1^6 x_i^2 - 6(\bar{x})^2\right\}$$

$$= \frac{1}{5}\{760.1787 - 760.0501\} = \frac{0.1286}{5}$$

$$= 0.0257$$

and

$$S_2^2 = \frac{1}{7-1}\sum_1^7 (y_i - \bar{y})^2 = \frac{1}{6}\left\{\sum y_i^2 - 7(\bar{y})^2\right\}$$

$$= \frac{1}{6}\{894.9240 - 894.7794\}$$

$$= \frac{0.1445}{6} = 0.0241$$

$$S^2 = \frac{(6-1)S_1^2 + (7-1)S_2^2}{6+7-2}$$

$$= \frac{0.1285 + 0.1446}{11}$$

$$= 0.0248$$

We form the test statistic

$$(t; 11) = \frac{\bar{x} - \bar{y}}{S\sqrt{(1/n_1) + (1/n_2)}} = \frac{-0.051}{0.0876}$$

$$= -0.582$$

We need what is called a "*two-tailed test*" since we don't care whether $(\bar{x} - \bar{y})$ is positive or negative. What matters is whether it differs from zero.

$$\Pr(|t| > 0.582) = 2[1 - P(0.58; 11)]$$

$$\approx 2[1 - 0.72]$$

$$= 0.56$$

The manager has not proved his point. ◄

28

MISCELLANEOUS OTHER PROBABILITY DISTRIBUTIONS AND SOME EXAMPLES

We shall briefly review the properties of some other distributions that may be useful.

The Negative Binomial Distribution

We have already considered the expectation value of a related problem: given our usual dichotomous situation in which an event has a constant probability, p, of success, and, $q = 1 - p$, of failure. We now suppose that the trials are continued until the event has occurred exactly r times. We want to determine the probability, $\Pr(N = n)$, that this will require **exactly** n trials, where N is the number of trials.

$$\Pr(N - n) = \Pr\{[(r-1) \text{ events in the first } (n-1) \text{ trials}] \textit{ and } (\text{one event on the } n\text{th trial})\}$$
$$= \Pr\{(r-1) \text{ events in } (n-1) \text{ trials}\} \Pr\{1 \text{ on } n\text{th}\}$$
$$= \binom{n-1}{r-1} p^{r-1} q^{n-r} p = \mathbf{b}_{\text{neg}}(n)$$

$$\therefore \mathbf{b}_{\text{neg}}(n) = \binom{n-1}{r-1} p^r q^{n-r}, \quad n = r, r+1, \ldots \qquad (28.1)$$

This distribution is called the *negative binomial* since the probabilities may be obtained from successive terms of the expansion of a negative binomial, for example,

$$\sum_{n=r}^{\infty} \binom{n-1}{r-1} p^r q^{n-r} = p^r (1-q)^{-r}$$

which may be written $= (Q - P)^{-r}$ where $Q = 1/p$, $P = q/p$ and $Q - P = \frac{1}{p}(1-q) = 1$.

We may rewrite $\mathbf{b}_{\text{neg}}(n)$ by substituting $s = n - r$ whence

$$\Pr(s) = \binom{r+s-1}{r-1} p^r q^s \qquad (28.2)$$

The *expectation value* is

$$E(s) = \frac{r}{p} \qquad (28.3)$$

and the *variance*

$$\text{Var}(s) = \frac{rq}{p}\left(1 + \frac{q}{p}\right) = \frac{rq}{p^2} \qquad (28.4)$$

The Multinomial Distribution

This is a generalization of the binomial distribution to the case of k-fold mutually exclusive events each of which occurs with constant probability p_i, $i = 1, \ldots, k$, but where

$$\sum_{i=1}^{k} p_i = 1 \qquad (28.5)$$

If n is the total number of trials, we have the *joint probability* that there are x_1 events of the first kind, x_2 events of the second kind, etc:

$$\Pr(x_1, x_2, \ldots, x_k) \equiv \mathbf{m}(x_1, x_2, \ldots, x_k; p_1, \ldots, p_{k-1}, n)$$

$$= \frac{n!}{x_1!x_2!\cdots x_k!} p_1^{x_1} p_2^{x_2} \cdots p_k^{x_k} \qquad (28.6)$$

where the parameters are only enumerated up to $k-1$ since one parameter is fixed by (28.5).

The *expectation values* and *variances* are k-fold:

$$E(x_i) = np_i \qquad (28.7)$$

and

$$\mathrm{Var}(x_i) = \sigma_{x_i}^2 = np_i(1-p_i) \qquad (28.8)$$

The Exponential Distribution

We have encountered this in considering radioactive decay. In general, it has application to life-testing problems and also to problems of atomic and nuclear physics. It is defined by the p.d.f.

$$\mathbf{e}(x; \lambda) \equiv \lambda e^{-\lambda x} \quad \text{for} \quad 0 \leqslant x \quad 0 < \lambda \qquad (28.9)$$

The *mean* and *variance* are given by

$$E(x) = \frac{1}{\lambda} \qquad (28.10)$$

and

$$\sigma^2(x) = \mathrm{Var}(x) = \frac{1}{\lambda^2} \qquad (28.11)$$

The Weibull Distribution

This distribution is a generalization of the exponential, having *three* parameters instead of one. It is also used in life-testing and reliability studies where its parameters are fitted to particular situations.

The Weibull p.d.f. is

$$\mathbf{w}(x; a,b,c) \equiv \frac{c}{b}\left(\frac{x-a}{b}\right)^{c-1} e^{-[(x-a)/b]^c} \qquad (28.12)$$

where $a \leqslant x$ and $0 < b,c$.

As with the Gaussian distribution, the Weibull distribution may be transformed into a *one-parameter distribution* for purposes of tabulation and graphing.

$$y = \frac{x-a}{b} \qquad (28.13)$$

yields

$$\mathbf{w}(y; c) = cy^{c-1}e^{-y^c} \qquad y \geqslant 0 \qquad c > 0 \qquad (28.14)$$

This has the c.d.f.

$$\mathbf{W}(y; c) = \int_0^y \mathbf{w}(y; c)\,dy = 1 - e^{-y^c} \qquad (28.15)$$

The *expectation value* is

$$E(y) = \Gamma(1 + c^{-1}) \qquad (28.16)$$

where

$$\Gamma(z) = \int_0^\infty e^{-y} y^{z-1}\,dy \qquad z > 0$$

is the Gamma function discussed in Appendix III.

The *variance* is

$$\mathrm{Var}(y) = \Gamma\left(1 + \frac{2}{c}\right) - \Gamma^2\left(1 + \frac{1}{c}\right) \qquad (28.17)$$

The Log-Normal Distribution

This is the distribution of a random variable, x, whose **logarithm** is *normally distributed,* that is,

$$y = \ln x \quad \text{is} \quad N(\mu, \sigma^2) \quad (28.18)$$

The log-normal distribution* has been found useful in describing the *distribution of incomes* and other economic indices. In medicine it has been used to describe the *distribution of recovery times* from various illness, the *survival times* of bacteria in disinfectant solutions, and the weights and blood pressures of people.

We may relate the log-normal distribution to the *normal convergence theorem* by considering a variate X that is the result of a large number of independent influences each of which produces a small effect *proportional* to X. Each influence then, produces an effect such that

$$\frac{\Delta X}{X} = k = \Delta(\ln X)$$

The distribution resulting from a large number of independent influences, each producing a small constant effect, is normal, that is, $\ln X$ is normally distributed and X is log-normal.

The *cumulative distribution function* is

$$F_X(x) = \Pr[X \leq x] = \Pr[\ln X \leq \ln x]$$

$$= \frac{1}{\sqrt{2\pi}} \int_{-\infty}^{(\ln x - \mu)/\sigma} e^{-t^2/2}\, dt \quad (28.19)$$

*See, for example, J. Aitchison and J. A. C. Brown, *The Lognormal Distribution*, Cambridge University Press, New York, 1957.

The p.d.f. is obtained from

$$\mathbf{LN}(x) = F_X'(x) = \frac{1}{\sqrt{2\pi}\sigma}\frac{1}{x}\, e^{-(\ln x - \mu)^2/(2\sigma^2)} \qquad x > 0 \quad (28.20)$$

We may get the *expectation value*

$$E(x) = e^{\mu + \sigma^2/2} \quad (28.21)$$

and the *variance*

$$\operatorname{Var}(x) = e^{2\mu + \sigma^2}(e^{\sigma^2} - 1) \quad (28.22)$$

The expectation value is not so useful for an **asymmetric** distribution. Often, more significant are such measures as the *median, mode,* and the *geometric mean.*

We note that the *mode,* obtained from setting

$$\frac{d}{dx}[\mathbf{LN}(x)] = 0 \quad \text{is} \quad x_{\text{mode}} = e^{\mu - \sigma^2} \quad (28.23)$$

For discrete data believed to follow a logarithmic distribution, $x_1, x_2, \ldots, x_n$, it is often useful to define the *geometric mean*

$$x_{\text{geometric mean}} = \left(\prod_{x=1}^{n} x_i\right)^{1/n}$$

$$= \exp\left[\left(\frac{1}{n}\sum_{i=1}^{n} \ln x_i\right)\right] \quad (28.24)$$

(See the end of Chapter 7.)

The *median* is

$$x_{\text{med}} = e^{\mu} \quad (28.24a)$$

The F-Distribution

Consider that x_1 and x_2 are independent, and that each is chi-square distributed with n_1 and n_2 d.o.f., respectively. Then

$$F(n_1, n_2) = \frac{x_1/n_1}{x_2/n_2} \quad (28.25)$$

has the *p.d.f.*

$$\mathbf{f}(F; n_1, n_2) = \frac{\Gamma[(n_1 + n_2)/2]}{\Gamma(n_1/2)\Gamma(n_2/2)} \frac{(n_1/n_2)^{n_1/2} F^{(n_1-2)/2}}{(1 + n_1 F/n_2)^{(n_1+n_2)/2}} \qquad F > 0 \quad (28.26)$$

The *mean* of the F-distribution is

$$E(F; n_1,n_2) = \frac{n_2}{n_2 - 2} \qquad n_2 > 2 \qquad (28.27)$$

and the *variance* is

$$\text{Var}(F; n_1,n_2) = \frac{2n_2^2(n_1 + n_2 - 2)}{n_1(n_2 - 2)^2(n_2 - 4)} \qquad n_2 > 4 \qquad (28.28)$$

The *mode* is that value of $F(n_1, n_2)$ for which (28.26) is maximum:

$$\text{mode}\,(F; n_1,n_2) = \frac{n_2(n_1 - 2)}{n_1(n_2 + 2)} \qquad (28.29)$$

Suppose we have random samples of sizes N_1, N_2 from *normal* populations that have variances σ_1^2, σ_2^2, respectively. Let s_1^2, s_2^2 be the *sample variances.* We have seen in Chapter 26 that

$s_1^2(N_1 - 1)/\sigma_1^2$ is $\chi^2(N_1 - 1)$-distributed

and

$s_2^2(N_2 - 1)/\sigma_2^2$ is $\chi^2(N_2 - 1)$-distributed.

Then

$$F(N_1 - 1, N_2 - 1) = \frac{\left[\dfrac{s_1^2(N_1 - 1)}{\sigma_1^2}\right]\Big/(N_1 - 1)}{\left[\dfrac{s_2^2(N_2 - 1)}{\sigma_2^2}\right]\Big/(N_2 - 1)} = \frac{\sigma_2^2}{\sigma_1^2}\frac{s_1^2}{s_2^2}$$

and the ratio of sample variances

$$\frac{s_1^2}{s_2^2} = \frac{\sigma_1^2}{\sigma_2^2}F(N_1 - 1, N_2 - 1) \qquad (28.30)$$

F-TEST

Equation (28.30) underlies the major use of the F-distribution. Just as Student's t distribution can be used to test whether there is a statistically significant difference between the means of samples drawn from two normally distributed populations so **the F-distribution can be used to test whether there is a significant difference between the two sample variances** in the same case of two normal populations. In other words, we can test the hypothesis that $\sigma_1 = \sigma_2 = \sigma$ by seeing how often it will happen that $F(N_1 - 1, N_2 - 1)$ will take on values $\geq s_1^2/s_2^2$ because of random fluctuations.

There are tables of the cumulative F-distribution* tabulating against n_1, n_2 the values of $F(n_1, n_2)$ such that

$$P = \int_0^F f(F; n_1, n_2)dF$$

that is, $100P\%$ of the time the observed values s_1^2/s_2^2 will be $\leq F$.

► **Example 28.1**

In our example of coin weighing, we noted that the 1964 sample of 21 measurements taken with a beam balance yielded $S_{64}^2 = (0.0939)^2$ while 33 measurements of the 1970 population yielded $S_{70}^2 = (0.1066)^2$. Query: Are the variances consistent with being equal in the two samples?

$$\left(\frac{S_{64}}{S_{70}}\right)^2 = \frac{(0.0939)^2}{(0.1066)^2} = F(20, 32) = 0.773$$

The tables are presented so that $F \geq 1$, but we use the relation

$$\Pr[F(n_1, n_2) < F_0] = \Pr[F(n_2, n_1) > 1/F_0] \qquad (28.31)$$

Therefore we consider

$$F(32, 20) = 1.29$$

and note that it is 95% probable that $F > 2.04$.

Therefore, the two sample variances are consistent. ◄

*See Pearson and Hartley, *Biometrika Tables for Statisticians*, Vol. I, 1958, pp. 159–163.

Folded Distributions

It sometimes happens that a random variable that is distributed about some mean is recorded so that only the *absolute values* of the deviations from the mean are noted. We may inquire about the distribution of this "*folded variable.*"

If the original variable is X with p.d.f. $f_X(x)$ and we are interested in the variable folded about 0, we note that

$$f_{Z=X^2}(z) = \frac{1}{2\sqrt{z}}\{f_X(\sqrt{z}) + f_X(-\sqrt{z})\} \qquad z > 0$$

Furthermore, we have that

$$f_{X=\sqrt{Z}}(x) = 2xf_Z(x^2)$$

$$\therefore f_{|X|=\sqrt{Z}}(|x|) = 2|x|f_Z(|x|^2) = f_X(|x|) + f_X(-|x|) \tag{28.32}$$

Folded Normal Distribution

For X a normally distributed random variable with mean μ and variance σ^2, the *folded distribution* has a p.d.f.

$$f_{|X|}(x) = \frac{1}{\sqrt{2\pi}\sigma}[e^{-(x-\mu)^2/2\sigma^2} + e^{-(x+\mu)^2/2\sigma^2}] \quad \text{for } x \geqslant 0 \tag{28.33}$$

The *mean* for the folded normal distribution is

$$\mu_f = \frac{1}{\sqrt{2\pi}}\int_{-\mu/\sigma}^{\mu/\sigma} e^{-u^2/2}\,du + \sigma\sqrt{\frac{2}{\pi}}\,e^{-(1/2)(\mu/\sigma)^2} \tag{28.34}$$

and the *variance* is

$$\sigma_f^2 = \sigma^2 + \mu^2 - \mu_f^2 \tag{28.35}$$

Truncated Distributions

It sometime happens that we are concerned with only a *restricted range* of values of a random variable, x. If we observe only values of $x < x_0$, then we say that the x-distribution is **truncated at x_0**. In particular, we must *normalize* the truncated p.d.f. over the restricted range $-\infty$ to x_0.

If $f_X(x)$ is the p.d.f., and $F_X(x)$ the c.d.f., of the unrestricted variate X, the *truncated p.d.f.* is

$$f_{\mathrm{Tr}}(x; x_0) = \frac{f_X(x)}{\int_{-\infty}^{x_0} f_X(x)\,dx} = \frac{f_X(x)}{F_X(x_0)} \quad \text{for} \quad x < x_0$$

$$= 0 \quad \text{for} \quad x \geq x_0 \tag{28.36}$$

Truncated Normal Distribution

If x is $N(\mu, \sigma^2)$ and its distribution is truncated to x_0, then it may be shown* that the *mean of the truncated distribution* is

$$\mu_{\mathrm{Tr}} = \mu - \sigma\frac{\varphi(z_0)}{\Phi(z_0)} \tag{28.37}$$

and the *variance* is

$$\sigma_{\mathrm{Tr}}^2 = \sigma^2\left\{1 - \left[\frac{\varphi(z_0)}{\Phi(z_0)}\right]^2 - z_0\left[\frac{\varphi(z_0)}{\Phi(z_0)}\right]\right\} \tag{28.38}$$

where

$$z_0 = \frac{x_0 - \mu}{\sigma}$$

$$\varphi(z_0) = \frac{1}{\sqrt{2\pi}}e^{-(1/2)(x_0-\mu)^2/\sigma^2}$$

and

$$\Phi(z_0) = \frac{1}{\sqrt{2\pi}}\int_{-\infty}^{[(x_0-\mu)/\sigma]} e^{-t^2/2}\,dt$$

*See N. L. Johnson and F. C. Leone, *Statistics and Experimental Design*, John Wiley, 1964, Volume I.

The Bivariate Normal Distribution

Two random variables are said to have a *bivariate normal distribution* if their **joint** probability distribution is

$$\mathbf{bn}(x, y) = \frac{1}{2\pi\sigma_x\sigma_y\sqrt{1-\rho^2}} e^{-(1/2)G(x,y;\mu_x,\mu_y,\sigma_x,\sigma_y)} \qquad (28.39)$$

where

$$G(x,y;\mu_x,\mu_y,\sigma_x,\sigma_y) = \frac{1}{(1-\rho^2)}\left\{\left[\frac{(x-\mu_x)}{\sigma_x}\right]^2 - 2\rho\frac{(x-\mu_x)(y-\mu_y)}{\sigma_x\sigma_y} + \frac{(y-\mu_y)^2}{\sigma_y^2}\right\} \qquad (28.40)$$

This is shown as a three-dimensional plot in Figure 28.1.

Above, ρ is the *correlation coefficient* between x, y, and the **marginal distributions** for x, y are **normal**:

$$\mathbf{bn}(x) = \int_{-\infty}^{\infty} \mathbf{bn}(x, y)\, dy = \frac{1}{\sqrt{2\pi}\sigma_x} e^{-(x-\mu_x)^2/(2\sigma_x^2)}$$

and

$$\mathbf{bn}(y) = \int_{-\infty}^{\infty} \mathbf{bn}(x, y)\, dx = \frac{1}{\sqrt{2\pi}\sigma_y} e^{-(y-\mu_y)^2/(2\sigma_y^2)}$$

If $\rho(x, y) = 0$, that is, x and y *uncorrelated*, then,

$$\mathbf{bn}(x, y) = \mathbf{bn}(x)\mathbf{bn}(y)$$

and x and y are *independent*. We emphasize that the statement, $\rho(x, y) = 0$, *implies independence* ***only*** *if we require that the joint p.d.f. is* ***bivariate normal.***

For $\rho = 0$ we have the *product* of two normal distributions

$$\mathbf{bn}(x, y) = \frac{1}{2\pi\sigma_x\sigma_y}\exp\left[-\frac{1}{2}\left(\frac{x-\mu_x}{\sigma_x}\right)^2 + \left(\frac{y-\mu_y}{\sigma_y}\right)^2\right] = \frac{1}{2\pi\sigma_x\sigma_y} e^{-(1/2)G}$$

We may define a *contour of equal probability* by setting

$$G = \text{constant} = \left(\frac{x-\mu_x}{\sigma_x}\right)^2 + \left(\frac{x-\mu_y}{\sigma_y}\right)^2$$

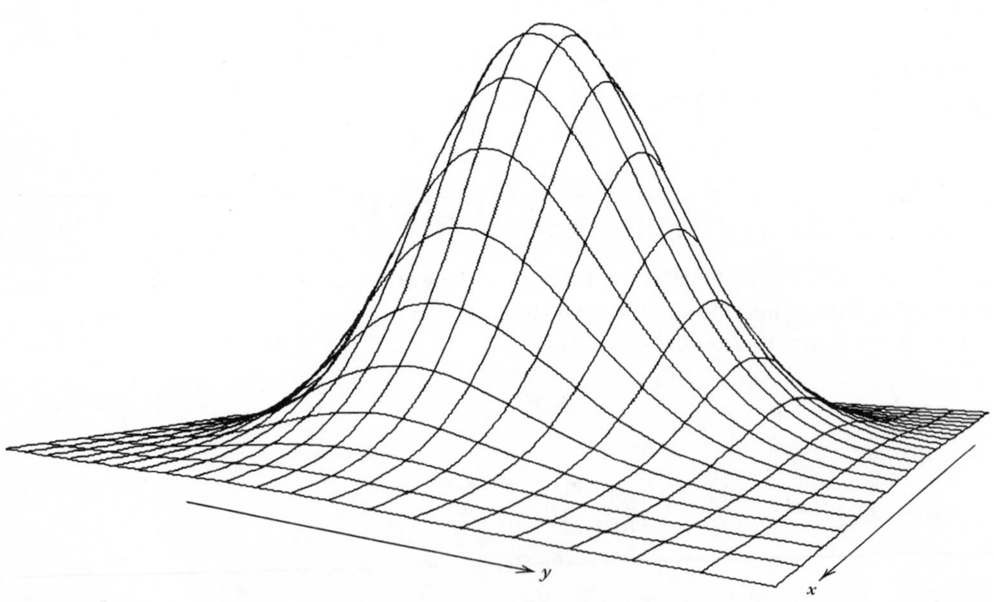

Figure 28.1
Three-dimensional plot of the joint probability distribution of the bivariate normal distribution.

which is an ellipse in the x, y plane centered on (μ_x, μ_y) and with principal axes parallel to the coordinate axes. For $G = 1$ the semimajor and semiminor axes of the ellipse are σ_x, σ_y, respectively, if $\sigma_x > \sigma_y$.

By definition all points (x, y) on the contour, G = constant, have equal *probability density*. Points **inside** the ellipse lie on contours that have probability density **greater** than this while points *outside* lie on contours with *lesser* probability.

The probability that a point (x, y) will lie anywhere *within* the ellipse is given by the integral

$$\frac{1}{2\pi\sigma_x\sigma_y}\iint_{G\leq k} e^{-(1/2)G}\,dx\,dy = \Pr[(x, y) \text{ inside } (G = k)]$$

where $G \leq k$ defines the area of integration.

We need not do this integral explicitly if we note that $G = (\chi^2; 2)$. Points outside the ellipse $G = k$ lie on contours $G = k'$ such that $k' > k$.

$$\begin{aligned}\therefore \Pr[(x, y) \text{ inside } (G = k)] &= \Pr[(\chi^2; 2) \leq k] \\ &= 1 - \Pr[(\chi^2; 2) > k]\end{aligned}$$

We note from Table A.VIII.1 the probabilities for χ^2 with two d.o.f. (Table 28.1).

Table ***28.1***

$\Pr[(x, y) \text{ inside } (G = k)]$	k
0.3935	1
0.6321	2
0.7769	3
0.8647	4
0.9179	5
0.9502	6

The univariate normal distribution allowed us to calculate the probability that the variate x was within a distance $t = (x - \mu)/\sigma$ of the mean μ. For the *bivariate normal distribution* we calculate the probability that the point (x, y) is **within an *area*** centered on the mean (μ_x, μ_y). We expect that 39.4% of the time (x, y) is contained within the ellipse $[(x - \mu_x)/\sigma_x]^2 + [(y - \mu_y)/\sigma_y]^2 = 1$, 63.2% of the time within the ellipse $[(x - \mu_x)/\sigma_x]^2 + [(y - \mu_y)/\sigma_y]^2 = 2$, etc. Equivalently, we might say that (x, y) is contained within the region defined by $G = 1$ *with 39.4% confidence.*

If $\rho \neq 0$, we have from (28.40) the general equation of an ellipse, called the ***covariance ellipse***:

$$\left(\frac{x - \mu_x}{\sigma_x}\right)^2 - 2\rho\frac{(x - \mu_x)(y - \mu_y)}{\sigma_x\sigma_y} + \left(\frac{y - \mu_y}{\sigma_y}\right)^2 = (1 - \rho^2)k$$

This ellipse has a center of symmetry at (μ_x, μ_y) and its principal axes x', y' make an angle θ with the x, y axes. The angle is

$$\theta = \frac{1}{2}\tan^{-1}\left\{\frac{2\rho\sigma_x\sigma_y}{\sigma_x^2 - \sigma_y^2}\right\}$$

The semidiameters along the principal axes may be shown to be r_x, r_y, where (for $k = 1$)

$$r_{x'}^2 = \frac{\sigma_x^2\sigma_y^2(1 - \rho^2)}{\sigma_y^2\cos^2\theta - \rho\sigma_x\sigma_y\sin 2\theta + \sigma_x^2\sin^2\theta}$$

$$r_{y'}^2 = \frac{\sigma_x^2\sigma_y^2(1 - \rho^2)}{\sigma_y^2\sin^2\theta + \rho\sigma_x\sigma_y\sin 2\theta + \sigma_x^2\cos^2\theta}$$

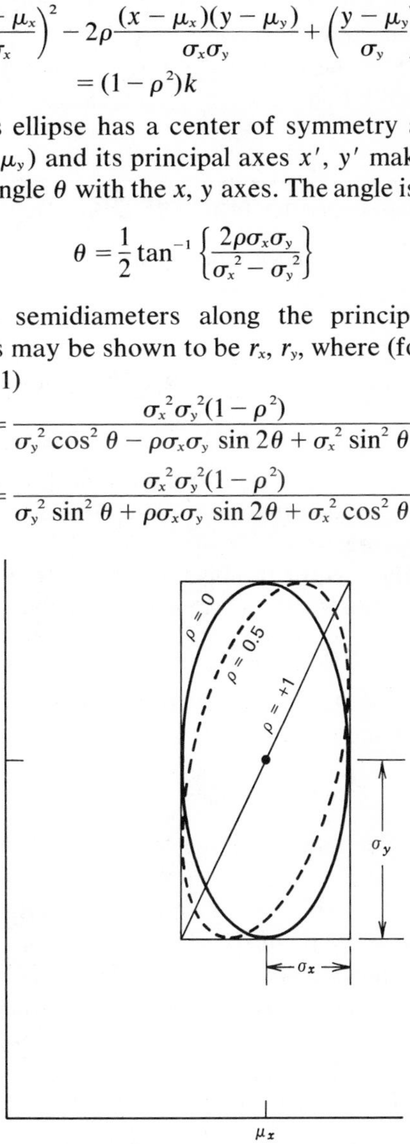

Figure 28.2
Covariance ellipses showing that sections of the three-dimensional surface are plotted versus x, y for different values of ρ, the *correlation coefficient* between x, y: $\rho = 0, +0.5$, and $+1.0$.

If we consider $\sigma_x, \sigma_y, \mu_x, \mu_y$ fixed and let ρ vary, we get different covariance ellipses, but all are *inscribed within a rectangle* bounded by the lines

$$y = \mu_y \pm \sigma_y$$
$$x = \mu_x \pm \sigma_x$$

The covariance ellipses represent *horizontal sections* of the three-dimensional plot shown in Figure 28.1.

For the case $\rho = \pm 1$ we have perfect correlation between x and y:

$$\left(\frac{x-\mu_x}{\sigma_x}\right)^2 \mp 2\left(\frac{x-\mu_x}{\sigma_x}\right)\left(\frac{y-\mu_y}{\sigma_x}\right) + \left(\frac{y-\mu_y}{\sigma_y}\right)^2 = 0 = \left[\left(\frac{x-\mu_x}{\sigma_x}\right) \mp \left(\frac{y-\mu_y}{\sigma_y}\right)\right]^2 = 0$$

or

$$\left(\frac{x-\mu_x}{\sigma_x}\right) = \pm\left(\frac{y-\mu_y}{\sigma_y}\right)$$

which is the equation of a straight line along one of the two diagonals of the rectangle.

The cases $\rho = 0$, $\rho = +0.5$, and $\rho = +1.0$ are shown in Figure 28.2.

Multivariate Normal Distribution

If n random variables $x_1, x_2, \ldots, x_n$ have the joint p.d.f.

$$\mathbf{m}_n(x_1, \ldots, x_n) = C \exp\left[-\frac{1}{2}\sum_{i=1}^{n}\sum_{j=1}^{n} A_{ij}(x_i-\mu_i)(x_j-\mu_j)\right] \tag{28.41}$$

then they are said to have a *multivariate normal distribution.* C is a normalization constant.

Just as for the bivariate case, the marginal distribution for each x_i is normal and

$$E(x_i) = \mu_i \qquad i = 1, \ldots, n$$

The A_{ij} form a matrix $\underline{\mathbf{A}}$ which is $n \times n$. We see that the matrix is symmetric, that is,

$$A_{ij} = A_{ji}$$

We may write the variances and covariances as $\sigma_{ii} = \sigma_i^2$ and $\sigma_{ij} = \text{Covar}(x_i, x_j)$. We then have a *variance-covariance matrix* of elements $(\underline{\sigma})_{ij}$.

The symmetric matrix $\underline{\mathbf{A}}$ has an inverse $\underline{\mathbf{A}}^{-1}$, and it may be demonstrated that

$$\sigma_{ij} = (\underline{\mathbf{A}}^{-1})_{ij} \tag{28.42}$$

The normalization constant is

$$C = [(2\pi)^n(\det \underline{\mathbf{A}})]^{-1/2} \tag{28.43}$$

where $(\det \underline{\mathbf{A}})$ is the determinant of the square matrix $\underline{\mathbf{A}}$.

For $n = 2$, the *exponent* in (28.41) is

$$\sum_{i,j=1}^{2} A_{ij}(x_i-\mu_i)(x_j-\mu_j) = A_{11}(x_1-\mu_1)^2 + (A_{12}+A_{21})(x_1-\mu_1) \times (x_2-\mu_2) + A_{22}(x_2-\mu_2)^2$$

where

$$\underline{\mathbf{A}}_{ij} = \frac{1}{1-\rho^2}\begin{pmatrix} \frac{1}{\sigma_{x_1}^2} & \frac{-\rho}{\sigma_{x_1}\sigma_{x_2}} \\ \frac{-\rho}{\sigma_{x_1}\sigma_{x_2}} & \frac{1}{\sigma_{x_2}^2} \end{pmatrix} \text{ by (28.40)}$$
$$= \frac{1}{(1-\rho^2)\sigma_{x_1}^2\sigma_{x_2}^2}\begin{pmatrix} \sigma_{x_2}^2 & -\rho\sigma_{x_1}\sigma_{x_2} \\ -\rho\sigma_{x_1}\sigma_{x_2} & \sigma_{x_1}^2 \end{pmatrix} \tag{28.44}$$

We get

$$(\underline{\mathbf{A}}^{-1})_{ij} = \frac{\text{Cofactor } A_{ij}}{|\mathbf{A}|}$$

$$\underline{\mathbf{A}}^{-1} = \frac{\begin{pmatrix} \sigma_{x_1}^2 & -\rho\sigma_{x_1}\sigma_{x_2} \\ -\rho\sigma_{x_1}\sigma_{x_2} & \sigma_{x_2}^2 \end{pmatrix}}{\frac{1}{(1-\rho^2)\sigma_{x_1}^2\sigma_{x_2}^2}(\sigma_{x_1}^2\sigma_{x_2}^2 - \rho^2\sigma_{x_1}^2\sigma_{x_2}^2)}$$
$$= \begin{pmatrix} \sigma_{x_1}^2 & -\rho\sigma_{x_1}\sigma_{x_2} \\ -\rho\sigma_{x_1}\sigma_{x_2} & \sigma_{x_2}^2 \end{pmatrix} \tag{28.45}$$
$$= \begin{pmatrix} \sigma_1^2 & -\sigma_{12} \\ -\sigma_{12} & \sigma_2^2 \end{pmatrix} = \underline{\boldsymbol{\sigma}}_{ij}$$

PART V

Statistical Inference

Since we have developed some of the necessary *tools*, we may now examine in greater detail some of their *uses* in data analysis.

In Chapter 17 we first mentioned some of the *areas* of statistical inference. We have already considered several of them in discussions of inverse probability and hypothesis testing (e.g., normal deviate test and Student's t test). We recognize the following broad headings of statistical inference, but they are overlapping, not disparate, and we make the divisions only for convenience in emphasizing different points and procedures. We shall consider all too briefly the problems of *estimation, hypothesis testing*, and *curve fitting*. We stress the point that they are all of a piece and completely interrelated. The distinctions are quite artificial and are made only to facilitate the cataloging of methods.

For example, in a gambling situation we might wish to determine from the data the probability of getting a certain outcome (*estimation*). We might wish to test the hypothesis that the game is fair (*hypothesis testing*). We might wish to adjust our estimates of the probabilities of all the outcomes to accord with the observed frequency distribution (*curve fitting*). We might wish to spend a limited amount of time at the gaming table to use in our study of the situation and to choose our observations to yield as much information as possible (*experimental design*). Finally, we might wish to use our study to determine the best way to play the game, that is, to maximize our expected return, or to minimize our expected loss (*decision theory*).

In the scope of this book we can only indicate *some* of the elementary considerations involved in such problems.

29

ESTIMATION

In problems of estimation we wish to make the *best determination we can*, within the constraints of the experimental situation, of certain features of a real (finite) or conceptual (infinite) population (Chapter 7).

We discussed the idea of a *statistic*, $p_{est} = \hat{p}$, as an *estimator* of a population parameter, p. The statistic is usually related to a **sample** of size n.

If, **for all** n, $E(p_{est}) = p$, then the statistic is an *unbiased* estimator of p. Figure 29.1*a* shows a plot of the p.d.f. of an unbiased statistic for p while Figure 29.1*b* indicates a biased statistic for p.

Furthermore, if

$$\text{Var}(p_{est}) \to 0 \quad \text{as} \quad n \to \infty$$

then the statistic is *consistent*, that is, the result becomes more precise as the sample size increases.

We showed earlier that the *sample mean* is both **unbiased** and **consistent** and that the *simple definition* of the variance of a sample

$$\sigma^2_{est} = \frac{1}{n} \sum_{i=1}^{n} (x_i - \bar{x})^2$$

is *biased* but **consistent**. In this last case we saw that the bias could be **corrected** by multiplying by $n/(n-1)$ so that the *sample variance* is both **consistent** and **unbiased**.

In general, there is *not* necessarily one *unique* statistic to use in estimating a parameter although frequently one will

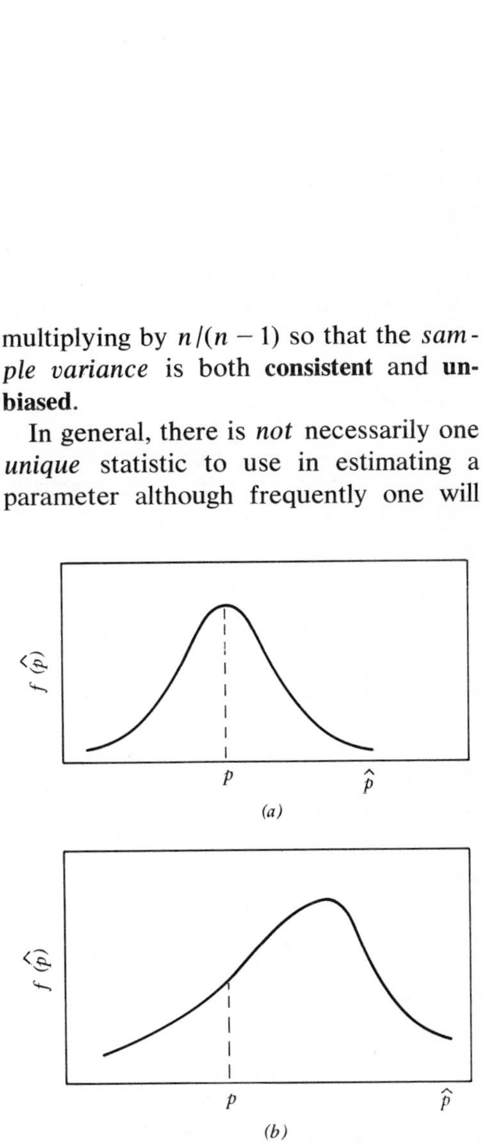

Figure 29.1
(*a*) Plot of the p.d.f. of an *unbiased* statistic for p versus p. (*b*) Plot of the p.d.f. of a *biased* statistic for p versus p.

appear to be "*natural*." We usually wish our statistics to be unbiased and consistent. If we have more than one statistic fulfilling these requirements, we would *prefer* the statistic that has the **smallest variance** for a given sample size. For example, the statistic in Figure 29.2*a* is to be preferred to that in Figure 29.2*b*.

To compare the *relative efficiencies* of two estimators, $\hat{p}_1$ and $\hat{p}_2$, we sometimes use the ratio

$$e_{12} = \sigma^2(\hat{p}_1)/\sigma^2(\hat{p}_2)$$

We may also ask the question: Does there exist within the category of unbiased consistent statistics one which has a **minimum** variance? We seek to find a "*most efficient statistic*." It may not be convenient to use the most efficient statistic, but we should like to know if such an estimator *exists* so that, at least, we may evaluate our practical estimators with respect to it.

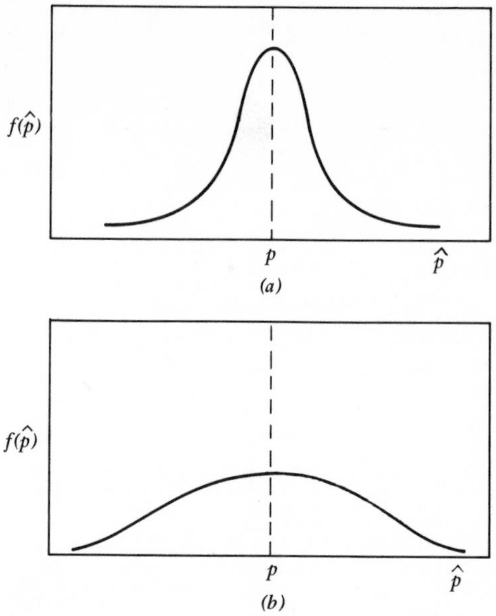

Figure 29.2
Examples of p.d.f.'s for unbiased, consistent statistics in p. (a) Plot shows small variance. (b) Plot shows larger variance.

Confidence Intervals

We may express an *estimator*, $\hat{p}$, of a **parameter** p in the following terms:

$$\Pr\{|\hat{p} - p| \geq \delta\} \leq a$$

or, equivalently,

$$\Pr\{|\hat{p} - p| \leq \delta\} \geq 1 - a$$

This may be stated in words, "The probability is greater than or equal to $(1 - a)$ that the parameter p is included in the interval

$$[\hat{p} - \delta, \hat{p} + \delta]\text{"}$$

We may also say, "With $100(1 - a)$ percent confidence, the interval includes p." We mean by this that a large number of repetitions of the observation that yielded $\hat{p}$ will result in the interval including p about $100(1 - a)$ percent of the time. We call the interval the *confidence interval* at the **level** of $(100)(1 - a)$ percent confidence.

Experimental scientists frequently quote their estimates of parameters in the form

$$p = \hat{p} \pm \sigma(\hat{p}) = \hat{p} \pm \sigma$$

where σ is a *standard deviation*. This usually means that the interval $[\hat{p} - \sigma, \hat{p} + \sigma]$ represents a 68% *confidence interval* if the estimator is assumed to be normally distributed. Sometimes, in an effort to be conservative, a scientist will quote his results

$$p = \hat{p} \pm (1.96\sigma) \qquad \text{"to 95\% confidence"}$$

This is actually no more conservative than quoting one standard deviation if one understands what is meant. If no confidence interval is specified, the *one-standard-deviation interval is generally assumed*. It is necessary to specify the confidence level if one uses something other than one standard deviation. It is important in reporting the results of an investigation to specify *exactly* what the results are. It is reprehensible to *understate* the precision (proportional to $1/\sigma$) of an estimate as well as to **overstate** it.

We emphasize that the confidence interval does *not* imply that the *parameter* in question has a probability distribution such that the probability is $\geq 1 - a$ that it is in a certain interval. The parameter is *fixed* and is **not** distributed. *Our* **estimator** *is the random variable.* Here is an apt analogy:

"A confidence interval and statements concerning it are somewhat like the game of horseshoe tossing. The stake is the parameter in question. (It never moves, regardless of some sportsmen's misconceptions.) The horseshoe is the confidence interval. If out of 100 tosses of the horseshoe one rings the stake 90 times on the average, he has 90% assurance (confidence) of ringing the stake. The parameter, just like the stake, is the constant. At any *one* toss (or one interval estimation) the stake (or parameter) is either enclosed or not. We make a probability statement about the variable quantities represented by the positions of the 'arms' of the horseshoe."*

*N. L. Johnson and F. C. Leone, *Statistics and Experimental Design*, John Wiley, 1964 p. 188, Vol. I.

Estimation of a Population Mean for a Large, Homogeneous Population

A common problem is to estimate the *mean of a population.* We have seen that the arithmetic sample mean is a good statistic to use for a *homogeneous* population, that is, one in which there are not subdivisions (partitions) with differing means. For the case in which the parent population is infinite or effectively infinite with respect to any feasible sample size, the usual considerations of the sample variance obtain. One question that we might address ourselves to is how large a sample size, n, to choose to produce an estimate that satisfies a given *criterion.*

SAMPLE SIZE REQUIRED

Suppose the population is characterized by a mean μ and a variance σ^2. We wish to **estimate** μ by the **sample mean,** $(\bar{x})_n$, where n refers to the *sample size.* We wish to determine the smallest size n such that $(\bar{x})_n$ will deviate from the value μ by more than δ with a probability that is less than or equal to some probability, a. We may state this

$$\Pr(|(\bar{x})_n - \mu| \geq \delta) \leq a \qquad (29.1)$$

We make use of the fact that the sample mean, $(\bar{x})_n$, is **approximately** *normally distributed* with mean μ and variance σ^2/n.

$$\begin{aligned}
\Pr(|(\bar{x})_n - \mu| \geq \delta) &= \Pr\{(\bar{x})_n \leq \mu - \delta\} + \Pr\{(\bar{x})_n \geq \mu + \delta\} \\
&\approx \Phi\left(\frac{(\mu - \delta) - \mu}{\sigma/\sqrt{n}}\right) + \left[1 - \Phi\left(\frac{(\mu + \delta) - \mu}{\sigma/\sqrt{n}}\right)\right] \\
&= \Phi\left(\frac{-\delta\sqrt{n}}{\sigma}\right) - \Phi\left(\frac{\delta\sqrt{n}}{\sigma}\right) + 1 \\
&= \left[1 - \Phi\left(\frac{\delta\sqrt{n}}{\sigma}\right)\right] - \Phi\left(\frac{\delta\sqrt{n}}{\sigma}\right) + 1 \\
&= 2\left[1 - \Phi\left(\frac{\delta\sqrt{n}}{\sigma}\right)\right] \\
&= 1 - A\left(\frac{\delta\sqrt{n}}{\sigma}\right)
\end{aligned}$$

by (25.33).

For the smallest n we set this probability equal to a:

$$1 - A\left(\frac{\delta\sqrt{n}}{\sigma}\right) = 2\left[1 - \Phi\left(\frac{\delta\sqrt{n}}{\sigma}\right)\right] = a$$

We may use the tables of the normal distribution to determine a value Z_a such that

$$A(Z_a) = 1 - a \qquad \text{and} \qquad \Phi(Z_a) = 1 - \frac{a}{2} \qquad (29.2)$$

$$\therefore \frac{\delta\sqrt{n}}{\sigma} = Z_a$$

and

$$n = \frac{\sigma^2}{\delta^2}(Z_a)^2 \qquad (29.3)$$

Actually, n is the first *integer* larger than this quantity.

The required number in the sample increases as the *square* of the population variance and **inversely as the square** of the size of the interval. We have seen this common characteristic of measurements before: if we wish to *halve* the uncertainty by repetition, we need to *quadruple* the number of measurements. It is thus important to determine in advance how precise one wishes the estimate to be.

We stress that (29.3) requires *knowledge of the population variance, σ.* This must either be given *a priori* or estimated separately in another experiment. The normal approximation provides a less stringent limit than was obtained from Chebychev's inequality but one that is adequate for the stated objective.

► **Example 29.1**
Sampling photomultiplier tubes

One of the United States national laboratories has been using photomultiplier tubes obtained from one manufacturer. Another manufacturer offers to sell similar tubes at a lower price. The purchasing officer of the laboratory is afraid that the lower-priced tubes may have inferior characteristics, in this case the overall *amplification factor*. He wishes to order a sample of tubes to measure the **average** amplification factor. Assuming that the manufacturer really sends a random sample of tubes, how many should the purchasing officer order so that the probability that the average for the sample tubes differs from that of all the tubes from that manufacturer by more than 5% is less than 0.05? Suppose that the tubes have a distribution with a *standard deviation* that is 10% of the *mean amplification factor*.

Here $\sigma = 0.1\bar{A}$ and $\delta = 0.05\bar{A}$. From the tables of the normal distribution, $A(Z_{0.05}) = 0.95$ for $Z = 1.96$. Therefore,

$$n = \frac{(0.1\bar{A})^2(1.96)^2}{(0.05)^2} = 15.3$$

or 16 tubes are required. ◄

Estimation of Population Mean for a Finite Population

We should consider again the problem that arose in connection with the *hypergeometric distribution*: When the parent population is finite, the process of drawing samples without replacement is not independent. Suppose we consider a *finite* parent population with N elements: $x_1, \ldots, x_N$. We draw a sample of size n: $z_1, \ldots, z_n$. As usual we wish to estimate the **population mean,** $\bar{x}$,

$$\bar{x} = \frac{1}{N}\sum_{1=1}^{N} x_i \qquad (29.4)$$

and the **population variance,** σ^2,

$$\sigma^2 = \frac{1}{N}\sum_{i=1}^{N}(x_i - \bar{x})^2 \qquad (29.5)$$

by using the *sample mean*

$$\bar{z} = \frac{1}{n}\sum_{j=1}^{n} z_j \qquad (29.6)$$

and the *sample variance*

$$s^2 = \frac{1}{n-1}\sum_{j=1}^{n}(z_j - \bar{z})^2 \qquad (29.7)$$

This situation should be analyzed anew because the *sampling is not independent.* We may summarize the results of such an analysis as follows.

$$E(\bar{z}) = \bar{x} \qquad (29.8)$$

that is, the sample mean is an unbiased estimator of the population mean.

$$E(s^2) = \sigma^2 \qquad (29.9)$$

that is, the sample variance is an unbiased estimator of the population variance.

What is different in this case of a finite parent population is the *variance of the sample mean.* Here we do not simply divide the sample variance by n. Instead,

$$\text{Var}(\bar{x}) = \frac{\sigma^2}{n}\frac{(1 - n/N)}{(1 - 1/N)} = \frac{\sigma^2}{n}\frac{(N - n)}{(N - 1)} \qquad (29.10)$$

This is an important modification, but one that is reasonable. As $n \to N$, the variance of $\bar{x}$ must **go to zero**, since the sample mean is identical with the population mean and the uncertainty vanishes.

Stratified Sampling: Estimation of a Population Mean where the Population is Large and Partitioned into Strata

Suppose that we are interested in *estimating by sampling* the mean, μ, of a parent population. We are always interested in choosing a statistic with the smallest variance for a given sample size, n.

The straightforward and simplest procedure is to select a sample of size n at random and calculate

$$\bar{x} = \frac{1}{n}\sum_{i=1}^{n} x_i \qquad (29.11)$$

Suppose that there are m different *classes* or *strata* within the parent population and that each class i has *population mean* and *variance* μ_i and σ_i^2, respectively. We still wish to estimate the **overall population mean,** μ.

Let us assume that our sample of *total size n* consists of n_i items selected from each class, i. Then we see that

$$\sum_{i=1}^{m} n_i = n \qquad (29.12)$$

and we may define the *sample average* for the ith class

$$(\bar{x}_i) = \frac{1}{n_i}\sum_{i=1}^{n_i} x_i \qquad (29.13)$$

We further define that α_i = the *fraction* of the parent population in class i.

We may define a new statistic

$$\bar{u} = \sum_{i=1}^{m} \alpha_i(\bar{x}_i) \qquad (29.14)$$

where $(\bar{x}_i)$ is defined in (29.13).

We may see that $\bar{x}$ and $\bar{u}$ are both *unbiased statistics*:

$$\langle\bar{x}\rangle = \left\langle\frac{1}{n}\sum_{i=1}^{n} x_i\right\rangle = \mu \qquad (29.15)$$

and

$$\langle\bar{u}\rangle = \left\langle\sum_{i=1}^{m} \alpha_i(\bar{x}_i)\right\rangle$$

$$= \sum_{i=1}^{m} \alpha_i\langle(\bar{x}_i)\rangle$$

$$= \sum_{i=1}^{m} \alpha_i\mu_i = \mu \qquad (29.16)$$

We wish to show that $\text{Var}(\bar{u}) \le \text{Var}(\bar{x})$ for appropriately chosen n_i. By (20.17c) and (20.17b),

$$\text{Var}(\bar{u}) = \sum_{i=1}^{m} \alpha_i^2\,\text{Var}(\bar{x}_i)$$

$$= \sum_{i=1}^{m} \alpha_i^2\frac{\sigma_i^2}{n_i} \qquad (29.17)$$

since $(\bar{x}_i)$ is a *sample mean* for the class i.

We shall consider the *discrete* case for simplicity although the argument can be also made for the continuous case. If only k values of x are possible, then

$$\text{Var}(x) = \sum_{j=1}^{k} (x_j - \mu)^2 p(x_j) \qquad (29.18)$$

where $p(x_j)$ is the probability of the value x_j.

The probability of a given value **depends on the class** from which the sampled item came, that is, we may express $p(x_j)$ in terms of *conditional probabilities*:

$$p(x_j) = \sum_{i=1}^{m} \Pr_i(x_j)p(i) \qquad (29.19)$$

where $\Pr_i(x_j)$ is the *conditional probability* of the value x_j if the variate has been chosen from class i.

Here $p(i)$ is the probability that the variate was selected from class i, so

$$p(i) = \alpha_i \qquad (29.20)$$

Therefore

$$p(x_j) = \sum_{i=1}^{m} \alpha_i \Pr_i(x_j)$$

and

$$\mathrm{Var}(x) = \sum_{j=1}^{k} (x_j - \mu)^2 \sum_{i=1}^{m} \alpha_i \Pr_i(x_j)$$

$$= \sum_{j=1}^{k} \sum_{i=1}^{m} \alpha_i (x_j - \mu)^2 \Pr_i(x_j)$$

$$= \sum_{i=1}^{m} \alpha_i \left[\sum_{j=1}^{k} (x_j - \mu)^2 \Pr_i(x_j)\right] \qquad (29.21)$$

We note that

$$\left[\sum_{j=1}^{k} (x_j - \mu)^2 \Pr_i(x_j)\right]$$

$$= \sum_{j=1}^{k} [(x_j - \mu_i) + (\mu_i - \mu)]^2 \Pr_i(x_j)$$

$$= \sum_{j=1}^{k} (x_j - \mu_i)^2 \Pr_i(x_j)$$

$$+ 2(\mu_i - \mu)$$

$$\times \sum_{j=1}^{k} (x_j - \mu_i) \Pr_i(x_j)$$

$$+ (\mu_i - \mu)^2 \sum_{j=1}^{k} \Pr_i(x_j)$$

where

$$\sum_{j=1}^{k} \Pr_i(x_j) = 1$$

and

$$\sum_{j=1}^{k} (x_j - \mu_i)\Pr_i(x_j) = \sum_{j=1}^{k} x_j \Pr_i(x_j) - \mu_i = 0$$

since

$$\sum_{j=1}^{k} x_j \Pr_i(x_i) = \mu_i$$

$$\therefore \left[\sum_{j=1}^{k} (x_j - \mu)^2 \Pr_i(x_j)\right]$$

$$= \sum_{j=1}^{k} (x_j - \mu_i)^2 \Pr_i(x_j) + (\mu_i - \mu)^2$$

$$= \sigma_i^2 + (\mu_i - \mu)^2 \qquad (29.22)$$

Using (29.21) we get

$$\mathrm{Var}(x) = \sum_{i=1}^{m} \alpha_i [\sigma_i^2 + (\mu_i - \mu)^2] \qquad (29.23)$$

Thus,

$$\mathrm{Var}(\bar{x}) = \frac{\mathrm{Var}(x)}{n}$$

$$= \sum_{i=1}^{m} \frac{\alpha_i}{n} \sigma_i^2$$

$$+ \sum_{i=1}^{m} \frac{\alpha_i}{n} (\mu_i - \mu)^2 \qquad (29.24)$$

Recall (29.17)

$$\mathrm{Var}(\bar{u}) = \sum_{i=1}^{m} \alpha_i^2 \frac{\sigma_i^2}{n_i}$$

which is true for any n_i.

OPTIMAL ALLOCATION OF SAMPLE SIZES

If we choose a *proportional allocation*, we may set

$$n_i = \alpha_i n \qquad (29.25)$$

$$\mathrm{Var}(\bar{u}) = \sum_{i=1}^{m} \frac{\alpha_i^2 \sigma_i^2}{\alpha_i n}$$

$$= \sum_{i=1}^{m} \alpha_i \frac{\sigma_i^2}{n} \qquad (29.26)$$

and

$$\mathrm{Var}(\bar{x}) = \mathrm{Var}(\bar{u}) + \sum_{i=1}^{m} \frac{\alpha_i}{n} (\mu_i - \mu)^2 \qquad (29.27)$$

where the second term is necessarily ≥ 0. If any class exists such that $\mu_i \neq \mu$, then $\mathrm{Var}(\bar{u}) < \mathrm{Var}(\bar{x})$ and

$$\bar{u} = \sum_{i=1}^{m} \alpha_i (\bar{x}_i) = \sum_{i=1}^{m} \frac{n_i}{n} (\bar{x}_i) \qquad (29.28)$$

is a better statistic to use than is (29.11) for the same sample size, since its variance is smaller.

Still better in the general case where the σ_i are different (or, at least one σ_i is different) is to minimize (29.17) subject to the constraint of (29.12).

Here

$$X = \sum_{i=1}^{m} \frac{\alpha_i^2 \sigma_i^2}{n_i} = X(n_i)$$

and

$$\varphi(n_i) = \sum_{i=1}^{m} n_i - n = 0$$

We solve

$$\frac{\partial X}{\partial n_i} + \lambda \frac{\partial \varphi}{\partial n_i} = 0 = \frac{-\alpha_i^2 \sigma_i^2}{n_i^2} + \lambda$$

together with (29.12)

$$n_i^2 = \frac{\alpha_i^2 \sigma_i^2}{\lambda} \qquad \text{or} \qquad n_i = \frac{\alpha_i \sigma_i}{\sqrt{\lambda}}$$

By (29.12)

$$n = \frac{\sum_{i=1}^{m} \alpha_i \sigma_i}{\sqrt{\lambda}} \qquad \text{so} \qquad \sqrt{\lambda} = \frac{1}{n} \sum_{i=1}^{m} \alpha_i \sigma_i$$

$$\therefore n_i = \frac{\alpha_i \sigma_i n}{\sum_{i=1}^{m} \alpha_i \sigma_i} \qquad (29.29)$$

is the best choice of n_i to minimize the variance of the statistic. That is, we choose m values n_i according to (29.29) and use the appropriate statistic

$$\bar{u} = \sum_{i=1}^{m} \frac{\alpha_i}{n_i} \sum_{j=1}^{n_i} x_j \qquad (29.30)$$

Relation (29.29) is sometimes called the ***Chuprow-Neyman** allocation.*

Stratified sampling is very important in opinion polling where the response of an individual may depend (on average) on his socioeconomic stratum, on his ethnic background, on his geographical location, etc. At the very least, the population *proportions* must be determined so as not to bias the sample average. An infamous example where this was not correctly done was the United States presidential preference poll taken by the *Literary Digest* in 1936. A very large sample of people whose names were chosen at random from telephone directories was used to determine whether the Republican or the Democratic presidential candidate was favored. The results indicated a victory for the Republican, although the election proved to be an overwhelming victory for the Democrat. The problem was that possession of a telephone in 1936 was very strongly correlated with being in an *upper* socioeconomic class and that the preference of this group was not appropriately weighted with the preference of everyone else.

Stratified sampling is also of importance in the *Monte Carlo simulation* of problems. Usually Monte Carlo problems can be couched in terms of *evaluating an integral* (n-dimensional, for generality). Here, the mean value of the function to be integrated takes the place of the means, μ_i. It is not always desirable to choose the distribution of points so as to make the points evenly distributed over the space. This is also related to the Monte Carlo technique of "**importance sampling**" where the points are preferentially chosen in regions where the contributions to the integral are larger. It may be stated rather obviously that points should not be chosen completely at random in a Monte Carlo problem but, instead, more points should be chosen in "**important**" regions (and weighted appropriately).

Stratified sampling is one example of a *variance-reducing technique.* One of the proper objects of experimental design, in addition to avoidance of systematic errors (biases), is the **minimization of the variance** of the resulting estimator. For a given amount of experimental effort (e.g., a given sample size) we wish the variance to be as small as possible.

►Example 29.2

A variance-reducing technique in weighing

Let us consider the common problem of ascertaining the weight of four objects (e.g., coins,) with a balance and set of accurately calibrated weights known to arbitrary precision. Let us assume that the error in the balance is **random** and is describable by a **constant** standard deviation of σ. (We mean that σ represents the standard deviation of a distribution of repeated weighings of the same object. Typically, σ is of the order of 1 milligram for an analytical balance. Strictly speaking, the precision should be defined as $1/\sigma$ so that a smaller error implies a higher precision, but our meaning will be unambiguous here). Let us call the *actual values* of the four weights w_1, w_2, w_3, and w_4.

Direct method

The straightforward method is to weigh each coin separately, yielding the measurements $w_i' = w_i + \epsilon_i'$ where the ϵ_i' is a *random error* with the results

$$w_1' \pm \sigma,\ w_2' \pm \sigma,\ w_3' \pm \sigma,\ w_4' \pm \sigma$$

where we mean to imply that

$$\sigma_{w_i'} = \sigma \qquad \text{for each} \qquad w_i'$$

These weighings are each *independent*. The results w_i' are *uncorrelated* with each other, and we may determine the mean

$$\bar{w}' = \tfrac{1}{4} \sum_{i=1}^{4} w_i'$$

with a *standard deviation of the mean*

$$\sigma_{\bar{w}'} = \sqrt{\frac{\sigma^2}{4}} = \frac{\sigma}{2}$$

by (20.17b).

Naturally, we should prefer any method that yielded a smaller *standard deviation* (or variance) for the results.

Difference method

Let us consider that the weighings are made to involve all four coins at each of four steps, that is, we make the following measurements:

$$y_1' = (w_1 + w_2) - (w_3 + w_4) + \epsilon_1$$
$$y_2' = (w_1 + w_3) - (w_2 + w_4) + \epsilon_2$$
$$y_3' = (w_1 + w_4) - (w_2 + w_3) + \epsilon_3$$
$$y_4' = w_1 + w_2 + w_3 + w_4 + \epsilon_4$$

where the ϵ's are random errors that may be positive, negative, or zero. The measurements y_i' are *independent* and may be positive, negative, or zero, but each measurement has associated with it the standard deviation

$$\sigma_{y'} = \sigma$$

From these measurements we may calculate *estimates* of the *individual* weights:

$$\hat{w}_1 = \tfrac{1}{4}(y_1' + y_2' + y_3' + y_4')$$
$$\hat{w}_2 = \tfrac{1}{4}(y_1' - y_2' - y_3' + y_4')$$
$$\hat{w}_3 = \tfrac{1}{4}(-y_1' - y_2' - y_3' + y_4')$$
$$\hat{w}_4 = \tfrac{1}{4}(-y_1' - y_2' + y_3' + y_4')$$

We ask what is the **precision** of these estimates, and note that the y_i' are independent and, hence, $\text{Covar}\,(y_i', y_j') = 0$ for all i, j. Therefore,

$$\sigma_{\hat{w}_i}^2 = \frac{1}{4^2} \sum_{j=1}^{4} \sigma_{y_i'}^2 = \frac{\sigma^2}{4}$$

by (20.17c).

We note that

$$\sigma_{\hat{w}_i} = \tfrac{1}{2}\sigma_{w_i'}$$

That is, the estimates obtained by the difference method are *twice as precise* as those obtained directly *for the same number of weighings*.

However, the estimates $\hat{w}_i$ are no longer necessarily independent and we must calculate their *correlations* before we can estimate the precision of the *mean*,

$$\bar{\hat{w}} = \frac{1}{4} \sum_{i=1}^{4} \hat{w}_i$$

In particular, we must calculate the **covariances**,

$$\text{Covar}(\hat{w}_i, \hat{w}_j) = \langle [\hat{w}_i \hat{w}_j - \langle \hat{w}_i \rangle \langle \hat{w}_j \rangle] \rangle$$

for each i, j in order to use (20.17a).

We recall Example 20.3 in which the case was considered of $A_1 = x_1 + x_2$, $A_2 = x_1 - x_2$ with the x_1, x_2 independent. The result was (20.17d):

$$\text{Covar}(A_1, A_2) = \sigma_{x_1}^2 - \sigma_{x_2}^2$$

We note here that

$$4\hat{w}_1 = (y_1' + y_4') + (y_2' + y_3')$$
$$4\hat{w}_2 = (y_1' + y_4') - (y_2' + y_3')$$

which yields

$$\text{Covar}(\hat{w}_1, \hat{w}_2) = \frac{1}{4^2}\{\sigma_{y_1'+y_4'}^2 - \sigma_{y_2'+y_3'}^2\}$$

But all the y_i' are *independent* and, hence, in general

$$\sigma_{y_i' \pm y_j'}^2 = \sigma_{y_i'}^2 + \sigma_{y_j'}^2 \qquad \text{for} \qquad i \neq j$$

Therefore

$$\text{Covar}(\hat{w}_1, \hat{w}_2) = \frac{1}{4^2}\{\sigma_{y_1'}^2 + \sigma_{y_4'}^2) - (\sigma_{y_2'}^2 + \sigma_{y_3'}^2)\}$$

Also,

$$4\hat{w}_1 = (y_2' + y_4') + (y_1' + y_3')$$
$$4\hat{w}_3 = (y_2' + y_4') - (y_1' + y_3')$$

and

$$\text{Covar}(\hat{w}_1, \hat{w}_3) = \frac{1}{4^2}\{(\sigma_{y_2'}^2 + \sigma_{y_4'}^2) - (\sigma_{y_1'}^2 + \sigma_{y_2'}^2)\}$$

There are $\binom{4}{2} = 6$ covariances to consider. We write the various pairs in similar fashion:

$$(\hat{w}_1, \hat{w}_4)\begin{cases}4\hat{w}_1 = (y_3' + y_4') + (y_1' + y_2')\\ 4\hat{w}_4 = (y_3' + y_4') - (y_1' + y_2')\end{cases}$$

$$(\hat{w}_2, \hat{w}_3)\begin{cases}4\hat{w}_2 = (y_4' - y_3') + (y_1' - y_2')\\ 4\hat{w}_3 = (y_4' - y_3') - (y_1' - y_2')\end{cases}$$

$$(\hat{w}_2, \hat{w}_4)\begin{cases}4\hat{w}_2 = (y_4' - y_2') + (y_1' - y_3')\\ 4\hat{w}_4 = (y_4' - y_2') - (y_1' - y_3')\end{cases}$$

$$(\hat{w}_3, \hat{w}_4)\begin{cases}4\hat{w}_3 = (y_4' - y_1') + (y_2' - y_3')\\ 4\hat{w}_4 = (y_4' - y_1') - (y_2' - y_3')\end{cases}$$

The remaining covariances are seen to be:

$$\text{Covar}(\hat{w}_1, \hat{w}_4) = \frac{1}{4^2}\{(\sigma_{y_3'}^2 + \sigma_{y_4'}^2) - (\sigma_{y_1'}^2 + \sigma_{y_2'}^2)\}$$

$$\text{Covar}(\hat{w}_2, \hat{w}_3) = \frac{1}{4^2}\{(\sigma_{y_4'}^2 + \sigma_{y_3'}^2) - (\sigma_{y_1'}^2 + \sigma_{y_2'}^2)\}$$

$$\text{Covar}(\hat{w}_2, \hat{w}_4) = \frac{1}{4^2}\{(\sigma_{y_4'}^2 + \sigma_{y_2'}^2) - (\sigma_{y_1'}^2 + \sigma_{y_3'}^2)\}$$

$$\text{Covar}(\hat{w}_3, \hat{w}_4) = \frac{1}{4^2}\{(\sigma_{y_4'}^2 + \sigma_{y_1'}^2) - (\sigma_{y_2'}^2 + \sigma_{y_3'}^2)\}$$

Therefore, although the estimates $\hat{w}_1$, $\hat{w}_2$, $\hat{w}_3$, and $\hat{w}_4$ are *not independent*, the circumstance that each of the (independent) measurements, y_i', has the *same variance* ($\sigma_{y_i'}^2 = \sigma^2$) has the consequence that

$$\text{Covar}(\hat{w}_i, \hat{w}_j) = 0 \qquad \text{for each } i, j$$

We emphasize that the estimates $\hat{w}_i$ are ***uncorrelated*** because of the constancy of $\sigma_{y_i'}$.

Accordingly, the average

$$\bar{\hat{w}} = \frac{1}{4}\sum_{i=1}^{4} \hat{w}_i$$

has the variance

$$\text{Var}(\bar{\hat{w}}) = \frac{1}{4^2}\sum_{i=1}^{4} \text{Var}(\hat{w}_i) = \frac{1}{4^2}4\left\{\frac{\sigma^2}{4}\right\} = \frac{\sigma^2}{16}$$

That is, four weighings (to obtain y_1', y_2', y_3', and y_4') yield a standard deviation on the mean weight of $\sigma/4$. Recall that four separate weighings (to obtain w_1', w_2', w_3', and w_4') yield a standard deviation on the mean weight of $\sigma/2$.

The difference method is clearly preferable to maximize the precisions of both the individual weights and the mean. The generalization to n objects is to form n linear independent sums and differences, in which case the variance is reduced by $1/n$. ◄

Monte Carlo simulation of the previous example

To illustrate what actually happens in the laboratory we have simulated the preceding example. Let the reader be assured that the Monte Carlo procedure actually reflects the *real* laboratory situation, but the use of a computer to generate and to analyze the data was *convenient* for the purpose of displaying clearly all that is happening. These results were essentially *duplicated* by a freshman student working with four United States cents and a beam balance, an analytical balance, and an electronic (Mettler) balance.

We have assumed the weights of four coins to be, respectively, 3.1077 grams, 3.1122 grams, 3.1068 grams, and 3.0392 grams. We assumed the standard deviation characteristic of the balance used to be 0.001 grams. For any given weighing (whether of an individual coin or the differences or the sum of the coins) the experimental values were generated by taking the actual value and adding to it the product of the assumed standard deviation and a random Gaussian deviate (which may be *positive* or *negative*). Thus, for *individual weighings* w_i' the actual weight w_i of the coin was incremented by a random Gaussian deviate multiplied by 0.001 grams for each trial. For the *difference weighings* the actual central value of the difference (or sum) y_i was incremented in similar fashion to yield the **experimental** values y_i'. The quantities $\hat{w}_i$ were then calculated by formula from the y_i'. **The term "*difference weighings*" refers to the quantities** $\hat{w}_i$, and **the term "*single weighings*" refers to the** w_i'. The *same number* of weighings was compared for the "single" and "difference" cases. Histograms of the experimental quantities $\hat{w}_i$ and w'_i are shown in Figures 29.3*a*, and *b* to 29.6*a* and *b* together with superimposed normalized Gaussian curves of the form $\mathbf{n}(\bar{w}, s^2)$ where the parameters are the sample mean and sample variance, respectively. The curves for the difference weighings for 36 trials are clearly about a factor of two narrower in Gaussian width than are the curves for the individual weighings.

We indicate in Table 29.1 the summary of results for different numbers of trials. In all cases, the difference technique yields a *standard deviation on the estimator* that is approximately **half** that obtained by the direct method.

We may examine the correlations between the nonindependent estimators $\hat{w}_i$ and $\hat{w}_j$. This is shown as a scatter plot in Figures 29.7 to 29.12. Also indicated on the plots are the values of the empirical correlation coefficients.

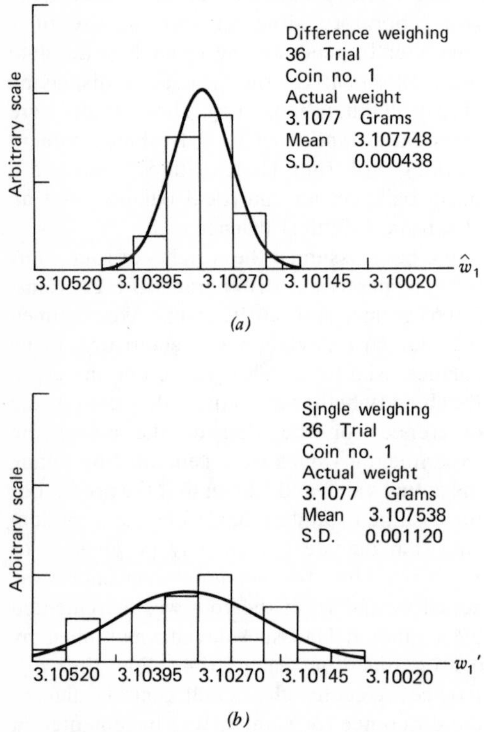

Figure 29.3
Plot of the frequency histograms for Coin 1 for "difference" weighings, $\hat{w}_i$, and "single" weighings, w'_i, with superimposed gaussian curves of the form $\mathbf{n}(\bar{w}, s^2)$ where the parameters are the sample mean and the sample variance. (*a*) Difference weighings. (*b*) Single weighings.

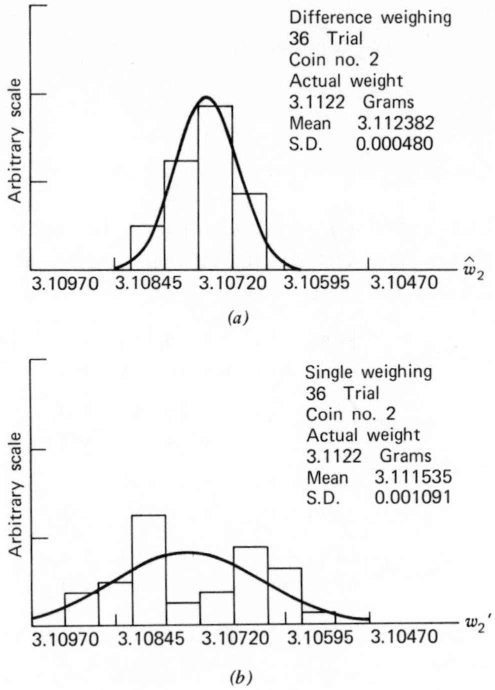

Figure 29.4
Same plot as in Figure 29.3 for Coin 2.

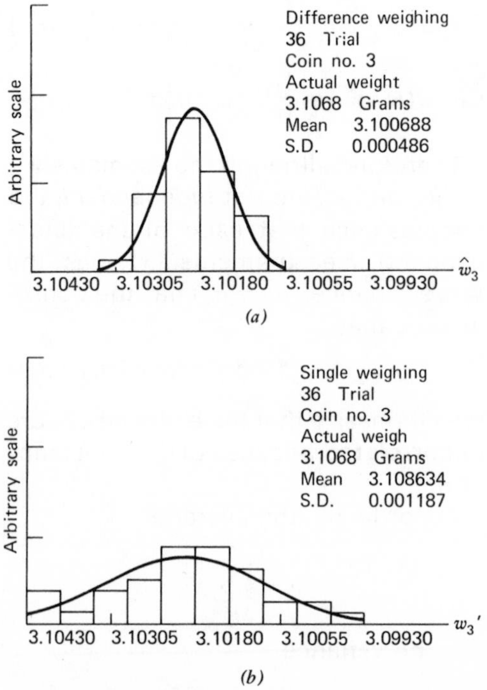

Figure 29.5
Same plot as in Figure 29.3 for Coin 3.

Table ***29.1*** Summary of Results for Different Numbers of Trials.

		Single Weighing			Difference Weighing		
No. of Trial	Coin	Mean	S.D.	S.D. of Mean	Mean	S.D.	S.D. of Mean
4	1	3.107039	.001305	.000653	3.107551	.000442	.000221
4	2	3.112651	.000897	.000449	3.112210	.000377	.000188
4	3	3.106875	.001570	.000785	3.106864	.000529	.000264
4	4	3.039831	.000580	.000290	3.039348	.000619	.000310
16	1	3.107810	.001223	.000306	3.107694	.000440	.000110
16	2	3.111964	.000892	.000223	3.112276	.000517	.000129
16	3	3.106608	.000915	.000229	3.106775	.000418	.000104
16	4	3.039216	.000844	.000211	3.039179	.000563	.000141
36	1	3.107538	.001120	.000187	3.107748	.000438	.000073
36	2	3.111995	.001091	.000182	3.112302	.000455	.000076
36	3	3.106634	.001187	.000198	3.106688	.000486	.000081
36	4	3.039212	.000901	.000150	3.039236	.000512	.000085
(S.D.) square	of Y(1, 2, 3, 4)	.000001	.000001	.000001	.000001		
S.D.	of Y(1, 2, 3, 4)	.000947	.000921	.001063	.000841		

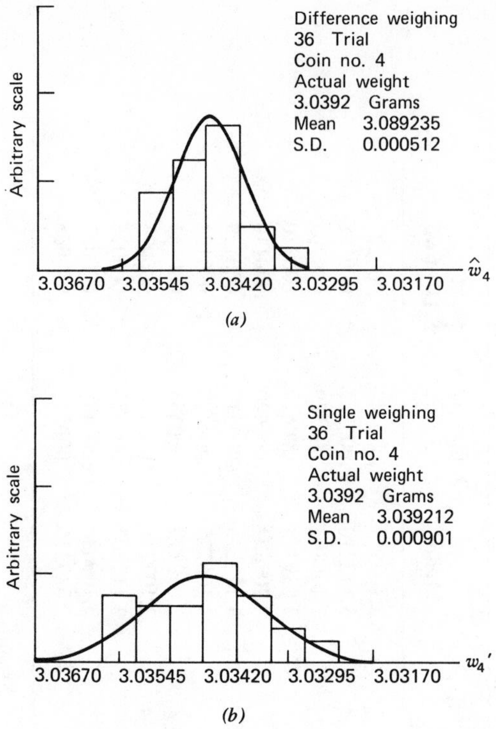

Figure 29.6
Same plot as in Figure 29.3 for Coin 4.

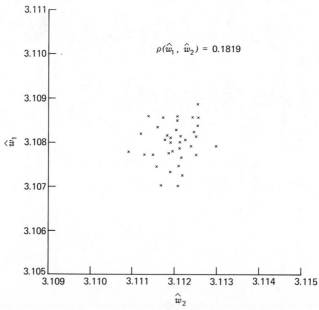

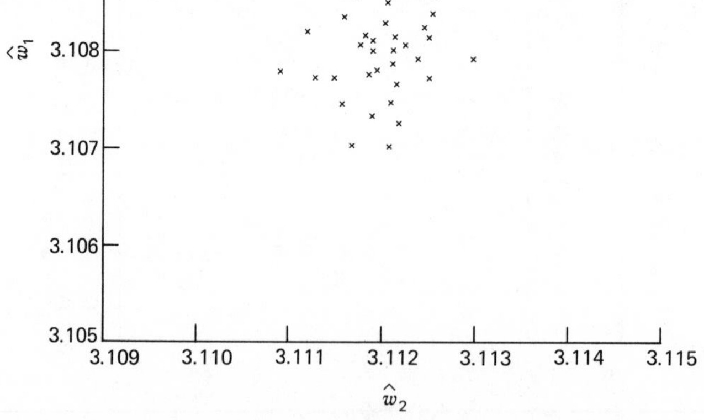

Figure 29.7
Examination of the correlations between the nonindependent estimators $\hat{w}_i$, $\hat{w}_j$. Shown is a scatter plot of the 36 values of $\hat{w}_1$, $\hat{w}_2$ with the value of the empirical correlation coefficient.

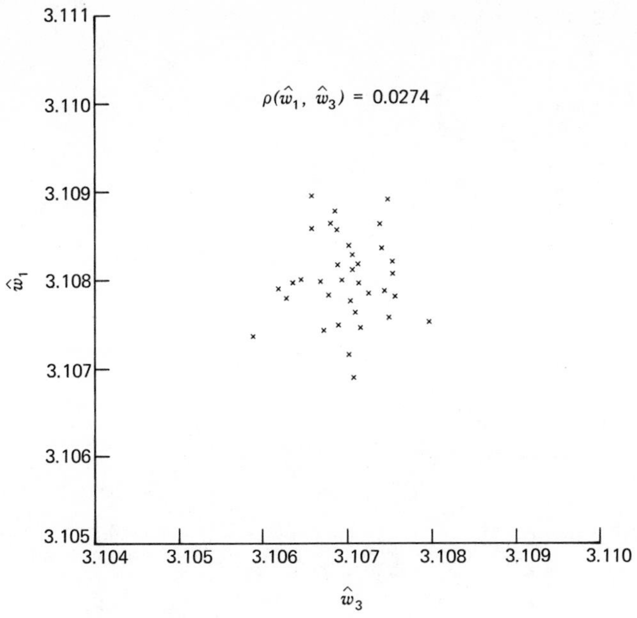

Figure 29.8
Same as Figure 29.7 for the estimators $\hat{w}_1$, $\hat{w}_3$.

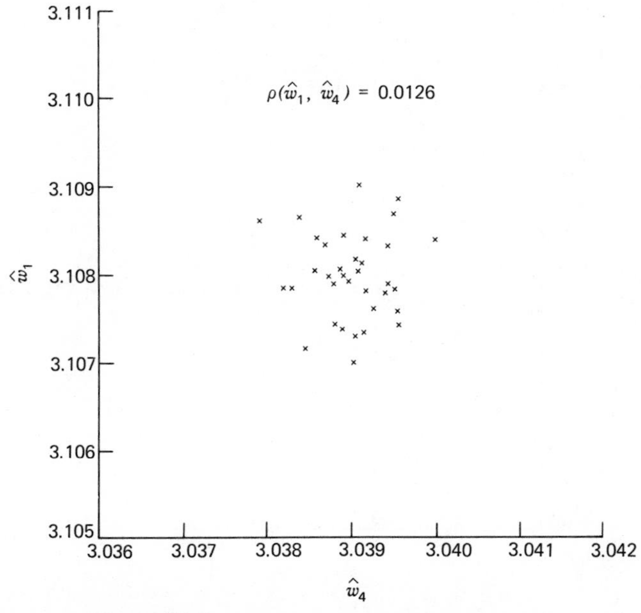

Figure 29.9
Same as Figure 29.7 for the estimators $\hat{w}_1$, $\hat{w}_4$.

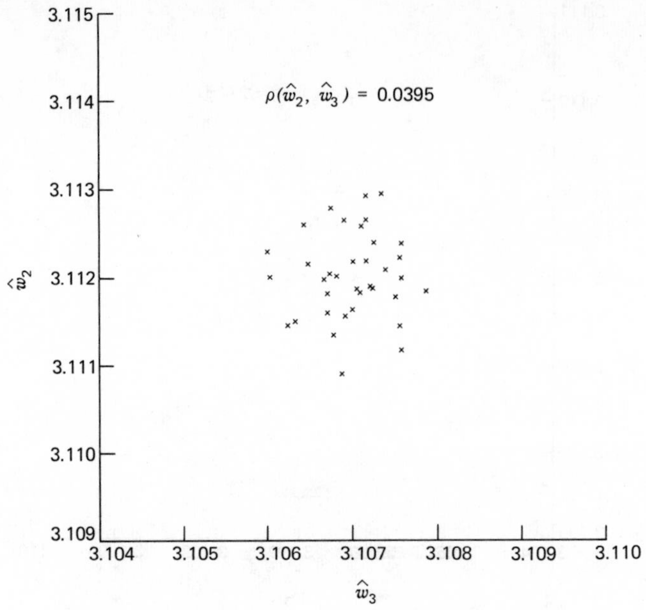

Figure 29.10
Same as Figure 29.7 for the estimators $\hat{w}_2$, $\hat{w}_3$.

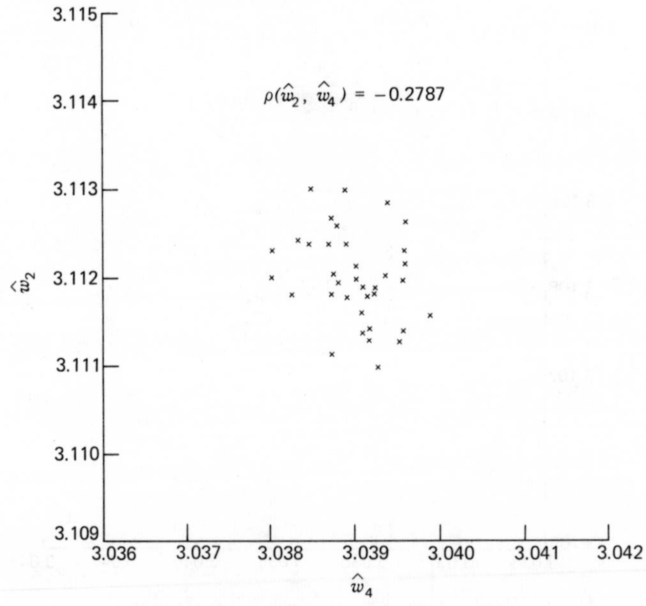

Figure 29.11
Same as Figure 29.7 for the estimators $\hat{w}_2$, $\hat{w}_4$.

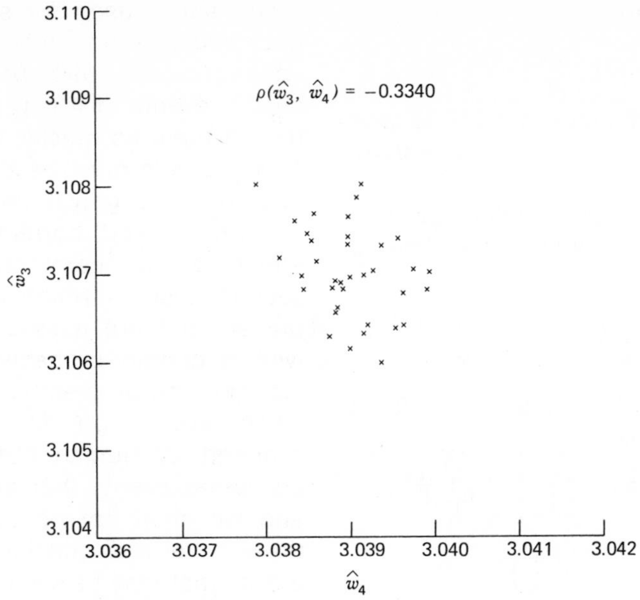

Figure 29.12
Same as Figure 29.7 for the estimators $\hat{w}_3$, $\hat{w}_4$.

We note that the *correlations appear to be absent* which is borne out by the correlation coefficients being close to zero. In the cases where the correlation coefficient is furthest from zero we note that the explanation comes from the fact that the sample variances of the individual y_i are not identical. For larger samples, the sample variances should be closer to being all equal and the correlation coefficients will be closer yet to zero. *For this quite realistic case, therefore, the estimates obtained by the difference method are* **uncorrelated.**

Estimation of a Probability (Binomial)

In Chapter 22 we considered the *inverse probability* problem of **estimating** a probability p for a binomial situation in which we count only successes and failures and where r successes have been observed in n trials.

We noted that

$$\hat{p} = \frac{r}{n}$$

is an *unbiased* and *consistent* estimator with variance

$$\sigma^2(\hat{p}) = \frac{p(1-p)}{n}$$

We note that the probability p must be known to evaluate the variance. When it is not known it is frequently *satisfactory* to use

$$\sigma^2(\hat{p}) = \frac{\hat{p}(1-\hat{p})}{n}$$

although it should be recognized that this is *not rigorous.*

We may consider two examples, the first where the probability p is known and the second where $\hat{p}$ must be used.

► **Example 29.3**
Buffon Needle

Recall the needle-tossing experiment of Chapter 21. Needles of length ℓ were tossed *at random* onto a ruled grid of spacing a, and the number of intersections M noted for N trials. For the case

$\ell = a$, we note that

$$\pi = E\left(\frac{2N}{M}\right)$$

We use the estimator $\hat{\pi} = 2N/M$ and ask how close $\hat{\pi}$ is to the "*true value*" π.

The probability of an intersection is $p = E(M/N) = 2/\pi$, which is estimated by $\hat{p} = M/N$ and

$$\sigma^2(\hat{p}) = \frac{p(1-p)}{N} = \frac{2}{\pi}\left(1 - \frac{2}{\pi}\right)\frac{1}{N}$$

But $\hat{\pi} = 2/\hat{p}$, and we may use (10.4b) and (10.4g) to get

$$\frac{\sigma^2(\hat{\pi})}{(\hat{\pi})^2} = \frac{\sigma^2(\hat{p})}{(\hat{p})^2} = \frac{2}{\pi}\left(1 - \frac{2}{\pi}\right)\frac{1}{N}\frac{\hat{\pi}^2}{4}$$

$$\approx \frac{2}{\pi}\left(1 - \frac{2}{\pi}\right)\left(\frac{1}{N}\right)\frac{\pi^2}{4}$$

$$= \left(\frac{\pi - 2}{2}\right)\frac{1}{N}$$

The percentage standard deviation of $\hat{\pi}$ is thus

$$\frac{\sigma(\hat{\pi})}{\hat{\pi}} = \frac{1}{\sqrt{N}}\left[\frac{\pi - 2}{2}\right]^{1/2} = \frac{0.7555}{\sqrt{N}}$$

We might compare $\sigma(\hat{\pi})$ with $\Delta\pi = |\pi - \pi_{\text{obs}}|$ from a Monte Carlo simulation of the Buffon Needle problem (Table 29.2).

Table ***29.2***

N	Observed	$\Delta\pi$	$\sigma(\hat{\pi})$
100	3.356	0.214	0.2373
1000	3.173	0.031	0.0751
10000	3.1447	0.003	0.0237
40000	3.1420	0.0004	0.0119
100000	3.1428	0.0016	0.0075

◄

► Example 29.4
Scanning efficiency

It frequently happens in experimental science that the observer does not observe all that there is to see. Two cases are important. There may be a *systematic* reason for some "events" not being seen, for example, the apparatus may not be sensitive to certain events or it may not be able to record them as quickly as they arrive. This problem must be attacked on the basis of thoroughly understanding your experiment and *correcting* for such effects. We are here concerned with a second case in which the events are present but are missed "*at random*" with a certain probability (the **inefficiency**). The problem is quite common when events on film are visually scanned by human observers. There are some events that are *never* seen and we shall not be concerned with these. We are considering, instead, events that may be seen on one trial by one scanner and missed on another trial by either the same observer or a different one. It is important to know the "*scanning efficiency*" before one can estimate how many events there really are in the sample. We emphasize that the probability of one scanner finding or not finding an event is *independent* of the probability of finding it or not finding it on another scan. There are always events for which this is *not* true, and these must be treated *differently*.

Suppose that the body of data has been scanned once and $N_1(x)$ events of the type x have been found. To determine the efficiency with which the $N_1(x)$ events were found, $E_1(x)$, it is necessary to perform a second (*independent*) scan of the same body of data. The *second scan* yields a total of $N_2(x)$ events that are not necessarily the same events found in the first scan although $N_{12}(x)$ events are *common* to *both*. The efficiency for the second scan, $E_2(x)$, may in general be different from that for the first scan. *It is necessary only that it be independent.* Since the same body of data that contained $N_1(x)$ events was scanned with efficiency $E_2(x)$, the result was $N_{12}(x)$, and we may write

$$E_2(x) = \frac{N_{12}(x)}{N_1(x)}$$

Also, we may consider that $N_2(x)$ events were scanned with efficiency $E_1(x)$ to yield $N_{12}(x)$. Therefore,

$$E_1(x) = \frac{N_{12}(x)}{N_2(x)}$$

We have here a *binomial* quantity, since there are only possibilities that an event is found or missed. We may consider that the total number of binomial trials is $N_2(x)$ and the number of successes is $N_{12}(x)$ which yields the probability

$$E_1(x) = N_{12}(x)/N_2(x) = \hat{p}$$

We may obtain an estimate for the *variance* of this estimator by using the binomial formula $\sigma^2(\hat{p}) = p(1-p)/N$. Here, we must approximate by replacing p by its estimator $\hat{p}$. We thus get

$$\sigma^2(E_1) \approx (1 - E_1)E_1/N_2(x) = \frac{[1-(N_{12}/N_2)](N_{12}/N_2)}{N_{12}}$$

We may write the result in the form

$$E_1(x) \pm \sigma[E_1(x)] = \frac{N_{12}(x)}{N_2(x)}\left\{1 \pm \left[\frac{1}{N_{12}(x)} - \frac{1}{N_2(x)}\right]^{1/2}\right\}$$

and

$$E_2(x) \pm \sigma[E_2(x)] = \frac{N_{12}(x)}{N_1(x)}\left\{1 \pm \left[\frac{1}{N_{12}(x)} - \frac{1}{N_1(x)}\right]^{1/2}\right\}$$

We may estimate the *total number of events* in the film

$$N(x) = \frac{N_1(x)}{E_1(x)} = \frac{N_2(x)}{E_2(x)} = \frac{N_1(x)N_2(x)}{N_{12}(x)}$$

We may also obtain this by considering the total number of *independent* events N_I identified in either scan 1 or scan 2 and using (19.4):

$$N(x) = N_I(x)/E_I(x) = \frac{N_1(x) + N_2(x) - N_{12}(x)}{E_1(x) + E_2(x) - E_1(x)E_2(x)} = \frac{N_1(x)N_2(x)}{N_{12}(x)}$$

since the scans are independent. ◄

Estimation of a Population Proportion

We may now return to the problem of *inverse probability* connected with the **hypergeometric distribution**. We consider a population that contains n members of which a *fraction p* has a certain property. We wish to estimate p by **sampling**, that is, without counting the entire population. We draw a random sample of size r and count the number, k, of members in this sample that have the property of interest.

In (23.20) and (23.22) we saw that the *natural estimator* is

$$\hat{p} = \frac{k}{r}$$

with

$$\sigma^2(\hat{p}) = \frac{p(1-p)}{r}\left(\frac{n-r}{n-1}\right)$$

The estimation problem is to determine *how large* a sample, r, to take to get an estimate $\hat{p}$ that is close to p, that is, we want the probability to be small ($\leq \alpha$) that $\hat{p}$ differs from p by more than an arbitrary amount (δ), which we may write as before.

$$\Pr\{|\hat{p} - p| \geq \delta\} \leq \alpha,$$

or

$$\Pr\{|\hat{p} - p| \leq \delta\} \geq 1 - \alpha$$

For practical problems it is usually a good approximation to consider that k is **normally distributed** and, hence, that $\hat{p}$ is *normally distributed* as well with mean p and variance $\sigma^2(\hat{p})$ as above.

$$\begin{aligned}\Pr\{|\hat{p} - p| \leq \delta\} &= \Pr\{p - \delta \leq \hat{p} \leq p + \delta\} \\ &= \Pr\{\hat{p} \leq p + \delta\} \\ &\quad - \Pr\{\hat{p} \leq p - \delta\} \\ &\approx \Phi\left\{\frac{(p+\delta) - \hat{p}}{\sigma(\hat{p})}\right\} \\ &\quad - \Phi\left\{\frac{(p-\delta) - \hat{p}}{\sigma(\hat{p})}\right\}\end{aligned}$$

in the normal approximation

$$= \Phi\left\{\frac{\delta}{\sigma(\hat{p})}\right\} - \Phi\left\{\frac{-\delta}{\sigma(\hat{p})}\right\}$$

$$= A\left\{\frac{\delta}{\sigma(\hat{p})}\right\}$$

$$= 2\Phi\left\{\frac{\delta}{\sigma(\hat{p})}\right\} - 1$$

For the **minimum** r we have

$$\Pr\{|\hat{p} - p| \leq \delta\} = 1 - \alpha$$

$$\approx 2\Phi\left\{\frac{\delta}{\sigma(\hat{p})}\right\} - 1$$

$$\therefore 1 - \frac{\alpha}{2} = \Phi\left\{\frac{\delta}{\sigma(\hat{p})}\right\}$$

As in our earlier uses of the normal distribution, we may use the tables to determine a value Z_α such that

$$\Phi(Z_\alpha) = 1 - \frac{\alpha}{2}$$

or

$$A(Z_\alpha) = 1 - \alpha$$

Having obtained Z_α we may calculate the desired sample size r, since

$$\left(\frac{\delta}{\sigma(\hat{p})}\right)^2 = (Z_\alpha)^2 = \frac{\delta^2}{p(1-p)}\frac{r(n-1)}{(n-r)}$$

This has the solution

$$r_{\min} = \frac{p(1-p)}{[(n-1)/n](\delta/Z_\alpha)^2 + p(1-p)/n}$$

To check some of the expected features of $r_{\min}$ we may write this as

$$r_{\min} = \frac{n/(n-1)}{1/[p(1-p)](\delta/Z_\alpha)^2 + 1/(n-1)}$$

The second term in the denominator decreases as n increases; thus $r_{\min}$ increases with n.

Also, $r_{\min}$ increases as δ decreases and increases with Z_α. That is, as the restriction on $\hat{p}$ becomes "tighter" either in magnitude or probability, the sample size must increase.

The probability product, for all p,

$$p(1-p) \leq \frac{1}{4}$$

Therefore, $r_{\min}$ is *largest* for $p(1-p) = 1/4$. We note the difficulty that the answer depends on knowledge of $p(1-p)$. If this is not known, a *conservative* value of $r_{\min}$ is obtained by setting $p(1-p) = 1/4$.

► **Example 29.5**

We are now able to give a better answer to Example 23.6 of Chapter 23.

Here $\delta = 0.01$, $\alpha = 0.25$, and $n = 40{,}000$ with $p \approx 0.5$.

From Table A.VI.2,

$$A(Z_\alpha) = 1 - \alpha = 0.75$$

for $Z_\alpha \approx 0.675$.

We note $n/(n-1) \approx 1$ for this case and

$$r_{\min} = \frac{1/4}{(0.01/0.675)^2 + 1/[4(40000)]} \approx 1108$$

This is significantly less than the estimate, previously obtained from the *Chebychev inequality*, of 8060.

We might ask what probability level, α, would correspond to a sample of 8060, and we obtain this from the relation

$$(Z_\alpha)^2 = \frac{\delta^2}{p(1-p)}\frac{r(n-1)}{(n-r)}$$

$$= \frac{(0.01)^2(8060)(40000)}{(1/4)(40000-8060)}$$

$$= 4.038$$

or

$$Z_\alpha = 2.010$$

This yields $1 - \alpha = 0.95$

or

$$\alpha = 0.05$$

That is, the probability is greater than 95% that the answer obtained by sampling 8060 persons in a random way will be within 0.01 of the true probability. ◄

30

ESTIMATION AND THE METHOD OF MAXIMUM LIKELIHOOD

In Chapter 19 we first considered the construction of likelihood functions and indicated how the *method of maximum likelihood* leads to **estimation of parameters** (we also used likelihood functions in our discussions of the normal and chi-square distributions).

Likelihood Estimators

Let us consider that a distribution is characterized by r **parameters** c_i, $i = 1, \ldots, r$, which we summarize as

$$\mathbf{c} = (c_1, \ldots, c_r)$$

The **observables** (measurable quantities for ***each*** experiment) may be called

$$\mathbf{x} = (x_1, \ldots, x_s)$$

We define the *a posteriori* probability, $f(\mathbf{x}; \mathbf{c})$, and the likelihood function as before:

$$\mathscr{L}(\mathbf{x}; \mathbf{c}) = \prod_{i=1}^{n} f_i(\mathbf{x}; \mathbf{c}) \qquad (19.25)$$

where n is the number of experiments or data points (each **data point** is a point in **s-dimensional space**) where it will frequently be true that the experiments are *independent* but **identical**

$$f_i(\mathbf{x}; \mathbf{c}) = f(\mathbf{x}; \mathbf{c})$$

LIKELIHOOD RATIO

Where we have no *a priori* reason to prefer one set of parameters $\mathbf{c}_A$, over another set $\mathbf{c}_B$, we may use the *likelihood ratio*, Q, obtained from (19.24) to measure our *a posteriori* belief in $\mathbf{c}_A$ over $\mathbf{c}_B$:

$$Q(\mathbf{c}_A; \mathbf{c}_B) = \frac{\prod_{i=1}^{n} f_i(\mathbf{x}; \mathbf{c}_A)}{\prod_{i=1}^{n} f_i(\mathbf{x}; \mathbf{c}_B)} \qquad (30.1)$$

► **Example 30.1**
It is known that the probability of having a son for any given birth is about 0.5 for the *general populace* (GP). For a certain *special group* (SG) of families, the probability of a son is 0.7. The Jones family has five sons and one daughter. We should like to investigate the probability that they are (SG) instead of (GP).

The *likelihood functions* are as follows:

$$\mathcal{L}_{GP} = (0.5)^5(0.5)^1 = 0.0156$$

and

$$\mathcal{L}_{SG} = (0.7)^5(1-0.7)^1 = 0.0504$$

The **likelihood ratio** is, therefore,

$$Q(SG; GP) = \frac{\mathcal{L}_{SG}}{\mathcal{L}_{GP}} = 3.23 \quad ◄$$

MAXIMUM LIKELIHOOD

As discussed in Chapter 19 we have the most confidence in the set of parameters for which the likelihood function is *maximum*. This will frequently occur at a local maximum that we find by taking derivatives as in (19.26). (See Example 30.6 for an exception.)

It is convenient to consider the *logarithm* of the likelihood function,

$$\ell = \ln \mathcal{L} = \ln \left[\prod_{i=1}^{n} f_i(\mathbf{x}; \mathbf{c})\right] = \sum_{i=1}^{n} \ln [f_i(\mathbf{x}; \mathbf{c})] \qquad (30.2)$$

called the "*log-likelihood function*" for short. The positions of the maxima of ℓ and $\mathcal{L}$ are identical, and we may locate $\mathbf{c}^*$, the "*most likely values*" of $\mathbf{c}$, by taking the derivatives and writing the r equations:

$$\left.\frac{\partial \ell(\mathbf{x}; \mathbf{c})}{\partial c_i}\right|_{c_i^*} = 0 \qquad i = 1 \cdots r \qquad (30.3)$$

These are called the **likelihood equations.**

We shall consider primarily the case in which we seek to estimate *one* parameter, c. For this case the likelihood equation becomes

$$\frac{d\ell(\mathbf{x}; c)}{dc} = 0 \qquad (30.3a)$$

Uniqueness

Maximum likelihood estimators have the property of being unique since we may choose to estimate any function of the parameter of interest,

$$\lambda = \lambda(c)$$

since the likelihood equation for λ

$$\frac{\partial \ell(\mathbf{x}; \lambda)}{\partial \lambda} = \frac{\partial \ell(\mathbf{x}; c)}{\partial c}\frac{\partial c}{\partial \lambda} = 0 \qquad (30.4)$$

is necessarily satisfied because of (30.3), and conversely.

$$\therefore \lambda^* = \lambda(c^*) \qquad (30.4a)$$

The logarithmic derivative

It is useful to define the "*logarithmic derivative*" to be

$$\varphi_i(\mathbf{x}; c) \equiv \frac{\partial}{\partial c}[\ln f_i(\mathbf{x}; c)] = \frac{1}{f_i(\mathbf{x}; c)}\frac{\partial}{\partial c} f_i(\mathbf{x}; c) \qquad (30.5)$$

We note that

$$\frac{\partial \ell}{\partial c} = \frac{\partial}{\partial c}\ln \mathcal{L} = \sum_{i=1}^{n} \varphi_i(\mathbf{x}; c) = \ell' \quad \text{for one } c \qquad (30.6)$$

► Example 30.2
Binomial distribution

Consider the case of n independent trials of an asymmetric coin. We wish to estimate the probability, p, of success (heads) by observing the results of the n trials.

We may regard this as one experiment with n observables in which case

$$\mathcal{L} = f(x; p) = \binom{n}{s(x)} p^{s(x)}(1-p)^{n-s(x)}$$

where $s(x)$ is the *observed number of successes* or, alternatively, as n trials of one observation

$$\mathcal{L} = \prod_{i=1}^{n} f_i(x; p)$$

where each

$$f_i(x; p) = p \quad \text{for a head}$$
$$= (1-p) \quad \text{for a tail}$$

so that

$$\mathcal{L} = p^{s(x)}(1-p)^{n-s(x)}$$

We note that *both of these likelihood functions yield the same estimator*. It is often a property of the maximum likelihood method that constants of the problem have no bearing on the solution. This is usually a great convenience.

We form

$$\ell = s(x) \ln p + [n - s(x)] \ln (1 - p)$$

and get

$$\ell' = \frac{1}{p} s(x) + [n - s(x)] \frac{1(-1)}{1-p}$$

which we set to zero for $p = \hat{p}$.

$$\therefore \frac{s(x)}{\hat{p}} = \frac{[n - s(x)]}{1 - \hat{p}}$$

and

$$\hat{p} = \frac{s(x)}{n} \tag{30.7}$$

which accords with the "*natural*" *statistic* discussed in Chapter 22. We shall verify later that this is a "**most efficient statistic**" *with variance* $p(1-p)/n$. ◄

► Example 30.3
Hypergeometric distribution ("tagging")

Suppose we have an urn containing an unknown number of white balls. The constraints of the situation prevent us from counting all the balls, but we may add n_1 white balls, mix the balls thoroughly, and then draw at random a sample of r balls. Let n now be the (unknown) total number of balls.

In a sample of r balls, the probability that there will be k white balls is by (23.1)

$$\Pr(k) = \frac{\binom{n_1}{k}\binom{n - n_1}{r - k}}{\binom{n}{r}}$$

We form the *likelihood function*

$$\mathscr{L}(k; n_1, r, n) = \Pr(k)$$

and we try to find the value of n for which $\mathscr{L}$ is *maximum*.

We may investigate the maximum by considering

$$\frac{\mathscr{L}(k; n_1, r, n)}{\mathscr{L}(k; n_1, r, n-1)} = \frac{\binom{n - n_1}{r - k}\binom{n-1}{r}}{\binom{n}{r}\binom{n - 1 - n_1}{r - k}}$$

$$= \frac{(n - n_1)(n - r)}{n(n - n_1 - r + k)}$$

$$= Q(n; n-1)$$

We note

$$Q > 1 \quad \text{if} \quad n_1 r > nk$$
$$Q < 1 \quad \text{if} \quad n_1 r < nk$$

$\mathscr{L}$ is maximum for the **integer** n closest to

$$n_1 r / k \tag{30.8}$$

This problem frequently arises in estimating wildlife populations. First, a sample of size n_1 of animals, fish, birds, etc. is caught *and tagged*. They are then released in such fashion as to insure that they *mix randomly* with the rest of the parent population (total n). At some later time another sample of size r is drawn and the number, k, of tagged specimens observed. It is desired to estimate n from this procedure, which is done as shown.

$$n \approx \frac{n_1 r}{k}$$

There are obvious extensions to other tagging problems, for example, radioactive tracers. ◄

► Example 30.4
Poisson distribution

Consider that we observe a series of n counts $\mathbf{x} = (x_1, \ldots, x_n)$ where $x_i = r_i$ and such that we believe the underlying variate to be Poisson-distributed with parameter μ, that is,

$$f_i(r_i; \mu) = \mathbf{p}(r_i; \mu) = \frac{\mu^{r_i} e^{-\mu}}{r_i!}$$

The *likelihood function* is

$$\mathscr{L} = \prod_{i=1}^{n} f_i(r_i; \mu)$$

and

$$\ell = \sum_{i=1}^{n} \{r_i \ln \mu - \mu) - \ln (r_i!)\}$$

We differentiate and obtain

$$\ell' = \sum_{i=1}^{n} \left\{\frac{r_i}{\mu} - 1\right\}$$

which we set to zero for $\hat{\mu}$:

$$\frac{1}{\hat{\mu}}\sum_{i=1}^{n} r_i = n$$

or

$$\hat{\mu} = \frac{1}{n}\sum_{i=1}^{n} r_i = \bar{r}$$

This is the *maximum likelihood estimate* which corresponds to the **most probable value** for $\hat{\mu}$ in (24.26). ◄

For later use we note that

$$\ell' = \frac{1}{\mu}\sum_{i=1}^{n}(r_i - \mu) = \frac{n}{\mu}(\bar{r} - \mu) \qquad (30.9)$$

► **Example 30.5**
n independent normally distributed measurements of the same quantity

We have already considered this problem by the *maximum likelihood method* in Chapter 26,

$$\mathcal{L} = e^{-\chi^2/2}\prod_{i=1}^{n}\frac{1}{\sqrt{2\pi\sigma_i^2}} \qquad \text{by (25.4)}$$

$$\ell = -\frac{\chi^2}{2} + \sum_{i=1}^{n}\frac{1}{\sqrt{2\pi\sigma_i^2}}$$

where

$$\chi^2 = \sum_{i=1}^{n}\left(\frac{x_i - \bar{x}}{\sigma_i}\right)^2$$

and

$$\ell' = -\frac{1}{2}\frac{d\chi^2}{d\bar{x}} = \sum_{i=1}^{n}\frac{(x_i - \bar{x})}{\sigma_i^2}$$

Setting $\ell' = 0$, we obtained

$$(\bar{x})_{\text{est}} = \frac{\sum_{i=1}^{n}\frac{x_i}{\sigma_i^2}}{\sum_{i=1}^{n}\frac{1}{\sigma_i^2}} \qquad (30.10)$$

◄

All estimators obtained by **minimizing χ^2** are also *maximum likelihood* estimators as well (with the additional assumption of *normality*).

► **Example 30.6**
The ski tow problem

We may also use the *method of maximum likelihood* to consider problems where we cannot readily obtain an estimator by using a distribution in the sense of inverse probability.

Suppose a skier goes up a mountain on a ski tow but is able to read only the number of the ski tow seat he is on. The first time he goes up the mountain he notes the number is n_1. The second trip he notes the number is n_2, and so on. He does this s times and records the numbers n_i where $i = 1 \cdots s$.

Query: How many ski tow seats, N, are there?

We first state that the ski tow seat numbers have significance and remark that $N \geq \text{maximum}\,(n_1, \ldots, n_s)$. Furthermore, we assume that he takes a seat *at random*, that is, he is equally likely to get any number at any given time. If there are N seats, the probability of getting n_i is just $1/N$.

$$\therefore \mathcal{L}(\mathbf{n}; N) = \prod_{i=1}^{s}\left(\frac{1}{N}\right) = \left(\frac{1}{N}\right)^s$$

with the constraint that $N \geq \max\,(n_1, \ldots, n_s)$.

This is a monotonically decreasing function which starts at $N = (n_i)_{\max}$. The likelihood function has no *local* maximum but does have an **absolute** maximum at $N = (n_i)_{\max}$. As s increases, the likelihood function gets sharper (more peaked). The situation is indicated in Figures 30.1 and 30.2. ◄

► **Example 30.7**
Tank problem

A similar problem had a military application in World War II when the numbers on captured German tanks were decoded and an estimate made of the number of tanks produced.

The tank problem is somewhat different from the ski tow situation since tank numbers *cannot repeat*. In the ski tow problem the observations were truly **independent** since the skier had the same chance of seeing any number, even one he had seen before, that is, the sampling was performed *with re-*

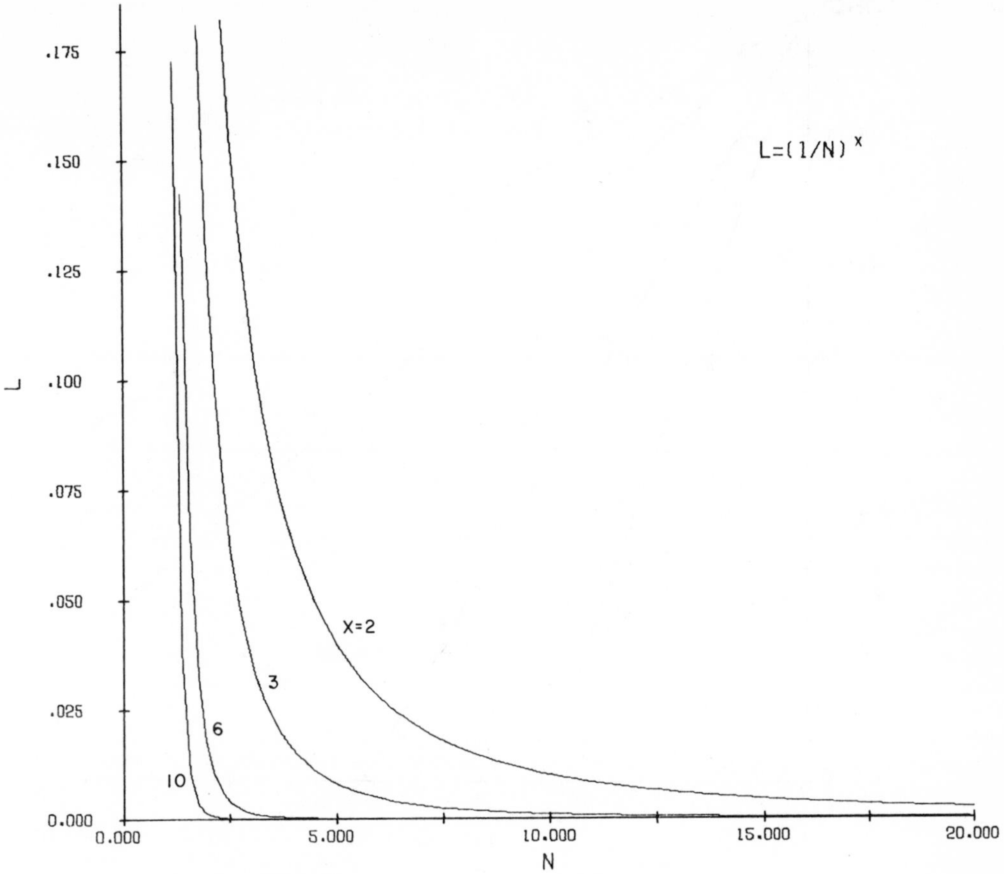

Figure 30.1
Plot of the likelihood function of Example 30.6 on linear scale.

placement. The tank problem represented a situation in which sampling occurred *without replacement*. This problem yields to solution by the following argument.

Let X_i be the number on a tank assuming the tanks were labeled consecutively. Each tank had probability $1/N$ of being observed where N is the total number that we wish to estimate

$$E(X) = \sum_{i=1}^{N} i \frac{1}{N} = \frac{1}{N} \frac{N(N+1)}{2}$$

since it is the summation of an arithmetic progression.

$$\therefore\ E(x) = \frac{(N+1)}{2}$$

We may estimate $E(x)$ by $\bar{x}$, the *sample mean of the numbers observed.*

$$\therefore\ \hat{N} = 2\bar{x} - 1$$

is an estimator of N. ◄

We should note that estimators obtained by *maximum likelihood* are usually **efficient** estimators (see later discussion) but are *not always* **unbiased**.

► **Example 30.8**

The *maximum likelihood* estimator for the Poisson case was $\hat{\mu} = \bar{r}$ whereas the *unbiased* estimator is $(\bar{r}+1)$ as shown in Chapter 25. ◄

► **Example 30.9**

Suppose we have a sample of size n drawn from a *normal* distribution $N(\mu, \sigma^2)$. We wish to estimate μ and σ^2.

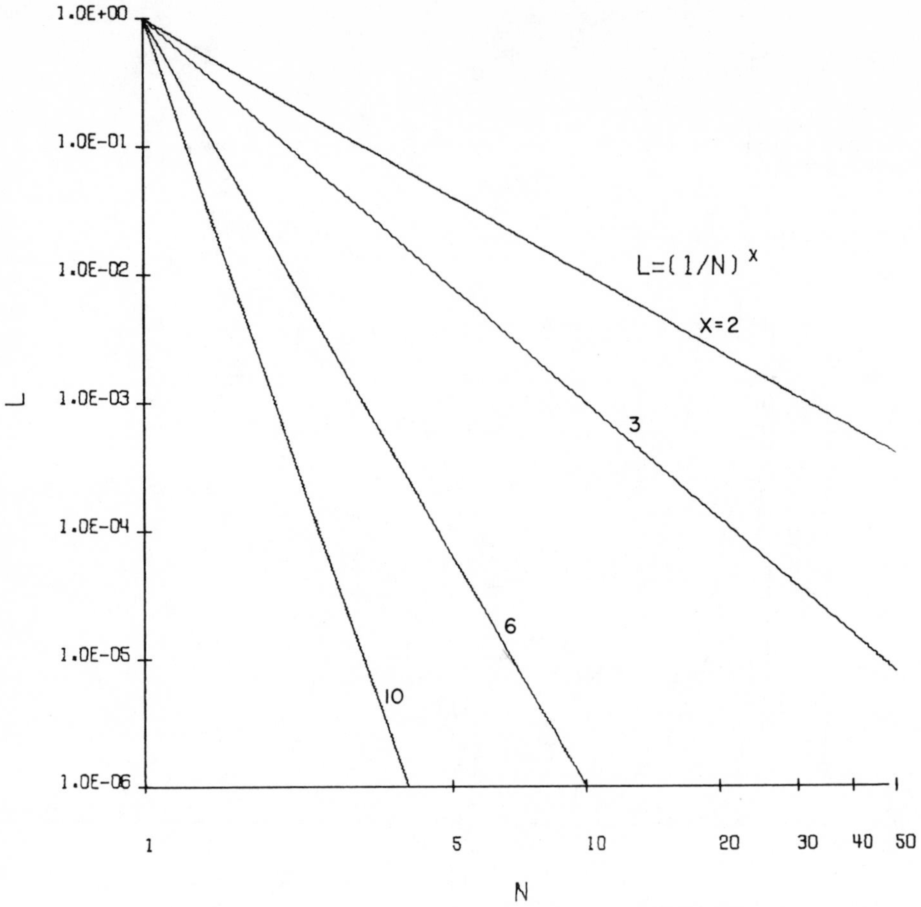

Figure 30.2
Plot of the likelihood function of Example 30.6 on log scale.

Here

$$f(\mathbf{x};\,\mu,\,\sigma^2) = \frac{1}{\sqrt{2\pi\sigma^2}}\,e^{-(x-\mu)^2/2\sigma^2}$$

and

$$\mathcal{L}(\mathbf{x};\,\mu,\,\sigma^2)$$

$$= \prod_{i=1}^{n} \frac{1}{\sqrt{2\pi\sigma^2}}\,e^{-(x_i-\mu)^2/2\sigma^2}$$

$$= (2\pi\sigma^2)^{-n/2}\exp\left[-\sum_{i=1}^{n}\frac{(x_i-\mu)^2}{2\sigma^2}\right]$$

$$\therefore\ \ell(\mathbf{x};\,\mu,\,\sigma^2) = \frac{-n}{2}\ln(2\pi) + \frac{-n}{2}\ln\sigma^2 - \frac{1}{2\sigma^2}\sum_{i=1}^{n}(x_i-\mu)^2 \qquad (30.11)$$

Here

$$\frac{\partial\ell}{\partial\mu} = \frac{+2}{2\sigma^2}\sum_{i=1}^{n}(x_i-\mu)$$

and

$$\frac{\partial\ell}{\partial(\sigma^2)} = -\frac{n}{2\sigma^2} - \frac{1}{2}\sum_{i=1}^{n}(x_i-\mu)^2\left[\frac{-1}{(\sigma^2)^2}\right]$$

The *likelihood equations* are

$$\left.\frac{\partial\ell(\mathbf{x};\,\mu,\,\sigma)}{\partial\mu}\right|_{\hat\mu} = 0 = \frac{1}{\sigma^2}\left[\sum_{i=1}^{n}x_i - n_{\hat\mu}\right]$$

and

$$\left.\frac{\partial\ell(\mathbf{x};\,\mu,\,\sigma)}{\partial\sigma^2}\right|_{\widehat{\sigma^2}} = 0$$

$$= -\frac{n}{2\widehat{\sigma^2}} + \frac{1}{2(\widehat{\sigma^2})^2}\sum(x_i-\hat\mu)$$

These yield $\hat{\mu} = \bar{x}$ as usual and

$$\widehat{(\sigma^2)} = \frac{1}{n}\sum_{i=1}^{n}(x_i - \bar{x})^2$$

This latter is a **biased** *estimator* since

$$E\widehat{(\sigma^2)} = \left(\frac{n-1}{n}\right)\sigma^2$$ ◄

General Properties of Estimators

We have already stated that estimators should be unbiased and have a variance "as small as possible." When dealing with the real world, compromises are usually required when trying to utilize such criteria. We frequently find that we must *compromise* the desiderata of minimum *bias* and minimum *variance* for an estimator. For example, an estimator that is a single (guessed) *constant* has the characteristic of having *zero variance.* The **bias**, of course, may be enormous.

UNBIASED ESTIMATORS

Consider that we wish to investigate possible **limits** on the *efficiency* of an *unbiased estimator,* that is, we seek to find either a *minimum variance* unbiased statistic or a *lower bound* on the variance of an unbiased estimator.

If $\hat{c}$ is an *unbiased estimator* of c, then we wish to show

$$\text{Var}(\hat{c}) \geq \frac{1}{E\left[\left\{\frac{\partial}{\partial c}\ln f(\mathbf{x}; c)\right\}^2\right]} \tag{30.12}$$

where $f(\mathbf{x}; c)$ is the *a posteriori* probability as before.

This is one form of the *Cramer-Rao-Frechet inequality.*

Since $\hat{c}$ is an unbiased estimator of c,

$$\int \hat{c}(\mathbf{x})f(\mathbf{x}; c)d\mathbf{x} = c$$

and

$$\frac{\partial}{\partial c}\int \hat{c}(\mathbf{x})f(\mathbf{x}; c)d\mathbf{x} = 1$$

We assume that we can differentiate under the integral sign and get

$$1 = \int \hat{c}(\mathbf{x})\frac{\partial f(\mathbf{x}; c)}{\partial c}d\mathbf{x}$$

$$= \int \frac{\hat{c}(\mathbf{x})}{f(\mathbf{x}; c)}\frac{\partial f(\mathbf{x}; c)}{\partial c}f(\mathbf{x}; c)d\mathbf{x}$$

$$= \int \left\{\hat{c}(\mathbf{x})\left[\frac{\partial}{\partial c}\ln f(\mathbf{x}; c)\right]\right\}f(\mathbf{x}; c)d\mathbf{x}$$

$$= E\left\{\hat{c}(\mathbf{x})\left[\frac{\partial}{\partial c}\ln f(\mathbf{x}; c)\right]\right\}$$ by definition of *expectation value*

Consider the quantities

$$y = \hat{c}(\mathbf{x}) \tag{30.13}$$

and

$$z = \left[\frac{\partial}{\partial c}\ln f(\mathbf{x}; c)\right] \tag{30.14}$$

$$\therefore 1 = E[yz]$$

We note that

$$E(z) = \int \left[\frac{\partial}{\partial c}\ln f(\mathbf{x}; c)\right]f(\mathbf{x}; c)d\mathbf{x}$$

$$= \int \frac{\partial}{\partial c}f(\mathbf{x}; c)d\mathbf{x}$$

$$= \frac{\partial}{\partial c}\int f(\mathbf{x}; c)d\mathbf{x}$$

$$= \frac{\partial}{\partial c}\{1\} = 0 \tag{30.15}$$

$\text{Covar}(y, z) = E(yz) - E(y)E(z)$ by (20.12).

$$\therefore E(yz) = \text{Covar}(y,z) = 1$$

Recall that

$$\frac{[\text{Covar}(y, z)]^2}{\text{Var}(y)\,\text{Var}(z)} \leq 1 \quad \text{by (20.13a)}$$

$$[\text{Covar}(y, z)]^2 = 1 \leq \text{Var}(y)\,\text{Var}(z)$$

and

$$\text{Var}(y) \geq \frac{1}{\text{Var}(z)}$$

Since $E(z) = 0$, $\text{Var}(z) = E(z^2)$

$$\therefore \text{Var}(\hat{c}) \geq \frac{1}{E\left[\left\{\frac{\partial}{\partial c}\ln f(\mathbf{x}; c)\right\}^2\right]}$$

which is (30.12). **Q.E.D.**

INFORMATION

The quantity,

$$I(c) = E\left[\left\{\frac{\partial}{\partial c}\ln f(\mathbf{x}; c)\right\}^2\right] \tag{30.16}$$

is called the "**information about** c," which is contained in one observation of $\mathbf{x}$.

$$\mathrm{Var}(\hat{c}) \geq \frac{1}{I(c)} \tag{30.17}$$

The appellation is self-explanatory. The more information about c provided by an observation, the smaller we might expect the variance (of a *most efficient statistic*) to be. If there are n **identical** but *independent* observations, then the set of n data points provides n times the information provided by each.

We may see this as follows:

$$\begin{aligned}
I_n(c) &= E\left[\left\{\frac{\partial}{\partial c}\ln\left[\prod_{i=1}^{n} f_i(\mathbf{x}; c)\right]\right\}^2\right] \\
&= E\left[\left\{\frac{\partial}{\partial c}\sum_{i=1}^{n}\ln f_i(\mathbf{x}; c)\right\}^2\right] \\
&= E\left[\left\{\sum_{i=1}^{n}\frac{\partial}{\partial c}\ln f_i(\mathbf{x}; c)\right\}^2\right] \\
&= E\left[\left\{\sum_{i=1}^{n}\varphi_i(\mathbf{x}; c)\right\}^2\right] \\
&= E\left[\sum_{i=1}^{n}\{\varphi_i(\mathbf{x}; c)\}^2 + \sum_{i\neq j}^{n}\{\varphi_i(\mathbf{x}; c)\varphi_j(\mathbf{x}; c)\}\right] \\
&= \sum_{i=1}^{n} E\{[\varphi_i(\mathbf{x}; c)]^2\} \\
&\quad + \sum_{i=j} E\{\varphi_i(\mathbf{x}; c)\varphi_j(\mathbf{x}; c)\}
\end{aligned}$$

The various measurements are *independent* for $i \neq j$. Hence the $\varphi_i(\mathbf{x}; c)$ and $\varphi_j(\mathbf{x}; c)$ are *independent* and

$$\begin{aligned}
&\sum_{i\neq j} E\{\varphi_i(\mathbf{x}; c)\varphi_j(\mathbf{x}; c)\} \\
&\quad = \sum_{i\neq j} E\{\varphi_i(\mathbf{x}; c)\}E\{\varphi_j(\mathbf{x}; c)\}
\end{aligned}$$

by (20.9).

But $E\{\varphi_i(\mathbf{x}; c)\} = E\{z\} = 0$ for each i since the measurements are **identical**.

$$\begin{aligned}
\therefore I_n(c) &= \sum_{i=1}^{n} E\{[\varphi_i(\mathbf{x}; c)]^2\} = \sum_{i=1}^{n} E\{z^2\} \\
&= nE(z^2) = nI(c)
\end{aligned} \tag{30.18}$$

The "**information about** c" scales linearly with the number of independent, identical measurements.

Because of the inequality the "**minimum-variance** (or *Cramer-Rao*) **bound**" for n independent measurements determining an unbiased estimator $\hat{c}$ is

$$\{I_n(c)\}^{-1} \tag{30.19}$$

where

$$I_n(c) = nE\left\{\left[\frac{\partial}{\partial c}\ln f(\mathbf{x}; c)\right]^2\right\} \tag{30.20}$$

The *information* may be written in alternative form for most likelihood functions:

$$\begin{aligned}
I(c) &= E\left\{\left[\frac{\partial}{\partial c}\ln f(\mathbf{x}; c)\right]^2\right\} \\
&= E\left\{\left[\frac{-\partial^2}{\partial c^2}\ln f(\mathbf{x}; c)\right]\right\}
\end{aligned} \tag{30.21}$$

We may show this as follows:

$$\begin{aligned}
\frac{\partial^2}{\partial c^2}\ln f(\mathbf{x}; c) &= \frac{\partial}{\partial c}\left[\frac{1}{f}\frac{\partial f}{\partial c}\right] \\
&= \frac{-1}{f^2}\left(\frac{\partial f}{\partial c}\right)^2 + \frac{1}{f}\frac{\partial^2 f}{\partial c^2} \\
&= -\left[\frac{\partial \ln f}{\partial c}\right]^2 + \frac{1}{f}\frac{\partial^2 f}{\partial c^2}
\end{aligned}$$

Taking the *expectation value* of both sides, we get

$$\begin{aligned}
E\left[\frac{\partial^2}{\partial c^2}\ln f(\mathbf{x}; c)\right] &= -I(c) \\
&\quad + E\left\{\frac{1}{f(\mathbf{x}; c)}\frac{\partial^2 f(\mathbf{x}; c)}{\partial c^2}\right\}
\end{aligned}$$

where the last term is

$$\begin{aligned}
E\left\{\frac{1}{f}\frac{\partial^2 f}{\partial c^2}\right\} &= \int \frac{\partial^2 f(\mathbf{x}; c)}{\partial c^2}\,d\mathbf{x} \\
&= \frac{\partial^2}{\partial c^2}\int f(\mathbf{x}; c)\,d\mathbf{x}
\end{aligned}$$

assuming we can differentiate inside or outside the integral.

$$\therefore E\left\{\frac{1}{f}\frac{\partial^2 f}{\partial c^2}\right\} = \frac{\partial^2}{\partial c^2}(1) = 0$$

which yields

$$E\left[-\frac{\partial^2}{\partial c^2}\ln f(\mathbf{x}; c)\right] = I(c) \qquad \textbf{Q.E.D.}$$

MINIMUM VARIANCE ESTIMATORS

We want to inquire about the conditions for a *minimum variance estimator* or **most efficient statistic**, that is, conditions for an estimator such that

$$\sigma^2(\hat{c}) = \frac{1}{I_n(c)} \tag{30.22}$$

Recall that the **in***equality* arose because of the *Cauchy-Schwartz* relation (20.13c). The *minimum bound occurs when the equality sign obtains,* which happens when $(ax + y) = 0$ in (20.13f) for all a, x and y, that is, the variables must be *linearly related* where a must be independent of the observables $\mathbf{x}$ but may, in general, depend on c.

$$\therefore a(c)[y - \langle y \rangle] = [z - \langle z \rangle] = z$$

since $\langle z \rangle = E(z) = 0$, where y, z are given by (30.13) and (30.14).

The *requirement* is

$$a(c)\{\hat{c}(\mathbf{x}) - c\} = \frac{\partial}{\partial c} \ln f(\mathbf{x}; c) = \ell' \tag{30.23}$$

and

$$\begin{aligned}\text{Covar}(y, z) = 1 &= \text{Var}(y)\,\text{Var}(z)\\ &= \text{Var}(\hat{c}) I(c)\end{aligned}$$

$$\therefore \text{Var}(\hat{c}) = \{I(c)\}^{-1}$$

and $\hat{c}$ is a *minimum variance estimator* when (30.23) holds. Furthermore,

$$\begin{aligned}\text{Var}(z) &= [a(c)]^2 \,\text{Var}[\hat{c} - c]\\ &= [a(c)]^2 \sigma^2(\hat{c})\\ &= I(c)\end{aligned}$$

or

$$[a(c)]^2 = \frac{I(c)}{\sigma^2(\hat{c})} = [I(c)]^2$$

$$\therefore a(c) = I(c) \tag{30.24}$$

We have shown that $\hat{c}$ is a *minimum variance unbiased estimator* if (and only if)

$$\frac{\partial}{\partial c} \ln f(\mathbf{x}; c) = I(c)[\hat{c}(\mathbf{x}) - c] \tag{30.25}$$

► **Example 30.10**

In coin tossing consider when the coin may be *asymmetrical*. We wish to estimate the probability, p, of getting a head by observing the results $x = (x_1, \ldots, x_n)$ of n independent trials with the same coin.

$$f(x, p) = \binom{n}{s(x)} p^{s(x)} (1 - p)^{n - s(x)}$$

where $s(x)$ is the number of successes (heads) in n trials.

$$\begin{aligned}\ell = \ln f_n(x, p) &= \text{constant} + s(x) \ln p\\ &\quad + [n - s(x)] \ln (1 - p)\end{aligned}$$

$$\begin{aligned}\ell' &= 0 + s(x)\frac{1}{p} + [n - s(x)]\frac{1}{1 - p}(-1)\\ &= 0\end{aligned}$$

$$\frac{s(x)}{p} = \frac{n - s(x)}{1 - p}$$

and

$$\hat{p} = \frac{s(x)}{n} \tag{30.7}$$

the *fraction of successes* in n trials, is the estimator.

We see that

$$\begin{aligned}\ell' &= \frac{\partial}{\partial c} \ln f(\mathbf{x}; c) = \frac{s(x)}{p} - \frac{[n - s(x)]}{1 - p}\\ &= \frac{s(x)(1 - p) - [n - s(x)]p}{p(1 - p)}\\ &= \frac{n\hat{p}(1 - p) - (n - n\hat{p})p}{p(1 - p)}\\ &= \frac{n}{p(1 - p)}[\hat{p} - p]\end{aligned}$$

is of the required form (30.25).

Equation (30.7) is a *minimum variance estimator* and

$$\sigma^2(\hat{p}) = \frac{1}{I(p)} = \frac{1}{a(p)} = \frac{p(1 - p)}{n} \tag{30.26}$$

◄

► **Example 30.11**

Poisson distribution

Considering again our *likelihood estimation* of the Poisson parameter μ in Example 30.4, we noted that

$$\hat{\mu} = \bar{r}$$

and

$$\ell' = \frac{n}{\mu}(\bar{r} - \mu) \tag{30.9}$$

This satisfies the requirement (30.25) and $\hat{\mu}$ is a *minimum variance estimator* with

$$\sigma^2(\hat{\mu}) = \frac{1}{I(\mu)} = \frac{1}{a(\mu)} = \frac{\mu}{n} \tag{30.27}$$

◄

► **Example 30.12**
n independent normally distributed measurements of the same quantity

We recall from Example 30.5 that

$$(\bar{x})_{\text{est}} = \frac{\sum_{i=1}^{n} (x_i/\sigma_i^2)}{\sum_{i=1}^{n} (1/\sigma_i^2)} \qquad (30.10)$$

with

$$\ell' = \sum_{i=1}^{n} \frac{x_i}{\sigma_i^2} - (\bar{x}) \sum_{i=1}^{n} \frac{1}{\sigma_i^2}$$
$$= \sum_{i=1}^{n} \frac{1}{\sigma_i^2} \left\{ \frac{\sum_{i=1}^{n} (x_i/\sigma_i^2)}{\sum_{i=1}^{n} (1/\sigma_i^2)} - (\bar{x}) \right\}$$
$$= \left(\sum_{i=1}^{n} \frac{1}{\sigma_i^2} \right) \{(\bar{x})_{\text{est}} - (\bar{x})\}$$

ℓ' is of the form (30.25) and $(\bar{x})_{\text{est}}$ is a *minimum variance estimator* with

$$\sigma^2[(\bar{x})_{\text{est}}] = \frac{1}{I(\bar{x})} = \left\{ \sum_{i=1}^{n} \left(\frac{1}{\sigma_i^2} \right) \right\}^{-1} \qquad (30.28)$$

If all the measurements are of *equal precision* $\sigma_i = \sigma$ for all i and

$$(\bar{x})_{\text{est}} = \frac{(1/\sigma^2) \sum_{i=1}^{n} x_i}{(1/\sigma^2) \sum_{i=1}^{n} 1} = \frac{\sum_{i=1}^{n} x_i}{n}$$

the *arithmetic mean*. The *variance*

$$\sigma^2[(\bar{x})_{\text{est}}] = \left\{ \frac{1}{\sigma^2} \sum_{i=1}^{n} 1 \right\}^{-1} = \left\{ \frac{n}{\sigma^2} \right\}^{-1}$$
$$= \frac{\sigma^2}{n}$$

as before. ◄

BIASED ESTIMATORS

We can extend this analysis to *biased estimators* by defining the **bias** to be

$$b(\hat{c}\,; c) = E(\hat{c}) - c \qquad (30.29)$$

where $b(\hat{c}\,; c)$ may be a function of c. (If the bias is *not* a function of c but, instead, a constant c_0, then we may trivially produce an unbiased estimator by considering $\hat{c} - c_0$.)

The analysis proceeds as before with

$$\int \hat{c}(\mathbf{x}) \left\{ \prod_{i=1}^{n} f_i(\mathbf{x}; c) \right\} d\mathbf{x} = c + b(\hat{c}\,; c)$$

and

$$\frac{\partial}{\partial c} \int \hat{c}(\mathbf{x}) \left\{ \prod_{i=1}^{n} f_i(\mathbf{x}; c) \right\} d\mathbf{x} = 1 + b'(\hat{c}\,; c)$$

Again we *differentiate under the* integral sign to get

$$1 + b'(\hat{c}\,; c) = \int \hat{c}(\mathbf{x}) \times \left\{ \sum_{i=1}^{n} \varphi_i(\mathbf{x}; c) \right\} \prod_{i=1}^{n} f_i(\mathbf{x}; c)\, d\mathbf{x}$$
$$= E\left\{ \hat{c}(\mathbf{x}) \left[\sum_{i=1}^{n} \varphi_i(\mathbf{x}; c) \right] \right\}$$
$$= E\{\hat{c}(\mathbf{x}) \ell'(\mathbf{x}; c)\}$$

using (30.6).

As before we use

$$\int \prod_{i=1}^{n} f_i(\mathbf{x}; c)\, d\mathbf{x} = 1$$

and *differentiate* to get

$$\int \sum_{i=1}^{n} \frac{f_i'(\mathbf{x}; c)}{f_i(\mathbf{x}; c)} \prod_{i=1}^{n} f_i(\mathbf{x}; c)\, d\mathbf{x} = 0$$
$$= E\left\{ \sum_{i=1}^{n} \frac{f_i'(\mathbf{x}; c)}{f_i(\mathbf{x}; c)} \right\} = E(\ell')$$

$$\therefore 1 + b'(\hat{c}\,; c) = E\{\hat{c}(\mathbf{x}) \ell'(\mathbf{x}; c)\} - E(\hat{c}) E(\ell')$$
$$= E\{[\hat{c}(\mathbf{x}) - E(\hat{c})] \ell'\}$$

Using the *Cauchy-Schwartz* relation (20.13c) again, we get

$$[1 + b'(\hat{c}\,; c)]^2 \le E\{[\hat{c}(\mathbf{x}) - E(\hat{c})]^2\} E(\ell'^2)$$

where

$$E(\ell'^2) = E\left\{ \left[\sum \varphi_i(\mathbf{x}; c) \right]^2 \right\}$$
$$= E\left\{ \sum_{i=1}^{n} [\varphi_i(\mathbf{x}; c)]^2 \right\}$$

because of the *independence* of the n measurements,

$$= I_n(c) = nI(c)$$

From the *definition*

$$\sigma^2(\hat{c}) = E\{[\hat{c}(\mathbf{x}) - E(\hat{c})]^2\}$$

and so we get the general inequality

$$[1 + b'(\hat{c}\,; c)]^2 \le \sigma^2(\hat{c}) I_n(c)$$

or

$$\sigma^2(\hat{c}) \ge \frac{[1 + b'(\hat{c}\,; c)]^2}{I_n(c)} \qquad (30.30)$$

as the ***Cramer-Rao-Frechet* inequality for *biased* estimators.**

This relation tells us that we may be able to *trade off* bias for smaller variance; that is, it may happen that we wish to choose a biased estimator that has a smaller variance than one that is unbiased. Also, we might prefer an unbiased estimator with a larger variance.

► Example 30.13

If we wish to estimate the **variance** of a normal population $N(\mu, \sigma^2)$ by considering the sample $x_1, \ldots, x_n$, we saw in (26.55a) that the *sample variance statistic*, s^2, has variance

$$\text{Var}(s^2) = \text{Var}\left\{\frac{1}{n-1}\sum_{i=1}^{n}(x_i - \bar{x})^2\right\} = \frac{2(\sigma^2)^2}{n-1}$$

The statistic $(1/n)\Sigma_{i=1}^{n}(x_i - \bar{x})^2$ is *biased* but has the variance

$$\text{Var}\left\{\left(\frac{n-1}{n}\right)s^2\right\} = \left(\frac{n-1}{n}\right)^2 \text{Var}(s^2)$$

The biased statistic has a smaller variance than the unbiased statistic, s^2. ◄

EFFICIENCY OF UNBIASED ESTIMATORS

Even when we are not able to express ℓ' in the required form to produce a minimum variance estimator, we may nevertheless use the *Cramer-Rao* inequality to define the ***efficiency*** of a statistic $\hat{c}$:

$$e(\hat{c}) = \frac{[I(c)]^{-1}}{\sigma^2(\hat{c})} \qquad (30.31)$$

where $[I(c)]^{-1} = \sigma^2_{\min}$, the *minimum* possible variance.

Note that we should like $e(\hat{c})$ to be close to unity, but *we should not always reject a statistic with low efficiency*. It may be the only one available, or it may be more *convenient* than more efficient estimators.

SUFFICIENCY

An important idea is that of *sufficient* statistics. As usual, our discussion of this will be heuristic since we are more concerned with plausibility than mathematical rigor. It is intuitively clear that there are usually certain features of a set of observations, **x**, which provide *no information* about a parameter c that we wish to estimate. For example, the assumption of independence in a sequence of binomial trials that we wish to use to estimate the probability p means that we may safely neglect the **order** in which the observations appeared.

A *sufficient statistic* is one that contains **all** the information *relevant* to the parameter of interest. A *minimal-sufficient statistic* corresponds to the **greatest possible reduction** of a set of data without losing any information relevant to the parameter of interest. For example, if we have a set of observations **x** that we know comes from a normal population, sufficient statistics are Σx_i and Σx_i^2. No other information need be kept to determine μ and σ^2. Actually, we may discard the original **x** without loss of information once we have the sufficient statistics.

The Factorization Theorem: Recognition of Sufficient Statistics

Suppose that **x** is a sample from a population with the *a posteriori* probabilities $f(\mathbf{x}; c)$. The likelihood function is $\Pi_{i=1}^{n} f(x_i; c)$. Suppose that this can be *factorized* such that

$$\prod_{i=1}^{n} f(x_i; c) = g(\mathbf{x}; \hat{c})h(\hat{c}; c) \qquad (30.32)$$

that is, factorized into a function g *that does not depend on the* ***parameter*** c (although g may depend on the **estimator** $\hat{c}$) and a function h *that does depend on* c (also on $\hat{c}$). The function $g(\mathbf{x}; \hat{c})$ may be regarded as the *conditional probability* distribution of **x** given the specific value of $\hat{c}$. Since $g(\mathbf{x}; \hat{c})$ does not depend in any way on c, it cannot be used to provide any information about c. Also, $h(\hat{c}; c)$ does not depend on **x** except through the estimator $\hat{c}(\mathbf{x})$; therefore, *$\hat{c}$ must contain all the information pertinent to the estimation of c.* This argument says that *the* **factorizability** *implies that $\hat{c}$ is a* **sufficient estimator.**

Conversely, if $\hat{c}$ is a sufficient estimator, then

$$\mathscr{L}(\mathbf{x}; c) = \Pr(c = \hat{c}) \Pr_{c=\hat{c}}(\mathbf{x}) \quad (30.33)$$

where the second term is the *conditional probability* of getting **x** given that the parameter $c = \hat{c}$. We may write $h(\hat{c}; c) = \Pr(c = \hat{c})$ and $g(\mathbf{x}; \hat{c}) = \Pr_{c=\hat{c}}(\mathbf{x})$, which shows the *converse.* Therefore, **c is a *sufficient statistic* if and only if the factorization is possible.**

► Example 30.14

Sampling with the geometric distribution

Suppose we consider a sequence of *independent dichotomous* trials, each with only two possible outcomes, success or failure, with constant probabilities p and $(1-p)$, respectively, where $0 < p \le 1$. If we ask how many trials, X, are required to achieve a first success, the (discrete) distribution of X is given by

$$\Pr(X = x; p) = (1-p)^{x-1} p = f(x; p)$$

This is called the *geometric distribution* with parameter p.

Suppose we wish to estimate the parameter p, and we perform this random measurement N times noting the data $\mathbf{x} = x_1, \ldots, x_N$.

The *likelihood function* is

$$\mathscr{L}(\mathbf{x}; p) = \prod_{i=1}^{N} [(1-p)^{x_i - 1} p]$$

$$= p^N (1-p)^{\left[\sum_{i=1}^{N}(x_i - 1)\right]}$$

$$= p^N (1-p)^{-N} (1-p)^{\left[\sum_{i=1}^{N} x_i\right]}$$

If we set

$$g(\mathbf{x}; \hat{c}) = 1$$

and

$$h(\hat{c}; c) = \left(\frac{p}{1-p}\right)^N (1-p)^{\left[\sum_{i=1}^{N} x_i\right]}$$

we satisfy (30.32) and conclude that

$$\hat{c}(\mathbf{x}) = \sum_{i=1}^{N} x_i$$

is a *sufficient statistic* to estimate p ($\hat{p}$ is a function of Σx_i). ◄

► Example 30.15

Sampling from a normal distribution with unknown mean and variance

If $\mathbf{x} = x_1, x_2, \ldots, x_n$ is a random sample of size n drawn from a normal population $N(\mu, \sigma^2)$ with $\mathbf{c} = (\mu, \sigma^2)$ the parameters we wish to estimate, then

$$\mathscr{L}(x; \mu, \sigma^2) = (2\pi\sigma^2)^{-n/2} \times \left\{\exp\left[-\frac{1}{2\sigma^2}\sum_{i=1}^{n}(x_i - \bar{x})^2 + n(\bar{x} - \mu)^2\right]\right\}.$$

The requirements of the *factorization theorem* are met where $h(\hat{c}; c)$ depends on the data only through the statistics, $[\bar{x}, \Sigma(x_i - \bar{x})^2] = \mathbf{t}_1$ or, equivalently, through the statistics, $(\Sigma_{i=1}^n x_i, \Sigma_{i=1}^n x_i^2) = \mathbf{t}_2$. Either of these statistics, $\mathbf{t}_i$, is sufficient. *Any* **function** *of these statistics is a* **sufficient statistic**. So we also have $(\bar{x}, s^2)$, the *sample mean* and the *sample variance*, as sufficient statistics. For a sample drawn from a normal distribution, therefore, we need keep only the *sample mean* and the *sample variance.* This applies only to *normally* distributed samples, however. ◄

► Example 30.16

Sampling from a normal distribution with unit variance

If $\mathbf{x} = x_1, \ldots, x_n$ is a random sample drawn from a normal distribution with (*known*) unit variance, $N(\mu, 1)$, the parameter we wish to estimate is $c = \mu$.

Here

$$f(x_i; \mu) = \frac{1}{\sqrt{2\pi}} e^{-(x_i - \mu)^2/2}$$

and

$$\mathscr{L}(\mu ; \mathbf{x}) = \prod_{i=1}^{n} f(x_i ; \mu)$$
$$= (2\pi)^{-n/2} e^{-\Sigma(x_i-\mu)^2/2}$$
$$= [(2\pi)^{-n/2} e^{-\Sigma x_i^2/2}]\{e^{-1/2[n\mu^2-2\mu\Sigma x_i]}\}$$

Here $h(\hat{c}; c) = \{e^{-1/2[n\mu^2-2\mu\Sigma x_i]}\}$ and a *sufficient statistic* for μ is $\Sigma_{i=1}^{n} x_i$ or, equivalently, the *sample mean*

$$\bar{x} = \frac{1}{n}\sum_{i=1}^{n} x_i$$

is a sufficient estimator for μ. The statistic $\Sigma_{i=1}^{n} x_i^2$ is not relevant to the estimation of μ. ◄

A *minimal-sufficient statistic* has the property that it is a function of all possible *sufficient statistics.*

Note that we have not proved that a sufficient statistic always exists or that a minimum variance estimator always exists. If either of these does exist, however, *the maximum likelihood method will find it.*

It is usually harder to show that a statistic T is *not* sufficient since it is necessary to show that $f(\mathbf{x}; c)$ *cannot* be factored according to (30.32). It is frequently easier to show that T is not sufficient by showing that the *conditional distribution* of $\mathbf{X}$, given $T = t$, **depends** on c.

► Example 30.17
Excluding data

Consider a random sample $(x_1, \ldots, x_n)$ of independent points from a discrete population with frequency function $p(\mathbf{x}; c)$.

Consider the statistic $\mathbf{T} = (X_1, \ldots, X_{n-1})$, that is, exclude the nth data point. The *conditional probability* of $\mathbf{X} = \mathbf{x}$ given that $\mathbf{T} = \mathbf{t} = (t_1, \ldots, t_{n-1})$ is

$$\Pr_{\mathbf{T}=\mathbf{t}}(\mathbf{X} = \mathbf{x}) = \frac{\Pr(\mathbf{X} = \mathbf{x})}{\Pr(\mathbf{T} = x_1, \ldots, x_{n-1})}$$
$$= \Pr(X_n = x_n)$$

since the observations are independent.

But $\Pr(X_n = x_n) = p(x_n; c)$, which **depends** on the parameter c. Therefore, the factorization (30.33) fails and the statistic T is *not* a sufficient statistic. Also, any function of T alone is not sufficient. In general, any time an observation is discarded what remains is not sufficient, that is, something has been lost. ◄

INFORMATION FOR r PARAMETERS $\mathbf{c} = c_1, \ldots, c_r$

We have defined the "*information*" for one parameter c

$$I(c) = E\left[\left\{\frac{\partial}{\partial c}\ln f(\mathbf{x}; c)\right\}^2\right] \tag{30.16}$$

$$= E\left[\left\{\frac{-\partial^2}{\partial c^2}\ln f(\mathbf{x}; c)\right\}\right] \tag{30.21}$$

where

$$\mathrm{Var}(\hat{c}) \geq \frac{1}{I(c)} \tag{30.17}$$

If we have r parameters $\mathbf{c} = c_1, \ldots, c_r$, we are interested in the r^2 quantities

$$\sigma_{ij} \qquad i,j = 1, \ldots, r$$

where

$$\sigma_{ii} = \sigma_i^2 = \mathrm{Var}(\hat{c}_i)$$

and

$$\sigma_{ij} = \sigma_{ji} = \mathrm{Covar}(\hat{c}_i, \hat{c}_j)$$

and

$$\hat{\mathbf{c}} = \hat{c}_1, \ldots, \hat{c}_r$$

are the unbiased estimators of the parameters $\mathbf{c}$.

The σ_{ij} form the $(r \times r)$ **covariance matrix** $(\boldsymbol{\sigma})$.

We may define the elements of the $(r \times r)$ **information matrix** $(\underline{\mathbf{A}})$:

$$(\underline{\mathbf{A}})_{ij} = E\left[\left\{\frac{\partial}{\partial c_i}\ln f(\mathbf{x}; \mathbf{c})\right\}\left\{\frac{\partial}{\partial c_j}\ln f(\mathbf{x}; \mathbf{c})\right\}\right] \tag{30.34}$$

$$= -E\left[\frac{\partial^2 \ln f(\mathbf{x}; \mathbf{c})}{\partial c_i\, \partial c_j}\right] \tag{30.35}$$

for most likelihood functions with the features used in the derivation of (30.21).

The *generalization of the Cramer-Rao-Frechet inequality* is in terms of the covariance matrix and the information matrix:

$$(\underline{\sigma}) \geq (\underline{A}^{-1}) \qquad (30.36)$$

that is, each element of the *inverse* of the information matrix acts as a lower bound for the corresponding element of the covariance matrix for the estimators $\hat{\mathbf{c}}$.

If we can find a set of unbiased estimators $\hat{\mathbf{c}}$ such that

$$(\underline{\sigma}) = (\underline{A}^{-1}) \qquad (30.37)$$

then these are *most efficient* statistics.

►Example 30.18
Sampling from a normal distribution

Let us continue the discussion of Examples 30.9 and 30.15 for a sample of size n drawn from a normal distribution where it is required to estimate μ and σ^2.

We recall

$$\frac{\partial \ell}{\partial \mu} = \sum_{i=1}^{n} \frac{(x_i - \mu)}{\sigma^2}$$

and

$$\frac{\partial \ell}{\partial (\sigma^2)} = \frac{-n}{2\sigma^2} - \frac{1}{2}\sum_{i=1}^{n} (x_i - \mu)^2 \left[\frac{-1}{(\sigma^2)^2}\right]$$

yielding

$$\hat{\mu} = \bar{x} = \frac{1}{n}\sum_{i=1}^{n} x_i$$

and

$$\widehat{\sigma^2} = \frac{1}{n}\sum_{i=1}^{n} (x_i - \bar{x})^2$$

We calculate

$$\frac{\partial^2 \ell}{\partial \mu^2} = -\sum_{i=1}^{n} \frac{1}{\sigma^2} = -\frac{n}{\sigma^2}$$

$$= \left(\frac{\partial \ell^2}{\partial \mu^2}\bigg|_{\hat{\mu}}\right) = E\left(\frac{\partial^2 \ell}{\partial \mu^2}\right)$$

$$\frac{\partial^2 \ell}{\partial \mu \partial (\sigma^2)} = \frac{-1}{(\sigma^2)^2}\sum_{i=1}^{n} (x_i - \mu)$$

$$= \frac{-1}{(\sigma^2)^2}\left(\sum_{i=1}^{n} x_i - n\mu\right)$$

$$\frac{\partial^2 \ell}{\partial (\sigma^2)^2} = \frac{+n}{2(\sigma^2)^2} - \sum \frac{(x_i - \mu)^2}{(\sigma^2)^3}$$

The ***information matrix*** uses the elements

$$\underline{A} = \begin{pmatrix} -E\left(\dfrac{\partial \ell^2}{\partial \mu^2}\right) & -E\left(\dfrac{\partial^2 \ell}{\partial \mu\, \partial \sigma^2}\right) \\ -E\left(\dfrac{\partial^2 \ell}{\partial \mu\, \partial \sigma^2}\right) & -E\left(\dfrac{\partial^2 \ell}{\partial (\sigma^2)^2}\right) \end{pmatrix}$$

If we assume that the distribution $\ell(\mu, \sigma^2)$ is *symmetrical*, then

$$E\left(\frac{\partial^2 \ell}{\partial \mu^2}\right) \approx \left(\frac{\partial^2 \ell}{\partial \mu^2}\right)\bigg|_{\sigma^2=\widehat{\sigma^2}} = -\frac{n}{\widehat{\sigma^2}}$$

$$E\left(\frac{\partial^2 \ell}{\partial \mu \partial \sigma^2}\right) \approx \left(\frac{\partial^2 \ell}{\partial \mu \partial \sigma^2}\right)\bigg|_{\sigma^2=\widehat{\sigma^2}} = 0$$

$$E\left(\frac{\partial^2 \ell}{\partial (\sigma^2)^2}\right) \approx \left(\frac{\partial^2 \ell}{\partial (\sigma^2)^2}\right)\bigg|_{\sigma^2=\widehat{\sigma^2}}$$

$$= \frac{n}{2(\widehat{\sigma^2})^2} - \frac{n\widehat{\sigma^2}}{(\widehat{\sigma^2})^3}$$

$$= \frac{n}{2(\widehat{\sigma^2})^2}$$

We shall see that the *asymptotic properties* of the likelihood function are such that a large n guarantees that

$$E\left(\frac{\partial^2 \ell}{\partial c_1 \partial c_2}\right) \approx \left(\frac{\partial^2 \ell}{\partial c_1 \partial c_2}\right)\bigg|_{\hat{c}_1, \hat{c}_2}$$

(Plotting the likelihood function when n is **not** clearly large may reveal symmetry *anyway*. If not, then the information matrix should use the appropriate *average* of each of the second derivatives.)

The information matrix is then

$$\underline{A} = \begin{pmatrix} \dfrac{n}{\widehat{\sigma^2}} & 0 \\ 0 & \dfrac{n}{2(\widehat{\sigma^2})^2} \end{pmatrix}$$

with

$$\det \underline{A} = \frac{n^2}{2(\widehat{\sigma^2})^3}$$

The *inverse* of the information matrix is

$$\underline{A}^{-1} = \frac{2(\widehat{\sigma^2})^3}{n^2}\begin{pmatrix} \dfrac{n}{2(\widehat{\sigma^2})^2} & 0 \\ 0 & \dfrac{n}{\widehat{\sigma^2}} \end{pmatrix} = \begin{pmatrix} \dfrac{\widehat{\sigma^2}}{n} & 0 \\ 0 & \dfrac{2(\widehat{\sigma^2})^2}{n} \end{pmatrix}$$

and

$$\operatorname{Var}(\hat{\mu}) \geq \frac{\widehat{\sigma^2}}{n}$$

$$\operatorname{Var}(\widehat{\sigma^2}) \geq \frac{2(\widehat{\sigma^2})^2}{n}$$

and

$$\mathrm{Covar}(\hat{\mu}, \widehat{\sigma^2}) \geq 0$$

Because $\partial\ell/\partial\mu$ may be cast into the form of (30.23),

$$\mathrm{Var}(\hat{\mu}) = \frac{\widehat{\sigma^2}}{n}$$

For the normal distribution,

$$\mathrm{Var}(\sigma^2) = \frac{2(\sigma^2)^2}{n-1}$$

and so the *inequality holds.* ◄

► **Example 30.19**
Tossing a fat coin

Consider a "fat coin" that is sufficiently like a cylinder that we must consider the appreciable probability that it lands "*on edge*" (E) in addition to landing "*heads*" (H) or "*tails*" (T). Consider a large number N of independent trials in which the coin is tossed *randomly* with the results $(n_E, n_H, n_T) = \mathbf{n}$ being the number of E's, H's, and T's, respectively, where

$$n_E + n_H + n_T = N$$

The probability of getting a head is p_H, that of getting a tail is p_T and that of landing on edge p_E, where

$$p_E = 1 - p_H - p_T$$

and we wish to estimate

$$\mathbf{p} = (p_H, p_T)$$

The *likelihood function* is given by the *multinomial probability* (28.6):

$$\mathscr{L}(\mathbf{n}; \mathbf{p}) = \frac{N!}{n_E!n_H!n_T!} p_H^{n_H} p_T^{n_T} \times (1 - p_H - p_T)^{n_E},$$

and the *log likelihood* by

$$\ell(\mathbf{n}; \mathbf{p}) = \ln\left(\frac{N!}{n_E!n_H!n_T!}\right) + n_H \ln p_H + n_T \ln p_T + n_E \ln(1 - p_H - p_T)$$

The *first derivatives*

$$\frac{\partial\ell}{\partial p_H}(\mathbf{n}; \mathbf{p}) = \frac{n_H}{p_H} - \frac{n_E}{1 - p_H - p_T}$$

$$\frac{\partial\ell}{\partial p_T}(\mathbf{n}; \mathbf{p}) = \frac{n_T}{p_T} - \frac{n_E}{1 - p_H - p_T}$$

may be set to zero and solved for the *natural estimators*

$$\hat{p}_H = \frac{n_H}{N}$$

and

$$\hat{p}_T = \frac{n_T}{N}$$

The *second derivatives* are

$$\frac{\partial^2\ell}{\partial p_H^2} = \frac{-n_H}{p_H^2} - \frac{n_E}{(1 - p_H - p_T)^2}$$

$$\frac{\partial^2\ell}{\partial p_H \partial p_T} = \frac{-n_E}{(1 - p_H - p_T)^2}$$

and

$$\frac{\partial^2\ell}{\partial p_T^2} = \frac{-n_T}{p_T} - \frac{n_E}{(1 - p_H - p_T)^2}$$

To find the **information matrix** we must take the *expectation value* of the derivatives noting that

$$E(n_H) = Np_H$$

and

$$E(n_E) = N(1 - p_H - p_T)$$

Therefore,

$$-E\left[\frac{\partial^2\ell}{\partial p_H^2}\right] = N\left[\frac{1}{p_H} + \frac{1}{1 - p_H - p_T}\right]$$

$$-E\left[\frac{\partial^2\ell}{\partial p_H \partial p_T}\right] = \frac{N}{(1 - p_H - p_T)}$$

and

$$-E\left[\frac{\partial^2\ell}{\partial p_T^2}\right] = N\left[\frac{1}{p_T} + \frac{1}{1 - p_H - p_T}\right]$$

The **information matrix** is

$$\underline{\mathbf{A}} = N\begin{pmatrix} \frac{1}{p_H} + \frac{1}{1 - p_H - p_T} & \frac{1}{1 - p_H - p_T} \\ \frac{1}{1 - p_H - p_T} & \frac{1}{p_T} + \frac{1}{1 - p_H - p_T} \end{pmatrix}$$

and its *inverse matrix* is

$$(\underline{\mathbf{A}}^{-1}) = \frac{1}{N}\begin{pmatrix} p_H(1 - p_H) & -p_H p_T \\ -p_H p_T & p_T(1 - p_T) \end{pmatrix}$$

so that

$$\text{Var}(\hat{p}_H) \geq p_H(1-p_H)$$
$$\text{Var}(\hat{p}_T) \geq p_T(1-p_T)$$

and

$$\text{Covar}(\hat{p}_H, p_T) \geq -p_H p_T$$

In this case, the estimators $\hat{p}_H$, $\hat{p}_T$ are *minimum variance estimators* and the equality sign obtains,

$$\underline{\sigma} = \underline{\mathbf{A}}^{-1}$$ ◄

Nonanalytical Solution of the Likelihood Equation

It frequently happens that the likelihood equation is not solvable **analytically**. In such cases we use *numerical* or *graphical* methods to obtain the maximum likelihood estimator. The method is illustrated in the following example, which also shows how to fold in such experimental effects as *finite resolution.*

► **Example 30.20**
Analysis of Lifetime Data Taken with Finite Resolution

We may consider the measurement of a system undergoing exponential decay so as to yield a measurement of the *lifetime*, or, equivalently the *decay rate*, λ. The analysis is similar to one invented by Peierls.*

Suppose that we wish to find λ where

λ = decay rate
N = the number of systems decaying

We record the times of decay in discrete intervals of width Δ centered on t_i, that is, if a decay occurs at a time t such that

$$t_i - \frac{\Delta}{2} \leq t \leq t_i + \frac{\Delta}{2}$$

then the count goes into the ith channel. The data consist of a series of numbers n_i where each i corresponds to a time, t_i. Clearly an exponential decay could be observed forever so that, practically speaking, there is a maximum time of observation for any finite experiment. We shall call this T. This means that we do not observe all the N systems to decay. We record only those that decay in $t \leq T$.

The expected value for the number of counts in the ith channel is

$$\langle n_i \rangle = \lambda N \int_{t_i-\Delta/2}^{t_i+\Delta/2} e^{-\lambda t_i}\, dt_i$$
$$= -N[e^{-\lambda t_i - \lambda\Delta/2} - e^{-\lambda t_i + \lambda\Delta/2}]$$

which we may rewrite, using the definition of the *hyperbolic function*, as

$$\langle n_i \rangle = 2Ne^{-\lambda t_i} \sinh \frac{\lambda\Delta}{2} \qquad (30.38a)$$

The probability of getting n_i counts in the ith channel, where the *expected number* is (30.38a), is given by the Poisson distribution

$$P(n_i) = \frac{\langle n_i \rangle^{n_i} e^{-\langle n_i \rangle}}{n_i!}$$

$$\therefore P(n_i; \lambda) = \frac{[2Ne^{-\lambda t_i} \sinh \lambda\Delta/2]^{n_i} e^{-(2Ne^{-\lambda t_i}) \sinh \lambda\Delta/2)}}{n_i!}$$

We may form the likelihood function, the *a posteriori* probability of a specific sequence of n_i in k successive channels i:

$$\mathcal{L}(\lambda) = \prod_{i=1}^{k} P(n_i; \lambda)$$
$$= \left[\prod_{i=1}^{k} \frac{(2N)^{n_i} \sinh^{n_i} \lambda\Delta/2}{n_i!}\right]$$
$$\times \left[\exp\left(-\lambda \sum_{i=1}^{k} n_i t_i\right)\right]$$
$$\times \exp\left(-2N \sinh \frac{\lambda\Delta}{2} \sum_{i=1}^{k} e^{-\lambda t_i}\right)$$
$$= \left[\prod_{i=1}^{k} \left(2N \sinh \frac{\lambda\Delta}{2}\right)^{n_i}\right]$$
$$\times \exp\left(-\lambda \sum_{i=1}^{k} n_i t_i\right) \exp\left(-\sum_{i=1}^{k} \langle n_i \rangle\right)$$

Proceedings of the Royal Society* (London) **A149, 467 (1935).

We note that

$$\sum_{i=1}^{k} \langle n_i \rangle = \sum_{i=1}^{k} \lambda N \int_{t_i-\Delta/2}^{t_i+\Delta/2} e^{-\lambda t_i} dt_i$$
$$= \lambda N \int_0^T e^{-\lambda t} dt$$
$$= N(1-e^{-\lambda T})$$

where T is the *largest time considered.*

We may define

$$s = \frac{\sum_{i=1}^{k} n_i t_i}{\sum_{i=1}^{k} n_i} = \frac{\left(\sum_{i=1}^{k} n_i t_i\right)}{m}$$

where

$$m = \sum_{i=1}^{k} n_i$$

is the *total number* of decays observed. Note that $m \neq N$ since we *stop* observing at $t = T$. Actually,

$$m = \sum_{i=1}^{k} n_i = N(1-e^{-\lambda T}) \quad (30.38b)$$

$$\therefore \mathscr{L}(\lambda) = \frac{\left[\prod_{i=1}^{k} (2N \sinh \lambda\Delta/2)^{n_i}\right] e^{-\lambda m s} e^{-N(1-e^{-\lambda T})}}{\prod_i^k n_i!}$$

and

$$\ln \mathscr{L} = \ell(\lambda) = \sum_{i=1}^{k} n_i \ln\left[2N \sinh \lambda\Delta/2\right]$$
$$- \lambda m s - N(1-e^{-\lambda T}) - \sum_{i=1}^{k} \ln(n_i!)$$
$$= m \ln 2N + m \ln \sinh \frac{\lambda\Delta}{2}$$
$$- \lambda m s - N(1-e^{-\lambda T}) - \sum_{i=1}^{k} \ln(n_i!)$$

We maximize $\ell(\lambda)$ with respect to λ:

$$\frac{d}{d\lambda}[\ell(\lambda)] = 0 = \frac{m \cosh(\lambda\Delta/2)}{\sinh \lambda\Delta/2}(\Delta/2) - ms - NTe^{-\lambda T}$$

$$\therefore \frac{\Delta}{2} \coth \frac{\lambda\Delta}{2} = s + \frac{NT}{m} e^{-\lambda T} = s + \frac{Te^{-\lambda T}}{1-e^{-\lambda T}}$$

using (30.38b).

T is a parameter chosen *a priori* while only s depends on the data. It suffices to *solve the equation*

$$\frac{\Delta}{2} \coth \frac{(\lambda\Delta)}{2} = s + \frac{T}{e^{\lambda T}-1} \quad (30.38c)$$

to get the *best estimate* λ^*.

We may cast this into a form suitable for *graphical* solution.

Express all times in units of Δ:

$$t_i = (i - 1/2)\Delta$$
$$T = k\Delta$$
$$s = \frac{\sum n_i (i-1/2)\Delta}{\sum n_i} = \frac{\sum i n_i \Delta}{\sum n_i} - \frac{\Delta}{2}$$

(30.38c) becomes

$$\frac{1}{2}\frac{e^{\lambda/2}+e^{-\lambda/2}}{e^{\lambda/2}-e^{-\lambda/2}} = \frac{\sum_{i=1}^{k} i n_i}{\sum_{i=1}^{k} n_i} - \frac{1}{2} + \frac{k}{e^{\lambda k}-1}$$

If we let

$$y = e^{\lambda}$$

we get

$$\frac{1}{2}\frac{y+1}{y+1} = \frac{\sum_{i=1}^{k} i n_i}{\sum_{i=1}^{k} n_i} - \frac{1}{2} + \frac{k}{y^k - 1}$$

or

$$\frac{1}{y-1} - \frac{k}{y^k-1} = \frac{\sum_{i=1}^{k} i n_i}{\sum_{i=1}^{k} n_i} - 1 \quad (30.38d)$$

Therefore, we need only plot

$$\frac{1}{y-1} - \frac{k}{y^k-1} \quad \text{against} \quad \lambda$$

and, from the data, we calculate

$$\frac{\sum_{i=1}^{k} i n_i}{\sum_{i=1}^{k} n_i} - 1$$

We read off the value of $\lambda = \lambda^*$, which is the maximum likelihood estimate from the data.

A general plot is shown in Figure 30.3. When looked at in detail, the plot becomes a straight line over a localized region and becomes independent of k for k large enough. Figure 30.4 shows a plot used in the analysis of a measurement of the lifetime of a particle called the negative muon. ◄

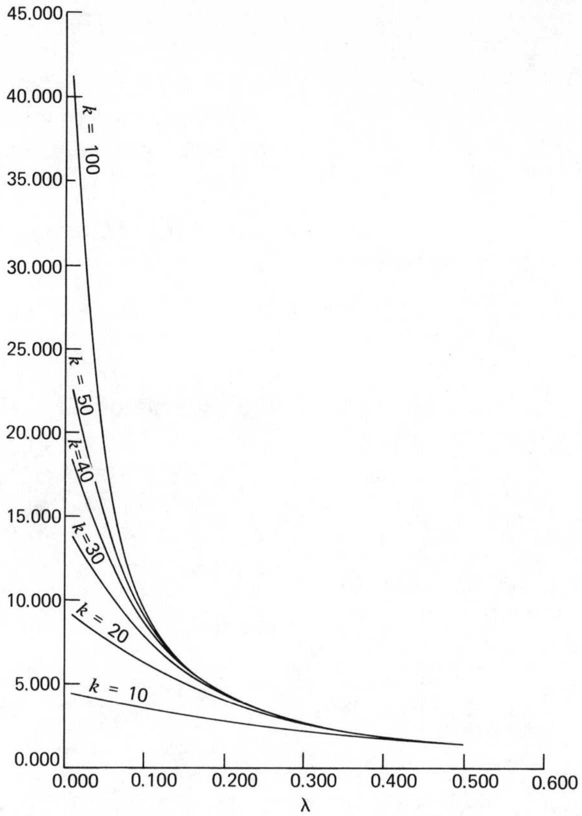

Figure 30.3
Plot of $1/(e^{\lambda}-1)-k/(e^{k\lambda}-1)$ versus λ.

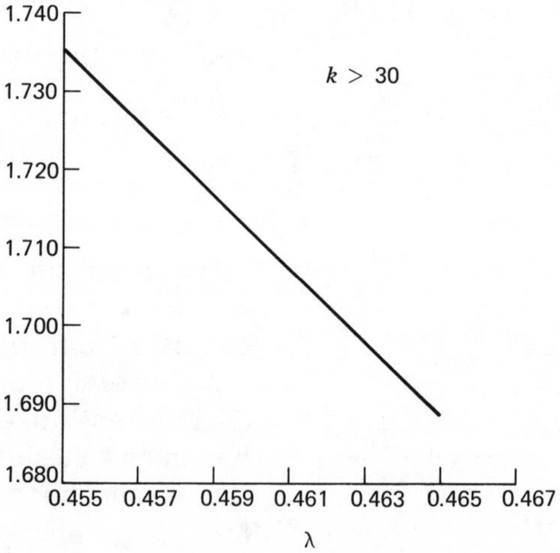

Figure 30.4
Enlarged view over a localized region of the plot shown in Figure 30.3. This plot was used in a measurement of the lifetime of the muon.

Asymptotic Properties of the Likelihood Function

Among the most important features of the method of maximum likelihood is the behavior of the likelihood function as the number of data points, n, becomes large.

ONE PARAMETER

We have

$$\ell(\mathbf{x}; c) = \sum_{i=1}^{n} \ln f_i(\mathbf{x}; c)$$

and set

$$\ell'(\mathbf{x}; \hat{c}) = \sum_{i=1}^{n} \varphi_i(\mathbf{x}; \hat{c}) = 0 \quad (30.3a)$$

to define the maximum likelihood estimator $\hat{c}$ by (30.4) to (30.6).

Assuming the existence of higher derivatives we may *expand* $\ell'(\mathbf{x}; c)$ about $c = \hat{c}$:

$$\begin{aligned}\ell'(\mathbf{x}; c) &= \ell'(\hat{c}) + (c - \hat{c})\ell''(\hat{c}) + \cdots \\ &= (c - \hat{c})\ell''(\hat{c}) + \cdots\end{aligned}$$

since the first term vanishes by (30.3a).

Now

$$\ell''(\hat{c}) = \sum_{i=1}^{n} \left[\frac{\partial}{\partial c} \varphi_i(\mathbf{x}; c)\right]_{c=\hat{c}}$$

Consider the expectation value

$$\begin{aligned}E\left[\frac{\partial}{\partial c} \varphi_i(\mathbf{x}; c)\right] &= E\left[\frac{\partial^2}{\partial c^2} \ln f(\mathbf{x}; c)\right] \\ &= E[\ell''(\mathbf{x}; c)] \\ &= -E\left\{\left[\frac{\partial}{\partial c} \ln f(\mathbf{x}; c)\right]^2\right\} \\ &= -I(c)\end{aligned}$$

by (30.21).

For large n we may use the *sample mean* for the *expectation value*

$$E\left[\frac{\partial}{\partial c} \varphi(\mathbf{x}; c)\right] \approx \frac{1}{n} \sum_{i=1}^{n} \left(\frac{\partial}{\partial c} \varphi_i(\mathbf{x}; c)\bigg|_{\hat{c}}\right) \quad (30.39)$$

$$\begin{aligned}\therefore \ell''(\hat{c}) &\approx -nI(c) \\ &= -I_n(c)\end{aligned}$$

by (30.18).

The Taylor series expansion of $\ell'(c)$, neglecting higher order terms, then yields

$$\begin{aligned}\ell'(c) &\approx (c - \hat{c})\ell''(\hat{c}) \\ &= (\hat{c} - c)I_n(c) \quad (30.40)\end{aligned}$$

which shows that $\hat{c}$, **for large n**, is a *minimum-variance unbiased estimator* since it fulfills the condition (30.25). We emphasize that $I_n(c)$ does not depend on the data but only on the form of the posterior probability distribution f and the estimator $\hat{c}$. This is a most important result. Although the maximum likelihood estimator may be biased for finite n, we now see that the *estimator is consistent*, that is, $\hat{c} = E(c)$ for large n.

If we set

$$I_n(c) = \frac{1}{\sigma_0^2} = \text{constant}$$

then, for n large, we have

$$\ell'(c) \approx -(c - \hat{c})\frac{1}{\sigma_0^2}$$

which may be integrated to yield

$$\ell(c) = -\frac{1}{2\sigma_0^2}(c - \hat{c}) + \ln K$$

or

$$\mathcal{L}(c) \approx K e^{-(c-\hat{c})^2/2\sigma_0^2} \quad (30.41)$$

which implies that, **for large n,** *the likelihood function approaches a normal distribution* with **mean** $\hat{c}$ and **variance**

$$\sigma_0^2 = [I_n(c)]^{-1} = -\left(\frac{d^2\ell}{dc^2}\bigg|_{\hat{c}}\right)^{-1} \quad (30.42)$$

Note that $\sigma^2(\hat{c}) = [I_n(c)]^{-1}$ holds for all n if $\hat{c}$ is a minimum variance estimator satisfying (30.25) for all n. What is added here is that (30.42) holds for *any* maximum likelihood estimator, provided n is large enough. We say that the likelihood function is *asymptotically normal* in the large sample limit. The estimator is also *asymptotically efficient* because of (30.40).

SEVERAL PARAMETERS

For more than one parameter we have the *log likelihood function*

$$\ell(\mathbf{x}; \mathbf{c}) = \sum_{i=1}^{n} \ln f_i(\mathbf{x}; \mathbf{c})$$

where there are r parameters of interest

$$\mathbf{c} = (c_1, \ldots, c_r)$$

Again assuming the existence of first and higher derivatives we have the r equations

$$\left(\frac{\partial \ell}{\partial c_j}\right)\bigg|_{c_j=\hat{c}_j} = 0 \qquad j = 1, \ldots, r$$

to define the likelihood estimators

$$\hat{\mathbf{c}} = \hat{c}_1, \ldots, \hat{c}_r$$

and the Taylor series expansion about $\hat{\mathbf{c}}$:

$$\begin{aligned}\ell(\mathbf{x}; \mathbf{c}) &= \ell(\hat{\mathbf{c}}) + \sum_{j=1}^{r} \left(\frac{\partial \ell}{\partial c_j}\bigg|_{c_j=\hat{c}_j}\right)(c_j - \hat{c}_j) \\ &\quad - \frac{1}{2}\sum_{k,\ell=1}^{r} \left(\frac{\partial^2 \ell}{\partial c_k\,\partial c_\ell}\bigg|_{\hat{c}_k,\hat{c}_\ell}\right)(c_k - \hat{c}_k) \\ &\quad \times (c_\ell - \hat{c}_\ell) + \cdots \\ &= \ell(\hat{\mathbf{c}}) - \frac{1}{2}\sum_{k,\ell=1}^{r} \left(\frac{\partial^2 \ell}{\partial c_k\,\partial c_\ell}\bigg|_{\hat{c}_k,\hat{c}_\ell}\right) \\ &\quad \times (c_k - \hat{c}_k)(c_\ell - \hat{c}_\ell) + \cdots\end{aligned}$$

Again we consider that, for n large, we may make the replacement

$$\frac{\partial^2 \ell}{\partial c_k\,\partial c_\ell}\bigg|_{\hat{c}_k,\hat{c}_\ell} \approx E\left(\frac{\partial^2 \ell}{\partial c_k\,\partial c_\ell}\right) \tag{30.43}$$

The right-hand side of (30.43) does not formally depend on the data, representing the appropriate *average over all possible data.*

$$\therefore E\left(\frac{\partial^2 \ell}{\partial c_k\,\partial c_\ell}\right) = A_{k\ell} \tag{30.44}$$

and we may write

$$\ell(\mathbf{x}; \mathbf{c}) \approx \ell(\hat{c}) - \frac{1}{2}\sum_{k=1}^{r}\sum_{\ell=1}^{r} A_{k\ell}(c_k - \hat{c}_k) \times (c_\ell - \hat{c}_\ell)$$

or

$$\mathcal{L}(\mathbf{x}; c) \approx K \exp\left[-\frac{1}{2}\sum_{k=1}^{r}\sum_{\ell=1}^{r} A_{k\ell}(c_k - \hat{c}_k) \times (c_\ell - \hat{c}_\ell)\right] \tag{30.45}$$

where $K = e^{\ell(\hat{c})}$, which has the form of (28.41), the *multivariate normal distribution.* The $A_{k\ell}$ form a square matrix $\underline{\mathbf{A}}$ which is $(r \times r)$. The *inverse* of this matrix is the *covariance matrix* by (28.42):

$$(\underline{\boldsymbol{\sigma}})_{k\ell} = (\underline{\mathbf{A}}^{-1})_{k\ell} \tag{30.46}$$

Finite Data

In the *asymptotic limit* of a Gaussian likelihood function,

$$-\ell''(\mathbf{x}; c)$$

is a constant, independent of c. The various alternative formulas for $\sigma^2(\hat{c})$ are then all equivalent

$$\sigma^2(\hat{c}) \approx -[\ell''(\mathbf{x}; \hat{c})]^{-1} \tag{30.42}$$

$$\approx \frac{\displaystyle\int_{c_{\min}}^{c_{\max}} (c - \hat{c})^2 \mathcal{L}(\mathbf{x}; c)\, dc}{\displaystyle\int_{c_{\min}}^{c_{\max}} \mathcal{L}(\mathbf{x}; c)\, dc} \tag{30.47}$$

$$\approx \left\{-\left[\frac{d^2\ell}{dc^2}\right]_{\text{avg}}\right\}^{-1} \tag{30.48}$$

where

$$\left[\frac{d^2\ell}{dc^2}\right]_{\text{avg}} = \frac{\displaystyle\int_{c_{\min}}^{c_{\max}} \left(\frac{d^2\ell}{dc^2}\right) \mathcal{L}(\mathbf{x}; c)dc}{\displaystyle\int_{c_{\min}}^{c_{\max}} \mathcal{L}(\mathbf{x}; c)dc} \tag{30.49}$$

All of these may be read off from the data, that is, from a plot of the *experimental* likelihood function. If the likelihood function does not look Gaussian, then it is somewhat unclear how to interpret the variance of the estimator $\hat{c}$. However, a general "rule of thumb" is that (30.48) and (30.49) may still be used when (30.42) should be avoided.

CONFIDENCE INTERVALS

We should note the applicability of *confidence intervals* (Chapter 29) to *maximum likelihood estimators.* The probability that an interval between estimates c_1 and c_2 includes the true value of the parameter c is

$$\Pr\{c_1 \le c \le c_2\}$$

$$= \frac{\displaystyle\int_{c_1}^{c_2} \mathcal{L}(c)dc}{\displaystyle\int_{-\infty}^{\infty} \mathcal{L}(c)dc} = \frac{\displaystyle\int_{c_1}^{c_2} \mathcal{L}(c)dc}{\displaystyle\int_{c_{\min}}^{c_{\max}} \mathcal{L}(c)dc} \tag{30.50}$$

where $\mathscr{L}(c < c_{min}) = \mathscr{L}(c > c_{max}) = 0$. If $\Pr\{c_1 \le c \le c_2\}$ is chosen equal to 0.68, then the limits c_1, c_2 correspond to *one standard deviation* in the usual sense. If the likelihood function is *symmetrical*, we may write the maximum likelihood estimate and its "error" as

$$\hat{c} \pm \Delta$$

where $c_1 = \hat{c} - \Delta$ and $c_2 = \hat{c} + \Delta$. Although any function $\lambda = \lambda(c)$ of the parameter c will have its maximum likelihood at the same point by (30.4a).

$$\hat{\lambda} = \lambda(\hat{c})$$

the confidence interval will be different since

$$\Pr(\lambda_1 \le \lambda \le \lambda_2) = \frac{\int_{\lambda_1}^{\lambda_2} \mathscr{L}(\lambda)d\lambda}{\int_{-\infty}^{\infty} \mathscr{L}(\lambda)d\lambda} = \frac{\int_{c_1}^{c_2} \mathscr{L}(c)\frac{\partial\lambda}{\partial c}\,dc}{\int_{-\infty}^{\infty} \mathscr{L}(\lambda)d\lambda}$$

and this, in general, is not equal to (30.50). In particular, if the parameter c has a symmetrical confidence interval, the parameter $\lambda = \lambda(c)$ in general does not, and we write the one standard deviation limits as

$$\hat{\lambda}^{+\Delta_+}_{-\Delta_-}$$

where Δ_+, Δ_- *need not be equal.*

Error Expected Prior to a Measurement

We may write an *a priori* estimate of the "**error**" to be expected in a proposed measurement of $\hat{c}$ by generating n data points using *Monte Carlo simulation* for various values of $\hat{c}$ and applying formulas (30.42), (30.47), or (30.48). This is frequently desirable in any event because it is the most straightforward means of including such effects as *finite resolution* and *experimental efficiency* functions.

Another means of estimating the **error** to be expected *a priori* from an experimental measurement may be derived if the **probability density function** $f(\mathbf{x}; c)$ is known (in principle including any significant effects of resolution or efficiency). We wish to estimate

$$E[\ell''(\mathbf{x}; c)]$$

by averaging over all possible data.

Assume $\mathbf{x} = x$ for simplicity. For one measurement

$$E\left[\frac{\partial^2 \ell}{\partial c^2}\right] = \int_{x_{min}}^{x_{max}} \left[\frac{\partial^2 \ln f}{\partial c^2}\right] f(x; c)dx$$

while for n events we have

$$E\left[\frac{\partial^2 \ell}{\partial c^2}\right] = \sum_{i=1}^{n} \int_{x_{min}}^{x_{max}} \left[\frac{\partial^2 \ln f_i}{\partial c^2}\right] f_i(x; c)dx$$

$$= n \int_{x_{min}}^{x_{max}} \left[\frac{\partial^2 \ln f}{\partial c^2}\right] f(x; c)dx$$

We recall relation (30.21) for

$$\int_{x_{min}}^{x_{max}} f(x; c)dx = 1$$

$$E\left[\frac{\partial^2}{\partial c^2}\ln f(x; c)\right] = -E\left[\left(\frac{\partial}{\partial c}\ln f(x; c)\right)^2\right] = -E\left[\left(\frac{1}{f(x; c)}\frac{\partial}{\partial c}f(x; c)\right)^2\right]$$

$$\therefore E\left[\frac{\partial^2 \ell}{\partial c^2}\right] = -\int_{x_{min}}^{x_{max}} \frac{1}{f(x; c)}\left(\frac{\partial f(x; c)}{\partial c}\right)^2 dx$$

for *one* measurement or

$$E\left[\frac{\partial^2 \ell}{\partial c^2}\right] = -n \int_{x_{min}}^{x_{max}} \frac{1}{f(x; c)}\left(\frac{\partial f(x; c)}{\partial c}\right)^2 dx$$

for n measurements.

$$\therefore \sigma^2(\hat{c}) = \frac{1}{n}\left\{\int_{x_{min}}^{x_{max}} \frac{1}{f(x; c)}\left[\frac{\partial f(x; c)}{\partial c}\right]^2 dx\right\}^{-1} \tag{30.51}$$

which yields an *a priori* estimate of the **variance** of $\hat{c}$ to be *expected* from a measurement involving n data points.

► **Example 30.21**

Suppose we have a population described by

$$f(x;a)=(1+ax)k; \qquad -1\le x\le +1$$
$$-1\le a\le +1$$

and it is required to estimate a with n measurements. How many measurements should be made in order that

$$\frac{\sigma(\hat{a})}{a}\approx \epsilon\%?$$

We have

$$f(x;a)=\tfrac{1}{2}(1+ax)$$

since

$$k\int_{-1}^{+1}(1+ax)\,dx=1$$

$$\frac{\partial f}{\partial a}=\frac{x}{2}$$

and

$$\int_{-1}^{+1}\frac{1}{f}\left(\frac{\partial f}{\partial a}\right)^2 dx=\int_{-1}^{+1}\frac{1}{1/2(1+ax)}\frac{x^2}{4}\,dx$$
$$=\frac{1}{2}\int_{-1}^{+1}\frac{x^2}{1+ax}\,dx$$
$$=\frac{1}{2a^3}[1/2(1+ax)^2-2(1+ax)$$
$$+\ln(1+ax)]_{-1}^{+1}$$
$$=\frac{1}{2a^3}\left[-2(1+a)+2(1-a)+\ln\left(\frac{1+a}{1-a}\right)\right]$$
$$=\frac{1}{2a^3}\left[\ln\left(\frac{1+a}{1-a}\right)-2a\right]$$
$$\therefore \sigma^2(\hat{a})=\frac{1}{n}\left\{\frac{1}{2a^3}\left[\ln\left(\frac{1+a}{1-a}\right)-2a\right]\right\}^{-1}$$

or

$$\frac{\sigma(\hat{a})}{a}=\frac{1}{\sqrt{n}}\left[\frac{2a}{\ln\left(\frac{1+a}{1-a}\right)-2a}\right]^{1/2}$$
$$=\frac{\epsilon}{100}$$
$$\therefore n=\frac{10^4}{\epsilon^2}\left[\frac{2a}{\ln\left(\frac{1+a}{1-a}\right)-2a}\right]^{1/2}$$

The *required number* of measurements, n, depends on the *state of nature*, that is, the value of a. It is, therefore, important to *guess* the magnitude of a to design the experiment effectively.

Since $a^2<1$ we may expand using

$$\ln\left(\frac{1+a}{1-a}\right)=2\left(a+\frac{a^3}{3}+\frac{a^5}{5}+\frac{a^7}{7}+\cdots\right)$$

which yields

$$\sigma^2(\hat{a})\approx\frac{1}{n}\left[\frac{1}{3}+\frac{a^2}{5}+\frac{a^4}{7}+\cdots\right]^{-1} \qquad (30.52)$$

for the *maximum likelihood estimator*. ◄

Graphical Methods: The Score Function

It is often convenient to use a statistic other than the *maximum likelihood estimator*. When this is the case it is desirable to know how inefficient the convenient statistic is.

► **Example 30.22**

Convenient statistics

Consider the case of Example 30.21 where $x=\cos\theta$ and the maximum likelihood estimator $\hat{a}$ has variance given by (30.52). Define "*up*" to be all events such that $\cos\theta>0$, and "*down*" to be all events such that $\cos\theta<0$.

We may consider the statistic (the "*up-down asymmetry*")

$$\bar{\bar{a}}=\frac{n_{\text{up}}-n_{\text{down}}}{\frac{1}{2}(n_{\text{up}}+n_{\text{down}})}$$

as an estimator of a. *Experimentally* this is simpler than measuring all the x_i. We need only record if x_i is positive or negative.

We note that $\bar{\bar{a}}$ is an *unbiased* estimator of a:

$$E(\bar{\bar{a}})=\frac{E(n_{\text{up}})-E(n_{\text{down}})}{\frac{1}{2}n}$$
$$=\frac{n\,\Pr(x>0)-n\,\Pr(x<0)}{\frac{1}{2}n}$$
$$=2[\Pr(x>0)-\Pr(x<0)]$$

where

$$\Pr(x>0)=\int_0^{+1}\left(\frac{1+ax}{2}\right)dx=\frac{1}{4}(2+a)$$

and

$$\Pr(x < 0) = \int_{-1}^{0} \left(\frac{1+ax}{2}\right) dx = \frac{1}{4}(2-a)$$

$$\therefore E(\bar{\bar{a}}) = 2\left[\left(\frac{2+a}{4}\right) - \left(\frac{2-a}{4}\right)\right] = a$$

We may show that

$$\sigma^2(\bar{\bar{a}}) = \frac{1}{n}(4-a^2)$$

which may be compared with (30.52)

$$e_{\hat{a}\bar{\bar{a}}} = \frac{\sigma^2(\hat{a})}{\sigma^2(\bar{\bar{a}})}$$

for different a.

Another convenient statistic is obtained by noting the relation to the "*average value of* x".

$$E(x) = \int_{-1}^{+1} \left(\frac{1+ax}{2}\right) dx = \frac{a}{3}$$

Therefore, we may define the statistic

$$\tilde{a} = 3\bar{x}$$

where $\bar{x}$ is the *sample average* of x.

$$\therefore E(\tilde{a}) = 3E(\bar{x}) = a$$

and $\tilde{a}$ is an *unbiased* estimator of a.

We may show that

$$\sigma^2(\tilde{a}) = \frac{1}{n}(3-a^2)$$

and compare this with (30.52).

We show the efficiencies of $\tilde{a}$ and $\bar{\bar{a}}$ relative to $\hat{a}$ in Figure 30.5. ◄

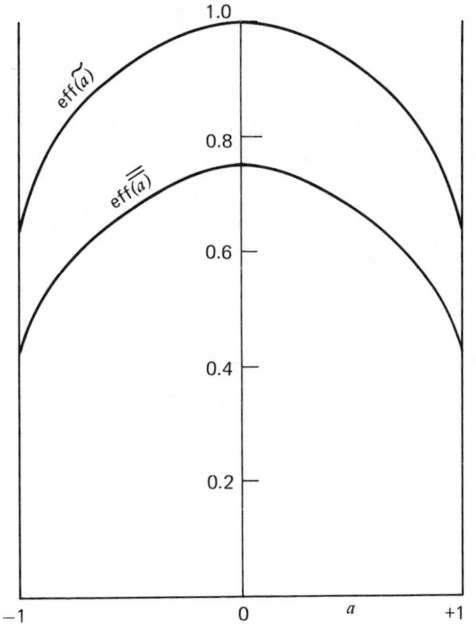

Figure 30.5
Plot of the statistical efficiencies of the "convenient statistics" $\tilde{a}$ and $\bar{\bar{a}}$ relative to that of $\hat{a}$. The plot is against different values of a.

Graphical Methods: The Score Function

It is frequently useful to look at what has been called the "**score function**":

$$\mathcal{S}(c) = \frac{\partial}{\partial c} \ln \mathcal{L}(c) = \frac{\partial \ell(c)}{\partial c} \quad (30.53)$$

and consider it plotted as the ordinate against the abscissa c. The score has some useful features when considered in this way.

1. The maximum likelihood estimate c^* is obtained when the score function passes through zero, that is, $\mathcal{S}(c^*) = 0$ defines c^*. This is indicated schematically in Figure 30.6*a*.
2. In the vicinity of a likelihood *maximum* the **slope** of the score function must be *negative* and reflects the **error** in the estimate, c^*: that is,

$$\frac{\partial}{\partial c}\mathcal{S}(c) = \ell''(c) = -I_n(c) \quad (30.54)$$

The larger the magnitude of the (negative) slope, the more precise is the estimate of c. A theoretical number c_0 with no error might be represented by a score function that is a vertical line at $c = c_0$. These ideas are indicated in Figure 30.6*b*.
3. If we wish to combine two sets of data with score functions $\mathcal{S}_1(c)$ and $\mathcal{S}_2(c)$, respectively, we get the score function for the combined data by *adding*

$$\mathcal{S}_{1+2}(c) = \mathcal{S}_1(c) + \mathcal{S}_2(c) \quad (30.55)$$

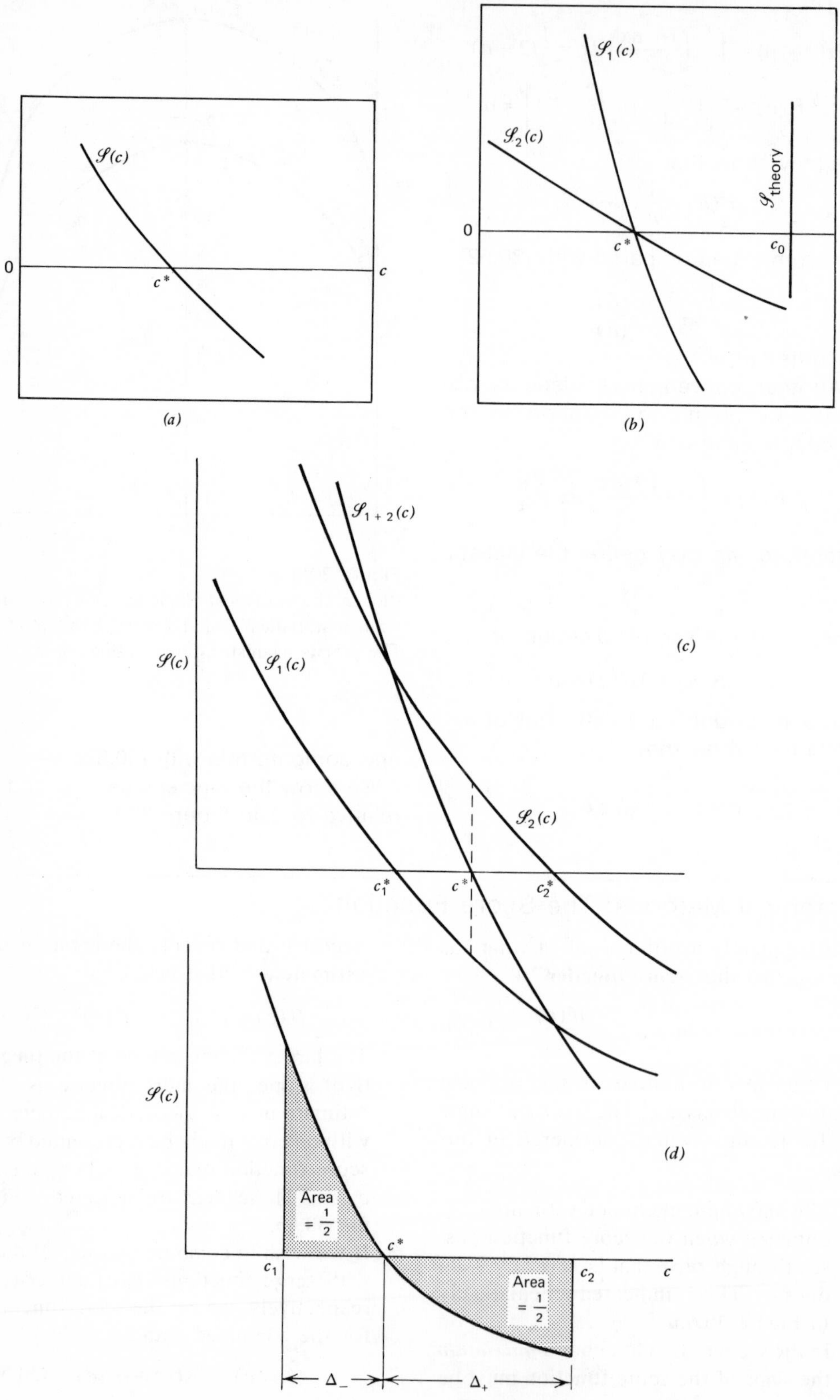

𝒮(c)
0
c*
c
(a)
𝒮₁(c)
𝒮₂(c)
𝒮theory
0
c*
c₀
c
(b)
𝒮₁₊₂(c)
𝒮(c)
𝒮₁(c)
𝒮₂(c)
c₁*
c*
c₂*
(c)
𝒮(c)
Area = 1/2
c*
c₁
c₂
c
Area = 1/2
Δ₋
Δ₊
(d)

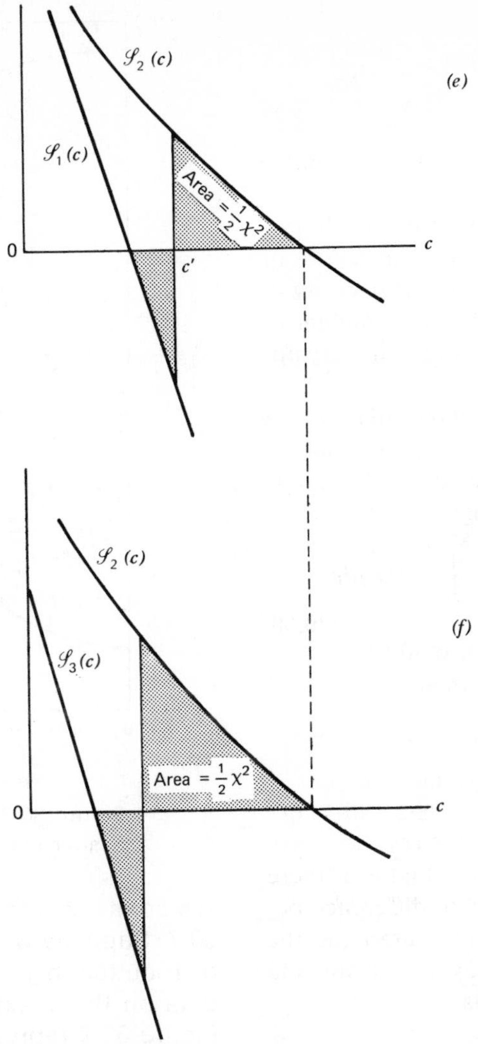

Figure 30.6
Plots of score functions $\mathcal{S}(c)$ versus the parameter c. (*a*) The score function passes through zero (x-axis intercept) at the value c^*, the maximum likelihood estimate of c. (*b*) For two score functions, the one with the more negative slope determines the estimate of c with greater precision. A determination with no error is represented by a vertical score function. (*c*) If two experiments have produced respective score functions, the combined data may be represented by simple addition of the score functions. (*d*) Deviation from linearity of the score function indicates the *nongaussian* nature of the likelihood function. One-standard-deviation limits on c^* may be obtained graphically. (*e*) The consistency of two sets of data may be checked by measuring areas on the score plots. (*f*) Shows a larger difference between measurement 3 and measurement 2 than between measurement 1 and measurement 2.

This follows from the definitions and is shown in Figure 30.6*c*.

4. The *weighted mean* of the *combined data* is obtained by choosing that point c' such that $|\mathcal{S}_1(c')| = |\mathcal{S}_2(c')|$, that is, equal heights above and below the axis. This, of course, is the point $\mathcal{S}_{1+2}(c') = 0$ or

$$\mathcal{S}_1(c') = -\mathcal{S}_2(c') \qquad (30.56)$$

5. The score function is a *straight line* if the likelihood function is *gaussian*, that

is,

$$\mathcal{L} \approx Ke^{-(c-c^*)^2/2\sigma_0^2};\ \ell(c) = -\frac{(c-c^*)^2}{2\sigma_0^2}$$

and

$$\ell'(c) = -\frac{(c-c^*)}{\sigma_0^2} = \mathcal{S}(c) \qquad (30.57)$$

Deviations from score-function linearity indicate the nongaussian nature of the likelihood function. The *linearity* of the score function is thus a qualitative check of the "*approach to asymptopia.*"

6. We may get the *one-standard deviation* limits on c^* from the score plot by considering the points $c_2 = c^* + \Delta_+$ and $c_1 = c^* - \Delta_-$ such that

$$\int_{c^*}^{c_2} \mathcal{S}(c)dc = \frac{1}{2} = \int_{c_1}^{c^*} \mathcal{S}(c)dc \qquad (30.58)$$

This is shown in Figure 30.6*d*. If $\mathcal{L}$(c) is *Gaussian*, then

$$c_1 = c^* - \sigma_0 \qquad \text{and} \qquad \sigma_2 = c^* + \sigma_0$$

and the errors are *symmetrical.*

7. If we have two sets of data with score functions $\mathcal{S}_1(c)$ and $\mathcal{S}_2(c)$, we may check their *consistency*, that is, is there a statistically significant difference between c_1^* and c_2^*? We determine the weighted mean c' as in (30.56) and consider the integrals

$$\int_{c_1^*}^{c'} \mathcal{S}_1(c)dc + \int_{c'}^{c_2^*} \mathcal{S}_2(c)dc = \chi^2(1) \qquad (30.59)$$

For Gaussian likelihood functions each integral should separately be $(1/2)\chi^2(1)$. If the measured areas are summed and compared with a table of χ^2 probabilities, one can calculate the probability of obtaining by random sampling a χ^2 as large as that observed. This measures the probability that a difference as large as $(c_1^* - c_2^*)$ could have arisen by chance. *This is indicated schematically in Figures* 30.6*e and* 30.6*f.*

► **Example 30.23**
As measured in an actual experiment,

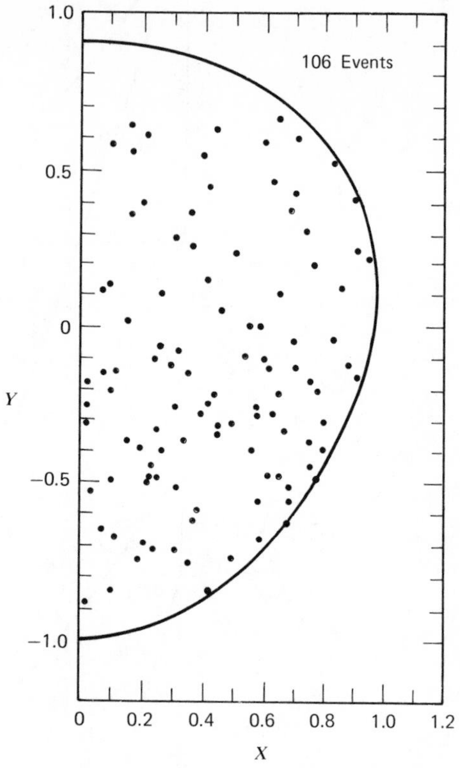

Figure 30.7
Scatter plot of actual data.

106 points (x, y) are displayed in Figure 30.7. Suppose we are interested in the distribution in y, that is, we project the data on the y axis. The solid curve in Figure 30.7 represents a constraint on the data, that is, the physically allowed region is bounded by the solid curve. For any given region of y, for example, Δy at y, the probability may be factored into a "*geometrical*" quantity $\Phi(y)$ and the **residual probability** $g(y; c)$. The quantity $\Phi(y)$ depends only on known constants and may be calculated *a priori*. It measures the "*available space*" at y. If the residual probability were $g(y; c) = 1$, then the probability of y being in the interval 0.8 to 0.9 would be less than the probability of being in the interval 0.0 to 0.1, that is, $\Pr(0.8 \le y \le 0.9) < \Pr(0.0 \le y \le 0.1)$. The physically unallowed region may be described by $\Phi(y) = 0$. In physics, the

function $\Phi(y)$ is called the "*phase space factor*" while the information bearing function, $g(y;c)$, is called the "**matrix element squared**."

The separation into $\Phi(y)$ and $g(y;c)$ is convenient since the $\Phi(y)$ is *independent* of the data and is known *a priori*.

In this example, it was required to fit the data to $g(y;c) \propto (1+ay)$. We, therefore, write the likelihood function for N events:

$$\mathscr{L}(a) = \prod_{i=1}^{N} K(a)[1+ay_i]\Phi(y_i)$$

We may *normalize* to the total number of events, N:

$$N = \int_{y_{\min}}^{y_{\max}} K(a)(1+ay)\Phi(y)dy$$

$$= K(a)\left\{\int_{y_{\min}}^{y_{\max}} \Phi(y)dy + a\int_{y_{\min}}^{y_{\max}} y\Phi(y)dy\right\}$$

$$= K(a)\{1+a\langle y\rangle\}$$

where we have used the *normalized geometrical factor*

$$\int_{y_{\min}}^{y_{\max}} \Phi(y)dy = 1$$

and the quantity

$$\langle y\rangle = \int_{y_{\min}}^{y_{\max}} y\Phi(y)dy$$

The quantity $\langle y\rangle$ is a known constant, since it is calculable *independent of performing the experiment.*

$$\therefore \mathscr{L}(\mathbf{y};a) = \left(\frac{N}{1+a\langle y\rangle}\right)^N \prod_{i=1}^{N}(1+ay_i)\Phi(y_i)$$

$$\ell(\mathbf{y};a) = \ln\mathscr{L} = N\ln N - N\ln(1+a\langle y\rangle) + \sum_{i=1}^{N}[\ln(1+ay_i) + \ln\Phi(y_i)]$$

and

$$\frac{d}{da}\ell(y;a) = \mathscr{S}(a) = \frac{-N\langle y\rangle}{1+a\langle y\rangle} + \sum_{i=1}^{N}\left[\frac{y_i}{1+ay_i}\right]$$

The *score function* $\mathscr{S}(a)$ is plotted against a in Figure 30.8 for the data of Figure 30.7 with the confidence interval indicated. Figure 30.9 shows the *likelihood plot* $\mathscr{L}(a)$ versus a for these data where the confidence interval is indicated for the points $e^{-1/2}\mathscr{L}_{\max}$.

Another way of treating these data is to group them into "*bins*." This grouping lends itself to treatment by

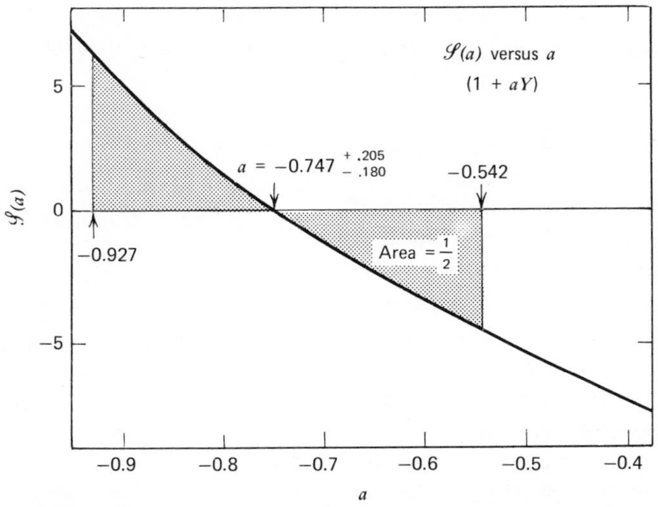

Figure 30.8
Score function plotted against a for the data of Figure 30.7.

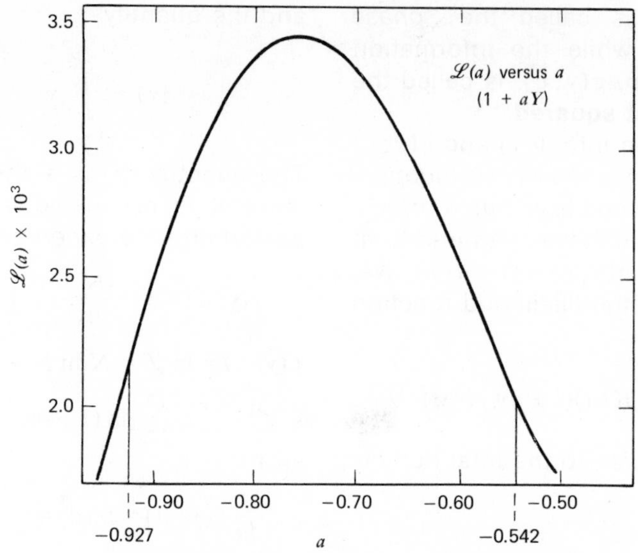

Figure 30.9
Likelihood plot for the data of Figure 30.7.

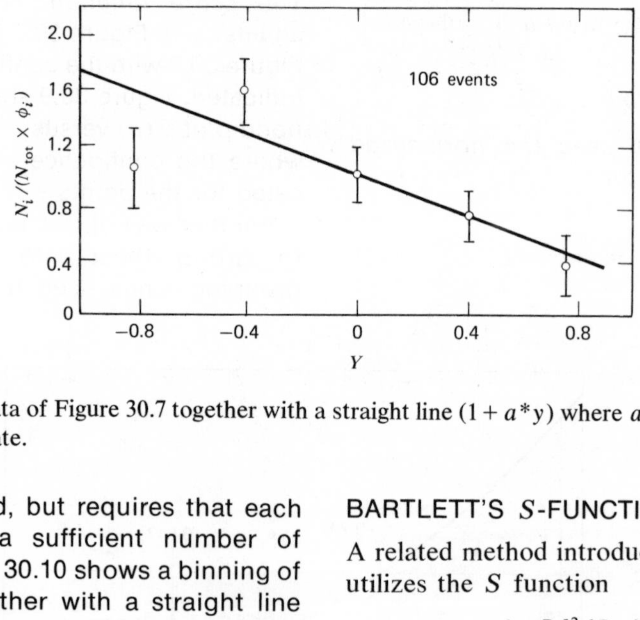

Figure 30.10
Binning of the data of Figure 30.7 together with a straight line $(1 + a^*y)$ where a^* is the maximum likelihood estimate.

the χ^2 method, but requires that each bin contain a sufficient number of events. Figure 30.10 shows a binning of the data together with a straight line $(1 + a^*y)$ where a^* is the *maximum likelihood estimate*. In this case the binning was not used for data analysis but only to see *ex post facto* how well the function $1 + a^*y$ fitted the data. When the experimenter has only 100 events, he is well-advised to use the *maximum likelihood method* since *it utilizes the maximum amount of information* contained in the data. ◄

BARTLETT'S *S*-FUNCTION

A related method introduced by Bartlett* utilizes the S function

$$S(c) = \left\{ -\left[\frac{d^2\ell}{dc^2}\right]_{\text{avg}} \right\}^{-1/2} \frac{\partial \ell}{\partial c} \qquad (30.60)$$

where $[d^2\ell/dc^2]_{\text{avg}}$ is given by (30.49).

If $S(c)$ is regarded as a variate, it may be seen to have a *mean of zero* and a *variance of unity*, that is,

$$E(S) = 0 \qquad (30.61)$$

$$\text{Var}(S) = 1 \qquad (30.62)$$

*Bartlett, Philosophical Magazine *44* (1953), p. 249.

When the likelihood function is Gaussian, the S-function plotted against c is a straight line.

When the S-function passes through zero, we get the maximum likelihood estimate

$$S(c^*) = 0 \tag{30.63}$$

The one-standard-deviation confidence limits c_1, c_2 are easily read off since

$$S(c_1) = -1 \quad \text{and} \quad S(c_2) = +1 \tag{30.64}$$

where

$$c_1 = c^* - \Delta_- \quad \text{and} \quad c_2 = c^* + \Delta_+$$

31

HYPOTHESIS TESTING AND SIGNIFICANCE

We have previously used the idea of testing a "*null hypothesis,*" and we have considered such procedures as the **normal deviate** test, **chi-square** test, **Student's** *t* test, and the ***F*-test**. In this section we wish to restate the procedure of *hypothesis testing* in more precise terms to illustrate certain general features.

Kinds of Hypotheses

An hypothesis can usually be cast into a form such that it makes a statement about a probability function $f(\mathbf{X};\mathbf{k})$ where we wish to include both discrete and continuous cases. A *parametric hypothesis* concerns a statement about a certain set of parameters, **k**. A *nonparametric hypothesis* makes a statement about the **form** of a distribution rather than about a **parameter**. We may make further distinctions: in a *simple* parametric hypothesis a parameter has a specific value, while in a *composite* hypothesis a distribution has a certain *form* involving one or more parameters **k** and leaves the parameters *to be determined.*

The hypothesis to be tested is called the null hypothesis, $\mathscr{H}_0$. If the null hypothesis is false, then one or more of the *alternative hypotheses,* H_k, will be true although we are frequently not concerned with specifying the alternative hypotheses.

Consistency and Proof

There is a fundamental asymmetry in hypothesis testing. Normally, one can never prove *without deductive reasoning* that a given hypothesis about an infinite population is *true,* although **it can be proved false**. On the affirmative side, the most that can be done is to determine that the hypothesis is *consistent* with all the data available **thus far**. This may be illustrated by an example based on one due to E. B. Wilson.*

Suppose one is investigating the

*E. B. Wilson, *An Introduction to Scientific Research,* McGraw-Hill, 1952.

hypothesis that the seat numbers on a ski tow, N_i, are given by the simple formula

$$\mathcal{H}_0: N_i = (i + 1)$$

One observes the first seat to be 2, the second to be 3, and the entire sequence of the first 6 seats to be

$$2,3,4,5,6,7$$

These data are certainly *consistent* with the hypothesis, $\mathcal{H}_0$. However, we have not *excluded* the hypothesis:

$$H_1: N_i = (i + 1) + (i - 1)(i - 2)(i - 3) \times (i - 4)(i - 5)(i - 6)$$

Suppose that a stubborn observer records the *next two* seats and notes

$$N_7 = 728 \qquad \text{and} \qquad N_8 = 5049$$

Hypothesis $\mathcal{H}_0$ is clearly inconsistent with *these* data and hence is disproved (*rejected*) and *false.* All we can say positively, however, is that H_1 is *consistent* with the data **thus far available**.

Sometimes we cannot even *disprove* an hypothesis by statistical reasoning alone. For example, the hypothesis "The Democratic candidate for the U.S. Presidency will win the election on November 7, 1972" cannot be proved or disproved by statistical sampling *before the fact* although strong indications may result from careful surveys prior to Election Day. However, the truth or falsity of the hypothesis is usually clear by the following day.

For a *finite* population, a clear test can be made although it is not always convenient to do so. For example, one could look at *all* the seats on a finite ski tow and verify that hypothesis H_1 describes *all* the possible data. Also, a prediction about the outcome of a finite experiment can be checked against the finite result.

A more general hypothesis, however, can never be proved by statistical reasoning alone. An "*accepted*" hypothesis should always be interpreted to mean that it is consistent with all data *thus far observed,* that is, it has not yet been rejected.

All the experimental scientist can do is to reject an hypothesis or not reject it. Certainty about the "*truth*" of an hypothesis is usually reserved to logicians, mathematicians, and God. For most of us, the idea of a "*working hypothesis*" is good enough. This is the one that we accept as true (i.e., not rejected) until shown otherwise. Quite often, science advances by a large step forward when a long-standing hypothesis *falls* in the face of the data from a new experiment. Conversely, much effort is wasted when an important hypothesis is thought to be inconsistent with the data from an experiment, and the experiment turns out subsequently to be wrong or the inconsistency is due to a random fluctuation!

One should also be prepared to replace a working hypothesis with another *that is equally consistent with all available data* and that is *simpler* or *more general* or has *other superior features.* We must never rest too secure in the belief that the current hypothesis is the "last word." The situation may be described by the following paraphrase of Edna St. Vincent Millay:

How can I tell, unless I smell
  The Carthaginian rose,
If the best flower be the one I see
  Here beneath my nose!?

Two Kinds of Error and the Cost of Being Wrong

In subjecting an hypothesis $\mathcal{H}_0$ to test, we have two errors that we can make. We can reject $\mathcal{H}_0$ when it is true, an "*error of the first kind*" or "error *I.*" Conversely, we can accept the hypothesis $\mathcal{H}_0$ when it is false, producing an "*error of the second kind*", or "error II." We may represent the situation as in Table 31.1.

Table 31.1

Truth \ Decision	Accept $\mathcal{H}_0$	Reject $\mathcal{H}_0$
$\mathcal{H}_0$	Correct	Error I
not-$\mathcal{H}_0$	Error II	Correct

In the judicial system in English-speaking countries, a possible hypothesis in a criminal proceeding is "*The accused is innocent.*" The system must decide whether the hypothesis is rejected (finding of "guilty") or not rejected (finding of "acquittal"). An error of the first kind would be to convict (find guilty) an innocent defendant while an error of the second kind would be to acquit a guilty defendant. Clearly, the system could totally avoid errors of the first kind by acquitting *all* defendants. Also, errors of the second kind could be eliminated by convicting *everyone who is accused.* We must choose a strategy that steers a course between these two extreme cases basing the procedure on the probability of each of the two kinds of error and the weight or importance we attach to each. In quantitative terms, we must ask what "*loss*" attaches to each kind of error. The judicial system following the Judaeo-Christian tradition considers errors of the first kind much more serious. ("It is better that 99 guilty men should go free than that one innocent man should be convicted.") The very premise of the system (the accused is innocent until proven guilty beyond a reasonable doubt) reflects this concern to avoid errors of the first kind, even at the expense of producing errors of the second kind.

On the other hand, Artemus Ward ascribed much of the troubles of the world to the other source: "It ain't so much the things we don't know that get us into trouble. It's the things we know that ain't so."

The *loss* associated with each kind of error becomes particularly important when we are concerned with testing hypotheses as a *guide to making decisions.* The distinction between decision problems and others is admittedly *artificial* since we are always making decisions of some kind in analyzing data (Should we design a new experiment? Should we conclude the current measurement? Should we publish our results? Should we quote number x_1 or x_2?). Nevertheless, we shall assume for the present that we are more concerned with the *probabilities* of the two kinds of error than with the *importance to us* of each.

Concepts in Hypothesis Testing

In its most basic form, the **test** of an hypothesis $\mathcal{H}_0$ consists of performing a *random experiment* that yields values $\mathbf{x}_i$ for the observables $\mathbf{X}$ from which we may compute values $\mathbf{t}_i$ for some variate $\mathbf{T} = \mathbf{T}(\mathbf{X})$, a *function of the observables.* Based on the values $\mathbf{x}_i$ observed, we either reject the hypothesis $\mathcal{H}_0$ or we accept it (in the sense of *not rejecting it*). The variate $\mathbf{T}$ is called the "**statistic**" used to test the hypothesis.

We are concerned with the problem of **choosing** the statistic $\mathbf{T}$ and **formulating** a *strategy* (or *rule, procedure,* or *criterion*) based on the range of observed values $\mathbf{x}_i$ that causes us to make a decision to reject or to accept the hypothesis. In general, we should insist that the strategy be decided upon *before* the experiment is performed and that *ambiguities should be eliminated.* We then have a rule that assigns each possible observation to *one and only one* of two categories: *consistent* with $\mathcal{H}_0$ or *not consistent* with $\mathcal{H}_0$.

We may formulate this rule by considering points $\mathbf{x}$ in the n-dimensional space of possible observations. The dimensionality of the space is that of the observable $\mathbf{X}$ (e.g., if $\mathbf{X}$ has four components $X_1, \ldots, X_4$, the space is four-dimensional). The rule will choose a region $R_c(\mathbf{X})$ in this space, called the *critical region,* and decide that $\mathcal{H}_0$ will be rejected if $\mathbf{x}$ is in R_c and accepted if $\mathbf{x}$ is not in R_c:

$$\begin{array}{lll} \text{Reject } \mathcal{H}_0 & \text{if} & (\mathbf{x} \in R_c) \\ \text{Accept } \mathcal{H}_0 & \text{if} & (\mathbf{x} \notin R_c) \end{array} \qquad (31.1\text{a})$$

We may cast this rule *in terms of the test statistic* $\mathbf{T}(\mathbf{X})$ where the region $R_c(\mathbf{X})$ becomes the region $\mathcal{R}_c(\mathbf{T})$:

$$\begin{array}{lll} \text{Reject } \mathcal{H}_0 & \text{if} & (\mathbf{t} \in \mathcal{R}_c) \\ \text{Accept } \mathcal{H}_0 & \text{if} & (\mathbf{t} \notin \mathcal{R}_c) \end{array} \qquad (31.1\text{b})$$

It will often be true that the dimensionality of the statistic $\mathbf{T}(\mathbf{X})$ is *less* than that of $\mathbf{X}$. If this is true, the critical region $\mathcal{R}_c$ is more *convenient* to use than R_c. If the statistic $\mathbf{T}(\mathbf{X})$ can be reduced to one dimension (*one* test parameter) the situation is particularly convenient.

If we assume that $\mathcal{H}_0$ is true, there is nevertheless usually a nonvanishing probability that $\mathbf{x} \in R_c$ and, hence, that $\mathcal{H}_0$ will be rejected. We may write this *conditional probability* (assuming $\mathcal{H}_0$ true) as in Chapter 19:

$$\Pr_{\mathcal{H}_0}(\mathbf{x} \in R_c) = \alpha = \Pr_{\mathcal{H}_0}(\mathbf{t} \in \mathcal{R}_c) \qquad (31.2)$$

$$= \Pr(\text{error I}) \qquad (31.2\text{a})$$

because of the *rule* (31.1a,b). The probability α is sometimes called the "*significance level*" of the test. α is also referred to as the *size* of the critical region R_c (or $\mathcal{R}_c$).

Suppose we consider the null hypothesis $\mathcal{H}_0$, and the *alternative hypothesis,* H_1. On the assumption of H_1, ($\mathcal{H}_0$ is false) we may still have a nonvanishing probability that $\mathbf{x}$ is *not* in R_c

$$\Pr_{H_1}(\mathbf{x} \notin R_c) = \beta = \Pr_{H_1}(\mathbf{t} \notin \mathcal{R}_c) \qquad (31.3)$$

$$= \Pr(\text{error II}) \qquad (31.3\text{a})$$

since this constitutes accepting $\mathcal{H}_0$ when it is false.

Choosing the statistical *test* or rule involves *choosing the critical region* $\mathcal{R}_c$ with regard to the probabilities α and β of errors of the first and second kind. For a *given* significance level, $\alpha = \Pr(\text{error I})$, we should choose the critical region so that $\beta = \Pr(\text{error II})$ is *minimized.* We thus see that the choice of the critical region depends on the alternative hypotheses.

Also, note the probability of rejecting the null hypothesis $\mathcal{H}_0$ when an alternative hypothesis H_1 is true:

$$\Pr_{H_1}(\mathbf{x} \in R_c) = \Pr_{H_1}(\mathbf{t} \in \mathcal{R}_c) = 1 - \beta \qquad (31.4)$$

This is said to measure the *power* of the critical region with respect to the alternative hypothesis H_1. We may consider this a measure of the power to reject the null hypothesis. We see that among different possible critical regions *of the same size* ($\boldsymbol{\alpha}$) we choose the critical region to *minimize* β, and **maximize the power**.

The *most powerful test* of a null hypothesis $\mathcal{H}_0$ with respect to an alternative H_1 is a critical region for which $(1-\beta)$ is maximum when compared with *all* critical regions of the same size, α.

A *uniformly most powerful test* is most powerful with respect to *any possible* alternative hypothesis.

If we can describe the alternative hypotheses H_k by the parameter k, then we may define the *power function* $\Pi(k)$:

$$\Pr_{H_k}(\mathbf{t} \in \mathcal{R}_c) = \Pi(\mathcal{R}_c; k) \qquad (31.5)$$

This is the probability of rejecting the null hypothesis as a function of k.

We may also define the "*operating characteristic*"

$$OC(k) = \Pr_{H_k}(\mathbf{t} \notin \mathcal{R}_c) = 1 - \Pi(\mathcal{R}_c; k) \qquad (31.6)$$

which is the probability of accepting $\mathcal{H}_0$ when H_k is "true."

If k_0 is the value of the parameter corresponding to $\mathcal{H}_0$ then

$$\Pi(\mathcal{R}_c; k_0) = \alpha \qquad (31.7)$$

A test is said to be *unbiased* if its power is $\geq \alpha$ for *every possible* hypothesis

$$\Pi(\mathcal{R}_c; k) \geq \alpha \qquad \text{for all } k \qquad (31.8)$$

This reflects the fact that a good test should be such that the probability for

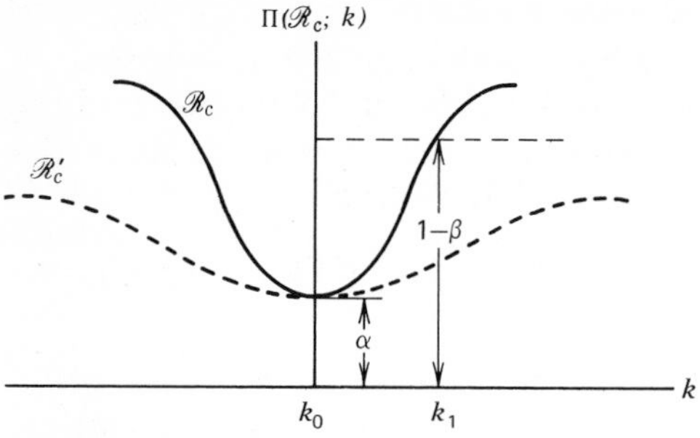

Figure 31.1
Plot of the power functions of two tests versus the parameter. Both tests are unbiased and have the same size α. The test indicated by the solid curve ($\mathcal{R}_c$) *dominates* the test indicated by the dashed curve ($\mathcal{R}'_c$). Hence, the test $\mathcal{R}'_c$ is *inadmissible.*

rejecting $\mathcal{H}_0$ should be smallest when $\mathcal{H}_0$ is true.

The power function of a good test is shown schematically in Figure 31.1 where k is the assumed true value (hypothesis H_k).

With respect to the null hypothesis $\mathcal{H}_0$ and the alternative H_1, we should like to minimize α and maximize $(1-\beta)$. If we have two **tests** (also called *critical regions, procedures, rules*) $\mathcal{R}_c$ and $\mathcal{R}'_c$ such that

$$\Pi(\mathcal{R}_c; k) \geq \Pi(\mathcal{R}'_c; k) \qquad \text{for all } k \tag{31.9}$$

where $\mathcal{R}_c$ and $\mathcal{R}'_c$ have the same size α, then we say that procedure $\mathcal{R}_c$ *dominates* $\mathcal{R}'_c$ and $\mathcal{R}'_c$ is *inadmissible.* We mean that there is no reason to use $\mathcal{R}'_c$ when $\mathcal{R}_c$ is better.

For discrete hypotheses it might happen that $\Pi(\mathcal{R}_c; c) \geq \Pi(\mathcal{R}'_c; c)$ for all $c \neq c_0$ while

$$\Pi(\mathcal{R}_c; c_0) < \Pi(\mathcal{R}'_c; c_0) \tag{31.9a}$$

This also means that $\mathcal{R}'_c$ is inadmissible.

We may illustrate some of these ideas by considering the following examples.

► **Example 31.1**
A coin-tossing problem

Suppose we are presented with a coin such that the possible hypotheses, H_k, are:

H_0: the coin has two tails, that is, no heads — $k = 0$
H_1: the coin is fair, that is, 1 head — $k = 1$
H_2: the coin has two heads — $k = 2$

We are not allowed to examine both sides of the coin but may note and record the results of two tosses. We write the possible data points as **x** and calculate their probabilities on the basis of the three hypotheses (Table 31.2).

Table 31.2

x	$\Pr_{H_0}(\mathbf{x})$	$\Pr_{H_1}(\mathbf{x})$	$\Pr_{H_2}(\mathbf{x})$
TT	1	1/4	0
HT	0	1/4	0
TH	0	1/4	0
HH	0	1/4	1

Suppose we wish to test the *null hypothesis*

$$\mathcal{H}_0 = H_1$$

and we wish to consider the following possible critical regions $R_c^{(i)}(\mathbf{x})$:

$R_c^{(1)}$: $\{TT\} + \{HH\}$, that is, reject $\mathcal{H}_0$ if we do not get one of each
$R_c^{(2)}$: $\{HT\} + \{TH\}$, that is, reject $\mathcal{H}_0$ if we do get one of each
$R_c^{(3)}$: $\{TT\} + \{HT\}$
$R_c^{(4)}$: $\{HH\} + \{TH\}$

We may easily calculate the power functions of each critical region $\Pr_{H_K}(\mathbf{x} \in R_c) = \Pi(R_c; k)$ (Table 31.3).

Table 31.3

Rule	$\Pr_{H_0}(\mathbf{x} \in R_c)$	$\Pr_{H_1}(\mathbf{x} \in R_c)$	$\Pr_{H_2}(\mathbf{x} \in R_c)$
$R_c^{(1)}$	1	1/2	1
$R_c^{(2)}$	0	1/2	0
$R_c^{(3)}$	1	1/2	0
$R_c^{(4)}$	0	1/2	1

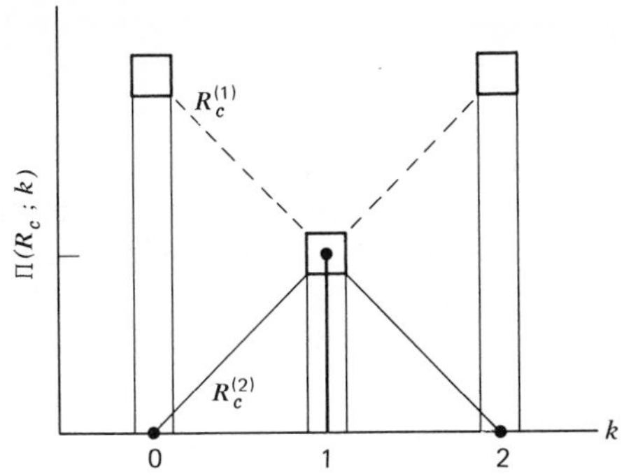

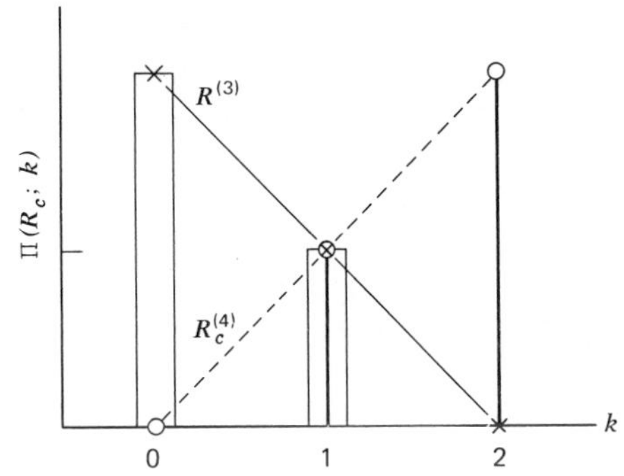

Figure 31.2
For Example 31.1, plot of the (discrete) power functions of the tests $R_c^{(1)}$, $R_c^{(2)}$, $R_c^{(3)}$ and $R_c^{(4)}$ versus k. $R_c^{(1)}$ is seen to be unbiased and uniformly most powerful.

The (discrete) power functions are indicated schematically in Figure 31.2. All the regions have the same size (significance level)

$$\Pr_{\mathcal{H}_0}(\mathbf{x} \in R_c) = \alpha = \frac{1}{2}$$

We note that $R_c^{(2)}$ is dominated by $R_c^{(1)}$ and, hence, is *inadmissible* by (31.9).

Also, $R_c^{(3)}$ and $R_c^{(4)}$ are dominated by $R_c^{(1)}$. Therefore, $R_c^{(1)}$ is a *uniformly most powerful test.* We also note that $R_c^{(1)}$ is *unbiased* by (31.8). $R_c^{(1)}$ is clearly the test of choice. ◄

It is not usually true that a unique test or critical region is the most powerful. The situation is somewhat more complex in the following example.

► **Example 31.2**
An urn problem

Consider an urn containing four balls that are identical except for color, which may be white (W) or black (B). We wish to consider the hypotheses of interest to involve the number of black balls in the urn

$$H_k: \text{ urn contains } k \text{ black balls}$$

where $k = 0, 1, 2, 3, 4$.

An observation consists of drawing a ball at random, noting its color, and replacing the ball. We consider the measurement to consist of *two* observations comprising possible data points $\mathbf{x}$.

We summarize the possible data points and their probabilities under the different possible hypotheses in Table 31.4.

Table 31.4

$\mathbf{x}$	$\Pr_{H_0}(\mathbf{x})$	$\Pr_{H_1}(\mathbf{x})$	$\Pr_{H_2}(\mathbf{x})$	$\Pr_{H_3}(\mathbf{x})$	$\Pr_{H_4}(\mathbf{x})$
WW	1	9/16	1/4	1/16	0
WB	0	3/16	1/4	3/16	0
BW	0	3/16	1/4	3/16	0
BB	0	1/16	1/4	9/16	1

Suppose we wish to test the hypothesis

$$\mathcal{H}_0 = H_2\text{: there are two black balls.}$$

We may consider several tests (critical regions):

$R_c^{(1)}$: $\{WB\} \cup \{BW\}$, reject $\mathcal{H}_0$ if one of each color appears.

$R_c^{(2)}$: $\{WB\} \cup \{WW\}$, reject $\mathcal{H}_0$ if at least one W appears.

$R_c^{(3)}$: $\{WW\} \cup \{BB\}$, reject H_0 if one of each is not observed.

We calculate the power functions of each critical region (Table 31.5)

$$\Pr_{H_k}(\mathbf{x} \in R_c) = \Pi(R_c; k)$$

The (discrete) power functions are

Table **31.5**

Rule	$\Pi(R_c;0)$	$\Pi(R_c;1)$	$\Pi(R_c;2)$	$\Pi(R_c;3)$	$\Pi(R_c;4)$
$R_c^{(1)}$	0	6/16	8/16	6/16	0
$R_c^{(2)}$	1	12/16	8/16	4/16	0
$R_c^{(3)}$	1	10/16	8/16	10/16	1

shown in Figure 31.3. We see that all regions have the same size or significance level: $\alpha = 1/2$.

We note that $R_c^{(1)}$ is dominated by $R_c^{(3)}$ by (31.9). Hence, $R_c^{(1)}$ is an inadmissible test that accords with our intuitive ideas about the situation. In this case, however, $R_c^{(2)}$ cannot be rejected as inadmissible since its power with respect to hypothesis H_1 is greater than that of $R_c^{(3)}$. If we were interested *only in the two hypotheses*, $\mathscr{H}_0$ and H_1, we would prefer $R_c^{(2)}$ to $R_c^{(3)}$; this should be kept in mind. In the absence of such information we note that $R_c^{(3)}$ is unbiased by (31.8) whereas $R_c^{(2)}$ is biased. On the grounds of *unbiasedness*, therefore, we might prefer $R_c^{(3)}$. ◄

► **Example 31.3**
Sampling from a normal population with unit variance and unknown mean [X *is* $N(\mu, 1)$]

Suppose we wish to test the null hypothesis

$$\mathscr{H}_0: \mu = \mu_0$$

by sampling n values of x and calculating the **sample mean**, $\bar{x}$, as the **test statistic**. We must choose the critical region $\mathscr{R}_c(\bar{X})$.

Suppose we set the significance level at $\alpha = 0.05$. We might choose the region

$$\mathscr{R}_c(\bar{X}): \quad \begin{array}{l} \bar{x} \le \mu_0 - 1.96\sigma_n(\bar{x}) \\ \bar{x} \ge \mu_0 + 1.96\sigma_n(\bar{x}) \end{array}$$

where $\sigma_n(\bar{x}) = 1/\sqrt{n}$, indicated as the shaded region in Figure 31.4*a*. From Chapter 25 we know that

$$\Pr_{\mathscr{H}_0}(\bar{x} \in \mathscr{R}_c) = 0.05$$

that is, assuming the mean is μ_0, the probability is 0.05 that the sample mean will differ from μ_0 by more than $1.96\sigma_n(\bar{x})$.

The significance level is not sufficient to define $\mathscr{R}_c$ uniquely. For example,

$$\mathscr{R}_c'(\bar{X}).\ \bar{x} \ge \mu_0 + 1.65\sigma_n(\bar{x})$$

may be seen from Table A.VI.2 to have the property that

$$\Pr_{\mathscr{H}_0}(\bar{x} \in \mathscr{R}_c') = 0.05$$

The critical region $\mathscr{R}_c'$ is shown as the shaded area in Figure 31.4*b*.

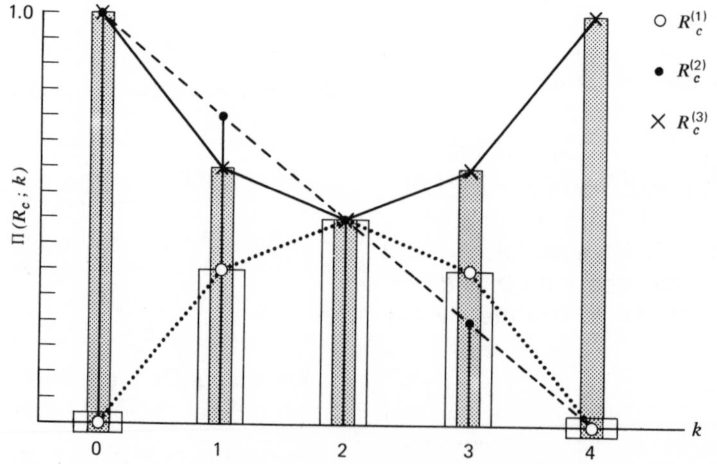

Figure 31.3
For Example 31.2, plot of the (discrete) power functions for the tests $R_c^{(1)}$, $R_c^{(2)}$, and $R_c^{(3)}$ versus k.

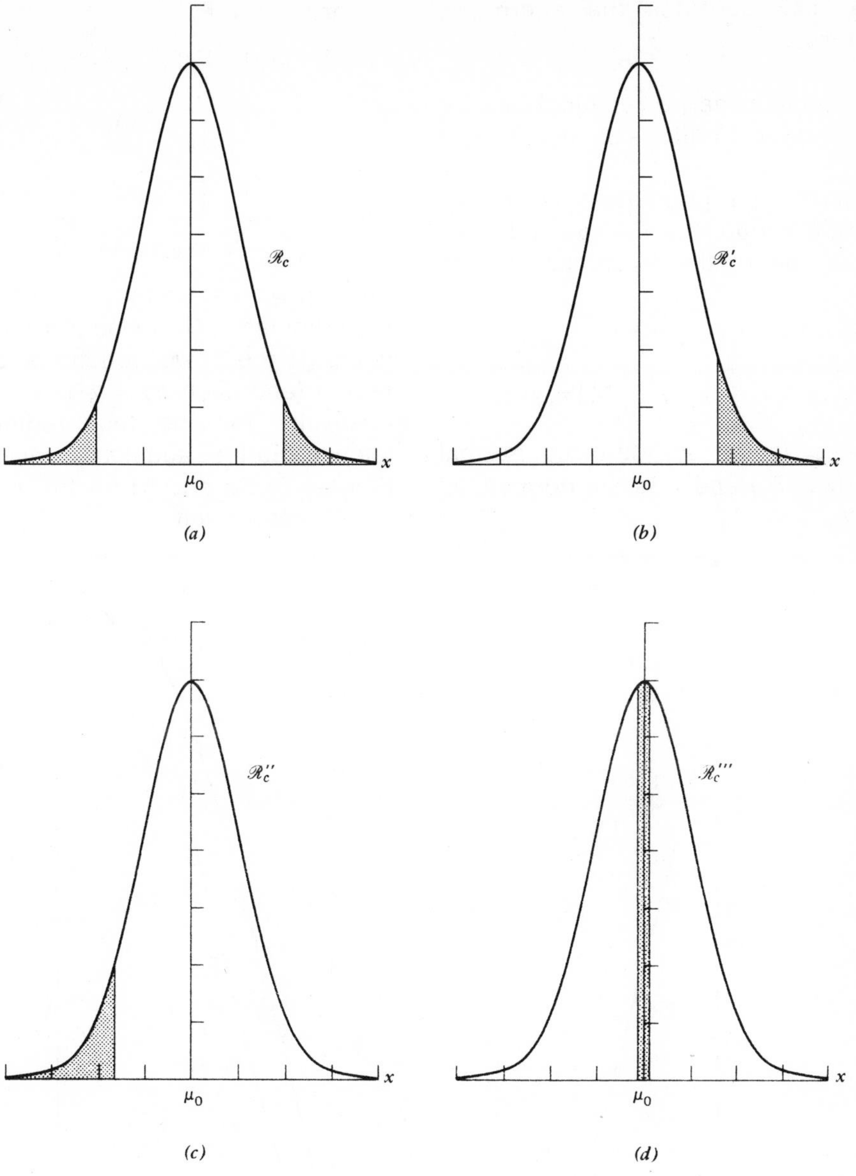

Figure 31.4
For Example 31.3, plot of the p.d.f. of the normal distribution with the critical regions for four different tests indicated as the shaded region of the plot. (*a*) $\mathcal{R}_c(\bar{X})$: $\mu_0 + 1.96\sigma_n(\bar{x}) \le \bar{x} \le \mu_0 - 1.96\sigma_n(\bar{x})$. (*b*) $\mathcal{R}'_c(\bar{X})$: $\bar{x} \ge \mu_0 + 1.65\sigma_n(\bar{x})$. (*c*) $\mathcal{R}''_c(\bar{X})$: $\bar{x} \le \mu_0 - 1.65\sigma_n(\bar{x})$. (*d*) $\mathcal{R}'''_c(\bar{X})$: $\mu_0 - 0.065\sigma_n(\bar{x}) \le \bar{x} \le \mu_0 + 0.065\sigma_n(\bar{x})$.

Other possible critical regions for which the significance level is the same are $\mathcal{R}''_c$

$$\mathcal{R}''_c(\bar{X}):\ \bar{x} \le \mu_0 - 1.65\sigma_n(\bar{x})$$

shown in Figure 31.4*c* and

$$\mathcal{R}'''_c(\bar{X}):\ \mu_0 - 0.065\sigma_n(\bar{x}) \le \bar{x} \le \mu_0 + 0.065\sigma_n(\bar{x})$$

shown in Figure 31.4*d*.

For all of these

$$\Pr_{\mathcal{H}_0}(\bar{x} \in \mathcal{R}''_c) = \Pr_{\mathcal{H}_0}(\bar{x} \in \mathcal{R}'''_c) = 0.05$$

We may consider the alternative hypotheses

$$H_k: \mu = \mu_k$$

and calculate the power functions for the four critical regions $\mathcal{R}_c$, $\mathcal{R}_c'$, $\mathcal{R}_c''$, and $\mathcal{R}_c'''$.

Because the population is *normal* with *unit variance*, we know the p.d.f. for each alternative hypothesis where $t = \bar{x}$

$$H_k: f(t; k, n) = \frac{n}{\sqrt{2\pi}} e^{-n(t-\mu_k)^2/2}$$
$$= \varphi\left(\frac{t-\mu_k}{\sigma_n}\right) = \varphi(z)$$

where φ is the standardized normal p.d.f. (25.17b) and z is the normalized variate.

For region $\mathcal{R}_c$,

$$\Pr_{H_k}(\bar{x} \in \mathcal{R}_c) = \Pi_n(\mathcal{R}_c; k)$$
$$= \int_{\mathcal{R}_c} \varphi(z)dz$$
$$= \int_{-\infty}^{z_1} \varphi(z)dz - \int_{z_2}^{\infty} \varphi(z)dz$$
$$= \Phi(z_1) - \Phi(z_2)$$

This shows that we may use the table of the *cumulative Gaussian distribution* to compute the power functions for each region (and each n).

Results for the four regions are shown plotted against $k = \mu_0 - \mu_k$ in Figures 31.5*a* and 31.5*b* for $n = 1$ and $n = 4$, respectively.

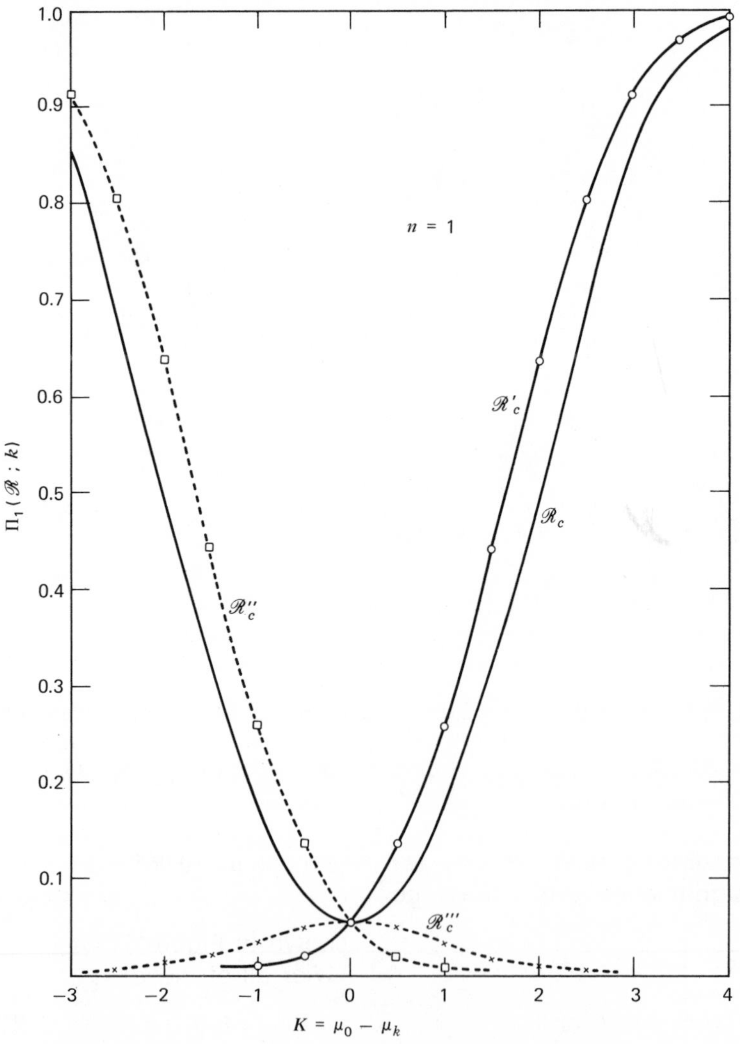

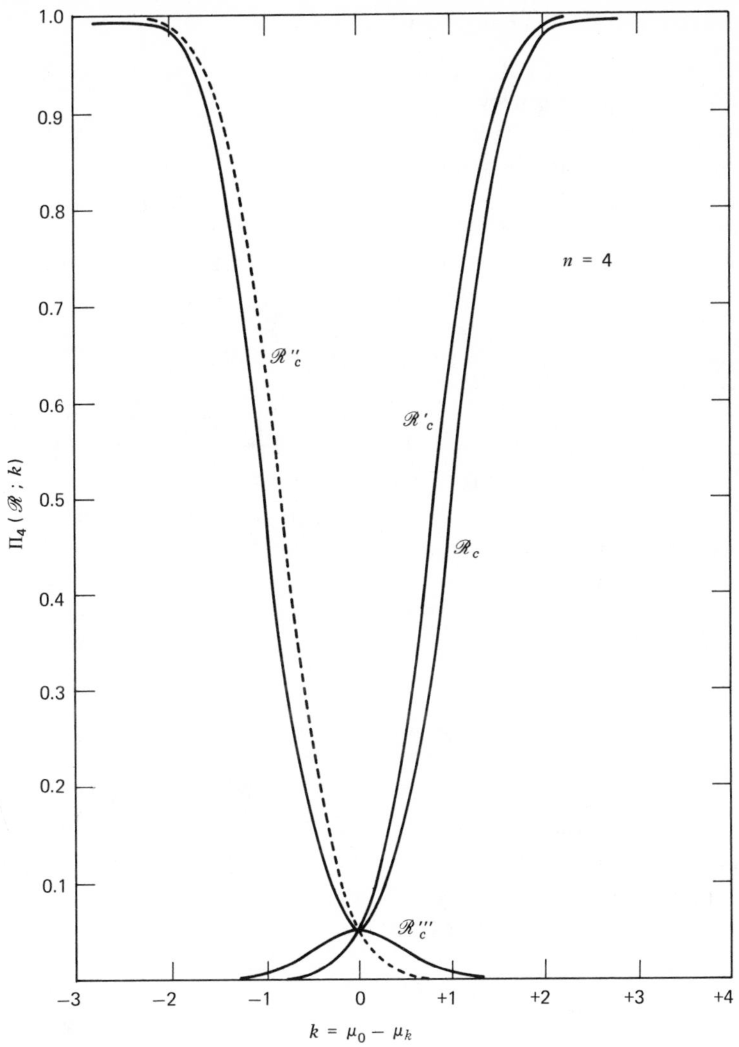

Figure 31.5
For Example 31.3, plot of the power functions of the four critical regions versus $k = \mu_0 - \mu_k$. (a) $n = 1$. (b) $n = 4$.

Specification of the *range of possible hypotheses* is very important. If the range includes all k, then only region $\mathcal{R}_c$ provides an *unbiased* test. In any case, the test $\mathcal{R}'''_c$ is dominated by $\mathcal{R}_c$ for all k and, hence, is inadmissible, which accords with our intuitive ideas on the matter. If we were to specify that $\mu \geq \mu_0$, then test $\mathcal{R}'_c$ would dominate $\mathcal{R}''_c$ and $\mathcal{R}_c$. Conversely, if $\mu \leq \mu_0$ then $\mathcal{R}''_c$ would dominate both $\mathcal{R}_c$ and $\mathcal{R}''_c$. We note that a "*two-tailed test*" has the alternative hypotheses of all k. A "*one-tailed test*" restricts the alternative hypotheses *either* to $k \geq 0$ *or* to $k \leq 0$. Any test becomes more powerful as n increases. (This is generally true except as noted earlier for the Cauchy distribution.) ◄

The Neyman-Pearson Theorem

Suppose we seek a test R_c for a null hypothesis $\mathcal{H}_0$ that is most powerful with respect to an alternative hypothesis H_1. If we restrict ourselves to a simple null and a simple alternative hypothesis with associated probability functions $f(\mathbf{X}; k_0)$, $f(\mathbf{X}; k_1)$, respectively, we may show that R_c is *most powerful* with respect to H_1 if a constant $c > 0$ exists such that

$$\frac{f(\mathbf{X}; k_0)}{f(\mathbf{X}; k_1)} \le c \quad \text{for} \quad \mathbf{X} \in R_c$$

$$\frac{f(\mathbf{X}; k_0)}{f(\mathbf{X}; k_1)} \ge c \quad \text{for} \quad \mathbf{X} \notin R_c \tag{31.10}$$

c may depend on α, the level of significance.

The demonstration of this follows from considering *another* region R'_c with the *same size* α:

$$\int_{R_c} f(\mathbf{X}; k_0)d\mathbf{X} = \alpha = \int_{R'_c} f(\mathbf{X}; k_0)d\mathbf{X}$$

In general, R'_c may overlap R_c, that is, $R'_c \cap R_c = I$ need *not* be the null set, $\emptyset$, in the language of Chapter 19. The general situation is indicated schematically in Figure 31.6 where $R_c = A \cup I, R'_c = B \cup I$.

We see that

$$\int_A f(\mathbf{X}; k_0)d\mathbf{X} = \int_{R_c} f(\mathbf{X}; k_0)d\mathbf{X} - \int_I f(\mathbf{X}; k_0)d\mathbf{X}$$

$$= \int_{R'_c} f(\mathbf{X}; k_0)d\mathbf{X} - \int_I f(\mathbf{X}; k_0)d\mathbf{X}$$

$$= \int_B f(\mathbf{X}; k_0)d\mathbf{X} \tag{31.11a}$$

Since $A \subset R_c$, we may use (31.10)

$$f(\mathbf{X}; k_0) \le cf(\mathbf{X}; k_1) \quad \text{for} \quad \mathbf{X} \in A$$

which yields

$$\frac{1}{c}\int_A f(\mathbf{X}; k_0)d\mathbf{X} \le \int_A f(\mathbf{X}; k_1)d\mathbf{X} \tag{31.11b}$$

Since B is strictly outside R_c, we use the second part of (31.10) to get

$$\int_B f(\mathbf{X}; k_0)d\mathbf{X} \ge c\int_B f(\mathbf{X}; k_1)d\mathbf{X} \tag{31.11c}$$

The power function for R_c with respect to the hypothesis H_1 is

$$\Pi(R_c; k_1) = \Pr_{H_1}(\mathbf{x} \in R_c) = \int_{R_c} f(\mathbf{X}; k_1)d\mathbf{X}$$

$$= \int_A f(\mathbf{X}; k_1)d\mathbf{X} + \int_I f(\mathbf{X}; k_1)d\mathbf{X}$$

$$\ge \frac{1}{c}\int_A f(\mathbf{X}; k_0)d\mathbf{X} + \int_I f(\mathbf{X}; k_1)d\mathbf{X}$$

using (31.11b). We substitute using (31.11a) to get

$$\Pi(R_c; k_1) \ge \frac{1}{c}\int_B f(\mathbf{X}; k_0)d\mathbf{X} + \int_I f(\mathbf{X}; k_1)d\mathbf{X}$$

and apply to inequality (31.11c)

$$\Pi(R_c; k_1) \ge \frac{c}{c}\int_B f(\mathbf{X}; k_1)d\mathbf{X} + \int_I f(\mathbf{X}; k_1)d\mathbf{X}$$

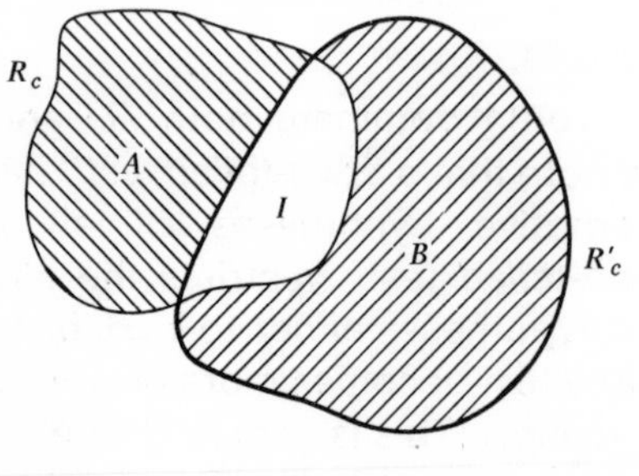

Figure 31.6
Schematic of critical regions $R_c = A \cup I$, $R'_c = B \cup I$.

or

$$\Pi(R_c; k_1) \geq \int_{R'_c} f(\mathbf{X}; k_1) d\mathbf{X} = \Pi(R'_c; k_1)$$

$$\therefore \Pi(R_c; k_1) \geq \Pi(R'_c; k_1)$$

where R'_c is any critical region of the same size as R_c. Therefore, R_c is *most powerful* with respect to any hypothesis H_1 for which (31.10) holds. **Q.E.D.**

► **Example 31.4**
Neyman-Pearson Theorem applied to Example 31.3

As before we consider the normal population

$$X \quad \text{is} \quad N(\mu, 1)$$

where μ is unknown, and we wish to test the null hypothesis

$$\mathcal{H}_0: \mu = \mu_0$$

with respect to the single alternative hypothesis

$$H_1: \mu = \mu_1$$

We wish to find the *most powerful test*, critical region R_c, for data which consist of a sample of size n: $x_1, \ldots, x_n$.

We have the (conditional) p.d.f.'s for the two hypotheses:

$$\mathcal{H}_0: f(\mathbf{X}; \mu_0) = \left(\frac{1}{2\pi}\right)^{n/2} \times \exp\left[-\frac{1}{2}\sum_{i=1}^{n}(x_i - \mu_0)^2\right] = \mathcal{L}(\mathbf{x}; \mu_0)$$

$$H_1: f(\mathbf{X}; \mu_1) = \left(\frac{1}{2\pi}\right)^{n/2} \times \exp\left[-\frac{1}{2}\sum_{i=1}^{n}(x_i - \mu_1)^2\right] = \mathcal{L}(\mathbf{x}; \mu_1)$$

We consider the ratio

$$Q(\mathbf{x}; \mu_0, \mu_1) = \frac{f(\mathbf{X}; \mu_0)}{f(\mathbf{X}; \mu_1)} = \frac{\mathcal{L}(\mathbf{x}; \mu_0)}{\mathcal{L}(\mathbf{x}; \mu_1)}$$

$$= \exp\left\{-\frac{1}{2}\left[\sum_{i=1}^{n}(x_i - \mu_0)^2 - \sum_{i=1}^{n}(x_i - \mu_1)^2\right]\right\}$$

$$= \exp\left\{-\frac{n}{2}(\mu_0^2 - \mu_1^2) + (\mu_0 - \mu_1)\sum_{i=1}^{n} x_i\right\}$$

$$= (\text{constant}) \exp\left\{(\mu_0 - \mu_1)\sum_{i=1}^{n} x_i\right\}$$

since $e^{-(n/2)(\mu_0^2 - \mu_1^2)}$ is a positive constant, independent of the data.

If we can demonstrate that a region R_c and a positive constant c both exist such that (31.10) holds, then we have R_c as a most powerful test. That is, we wish to show

$$\begin{aligned} Q(\mathbf{x}; \mu_0, \mu_1) &\leq c \quad \text{for} \quad \mathbf{x} \in R_c \\ &\geq c \quad \text{for} \quad \mathbf{x} \text{ outside } R_c \end{aligned} \tag{31.12a}$$

We need

$$\begin{aligned} (\mu_0 - \mu_1)\sum_{i=1}^{n} x_i &\leq c' \quad \text{for} \quad \mathbf{x} \in R_c \\ &\geq c' \quad \text{for} \quad \mathbf{x} \notin R_c \end{aligned}$$

or

$$\begin{aligned} (\mu_0 - \mu_1) n\bar{x} &\leq c_1 \quad \text{for} \quad \bar{x} \in R_c \\ &\geq c_1 \quad \text{for} \quad \bar{x} \notin R_c \end{aligned} \tag{31.12b}$$

Unlike c, *which must be positive*, the constants c' and c_1 may be either positive or negative since they relate to an exponent.

We see first of all that $\bar{x}$ is the *natural statistic* for finding a most powerful test.

We see also that there is no *one* most powerful region R_c against *all* alternative hypotheses to $\mathcal{H}_0$.

If $\mu_1 < \mu_0$, then (31.12b) is satisfied if

$$\bar{x} \leq c_2 \quad \text{for} \quad \bar{x} \in R_c$$

and

$$\bar{x} \geq c_2 \quad \text{for} \quad \bar{x} \notin R_c$$

so that R_c is the region: $\bar{x} \leq c_2$.

If $\mu_1 > \mu_0$ then (31.12b) is satisfied if

$$\bar{x} \geq c_3 \quad \text{for} \quad \bar{x} \in R_c$$

and

$$\bar{x} \leq c_3 \quad \text{for} \quad \bar{x} \notin R_c$$

so that R_c is the region: $\bar{x} \geq c_3$.

We stress our earlier remark that the constants c', c_1, c_2, and c_3 may be either positive or negative and still satisfy (31.11a) for $c > 0$.

The constant c_2 for the case $\mu_1 < \mu_0$ (and the constant c_3 for $\mu_1 > \mu_0$) may be chosen so as to fix α, the size of the critical region.

In general, there is no one critical region that is uniformly most powerful with respect to hypotheses $H_k: k = \mu_0 - \mu_k$ where k may be *both* positive and negative. ◄

The Likelihood Ratio

The Neyman-Pearson theorem deals with the *likelihood ratio statistic*

$$Q(\mathbf{x}; c_0, c_1) = \frac{\mathcal{L}(\mathbf{x}; c_0)}{\mathcal{L}(\mathbf{x}; c_1)}. \qquad (31.13)$$

We saw in Example 31.4 that the likelihood ratio led to the statistic $\bar{x}$, which was shown to be a *sufficient statistic* in Example 30.15). This is true in general; **the likelihood ratio leads to a sufficient statistic** *when one exists.*

A statistic is sufficient if and only if the factorization criterion (30.32) is met. Suppose we consider only two possible hypotheses, $\mathcal{H}_0$ and H_1, which correspond to $c = c_0$ and $c = c_1$, respectively. Let us consider the data $\mathbf{x}$ and suppose that the p.d.f. of $\mathbf{X}$ is $f(\mathbf{x}; c_0)$ and $f(\mathbf{x}; c_1)$ for the respective hypotheses. We have the likelihood ratio statistic

$$Q(\mathbf{x}) = \frac{f(\mathbf{x}; c_0)}{f(\mathbf{x}; c_1)}$$

We define

$$\begin{aligned} h[Q(\mathbf{x}), c] &= [Q(\mathbf{x})]^{1/2} \quad &\text{if} \quad c = c_0 \\ &= [Q(\mathbf{x})]^{-1/2} \quad &\text{if} \quad c = c_1 \end{aligned}$$

and

$$g(\mathbf{x}) = [f(\mathbf{x}; c_0)\, f(\mathbf{x}; c_1)]^{1/2}$$

$$\begin{aligned} \therefore h[Q(\mathbf{x}), c]g(\mathbf{x}) &= f(\mathbf{x}; c_0) \quad \text{if} \quad c = c_0 \\ &= f(\mathbf{x}; c_1) \quad \text{if} \quad c = c_1 \end{aligned}$$

However, the likelihood function is

$$\begin{aligned} \mathcal{L}(\mathbf{x}; c) &= f(\mathbf{x}; c_0) \quad &\text{if} \quad c = c_0 \\ &= f(\mathbf{x}; c_1) \quad &\text{if} \quad c = c_1 \end{aligned}$$

$$\therefore \mathcal{L}(\mathbf{x}; c) = h[Q(\mathbf{x}), c]g(\mathbf{x})$$

and the factorization criterion (30.32) is satisfied. Therefore, the statistic $Q(\mathbf{x})$ is a sufficient statistic.

The Generalized Likelihood Ratio

Suppose we may describe the null hypothesis $\mathcal{H}_0$ by specifying that the parameters $\mathbf{c}$, which determine the p.d.f. $f(\mathbf{X}; \mathbf{c})$ lie in a *restricted region* ω, of the parameter space, that is,

$$\mathcal{H}_0: f(\mathbf{X}; \mathbf{c}) \qquad \mathbf{c} \in \omega$$

The entire possible range of the parameters $\mathbf{c}$ is the space Ω, that is, the hypothesis *restricts* the possible values of $\mathbf{c}$. If

$$\mathbf{c} = (c_1, \ldots, c_r)$$

the *dimensionality* of Ω is r. It may be true that the hypothesis $\mathcal{H}_0$ restricts only *some* of the parameters (e.g., $c_1, \ldots, c_u$, $u \le r$) and, hence, the dimensionality of ω is u where

$$\text{Dim}(\omega) = u \le \text{Dim}(\Omega) = r$$

The alternative hypothesis H_a is

$$H_a: f(\mathbf{X}; \mathbf{c}) \qquad \mathbf{c} \in (\Omega - \omega)$$

where we have those values of the parameters $\mathbf{c}$ not in the region, ω, specified by $\mathcal{H}_0$.

As usual, we observe a sample of N values $\mathbf{x}$ and seek to test $\mathcal{H}_0$ on the basis of these results.

The likelihood of the sample may be written $\mathcal{L}(\mathbf{x}_N; \mathbf{c})$ and we distinguish $\mathcal{L}(\mathbf{x}_N; \mathbf{c} \in \omega)$ and $\mathcal{L}(\mathbf{x}_N; \mathbf{c} \in \Omega)$ where the likelihood functions are restricted to the regions ω and Ω, respectively. We seek the *maximum values*

$$\mathcal{L}_{max}(\mathbf{x}_N;\mathbf{c}\in\omega) \quad \text{and} \quad \mathcal{L}_{max}(\mathbf{x}_N;\mathbf{c}\in\Omega)$$

Here the *former* is the maximum value that $\mathcal{L}(\mathbf{x}_N;\mathbf{c}\in\omega)$ can assume for $\mathbf{c}\in\omega$; assume it occurs at $\mathbf{c}=\mathbf{c}^*_\omega$. The *latter* is the maximum value that $\mathcal{L}(\mathbf{x}_N; c\in\Omega)$ can assume over the entire unrestricted space Ω. Assume $\mathcal{L}_{max}(\mathbf{x}_N;\mathbf{c}\in\Omega)$ occurs at $\mathbf{c}=\mathbf{c}^*_\Omega$. In a condensed notation

$$\mathcal{L}_{max}(\mathbf{x}_N;\mathbf{c}\in\omega)=\mathcal{L}(\mathbf{c}^*_\omega)$$

and

$$\mathcal{L}_{max}(\mathbf{x}_N;\mathbf{c}\in\Omega)=\mathcal{L}(\mathbf{c}^*_\Omega)$$

We define the *generalized likelihood ratio*

$$Q(\omega^*,\Omega^*)=\frac{\mathcal{L}(\mathbf{c}^*_\omega)}{\mathcal{L}(\mathbf{c}^*_\Omega)} \tag{31.14}$$

We may regard (31.14) as an extension of (31.13) where we consider H_1 to be any hypothesis that has a likelihood greater than or equal to that of the null hypothesis.

Since $\mathbf{X}=x_1,\ldots,x_s$ is a random variable, so is $Q(\mathbf{x};\omega^*,\Omega^*)$. We note that

$$0\le Q\le 1$$

and would expect to believe that $\mathcal{H}_0$ is true if Q is close to 1 and reject $\mathcal{H}_0$ if Q is close to 0.

We might expect the p.d.f. of Q, $f_{\mathcal{H}_0}(Q)$, *under the assumption that $\mathcal{H}_0$ is true* to look something like Figure 31.7.

The probability of an error of the first kind is

$$\int_0^{Q_\alpha} f_{\mathcal{H}_0}(Q)\,dQ=\alpha=\Pr(\text{error I}) \tag{31.15}$$

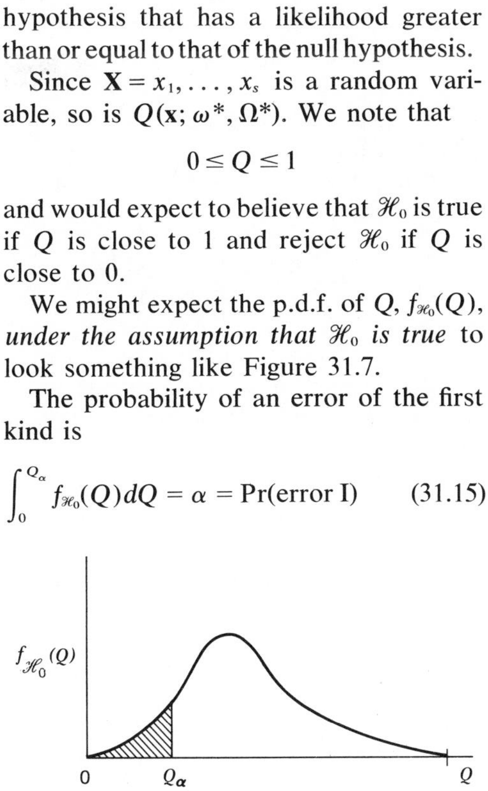

Figure 31.7
Expected appearance of the p.d.f. of Q, under the assumption that the null hypothesis is true (see text).

where the critical region for Q is $0\le Q\le Q_\alpha=\mathcal{R}_c(\alpha)$.

If we knew the p.d.f. of Q, $f_{H_a}(Q)$, under *the assumption that $\mathcal{H}_0$ is false* (H_a is true), then we could write the *power* of the Q test for the critical region $\mathcal{R}_c(\alpha)$ as:

$$\Pi[\mathcal{R}_c(\alpha)]=\int_0^{Q_a} f_{H_a}(Q)\,dQ \tag{31.16}$$

which depends on the possible alternate hypotheses.

► **Example 31.5**
Sampling from a normal distribution with unknown mean and unknown variance

Consider that $\mathbf{X}$ is $N(\mu,\sigma^2)$, and observe a sample of N measurements x. We wish to test the hypothesis, $\mathcal{H}_0$: $\mu=\mu_0$. This is an instance of a *composite* hypothesis since σ^2 may be *anything*. We form the likelihood ratio

$$Q=\frac{\mathcal{L}_{max}(\mu_0,\sigma^2)}{\mathcal{L}_{max}(\mu,\sigma^2)} \tag{31.17a}$$

where the numerator must be maximized with respect to σ^2 (μ is fixed at μ_0), but the denominator must be maximized with respect to *both* μ and σ^2.

For the numerator, we start with the likelihood function

$$\mathcal{L}(\mu_0,\sigma^2)=\left(\frac{1}{2\pi\sigma^2}\right)^{N/2}\exp\left[-\sum_{i=1}\frac{(x_i-\mu_0)^2}{2\sigma^2}\right]$$

and get the maximum

$$\mathcal{L}_{max}(\mu_0,\sigma^2)=\left(\frac{1}{2\pi(\sigma^2)^*}\right)^{N/2}\times\exp\left[-\sum_{i=1}^{N}\frac{(x_i-\mu_0)^2}{2(\sigma^2)^*}\right]$$

where the maximum value is obtained from differentiating

$$\ell(\mu_0,\sigma^2)=\ln\mathcal{L}(\mu_0,\sigma^2)=\frac{-N}{2}\ln 2\pi-N\ln\sigma-\frac{1}{2}\sum_{i=1}^{N}(x_i-\mu_0)^2\frac{1}{\sigma^2}$$

with respect to σ^2; which yields (after

setting the derivative equal to zero)

$$(\sigma^2)^* = \frac{1}{N}\sum_{i=1}^{N}(x_i-\mu_0)^2$$

$$\therefore \mathcal{L}_{\max}(\mu_0,\sigma^2) = \left(\frac{N}{2\pi\sum_{i=1}^{N}(x_i-\mu_0)^2}\right)^{N/2} e^{-N/2}$$

The denominator requires maximizing

$$\mathcal{L}(\mu,\sigma^2) = \left(\frac{1}{2\pi\sigma^2}\right)^{N/2}\exp\left[-\sum_{i=1}^{N}\frac{(x_i-\mu)^2}{2\sigma^2}\right]$$

with respect to *both* μ *and* σ^2. That is,

$$\mathcal{L}_{\max}(\mu,\sigma^2) = \left[\frac{1}{2\pi(\sigma^2)^*}\right]^{N/2} \times \exp\left[-\sum_{i=1}^{N}\frac{(x_i-\mu^*)^2}{2(\sigma^2)^*}\right]$$

We solve the two likelihood equations

$$\frac{\partial}{\partial\mu}\{\ln \mathcal{L}(\mu,\sigma^2)\} = 0$$

and

$$\frac{\partial}{\partial\sigma^2}\{\ln \mathcal{L}(\mu,\sigma^2)\} = 0$$

as in Example 30.9. We obtain

$$\mu^* = \bar{x} = \frac{1}{N}\sum_{i=1}^{N}x_i$$

and

$$(\sigma^2)^* = \frac{1}{N}\sum_{i=1}^{N}(x_i-\bar{x})^2$$

$$\therefore \mathcal{L}_{\max}(\mu,\sigma^2) = \left[\frac{N}{2\pi\sum_{i=1}^{N}(x_i-\bar{x})^2}\right]^{N/2} e^{-N/2}$$

The *likelihood ratio* is, therefore,

$$Q = \frac{\mathcal{L}_{\max}(\mu_0,\sigma^2)}{\mathcal{L}_{\max}(\mu,\sigma^2)} = \left\{\frac{\sum_{i=1}^{N}(x_i-\bar{x})^2}{\sum_{i=1}^{N}(x_i-\mu_0)^2}\right\}^{N/2}$$

$$= \frac{\sum_{i=1}^{N}(x_i-\bar{x})^2}{\sum_{i=1}^{N}(x_i-\bar{x})^2 + N(\bar{x}-\mu_0)^2}$$

$$= \frac{1}{1+\dfrac{(\bar{x}-\mu_0)^2}{\sum_{i=1}^{N}\dfrac{(x_i-\bar{x})^2}{N}}}$$

We recall (27.6a) and (27.3):

$$t = \left(\frac{\bar{x}-\mu_0}{\sigma\sqrt{N}}\right)\Big/\left(\frac{1}{\sigma}\sqrt{\sum_{i=1}^{N}\frac{(x_i-\bar{x})^2}{N-1}}\right)$$

$$= \frac{\bar{x}-\mu_0}{\sqrt{\sum_{i=1}^{N}\dfrac{(x_i-\bar{x})^2}{N(N-1)}}}$$

has the Student's t-distribution with $(N-1)$ degrees of freedom when X is normally distributed with unknown variance, σ^2.

$$\therefore Q = \left[\frac{1}{1+t^2/(N-1)}\right]^{N/2} \tag{31.17b}$$

that is, the likelihood ratio is a function only of Student's t, for any N. Since Q depends on the square of t, the critical value, Q_α, depends on *both tails* of the t-distribution for $(N-1)$ degrees of freedom, that is, we reject $\mathcal{H}_0$ when $Q \le Q_\alpha$ where

$$Q_\alpha = \left[\frac{1}{1+t_{\alpha/2}^2/(N-1)}\right]^{N/2}$$

or $t^2 \ge t_{\alpha/2}^2$. The significance level α may be read from the appropriate table for Student's t. The critical region for significance level α is

$$t < -t_{\alpha/2} \quad \text{and} \quad t > t_{\alpha/2}$$

where

$$\int_{t_{\alpha/2}}^{\infty} \mathbf{St}(t; N-1)\,dt = \alpha/2$$

as in (27.8). ◄

Large-Sample Properties of the Likelihood Ratio

As usual we consider a sample of N values of $\mathbf{x}$ and seek to test $\mathcal{H}_0$: $\mathbf{c} \in \omega$ on the basis of these observations. We note that the null hypothesis $\mathcal{H}_0$ restricts u

components of $\mathbf{c} = (c_1, \ldots, c_r)$ whereas the unrestricted parameter space, Ω, is r-dimensional.

We form the generalized likelihood ratio (31.14) and note a theorem due to S. S. Wilks:

For N large, the quantity

$$-2 \ln Q \approx \chi^2(r-u) \qquad \text{for } \mathcal{H}_0 \text{ true} \tag{31.18}$$

That is, if we assume the null hypothesis is true then, for large N, the logarithm of the likelihood ratio is distributed as χ^2 with n degrees of freedom where n equals the number of parameters not restricted by the null hypothesis. If the null hypothesis is simple, that is, $r = u$, then the number of degrees of freedom is *one*.

There are some modest requirements on the p.d.f.'s $f_{\mathcal{H}_0}(\mathbf{x}; \mathbf{c})$ and $f_{H_a}(\mathbf{x}; \mathbf{c})$ for (31.18) to hold true, but they need not concern us here.

We have considered examples in which the likelihood ratio Q could be calculated for all N. We now have the distribution for large N in all cases where the null hypothesis is assumed true. The critical region for the statistic $-2 \ln Q$ is the *right-hand tail* of the χ^2 distribution.

For example, if we wish to test $\mathcal{H}_0$ at the $\alpha = 0.05$ significance level with a large sample, we compute $-2 \ln Q$. If $-2 \ln Q > \chi^2(r-u)$ for $\alpha = 0.05$ we reject $\mathcal{H}_0$; otherwise, we accept it.

Except in special cases, it is difficult to calculate the *power* of the likelihood ratio test, since we usually do not know the form of the distribution for $\mathcal{H}_0$ *not true.* For some classes of alternative hypotheses it is possible to calculate "*noncentral*" distributions for the case, $\mathcal{H}_0$ *false.*

The Generalized χ^2 Test for Goodness-of-Fit

We have thus far considered tests wherein a test statistic is set up and evaluated on the basis of the sample. The result of this evaluation is then compared with an expected value based on the significance level and the null hypothesis accepted or rejected.

The important χ^2 test compares the **observed** *distribution of the sample* with the **(hypothesized)** *distribution of the population.*

Samples inherently involve sets of discrete numbers, that is, *observed frequencies* in *intervals* of the variate x. Suppose we have the hypothesized population p.d.f. $f(\mathbf{x})$ where we assume for the moment that all parameters are specified (simple hypothesis). Suppose we divide the entire range of x into n intervals $k = 1, \ldots, n$. The probability of observing x in interval k is

$$p_k = \Pr(x \in k) = \int_k f(x)\,dx \tag{31.19}$$

We know that each observation must fall into *some* interval; hence,

$$\sum_{k=1}^{n} p_k = 1 \tag{31.20}$$

If we observe a sample of N data points and call f_k = number (frequency) of sample observations in the kth interval, we have

$$N = \sum_{k=1}^{n} f_k \tag{31.21}$$

On the basis of the hypothesized p.d.f. $f(x)$ we would expect to find Np_k in the kth interval.

We define the *generalized chi-squared statistic*

$$X^2 = \sum_{k=1}^{n} \frac{(f_k - Np_k)^2}{Np_k} = \left(\frac{1}{N} \sum_{k=1}^{n} \frac{f_k^2}{p_k} \right) - N \tag{31.22}$$

to measure the deviation of the observed sample frequency distribution $(f_1, \ldots, f_n)$ from that expected from the hypothetical distribution.

As N becomes large, the statistic X^2 becomes (asymptotically) χ^2-distributed with $(n-1)$ degrees of freedom:

$$X^2 \to \chi^2(n-1) \qquad \text{as} \qquad N \to \infty \tag{31.23}$$

We may show this as follows. The joint probability $\Pr(f_1, \ldots, f_n)$ for observing the frequencies $(f_1, \ldots, f_n)$ is given by the *multinomial distribution* (28.6):

$$\Pr(f_1, \ldots, f_n) = \frac{N!}{f_1!f_2!\cdots f_n!}(p_1^{f_1}\cdots p_n^{f_n}) = \frac{N!}{\prod_{k=1}^{n} f_k!}\prod_{k=1}^{n} p_k^{f_k} \tag{31.24}$$

We may consider the f_k to be n independent *Poisson-distributed* variables, each with Poisson parameter Np_k. The n variables are linked by the requirement (31.21) and so (31.24) may be regarded as the *conditional* probability that n independent Poisson variables assume the values $f_1, \ldots, f_n$ *on the assumption that* (31.21) holds.

The probability $\Pr(f_1, \ldots, f_n)$ may be written

$$\Pr(f_1, \ldots, f_n) = \prod_{k=1}^{n} \frac{(Np_k)^{f_k}}{f_k!} e^{-Np_k} = N^{\Sigma f_k} \exp\left(-N\sum p_k\right)\prod_{k=1}^{n}\frac{p_k f_n}{f_k!} = N^N e^{-N}\prod_{k=1}^{n}\frac{p_k^{f_k}}{f_k!} \tag{31.25}$$

By the *reproductive property of the Poisson distribution* discussed in Chapter 24, the sum of n-independent Poisson variables is also a Poisson variable with Poisson parameter equal to the sum of the individual Poisson parameters, that is,

$$\Pr(N) = \frac{N^N}{N!}\exp\left(-\sum_{k=1}^{n} Np_k\right) = \frac{N^N}{N!}e^{-N} \tag{31.26}$$

The conditional probability needed (i.e., the probability of the sequence $f_1, \ldots, f_n$ given that $\Sigma_{k=1}^{n} f_k = N$) is the quotient of (31.25) and (31.26):

$$\Pr_{\Sigma f_k = N}(f_1, \ldots, f_n) = \frac{N!\prod_{k=1}^{n} p_k^{f_k}}{\prod_{k=1}^{n} f_k!}$$

which is (31.24). We now have, however, n independent variables each with mean and variance Np_k.

We now introduce the standardized variables

$$Z_k = \frac{(f_k - Np_k)}{\sqrt{Np_k}} \tag{31.27}$$

which are independent and *asymptotically normal* by (25.51).

$$\therefore X^2 = \sum_{k=1}^{n}\frac{(f_k - Np_k)^2}{Np_k} = \sum_{k=1}^{n} Z_k^2$$

is asymptotically $(N \to \infty)$ the sum of n squares of standard normal variables. The conditional distribution of such a sum, given

$$Z_1 + \cdots + Z_n = 0 \tag{31.28}$$

is the distribution $\chi^2(n-1)$ where the linear restriction (31.28) has reduced the number of degrees of freedom by 1.

For composite hypotheses, we wish to test the goodness of fit of the hypothesized p.d.f. $f(x;\mathbf{c})$ where we must determine the parameters $\mathbf{c} = c_1, \ldots, c_r$ from the data.

We again consider n intervals where

$$r < n$$

since otherwise the data cannot be used to determine the parameters. Also, the case $r = n$ may be used only to *solve* for a set of parameters without the "overdetermination" necessary to check *goodness-of-fit*.

In treating such composite hypotheses we assume that the parameters are evaluated by the *method of maximum likelihood.* We again write (31.19) using $f(x;\mathbf{c})$ and changing only the form of p_k to indicate that it depends on $\mathbf{c}$: $p_k(\mathbf{c})$.

The *log-likelihood function* is obtained from (31.24):

$$\ell(f_1, \ldots, f_n; \mathbf{c}) = \ln N! + \sum_{k=1}^{n} f_k \ln[p_k(\mathbf{c})] - \sum_{i=1}^{n} \ln(f_k!)$$

and we write the r likelihood equations

$$\frac{\partial \ell}{\partial c_i} = \sum_{k=1}^{n}\frac{f_k}{p_k(\mathbf{c})}\frac{\partial p_k}{\partial c_i} = 0 \qquad i = 1, \ldots, r$$

as *constraint equations* on the sum

$$X^2 = \sum_{k=1}^{n} \frac{[f_k - Np_k(\mathbf{c}^*)]^2}{Np_k(\mathbf{c}^*)} \tag{31.29}$$

The number of degrees of freedom is reduced by r, the number of constraint equations, and X^2 is asymptotically $\chi^2(n-r-1)$:

$$X^2 \xrightarrow[N\to\infty]{} \chi^2(n-r-1) \tag{31.30}$$

Use of the χ^2 Test for Goodness-of-Fit

We may now summarize the **generalized χ^2 test.** If we start with N observations of x, we divide the data into n intervals noting f_k points in each interval. This is what we have previously called *histogramming* or *binning the data.* We are interested in the asymptotic distribution of the quantity (31.29) and we must choose the intervals to contain f_k data points where f_k is *large enough* so that (31.30) is satisfied approximately. We recall that the normal approximation to the Poisson distribution is excellent for $N_p \approx f_k \geq 5$ (see Figure 25.8). Although the approximation may be adequate for smaller numbers, we usually take 4 to 5 as a lower limit: that is, we should choose our intervals to get *at least* 4 to 5 points in each interval. The intervals need not be of the same size and, in fact, intervals containing rare occurrences should be *summed together* to get the minimum number. Conversely, we learn more about the shape of the distribution the more intervals there are.

We first determine estimates $\mathbf{c}^*$ of the r parameters $\mathbf{c}$ by the maximum likelihood method (or the χ^2 minimization method). Having divided the N points x_i into n intervals, we calculate the $p_k(\mathbf{c}^*)$ and compute X^2 using (31.29).

We choose a level of significance α, and use the table of cumulative χ^2 for $(n-r-1)$ degrees of freedom to compare X^2 with $\chi^2_{1-\alpha}(n-r-1)$.

If $X^2 > \chi^2_{1-\alpha}(n-r-1)$, then the hypothesized $f(x;\mathbf{c})$ is *rejected* at the significance level α, that is, the fit is not good at the level α. Otherwise the fit [i.e., $f(x;\mathbf{c}^*)$] is accepted as being consistent with the data.

► **Example 31.6**

Prussian cavalrymen killed by horse kicks

A famous example of the Poisson distribution first discussed by L. von Bortkiewicz* concerns records kept over a period of 20 years of the number of cavalrymen in the Prussian army who died of horse kicks. The data consist of 200 numbers representing the number of deaths per year for each of 10 army corps for each of the 20 years. The data are summarized in Table 31.6.

We note the total number of observations

$$\sum_{k=0}^{7} f_k = 200$$

the total number of deaths

$$\sum_{k=0}^{7} kf_k = 122$$

and the *average* number of deaths per corps-year,

$$\bar{k} = \frac{\sum kf_k}{\sum f_k} = \frac{122}{200} = 0.61$$

*L. von Bortkiewicz, *Das Gesetz der Kleinen Zahlen,* Teubner, Leipzig, 1898.

Table ***31.6***

Number of deaths, k:	0	1	2	3	4	5	6	7
Number of observations of k, f_k:	109	65	22	3	1	0	0	0

Suppose we wish to test the hypothesis

$$\mathcal{H}_0:\ p_k = \frac{\mu^k e^{-\mu}}{k!}$$

that is, that the number of deaths is *Poisson-distributed.* We note that this is a *composite* hypothesis, since $p_k(\mu)$ depends on μ, which is not specified.

We determine $\hat{\mu}$ from the data by the maximum likelihood method. As we saw in Example 30.4, the estimate

$$\mu^* = \bar{k} = 0.61$$

is the maximum likelihood estimator.

The *expected* frequencies are then

$$Np_k = N\frac{(\bar{k})^k e^{-\bar{k}}}{k!} = \frac{(200)(0.61)^k\, e^{-0.61}}{k!}$$

The intervals $k = 3, 4, 5, 6, 7$ each contains fewer than 4 events. Accordingly, we sum the last 5 intervals and compute Np_k for the four intervals $k = 0, 1, 2, \geq 3$.

k:	0	1	2	≥ 3
f_k:	109	65	22	4
Np_k:	108.67	66.29	20.22	4.82

We calculate

$$X^2 = \sum_{k=0}^{3} \frac{(f_k - Np_k)^2}{Np_k}$$

$$= \sum_{k=0}^{3} \frac{f_k^2}{Np_k} - N$$

$$= 109.33 + 63.74 + 23.94 + 3.32 - 200 = 0.32$$

There are two constraints on the sum due to $\Sigma p_k = 1$ and $\mu = \bar{k}$. There are $n = 4$ data intervals; hence, the number of degrees of freedom is $\chi^2(4-1-1) = \chi^2(2)$.

Suppose we choose a significance level $\alpha = 0.05$, that is, there is a probability of 0.05 that the true hypothesis $\mathcal{H}_0$ will be rejected. For $\chi^2(2)$, the $\alpha = 0.05$ level corresponds to $\chi^2(2) = 5.99$, that is, 5% of the time random sampling will cause $\chi^2(2)$ to be greater than 5.99. This choice of α means that we reject the fit if $X^2 \geq 6$. The observed value 0.32 is clearly acceptable, and $\mathcal{H}_0$ is not inconsistent with the data. Actually, 0.32 is such that 85% of random $\chi^2(2)$ values are expected to be larger.

The data are clearly consistent with the hypothesis. ◄

32

CHI-SQUARE MINIMIZATION METHODS

We shall again consider a special class of the *maximum likelihood method* for estimating parameters of a probability distribution function where the data points may be assumed to be *normally distributed* about some mean value. Even in cases where the underlying distribution is not strictly normal, the central limit theorem may be invoked; therefore, the method of χ^2 minimization has quite broad utility. Once the "best estimate" parameters have been obtained, we may consider this set to represent an hypothesis, and we may use one or another version of the $\boldsymbol{\chi}^2$ **test** to test the hypothesis or goodness of fit to the observed data of the estimated probability function. In the next chapter we shall show the connection with the *method of least squares* which is computationally quite similar but even broader in applicability since it does not require the assumption of normality.

Review of χ^2

The χ^2 distribution was introduced in Chapter 25, and we shall use all of the results obtained there.

We shall consider the paradigm of the situations to be treated to consist of N data points of the form (x_i, y_i, σ_i) where the x_i are values of the independent variable X assumed known with arbitrary precision, and y_i is the *measured quantity* associated with x_i, with σ_i being the standard deviation of the y_i. We shall initially assume the y_i are *normally distributed* about some value $p(x_i; \mathbf{c})$ with variance σ_i^2.

In common laboratory usage we measure Y for different X, and σ_i represents the standard error associated with the *measurement* y_i.

Given the N data points (x_i, y_i, σ_i) we define

$$\chi^2 = \sum_{i=1}^{N} \frac{[y_i - p(x_i; \mathbf{c})]^2}{\sigma_i^2} = \chi^2(\mathbf{y}; \mathbf{x}, \boldsymbol{\sigma}, \mathbf{c}) \tag{32.1}$$

where the likelihood function is

$$\mathscr{L}(\mathbf{y}; \mathbf{x}, \boldsymbol{\sigma}, \mathbf{c}) = e^{-(1/2)\chi^2} \prod_{i=1}^{N} (2\pi\sigma_i^2)^{-1/2} \tag{32.2}$$

and

$$\ell(\mathbf{y}; \mathbf{x}, \boldsymbol{\sigma}, \mathbf{c}) = \ln \mathcal{L}$$
$$= -\frac{1}{2}\chi^2 - \frac{1}{2}\sum \ln \sigma_i^2 - \frac{N}{2}\ln 2\pi \qquad (32.3)$$

We assume the composite null hypothesis $\mathcal{H}_0$ to be that the expectation values of the y_i are given by $p(x_i; \mathbf{c})$. We seek to *determine the best estimates* of the parameters $\mathbf{c}$ and to *test* the null hypothesis at significance level α.

Because we assume the y_i are normally distributed, (32.1) follows the χ^2 distribution with the appropriate number, n, of *independent* degrees of freedom. For N data points we start with N degrees of freedom. We must reduce this number by the *number of parameters* to be determined and by the *number of constraints* on the parameters. If $\mathbf{c} = c_1, \ldots, c_r$, which are all to be determined from the data, the number of independent degrees of freedom is $N - r$, since we would use the r equations

$$\left.\frac{\partial\chi^2(\mathbf{y}; \mathbf{c})}{\partial c_i}\right|_{\mathbf{c}^*} = 0 \qquad i = 1, \ldots, r \qquad (32.4)$$

to determine $\mathbf{c}^* = c_1^*, \ldots, c_r^*$.

Also, if we added m constraint equations,

$$\varphi_j(\mathbf{c}) = 0 \qquad j = 1, \ldots, m \qquad (32.5)$$

as in (26.6), we would have

$$n = N - r - m$$

degrees of freedom.

If we expand χ^2 about its minimum value,

$$\chi^2_{\min} = \chi_0^2(\mathbf{c}^*)$$

We note that

$$\chi^2(\mathbf{y}; \mathbf{c}) = \chi_0^2(\mathbf{c}^*) + \sum_{i=1}^{r} \left.\frac{\partial\chi^2}{\partial c_i}\right|_{\mathbf{c}^*} (c_i - c_i^*)$$
$$+ \frac{1}{2}\sum_{\substack{i=1\\j=1}}^{r} \left.\frac{\partial^2\chi^2}{\partial c_i\,\partial c_j}\right|_{\mathbf{c}^*} (c_i - c_i^*)(c_j - c_j^*) + \cdots$$
$$= \chi_0^2(c^*) + \frac{1}{2}\sum_{\substack{i=1\\j=1}}^{r} \left.\frac{\partial^2\chi^2}{\partial c_i \partial c_j}\right|_{\mathbf{c}^*} \times (c_i - c_i^*)(c_j - c_j^*) + \cdots \qquad (32.6)$$

using (32.4). This represents a *quadratic surface* or *hyperparaboloid* in $(r+1)$-dimensional space.

We seek the $\chi^2_{\min}$ point $\mathbf{c}^* = (c_1^*, \ldots, c_r^*)$. If we have m constraint equations of the form (32.5), we realize that these define surfaces that intersect the hyperparaboloid and define a new surface whose $\chi^2_{\min}$ point we must determine. In this geometrical picture we see that *nonholonomic* constraints of the form

$$\Psi(\mathbf{c}) \geq 0$$

also define regions in the $(r+1)$-dimensional space that restrict the χ^2 surface. We always seek the $\chi^2_{\min}$ points on the *resulting surface*.

Having obtained the $\chi^2_{\min}(\mathbf{c}^*)$ point *by some means*, we have $\mathbf{c}^*$ as our best estimates of the parameters. We compute $\chi^2_{\min}(\mathbf{c}^*)$ and compare with $\chi^2(n)$ to determine the goodness of fit where n is the number of independent degrees of freedom.

We note that we have assumed the underlying distribution to be normal. If the constraint equations have not selected out a surface too different from the hyperparaboloid, we may use (30.45) in the form

$$\mathcal{L}(\mathbf{y}; \mathbf{x}, \boldsymbol{\sigma}, \mathbf{c})$$
$$= K \exp\left[-\frac{1}{2}\sum_{\substack{k=1\\l=1}}^{r} A_{kl}(c_k - c_k^*)(c_l - c_l^*)\right]$$

where

$$A_{kl} = \frac{1}{2}\left\{\left.\frac{\partial^2\chi^2}{\partial c_k \partial c_l}\right|_{\mathbf{c}^*}\right\} \qquad (32.7)$$

We use relation (30.46) to define the *covariance matrix* of the c_k^*, c_l^*, that is,

$$\mathrm{Covar}(c_k^*, c_l^*) = (\underline{\boldsymbol{\sigma}})_{kl} = (\underline{\mathbf{A}})_{kl}^{-1} \qquad (32.8)$$

I.e. we form the *matrix*

$$(\underline{\mathbf{A}})_{kl} = \left\{\frac{1}{2}\left.\frac{\partial^2\chi^2}{\partial c_k \partial c_l}\right|_{\mathbf{c}^*}\right\}$$

its *inverse*, also called the **error matrix**, specifies the variances and covariances of the (c_k^*, c_l^*). The *diagonal* elements of the error matrix represent the *variances* of the estimators while the **off-diagonal** elements represent the **covariances.** The **error matrix** (*covariance matrix*), of course, is symmetric: $\sigma_{ij} = \sigma_{ji}$.

If the constrained surface is too distorted from the hyperparaboloidal shape by the constraint equations, some appropriate average of

$$\left\{\frac{1}{2}\frac{\partial^2\chi^2}{\partial c_k \partial c_l}\right\}$$

should be employed in place of

$$\left\{\frac{1}{2}\frac{\partial^2\chi^2}{\partial c_k \partial c_l}\bigg|_{c*}\right\}$$

similar to (30.49).

One parameter χ^2: $\mathbf{c} = c_1$ $(r = 1)$

Here we have

$$\chi^2 = \sum_{i=1}^{N} \frac{[y_i - p(x_i; c_1)]^2}{\sigma_i^2}$$

which we minimize by setting $c_1 = c_1^*$, where

$$\frac{\partial\chi^2}{\partial c_1}\bigg|_{c_1^*} = \frac{d\chi^2}{dc_1}\bigg|_{c_1^*} = 0$$

$$= -2\sum_{i=1}^{N}[y_i - p(x_i; c_1^*)] \times \left(\frac{1}{\sigma_i^2}\right)\frac{dp(x_i; c_1}{dc_1}$$

We note that

$$-\frac{d^2\ell}{dc_1^2} = \frac{1}{2}\frac{d^2\chi^2}{dc_1^2}$$

and the assumption of *normality* permits us to use (30.42).

$$\therefore \mathrm{Var}(c_1^*) = \sigma^2(c_1^*) = -\left(\frac{d^2\ell}{dc^2}\bigg|_{c*}\right)^{-1}$$

$$= \left\{\frac{1}{2}\frac{d^2\chi^2}{dc_1^2}\bigg|_{c_1 = c_1^*}\right\}^{-1} \qquad (32.9)$$

We also note that

$$\chi^2(c) - \chi_0^2(c*) = \left\{\frac{1}{2}\frac{\partial^2\chi^2(c)}{\partial c_1^2}\bigg|_{c_1^*}\right\}(c_1 - c_1^*)^2$$

may be evaluated at $c_1 = c_1^* \pm \sigma(c_1^*)$:

$$\chi^2[c_1^* \pm \sigma(c_1^*)] - \chi_0^2(c_1^*) = 1 \qquad (32.10)$$

that is, we may read c_1^* and $\sigma(c_1^*)$ from the graph of $\chi^2(c)$ versus c. c_1^* occurs at the *minimum point* χ_0^2 while $c_1^* \pm \sigma(c_1^*)$ occurs at $\chi^2_{\min} + 1$.

► **Example 32.1**

***N* measurements of the same quantity**

Suppose we have N measurements (y_i, σ_i) that purport to be of the *same* quantity, $\bar{y}$. We have treated this problem in Example 26.4. We specify the hypothesis $\mathcal{H}_0$: $y_i = \bar{y}$ where we wish to estimate the *single* parameter, $\bar{y}$. As in (26.12) we note that

$$\chi^2 = \sum_{i=1}^{N}\frac{(y_i - \bar{y})^2}{\sigma_i^2}$$

is minimized by

$$\bar{y} = \frac{\sum_{i=1}^{N}(y_i/\sigma_i^2)}{\sum_{i=1}^{N}(1/\sigma_i^2)} \qquad (26.12)$$

and that

$$\mathrm{Var}(\bar{y}) = \left(\sum_{i=1}^{N}\frac{1}{\sigma_i^2}\right)^{-1} \qquad (26.13)$$

which may be written

$$\sigma^2(c_1^*) = \left[\frac{1}{2}\frac{\partial^2\chi^2}{\partial c_1^2}\bigg|_{c_1^*}\right]^{-1}$$

This is an instance of (32.9), which we emphasize is true for *all* one-parameter problems.

Numerical example (Example 25.1) coin weighing

Let us consider the situation in which we have compared the weights of groups of 1964 coins and 1970 coins using a beam balance (BB) and also an analytical balance (AB), which is more precise. The results are summarized below where we write: $\bar{w} \pm \Delta\bar{w}$ to connote the mean and the standard deviation of the mean

	BB	AB
1964:	3.0733 ± 0.0205	3.0806 ± 0.0080
1970:	3.0968 ± 0.0183	3.1045 ± 0.0059

We wish to consider the hypothesis that the means are equal, that is, $\mathcal{H}_0$: $\bar{w}_{1964} = \bar{w}_{1970} = \mu$.

Using the *BB* results we form

$$\chi^2(\mu) = \sum_{i=1}^{2} \frac{(\bar{w}_i - \mu)^2}{\sigma_i^2} = \frac{(3.0733 - \mu)^2}{(0.0205)^2} + \frac{(3.0968 - \mu)^2}{(0.0183)^2}$$

We may plot $\chi^2(\mu)$ versus μ in Figure 32.1. We may observe several features of this plot:

(i) The curve is *parabolic.*

(ii) The minimum χ^2 occurs at a point given by (26.12):

$$\mu^* = \left\{\frac{3.0733}{(0.0205)^2} + \frac{3.0968}{(0.0183)^2}\right\} \Big/ \left\{\left(\frac{1}{0.0205}\right)^2 + \left(\frac{1}{0.0183}\right)^2\right\}$$
$$= 3.0864$$

(iii) The *minimum value* is

$$\chi^2_{\min} = \chi^2(\mu^*) = 0.731$$

(iv) $\chi^2(\mu)$ rises to $\chi^2_{\min} + 1 = 1.731$ at points

$$\mu = \mu^* \pm \Delta\mu^*$$

where $\Delta\mu^*$ is given by (26.13):

$$\Delta\mu^* = \left\{\left(\frac{1}{0.0205}\right)^2 + \left(\frac{1}{0.0183}\right)^2\right\}^{-1/2}$$
$$= 0.0136$$

We have two data points; hence $N = 2$. Since we determine μ^* from the data, there are $N - r = 2 - 1 = 1$ independent degree of freedom. We must, therefore, compare $\chi^2_{\min}$ with $\chi^2(1)$. By linear interpolation in the table of percentage points of the χ^2 distribution, we get

$$\Pr\{\chi^2(1) \geq 0.731\} = 0.38$$

We usually set the level of significance at $\alpha = 0.05$, which corresponds for $\chi^2(1)$ to $\chi^2 \geq 3.84$, that is, we reject the null hypothesis if $\chi^2_{\min} \geq 3.84$. We see that the beam balance data are *consistent* with the hypothesis that the two populations have equal means, that is, $\bar{w}_{1964} = \bar{w}_{1970} = \mu^*$, where $\mu = 3.0864 \pm 0.0136$.

This is shown in Figure 32.2*a*, where the two data points are shown with their (*one-standard-deviation*) **error bars**, and the value $\mu^* = \bar{w}$ is shown as a *horizontal line* at $w = 3.0864$. Graphically, the data are seen to be consistent with the hypothesis.

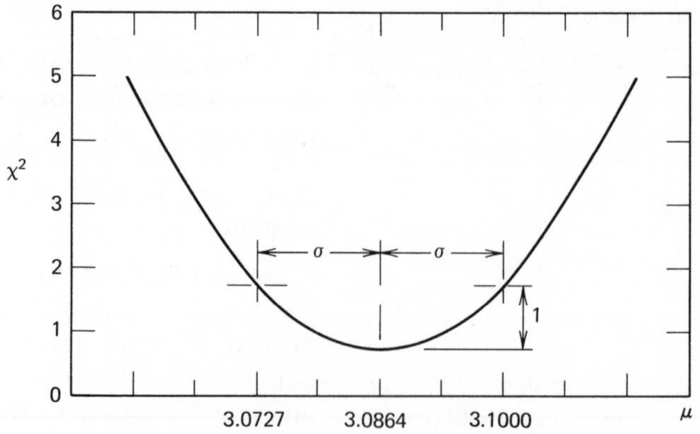

Figure 32.1
Plot of $\chi^2(\mu)$ versus μ (beam balance data).

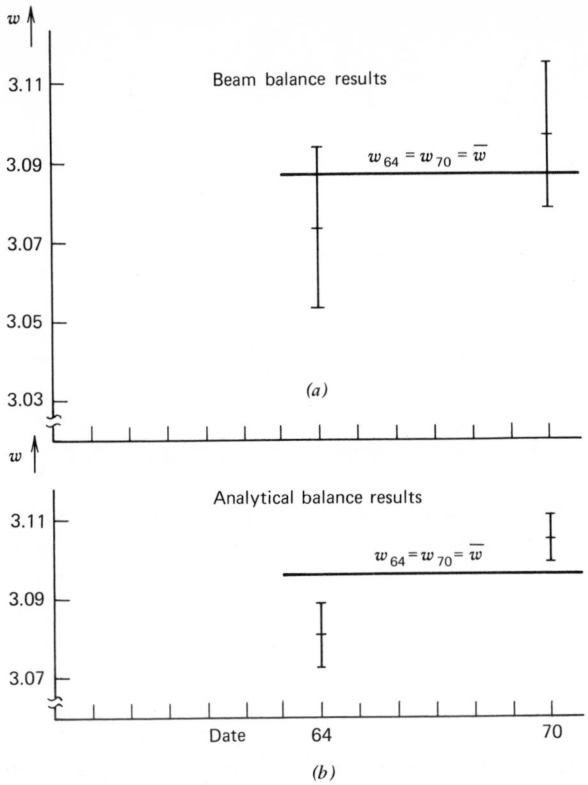

Figure 31.2
Plot of two data points with one-standard-deviation error bars and a straight line indicating the means are equal. (*a*) Beam balance data. (*b*) Analytical balance data.

We note that the *probability* (0.38) of a difference arising by chance as large as (or larger than) that observed *given that the hypothesis is true* agrees with what was obtained by the *normal deviate test* in Example 25.1.

We must emphasize that we have *not proven* the null hypothesis to be true. We have only shown that the beam balance data are *consistent* with the hypothesis. A more stringent test of any hypothesis is always valuable, and we turn to the *analytical balance data*.

Using the AB results, we form

$$\chi^2(\mu) = \sum_{i=1}^{2} \frac{(\bar{w}_i - \mu)^2}{\sigma_i^2} = \frac{(3.1045 - \mu)^2}{(0.0059)^2} + \frac{(3.0806 - \mu)^2}{(0.0080)^2}$$

We plot $\chi^2(\mu)$ versus μ in Figure 32.3.

We note again that

(i) The curve is *parabolic.*

(ii) The minimum $\chi^2(\mu)$ occurs at μ^* given by (26.12)

$$\mu^* = \left\{\frac{3.1045}{(0.0059)^2} + \frac{3.0806}{(0.0080)^2}\right\} \Big/ \left\{\left(\frac{1}{0.0059}\right)^2 + \left(\frac{1}{0.0080}\right)^2\right\}$$
$$= 3.0961$$

(iii) The *minimum value* is

$$\chi^2_{\min} = \chi^2(\mu^*) = 5.781$$

(iv) $\chi^2(\mu)$ rises to $\chi^2_{\min} + 1 = 6.781$ at

$$\mu = \mu^* \pm \Delta\mu^*$$

where $\Delta\mu^*$ is given by (26.13):

$$\Delta\mu^* = \left\{\left(\frac{1}{0.0059}\right)^2 + \left(\frac{1}{0.0080}\right)^2\right\}^{-1/2}$$
$$= 0.00475$$

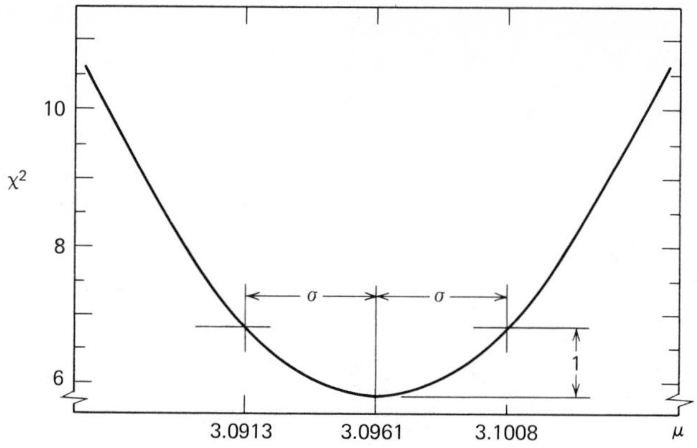

Figure 32.3
Plot of $\chi^2(u)$ versus u (analytical balance data).

We have this as our best estimate of μ, assuming the null hypothesis

$$\mu^* = 3.0961 \pm 0.0048$$

We compare $\chi^2_{\min} = 5.781$ with $\chi^2(1)$. By linear interpolation in the χ^2 tables we note

$$\Pr\{\chi^2(1) \geq 5.781\} = 0.017$$

That is, there is only a 1.7% chance that $\chi^2(1)$ can assume a value as large as 5.781, *provided the hypothesis is true*. Setting our critical region at $\alpha = 0.05$, we *reject* the null hypothesis, that is, the analytical balance data are *not* consistent with the null hypothesis at the 5% level of significance. We plot the data points in Figure 32.2*b* together with the "best-fit curve" (here the horizontal line $w = 3.0961$) and note that the fit is not terribly good.

Again we remark that the fact that the analytical balance data are only consistent with the hypothesis at the 1.7% confidence level agrees with the result of the *normal deviate test* in Example 25.1. ◄

Multiparameter χ^2: $\mathbf{c} = (c_1, \ldots, c_r)$

For this general case we use (32.1) as the definition of χ^2. After obtaining the minimum point by some means, we notice that the *best estimators* are, in general, **correlated**. An *arbitrary* χ^2 problem must usually be attacked by special or numerical methods. An interesting class of problems that is more tractable than most is that in which the hypothetical frequency distribution is *linear* in the parameters to be estimated.

LINEAR CASE

Suppose

$$p(x_i; \mathbf{c}) = \sum_{k=1}^{r} c_k g_k(x_i) \qquad (32.11)$$

where the $g_k(x_i)$ are *known functions* of only the known quantities x_i. Therefore,

$$\chi^2 = \sum_{i=1}^{N} \left\{ \left[y_i - \sum_{k=1}^{r} c_k g_k(x_i) \right]^2 \Big/ \sigma_i^2 \right\} \qquad (32.12)$$

This case includes any series expansion

$$y_i = c_1 g_1(x_i) + c_2 g_2(x_i) + \cdots + c_r g_r(x_i) \qquad (32.13)$$

where we wish to determine the *coefficients* or *expansion parameters.*

We may take derivatives:

$$\begin{aligned} \frac{\partial \chi^2}{\partial c_l} &= -2 \sum_{i=1}^{N} \left\{ \left[y_i - \sum_{k=1}^{r} c_k g_k(x_i) \right] \left(\frac{1}{\sigma_i^2} \right) \left(\frac{\partial p}{\partial c_l} \right) \right\} \\ &= -2 \sum_{i=1}^{N} \left\{ \left[y_i - \sum_{k=1}^{r} c_k g_k(x_i) \right] \left(\frac{1}{\sigma_i^2} \right) g_l(x_i) \right\} \end{aligned} \qquad (32.14)$$

The second derivative is

$$\frac{\partial^2 \chi^2}{\partial c_l \, \partial c_m} = +2 \sum_{i=1}^{N} \left(\frac{1}{\sigma_i^2}\right) g_l(x_i) g_m(x_i) \tag{32.15}$$

Let us call

$$\frac{1}{2} \frac{\partial^2 \chi^2}{\partial c_l \, \partial c_m} = \sum_{i=1}^{N} g_l(x_i) g_m(x_i) \frac{1}{\sigma_i^2} \equiv A_{lm}, \; l, m = 1, \ldots, r \tag{32.16}$$

which are the elements of an $(r \times r)$ symmetric matrix $\underline{\mathbf{A}}$, which depends only on the x_i and the (*inverse weights*) σ_i.

We use (32.4) and (32.14) to get the r equations (called the *normal equations*)

$$\left.\frac{\partial \chi^2}{\partial c_l}\right|_{c^*} = 0 = \sum_{i=1}^{N} \frac{y_i g_l(x_i)}{\sigma_i^2} - \sum_{i=1}^{N} \left[\frac{g_l(x_i)}{\sigma_i^2} \sum_{k=1}^{r} c_k^* g_k(x_i)\right] = u_l - \sum_{k=1}^{r} c_k^* \left\{\sum_{i=1}^{N} g_k(x_i) \, g_l(x_i) \left(\frac{1}{\sigma_i^2}\right)\right\}$$

where we have introduced the r-dimensional *data vector* $\mathbf{u}$:

$$u_l = \sum_{i=1}^{N} \frac{y_i g_l(x_i)}{\sigma_i^2} \qquad l = 1, \ldots, r \tag{32.17}$$

which depends on the measured values, y_i. We therefore have

$$u_l = \sum_{k=1}^{r} c_k^* \left\{\sum_{i=1}^{N} g_k(x_i) \, g_l(x_i) \frac{1}{\sigma_i^2}\right\} = \sum_{k=1}^{r} c_k^* A_{kl}$$

using (32.16). Since the matrix $\underline{\mathbf{A}}$ is symmetric ($A_{kl} = A_{lk}$), we may write this as

$$u_l = \sum_{k=1}^{r} A_{lk} c_k^* \tag{32.18}$$

which is the matrix equation

$$\mathbf{u} = \underline{\mathbf{A}}\mathbf{c}^* \tag{32.19}$$

We may solve this matrix equation for the *vector of best estimates*, $\mathbf{c}^*$:

$$\mathbf{c}^* = (\underline{\mathbf{A}})^{-1} \mathbf{u} \tag{32.20}$$

where $\underline{\mathbf{A}}^{-1}$ is the matrix inverse to $\underline{\mathbf{A}}$:

$$(\underline{\mathbf{A}}^{-1})_{ij} = \frac{\text{Cofactor } A_{ji}}{|\underline{\mathbf{A}}|} = \frac{\text{Cof } A_{ij}}{|\underline{\mathbf{A}}|}. \tag{32.21}$$

Here $|\underline{\mathbf{A}}|$ is the determinant of the matrix $\underline{\mathbf{A}}$. We note that the formula (32.21), while true in general, is not always the best way to evaluate the *inverse matrix*, especially when $r \geq 4$. One of the nice features of a matrix formulation is that digital computer installations are usually set up to handle matrix manipulations in a convenient fashion.

FITTING N MEASUREMENTS TO A STRAIGHT LINE: $y = ax + b$

We shall discuss straight-line fitting in detail. This case is important because it arises directly in many situations of interest. Also, as we have seen in Part II, the resulting graph is easy to plot. Furthermore, we showed in Part II that if a *relationship* between $\mathbf{Y}$ and $\mathbf{X}$ exists, it can often be transformed to new variables $\mathbf{Y}'$ and $\mathbf{X}'$ that do have a *linear relationship.* (We should caution the reader that the *relative weights* of the points must be carefully tracked through the transformation.) Finally, linear fitting involving only *two* parameters enables us to visualize the geometry of the χ^2 minimization method, since the χ^2 surface exists in *three* dimensions.

Consider that we have N measurements of the form (x_i, y_i, σ_i) where x_i is the independent mathematical (*not random*) variable assumed *known* with arbitrary precision and (y_i, σ_i) is the *measured* quantity with its given *error* (in the sense of a *standard deviation*). We assume that the y_i are *random variables* that are *normally* distributed about values $ax_i + b$ with variance σ_i^2, that is, y_i is $N(ax_i + b, \sigma_i^2)$. We wish to determine the parameters of a straight line $y = ax + b$ that "*fits*" the data. This is called the "*regression line.*"

We form

$$\chi^2 = \sum_{i=1}^{N} \frac{[y_i - (ax_i + b)]^2}{\sigma_i^2} \tag{32.22}$$

We expect that $\chi^2(a^*, b^*)$, where a^*, b^* are determined from the data, is distributed as $\chi^2(N-2)$ since two parameters are determined.

We note that

$$\left.\frac{\partial\chi^2}{\partial a}\right|_{a^*} = 0 = \sum_{i=1}^{N} 2\frac{[y_i - (a^*x_i + b^*)]}{\sigma_i^2}(-x_i) \tag{32.23}$$

and

$$\left.\frac{\partial\chi^2}{\partial b}\right|_{b^*} = 0 = \sum_{i=1}^{N} 2\frac{[y_i - (a^*x_i + b^*)]}{\sigma_i^2}(-1) \tag{32.24}$$

If we define

$$A = \sum_{i=1}^{N}\frac{x_i}{\sigma_i^2} \qquad B = \sum_{i=1}^{N}\frac{1}{\sigma_i^2}$$

$$C = \sum_{i=1}^{N}\frac{y_i}{\sigma_i^2} \qquad D = \sum_{i=1}^{N}\frac{x_i^2}{\sigma_i^2} \tag{32.25}$$

$$E = \sum_{i=1}^{N}\frac{x_iy_i}{\sigma_i^2} \qquad F = \sum_{i=1}^{N}\frac{y_i^2}{\sigma_i^2}$$

then (32.23) yields

$$0 = E - a^*D - b^*D$$

and (32.24) gives

$$0 = C - a^*A - b^*B$$

These are *linear equations* in a^*, b^*:

$$E = a^*D + b^*A$$
$$C = a^*A + b^*B$$

which may be solved directly, since there are only two or by *Cramer's rule* (Appendix B) using determinants:

$$a^* = \frac{\begin{vmatrix} E & A \\ C & B \end{vmatrix}}{\begin{vmatrix} D & A \\ A & B \end{vmatrix}} = \frac{EB - CA}{DB - A^2} \tag{32.26}$$

and

$$b^* = \frac{\begin{vmatrix} D & E \\ A & C \end{vmatrix}}{\begin{vmatrix} D & A \\ A & B \end{vmatrix}} = \frac{DC - EA}{DB - A^2} \tag{32.27}$$

The parameters of the "best-fit" straight line are thus given by formulas that are easily derived.

We note that (32.26) and (32.27) could also have been obtained using (32.20).

For this problem, [see (32.13)]

$$g_1(x) = x \qquad c_1 = a$$

and

$$g_2(x) = 1 \qquad c_2 = b$$

From (32.17)

$$u_1 = \sum_{i=1}^{N}\frac{y_ig_1(x_i)}{\sigma_i^2} = \sum_{i=1}^{N}\frac{y_ix_i}{\sigma_i^2} = E$$

$$u_2 = \sum_{i=1}^{N}\frac{y_ig_2(x_i)}{\sigma_i^2} = \sum_{i=1}^{N}\frac{y_i}{\sigma_i^2} = C$$

and

$$\mathbf{u} = \begin{pmatrix} E \\ C \end{pmatrix}$$

$$\therefore \mathbf{c}^* = \begin{pmatrix} c_1^* \\ c_2^* \end{pmatrix} = \begin{pmatrix} a^* \\ b^* \end{pmatrix} = (\underline{\mathbf{A}}^{-1})\mathbf{u}$$

$$= \frac{1}{BD - A^2}\begin{pmatrix} B & -A \\ -A & D \end{pmatrix}\begin{pmatrix} E \\ C \end{pmatrix}$$

$$= \frac{1}{BD - A^2}\begin{pmatrix} EB - AC \\ -AE + DC \end{pmatrix}$$

which agrees with (32.26) and (32.27).

Goodness of fit

The quality of the fit may be studied by calculating

$$\chi^2(a^*, b^*) = \sum_{i=1}^{N}\frac{[y_i - (a^*x_i + b^*)]^2}{\sigma_i^2}$$

and comparing the observed value with $Q(\chi^2; N-2)$ from the χ^2 tables.

Error matrix

We must also inquire into the errors that should be assigned to the estimates a^*, b^*. We must calculate the matrix (32.16), which is (2×2) in this case:

$$\begin{pmatrix} A_{11} & A_{12} \\ A_{21} & A_{22} \end{pmatrix}$$

where

$$A_{11} = \left.\frac{1}{2}\frac{\partial^2\chi^2}{\partial a^2}\right|_{a^*,b^*}$$

$$A_{12} = A_{21} = \left.\frac{1}{2}\frac{\partial^2\chi^2}{\partial b^2}\right|_{a^*,b^*}$$

$$A_{22} = \left.\frac{1}{2}\frac{\partial^2\chi^2}{\partial b^2}\right|_{a^*,b^*}$$

The inverse of $\underline{\mathbf{A}}$ is the *covariance* or *error matrix* for a^*, b^*.

Using (32.22) and (32.25) we get

$$\chi^2(a, b) = \sum_{i=1}^{N} \frac{1}{\sigma_i^2} [y_i^2 - 2ax_iy_i - 2by_i + a^2x_i^2 + 2abx_i + b^2]$$
$$= F - 2aE - 2bC + a^2D + 2abA + b^2B$$

Therefore,

$$A_{11} = \frac{1}{2}\frac{\partial^2\chi^2}{\partial a^2} = D$$

$$A_{12} = A_{21} = \frac{1}{2}\frac{\partial^2\chi^2}{\partial a\,\partial b} = A$$

$$A_{22} = \frac{1}{2}\frac{\partial^2\chi^2}{\partial b^2} = B$$

or

$$\underline{\mathbf{A}} = \begin{pmatrix} D & A \\ A & B \end{pmatrix} \qquad (32.28)$$

For the *inverse matrix* we may use (32.21) since the matrix is only (2 × 2).

$$|\underline{\mathbf{A}}| = \det(\underline{\mathbf{A}}) = BD - A^2$$

We get the elements of the *covariance matrix*

$$(A^{-1})_{11} = \frac{\text{Cof } A_{11}}{|\underline{\mathbf{A}}|} = \frac{B}{BD - A^2} = \text{Var}(a^*)$$

$$(A^{-1})_{12} = (A^{-1})_{21} = \frac{\text{Cof } A_{12}}{|\underline{\mathbf{A}}|} = \frac{-A}{BD - A^2} = \text{Covar}(a^*, b^*)$$

$$(A^{-1})_{22} = \frac{\text{Cof } A_{22}}{|\underline{\mathbf{A}}|} = \frac{D}{BD - A^2} = \text{Var}(b^*)$$

or

$$\underline{\mathbf{A}}^{-1} = \frac{1}{BD - A^2}\begin{pmatrix} B & -A \\ -A & D \end{pmatrix}$$
$$= \begin{pmatrix} \sigma_{a^*}^2 & \sigma_{a^*b^*} \\ \sigma_{a^*b^*} & \sigma_{b^*}^2 \end{pmatrix} \qquad (32.29)$$

We see that the parameters a^* and b^* are *correlated*, and a complete statement of the results of the analysis would quote the entire *error matrix*. It is common, if sometimes inadequate, to quote

$$a^* \pm \sigma_{a^*}$$
$$b^* \pm \sigma_{b^*}$$

as the "results," but we should keep the existence of *correlations* in mind.

Choice of points

The choice of points x_i affects the correlation of the parameters and the standard deviations of a^* and b^*. In particular, the quantities A, B, and D depend only on the x_i.

Since $\text{Covar}(a^*, b^*) \propto A = \sum_{i=1}^N x_i/\sigma_i^2$, we may insure that the estimates of a^* and b^* are *uncorrelated* by choosing the (weighted) x_i *symmetric* about $x = 0$, that is, choose $A = \sum_{i=1}^N x_i/\sigma_i^2 = 0$.

If we define

$$\bar{x} = \left(\sum_{i=1}^{N}\frac{x_i}{\sigma_i^2}\right)\left(\sum_{i=1}^{N}\frac{1}{\sigma_i^2}\right)^{-1} = A/B \qquad (32.30)$$

and

$$\overline{x^2} = \left(\sum_{i=1}^{N}\frac{x_i^2}{\sigma_i^2}\right)\left(\sum_{i=1}^{N}\frac{1}{\sigma_i^2}\right)^{-1} = D/B$$

the denominator of Var(a^*) and Var(b^*) becomes

$$DB - A^2 = [\overline{x^2} - (\bar{x})^2]\left(\sum_{i=1}^{N}\frac{1}{\sigma_i^2}\right)^2$$
$$= \text{Var}(x)\left(\sum_{i=1}^{N}\frac{1}{\sigma_i^2}\right)^2$$

Thus we may *minimize* the errors on a^* and b^* by choosing the x_i to have as *broad a spread* as possible. If we were *not* concerned about *testing* the goodness of fit of the linearity hypothesis, we would endeavor to take *all N* measurements at the two *most distant* values of x. This is intuitively clear from the graph of a straight line. One loses, of course, a *test* of linearity. A good compromise is to choose the points x_i/σ_i^2 *symmetrically* about the origin ($\bar{x} = 0$) and as *widely separated* as possible.

A feature of this method of straight-line fitting is that the regression line

$$y = a^*x + b^*$$

passes through the **weighted centroid** $(\bar{x}, \bar{y})$ of the data where $\bar{x}$ is given by (32.30) and

$$\bar{y} = \left(\sum_{i=1}^{N}\frac{y_i}{\sigma_i^2}\right)\left(\sum_{i=1}^{N}\frac{1}{\sigma_i^2}\right)^{-1} = C/B \qquad (32.31)$$

which can be shown by substitution, that is,

$$\bar{y} = a^*\bar{x} + b^* \qquad (32.32)$$

► **Example 32.2**
Straight-line fitting

Generation of random data

Let us take a specific relation $y = ax + b$ and *generate* normally distributed data points from this for six values x_i and σ_i. We do this by choosing a set of six random Gaussian deviates R_i and generating

$$y_i = ax_i + b + R_i\sigma_i$$

These correspond to measurements, y_i, with their "errors," σ_i. The generation of these data is shown in Table 32.1 for $a = 4$, $b = 1$.

$$b^* = \frac{CD - EA}{DB - A^2}$$

$$= \frac{(61.47)(40.98) - (184.40)(12.70)}{201.79}$$

$$= \frac{177.16}{201.79} = 0.878$$

The *best-fit straight line* is, therefore,

$$y = 4.23x + 0.88$$

This is plotted in Figure 32.4 together with the six data points y_i with their assigned standard deviations plotted as *"error flags"* ($\pm\sigma_i$).

A measure of the uncertainties in the estimates a^* and b^* is obtained from (32.29):

$$\overline{(\Delta a^*)^2} = \sigma_{a^*}^2 = \frac{B}{DB - A^2} = \frac{8.86}{201.79}$$
$$= 0.0439$$

$$\overline{(\Delta b^*)^2} = \sigma_{b^*}^2 = \frac{D}{DB - A^2} = \frac{40.98}{201.79}$$
$$= 0.2031$$

$$\overline{\Delta a^* \Delta b^*} = \sigma_{a^*b^*} = \frac{-A}{DB - A^2} = \frac{-12.70}{201.79}$$
$$= -0.0629$$

If we ignore the correlation between a^* and b^*, we may quote

$$a^* \pm \Delta a^* = 4.23 \pm 0.21$$

and

$$b^* \pm \Delta b^* = 0.88 \pm 0.45$$

Table ***32.1***

x_i	0	1	2	3	4	5
$4x_i + 1$	1	5	9	13	17	21
σ_i	1/2	1	3/4	5/4	1	3/2
Random Deviate	−0.158	−0.848	+1.041	+1.167	+0.261	0.601
$R_i\sigma_i$	−0.08	−0.85	+0.78	+1.46	+0.26	+0.90
y_i	0.92	4.15	9.78	14.46	17.26	21.90

Analysis

We now have six data points (y_i; x_i, σ_i) that we may treat as the results of an experiment and from which we may calculate the quantities (32.25) (Table 32.2).

We may calculate the best estimates of a^*, b^* using (32.26) and (32.27):

$$a^* = \frac{EB - CA}{DB - A^2}$$

$$= \frac{(184.40)(8.86) - (61.47)(12.70)}{(40.98)(8.86) - (12.70)^2}$$

$$= \frac{853.12}{201.79} = 4.227$$

Table **32.2**

x_i	0	1	2	3	4	5	
σ_i	1/2	1	3/4	5/4	1	3/2	
y_i	0.92	4.15	9.78	14.46	17.26	21.90	
σ_i^2	0.25	1.00	0.56	1.56	1.00	2.25	
$1/\sigma_i^2$	4.00	1.00	1.78	0.64	1.00	0.44	$\Sigma\,(1/\sigma_i^2) = B = 8.86$
x_i/σ_i^2	0	1.00	3.56	1.92	4.00	2.22	$\Sigma\,(x_i/\sigma_i^2) = A = 12.70$
y_i/σ_i^2	3.68	4.15	17.39	9.25	17.26	9.73	$\Sigma\,(y_i/\sigma_i^2) = C = 61.47$
x_iy_i/σ_i^2	0	4.15	34.78	27.76	69.04	48.67	$\Sigma\,(x_iy_i/\sigma_i^2) = E = 184.40$
y_i^2/σ_i^2	3.40	17.24	170.07	133.80	297.94	213.19	$\Sigma\,(y_i^2/\sigma_i^2) = F = 835.63$
x_i^2/σ_i^2	0	1.00	7.11	5.76	16.00	11.11	$\Sigma\,(x_i^2/\sigma_i^2) = D = 40.98$

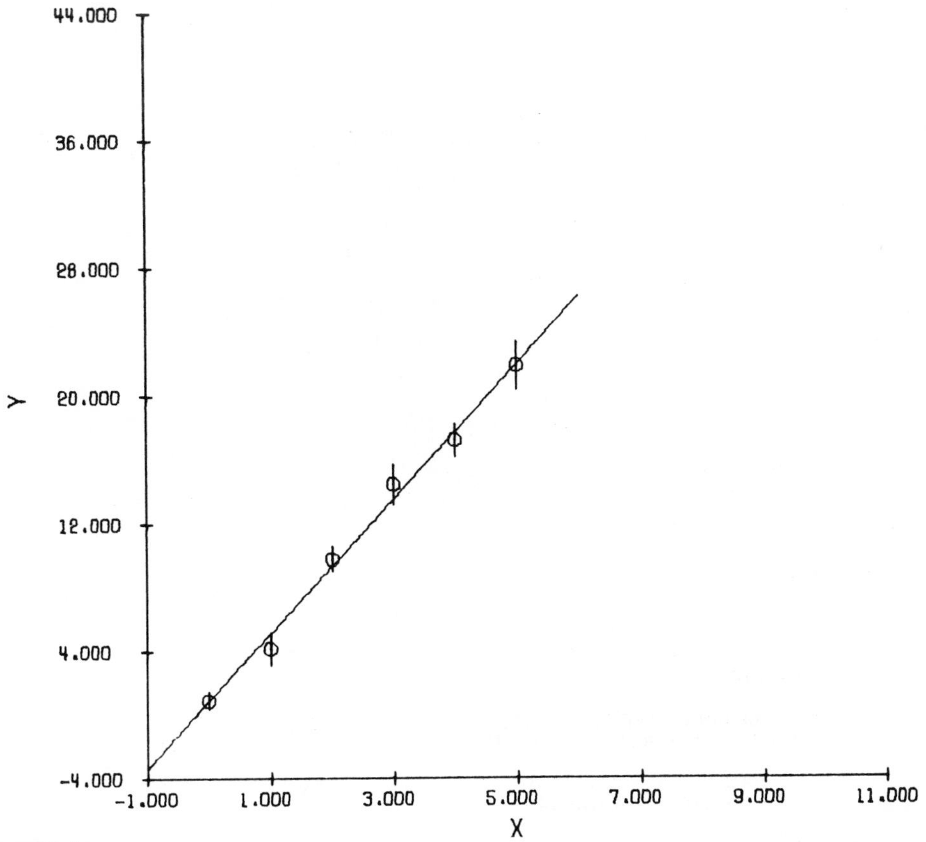

Figure 32.4
Plot of the six data points y_i with their assigned standard deviations as error flags together with the best-fit straight line.

However, our estimates are *negatively correlated:* smaller values for a^* require larger values for b^* and conversely.

The minimum value of χ^2 occurs at

$$\chi^2_{\min} = \chi^2(a^*, b^*) = 2.078$$

For $N - r = 6 - 2 = 4$ degrees of freedom

$$\Pr\{\chi^2(4) \geq 2.1\} = 0.72$$

and so the fit is "good" as visual inspection of Figure 32.4 will verify. ◄

Geometrical view

It is useful to look at this χ^2 minimization problem from a *geometrical viewpoint.* It is not always true that the problem can be solved analytically, and numerical techniques are usually best guided by some appreciation for what is happening on the χ^2 surface.

If we expand (32.22) using (32.25), we get

$$\chi^2(a, b) = a^2D + 2abA - 2aE + b^2B - 2bC + F$$

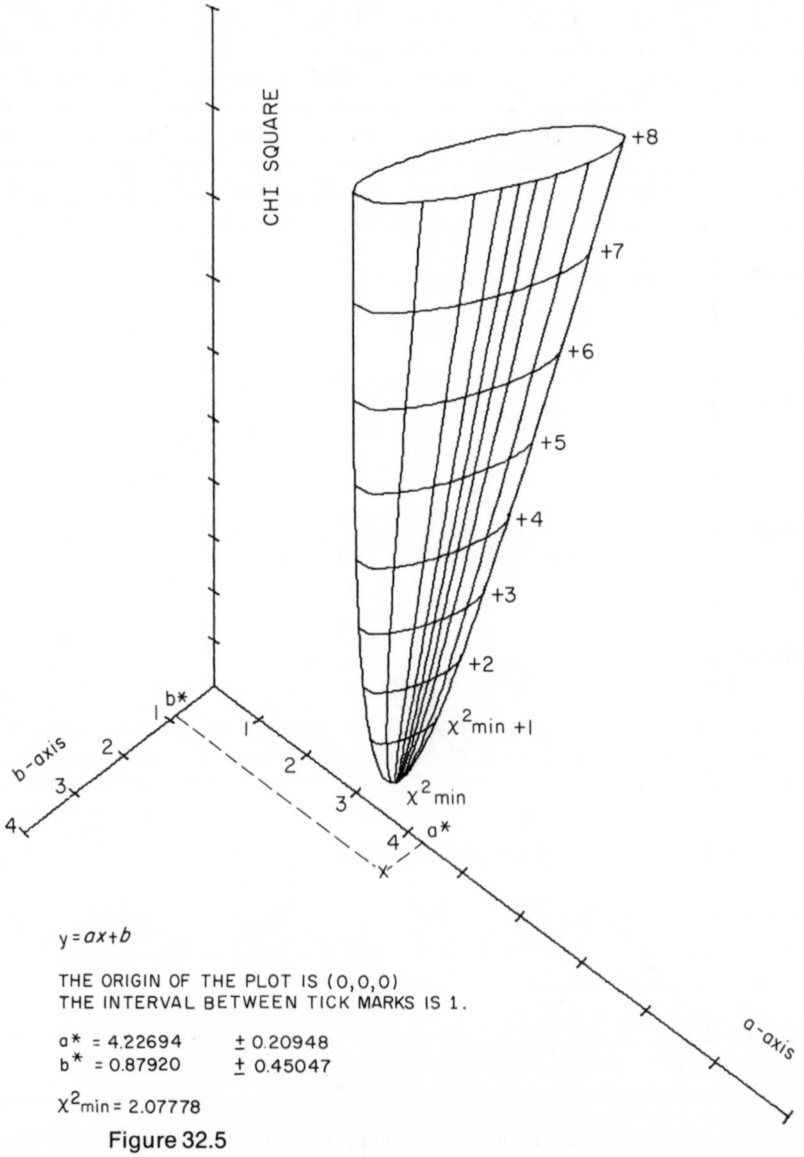

Figure 32.5
Plot of the three-dimensional surface, $\chi^2(a, b)$ versus a, b.

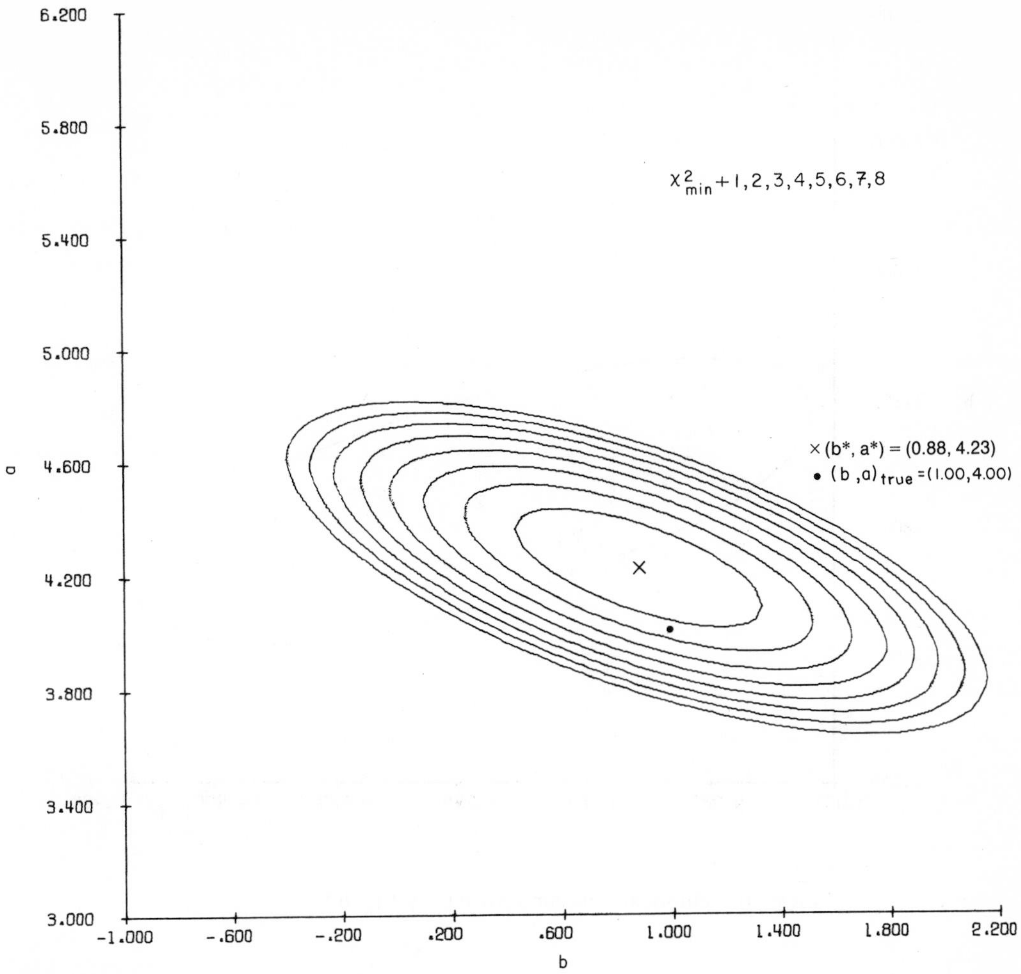

Figure 32.6
Plot of the contours of the three-dimensional surface projected onto the a, b plane. × equals "best-fit" value. • equals "true" values.

which is shown as a three-dimensional plot in Figure 32.5.

This is a general form of an ellipse for $\chi^2(a, b) = \text{constant}$. If we *project* the three-dimensional surface on the a, b plane, we get **contours of constant χ^2**. This is shown in Figure 32.6 where $\chi^2_{\min}$ projects to the point (a^*, b^*).

Let us focus our attention on the *elliptical contour*

$$\chi^2(a, b) = \chi^2(a^*, b^*) + 1$$

which is shown in Figure 32.7.

We wish to study some of the geometrical properties of the ellipse. We may ask for δa where

$$\chi^2(a^* \pm \delta a, b^*) = \chi^2(a^*, b^*) + 1$$

$$\begin{aligned}(a^* + \delta a)^2 D &+ 2(a^* + \delta a)b^* A \\ &- 2(a^* + \delta a)E + b^{*2}B - 2b^*C + F \\ &= a^{*2}D + 2a^*b^*A - 2a^*E + b^{*2}B \\ &\quad - 2b^*C + F + 1\end{aligned}$$

Therefore,

$$[2a^*(\delta a) + (\delta a)^2]D + 2(\delta a)b^*A - 2(\delta a)E = 1$$

and

$$(\delta a)^2 D + 2\delta a(a^*D + b^*A - E) - 1 = 0$$

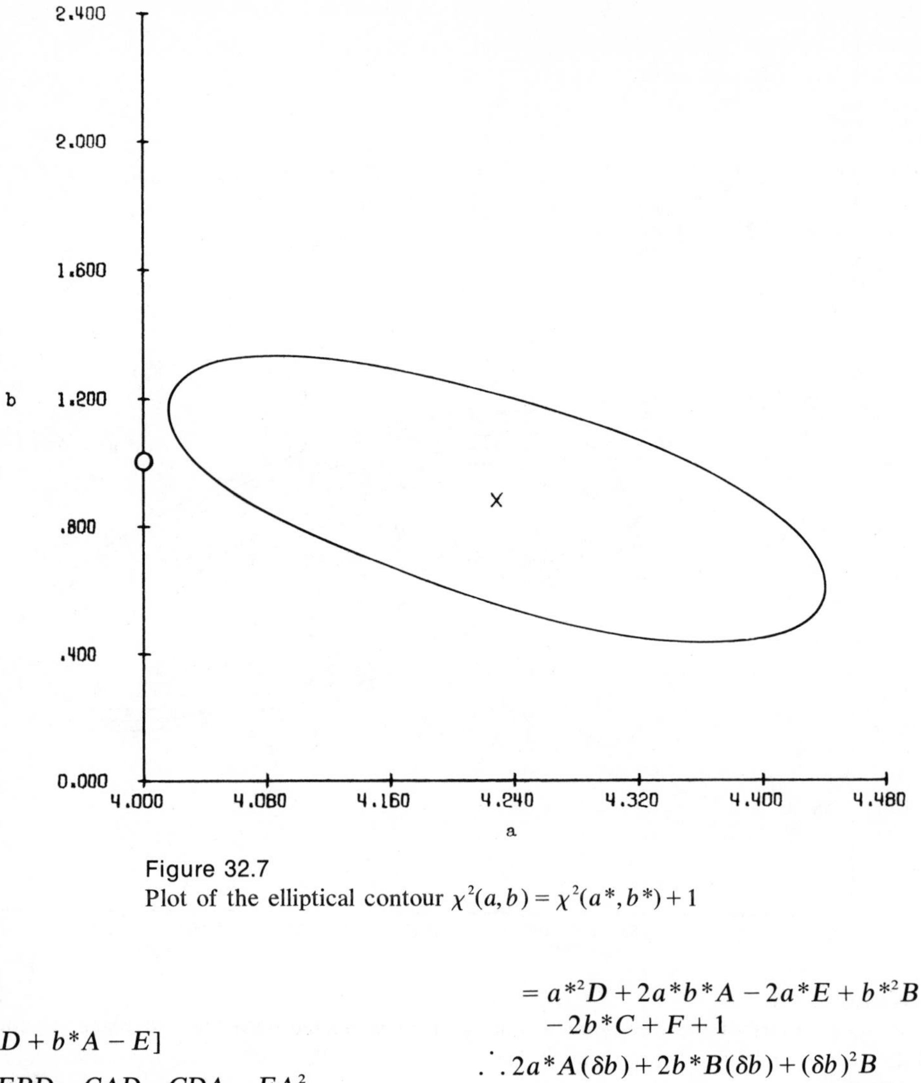

Figure 32.7
Plot of the elliptical contour $\chi^2(a,b) = \chi^2(a^*,b^*) + 1$

But

$[a^*D + b^*A - E]$

$$= \frac{EBD - CAD}{DB - A^2} + \frac{CDA - EA^2}{DB - A^2} - E$$

$$= \frac{EBD-CAD+CDA-EA^2-EDB+EA^2}{DB-A^2}$$

$= 0$

$$\therefore (\delta a)^2 = \frac{1}{D}$$

and

$$\delta a = \pm\frac{1}{\sqrt{D}}$$

Also,

$$\chi^2(a^*, b^* + \delta b) = \chi^2(a^*, b^*) + 1$$

and

$$a^{*2}D + 2a^*(b^* + \delta b)A - 2a^*E + (b^* + \delta b)^2B - 2(b^* + \delta b)C + F$$

$$= a^{*2}D + 2a^*b^*A - 2a^*E + b^{*2}B - 2b^*C + F + 1$$

$$\therefore 2a^*A(\delta b) + 2b^*B(\delta b) + (\delta b)^2B - 2C(\delta b) = 1$$

or

$$(\delta b)^2B + 2\delta b[a^*A + b^*B - C] - 1 = 0$$

But

$[a^*A + b^*B - C]$

$$= \frac{EBA-CA^2+CDB-EAB-CDB+CA^2}{DB-A^2}$$

$= 0$

$$\therefore \delta b = \pm\frac{1}{\sqrt{B}}$$

We note that δa, δb correspond to $\chi^2_{\min} + 1$ points on plots of $\chi^2(a, b^*)$ versus a and $\chi^2(a^*, b)$ versus b, respectively. These are shown in Figures 32.8a and

32.8*b*. *If* the statistics a^* and b^* were *uncorrelated,* these points would correspond to the one-standard-deviation interval for the estimators a^*, b^*. However, the one-standard deviation points may *not* be obtained in this way when the parameters are *correlated.*

When the parameters are *correlated* (as in this case), the curve $\chi^2(a, b) = \chi^2(a^*, b^*) + 1$ yields the *one-standard-deviation* points as follows.

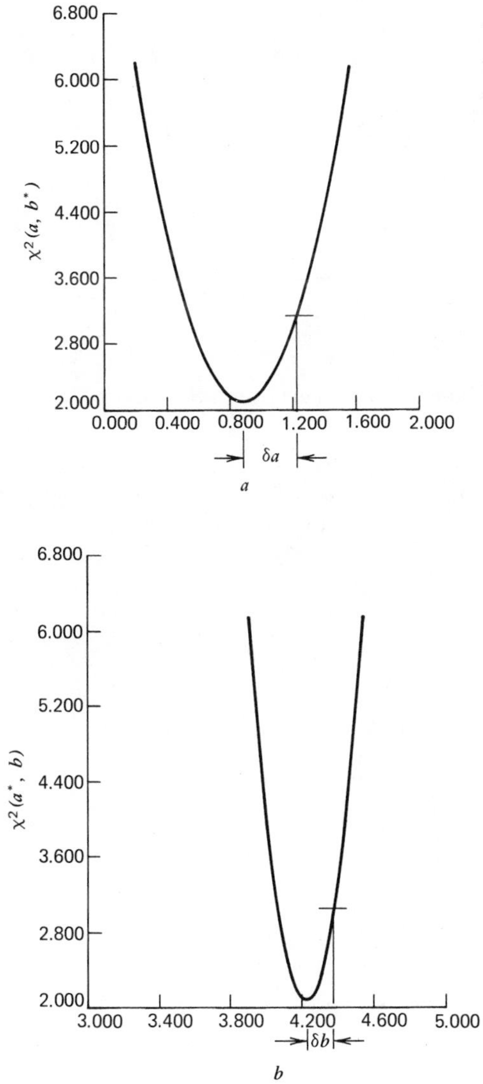

Figure 32.8
(*a*) Plot of $\chi^2(a, b^*)$ versus *a*. (*b*) Plot of $\chi^2(a^*, b)$ versus *b*.

Bounds on the ellipse

We are also interested in $a_{\max}$, $a_{\min}$, and $b_{\max}$, $b_{\min}$ that are on the curve. We may take the constraint equation

$$\chi^2(a, b) = \chi^2(a^*, b^*) + 1$$

and use it together with the derivative equation where

$$\frac{\partial a}{\partial b} = 0 \qquad \text{for} \qquad a_{\max}, a_{\min}$$

and

$$\frac{\partial b}{\partial a} = 0 \qquad \text{for} \qquad b_{\max}, b_{\min}$$

$$\begin{aligned}\chi^2(a, b) &= a^2 D + 2baA - 2aE \\ &\quad + b^2 B - 2bC + F \\ &= \chi^2(a^*, b^*) + 1\end{aligned}$$

yields the derivative equation

$$2aD\frac{\partial a}{\partial b} + 2bA\frac{\partial a}{\partial b} + 2aA - 2E\frac{\partial a}{\partial b} + 2bB - 2C = 0$$

Setting $\partial a/\partial b = 0$, we get

$$aA + bB = C$$

This, together with the constraint equation yields

$$a_{\substack{\max\\ \min}} = a^* \pm \frac{\sqrt{B}}{\sqrt{DB - A^2}}$$

Also

$$b_{\substack{\max\\ \min}} = b^* \pm \frac{\sqrt{D}}{\sqrt{DB - A^2}}$$

The standard deviations of a^*, b^* may thus be read from *the rectangle that bounds the ellipse,* as shown in Figure 32.9, and which is parallel to the *a*, *b* axes. In cases where the parameters a^*, b^* are uncorrelated, the principal axes of the χ^2 ellipse are parallel to the coordinate axes.

It is convenient if the estimators may be *chosen* to be *uncorrelated.* The estimators are then said to be *orthogonal.* Sometimes the values of x_i may be chosen to produce uncorrelated estimators, but this is not always possible.

Confidence regions

The regression line given by

$$y = a^* x + b^*$$

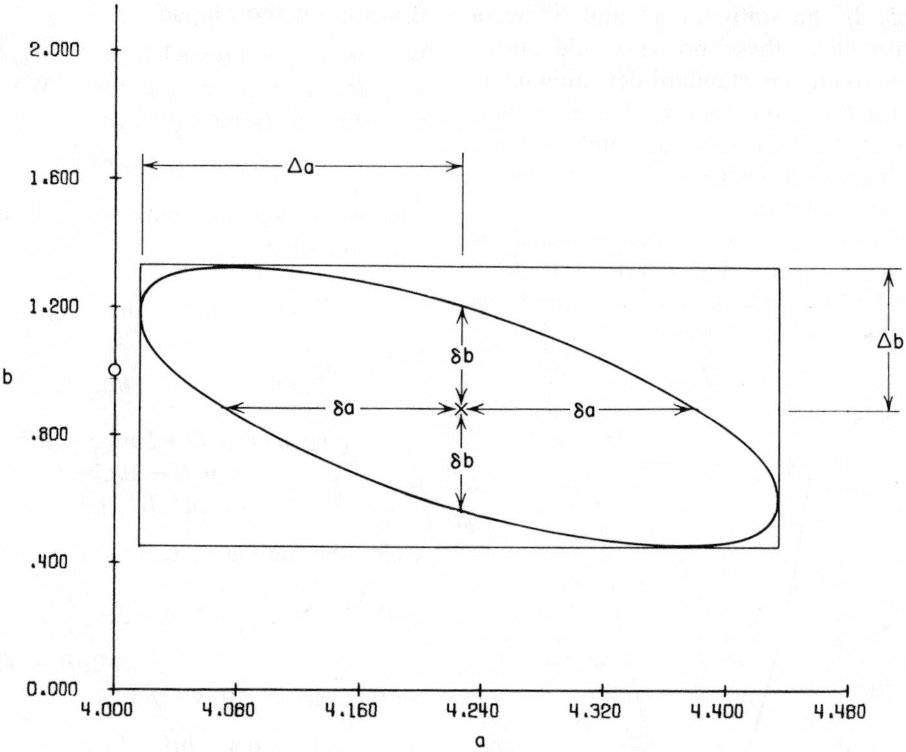

Figure 32.9
Plot of the contour of Figure 32.7 together with the bounding rectangle from which may be read the standard deviations of a^*, b^*.

we believe to have as its "*expected value*" the *true straight line*

$$y = ax + b$$

that is, if we *repeat* the random experiment and analysis a large number of times, $\mathcal{N}$, the average

$$\frac{1}{\mathcal{N}} \sum_{i=1}^{\mathcal{N}} a_i^* \approx a$$

and

$$\frac{1}{\mathcal{N}} \sum_{i=1}^{\mathcal{N}} b_i^* \approx b$$

where a and b are the *true values.*

The $(\chi^2_{\min})_i = \chi^2(a_i^*, b_i^*)$ for each experiment should follow the $\chi^2(N-2)$ distribution. We have repeated this experiment and analysis 100 times by choosing 100 sets of six random gaussian deviates and repeating the analysis described for each. A histogram of the frequencies of the observed $(\chi^2_{\min})_i$ is shown in Figure 32.10 together with a superimposed curve of **CS**(4). Also, the observed *mean*

$$\frac{1}{100} \sum_{i=1}^{100} (\chi^2_{\min})_i = 4.073 = \overline{(\chi^2_{\min})}$$

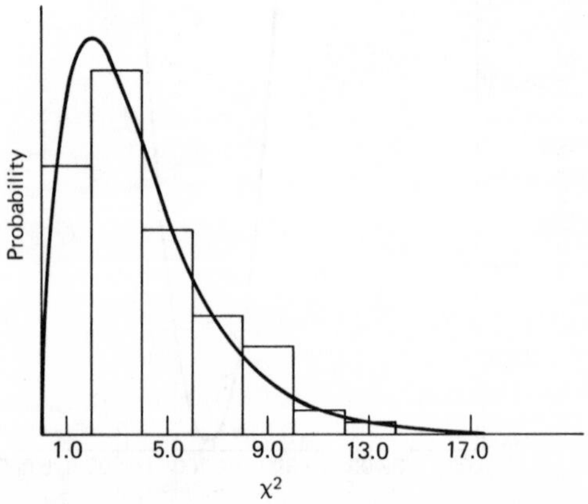

Figure 32.10
Histogram of the frequencies of the observed values $(\chi^2_{\min})_i$ with a superimposed curve of **CS**(4).

and *variance*

$$\sigma^2 = \frac{1}{100-1}\sum_{i=1}^{100}[(\chi^2_{\min})_i - \overline{(\chi^2_{\min})}]^2 = 7.725$$

This checks that $\chi^2_{\min}$ for this problem **follows the $\chi^2(4)$ distribution.**

The individual (a^*, b^*) will be *bivariate normally distributed* about the true values (a, b), according to (28.39) with

$$\mathbf{bn}(a^*, b^*; \mu_{a^*}, \mu_{b^*}, \sigma_{a^*}, \sigma_{b^*}, \rho_{a^*b^*})$$

where

$$\mu_{a^*} = a \qquad \mu_{b^*} = b \qquad \text{and} \qquad \sigma_{a^*}, \sigma_{b^*}$$

and

$$\rho_{a^*b^*} = \frac{\sigma_{a^*b^*}}{\sigma_{a^*}\sigma_{b^*}}$$

are given by (32.29).

The probability that the true value (a, b) will be found *within* the elliptical contour $\chi^2(a, b) = \chi^2_{\min} + k$, which we may call R_k, is given by

$$\Pr[(a, b) \text{ inside } R_k] = \Pr[\chi^2(2) \le k] \tag{32.33}$$

As in our discussion of the *bivariate normal distribution*, we may use the χ^2 table for $\chi^2(2)$ to get the results in Table 32.3.

Table ***32.3***

$\Pr(a, b)$ inside R_k	k
0.3935	1
0.6321	2
0.7769	3
0.8647	4
0.9179	5
0.9502	6
0.9698	7
0.9817	8

If we perform a single experiment (e.g., six data points) and get (a^*, b^*) and the ellipses $R_k: \chi^2(a, b) = \chi^2_{\min} + k$, we may assert that the true value of (a, b) lies within R_1 with 39% confidence, within R_2 with 63% confidence, etc. For our *detailed example*, the true value is outside R_1 but inside R_2.

For our 100 experiments we know the true value to be $(a, b) = (4, 1)$. For each experiment i we may compute $[\chi^2(4, 1)]_i$ and also $[\chi^2_{\min} + k]_i$. If

$$[\chi^2(4, 1)]_i \le [\chi^2_{\min} + k]_i$$

then the true point (4, 1) is contained within R_k. We have counted this for our 100 experiments and noted that $\chi^2(4, 1) \le \chi^2_{\min} + 1$ occurred 43 times and $\chi^2(4, 1) \le \chi^2_{\min} + 6$ occurred 97 times.

Confidence limits on the regression line

We have defined confidence regions R_k for the points (a, b). Each point (a_k, b_k) in R_k defines a straight line

$$y = a_k x + b_k \tag{32.34}$$

so that each *confidence interval* R_k corresponds to a *bundle of straight lines* (32.34). The confidence limits on the regression line may be described by the *envelope* of the bundle of straight lines. If the envelope has equation $g_k(x) = y$, then the **upper** envelope for region R_k is defined by

$$y = a_k x + b_k \le g_k^+(x) \tag{32.35}$$

and the **lower** envelope by

$$y = a_k x + b_k \ge g_k^-(x) \tag{32.36}$$

$g_k^+(x)$ is the equation of the maxima of (32.34) *such that* (a_k, b_k) are **inside** R_k. For each x the value (a_k, b_k) that maximizes (32.34) lies on the *boundary* of the region R_k. This may be solved as a *maximization* of (32.34) with respect to a_k, b_k and *subject to the constraint* that $\chi^2(a_k, b_k) = \chi^2_{\min} + k$.

We cite the result for the *envelope*

$$\begin{aligned} g_k(x) &= (a^*x + b^*) \\ &\quad \pm\left\{(\Delta\chi^2)\left[\frac{1}{\Sigma(1/\sigma_i^2)} + \frac{(x-\bar{x})^2}{\Sigma(x_i^2/\sigma_i^2)}\right]\right\}^{1/2} \\ &= (a^*x + b^*) \\ &\quad \pm\left\{k\left[\frac{1}{B} + \frac{(x-\bar{x})^2}{D}\right]\right\}^{1/2} \end{aligned} \tag{32.37}$$

where the $\pm$ refers to the *upper* and *lower* envelopes.

The envelopes are *hyperbolas* symmetric about the regression line. The hyperbolas are shown schematically in Figure

32.11*a* and in Figure 32.11*b* for the case at hand.

The envelopes also define *confidence limits* on y for each value of x. The uncertainty in y is least for $x = \bar{x}$, the *centroid* given by (32.30), that is, the "narrow waist" of the envelope occurs at the centroid $(\bar{x}, \bar{y})$. This emphasizes our previous assertion that the uncertainty in the y-intercept (parameter b) is least if the points are chosen symmetric about the origin.

FITTING A PARABOLA

Another example that illustrates the procedures, but yet is tractable enough so that we may follow a hand calculation is the case of fitting a parabola through a number of experimental points. Although we could derive formulas from the beginning we prefer to use the *matrix methods* of equations (32.11) to (32.21).

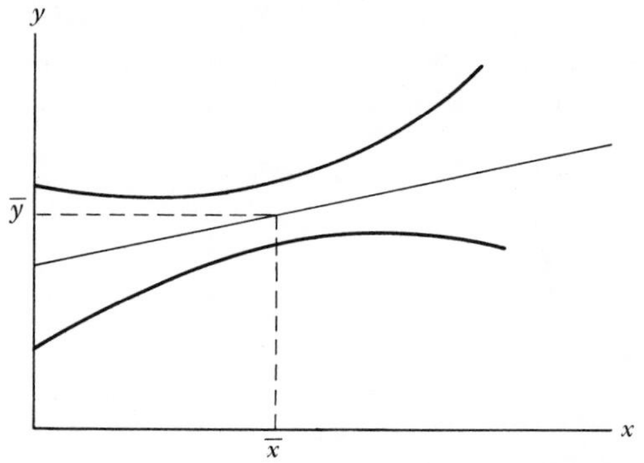

(*a*)

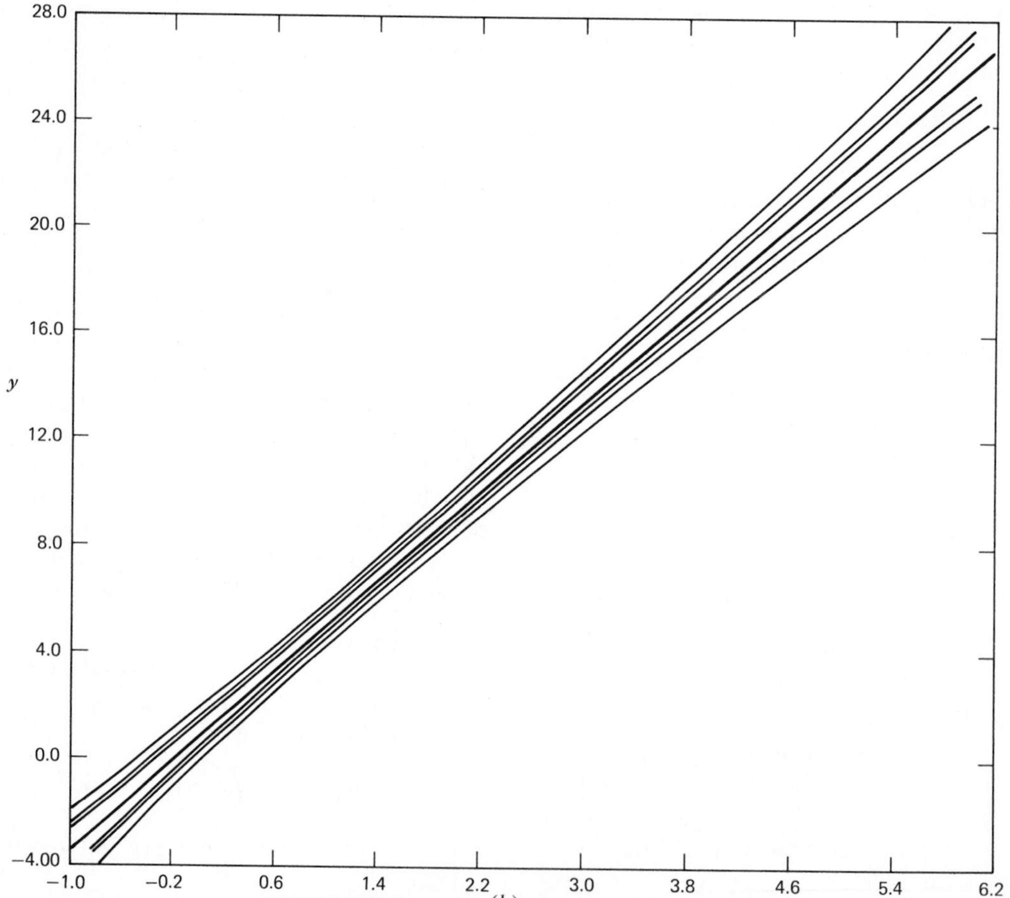

(*b*)

Figure 32.11
(*a*) Schematic picture of the hyperbolas representing the envelopes of the regression line. (*b*) Actual plot of the envelopes of the regression line for the example discussed in the text.

► **Example 32.3**
Fitting a parabola

Let us consider the parabola $y = 1 + 3x + 4x^2$ and *generate* five normally distributed data points y_i, each with assigned error σ_i, corresponding to the known points x_i. We shall then treat the five points $(y_i; x_i, \sigma_i)$ as given and get best estimates for the fit

$$y = c_1 + c_2 x + c_3 x^2 = a + bx + cx^2$$

We shall do this example **twice** to indicate the effects of choosing the data points *with* and *without* regard to *symmetry*.

I. Asymmetrical data

We generate a set of (asymmetrical) data as in Table 32.4.

Table **32.4**

x_i	−3.00	−2.00	0.00	1.00	2.00
$1+3x_i+4x_i^2$	28.00	11.00	1.00	8.00	23.00
σ_i	0.50	1.00	0.75	1.25	1.00
R_i	−0.158	−0.848	1.041	1.167	0.261
$R_i\sigma_i$	−0.08	−0.85	0.78	1.46	0.26
y_i	27.92	10.15	1.78	9.46	23.26

We now have as *data input* the five points $(y_i; x_i, \sigma_i)$. We are fitting y_i to

$$p(x_i; \mathbf{c}) = \sum_{k=1}^{r=3} c_k g_k(x_i)$$

from (32.11) where

$$g_1(x_i) = 1$$

$$g_2(x_i) = x_i$$

and

$$g_3(x_i) = x_i^2$$

We must form the (3×3) matrix $\underline{\mathbf{A}}$ where by (32.16)

$$A_{kl} = \sum_{i=1}^{N=5} g_k(x_i) g_l(x_i) \left(\frac{1}{\sigma_i^2}\right) \qquad k, l = 1, 2, 3$$

and the (3×1) vector $\mathbf{u}$ where by (32.17)

$$u_l = \sum_{i=1}^{N=5} y_i g_l(x_i) \left(\frac{1}{\sigma_i^2}\right) \qquad l = 1, 2, 3$$

We need, therefore, the quantities (32.25) and also

$$G = \sum_{i=1}^{N} \frac{x_i^2 y_i}{\sigma_i^2}$$

$$H = \sum_{i=1}^{N} \frac{x_i^3 y_i}{\sigma_i^2} \qquad (32.38)$$

and

$$I = \sum_{i=1}^{N} \frac{x_i^4}{\sigma_i^2}$$

We may summarize the data table and the needed quantities to be calculated from the data (Table 32.5).

Table **32.5**

						$\sum_{i=1}^{5}$
x_i	−3.00	−2.00	0.00	1.00	2.00	
y_i	27.92	10.15	1.78	9.46	23.26	
σ_i	0.50	1.00	0.75	1.25	1.00	
σ_i^2	0.25	1.00	0.56	1.56	1.00	
$1/\sigma_i^2$	4.00	1.00	1.78	0.64	1.00	$8.42 = B$
x_i/σ_i^2	−12.00	−2.00	0.00	0.64	2.00	$-11.36 = A$
y_i/σ_i^2	111.68	10.15	3.17	6.05	23.26	$154.32 = C$
$x_i y_i/\sigma_i^2$	−335.05	−20.30	0.00	6.05	46.52	$-302.78 = E$
y_i^2/σ_i^2	3118.33	103.06	5.64	57.26	541.07	$3825.36 = F$
x_i^2/σ_i^2	36.00	4.00	0.00	0.64	4.00	$44.64 = D$
$x_i^2 y_i/\sigma_i^2$	1005.16	40.61	0.00	6.05	93.04	$1144.86 = G$
$x_i^3 y_i/\sigma_i^2$	−108.00	−8.00	0.00	0.64	8.00	$-107.36 = H$
x_i^4/σ_i^2	324.00	16.00	0.00	0.64	16.00	$356.64 = I$

We may calculate the elements of the matrix $\underline{\mathbf{A}}$:

$$A_{11} = \sum_{i=1}^{5} g_1(x_i)g_1(x_i)\frac{1}{\sigma_i^2} = \sum_{i=1}^{5}\frac{1}{\sigma_i^2} = B = 8.42$$

$$A_{22} = \sum_{i=1}^{5} g_2(x_i)g_2(x_i)\frac{1}{\sigma_i^2} = \sum_{i=1}^{5}\frac{x_i^2}{\sigma_i^2} = D = 44.64$$

$$A_{33} = \sum_{i=1}^{5} g_3(x_i)g_3(x_i)\frac{1}{\sigma_i^2} = \sum_{i=1}^{5}\frac{x_i^4}{\sigma_i^2} = I = 356.64$$

$$A_{12} = A_{21} = \sum_{i=1}^{5} g_1(x_i)g_2(x_i)\frac{1}{\sigma_i^2} = \sum_{i=1}^{5}\frac{x_i}{\sigma_i^2} = A = -11.36$$

$$A_{13} = A_{31} = \sum_{i=1}^{5} g_1(x_i)g_3(x_i)\frac{1}{\sigma_i^2} = \sum_{i=1}^{5}\frac{x_i^2}{\sigma_i^2} = D = 44.64$$

$$A_{23} = A_{32} = \sum_{i=1}^{5} g_2(x_i)g_3(x_i)\frac{1}{\sigma_i^2} = \sum_{i=1}^{5}\frac{x_i^3}{\sigma_i^2} = H = -107.36$$

The *matrix* is, therefore,

$$\underline{\mathbf{A}} = \begin{bmatrix} 8.42 & -11.36 & 44.64 \\ -11.36 & 44.64 & -107.36 \\ 44.64 & -107.36 & 356.64 \end{bmatrix}$$

The *error matrix* is the *inverse*, $\underline{\mathbf{A}}^{-1}$, which is

$$\underline{\mathbf{A}}^{-1} = \begin{bmatrix} 0.4033 & -0.068 & -0.0710 \\ -0.068 & 0.0926 & 0.0364 \\ -0.0710 & 0.0364 & 0.0226 \end{bmatrix}$$

We form the *data vector* **u**.

$$u_1 = \sum_{i=1}^{5} y_i g_1(x_i)\left(\frac{1}{\sigma_i^2}\right) = \sum_{i=1}^{5} y_i\frac{1}{\sigma_i^2} = C = 154.32$$

$$u_2 = \sum_{i=1}^{5} y_i g_2(x_i)\frac{1}{\sigma_i^2} = \sum_{i=1}^{5} y_i x_i\frac{1}{\sigma_i^2} = E = -302.78$$

$$u_3 = \sum_{i=1}^{5} y_i g_3(x_i)\frac{1}{\sigma_i^2} = \sum_{i=1}^{5} y_i x_i^2\frac{1}{\sigma_i^2} = G = 1144.86$$

The data vector is, therefore,

$$\mathbf{u} = \begin{bmatrix} 154.32 \\ -302.78 \\ 1144.86 \end{bmatrix}$$

We may solve for the best estimates, $\mathbf{c}^*$, using (32.20):

$$\mathbf{c}^* = (\underline{\mathbf{A}}^{-1})\mathbf{u}$$

$$\begin{bmatrix} c_1^* \\ c_2^* \\ c_3^* \end{bmatrix} = \begin{bmatrix} 0.4033 & -0.068 & -0.0710 \\ -0.068 & 0.0926 & 0.0364 \\ -0.0710 & 0.0364 & 0.0226 \end{bmatrix} \times \begin{bmatrix} 154.32 \\ -302.78 \\ 1144.86 \end{bmatrix} = \begin{bmatrix} 1.54 \\ 3.13 \\ 3.95 \end{bmatrix}$$

The *best-fit parameters* are thus

$$c_1^* = a^* = 1.54 \pm 0.64$$
$$c_2^* = b^* = 3.13 \pm 0.30$$
$$c_3^* = c^* = 3.95 \pm 0.150$$

with the elements of the *error matrix* giving the (**correlated**) errors:

$$\sigma(c_1^*) = \sqrt{\overline{(\Delta a)^2}} = \sqrt{(\underline{\mathbf{A}}^{-1})_{11}} = 0.64$$

$$\sigma(c_2^*) = \sqrt{\overline{(\Delta b)^2}} = \sqrt{(\underline{\mathbf{A}}^{-1})_{22}} = 0.30$$

$$\sigma(c_3^*) = \sqrt{\overline{(\Delta c)^2}} = \sqrt{(\underline{\mathbf{A}}^{-1})_{33}} = 0.150$$

$$\sigma(c_1^*, c_2^*) = \overline{(\Delta a)(\Delta b)} = (\underline{\mathbf{A}}^{-1})_{12} = (\underline{\mathbf{A}}^{-1})_{21} = -0.068$$

$$\sigma(c_1^*, c_3^*) = \overline{(\Delta a)(\Delta b)} = (\underline{\mathbf{A}}^{-1})_{13} = (\underline{\mathbf{A}}^{-1})_{31} = -0.071$$

$$\sigma(c_2^*, c_3^*) = \overline{(\Delta b)(\Delta c)} = (\underline{\mathbf{A}}^{-1})_{23} = (\underline{\mathbf{A}}^{-1})_{32} = +0.0364$$

The minimum value of $\chi^2 = \chi^2(a^*, b^*, c^*) = 1.684$, which must be compared with χ^2 for $(N - r) = 5 - 3 = 2$ degrees of freedom. From the χ^2 tables for $\chi^2(2)$ we see that

$$\Pr(\chi^2(2) \geq 1.684) > \Pr(\chi^2(2) \geq 1.83) = 0.40$$

Thus the fit is acceptable. The data points and the fitted curve are plotted in Figure 32.12.

Let us *repeat* this example *choosing the data points symmetrically* about

$$\bar{x} = \frac{1}{N}\sum_{i=1}^{5}\frac{x_i}{\sigma_i^2} = 0.$$

The object of this choice is to reduce to zero some of the off-diagonal terms in the error matrix thereby eliminating some of the correlations.

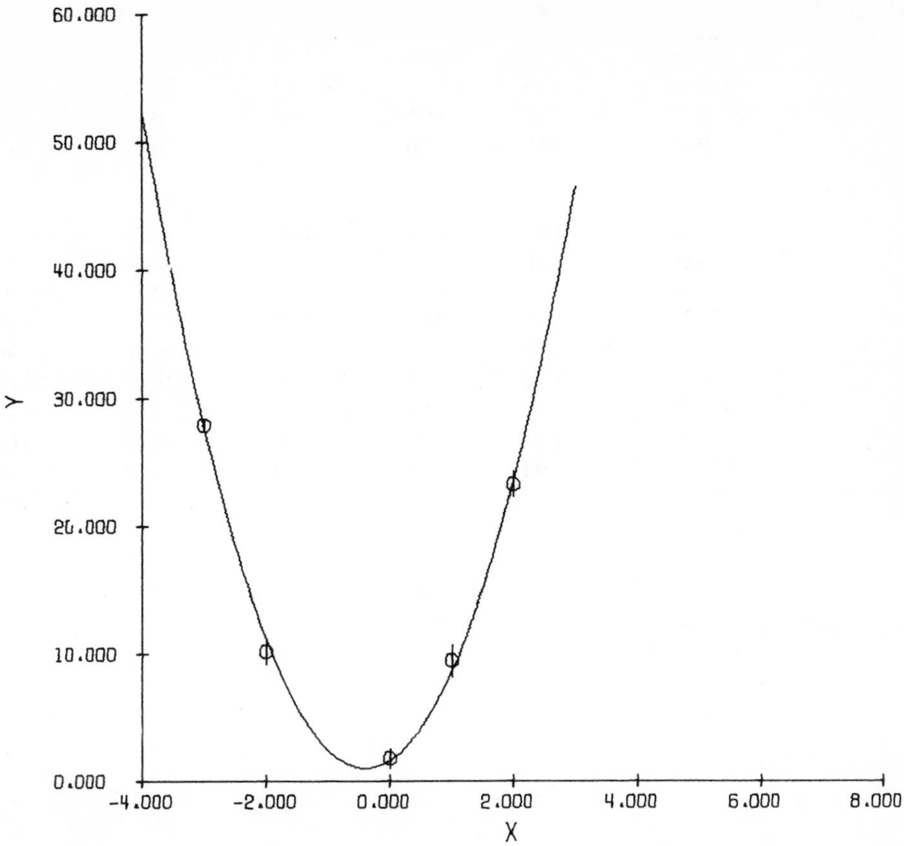

Figure 32.12
The data points (with error flags) and the fitted curve for the example discussed in the text.

II. Symmetrical data

Using the same equation as in (I), we generate a set of (symmetrical) data (Table 32.6).

Table ***32.6***

x_i	−2.00	−1.00	0.00	1.00	2.00
$1+3x_i+4x_i^2$	11.00	2.00	1.00	8.00	23.00
σ_i	0.50	1.00	1.25	1.00	0.50
R_i	−0.158	−0.848	1.041	1.167	0.261
$R_i\sigma_i$	−0.08	−0.85	1.30	1.17	0.13
y_i	10.92	1.15	2.30	9.17	23.13

The input data and needed quantities A to G, for the calculation of $\underline{\mathbf{A}}$, $\mathbf{u}$, and χ^2 are in Table 32.7.

The matrix $\underline{\mathbf{A}}$ may be written from (32.16)

$$\underline{\mathbf{A}} = \begin{bmatrix} B & A & D \\ A & D & H \\ D & H & I \end{bmatrix} = \begin{bmatrix} 10.64 & 0. & 34.00 \\ 0. & 34.0 & 0. \\ 34.00 & 0. & 130.00 \end{bmatrix}$$

where the terms that depend on an odd power of x_i vanish because of the symmetry *which we have chosen.*

This matrix may be *inverted* to get the *error matrix*

$$(\underline{\mathbf{A}}^{-1}) = \begin{bmatrix} 0.592 & 0. & -1.50 \\ 0.999 & 0.029 & 0. \\ -1.50 & 0. & 0.047 \end{bmatrix}$$

The vector $\mathbf{u}$ may be evaluated using (32.17):

$$\mathbf{u} = \begin{bmatrix} C \\ E \\ G \end{bmatrix} = \begin{bmatrix} 148.00 \\ 105.69 \\ 555.14 \end{bmatrix}$$

We solve for the estimates $\mathbf{c}^*$ using (32.20):

Table **32.7**

x_i	−2.00	−1.00	0.00	1.00	2.00	
y_i	10.92	1.15	2.30	9.17	23.13	
σ_i	0.50	1.00	1.25	1.00	0.50	
σ_i^2	0.25	1.00	1.56	1.00	0.25	
						$\sum_{i=1}^{5}$
$1/\sigma_i^2$	4.00	1.00	0.64	1.00	4.00	$10.64 = B$
x_i/σ_i^2	−8.00	−1.00	0.00	1.00	8.00	$0.00 = A$
y_i/σ_i^2	43.68	1.15	1.47	9.17	92.52	$148.00 = C$
x_iy_i/σ_i^2	−87.37	−1.15	0.00	9.17	185.04	$105.69 = E$
y_i^2/σ_i^2	477.07	1.33	3.39	84.03	2140.08	$2705.90 = F$
x_i^2/σ_i^2	16.00	1.00	0.00	1.00	16.00	$34.00 = D$
$x_i^2y_i/\sigma_i^2$	174.74	1.15	0.00	9.17	370.09	$555.14 = G$
$x_i^3y_i/\sigma_i^2$	−32.00	−1.00	0.00	1.00	32.00	$0.00 = H$
x_i^4/σ_i^2	64.00	1.00	0.00	1.00	64.00	$130.00 = I$

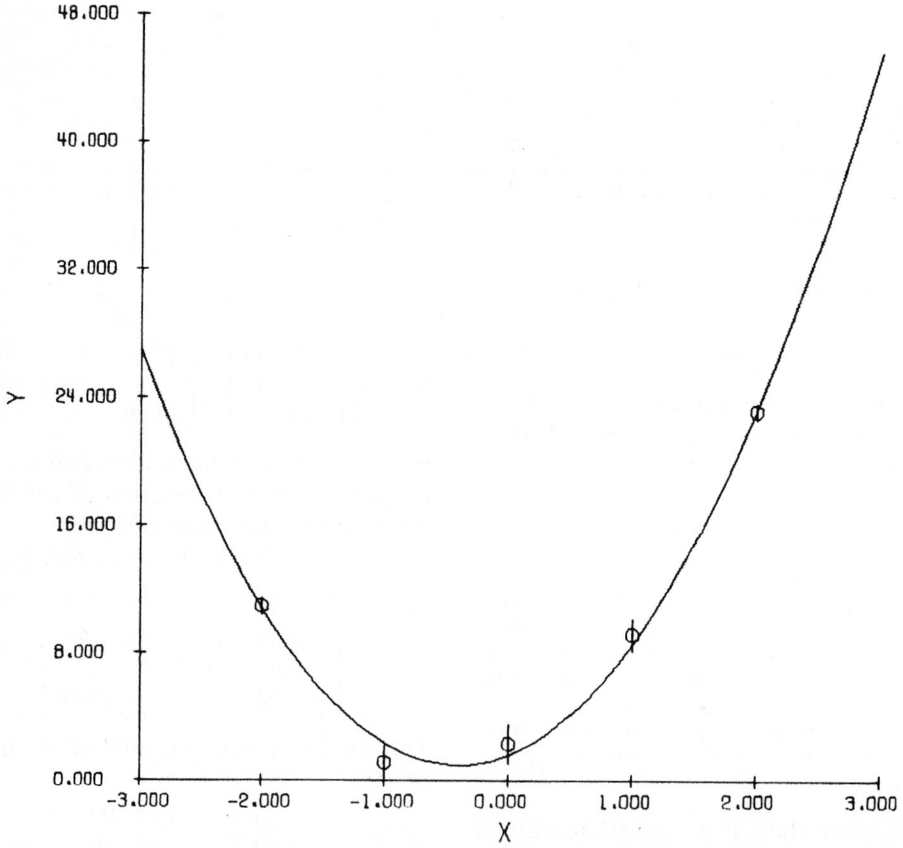

Figure 32.13
The data points (with error flags) and the fitted curve for the symmetric example discussed in the text.

$$\mathbf{c}^* = (\underline{\mathbf{A}}^{-1})\mathbf{u}$$

$$= \begin{bmatrix} c_1^* \\ c_2^* \\ c_3^* \end{bmatrix} = \begin{bmatrix} 0.572 & 0. & -1.50 \\ 0. & 0.029 & 0. \\ -1.50 & 0. & 0.047 \end{bmatrix} \times \begin{bmatrix} 148.00 \\ 105.69 \\ 555.14 \end{bmatrix} = \begin{bmatrix} 1.61 \\ 3.11 \\ 3.85 \end{bmatrix}$$

The best-fit parameters are

$$c_1^* = a^* \pm \sigma(a^*) = 1.61 \pm 0.76$$
$$c_2^* = b^* \pm \sigma(b^*) = 3.11 \pm 0.17$$
$$c_3^* = c^* \pm \sigma(c^*) = 3.85 \pm 0.22$$

with

$$\overline{(\Delta a^*)(\Delta b^*)} = 0 = \overline{(\Delta c^*)(\Delta b^*)}$$

and

$$\overline{(\Delta a^*)(\Delta c^*)} = -1.50$$

Here

$$\chi^2_{\min} = \chi^2(a^*, b^*, c^*) = 2.205$$

which is quite acceptable for $5-3=2$ degrees of freedom. This is visually verified as a good fit in Figure 32.13. ◄

Where lies the difference between Examples I and II? The difference is that our *choice* of data points has caused b^* to be uncorrelated with a^* and c^*, although a^* and c^* remain correlated with each other. In other words, b^* has been determined independently (since it is normally distributed and uncorrelated). This has some geometrical consequences that may be useful:

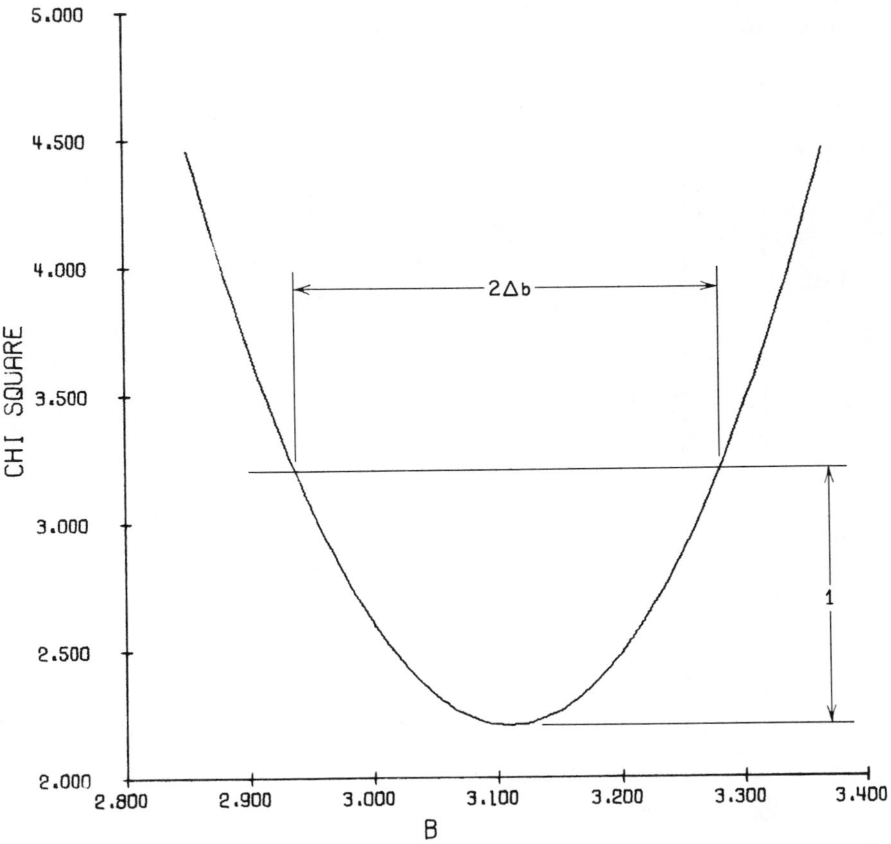

Figure 32.14
Plot of $\chi^2(a^*, b, c^*)$ versus b.

1. A plot of $\chi^2(a^*, b, c^*)$ versus b will yield the error limit on b^*: $\chi^2(a^*, b^* \pm \Delta b^*, c^*) = \chi^2(a^*, b^*, c^*) + 1$. That is, our graphical method for a one-parameter problem may be applied to b. This is shown in Figure 32.14. We stress that this may *not* be done for a^* and c^*.
2. The contour $\chi^2(a, b, c^*) = \chi^2(a^*, b^*, c^*) + 1$ will be an ellipse with principal axes *parallel* to the a and b axes, showing that a, b are *uncorrelated.* This is plotted in Figure 32.15. The standard deviation, σ_{b^*}, may be read from this plot. This may *not* be done for a, since a is correlated with c and depends on the specific value c^*. Also, the contour $\chi^2(a^*, b, c) = \chi^2(a^*, b^*, c^*) + 1$, plotted in Figure 32.16, may be used to estimate σ_{b^*} but *not* σ_{c^*}.
3. The contour $\chi^2(a, b^*, c) = \chi^2(a^*, b^*, c^*) + 1$ is plotted in Figure 32.17. Since a, c are independent of b, their entire variation is expressed by this ellipse. We may thus read off the standard deviations σ_{a^*}, σ_{c^*} from the plot.

UNCORRELATED AND TWOFOLD-CORRELATED PARAMETERS

The simplification exhibited by symmetrical data is a consequence of the relation

$$\chi^2(\mathbf{y}; c) - \chi_0^2(\mathbf{y}; c^*) \approx \frac{1}{2}\sum_{\substack{i=1\\j=1}}^{r} \left(\frac{\partial^2 \chi^2}{\partial c_i\, \partial c_j}\bigg|_{c^*}\right)(c_i - c_i^*)(c_j - c_j^*)$$

If a parameter c_m is *uncorrelated* with *all* the others [Covar$(c_m, c_i) = 0$, all $i \neq m$], then setting all $c_i = c_i^*$, $i \neq m$

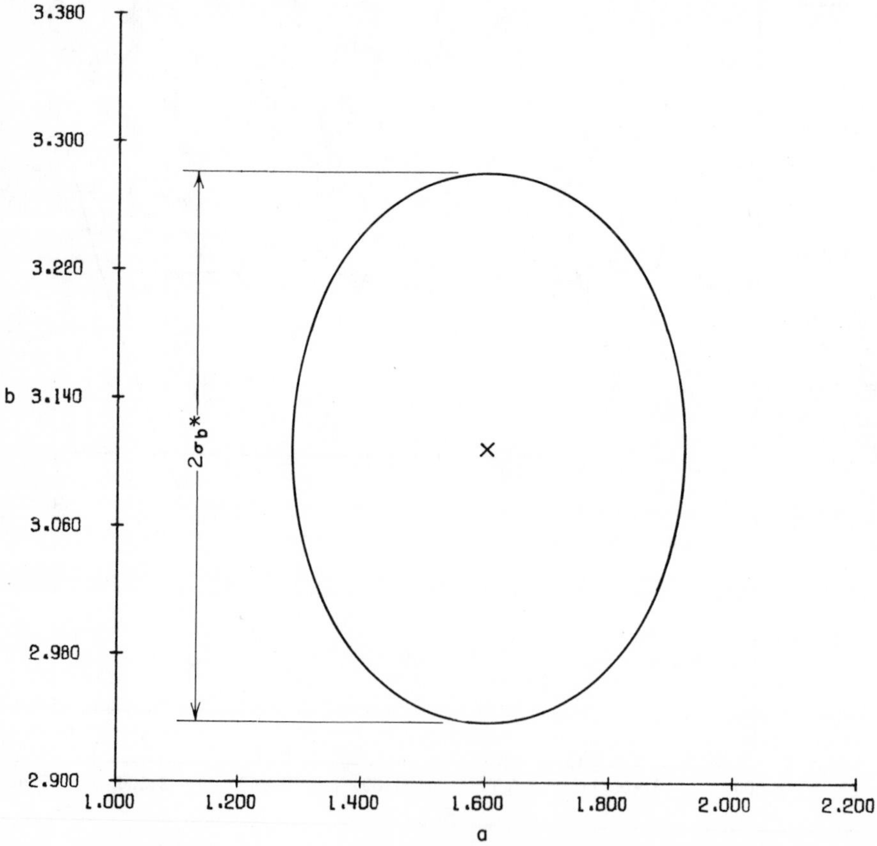

Figure 32.15
Plot of the contour $\chi^2(a, b, c^*) = \chi^2(a^*, b^*, c^*) + 1$ on the a, b plane.

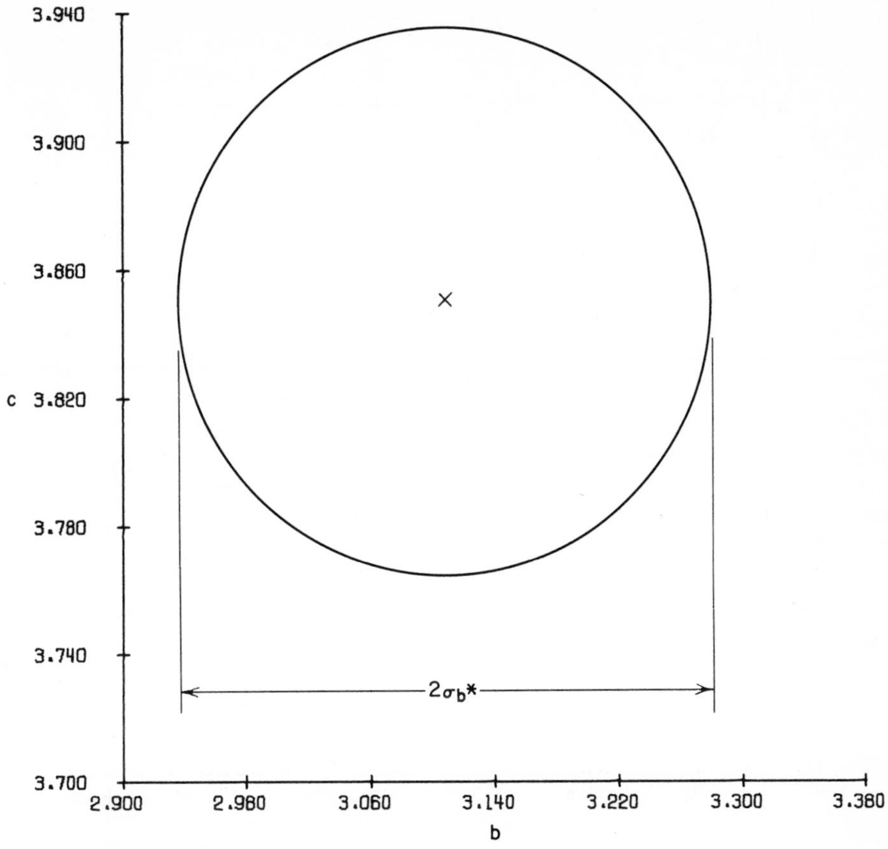

Figure 32.16
Plot of the contour $\chi^2(a^*,b,c) = \chi^2(a^*,b^*,c^*) + 1$ on the b,c plane.

yields

$$\chi^2(\mathbf{y};\mathbf{c}) - \chi_0^2(\mathbf{y};\mathbf{c}^*) = \left(\frac{1}{2}\frac{\partial^2\chi^2}{\partial c_m^{\,2}}\right)\bigg|_{c^*}(c_m - c_m^*)^2$$

which is a parabola, such as shown in Figure 32.14, from which $\sigma^2(c_m^*)$ may be obtained.

This results from a matrix of the form

that is, the *uncorrelated* parameter, c_m, has only a *diagonal* element nonzero

$$A_{mi} = A_{jm} = 0 \qquad \text{for } i, j \neq m$$

If two parameters c_k, c_l are *only correlated with each other* ($\mathrm{Covar}(c_k, c_i) = \mathrm{Covar}(c_l, c_i) = 0$ for $i \neq k, l$), then setting

$$\underline{\mathbf{A}} = \begin{bmatrix} A_1 & \cdots & A_{1,m-1} & 0 & A_{1,m+1} & \cdots & A_{1,r} \\ \vdots & & \vdots & \vdots & \vdots & & \\ A_{m-1,1} & \cdots & A_{m-1,m-1} & 0 & A_{m-1,m+1} & \cdots & A_{m-1,r} \\ 0 & \cdots & 0 & A_{m,m} & 0 & \cdots & 0 \\ A_{m+1,1} & \cdots & A_{m+1,m-1} & 0 & A_{m+1,m+1} & \cdots & A_{m+1,r} \\ \vdots & & \vdots & \vdots & \vdots & & \vdots \\ A_{r,1} & \cdots & A_{r,m-1} & 0 & A_{r,m+1} & \cdots & A_{r,r} \end{bmatrix}$$

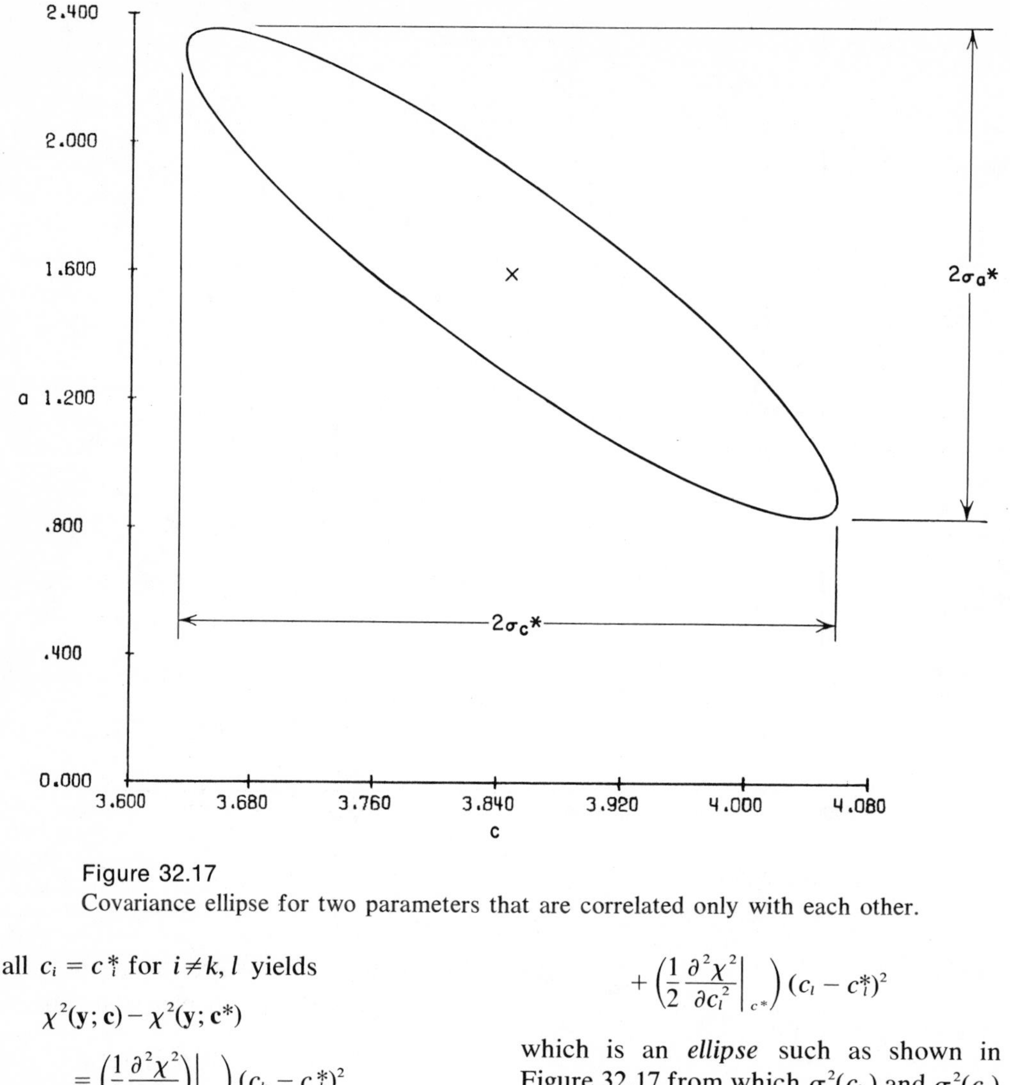

Figure 32.17
Covariance ellipse for two parameters that are correlated only with each other.

all $c_i = c_i^*$ for $i \neq k, l$ yields

$$\chi^2(\mathbf{y};\mathbf{c}) - \chi^2(\mathbf{y};\mathbf{c}^*) = \left(\frac{1}{2}\frac{\partial^2\chi^2}{\partial c_k^2}\bigg|_{\mathbf{c}^*}\right)(c_k - c_k^*)^2 + \left(\frac{\partial^2\chi^2}{\partial c_k\,\partial c_l}\bigg|_{\mathbf{c}^*}\right)(c_k - c_k^*)(c_l - c_l^*) + \left(\frac{1}{2}\frac{\partial^2\chi^2}{\partial c_l^2}\bigg|_{\mathbf{c}^*}\right)(c_l - c_l^*)^2$$

which is an *ellipse* such as shown in Figure 32.17 from which $\sigma^2(c_k)$ and $\sigma^2(c_l)$ may be obtained.

This arises from a matrix $\underline{\mathbf{A}}$ that may be cast into the form (for $l = k + 1$):

$$\underline{\mathbf{A}} = \begin{bmatrix} A_{11} & A_{12} & \cdots & 0 & 0 & A_{1,l+1} & \cdots & A_{1r} \\ A_{21} & A_{22} & \cdots & 0 & 0 & A_{2,l+1} & \cdots & A_{2r} \\ \vdots & \vdots & \ddots & \vdots & \vdots & \vdots & & \vdots \\ 0 & 0 & \cdots & \boxed{A_{kk}} & \boxed{A_{kl}} & 0 & \cdots & 0 \\ 0 & 0 & \cdots & \boxed{A_{lk}} & \boxed{A_{ll}} & 0 & \cdots & 0 \\ \vdots & \vdots & & \vdots & \vdots & \vdots & \ddots & \vdots \\ A_{r1} & A_{r2} & & 0 & 0 & A_{r,l+1} & \cdots & A_{rr} \end{bmatrix}$$

where there is only a 2×2 matrix on the diagonal at the positions k, l.

ORTHOGONAL POLYNOMIALS

It would clearly be convenient if the matrix $\underline{\mathbf{A}}$ were completely *diagonal*, that is, if

$$A_{ij} = 0 \qquad \text{for} \qquad i \neq j$$

Not only would the parameters c_i, $i = 1, \ldots, r$ be *uncorrelated*, but the matrix would be *easily inverted*, and one would avoid many of the computational difficulties encountered in inverting an arbitrary matrix.

Clearly one cannot do very much in this direction by choosing data points. However, it is sometimes useful to choose new *functions*

$$\Psi_k(x_i),\ k = 1, \ldots, r$$

[in place of the $g_k(x_i)$ in (32.11)] that have the desired property. That is, we have analogues of (32.11) to (32.21). In particular,

$$p(x_i; \mathbf{c}') = \sum_{k=1}^{r} c_k' \Psi_k(x_i)$$

and

$$A'_{lm} = \sum_{i=1}^{N} \Psi_l(x_i) \Psi_m(x_i) \frac{1}{\sigma_i^2}$$
$$= 0 \qquad \text{if} \qquad l \neq m$$

defines the choice of the *orthogonal polynomials*, $\Psi_k(x_i)$.

These functions may be found by a process analogous to diagonalizing any positive definite matrix.

NONLINEAR χ^2 MINIMIZATION PROBLEMS

When the regression function $p(\mathbf{x}; \mathbf{c})$ is not in the form of (32.11), we have the *nonlinear* case. In general, we must rely on numerical methods to locate the minimum of the χ^2 surface and to determine the "errors" on the parameter estimates. One feature shared by all the methods is that our geometrical ideas about the χ^2 surface may serve as a *guide*.

All the numerical procedures entail problems of finding *good approximations* (good first guesses on the final estimators) and *iterative procedures* with good convergence properties. It often happens that a central computational difficulty involves the *inversion of matrices* that may be large, "ill-conditioned," or both. Space does not permit discussion of the various procedures extant and, in any case, *availability* of a particular program at the local computer facility may dictate what is most convenient to use. References to some of the methods developed for digital computer installations are listed at the end of Appendix B.

The problem of χ^2 minimization with respect to n parameters is part of the general problem of *optimizing a multivariate function.* Again, this is too extensive a subject to summarize here, but some references are given at the end of Appendix A.

For purposes of illustration we indicate the outlines of *one* of the myriad procedures that have been developed. It is neither the most general nor the most efficient for all cases, but serves as a useful example for some cases.

Example of a computational procedure for χ^2 minimization

The general expression

$$\chi^2(\mathbf{c}) = \sum_{i=1}^{N} \frac{[y_i - p(x_i; \mathbf{c})]^2}{\sigma_i^2} \qquad (32.39)$$

may be used for any set of r values, $\mathbf{c} = (c_1, \ldots, c_r)$.

The expansion

$$\chi^2 = \sum_{j=1}^{r} \sum_{k=1}^{r} A_{jk} \Delta c_j \Delta c_k + 2 \sum_{j=1}^{r} B_j \Delta c_j + X_0 \qquad (32.40)$$

where

$$A_{jk} = A_{kj} \qquad \text{and} \qquad \Delta c_j = c_j^* - c_j$$

should be valid in the vicinity of $\chi^2_{\min} = \chi^2(\mathbf{c}^*)$.

If we knew the value of $\chi^2(\mathbf{c})$ at

$$n = \frac{r(r+1)}{2} + 2r + 1 \qquad (32.41)$$

points near $\mathbf{c}^*$, we could in principle solve the resulting equations for the A_{jk}, B_j, and X_0. With the (now-determined) expression (32.40), we may find the minimum point $\mathbf{c}^*$. We then have

$$\frac{1}{2}\left(\frac{\partial^2\chi^2}{\partial c_j\,\partial c_k}\bigg|_{\mathbf{c}^*}\right) = A_{jk} \qquad (32.42)$$

forming the symmetric matrix $\underline{\mathbf{A}}$ so that

$$\text{Covar}(c_j^*, c_k^*) = (\underline{\mathbf{A}}^{-1})_{jk} \qquad (32.43)$$

The computational procedure might be outlined as follows:

1. Guess a value for $\mathbf{c}^*$: $\mathbf{c}^{(1)}$. Calculate and store $\chi^2(\mathbf{c}^{(1)})$, using (32.39).
2. Step about the r-dimensional point $\mathbf{c}^{(1)}$ to get n points $\mathbf{c}_i^{(1)}$, $i = 1, \ldots, n$ where n is given by (32.41).
3. Evaluate $\chi^2(\mathbf{c})$ at the n points $\mathbf{c}_i^{(1)}$, setting up n equations of the form of (32.40).
4. Solve the n equations for the A_{jk}, B_j, X_0, getting the $(r \times r)$ values $A_{jk}^{(1)}$, the r values $B_j^{(1)}$ and $X_0^{(1)}$. (Recall that $A_{jk} = A_{kj}$.)
5. Minimize the function

$$\chi^2 = \sum_{j=1}^{r}\sum_{k=1}^{r} A_{jk}^{(1)}\Delta c_j\,\Delta c_k + 2\sum_{j=1}^{r} B_j^{(1)}\Delta c_j + X_0^{(1)}$$

with respect to $\mathbf{c}$, getting $\mathbf{c}^{(2)}$. We calculate and note $\chi^2(\mathbf{c}^{(2)})$, using (32.39). If the first guess was good enough, we should have

$$\chi^2(\mathbf{c}^{(2)}) \leqslant \chi^2(\mathbf{c}^{(1)})$$

If not, a first guess must be found such that this inequality holds, that is, such that the procedure starts close enough to the hyperparaboloidal region that it converges.

6. We go to (2) and continue the operations until

$$|\chi^2(\mathbf{c}^{(i)}) - \chi^2(\mathbf{c}^{(i-1)})| \leqslant \epsilon$$

where ϵ is some assigned (small) number. At this juncture, the best estimate is

$$\mathbf{c}^{(i)} \approx \mathbf{c}^*$$

and its covariance matrix is

$$[\underline{\mathbf{A}}^{(i)}]^{-1} \approx [\underline{\mathbf{Covar}}(\mathbf{c}^*)]$$

33

LEAST-SQUARES METHODS; CURVE FITTING

The *method of least squares* is computationally quite similar to the χ^2-minimization methods discussed. A general situation is to find the "best" values for the r parameters $\mathbf{c}$ so that the function $p(x_i\,;\mathbf{c})$ provides a "good fit" to the n measurements y_i. In our treatment of the χ^2 minimization method we took the x_i to be "known", that is, not random variables, whereas the y_i were assumed to be random variables that specifically were *normally distributed* about $p(x_i\,;\mathbf{c})$. We may use the method of least squares where the y_i are *not* normally distributed. We shall see that the least-squares method has certain useful *distribution-free* properties, that is, properties that do not depend on the distribution of the y_i being normal (or being *any* specific distribution).

If we consider the important case where we wish to fit the n measurements y_i to a function that is **linear** in the r parameters c_j, the *least-squares estimates* $\hat{c}_j$ may be shown to be **unbiased** and **minimum variance estimators** (*Gauss-Markov theorem*) irrespective of the distribution of y.

The least-squares method is even useful when the y_i are *not random variables.* An example is the construction of a *formula* to *describe* a table of precise values (x_i, y_i) when the formula is to be used for *interpolation, extrapolation,* or otherwise as a good *approximation.* In this sense, the method of least squares is generally suitable for *curve fitting.*

In this chapter we emphasize the *formulation* of the least-squares method especially in terms of *matrix algebra.* We shall, therefore, rely heavily on the summary in Appendix B. The application of the method is usually computationally identical to the χ^2 minimization methods previously discussed, and all the examples carry over.

General Formulation

The general case considers two random variables $\mathbf{X}$, $\mathbf{Y}$ as in the previous section and assumes that the *expectation values* of the $\mathbf{y}$ are given by a *function* of the $\mathbf{x}$ involving parameters $\mathbf{c}$ that must be estimated:

$$E\mathbf{y}(\mathbf{x}) = p(\mathbf{x};\mathbf{c}) \tag{33.1}$$

We shall take the $\mathbf{x}$ as *given* and the $\mathbf{y}$ as *measurements*, that is, the **x** are the independent mathematical variables whereas $p(\mathbf{x}; \mathbf{c})$ is the dependent variable. We shall in fact not treat **x** as random variables in our discussion. It is usually adequate to treat the *variability* of the **x** by considering them as *fixed* (known) and using the *phenomenological* variability of the **y**, which will then consist of the *inherent* variability of the **y** *augmented* by the variability of the **x** (*effective variance*).

In the χ^2 minimization method we have considered the case where **y** is normally distributed about $p(\mathbf{x}; \mathbf{c})$. The least-squares method assumes only that the *expectation value* of **y** is $p(\mathbf{x}; \mathbf{c})$ and that each y_i has a *constant variance*, σ_i^2, that may in general be *unknown* and *different* for each *i*. Furthermore, we shall start out with the general case that the y_i *need not be uncorrelated*, although we shall be primarily concerned with $\mathrm{Cov}(y_i, y_j) = 0$ for $i \neq j$. To be explicit, we need only assume for the least-squares method that, whatever the distribution of the population of the **y**, *a finite second moment exists*. When the distribution of the y_i is further specified to be normal with the y_i independent, the least-squares method reduces to the maximum likelihood or χ^2-minimum method.

The goal of all of these estimation methods is to provide estimators, $\hat{\mathbf{c}}$, of the parameters $\mathbf{c} = (c_1, \ldots, c_r)$ given the n data points (x_i, y_i), $i = 1, \ldots, n$. The relation

$$\hat{p} = p(\mathbf{x}; \hat{\mathbf{c}}) \tag{33.2}$$

is sometimes called the *best-fit regression curve*.

We shall start with the general case that the y_i have *different variances* and may be *correlated*. We suppose that the **covariance matrix** of the **y** is

$$[\underline{\mathbf{Covar}}(\mathbf{y})] = \sigma^2 \underline{\mathbf{W}}^{-1} \tag{33.3}$$

where σ^2 is a *scale factor*, often unknown *a priori*, and $\underline{\mathbf{W}}$ is the **weight matrix**. Because of the definition data points have variances that are *inversely* proportional to their weights, that is, for *uncorrelated* measurements y_i, the $\underline{\mathbf{W}}$ matrix is *diagonal* and

$$\mathrm{Var}(y_i) = \frac{\sigma^2}{W_{ii}}$$

For *correlated* measurements $\underline{\mathbf{W}}$ need *not* be diagonal but $\underline{\mathbf{W}}$ is *always symmetric*

$$W_{ij} = W_{ji} \tag{33.4}$$

Linear Case

Our primary concern is where the regression function, $p(\mathbf{x}; \mathbf{c})$, is a *linear* function of the parameters **c**, that is,

$$p(\mathbf{x}; \mathbf{c}) = c_1 g_1(\mathbf{x}) + c_2 g_2(\mathbf{x}) + \cdots + c_r g_r(\mathbf{x}) = \sum_{i=1}^{r} c_i g_i(\mathbf{x}) \tag{33.5}$$

where the $g_i(\mathbf{x})$ are *known* functions of **x**. The r functions $g_i(\mathbf{x})$ need only be *linearly independent*. Since there are n points x_i, we may define a **matrix** $(n \times r)$,

$$\underline{\boldsymbol{\alpha}} = \begin{bmatrix} g_1(x_1) & g_2(x_1) & \cdots & g_r(x_1) \\ g_1(x_2) & \cdot & \cdots & \cdot \\ \cdot & \cdot & \cdots & \cdot \\ g_1(x_n) & \cdot & \cdots & g_r(x_n) \end{bmatrix} \tag{33.6}$$

sometimes called the *design matrix*, with elements,

$$\alpha_{ij} = g_j(x_i)$$

In general,

$$y_i = p(x_i; \mathbf{c}) + \epsilon_i \qquad i = 1, \ldots, n \tag{33.7}$$

Here, ϵ_i is a *random variable with zero mean*

$$E(\epsilon_i) = 0$$

because of (33.1) and *constant variance* (but that may be different for different i). The distribution of the ϵ_i is *not* necessarily normal. We need only assume that the distribution of the ϵ_i has a *finite second moment*. The link to the **covariance matrix** of the **y** is clear:

$$\sigma_{ij} = \mathrm{Cov}(y_i, y_j) = E(\epsilon_i \epsilon_j) = \sigma^2 (\underline{\mathbf{W}}^{-1})_{ij}$$

The quantities, ϵ_i, are called the *errors*.

We may define the ($n \times 1$) column matrix of errors by the matrix equation

$$\boldsymbol{\epsilon} = \mathbf{y} - \underline{\boldsymbol{\alpha}}\mathbf{c} \tag{33.8}$$

We may form S, the *weighted sum of squares of the errors*. When the errors are *uncorrelated* and of *equal weight*

$$S = \sum_{i=1}^{n} \epsilon_i^2 \tag{33.9}$$

When the errors are *uncorrelated* but of *different weights*, W_{ii}, the sum is

$$S = \sum_{i=1}^{n} \epsilon_i^2 W_{ii} = \sum_{i=1}^{n} \frac{\epsilon_i^2}{\sigma_{ii}^2} \tag{33.10}$$

In the general case (*correlated* errors of *different* weights), the sum may be written as a *quadratic form*. In matrix notation,

$$S = \boldsymbol{\epsilon}^T \underline{\mathbf{W}} \boldsymbol{\epsilon} = (\mathbf{y} - \underline{\boldsymbol{\alpha}}\mathbf{c})^T \underline{\mathbf{W}} (\mathbf{y} - \underline{\boldsymbol{\alpha}}\mathbf{c}) \tag{33.11}$$

We define the *least-squares estimator*, $\hat{\mathbf{c}}$, to be the estimator that **minimizes** S. We therefore seek to minimize S with respect to each of the r quantities c_i, that is,

$$\left.\frac{\partial S}{\partial c_i}\right|_{\hat{c}} = 0 \qquad i = 1, \ldots, r \tag{33.12}$$

We may expand

$$\begin{aligned} S &= (\mathbf{y} - \underline{\boldsymbol{\alpha}}\mathbf{c})^T \underline{\mathbf{W}} (\mathbf{y} - \underline{\boldsymbol{\alpha}}\mathbf{c}) \\ &= (\mathbf{y}^T - \mathbf{c}^T \underline{\boldsymbol{\alpha}}^T) \underline{\mathbf{W}} (\mathbf{y} - \underline{\boldsymbol{\alpha}}\mathbf{c}) \\ &= (\mathbf{y}^T \underline{\mathbf{W}} \mathbf{y}) - \mathbf{c}^T \underline{\boldsymbol{\alpha}}^T \underline{\mathbf{W}} \mathbf{y} - \mathbf{y}^T \underline{\mathbf{W}} \underline{\boldsymbol{\alpha}} \mathbf{c} \\ &\quad + \mathbf{c}^T \underline{\boldsymbol{\alpha}}^T \underline{\mathbf{W}} \underline{\boldsymbol{\alpha}} \mathbf{c} \end{aligned}$$

The *middle two* terms may be shown to be *equal*:

$$\left[\begin{aligned} \mathbf{c}^T \underline{\boldsymbol{\alpha}}^T \underline{\mathbf{W}} \mathbf{y} &= \sum_{\sigma=1}^{r} \sum_{\rho=1}^{r} \sum_{k=1}^{n} c^T{}_\sigma (\underline{\boldsymbol{\alpha}}^T)_{\sigma\rho} W_{\rho k} y_k \\ &= \sum_{\sigma=1}^{r} \sum_{\rho=1}^{n} \sum_{k=1}^{n} c_\sigma \alpha_{\rho\sigma} W_{\rho k} y_k \\ \mathbf{y}^T \underline{\mathbf{W}} \underline{\boldsymbol{\alpha}} \mathbf{c} &= \sum_{k=1}^{n} \sum_{\rho=1}^{n} \sum_{\sigma=1}^{r} y^T{}_k W_{k\rho} \alpha_{\rho\sigma} c_\sigma ; \\ W_{k\rho} &= W_{\rho k} \end{aligned}\right]$$

Therefore,

$$S = \mathbf{y}^T \underline{\mathbf{W}} \mathbf{y} - 2\mathbf{c}^T \underline{\boldsymbol{\alpha}}^T \underline{\mathbf{W}} \mathbf{y} + \mathbf{c}^T \underline{\boldsymbol{\alpha}}^T \underline{\mathbf{W}} \underline{\boldsymbol{\alpha}} \mathbf{c} \tag{33.13}$$

We differentiate S with respect to $\mathbf{c}$. The first term yields zero, the second term is straightforward, while the third term yields the *derivative of a quadratic form* given by Section 30 of Appendix B, since

$$\underline{\boldsymbol{\alpha}}^T \underline{\mathbf{W}} \underline{\boldsymbol{\alpha}} = \underline{\boldsymbol{\alpha}}^T (r \times n) \underline{\mathbf{W}} (n \times n) \underline{\boldsymbol{\alpha}} (n \times r)$$

is a symmetric $r \times r$ matrix.

$$\therefore \left.\frac{\partial S}{\partial \mathbf{c}}\right|_{\mathbf{c}=\hat{\mathbf{c}}} = 0 = -2\underline{\boldsymbol{\alpha}}^T \underline{\mathbf{W}} \mathbf{y} + 2\underline{\boldsymbol{\alpha}}^T \underline{\mathbf{W}} \underline{\boldsymbol{\alpha}} \hat{\mathbf{c}} \tag{33.14}$$

which is equivalent to r equations (33.12).

This yields the matrix equation

$$\underline{\boldsymbol{\alpha}}^T \underline{\mathbf{W}} \mathbf{y} = \underline{\boldsymbol{\alpha}}^T \underline{\mathbf{W}} \underline{\boldsymbol{\alpha}} \hat{\mathbf{c}}$$

the so-called *normal equations for the parameters* $\mathbf{c}$. This may be *solved* for $\hat{\mathbf{c}}$:

$$\hat{\mathbf{c}} = (\underline{\boldsymbol{\alpha}}^T \underline{\mathbf{W}} \underline{\boldsymbol{\alpha}})^{-1} \underline{\boldsymbol{\alpha}}^T \underline{\mathbf{W}} \mathbf{y} \tag{33.15}$$

The *scale factor* σ^2 need *not* be known; only the *relative weight* matrix $\underline{\mathbf{W}}$ is required to make the least-squares estimate of the parameters.

We should stress that this expression *cannot* be reduced further since $\underline{\boldsymbol{\alpha}}$ is ($n \times r$), $\underline{\boldsymbol{\alpha}}^T$ is ($r \times n$), $\underline{\mathbf{W}}$ is ($n \times n$), $\mathbf{y}$ is ($n \times 1$), $\hat{\mathbf{c}}$ is ($r \times 1$), and $\underline{\boldsymbol{\alpha}}^T \underline{\mathbf{W}} \underline{\boldsymbol{\alpha}}$ is $r \times r$. The inverse is defined only for the *square* matrix $\underline{\boldsymbol{\alpha}}^T \underline{\mathbf{W}} \underline{\boldsymbol{\alpha}}$ and then only if $\underline{\boldsymbol{\alpha}}^T \underline{\mathbf{W}} \underline{\boldsymbol{\alpha}}$ is nonsingular. $\underline{\boldsymbol{\alpha}}^T \underline{\mathbf{W}} \underline{\boldsymbol{\alpha}}$ is nonsingular if and only if the columns of $\underline{\boldsymbol{\alpha}}$ are *linearly independent*. That is, we cannot use as design functions $g_l(\mathbf{x})$, $g_k(\mathbf{x})$ such that

$$g_l(\mathbf{x}) = K g_k(\mathbf{x}) + K^1$$

where K and K^1 are constants.

(If the $g_i(\mathbf{x})$ are chosen to be linearly independent *and* $r = n$, **then** $\hat{\mathbf{c}} = \underline{\boldsymbol{\alpha}}^{-1}\mathbf{y}$, which is a very special case.)

► **Example 33.1**
Uncorrelated measurements

If we specialize to the case of *uncorrelated* measurements with *unequal weights*, the matrix $\underline{\mathbf{W}}$ becomes diagonal with elements $W_{ij} = \delta_{ij}/\sigma_i^2$.

We therefore get from (33.15) the estimate

$$\hat{\mathbf{c}} = (\underline{\boldsymbol{\alpha}}^T \underline{\mathbf{W}} \underline{\boldsymbol{\alpha}})^{-1} \underline{\boldsymbol{\alpha}}^T \underline{\mathbf{W}} \mathbf{y} = \underline{\mathbf{M}}^{-1}(r \times r) \cdot \mathbf{v}(r \times 1)$$

where

$$\underline{\mathbf{M}} = (\underline{\boldsymbol{\alpha}}^T \underline{\mathbf{W}} \underline{\boldsymbol{\alpha}}), \ \mathbf{v} = \underline{\boldsymbol{\alpha}}^T \underline{\mathbf{W}} \mathbf{y}$$

Here

$$\begin{aligned} M_{lm} &= \sum_{i=1}^{n} \sum_{j=1}^{n} \alpha^T_{li} W_{ij} \alpha_{jm} \\ &= \sum_{i=1}^{n} \sum_{j=1}^{n} \alpha^T_{li} \frac{\delta_{ij}}{\sigma_i^2} \alpha_{jm} \\ &= \sum_{i=1}^{n} \alpha_{il} \alpha_{im} \frac{1}{\sigma_i^2} \\ &= \sum_{i=1}^{n} g_l(x_i) g_m(x_i) \frac{1}{\sigma_i^2} \end{aligned}$$

which is just (32.16), that is, $\underline{\mathbf{M}} = \underline{\mathbf{A}}$. Also,

$$\mathbf{v} = \underline{\boldsymbol{\alpha}}^T \underline{\mathbf{W}} \mathbf{y}$$

$$\begin{aligned} \therefore v_l &= \sum_{i=1}^{n} \sum_{j=1}^{n} \alpha^T_{li} \frac{\delta_{ij}}{\sigma_i^2} y_j \\ &= \sum_{i=1}^{n} \alpha_{il} \frac{y_i}{\sigma_i^2} \\ &= \sum_{i=1}^{n} g_l(x_i) y_i / \sigma_i^2 \qquad l = 1, \ldots, r \end{aligned}$$

this is u_l from (32.17).

We have therefore shown that the **χ^2-minimum estimate**, $\mathbf{c}^*$, from (32.20) is identical to the **least-squares estimate**, $\hat{\mathbf{c}}$, from (33.15). ◄

DISTRIBUTION-FREE PROPERTIES OF LEAST-SQUARES ESTIMATES

The χ^2-minimum estimate of $\mathbf{c}$ was obtained by assuming that the ϵ_i are normally distributed. When this is *not* true, the least-squares estimate $\hat{\mathbf{c}}$ still has certain important features that may make it the estimate of choice.

$\hat{\mathbf{c}}$ is an unbiased estimate of $\mathbf{c}$

We have

$$\hat{\mathbf{c}} = (\underline{\boldsymbol{\alpha}}^T \underline{\mathbf{W}} \underline{\boldsymbol{\alpha}})^{-1} \underline{\boldsymbol{\alpha}}^T \underline{\mathbf{W}} \mathbf{y}$$

and

$$E\mathbf{y}(\mathbf{x}) = p(\mathbf{x}; \mathbf{c}) = \underline{\boldsymbol{\alpha}} \mathbf{c}$$

in the linear case.

Therefore, the expectation value is

$$\begin{aligned} E(\hat{\mathbf{c}}) &= E[(\underline{\boldsymbol{\alpha}}^T \underline{\mathbf{W}} \underline{\boldsymbol{\alpha}})^{-1} \underline{\boldsymbol{\alpha}}^T \underline{\mathbf{W}} \mathbf{y}] \\ &= (\underline{\boldsymbol{\alpha}}^T \underline{\mathbf{W}} \underline{\boldsymbol{\alpha}})^{-1} \underline{\boldsymbol{\alpha}}^T \underline{\mathbf{W}} E(\mathbf{y}) \\ &= (\underline{\boldsymbol{\alpha}}^T \underline{\mathbf{W}} \underline{\boldsymbol{\alpha}})^{-1} (\underline{\boldsymbol{\alpha}}^T \underline{\mathbf{W}} \underline{\boldsymbol{\alpha}}) \mathbf{c} = \mathbf{c} \end{aligned}$$

therefore $\hat{\mathbf{c}}$ is unbiased.

$\hat{\mathbf{c}}$ is a minimum variance unbiased estimator

Gauss in 1821 and Markov in 1912 proved the *Gauss-Markov Theorem*: Among the class $\mathbf{c}_{\text{est}}$ of *unbiased* estimates of $\mathbf{c}$ that are *linear* functions of the measurements $\mathbf{y}$, the least-squares estimate $\hat{\mathbf{c}}$ has **minimum variance**, that is,

$$\mathrm{Var}(\hat{c}_i) \leq \mathrm{Var}((c_{\text{est}})_i) \qquad i = 1, \ldots, r$$

Case 1: We may show this first for the case that all components y_i are *independent* and have the *same variance*, σ^2, *which may be unknown*:

$$\mathrm{Var}(y_i) = \sigma^2$$

and

$$\mathrm{Covar}(y_i, y_j) = 0 \qquad i \neq j$$

As a **matrix**

$$[\underline{\mathbf{Covar}}(\mathbf{y})] = \sigma^2 \underline{\mathbf{I}}(n \times n) = \sigma^2 \underline{\mathbf{W}}^{-1}$$

where $\underline{\mathbf{I}}$ is the $(n \times n)$-**unit matrix** ($\underline{\mathbf{W}} = \underline{\mathbf{I}}$).

The least-squares estimate becomes

$$\hat{\mathbf{c}} = (\underline{\boldsymbol{\alpha}}^T \underline{\boldsymbol{\alpha}})^{-1} \underline{\boldsymbol{\alpha}}^T \mathbf{y} \tag{33.16}$$

which is a *linear function* of $\mathbf{y}$:

$$\hat{\mathbf{c}} = \underline{\mathbf{B}} \mathbf{y}$$

where $\underline{\mathbf{B}} = (\underline{\boldsymbol{\alpha}}^T \underline{\boldsymbol{\alpha}})^{-1} \underline{\boldsymbol{\alpha}}^T$.

Since $\hat{\mathbf{c}}$ is a *linear transformation* of $\mathbf{y}$, the covariance matrix of the $(\hat{c}_i, \hat{c}_j)$ may be obtained from the covariance matrix of the (y_i, y_j) by the *law of propagation of errors* discussed in Section 36 of Appendix B:

$$\begin{aligned} [\underline{\mathbf{Covar}}(\hat{\mathbf{c}})] &= \underline{\mathbf{B}} [\underline{\mathbf{Covar}}(\mathbf{y})] \underline{\mathbf{B}}^T \\ &= (\underline{\boldsymbol{\alpha}}^T \underline{\boldsymbol{\alpha}})^{-1} \underline{\boldsymbol{\alpha}}^T [\underline{\mathbf{Covar}}(\mathbf{y})] \underline{\boldsymbol{\alpha}} (\underline{\boldsymbol{\alpha}}^T \underline{\boldsymbol{\alpha}})^{-1} \end{aligned} \tag{33.17}$$

where we have used

$$[(\underline{\boldsymbol{\alpha}}^T \underline{\boldsymbol{\alpha}})^{-1}]^T = (\underline{\boldsymbol{\alpha}}^T \underline{\boldsymbol{\alpha}})^{-1}.$$

$$\therefore [\underline{\mathbf{Covar}}(\hat{\mathbf{c}})] = \sigma^2(\underline{\alpha}^T\underline{\alpha})^{-1}\underline{\alpha}^T\underline{\mathbf{I}}\underline{\alpha}(\underline{\alpha}^T\underline{\alpha})^{-1} = \sigma^2(\underline{\alpha}^T\underline{\alpha})^{-1} \quad (33.18)$$

Let $\mathbf{c}_{est}$ be *any other unbiased* estimate of $\mathbf{c}$ that is *linear* in the y_i:

$$\mathbf{c}_{est}(r \times 1) = \underline{\mathbf{L}}(r \times n)\mathbf{y}(n \times 1)$$

Since we assume $\mathbf{c}_{est}$ to be unbiased,

$$E(\mathbf{c}_{est}) = \underline{\mathbf{L}}E(\mathbf{y}) = \underline{\mathbf{L}}\underline{\alpha}\underline{\mathbf{c}} = \mathbf{c}$$

by assumption.

$$\therefore \underline{\mathbf{L}}\underline{\alpha} = \underline{\mathbf{I}} \quad (33.19)$$

where $\underline{\mathbf{L}}$ is $(r \times n)$, $\underline{\alpha}$ is $(n \times r)$ and $\underline{\mathbf{I}}$ is $(r \times r)$.

By *propagation of errors* (33.17),

$$[[\underline{\mathbf{Covar}}(\mathbf{c}_{est})] = \underline{\mathbf{L}}[\underline{\mathbf{Covar}}(\mathbf{y})]\underline{\mathbf{L}}^T = \sigma^2\underline{\mathbf{L}}\underline{\mathbf{I}}\underline{\mathbf{L}}^T = \sigma^2\underline{\mathbf{L}}\underline{\mathbf{L}}^T \quad (33.20)$$

The Gauss-Markov theorem is shown by demonstrating that

$$[\underline{\mathbf{Covar}}(\hat{\mathbf{c}})]_{ii} \le [\underline{\mathbf{Covar}}(\mathbf{c}_{est})]_{ii} \qquad \text{for all } i \quad (33.21)$$

Following Plackett,* we define

$$\underline{\mathbf{C}} = \underline{\alpha}^T\underline{\alpha} \quad (33.22)$$

which is $(r \times r)$. We write the identity

$$\underline{\mathbf{L}}\underline{\mathbf{L}}^T = (\underline{\alpha}^T\underline{\alpha})^{-1} + (\underline{\mathbf{L}} - \underline{\mathbf{C}}^{-1}\underline{\alpha}^T)(\underline{\mathbf{L}} - \underline{\mathbf{C}}^{-1}\underline{\alpha}^T)^T \quad (33.23)$$

which may be shown by expansion of terms, using

$$\underline{\mathbf{L}}\underline{\alpha} = \underline{\mathbf{I}} = \underline{\alpha}^T\underline{\mathbf{L}}^T$$

$$\begin{aligned}[(\underline{\alpha}^T\underline{\alpha})^{-1} &+ (\underline{\mathbf{L}} - \underline{\mathbf{C}}^{-1}\underline{\alpha}^T)(\underline{\mathbf{L}} - \underline{\mathbf{C}}^{-1}\underline{\alpha}^T)^T \\ &= (\underline{\alpha}^T\underline{\alpha})^{-1} + \underline{\mathbf{L}}\underline{\mathbf{L}}^T - \underline{\mathbf{C}}^{-1}\underline{\alpha}^T\underline{\mathbf{L}}^T \\ &\quad - \underline{\mathbf{L}}\underline{\alpha}\underline{\mathbf{C}}^{-1} + \underline{\mathbf{C}}^{-1}\underline{\alpha}^T\underline{\alpha}\underline{\mathbf{C}}^{-1} \\ &= (\underline{\alpha}^T\underline{\alpha})^{-1} + \underline{\mathbf{L}}\underline{\mathbf{L}}^T - (\underline{\alpha}^T\underline{\alpha})^{-1} \\ &\quad - (\underline{\alpha}^T\underline{\alpha})^{-1} + (\underline{\alpha}^T\underline{\alpha})^{-1} \\ &= \underline{\mathbf{L}}\underline{\mathbf{L}}^{-1}, \text{ verifying (33.23)}]\end{aligned}$$

$$\begin{aligned}\therefore [\underline{\mathbf{Covar}}(\mathbf{c}_{est})] &= \sigma^2\underline{\mathbf{L}}\underline{\mathbf{L}}^T \\ &= \sigma^2(\underline{\alpha}^T\underline{\alpha})^{-1} \\ &\quad + \sigma^2(\underline{\mathbf{L}} - \underline{\mathbf{C}}^{-1}\underline{\alpha}^T)(\underline{\mathbf{L}} - \underline{\mathbf{C}}^{-1}\underline{\alpha}^T)^T \\ &= [\underline{\mathbf{Covar}}(\hat{\mathbf{c}})] \\ &\quad + \sigma^2(\underline{\mathbf{L}} - \underline{\mathbf{C}}^{-1}\underline{\alpha}^T)(\underline{\mathbf{L}} - \underline{\mathbf{C}}^{-1}\underline{\alpha}^T)^T\end{aligned}$$

*R. L. Plackett, *Principles of Regression Analysis*, Oxford University Press, 1960, p. 37.

$$\begin{aligned}\therefore [\underline{\mathbf{Covar}}(\mathbf{c}_{est})]_{ij} &= [\underline{\mathbf{Covar}}(\hat{\mathbf{c}})]_{ij} \\ &\quad + \sigma^2[(\underline{\mathbf{L}} - \underline{\mathbf{C}}^{-1}\underline{\alpha}^T) \\ &\quad \times (\underline{\mathbf{L}} - \underline{\mathbf{C}}^{-1}\underline{\alpha}^T)^T]_{ij}\end{aligned}$$

The second term is a positive semidefinite quantity (i.e., ≥ 0) for $i = j$. It is *zero* for

$$\underline{\mathbf{L}} = \underline{\mathbf{C}}^{-1}\underline{\alpha}^T = (\underline{\alpha}^T\underline{\alpha})^{-1}\underline{\alpha}^T$$

the *least-squares* estimate.

$$\therefore [\underline{\mathbf{Covar}}(\mathbf{c}_{est})]_{ii} \ge [\underline{\mathbf{Covar}}(\hat{\mathbf{c}})]_{ii} \qquad \text{for all } i$$

showing (33.21) and proving the Gauss-Markov theorem. **Q.E.D.**

$$\therefore \mathrm{Var}(\mathbf{c}_{est}) \ge \mathrm{Var}(\hat{\mathbf{c}}) \quad (33.24)$$

The least-squares estimate, $\hat{\mathbf{c}}$, is the unbiased estimator linear in the *independent measurements* that has *minimum* variance.

Case 2: The Gauss-Markov theorem was extended to *correlated measurements* by Aitken.*

As before, we take (33.7) as the starting point

$$\mathbf{y} = \underline{\alpha}\mathbf{c} + \boldsymbol{\epsilon}$$

where now

$$[\underline{\mathbf{Covar}}(\boldsymbol{\epsilon})] = \sigma^2\underline{\mathbf{W}}^{-1} = \sigma^2\underline{\mathbf{V}} \quad (33.25)$$

Here $\underline{\mathbf{V}}$ is a *known, positive definite symmetric* **matrix** that need *not* be the unit matrix or even diagonal. However, since $\underline{\mathbf{V}}$ must be positive-definite, it may be expressed in the form

$$\underline{\mathbf{V}} = \underline{\mathbf{P}}\underline{\mathbf{P}}^T \quad (33.26)$$

where $\underline{\mathbf{P}}$ is a *nonsingular* matrix. We may define the linear transformations

$$\boldsymbol{\eta} = \underline{\mathbf{P}}^{-1}\boldsymbol{\epsilon}$$

and (33.27)

$$\mathbf{x} = \underline{\mathbf{P}}^{-1}\mathbf{y},$$

and rewrite (33.7):

$$\underline{\mathbf{P}}\mathbf{x} = \underline{\alpha}\mathbf{c} + \underline{\mathbf{P}}\boldsymbol{\eta}.$$

*A. C. Aitken, *On least squares and linear combinations of observations* Proceedings of the Royal Society Edinburgh A, *55*, 42 (1934).

We may solve this for **x**:

$$\mathbf{x} = (\underline{\mathbf{P}}^{-1}\underline{\boldsymbol{\alpha}})\mathbf{c} + \boldsymbol{\eta}$$
$$= \underline{\mathbf{B}}\mathbf{c} + \boldsymbol{\eta} \qquad (33.28)$$

where

$$\underline{\mathbf{B}} = \underline{\mathbf{P}}^{-1}\underline{\boldsymbol{\alpha}} \qquad (33.29)$$

But, by *propagation of errors,*

$$\underline{\mathbf{Covar}}(\boldsymbol{\eta}) = \underline{\mathbf{P}}^{-1}[\underline{\mathbf{Covar}}(\boldsymbol{\epsilon})](\underline{\mathbf{P}}^{T})^{-1}$$
$$= \sigma^2\underline{\mathbf{P}}^{-1}\underline{\mathbf{V}}(\underline{\mathbf{P}}^{T})^{-1}$$
$$= \sigma^2\underline{\mathbf{P}}^{-1}(\underline{\mathbf{P}}\underline{\mathbf{P}}^{T})(\underline{\mathbf{P}}^{T})^{-1}$$
$$= \sigma^2\underline{\mathbf{I}}$$

Therefore, the model (33.28) reduces to the case already proved, that is, the estimator $\hat{\mathbf{c}}$ obtained by minimizing

$$S = (\mathbf{x} - \underline{\mathbf{B}}\mathbf{c})^T(\mathbf{x} - \underline{\mathbf{B}}\mathbf{c})$$

has minimum variance.

But

$$S = (\mathbf{x}^T - \mathbf{c}^T\underline{\mathbf{B}}^T)(\mathbf{x} - \underline{\mathbf{B}}\mathbf{c})$$
$$= [(\underline{\mathbf{P}}^{-1}\mathbf{y})^T - \mathbf{c}^T(\underline{\mathbf{P}}^{-1}\underline{\boldsymbol{\alpha}})^T][\underline{\mathbf{P}}^{-1}\mathbf{y} - \underline{\mathbf{P}}^{-1}\underline{\boldsymbol{\alpha}}\mathbf{c}]$$
$$= (\mathbf{y}^T - \mathbf{c}^T\underline{\boldsymbol{\alpha}}^T)((\underline{\mathbf{P}}^{-1})^T(\underline{\mathbf{P}}^{-1}))(\mathbf{y} - \underline{\boldsymbol{\alpha}}\mathbf{c})$$
$$\therefore S = (\mathbf{y} - \underline{\boldsymbol{\alpha}}\mathbf{c})^T\underline{\mathbf{W}}(\mathbf{y} - \underline{\boldsymbol{\alpha}}\mathbf{c})$$

which is (33.11).

Thus, we have shown the Gauss-Markov theorem for the estimator obtained by minimizing the weighted least-squares sum (33.11). **Q.E.D.**

Covariance matrix

We have shown the general *least-squares estimator*

$$\hat{\mathbf{c}} = (\underline{\boldsymbol{\alpha}}^T\underline{\mathbf{W}}\underline{\boldsymbol{\alpha}})^{-1}\underline{\boldsymbol{\alpha}}^T\underline{\mathbf{W}}\mathbf{y} \qquad (33.15)$$

to be *unbiased* and of *minimum* variance with its **covariance matrix** given by the law of propagation of errors:

$$[\underline{\mathbf{Covar}}(\hat{\mathbf{c}})] = (\underline{\boldsymbol{\alpha}}^T\underline{\mathbf{W}}\underline{\boldsymbol{\alpha}})^{-1}\underline{\boldsymbol{\alpha}}^T\underline{\mathbf{W}}[\underline{\mathbf{Covar}}(\mathbf{y})]$$
$$\times [(\underline{\boldsymbol{\alpha}}^T\underline{\mathbf{W}}\underline{\boldsymbol{\alpha}})^{-1}\underline{\boldsymbol{\alpha}}^T\underline{\mathbf{W}}]^T$$
$$= (\underline{\boldsymbol{\alpha}}^T\underline{\mathbf{W}}\underline{\boldsymbol{\alpha}})^{-1}\underline{\boldsymbol{\alpha}}^T\underline{\mathbf{W}}$$
$$\times (\sigma^2\underline{\mathbf{W}}^{-1})\underline{\mathbf{W}}\underline{\boldsymbol{\alpha}}(\underline{\boldsymbol{\alpha}}^T\underline{\mathbf{W}}\underline{\boldsymbol{\alpha}})^{-1}$$
$$= \sigma^2(\underline{\boldsymbol{\alpha}}^T\underline{\mathbf{W}}\underline{\boldsymbol{\alpha}})^{-1} \qquad (33.30)$$

We need only the **matrix** of *relative weights,* $\underline{\mathbf{W}}$, to calculate the estimate (33.15), the scale factor σ^2 not entering into the calculations. For the *covariance matrix,* (33.30), however, the *scale factor* σ^2 *is* required. If this is *a priori* unknown, the least-squares method provides an *unbiased estimate* of σ^2 from the *sum of squared residuals.*

RESIDUALS

We define the *fitted* values

$$\hat{y}_i = \sum_{j=1}^{r} \hat{c}_j g_j(x_i) = \sum_{j=1}^{r} \alpha_{ij}\hat{c}_j \qquad (33.31)$$

which may be expressed as

$$\hat{\mathbf{y}} = \underline{\boldsymbol{\alpha}}\hat{\mathbf{c}} \qquad (33.32)$$

We may compare the *fitted* values $\hat{\mathbf{y}}$ with the *measured* values $\mathbf{y}$ and define **R**, the $(n \times 1)$ column vector of *residuals*

$$\mathbf{R} = \mathbf{y} - \hat{\mathbf{y}} = \mathbf{y} - \underline{\boldsymbol{\alpha}}\hat{\mathbf{c}} \qquad (33.33)$$

The ith residual is

$$R_i = y_i - \hat{y}_i \qquad (33.34)$$

Let us consider uncorrelated y_i with the *same* **unknown** variance σ^2. That is,

$$\underline{\mathbf{Covar}}(\mathbf{y}) = \sigma^2\underline{\mathbf{I}}(n \times n). \qquad (33.35)$$

The *sum of the squares of the residuals* here is

$$S_{\min} = (\mathbf{y} - \underline{\boldsymbol{\alpha}}\hat{\mathbf{c}})^T(\mathbf{y} - \underline{\boldsymbol{\alpha}}\hat{\mathbf{c}}) \qquad (33.36)$$

which is the minimum value of $S = (\mathbf{y} - \underline{\boldsymbol{\alpha}}\mathbf{c})^T(\mathbf{y} - \underline{\boldsymbol{\alpha}}\mathbf{c})$, where $\hat{\mathbf{c}}$ is given by (33.16). We may rewrite this:

$$S_{\min} = \mathbf{y}^T\mathbf{y} - \hat{\mathbf{c}}^T\underline{\boldsymbol{\alpha}}^T\mathbf{y} - \mathbf{y}^T\underline{\boldsymbol{\alpha}}\hat{\mathbf{c}} + \hat{\mathbf{c}}^T\underline{\boldsymbol{\alpha}}^T\underline{\boldsymbol{\alpha}}\hat{\mathbf{c}}$$
$$= \mathbf{y}^T\mathbf{y} - 2\mathbf{y}^T\underline{\boldsymbol{\alpha}}\hat{\mathbf{c}} + \hat{\mathbf{c}}^T\underline{\boldsymbol{\alpha}}^T\underline{\boldsymbol{\alpha}}\hat{\mathbf{c}}$$

We substitute for $\hat{\mathbf{c}}^T$ using (33.16) and get

$$S_{\min} = \mathbf{y}^T\mathbf{y} - 2\mathbf{y}^T\underline{\boldsymbol{\alpha}}\hat{\mathbf{c}}$$
$$+ [(\underline{\boldsymbol{\alpha}}^T\underline{\boldsymbol{\alpha}})^{-1}\underline{\boldsymbol{\alpha}}^T\mathbf{y}]^T(\underline{\boldsymbol{\alpha}}^T\underline{\boldsymbol{\alpha}})\hat{\mathbf{c}}$$
$$= \mathbf{y}^T\mathbf{y} - 2\mathbf{y}^T\underline{\boldsymbol{\alpha}}\hat{\mathbf{c}}$$
$$+ \mathbf{y}^T\underline{\boldsymbol{\alpha}}(\underline{\boldsymbol{\alpha}}^T\underline{\boldsymbol{\alpha}})^{-1}(\underline{\boldsymbol{\alpha}}^T\underline{\boldsymbol{\alpha}})\hat{\mathbf{c}}$$
$$= \mathbf{y}^T\mathbf{y} - \mathbf{y}^T\underline{\boldsymbol{\alpha}}\hat{\mathbf{c}} \qquad (33.37)$$

Computationally, this formula is often not as useful as (33.36) because of *rounding errors.*

ESTIMATOR OF σ^2

We have thus far assumed the *independent* y_i to have the *same* variance σ^2. If σ^2 is **unknown** *a priori*, we may **estimate** its value from the sum of squared residuals, $S_{\min}$.

We may rewrite this sum as

$$S_{\min} = (\mathbf{y} - \underline{\boldsymbol{\alpha}}\mathbf{c})^T(\underline{\mathbf{I}} - \underline{\boldsymbol{\alpha}}\underline{\mathbf{C}}^{-1}\underline{\boldsymbol{\alpha}}^T) \times (\mathbf{y} - \underline{\boldsymbol{\alpha}}\mathbf{c}) \qquad (33.38)$$

where $\underline{\mathbf{C}}(r \times r) = \underline{\boldsymbol{\alpha}}^T(r \times n) \cdot \underline{\boldsymbol{\alpha}}(n \times r)$ is defined as in (33.22), $\underline{\mathbf{I}}$ is the $(n \times n)$ **unit matrix**, and

$$\hat{\mathbf{c}} = (\underline{\boldsymbol{\alpha}}^T\underline{\boldsymbol{\alpha}})^{-1}\underline{\boldsymbol{\alpha}}^T\mathbf{y} = \underline{\mathbf{C}}^{-1}\underline{\boldsymbol{\alpha}}^T\mathbf{y}$$

We may check the identity (33.38) by expanding:

$$\begin{aligned}(\mathbf{y}^T - \mathbf{c}^T\underline{\boldsymbol{\alpha}}^T)(\mathbf{y} - \underline{\boldsymbol{\alpha}}\mathbf{c} - \underline{\boldsymbol{\alpha}}\underline{\mathbf{C}}^{-1}\underline{\boldsymbol{\alpha}}^T\mathbf{y} &+ \underline{\boldsymbol{\alpha}}\underline{\mathbf{C}}^{-1}\underline{\boldsymbol{\alpha}}^T\underline{\boldsymbol{\alpha}}\mathbf{c}) \\
= \mathbf{y}^T\mathbf{y} - \mathbf{y}^T\underline{\boldsymbol{\alpha}}\mathbf{c} - \mathbf{y}^T\underline{\boldsymbol{\alpha}}\underline{\mathbf{C}}^{-1}\underline{\boldsymbol{\alpha}}^T\mathbf{y} \\
+ \mathbf{y}^T\underline{\boldsymbol{\alpha}}\underline{\mathbf{C}}^{-1}\underline{\boldsymbol{\alpha}}^T\underline{\boldsymbol{\alpha}}\mathbf{c} - \mathbf{c}^T\underline{\boldsymbol{\alpha}}^T\mathbf{y} + \mathbf{c}^T\underline{\boldsymbol{\alpha}}^T\underline{\boldsymbol{\alpha}}\mathbf{c} \\
+ \mathbf{c}^T\underline{\boldsymbol{\alpha}}^T\underline{\boldsymbol{\alpha}}\underline{\mathbf{C}}^{-1}\underline{\boldsymbol{\alpha}}^T\mathbf{y} - \mathbf{c}^T\underline{\boldsymbol{\alpha}}^T\underline{\boldsymbol{\alpha}}\underline{\mathbf{C}}^{-1}\underline{\boldsymbol{\alpha}}^T\underline{\boldsymbol{\alpha}}\mathbf{c} \\
= \mathbf{y}^T\mathbf{y} - \mathbf{y}^T\underline{\boldsymbol{\alpha}}\underline{\mathbf{C}}^{-1}\underline{\boldsymbol{\alpha}}^T\mathbf{y} = \mathbf{y}^T\mathbf{y} - \mathbf{y}^T\underline{\boldsymbol{\alpha}}\hat{\mathbf{c}},\end{aligned}$$

which agrees with (33.37). We see that (33.38) has, in fact, *no dependence on* **c**, only on the *estimator* $\hat{\mathbf{c}}$.

$S_{\min}$ in (33.38) is a quadratic form of the type

$$\mathbf{x}^T\underline{\mathbf{A}}\mathbf{x} = Q \qquad (33.39)$$

which is discussed in Section 37 of Appendix B, where

$$\mathbf{x} = \mathbf{y} - \underline{\boldsymbol{\alpha}}\mathbf{c} \qquad (33.40)$$

$$\underline{\mathbf{A}} = \underline{\mathbf{I}} - \underline{\boldsymbol{\alpha}}\underline{\mathbf{C}}^{-1}\underline{\boldsymbol{\alpha}}^T \qquad (33.41)$$

We note that the *expectation value*

$$E(\mathbf{y} - \underline{\boldsymbol{\alpha}}\mathbf{c}) = \mathbf{0} = \langle\mathbf{x}\rangle \qquad (33.42)$$

and that the *covariance matrix* is

$$[\underline{\mathbf{Covar}}(\mathbf{x})] = [\underline{\mathbf{Covar}}(\mathbf{y} - \underline{\boldsymbol{\alpha}}\mathbf{c})] = [\underline{\mathbf{Covar}}(\mathbf{y})]$$

since $\underline{\boldsymbol{\alpha}}\mathbf{c}$ has no variance. Therefore,

$$\underline{\mathbf{Covar}}(\mathbf{x}) = \sigma^2\underline{\mathbf{I}}(n \times n) \qquad (33.43)$$

using (33.35).

We are now in a position to use the result derived in Section 37 of Appendix B for the *expectation value of a quadratic form* (33.39), given the expectation value $\langle\mathbf{x}\rangle$ and that the covariance matrix of the **x** is *diagonal* (33.43). We rewrite the result of Section B.37 here for convenience.

$$EQ = \sigma^2 Tr\underline{\mathbf{A}} + \langle\mathbf{x}\rangle^T\underline{\mathbf{A}}\langle\mathbf{x}\rangle \qquad (33.44)$$

We need the trace of the matrix $\underline{\mathbf{A}} = \underline{\mathbf{I}} - \underline{\boldsymbol{\alpha}}\underline{\mathbf{C}}^{-1}\underline{\boldsymbol{\alpha}}^T$.

$$\begin{aligned}Tr\,\underline{\mathbf{A}} &= Tr\,\underline{\mathbf{I}}(n \times n) - Tr\,(\underline{\boldsymbol{\alpha}}\underline{\mathbf{C}}^{-1}\underline{\boldsymbol{\alpha}}^T) \\ &= Tr\,\underline{\mathbf{I}}(n \times n) - Tr\,(\underline{\mathbf{C}}^{-1}\underline{\boldsymbol{\alpha}}^T\underline{\boldsymbol{\alpha}})\end{aligned}$$

using Section B.34. But by (33.22)

$$\underline{\mathbf{C}}^{-1}\underline{\boldsymbol{\alpha}}^T\underline{\boldsymbol{\alpha}} = \underline{\mathbf{I}}(r \times r)$$

Therefore,

$$\begin{aligned}Tr\,\underline{\mathbf{A}} &= Tr\,\underline{\mathbf{I}}(n \times n) - Tr\,\underline{\mathbf{I}}(r \times r) \\ &= n - r \qquad (33.45)\end{aligned}$$

The *expectation value* of the *sum of squared residuals* is

$$\begin{aligned}ES_{\min} &= \sigma^2\,Tr\,\underline{\mathbf{A}} + \langle\mathbf{x}\rangle^T\underline{\mathbf{A}}\langle\mathbf{x}\rangle \\ &= \sigma^2(n - r) \qquad (33.46)\end{aligned}$$

because of (33.42).

We may use (33.46) to provide an *unbiased estimate* for σ^2:

$$\widehat{\sigma^2} = \frac{S_{\min}}{n - r} = \frac{(\mathbf{y} - \underline{\boldsymbol{\alpha}}\hat{\mathbf{c}})^T(\mathbf{y} - \underline{\boldsymbol{\alpha}}\hat{\mathbf{c}})}{n - r} \qquad (33.47)$$

Procedurally, after solving for $\hat{\mathbf{c}} = \hat{c}_1, \ldots, \hat{c}_r$, we may compute

$$\widehat{\sigma^2} = \frac{1}{n - r}\sum_{i=1}^{n}\left[y_i - \sum_{j=1}^{r}\hat{c}_j g_j(x_i)\right]^2$$

Correlated measurements of unequal weight

This result may be generalized to the case of *correlated measurements* of *unequal weight* described by (33.25) with $\hat{\mathbf{c}}$ given by (33.15) and its **covariance matrix** given by (33.30). The **weighted** *sum of squared residuals* becomes

$$\begin{aligned}S_{\min} &= (\mathbf{y} - \underline{\boldsymbol{\alpha}}\hat{\mathbf{c}})^T\underline{\mathbf{W}}(\mathbf{y} - \underline{\boldsymbol{\alpha}}\hat{\mathbf{c}}) \\ &= \mathbf{y}^T\underline{\mathbf{W}}\mathbf{y} - \mathbf{y}^T\underline{\mathbf{W}}\underline{\boldsymbol{\alpha}}\hat{\mathbf{c}} \qquad (33.48)\end{aligned}$$

The *unbiased estimate of the scale factor*, σ^2, becomes

$$\widehat{\sigma^2} = \frac{S_{\min}}{n - r} = \frac{(\mathbf{y} - \underline{\boldsymbol{\alpha}}\hat{\mathbf{c}})^T\underline{\mathbf{W}}(\mathbf{y} - \underline{\boldsymbol{\alpha}}\hat{\mathbf{c}})}{n - r} \qquad (33.49)$$

and the *unbiased estimate* for the **covariance matrix** of ĉ is

$$[\underline{\mathbf{Covar}}(\hat{\mathbf{c}})]_{\text{est}} = \left(\frac{1}{n-r}\right)(\mathbf{y} - \underline{\boldsymbol{\alpha}}\,\hat{\mathbf{c}})^T$$
$$\times\, \underline{\mathbf{W}}(\mathbf{y} - \underline{\boldsymbol{\alpha}}\hat{\mathbf{c}})(\underline{\boldsymbol{\alpha}}^T\underline{\mathbf{W}}\underline{\boldsymbol{\alpha}})$$
$$= \underline{\mathbf{E}}(\hat{\mathbf{c}}) \qquad (33.50)$$

called the "**error matrix.**" We may show that $\underline{\mathbf{E}}(\hat{\mathbf{c}})$ is a *symmetric* matrix that may be made *diagonal* by a **unitary matrix** $\underline{\mathbf{U}}$:

$$\underline{\mathbf{U}}^{-1}\underline{\mathbf{E}}\underline{\mathbf{U}} = \underline{\mathbf{D}}$$

where the vector ĉ is linearly transformed by $\underline{\mathbf{U}}$ into a vector $\hat{\mathbf{k}}$ that has uncorrelated components. This is a way of finding *orthogonal functions,* $h_i(\mathbf{x})$, such that

$$p(\mathbf{x}; \mathbf{k}) = \sum_{i=1}^{r} k_i h_i(\mathbf{x})$$

The *fitted values* $\hat{y}(x)$ have uncertainties given by the law of propagation of errors applied to

$$\hat{\mathbf{y}}(x) = \underline{\boldsymbol{\alpha}}\,\hat{\mathbf{c}}$$

that is,

$$[\underline{\mathbf{Covar}}(\hat{\mathbf{y}})] = \underline{\boldsymbol{\alpha}}[\underline{\mathbf{Covar}}\ \hat{\mathbf{c}}]\underline{\boldsymbol{\alpha}}^T$$
$$= \underline{\boldsymbol{\alpha}}\,\underline{\mathbf{E}}\,\underline{\boldsymbol{\alpha}}^T$$
$$= \left(\frac{1}{n-r}\right)\underline{\boldsymbol{\alpha}}(\mathbf{y} - \underline{\boldsymbol{\alpha}}\hat{\mathbf{c}})^T\underline{\mathbf{W}}$$
$$\times\,(\mathbf{y} - \underline{\boldsymbol{\alpha}}\hat{\mathbf{c}})(\underline{\boldsymbol{\alpha}}^T\underline{\mathbf{W}}\underline{\boldsymbol{\alpha}})\underline{\boldsymbol{\alpha}}^T \qquad (33.51)$$

Obviously, the estimates $\hat{\mathbf{y}}(\mathbf{x})$ are *correlated,* that is, whether or not the original measurements **y** were independent, *the fitted values are in general correlated* with each other.

We may ignore this correlation and calculate the *variance* of $\hat{y}_i$:

$$[\underline{\mathbf{Covar}}(\hat{\mathbf{y}})]_{ii} = \sum_{\rho=1}^{r}\sum_{\sigma=1}^{r} \alpha_{i\rho}E_{\rho\sigma}\alpha^T_{\sigma i}$$
$$= \sum_{\rho=1}^{r}\sum_{\sigma=1}^{r} g_\rho(x_i)E_{\rho\sigma}g_\sigma(x_i) \qquad (33.52)$$

► Example 33.2
Combination of two experimental measurements of a single quantity

Suppose we have made two independent measurements of c:

c_{I} with variance σ_{I}^2
c_{II} with variance σ_{II}^2

We may regard this as *two measurements* y_i $(i = 1, \ldots, n;\ n = 2)$ of **one parameter** c_i, $(i = 1, \ldots, r;\ r = 1)$.

$$\therefore \mathbf{y} = \begin{pmatrix} c_{\text{I}} \\ c_{\text{II}} \end{pmatrix}$$

and

$$[\underline{\mathbf{Covar}}(\mathbf{y})] = \sigma^2\underline{\mathbf{W}}^{-1} = \begin{pmatrix} \sigma_{\text{I}}^2 & 0 \\ 0 & \sigma_{\text{II}}^2 \end{pmatrix}$$

$$\therefore \frac{1}{\sigma^2}\underline{\mathbf{W}} = \begin{pmatrix} \sigma_{\text{I}}^{-2} & 0 \\ 0 & \sigma_{\text{II}}^{-2} \end{pmatrix}$$

The *design matrix* $\underline{\boldsymbol{\alpha}}$ is $(n \times r) = (2 \times 1)$:

$$\underline{\boldsymbol{\alpha}} = \begin{pmatrix} 1 \\ 1 \end{pmatrix}$$
$$\therefore \underline{\boldsymbol{\alpha}}^T = (1 \quad 1)$$

The *best estimate* of c is, by (33.15)

$$\hat{\mathbf{c}}(1 \times 1) = \left[(1 \quad 1)\begin{pmatrix} \sigma_{\text{I}}^{-2} & 0 \\ 0 & \sigma_{\text{II}}^{-2} \end{pmatrix}\begin{pmatrix} 1 \\ 1 \end{pmatrix}\right]^{-1}$$
$$\times\,(1 \quad 1)\begin{pmatrix} \sigma_{\text{I}}^{-2} & 0 \\ 0 & \sigma_{\text{II}}^{-2} \end{pmatrix}\begin{pmatrix} c_{\text{I}} \\ c_{\text{II}} \end{pmatrix}$$
$$= [\sigma_{\text{I}}^{-2} + \sigma_{\text{II}}^{-2}]^{-1}\left(\frac{c_{\text{I}}}{\sigma_{\text{I}}^2} + \frac{c_{\text{II}}}{\sigma_{\text{II}}^2}\right)$$
$$= \frac{\sum_{i=\text{I}}^{\text{II}} \frac{c_i}{\sigma_i^2}}{\sum_{i=\text{I}}^{\text{II}} \frac{1}{\sigma_i^2}}$$

which agrees with previous results. As in Part I, we have *not* assumed a normal distribution.

The *variance* of $\hat{c}$ is given by (33.30) since the *scale factor* of the covariance matrix of the **y** is given:

$$\text{Var}\,\hat{c} = \sigma^2[\underline{\alpha}^T(\underline{\mathbf{W}})\underline{\alpha}]^{-1}\left[\sum_{i=\text{I}}^{\text{II}}\frac{1}{\sigma_i^2}\right]^{-1} \quad ◄$$

The situation in which we seek to combine measurements that are *correlated* is more complicated but still straightforward.

► Example 33.3
Combination of experimental results

Suppose we have performed two *independent experiments* to determine the parameters c_1 and c_2. Experiment I has yielded the result

$$c_1^{\mathrm{I}} = a$$
$$c_2^{\mathrm{I}} = b$$

with covariance matrix

$$\underline{\mathbf{E}}^{\mathrm{I}} = \begin{pmatrix} e_{11} & e_{12} \\ e_{21} & e_{22} \end{pmatrix}$$

Likewise experiment II has yielded the result

$$c_1^{\mathrm{II}} = a'$$
$$c_2^{\mathrm{II}} = b'$$

with covariance matrix

$$\underline{\mathbf{E}}^{\mathrm{II}} = \begin{pmatrix} e'_{11} & e'_{12} \\ e'_{21} & e'_{22} \end{pmatrix}$$

We wish to find the *best estimates* of c_1 and c_2 and their **covariance matrix**.

We may consider that we have four measurements $\mathbf{y}$

$$\mathbf{y} = \begin{bmatrix} c_1^{\mathrm{I}} \\ c_2^{\mathrm{I}} \\ c_1^{\mathrm{II}} \\ c_2^{\mathrm{II}} \end{bmatrix} = \begin{pmatrix} a \\ b \\ a' \\ b' \end{pmatrix}$$

with covariance matrix $[\underline{\mathbf{Covar}}(\mathbf{y})] = \underline{\mathbf{V}}$:

$$\underline{\mathbf{V}} = \begin{bmatrix} e_{11} & e_{12} & 0 & 0 \\ e_{21} & e_{22} & 0 & 0 \\ 0 & 0 & e'_{11} & e'_{12} \\ 0 & 0 & e'_{21} & e'_{22} \end{bmatrix}$$

The *inverse* is the **weight matrix**:

$$\underline{\mathbf{V}}^{-1} = \begin{pmatrix} \omega_{11} & \omega_{12} & \omega_{13} & \omega_{14} \\ \omega_{21} & \omega_{22} & \omega_{23} & \omega_{24} \\ \omega_{31} & \omega_{32} & \omega_{33} & \omega_{34} \\ \omega_{41} & \omega_{42} & \omega_{43} & \omega_{44} \end{pmatrix} = \underline{\mathbf{W}}$$

$$= \begin{bmatrix} \dfrac{e_{22}}{|e|} & \dfrac{-e_{12}}{|e|} & 0 & 0 \\ \dfrac{-e_{21}}{|e|} & \dfrac{e_{11}}{|e|} & 0 & 0 \\ 0 & 0 & \dfrac{e'_{22}}{|e'|} & \dfrac{-e'_{12}}{|e'|} \\ 0 & 0 & \dfrac{-e'_{21}}{|e'|} & \dfrac{e'_{11}}{|e'|} \end{bmatrix}$$

where $|e| = e_{11}e_{22} - e_{12}e_{21}$; $|e'| = e'_{11}e'_{22} - e'_{22}e'_{21}$.

We assume

$$\begin{aligned} Ey_1 &= c_1 = c_1 \cdot 1 + c_2 \cdot 0 \\ &= c_1 g_1(x_1) + c_1 g_2(x_1) \\ Ey_2 &= c_2 = c_1 \cdot 0 + c_2 \cdot 1 \\ &= c_1 g_1(x_2) + c_2 g_2(x_2) \\ Ey_3 &= c_1 = c_1 \cdot 1 + c_2 \cdot 0 \\ &= c_1 g_1(x_3) + c_2 g_2(x_3) \\ Ey_4 &= c_2 = c_1 \cdot 0 + c_2 \cdot 1 \\ &= c_1 g_1(x_4) + c_2 g_2(x_4) \end{aligned}$$

that is, the *design matrix* $\underline{\boldsymbol{\alpha}}(n \times r) = \underline{\boldsymbol{\alpha}}(4 \times 2)$ is from (33.6)

$$\underline{\boldsymbol{\alpha}} = \begin{pmatrix} 1 & 0 \\ 0 & 1 \\ 1 & 0 \\ 0 & 1 \end{pmatrix}$$

with

$$\underline{\boldsymbol{\alpha}}^T = \begin{pmatrix} 1 & 0 & 1 & 0 \\ 0 & 1 & 0 & 1 \end{pmatrix}$$

We want

$$\hat{\mathbf{c}} = (\underline{\boldsymbol{\alpha}}^T \underline{\mathbf{W}} \underline{\boldsymbol{\alpha}})^{-1} \underline{\boldsymbol{\alpha}}^T \underline{\mathbf{W}} \mathbf{y} \qquad \text{from (33.15)}$$

Now

$$\underline{\boldsymbol{\alpha}}^T \underline{\mathbf{W}} \mathbf{y} = \begin{pmatrix} 1 & 0 & 1 & 0 \\ 0 & 1 & 0 & 1 \end{pmatrix}$$

$$\times \begin{bmatrix} \dfrac{e_{22}}{|e|} & \dfrac{-e_{12}}{|e|} & 0 & 0 \\ \dfrac{-e_{21}}{|e|} & \dfrac{e_{11}}{|e|} & 0 & 0 \\ 0 & 0 & \dfrac{e'_{22}}{|e'|} & \dfrac{-e'_{12}}{|e'|} \\ 0 & 0 & \dfrac{-e'_{21}}{|e'|} & \dfrac{e'_{11}}{|e'|} \end{bmatrix} \begin{bmatrix} c_1^{\mathrm{I}} \\ c_2^{\mathrm{I}} \\ c_1^{\mathrm{II}} \\ c_2^{\mathrm{II}} \end{bmatrix}$$

$$= \begin{pmatrix} 1 & 0 & 1 & 0 \\ 0 & 1 & 0 & 1 \end{pmatrix} \begin{bmatrix} \dfrac{e_{22}}{|e|} a - \dfrac{-e_{12}}{|e|} b \\ \dfrac{-e_{21}}{|e|} a + \dfrac{e_{11}}{|e|} b \\ \dfrac{e'_{22}}{|e'|} a' - \dfrac{e'_{12}}{|e'|} b' \\ \dfrac{-e'_{21}}{|e'|} a' + \dfrac{e'_{11}}{|e'|} b' \end{bmatrix}$$

$$= \begin{pmatrix} \dfrac{e_{22}a}{|e|} - \dfrac{e_{12}b}{|e|} + \dfrac{e'_{22}a'}{|e'|} - \dfrac{e'_{22}b'}{|e'|} \\ \dfrac{-e_{21}a}{|e|} + \dfrac{e_{11}b}{|e|} - \dfrac{e'_{21}a'}{|e'|} + \dfrac{e'_{11}b'}{|e'|} \end{pmatrix}$$

$$= \underline{\boldsymbol{\alpha}}^T \underline{\mathbf{W}} \mathbf{y}$$

Also,

$$(\underline{\boldsymbol{\alpha}}^T\underline{\mathbf{W}}\underline{\boldsymbol{\alpha}}) = \begin{bmatrix} 1 & 0 & 1 & 0 \\ 0 & 1 & 0 & 1 \end{bmatrix} \times \begin{bmatrix} \frac{e_{22}}{|e|} & \frac{-e_{12}}{|e|} & 0 & 0 \\ \frac{-e_{21}}{|e|} & \frac{e_{11}}{|e|} & 0 & 0 \\ 0 & 0 & \frac{e'_{22}}{|e'|} & \frac{-e'_{12}}{|e'|} \\ 0 & 0 & \frac{-e'_{21}}{|e'|} & \frac{e'_{11}}{|e'|} \end{bmatrix} \times \begin{pmatrix} 1 & 0 \\ 0 & 1 \\ 1 & 0 \\ 0 & 1 \end{pmatrix}$$

$$= \begin{bmatrix} 1 & 0 & 1 & 0 \\ 0 & 1 & 0 & 1 \end{bmatrix} \begin{bmatrix} \frac{e_{22}}{|e|} & \frac{-e_{12}}{|e|} \\ \frac{-e_{21}}{|e|} & \frac{e_{11}}{|e|} \\ \frac{e'_{22}}{|e'|} & \frac{-e'_{12}}{|e'|} \\ \frac{-e'_{21}}{|e'|} & \frac{e'_{11}}{|e'|} \end{bmatrix}$$

$$= \begin{pmatrix} \left(\frac{e_{22}}{|e|} + \frac{e'_{22}}{|e'|}\right) & \left(\frac{-e_{12}}{|e|} - \frac{e'_{12}}{|e'|}\right) \\ \left(\frac{-e_{21}}{|e|} - \frac{e'_{21}}{|e'|}\right) & \left(\frac{e_{11}}{|e|} + \frac{e'_{11}}{|e'|}\right) \end{pmatrix}$$

$$\therefore [\underline{\boldsymbol{\alpha}}^T\underline{\mathbf{W}}\underline{\boldsymbol{\alpha}}]^{-1} = \begin{pmatrix} u_{11} & u_{12} \\ u_{21} & u_{22} \end{pmatrix}^{-1} = \frac{\begin{pmatrix} u_{22} & u_{12} \\ -u_{21} & u_{11} \end{pmatrix}}{u_{11}u_{22} - u_{12}u_{21}} = \frac{\begin{pmatrix} u_{22} & -u_{12} \\ -u_{21} & u_{11} \end{pmatrix}}{|u|}$$

where

$$u_{11} = \frac{e_{22}}{|e|} + \frac{e'_{22}}{|e'|}$$

$$u_{12} = -\left(\frac{e_{12}}{|e|} + \frac{e'_{12}}{|e'|}\right)$$

$$u_{21} = -\left(\frac{e_{21}}{|e|} + \frac{e'_{21}}{|e'|}\right)$$

$$u_{22} = \frac{e_{11}}{|e|} + \frac{e'_{11}}{|e'|}$$

and

$$|u| = u_{11}u_{22} - u_{12}u_{21}$$

The *best estimates* of c_1 and c_2 are

$$\hat{c}_1 = \frac{u_{22}}{|u|}\left[\frac{1}{|e|}(e_{22}a - e_{12}b) + \frac{1}{|e'|}(e'_{22}a' - e'_{12}b')\right] - \frac{u_{12}}{|u|}\left[\frac{(e_{21}a + e_{11}b)}{|e|} + \frac{(-e'_{21}a' + e'_{11}b')}{|e'|}\right]$$

$$\hat{c}_2 = -\frac{u_{21}}{|u|}\left[\frac{1}{|e|}(e_{22}a - e_{12}b) + \frac{1}{|e'|}(e'_{22}a' - e'_{12}b')\right] + \frac{u_{11}}{|u|}\left[\frac{1}{|e|}(-e_{21}a + e_{11}b) + \frac{1}{|e'|}(-e'_{21}a' + e'_{11}b')\right]$$

The **covariance matrix** of $\hat{c}_1$, $\hat{c}_2$ is given by (32.30):

$$[\underline{\mathbf{Covar}}(\hat{\mathbf{c}})] = \frac{1}{|u|}\begin{pmatrix} u_{22} & -u_{12} \\ -u_{21} & u_{22} \end{pmatrix}$$

Numerical example

Suppose we have two determinations of the parameters of a straight line, $y = ax + b$:

$$\text{I: } a_{\text{I}}^* = 4.2 \qquad b_{\text{I}}^* = 0.8$$

with covariance matrix $\begin{pmatrix} 0.04 & -0.06 \\ -0.06 & 0.20 \end{pmatrix}$;

$$\text{II: } a_{\text{I}}^* = 3.8 \qquad b_{\text{I}}^* = 1.2$$

with covariance matrix $\begin{pmatrix} 0.08 & -0.12 \\ -0.12 & 0.4 \end{pmatrix}$.

We form the *vector*

$$\mathbf{y} = \begin{pmatrix} 4.2 \\ 0.8 \\ 3.8 \\ 1.2 \end{pmatrix}$$

that has the *covariance matrix*

$$\sigma^2\underline{\mathbf{V}} = \begin{pmatrix} 0.04 & -0.06 & 0 & 0 \\ -0.06 & 0.20 & 0 & 0 \\ 0 & 0 & 0.08 & -0.12 \\ 0 & 0 & -0.12 & 0.4 \end{pmatrix}$$

The *design matrix* is

$$\underline{\alpha} = \begin{pmatrix} 1 & 0 \\ 0 & 1 \\ 1 & 0 \\ 0 & 1 \end{pmatrix}$$

with

$$\underline{\alpha}^T = \begin{pmatrix} 1 & 0 & 1 & 0 \\ 0 & 1 & 0 & 1 \end{pmatrix}$$

The *weight matrix* is

$$\underline{\mathbf{W}} = (\sigma^2)^{-1}\underline{\mathbf{V}}^{-1} = \begin{pmatrix} 45.45 & 13.64 & 0 & 0 \\ 13.64 & 9.09 & 0 & 0 \\ 0 & 0 & 22.73 & 6.82 \\ 0 & 0 & 6.82 & 4.54 \end{pmatrix}$$

We form

$$(\underline{\alpha}^T\underline{\mathbf{W}}\underline{\alpha}) = \begin{pmatrix} 1 & 0 & 1 & 0 \\ 0 & 1 & 0 & 1 \end{pmatrix} \times \begin{bmatrix} 45.45 & 13.64 & 0 & 0 \\ 13.64 & 9.09 & 0 & 0 \\ 0 & 0 & 22.73 & 6.82 \\ 0 & 0 & 6.82 & 4.54 \end{bmatrix} \begin{bmatrix} 1 & 0 \\ 0 & 1 \\ 1 & 0 \\ 0 & 1 \end{bmatrix}$$

$$= \begin{pmatrix} 1 & 0 & 1 & 0 \\ 0 & 1 & 0 & 1 \end{pmatrix} \begin{bmatrix} 45.45 & 13.64 \\ 13.64 & 9.09 \\ 22.73 & 6.82 \\ 6.82 & 4.45 \end{bmatrix}$$

$$= \begin{pmatrix} 68.18 & 20.46 \\ 20.46 & 13.63 \end{pmatrix}$$

whose *inverse* is

$$(\underline{\alpha}^T\underline{\mathbf{W}}\underline{\alpha})^{-1} = \frac{1}{510.68}\begin{pmatrix} 13.63 & -20.46 \\ -20.46 & 68.18 \end{pmatrix}$$

We may calculate the 2×1 *vector*

$$\underline{\alpha}^T\underline{\mathbf{W}}\mathbf{y} = \begin{pmatrix} 1 & 0 & 1 & 0 \\ 0 & 1 & 0 & 1 \end{pmatrix} \times \begin{bmatrix} 45.45 & 13.64 & 0 & 0 \\ 13.64 & 9.09 & 0 & 0 \\ 0 & 0 & 22.73 & 6.82 \\ 0 & 0 & 6.82 & 4.54 \end{bmatrix} \begin{pmatrix} 4.2 \\ 0.8 \\ 3.8 \\ 1.2 \end{pmatrix}$$

$$= \begin{pmatrix} 45.45 & 13.64 & 22.73 & 6.82 \\ 13.64 & 9.09 & 6.82 & 4.54 \end{pmatrix} \begin{pmatrix} 4.2 \\ 0.8 \\ 3.8 \\ 1.2 \end{pmatrix}$$

$$= \begin{pmatrix} 296.36 \\ 95.92 \end{pmatrix}$$

The best estimate $\hat{\mathbf{c}}$ is:

$$\hat{\mathbf{c}} = \frac{1}{510.68}\begin{pmatrix} 13.63 & -20.46 \\ -20.46 & 68.18 \end{pmatrix}\begin{pmatrix} 296.36 \\ 95.92 \end{pmatrix}$$

$$= \frac{1}{510.68}\begin{pmatrix} 2076.86 \\ 476.3 \end{pmatrix} = \begin{pmatrix} 4.067 \\ 0.933 \end{pmatrix}$$

with an error matrix given by

$$[\underline{\mathbf{Covar}}(\hat{\mathbf{c}})] = (\underline{\alpha}^T\underline{\mathbf{W}}\underline{\alpha})^{-1}$$

$$= \begin{pmatrix} 0.027 & -0.04 \\ -0.04 & 0.133 \end{pmatrix}$$ ◄

Goodness-of-Fit

To determine the quality or "goodness of fit" of the least-squares solution, we must *assume a distribution* for the data, y_i. If the y_i are taken to be normally distributed about their assumed expectation values, and the scale factor σ^2 is known, then

$$S_{\min} = \frac{1}{\sigma^2}(\mathbf{y} - \underline{\alpha}\hat{\mathbf{c}})^T\underline{\mathbf{W}}(\mathbf{y} - \underline{\alpha}\hat{\mathbf{c}})$$

is distributed as χ^2 with $(n - r)$ independent degrees of freedom. We reject the fit at the α level of significance if

$$\Pr\{\chi^2(n - r) \geq S_{\min}\} \leq \alpha$$

It frequently happens that certain requirements on (or knowledge about) the parameters exist.

Linear Least Squares with Linear Constraints

Suppose we have m conditions or *constraints* on the r parameters $\mathbf{c}$. An *equality* or *holonomic* constraint is of the form

$$\varphi_j(\mathbf{c}) = 0 \qquad (33.53)$$

An *inequality* or *nonholonomic* constraint is of the form

$$\psi(\mathbf{c}) \geq 0 \qquad (33.54)$$

It may happen that the m constraints

are holonomic and also *linear*:

$$\underline{\mathbf{K}}\mathbf{c} = \mathbf{d} \qquad (33.55)$$

where $\underline{\mathbf{K}}$ is $(m \times r)$ and $\underline{\mathbf{d}}$ is $(m \times 1)$.

The method of *Lagrange multipliers* may be profitably applied to this problem, that is, we minimize the matrix expression

$$\mathcal{S} = (\mathbf{y} - \underline{\boldsymbol{\alpha}}\mathbf{c})^T \underline{\mathbf{W}}(\mathbf{y} - \underline{\boldsymbol{\alpha}}\mathbf{c}) + 2\boldsymbol{\lambda}^T(\underline{\mathbf{K}}\mathbf{c} - \underline{\mathbf{d}}) \qquad (33.56)$$

with respect to $\mathbf{c}$ where $\boldsymbol{\lambda}$ is the $(m \times 1)$ vector of Lagrange multipliers. We set the derivative to zero and solve for the values, $\hat{\mathbf{c}}_c$, the estimates for the **constrained** problem.

$$\left.\frac{\partial \mathcal{S}}{\partial \mathbf{c}}\right|_{\mathbf{c}=\hat{\mathbf{c}}_c} = 0 = \frac{\partial}{\partial \mathbf{c}}\{(\mathbf{y}^T - \mathbf{c}^T\underline{\boldsymbol{\alpha}}^T)\underline{\mathbf{W}}(\mathbf{y} - \underline{\boldsymbol{\alpha}}\mathbf{c}) + 2\lambda^T\underline{\mathbf{K}}\mathbf{c} - 2\boldsymbol{\lambda}^T\mathbf{d}\}$$

$$0 = \frac{\partial}{\partial \mathbf{c}}\{\mathbf{y}^T\underline{\mathbf{W}}\mathbf{y} - \mathbf{c}^T\underline{\boldsymbol{\alpha}}^T\underline{\mathbf{W}}\mathbf{y} - \mathbf{y}^T\underline{\mathbf{W}}\underline{\boldsymbol{\alpha}}\mathbf{c} + \mathbf{c}^T\underline{\boldsymbol{\alpha}}^T\underline{\mathbf{W}}\underline{\boldsymbol{\alpha}}\mathbf{c} + 2\boldsymbol{\lambda}^T\underline{\mathbf{K}}\mathbf{c} - 2\boldsymbol{\lambda}^T\mathbf{d}\}$$

$$= -2\mathbf{y}^T\underline{\mathbf{W}}\underline{\boldsymbol{\alpha}} + 2\hat{\mathbf{c}}_c^T\underline{\boldsymbol{\alpha}}^T\underline{\mathbf{W}}\underline{\boldsymbol{\alpha}} - 2\boldsymbol{\lambda}^T\underline{\mathbf{K}}$$

We may solve for the Lagrange multipliers:

$$\boldsymbol{\lambda}^T\underline{\mathbf{K}} = \mathbf{c}_c^T(\underline{\boldsymbol{\alpha}}^T\underline{\mathbf{W}}\underline{\boldsymbol{\alpha}}) - \mathbf{y}^T\underline{\mathbf{W}}\underline{\boldsymbol{\alpha}} \qquad (33.57)$$

The least-squares solution for the problem *without* constraints (unconstrained) is given by (33.15):

$$\hat{\mathbf{c}}_{u.c.} = (\underline{\boldsymbol{\alpha}}^T\underline{\mathbf{W}}\underline{\boldsymbol{\alpha}})^{-1}\underline{\boldsymbol{\alpha}}^T\underline{\mathbf{W}}\mathbf{y}$$

which yields

$$\hat{\mathbf{c}}_{u.c.}^T(\underline{\boldsymbol{\alpha}}^T\underline{\mathbf{W}}\underline{\boldsymbol{\alpha}}) = \mathbf{y}^T\underline{\mathbf{W}}\underline{\boldsymbol{\alpha}}$$

$$\therefore \boldsymbol{\lambda}^T\underline{\mathbf{K}} = (\hat{\mathbf{c}}_c - \hat{\mathbf{c}}_{u.c.})^T(\underline{\boldsymbol{\alpha}}^T\underline{\mathbf{W}}\underline{\boldsymbol{\alpha}}) \qquad (33.58)$$

We have assumed all along that the $(r \times r)$ matrix

$$\underline{\mathbf{M}} = \underline{\boldsymbol{\alpha}}^T\underline{\mathbf{W}}\underline{\boldsymbol{\alpha}} \qquad (33.59)$$

is *nonsingular.*

We multiply (33.58) from the right by $\underline{\mathbf{M}}^{-1}\underline{\mathbf{K}}^T$ and get

$$\boldsymbol{\lambda}^T\underline{\mathbf{K}}\underline{\mathbf{M}}^{-1}\underline{\mathbf{K}}^T = (\hat{\mathbf{c}}_c - \hat{\mathbf{c}}_{u.c.}^T)^T\underline{\mathbf{K}}^T \qquad (33.60)$$

But (33.55) yields

$$\mathbf{d}^T = \hat{\mathbf{c}}_c\underline{\mathbf{K}}^T \qquad (33.61)$$

which we substitute in (33.60) to get

$$\boldsymbol{\lambda}^T\underline{\mathbf{K}}\underline{\mathbf{M}}^{-1}\underline{\mathbf{K}}^T = \mathbf{d}^T - \hat{\mathbf{c}}_{u.c.}^T\underline{\mathbf{K}}^T$$

Since $\underline{\mathbf{K}}\underline{\mathbf{M}}^{-1}\underline{\mathbf{K}}^T$ is $(m \times m)$, we multiply by its *inverse* and solve explicitly for the *Lagrange multipliers.*

$$\boldsymbol{\lambda}^T = (\mathbf{d}^T - \mathbf{c}_{u.c.}^T\underline{\mathbf{K}}^T)(\underline{\mathbf{K}}\underline{\mathbf{M}}^{-1}\underline{\mathbf{K}}^T)^{-1} \qquad (33.62)$$

We may substitute this in (33.58):

$$(\mathbf{d}^T - \hat{\mathbf{c}}_{u.c.}^T\underline{\mathbf{K}}^T)(\underline{\mathbf{K}}\underline{\mathbf{M}}^{-1}\underline{\mathbf{K}}^T)^{-1}\underline{\mathbf{K}} = (\hat{\mathbf{c}}_c - \hat{\mathbf{c}}_{u.c.})^T\underline{\mathbf{M}}$$

which we may solve for $\hat{\mathbf{c}}_c$, the least-squares estimator for the constrained problem:

$$\hat{\mathbf{c}}_c^T = \hat{\mathbf{c}}_{u.c.}^T + (\mathbf{d}^T - \hat{\mathbf{c}}_{u.c.}^T\underline{\mathbf{K}}^T) \times (\underline{\mathbf{K}}\underline{\mathbf{M}}^{-1}\underline{\mathbf{K}}^T)^{-1}\underline{\mathbf{K}}\underline{\mathbf{M}}^{-1} \qquad (33.63)$$

This is the solution for the least-squares estimate of $\mathbf{c}$, *under the constraints.* The first term on the right-hand side is the least-squares estimate for the *unconstrained* problem. The second term is linear in the amount by which the unconstrained solution misses satisfying the constraints. The procedure for this *explicit* solution of the problem with constraints is the same as for an *iterative* solution: *one first finds the solution to the unconstrained problem.*

We may obtain the **error matrix** for $\hat{\mathbf{c}}_c$:

$$\begin{aligned}[\underline{\mathbf{Covar}}(\hat{\mathbf{c}}_c)] &= [\underline{\mathbf{Covar}}(\hat{\mathbf{c}}_{u.c.})] \\ &\quad + \{\underline{\mathbf{Covar}}[\underline{\mathbf{M}}^{-1}\underline{\mathbf{K}}^T(\underline{\mathbf{K}}\underline{\mathbf{M}}^{-1}\underline{\mathbf{K}}^T)^{-1} \\ &\quad \times (\mathbf{d} - \underline{\mathbf{K}}\hat{\mathbf{c}}_{u.c.})]\} \\ &= [\underline{\mathbf{Covar}}(\hat{\mathbf{c}}_{u.c.})] \\ &\quad - \{\underline{\mathbf{Covar}}[\underline{\mathbf{M}}^{-1}\underline{\mathbf{K}}^T(\underline{\mathbf{K}}\underline{\mathbf{M}}^{-1}\underline{\mathbf{K}}^T)^{-1}(\underline{\mathbf{K}})\hat{\mathbf{c}}_{u.c.}]\}\end{aligned}$$

where, by (33.30) and (33.59),

$$[\underline{\mathbf{Covar}}(\hat{\mathbf{c}}_{u.c.})] = \sigma^2\underline{\mathbf{M}}^{-1}$$

The second term may be evaluated by the *law of propagation of errors*:

$$\begin{aligned}&\underline{\mathbf{M}}^{-1}\underline{\mathbf{K}}^T(\underline{\mathbf{K}}\underline{\mathbf{M}}^{-1}\underline{\mathbf{K}}^T)^{-1}\underline{\mathbf{K}}[\underline{\mathbf{Covar}}(\hat{\mathbf{c}}_{u.c.})] \times \underline{\mathbf{K}}^T(\underline{\mathbf{K}}\underline{\mathbf{M}}^{-1}\underline{\mathbf{K}}^T)^{-1}\underline{\mathbf{K}}\underline{\mathbf{M}}^{-1} \\ &= \underline{\mathbf{M}}^{-1}\underline{\mathbf{K}}^T(\underline{\mathbf{K}}\underline{\mathbf{M}}^{-1}\underline{\mathbf{K}}^T)^{-1}(\underline{\mathbf{K}}\sigma^2\underline{\mathbf{M}}^{-1}\underline{\mathbf{K}}^T) \times (\underline{\mathbf{K}}\underline{\mathbf{M}}^{-1}\underline{\mathbf{K}}^T)^{-1}\underline{\mathbf{K}}\underline{\mathbf{M}}^{-1} \\ &= \sigma^2\underline{\mathbf{M}}^{-1}\underline{\mathbf{K}}^T(\underline{\mathbf{K}}\underline{\mathbf{M}}^{-1}\underline{\mathbf{K}}^T)^{-1}\underline{\mathbf{K}}\underline{\mathbf{M}}^{-1}\end{aligned}$$

The **covariance matrix** is, therefore,

$$[\underline{\mathbf{Covar}}(\hat{\mathbf{c}}_c)] = \sigma^2\underline{\mathbf{M}}^{-1}[\underline{\mathbf{I}} + \underline{\mathbf{K}}^T(\underline{\mathbf{K}}\underline{\mathbf{M}}^{-1}\underline{\mathbf{K}}^T)^{-1}\underline{\mathbf{K}}\underline{\mathbf{M}}^{-1}] \qquad (33.64)$$

If the *scale factor* σ^2 is not given *a priori* with the weight matrix $\underline{\mathbf{W}}$, we may *estimate* it from the data in an *unbiased* way. Again we use $\hat{\mathbf{c}}_{u.c.}$ to be the estimators for the unconstrained problem:

$$\widehat{\sigma_c^2} = \frac{[(\mathbf{y}-\underline{\boldsymbol{\alpha}}\hat{\mathbf{c}}_{u.c.})^T\underline{\mathbf{W}}(\mathbf{y}-\underline{\boldsymbol{\alpha}}\hat{\mathbf{c}}_{u.c.}) + (\hat{\mathbf{c}}_c - \hat{\mathbf{c}}_{u.c.})^T(\underline{\boldsymbol{\alpha}}^T\underline{\mathbf{W}}\underline{\boldsymbol{\alpha}})(\hat{\mathbf{c}}_c - \hat{\mathbf{c}}_{u.c.})]}{n-r+m} \tag{33.65}$$

We may also substitute (33.63) in the above equation to eliminate $\hat{\mathbf{c}}_c$. Equation (33.65) may be substituted into (33.64) to yield the error matrix for $\hat{\mathbf{c}}_c$.

COMMENT ON THE CONSTRAINT EQUATIONS

We have assumed that there are m *independent* constraint equations expressed by the matrix

$$\underline{\mathbf{K}}(m \times r)$$

If the *rank* of the matrix $\underline{\mathbf{K}}$ (defined in Section B-17) is l where $l < \min(m, r)$, then some of the constraints are *redundant*, that is, are *not linearly independent.* In this case we should recast the constraint equations so that $\boldsymbol{\lambda}$ becomes an $(l \times 1)$ vector. Also, we should read l for m in all our formulas. The rank l is, therefore, a measure of the number of linearly independent constraint equations.

The goodness of fit of a constrained problem where the y_i are assumed to follow a normal distribution may be estimated by computing from (33.56)

$$\frac{\mathcal{S}_{\min}}{\sigma^2} = (\mathbf{y}-\underline{\boldsymbol{\alpha}}\hat{\mathbf{c}}_c)^T\frac{\underline{\mathbf{W}}}{\sigma^2}(\mathbf{y}-\underline{\boldsymbol{\alpha}}\hat{\mathbf{c}}_c)$$

Using the χ^2 tables, we may determine

$$\Pr\left\{\chi^2(n-r+l) \geq \frac{\mathcal{S}_{\min}}{\sigma^2}\right\} = \alpha$$

and reject the fit or not depending on α. We have taken the number of *independent degrees of freedom* to be $n-r+l$ where n is the number of data points, r is the number of independent parameters to be determined, and l is the number of independent constraint equations.

Nonlinear Least Squares

The most general approach to nonlinear problems is to attempt to get close enough to the best estimate, $\hat{\mathbf{c}}$, so that the nonlinear function may be approximated by the linear term in a Taylor series expansion. If this region can be entered, then the problem has been *linearized* and the methods of *linear* least squares may be used. The first approximation is usually quite important and can be made only by *ad hoc* procedures suited to the specific problem. One usually uses iteration to approach the final solution but, for an arbitrary starting point, one has no guarantee on the rapidity of a procedure's convergence or, even, that the procedure will converge at all.

We can provide only a sample of a procedure that is sometimes useful.

Suppose we have in general

$$p(\mathbf{x}; \mathbf{c}) = f(\mathbf{c}) \tag{33.66}$$

instead of (33.5). We have the $n \times 1$ vector $\mathbf{f}(\mathbf{x}; \mathbf{c})$, and we wish to minimize

$$\Sigma = \frac{1}{\sigma^2}[\mathbf{y}-\mathbf{f}(\mathbf{x};\mathbf{c})]^T\underline{\mathbf{W}}[\mathbf{y}-\mathbf{f}(\mathbf{x};\mathbf{c})] \tag{33.67}$$

We may differentiate Σ with respect to $\mathbf{c}$ as before, but the r simultaneous equations we get here are nonlinear and not easily soluble.

The usual method of attack is to start with a first approximation to $\mathbf{c}$, say $\mathbf{c}^{(1)}$, and minimize Σ by an iterative procedure.

We may expand

$$\mathbf{f}(\mathbf{c})$$

in a Taylor series about $\mathbf{c}^{(1)}$, keeping only the first term:

$$\mathbf{f}(\mathbf{c}) = \mathbf{f}(\mathbf{c}^{(1)}) + \left.\frac{\partial \mathbf{f}}{\partial \mathbf{c}}\right|_{\mathbf{c}^{(1)}} (\mathbf{c} - \mathbf{c}^{(1)}) + \cdots \qquad (33.68)$$

If $\mathbf{c}^{(1)}$ is sufficiently close to $\mathbf{c}$, this approximation is valid and we have the problem of *linear least squares* where

$$\underline{\mathbf{A}}^{(1)} = \left.\frac{\partial \mathbf{f}}{\partial \mathbf{c}}\right|_{\mathbf{c}^{(1)}} \qquad (33.69)$$

is the *design matrix.* This may be evaluated analytically or numerically, depending on $f(\mathbf{x}; \mathbf{c})$.

$$\begin{aligned} \Sigma &= [\mathbf{y} - \mathbf{f}(\mathbf{c}^{(1)}) + \underline{\mathbf{A}}^{(1)}\mathbf{c}^{(1)} - \underline{\mathbf{A}}^{(1)}\mathbf{c}]^T \underline{\mathbf{W}}[\mathbf{y} - \mathbf{f}(\mathbf{c}^{(1)}) \\ &\quad + \underline{\mathbf{A}}^{(1)}\mathbf{c}^{(1)} - \underline{\mathbf{A}}^{(1)}\mathbf{c}] \\ &= [\mathbf{y}^{(1)} - \underline{\mathbf{A}}^{(1)}\mathbf{c}]^T \underline{\mathbf{W}}[\mathbf{y}^{(1)} - \underline{\mathbf{A}}^{(1)}\mathbf{c}] \qquad (33.70) \end{aligned}$$

where

$$\mathbf{y}^{(1)} = \mathbf{y} - \mathbf{f}(\mathbf{c}^{(1)}) + \underline{\mathbf{A}}^{(1)}\mathbf{c}^{(1)} \qquad (33.71)$$

is now minimized with respect to $\mathbf{c}$, yielding new values $\mathbf{c}^{(2)}$ using (33.15). This procedure may be iterated until

$$\mathbf{c}^{(i)} - \mathbf{c}^{(i-1)} \leq \boldsymbol{\epsilon} \qquad (33.72)$$

where $\boldsymbol{\epsilon}$ is a preassigned *vector of small quantities.*

This is a heuristic example of an iterative method. The adequacy of any procedure depends on the "goodness" of the first approximation and the rapidity of convergence of the procedure. We are never guaranteed that a given procedure will converge or that it will converge to the desired solution. Among the myriad difficulties that are encountered in practical nonlinear problems several should be mentioned.

A given procedure starting at a specified point may converge to a *local* minimum that is not the *absolute* minimum sought. This can often be checked by *starting the procedure at different initial points to see if the same minimum is obtained.* The convergence procedure may not be monotonic, that is, the ith iteration may sometimes be inferior to the $(i-1)$th. This is often a problem of attempting to *invert* a singular or *nearly singular* matrix. This can sometimes be corrected by adopting a different method for matrix inversion or by going to double-precision computation or otherwise minimizing round-off errors. It sometimes happens that two *different* sets of parameter values, $\mathbf{c}_{\mathrm{I}}$ and $\mathbf{c}_{\mathrm{II}}$, will have the *same* (minimum) value for Σ. Mathematically, either of these is acceptable. Which solution is obtained often depends on the starting values. *One must be sure to get all acceptable solutions,* so that a selection can be made from among them on a basis other than a statistical one.

APPENDIX A

REVIEW OF NOTATION AND SOME ELEMENTARY MATHEMATICS

Notation

Greek Alphabet

A	α	Alpha	H	η	Eta	N	ν	Nu	T	τ	Tau
B	β	Beta	Θ	θ	Theta	Ξ	ξ	Xi	Υ	υ	Upsilon
Γ	γ	Gamma	I	ι	Iota	O	o	Omicron	Φ	ϕ, φ	Phi
Δ	δ	Delta	K	κ	Kappa	Π	π	Pi	X	χ	Chi
E	ϵ	Epsilon	Λ	λ	Lambda	P	ρ	Rho	Ψ	ψ	Psi
Z	ζ	Zeta	M	μ	Mu	Σ	σ	Sigma	Ω	ω	Omega

$+$	Plus or positive
$-$	Minus or negative
$\pm$	Plus or minus Positive or negative For example, $\pm a \mp b = (a-b)$ or $(-a+b)$.
$\mp$	Minus or Plus Negative or Positive
$=$	Equals
$\neq$, $\not\!\!\!+$	Does not equal
$\rightarrow$	Approaches as a limit
$\times$ or $\cdot$	Multiplied by For example, $a \times b = ab = a \cdot b = (a)(b)$.
$\div$ or : or / or —	Divided by For example, $a \div b = a : b = a/b = \frac{a}{b}$
$\dot{x}$	Repeated digit. For example, $1/3 = 0.\dot{3} = 0.3333\ldots$.
$\equiv$	Is identical to Is defined as
$\cong$ or $\approx$	Equals approximately
$>$	Greater than
$\geq$	Greater than *or* equal to
$<$	Less than
$\leq$	Less than or equal to
$\sim$	Similar to (geometry)
$\cong$	Is congruent to (geometry)
$\stackrel{D}{=}$	Has the dimensions of
$\therefore$	Therefore
$\sqrt[n]{\ }$	nth root
$\sqrt{\ }$	Square root
a^n	nth power of a
ln or $\log_e$	Natural logarithm
log or $\log_{10}$	Common logarithm, Briggsian
e	Base of natural system of logarithms $= \lim_{h \to 0} (1+h)^{1/h} = 2.7182818284\cdots$

Δx	Increment in x; *error* in x
∞	Infinity
$\angle$	Angle
$\parallel$	Parallel to
$\perp$	Perpendicular to
$\llcorner$	Right angle
Σ	Summation of, for example, $\Sigma_{i=0}^{n} x_i = x_0 + x_1 + \cdots + x_n$
Π	Product of, for example, $\Pi_{i=0}^{n} \; x_i = x_0 \cdot x_1 \cdot x_2 \cdot \cdots \cdot x_n$
$\mathbf{x}$ or $\vec{x}$	$\mathbf{x}$ is a vector or n-tuplet $(x_1, \ldots, x_n)$
$(\underline{C})_{ij}$	$\underline{C}$ is a matrix (not a single row or column); indicated is the i–jth element
$\sigma^2(x)$, $\mathrm{Var}(x)$	Variance of the variate x
π	Pi $(3.141592653589793 \cdots)$
$\hat{p}$, p_{est}	Statistic or estimator of the parameter p
$\bar{x}$, $\langle x \rangle$ $E(x)$, Ex	Mean, average or expected value of x
$A \cup B$	Union of the sets A, B
$A \cap B$	Intersection of the sets A, B
$\emptyset$	Null or empty set
cos	Cosine
cot or ctn	Cotangent
csc	Cosecant
sin	Sine
tan	Tangent
sec	Secant

$$\sin \theta = \frac{a}{c} = \frac{1}{\csc \theta}$$

$$\cos \theta = \frac{b}{c} = \frac{1}{\sec \theta}$$

$$\tan \theta = \frac{a}{b} = \frac{1}{\cot \theta}$$

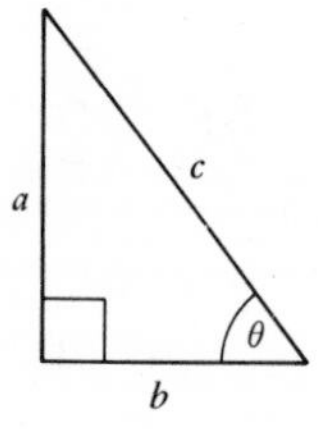

$\sin^{-1} a$ or $\arcsin a$	antisine a; angle whose sine is a; Inverse sine a
$\cos^{-1} a$ or	$\arccos a$
$\tan^{-1} a$ or	$\arctan a$

The complete solution of the equation $x = \sin y$ is $y = (-1)^n \sin^{-1} x + n(180°)$, $-\pi/2 \le \sin^{-1} x \le \pi/2$, where $\sin^{-1} x$ is the principal value of the angle whose sine is x.

Also, if $x = \cos y$, $y = \pm\cos^{-1} x + n(360°)$, $0 \le \cos^{-1} x \le \pi$.

If $x = \tan y$, $y = \tan^{-1} x + n(180°)$, $-\pi/2 \le \tan^{-1} x \le -\pi/2$.

δ_{ij}	Kronecker delta $= 1$ if $i = j$, $= 0$ if $i \ne j$
dy	Differential of y
$\dfrac{dy}{dx}$ or $f'(x)$	Derivative of $y = f(x)$ with respect to x
$\dfrac{d^2y}{dx^2}$ or $f''(x)$	Second derivative of $y = f(x)$ with respect to x
$\dfrac{d^n y}{dx^n}$ or $f^{(n)}(x)$	nth derivative of $y = f(x)$ with respect to x
$\dfrac{\partial z}{\partial x}$	Partial derivative of z with respect to x
$\dfrac{\partial^2 z}{\partial x \, \partial y}$	Second partial derivative of z with respect to y and z
j or i	Imaginary quantity $(\sqrt{-1})$, $i^2 = -1$
$\mathbf{z}$	$\mathbf{z}$ is a complex quantity $= z_R + iz_I$
$\mathbf{z}^*$	Complex conjugate of $\mathbf{z} = z_R - iz_I$.
$\mathrm{Re}\,\mathbf{z}$	Real part of $\mathbf{z} = z_R$
$\mathrm{Im}\,\mathbf{z}$	Imaginary part of $\mathbf{z} = z_I$
$\int$	Integral of
$\int_a^b$	Integral between the limits of a and b

Trigonometry

$\sin^2\theta + \cos^2\theta = 1,$

$1 + \tan^2\theta = \sec^2\theta,$

$1 + \mathrm{ctn}^2\,\theta = \csc^2\theta,$

$\sin 2\theta = 2\sin\theta\cos\theta,$

$$\cos 2\theta = 2\cos^2\theta - 1 = 1 - 2\sin^2\theta = \cos^2\theta - \sin^2\theta,$$

$$\tan 2\theta = \frac{2\tan\theta}{1-\tan^2\theta},$$

$\sin(\theta \pm \varphi) = \sin\theta\cos\varphi \pm \cos\theta\sin\varphi,$

$\cos(\theta \pm \varphi) = \cos\theta\cos\varphi \mp \sin\theta\sin\varphi,$

$$\tan(\theta \pm \varphi) = \frac{\tan\theta \pm \tan\varphi}{1 \mp \tan\theta\tan\varphi}.$$

$$\sin\theta + \sin\phi = 2\sin\frac{1}{2}(\theta+\phi)\cdot\cos\frac{1}{2}(\theta-\phi),$$

$$\sin\theta - \sin\phi = 2\cos\frac{1}{2}(\theta+\phi)\cdot\sin\frac{1}{2}(\theta-\phi),$$

$$\cos\theta + \cos\phi = 2\cos\frac{1}{2}(\theta+\phi)\cdot\cos\frac{1}{2}(\theta-\phi),$$

$$\cos\theta - \cos\phi = 2\sin\frac{1}{2}(\theta+\phi)\cdot\sin\frac{1}{2}(\theta-\phi).$$

Arithmetic Progression

$a,\ a+d,\ a+2d,\ a+3d, \ldots.$

n = number of terms

l = last term

S_n = sum of n terms

$l = a + (n-1)d$

$$S_n = \frac{n}{2}(a+l)$$

Geometric Progression

$a,\ ar,\ ar^2,\ ar^3, \ldots.$

n = number of terms

l = last term = ar^{n-1}

S_n = sum of n terms

$$S_n = \sum_{k=0}^{n-1} ar^k$$

$$S_n = a\left(\frac{r^n - 1}{r-1}\right) = \frac{rl - a}{r-1}$$

If $r^2 < 1$ $S_n \to S_\infty$ as $n \to \infty$

$$S_\infty = \frac{a}{1-r}$$

For example,

$$1 + x + x^2 + \cdots + x^n = \frac{1-x^{n+1}}{1-x} = \frac{1}{1-x} - \frac{x^{n+1}}{1-x}$$

Arithmetic mean of $a, b = \dfrac{a+b}{2}$

Arithmetic mean of $x_1, \ldots, x_n = \sum_{i=1}^{n} \dfrac{x_i}{n}$

Geometric mean of $a, b = \sqrt{ab}$

Geometric mean of $x_1, \ldots, x_n = \sqrt[n]{\prod_{i=1}^{n} x_i} = \left[\prod_{i=1}^{n} x_i\right]^{1/n}$

Harmonic Progression

A sequence of numbers whose reciprocals form an arithmetic progression is called an harmonic progression.

$$\frac{1}{a}, \frac{1}{a+d}, \frac{1}{a+2d}, \ldots.$$

The harmonic mean of a, $b = \left[\frac{1}{2}\left(\frac{1}{a}+\frac{1}{b}\right)\right]^{-1} = \left[\frac{a+b}{2ab}\right]^{-1} = \frac{2ab}{a+b}$

Taylor's Series

Any continuous, differentiable function may be expanded in a Taylor's series.

$$f(x) = f(a) + f'(a)\frac{(x-a)}{1!} + f''(a)\frac{(x-a)^2}{2!} + f'''(a)\frac{(x-a)^3}{3!} + \cdots + f^{(n-1)}(a)\frac{(x-a)^{n-1}}{(n-1)!} + R_n$$

$$R_n = f^{(n)}(x_1)\frac{(x-a)^n}{n!} \qquad \text{where} \qquad a < x_1 < x$$

Binomial Series

$$(a+x)^n = a^n + na^{n-1}x + \frac{n(n-1)}{2!}a^{n-2}x^2 + \cdots \qquad \text{for } x^2 < a^2$$

If n is a positive integer there are $n + 1$ terms; otherwise the number of terms is infinite.

Other Series

$$\frac{1}{a-bx} = \frac{1}{a}\left(1 + \frac{bx}{a} + \frac{b^2x^2}{a^2} + \frac{b^3x^3}{a^3} + \cdots\right) \qquad \text{if } b^2x^2 < a^2$$

$$e = \lim_{N\to\infty}\left(1+\frac{1}{N}\right)^N = \lim_{d\to 0}(1+d)^{1/d}$$

$$e^x = 1 + x + \frac{x^2}{2!} + \frac{x^3}{3!} + \frac{x^4}{4!} + \cdots$$

$$\sin x = x - \frac{x^3}{3!} + \frac{x^5}{5!} - \frac{x^7}{7!} + \cdots$$

$$\cos x = 1 - \frac{x^2}{2!} + \frac{x^4}{4!} + \frac{x^6}{6!} + \cdots$$

$$\tan x = x + \frac{x^3}{3} + \frac{2x^5}{15} + \frac{17x^7}{315} + \frac{62x^9}{2835} + \cdots \qquad \text{for } x^2 < \frac{\pi^2}{4}$$

$$\sin^{-1} x = x + \frac{x^3}{6} + \frac{1}{2}\cdot\frac{3}{4}\cdot\frac{x^5}{5} + \frac{1}{2}\cdot\frac{3}{4}\cdot\frac{5}{6}\cdot\frac{x^7}{7} + \cdots \qquad \text{for } x^2 < 1$$

$$\tan^{-1} x = x - \frac{1}{3}x^3 + \frac{x^5}{5} - \frac{x^7}{7} + \cdots \qquad \text{for } x^2 < 1$$

$$= \frac{\pi}{2} - \frac{1}{x} + \frac{1}{3x^3} - \frac{1}{5x^5} + \cdots \qquad \text{for } x^2 > 1$$

$$\ln(1+x) = x - \frac{x^2}{2} + \frac{x^3}{3} - \frac{x^4}{4} + \cdots \qquad \text{for } -1 < x \leq +1$$

Maxima and Minima of Functions

UNIVARIATE

Consider $f(x)$ over an interval $a \le x \le b$. A relative maximum (minimum) exists at x_0 if $f(x) \ge f(x_0)$ $[f(x) \le f(x_0)]$ for x near x_0.

At a relative maximum or minimum

$$f'(x_0) = 0$$

A point x_0 for which $f'(x_0) = 0$ is called a critical point. A critical point x_0 is a *maximum* if

$$f''(x_0) < 0$$

A critical point x_0 is a *minimum* if

$$f''(x_0) > 0$$

If $f''(x_0) = 0$, the higher derivatives must be considered. If $f'(x_0) = 0 = f''(x_0) = \cdots = f^{(n-1)}(x_0)$, but $f^{(n)}(x_0) \neq 0$, then $f(x)$ has a relative extremum (maximum or minimum) if n is even. Furthermore, the critical point is a maximum if $f^{(n)}(x_0) < 0$ and is a minimum if $f^{(n)}(x_0) > 0$. If n is odd for the first nonvanishing derivative, then the critical point is a horizontal inflection point.

An absolute maximum (minimum) is a point x_0 such that $f(x_0) \ge f(x)$ $[f(x_0) \le f(x)]$ for $a \le x \le b$. The absolute maximum (minimum) occurs either at a critical point or at a boundary point, $x = a$ or $x = b$.

MULTIVARIATE

Consider function $f(x_1, x_2, \ldots, x_n) = z$. Relative maxima (minima) are similarly defined to be points $(x_{10}, x_{20}, \ldots, x_{n0})$ such that $f(x_{10}, x_{20}, \ldots, x_{n0}) \ge f(x_1, x_2, \ldots, x_n)$ $[f(x_{10}, x_{20}, \ldots, x_{n0}) \le f(x_1, x_2, \ldots, x_n)]$ for all points $(x_1, x_2, \ldots, x_n)$ near $(x_{10}, x_{20}, \ldots, x_{n0})$. Critical points are points $(x_{10}, x_{20}, \ldots, x_{n0})$ such that all the first partial derivatives vanish:

$$\left.\frac{\partial f}{\partial x_1}\right|_{\mathbf{x}_0} = 0 = \left.\frac{\partial f}{\partial x_2}\right|_{\mathbf{x}_0} = \cdots = \left.\frac{\partial f}{\partial x_n}\right|_{\mathbf{x}_0}$$

Relative maxima and minima occur only at critical points.

SADDLE POINTS

It may happen that $f(x_1, x_{20}, \ldots, x_{n0})$ has a relative maximum at $\mathbf{x}_0$, while the function $f(x_{10}, x_2, x_{30}, \ldots, x_{n0})$ has a relative minimum at $\mathbf{x}_0$. For $n = 2$, the graph looks like a saddle; hence the term saddle point is applied to such points.

FUNCTIONS OF TWO VARIABLES

If we define

$$A = \left.\frac{\partial^2 f}{\partial x_1^2}\right|_{\mathbf{x}_0}$$

$$B = \left.\frac{\partial^2 f}{\partial x_1 x_2}\right|_{\mathbf{x}_0}$$

$$C = \left.\frac{\partial^2 f}{\partial x_2^2}\right|_{\mathbf{x}_0}$$

a relative extremum exists at $\mathbf{x}_0$ if $B^2 - AC < 0$. Furthermore, the extremum is a relative maximum if $A + C < 0$ and is a relative minimum if $A + C > 0$.

If $B^2 - AC > 0$, a saddle point exists at $\mathbf{x}_0$.

If $B^2 - AC = 0$, the nature of the critical point must be examined further.

FUNCTIONS WITH CONSTRAINTS

It is often necessary to extremize a function $f(x_1, x_2, \ldots, x_n)$ subject to the constraints

$$g_1(x_1, x_2, \ldots, x_n) = 0$$
$$g_2(x_1, x_2, \ldots, x_n) = 0$$
$$\vdots$$
$$g_m(x_1, x_2, \ldots, x_n) = 0$$

where $m < n$.

Example

$f(x, y, z) = w$ is to be maximized subject to the constraints

$$g_1(x, y, z) = 0$$

and

$$g_2(x, y, z) = 0$$

$g_1(x, y, z) = 0$ describes a surface in three-dimensional space as does $g_2(x, y, z) = 0$. Both equations together define their intersection which is a curve in space. It is thus required to maximize $f(x, y, z)$ as (x, y, z) varies along the curve of intersection. It is necessary that $\nabla f = (\partial f/\partial x)\mathbf{i} + (\partial f/\partial y)\mathbf{j} + (\partial f/\partial z)\mathbf{k}$ must lie in a plane normal to the curve at a maximum. Also, ∇g_1 and ∇g_2 must lie in the same plane. If three vectors are coplanar, there must exist scalar numbers λ_1, λ_2 such that

$$\nabla f + \lambda_1 \nabla g_1 + \lambda_2 \nabla g_2 = 0$$

In three-dimensional space, this represents three scalar equations:

$$\frac{\partial f}{\partial x} + \lambda_1 \frac{\partial g_1}{\partial x} + \lambda_2 \frac{\partial g_2}{\partial x} = 0$$

$$\frac{\partial f}{\partial y} + \lambda_1 \frac{\partial g_1}{\partial y} + \lambda_2 \frac{\partial g_2}{\partial y} = 0$$

and

$$\frac{\partial f}{\partial z} + \lambda_1 \frac{\partial g_1}{\partial z} + \lambda_2 \frac{\partial g_2}{\partial z} = 0$$

These three equations, together with

$$g_1(x, y, z) = 0$$

and

$$g_2(x, y, z) = 0$$

represent five equations that may be used to solve for the five unknowns: x_0, y_0, z_0, λ_1, λ_2. The λ's are called *Lagrange multipliers.*

In general, we form the n equations

$$\frac{\partial}{\partial x_i}\{f(\mathbf{x}) + \lambda_1 g_1(\mathbf{x}) + \lambda_2 g_2(\mathbf{x}) + \cdots + \lambda_m g_m(\mathbf{x})\} = 0 \qquad \text{for } i = 1, \ldots, n$$

and the m equations of constraint

$$g_j(\mathbf{x}) = 0 \qquad j = 1, \ldots, m$$

to solve for the $(n + m)$ unknowns: $\mathbf{x}_0$, λ_j $(j = 1, \ldots, m)$. This is the **method of Lagrange multipliers**.

NUMERICAL OPTIMIZATION OF GENERAL (NONLINEAR) MULTIPARAMETER FUNCTIONS

The *minimization* of an arbitrary chi-square function (or the *maximization* of a general likelihood function) may be effected through a variety of *optimization* procedures. These are largely numerical techniques that have been embodied in fairly general computer programs available in the scientific subroutine libraries of most digital computer installations. The techniques are described under such headings as, for example, *direct search methods, ravine stepping, grid search for independent parameters, gradient search, steepest descent, linear minimization in orthogonal directions, method of conjugate directions, variable metric, parabolic extrapolation,* etc. Which technique to choose in a given situation depends on the specific nature of the problem, the nature of the "first approximation" and, not least of all, the *availability* of particular routines at your local computer facility. Space permits only the enumeration of some references sufficient to explain the procedures most widely available at major computer centers.

REFERENCES

Lavi, A. and T. P. Vogl (eds.), *Recent Advances in Optimization Techniques,* Wiley, New York, 1966.

Kowalik, J. and M. R. Osborne, *Methods for Unconstrained Optimization Problems,* American Elsevier, 1968.

Box, M. J., D. Davies, and W. H. Swann, *Nonlinear Optimization Techniques,* Oliver and Boyd, London, 1969.

Fletcher, R. (ed.), *Optimization,* Academic Press, New York, 1969.

We shall cite only three references in the professional journals, all from Computer Journal:

Fletcher, R., *Computer J. 8,* 33 (1965).

Powell, M. J. D., *Computer J. 7,* 155 (1964).

Rosenbrock, H. H., *Computer J. 3,* 175 (1960).

APPENDIX B

MATRICES, DETERMINANTS, AND LINEAR EQUATIONS

1. DEFINITION OF A MATRIX

A matrix $\underline{\mathbf{A}}$ is a rectangular, two-dimensional array of numbers, called the *elements* of the matrix, arranged in rows and columns. The elements of the matrix $\underline{\mathbf{A}}$ may be written A_{ij} or $(\underline{\mathbf{A}})_{ij}$ where the first subscript, i, denotes the row and the second subscript, j, denotes the column. If there are m rows and n columns, we may write the matrix $\underline{\mathbf{A}}$ as

$$\underline{\mathbf{A}} = \begin{bmatrix} A_{11} & A_{12} & A_{13} & \cdots & A_{1n} \\ A_{21} & A_{22} & & \cdots & A_{2n} \\ \vdots & & & & \vdots \\ A_{m1} & A_{m2} & & \cdots & A_{mn} \end{bmatrix}$$

where

$$[A_{11} \quad A_{12} \quad A_{13} \quad \cdots \quad A_{1n}]$$

is the first row, etc. and

$$\begin{bmatrix} A_{11} \\ A_{21} \\ \vdots \\ A_{m1} \end{bmatrix}$$

is the first column, etc.

We say that $\underline{\mathbf{A}}$ is an $(m \times n)$ matrix, sometimes written $\underline{\mathbf{A}}(m \times n)$. Note. We may use various kinds of brackets for the array without changing the significance, for example, ( ), [ ], { }, etc.

2. VECTOR

A *vector* with m components is a *column* matrix that is $(m \times 1)$. We specify this particular matrix because it has such general usefulness. We have elsewhere defined a vector $\mathbf{v}$ with m components and sometimes written it as the m-tuplet

$$\mathbf{v} = (v_1, v_2, \ldots, v_m)$$

with commas between the components. We shall, however, whenever the matrix nature of a vector is relevant, regard a *vector* as a *column* matrix written

$$\mathbf{v} = \begin{bmatrix} v_1 \\ v_2 \\ \vdots \\ v_m \end{bmatrix}$$

3. SCALAR

A matrix, $\underline{\mathbf{A}}(m \times n)$, with $m = n = 1$ is just a number, or *scalar* quantity.

4. EQUALITY

We say that two matrices $\underline{\mathbf{A}}$ and $\underline{\mathbf{B}}$ are *equal*,

$$\underline{\mathbf{A}} = \underline{\mathbf{B}}$$

if they have the same number of rows and columns and, for all i and j.

$$A_{ij} = B_{ij}$$

5. ADDITION

If two matrices $\underline{\mathbf{A}}$ and $\underline{\mathbf{B}}$ have the same number of rows and columns their sum or difference is defined by $\underline{\mathbf{A}} \pm \underline{\mathbf{B}}$ where

$$(\underline{\mathbf{A}} + \underline{\mathbf{B}})_{ij} = A_{ij} \pm B_{ij}$$

Example:

Suppose $\underline{\mathbf{A}} = \begin{pmatrix} +4 & -1 \\ -2 & +2 \end{pmatrix}$ and $\underline{\mathbf{B}} = \begin{pmatrix} +2 & -5 \\ +2 & -6 \end{pmatrix}$ and we wish to find $\underline{\mathbf{A}} - \underline{\mathbf{B}}$.

We get

$$\underline{\mathbf{A}} - \underline{\mathbf{B}} = \begin{pmatrix} +2 & +4 \\ -4 & +8 \end{pmatrix}$$

Addition obeys the *associative law*:

$$\underline{\mathbf{A}} + (\underline{\mathbf{B}} + \underline{\mathbf{C}}) = (\underline{\mathbf{A}} + \underline{\mathbf{B}}) + \underline{\mathbf{C}} = \underline{\mathbf{A}} + \underline{\mathbf{B}} + \underline{\mathbf{C}}$$

Addition obeys the *commutative law*:

$$\underline{\mathbf{A}} + \underline{\mathbf{B}} = \underline{\mathbf{B}} + \underline{\mathbf{A}}$$

6. MULTIPLICATION BY A CONSTANT

If a matrix $\underline{\mathbf{A}}$ is multiplied by a constant, k, the product is written $k\underline{\mathbf{A}}$, which is a matrix of the same $(m \times n)$ *as* $\underline{\mathbf{A}}$ with elements

$$(k\underline{\mathbf{A}})_{ij} = kA_{ij}$$

7. ZERO MATRIX

A *zero* or *null* matrix, $\underline{\mathbf{0}}$, is a matrix whose elements are all zero:

$$(\underline{\mathbf{0}})_{ij} = 0 \qquad \text{for all } i, j$$

For each value $(m \times n)$, there is a zero matrix.

8. TRANSPOSE MATRIX

If we have a matrix $\underline{\mathbf{A}}(m \times n)$, we may define its *transpose matrix* $\underline{\mathbf{A}}^T(n \times m)$, which is a matrix with n rows and m columns obtained by *interchanging* rows and columns of $\underline{\mathbf{A}}$:

$$[\underline{\mathbf{A}}^T(n \times m)]_{ij} = [\underline{\mathbf{A}}(m \times n)]_{ji}$$
$$= [\underline{\mathbf{A}}(m \times n)]^T_{ij}$$

Example:

$$\underline{\mathbf{A}} = \begin{pmatrix} 2 & 4 \\ 3 & 5 \end{pmatrix} \qquad \underline{\mathbf{A}}^T = \begin{pmatrix} 2 & 3 \\ 4 & 5 \end{pmatrix}$$

This has the obvious consequence that

$$(\underline{\mathbf{A}}^T)^T = \underline{\mathbf{A}}$$

The transpose of the (column) vector

$$\mathbf{v} = \begin{pmatrix} v_1 \\ \vdots \\ v_m \end{pmatrix}$$

is the *row*

$$\mathbf{v}^T = (v_1 \cdots v_m)$$

and conversely.

9. COMPLEX CONJUGATION

If the elements of $\underline{\mathbf{A}}$ are complex, we may consider the *complex conjugate* matrix $\underline{\mathbf{A}}^*$ to be

$$(\underline{\mathbf{A}}^*)_{mn} = A^*_{mn}$$

That is, if

$$A_{mn} = a_{mn} + ib_{mn}$$

where a_{mn}, b_{mn} are real and i is $\sqrt{-1}$, then

$$A^*_{mn} = a_{mn} - ib_{mn}$$

10. HERMITIAN CONJUGATE

The *associate matrix* to $\underline{\mathbf{A}}$, also called the *hermitian conjugate* of $\underline{\mathbf{A}}$, is denoted

$$\underline{\mathbf{A}}^\dagger = (\underline{\mathbf{A}}^*)^T = (\underline{\mathbf{A}}^T)^*$$

A matrix is said to be *hermitian* if

$$\underline{\mathbf{A}}^\dagger = \underline{\mathbf{A}}$$

11. MATRIX MULTIPLICATION

If the number of *columns* of a matrix $\underline{\mathbf{A}}(r \times s)$ equals the number of *rows* of a matrix $\underline{\mathbf{B}}(s \times t)$, they are *conformable* in the order $\underline{\mathbf{A}}\,\underline{\mathbf{B}}$, and we may define their *product*

$$[\underline{\mathbf{A}}(r \times s)] \cdot [\underline{\mathbf{B}}(s \times t)] = \underline{\mathbf{C}}(r \times t)$$

with elements

$$(\underline{\mathbf{C}})_{ij} = \sum_{\rho=1}^{s} A_{i\rho} B_{\rho j}$$

Example

$$\underline{\mathbf{A}} = \begin{pmatrix} A_{11} & A_{12} \\ A_{21} & A_{22} \\ A_{31} & A_{32} \end{pmatrix}$$

$$\underline{\mathbf{B}} = \begin{pmatrix} B_{11} & B_{12} & B_{13} & B_{14} \\ B_{21} & B_{22} & B_{23} & B_{24} \end{pmatrix}$$

$$\underline{\mathbf{A}} \cdot \underline{\mathbf{B}} = \underline{\mathbf{C}} = \begin{pmatrix} C_{11} & C_{12} & C_{13} & C_{14} \\ C_{21} & C_{22} & C_{23} & C_{24} \\ C_{31} & C_{32} & C_{33} & C_{34} \end{pmatrix}$$

where

$$\begin{aligned} C_{11} &= A_{11}B_{11} + A_{12}B_{21} & C_{12} &= A_{11}B_{12} + A_{12}B_{22} \\ C_{13} &= A_{11}B_{13} + A_{12}B_{23} & C_{14} &= A_{11}B_{14} + A_{12}B_{24} \\ C_{21} &= A_{21}B_{11} + A_{22}B_{21} & C_{22} &= A_{21}B_{12} + A_{22}B_{22} \\ C_{23} &= A_{21}B_{13} + A_{22}B_{23} & C_{24} &= A_{21}B_{14} + A_{22}B_{24} \\ C_{31} &= A_{31}B_{11} + A_{32}B_{21} & C_{32} &= A_{31}B_{12} + A_{32}B_{22} \\ C_{33} &= A_{31}B_{13} + A_{32}B_{23} & C_{34} &= A_{31}B_{14} + A_{32}B_{24} \end{aligned}$$

We see that a product has the same number of rows as the first factor and the same number of columns as the second factor.

If multiplication is defined (i.e., the number of columns of the first factor is equal to the number of rows of the second), it obeys the *associative law*:

$$\begin{aligned} &[(\underline{\mathbf{A}}(r \times s)) \cdot (\underline{\mathbf{B}}(s \times t))] \cdot \underline{\mathbf{C}}(t \times u) \\ &\quad = \underline{\mathbf{A}}(r \times s) \cdot [\underline{\mathbf{B}}(s \times t) \cdot \underline{\mathbf{C}}(t \times u)] \end{aligned}$$

the result being an $(r \times u)$ matrix.

Multiplication also obeys the *distributive law*:

$$\begin{aligned} &\underline{\mathbf{A}}(r \times s) \cdot [\underline{\mathbf{B}}(s \times t) + \underline{\mathbf{D}}(s \times t)] \\ &\quad = \underline{\mathbf{A}}(r \times s) \cdot \underline{\mathbf{B}}(s \times t) \\ &\qquad + \underline{\mathbf{A}}(r \times s) \cdot \underline{\mathbf{D}}(s \times t) \end{aligned}$$

In general, multiplication does *not* obey the *commutative law*, that is,

$$\underline{\mathbf{A}} \cdot \underline{\mathbf{B}} \neq \underline{\mathbf{B}} \cdot \underline{\mathbf{A}}$$

In fact, the commuted product $\underline{\mathbf{B}} \cdot \underline{\mathbf{A}}$ may not even be defined when $\underline{\mathbf{A}} \cdot \underline{\mathbf{B}}$ is. Matrices $\underline{\mathbf{A}}$ and $\underline{\mathbf{B}}$ that do obey the relation

$$\underline{\mathbf{A}} \cdot \underline{\mathbf{B}} = \underline{\mathbf{B}} \cdot \underline{\mathbf{A}}$$

are said to *commute*.

Example

Suppose $\underline{\mathbf{A}} = (A_{11}A_{12}A_{13}) = (\mathbf{a})^T = (a_1a_2a_3)$

and $$\underline{\mathbf{B}} = \begin{pmatrix} B_{11} \\ B_{21} \\ B_{31} \end{pmatrix} = \mathbf{b} = \begin{pmatrix} b_1 \\ b_2 \\ b_3 \end{pmatrix}$$

Here, $\underline{\mathbf{A}} \cdot \underline{\mathbf{B}} = \mathbf{a}^T \cdot \mathbf{b} = a_1b_1 + a_2b_2 + a_3b_3$, a *number*, since

$$\underline{\mathbf{A}}(1 \times 3) \cdot \underline{\mathbf{B}}(3 \times 1) = \underline{\mathbf{C}}(1 \times 1)$$

This, of course, is the *scalar product* of two vectors.

However,

$$\underline{\mathbf{B}} \cdot \underline{\mathbf{A}} = \begin{pmatrix} B_{11}A_{11} & B_{11}A_{12} & B_{11}A_{13} \\ B_{21}A_{11} & B_{21}A_{12} & B_{21}A_{13} \\ B_{31}A_{11} & B_{31}A_{12} & B_{31}A_{13} \end{pmatrix}$$

$$= \begin{pmatrix} b_1a_1 & b_1a_2 & b_1a_3 \\ b_2a_1 & b_2a_2 & b_2a_3 \\ b_3a_1 & b_3a_2 & b_3a_3 \end{pmatrix}$$

which is a (3×3) matrix, since

$$\underline{\mathbf{B}}(3 \times 1) \cdot \underline{\mathbf{A}}(1 \times 3) = \underline{\mathbf{C}}(3 \times 3)$$

This is an example of the *dyadic product* of two vectors.

12. TRANSPOSITION AND MULTIPLICATION

$$(\underline{\mathbf{A}}(r \times s) \cdot \underline{\mathbf{B}}(s \times t))^T = (\underline{\mathbf{B}}(s \times t))^T \cdot (\underline{\mathbf{A}}(r \times s))^T$$

or

$$(\underline{\mathbf{A}} \cdot \underline{\mathbf{B}})^T = (\underline{\mathbf{B}}^T) \cdot (\underline{\mathbf{A}}^T)$$

since multiplication is automatically defined if $\underline{\mathbf{A}} \cdot \underline{\mathbf{B}}$ is defined.

This generalizes to

$$(\underline{\mathbf{A}} \cdot \underline{\mathbf{B}} \cdots \underline{\mathbf{Z}})^T = (\underline{\mathbf{Z}}^T) \cdots (\underline{\mathbf{B}}^T) \cdot (\underline{\mathbf{A}}^T)$$

that is, the transpose of a product is equal to the product of the transposes, *in reverse order.*

The product of a matrix and its transpose is *always* defined:

$$\underline{\mathbf{A}}(m \times n) \cdot \underline{\mathbf{A}}^T(n \times m) = (\underline{\mathbf{A}} \cdot \underline{\mathbf{A}}^T)$$

which is $(m \times m)$ and

$$\underline{\mathbf{A}}^T(n \times m) \cdot \underline{\mathbf{A}}(m \times n) = (\underline{\mathbf{A}}^T \cdot \underline{\mathbf{A}})$$

which is $(n \times n)$.

The product of the transpose of a vector with the vector is a *scalar* called the *square of the vector*:

$$\mathbf{v}^T \cdot \mathbf{v} = (v_1 v_2 \cdots v_n) \begin{pmatrix} v_1 \\ v_2 \\ \vdots \\ v_n \end{pmatrix} = \sum_{i=1}^{n} v_i^2$$

13. SQUARE MATRIX

A *square* or *quadratic* matrix, $\underline{\mathbf{A}}(m \times m)$, is one for which the number of rows equals the number of columns, that is, $\underline{\mathbf{A}}(m \times n)$ is square if $m = n$. The product of a matrix and its transpose is always a square matrix. The *order* of a square matrix is the number of rows or columns.

14. SYMMETRIC MATRIX

A square matrix $\underline{\mathbf{A}}(m \times m)$ is *symmetric* if

$$\underline{\mathbf{A}}^T = \underline{\mathbf{A}}$$

that is,

$$(\underline{\mathbf{A}})_{ij} = (\underline{\mathbf{A}})_{ji} \qquad \text{for all } i \text{ and } j$$

The *product* of a matrix and its transpose is a symmetric matrix since

$$(\underline{\mathbf{A}} \cdot \underline{\mathbf{A}}^T)^T = (\underline{\mathbf{A}}^T)^T \cdot \underline{\mathbf{A}}^T = \underline{\mathbf{A}} \cdot \underline{\mathbf{A}}^T$$

Also,

$$(\underline{\mathbf{A}}^T \cdot \underline{\mathbf{A}})^T = \underline{\mathbf{A}}^T \cdot \underline{\mathbf{A}}$$

15. DIAGONAL MATRIX

A square matrix, $\underline{\mathbf{D}}(m \times m)$, for which

$$(\underline{\mathbf{D}})_{ij} = 0 \qquad \text{if} \qquad i \neq j$$

is said to be *diagonal* and to have only diagonal elements $(\underline{\mathbf{D}})_{ii}$.

All diagonal matrices $\underline{\mathbf{D}}$, $\underline{\mathbf{D}}'$ commute:

$$\underline{\mathbf{D}} \cdot \underline{\mathbf{D}}' = \underline{\mathbf{D}}' \cdot \underline{\mathbf{D}}$$

16. UNIT MATRIX

A diagonal matrix $\underline{\mathbf{A}}(m \times m)$ for which

$$(\underline{\mathbf{A}})_{ii} = 1 \qquad \text{for all } i$$

is called the $(m \times m)$ *unit matrix* $\underline{\mathbf{I}}(m \times m)$:

$$\underline{\mathbf{I}} = \begin{bmatrix} 1 & 0 & 0 & \cdots & 0 \\ 0 & 1 & 0 & \cdots & 0 \\ 0 & 0 & 1 & \cdots & 0 \\ \vdots & & & & \vdots \\ 0 & 0 & 0 & \cdots & 1 \end{bmatrix}$$

or *unit matrix of order m.*

For all matrices $\underline{\mathbf{A}}(m \times n)$, $\underline{\mathbf{B}}(n \times m)$ we have

$$\underline{\mathbf{I}}(m \times m) \cdot \underline{\mathbf{A}}(m \times n) = \underline{\mathbf{A}}(m \times n)$$

and

$$\underline{\mathbf{B}}(n \times m) \cdot \underline{\mathbf{I}}(m \times m) = \underline{\mathbf{B}}(n \times m)$$

If we understand that the appropriate $\underline{\mathbf{I}}$ is always chosen so that the product is defined, we have

$$\underline{\mathbf{A}} \cdot \underline{\mathbf{I}} = \underline{\mathbf{I}} \cdot \underline{\mathbf{A}} = \underline{\mathbf{A}}$$

for *any* $\underline{\mathbf{A}}$.

The elements of the unit matrix are given by the *Kronecker delta*:

$$\delta_{ij} = 1 \qquad i = j$$
$$= 0 \qquad i \neq j$$

17. DETERMINANT: RANK

We may define the *determinant* $\underline{\mathbf{A}}$ of order m of a square matrix, $\underline{\mathbf{A}}(m \times m)$, to be a scalar quantity (also called the *value of the determinant*) obtained from a calculation involving the elements of the matrix. We write

$$\det \underline{\mathbf{A}} = |\underline{\mathbf{A}}| = \begin{vmatrix} A_{11} & A_{12} & \cdots & A_{1m} \\ A_{21} & A_{22} & \cdots & A_{2m} \\ \vdots & & & \vdots \\ A_{m1} & & \cdots & A_{mm} \end{vmatrix}$$

One definition of the determinant is

$$|\underline{\mathbf{A}}| = \sum (-1)^p A_{1a_1} A_{2a_2} A_{3a_3} \cdots A_{ma_m}$$

where we take the sum of all (signed) possible products of elements in each of which appears one and only one element from each row and each column. There are $m!$ such products. The sign of each product is determined by the number of *inversions* or *permutations*, p, of the sequence

$$a_1, a_2, a_3, \ldots, a_n$$

from the sequence $1, 2, 3, \ldots, m$. If the number of permutations is even, the + sign is taken $[(-1)^p = +1]$ while the − sign is taken for an odd number of permutations $[(-1)^p = -1]$.

Another way of defining the rule for assigning signs to the product comes from Hildebrand.* Join the elements in a given product in pairs by line segments. Let p in the above formula be the total number of such segments *sloping upward to the right.*

Example

$$\det \underline{\mathbf{A}} = \begin{vmatrix} A_{11} & A_{12} & A_{13} \\ A_{21} & A_{22} & A_{23} \\ A_{31} & A_{32} & A_{33} \end{vmatrix}$$

The possible products, P_i, are $3! = 6$ in number:

$$P_i = A_{11}A_{22}, A_{33}, A_{12}A_{23}A_{31}, A_{13}A_{21}A_{32},$$
$$A_{11}A_{23}A_{32},\ A_{12}A_{21}A_{33},\ A_{13}A_{22}A_{31}$$
$$i = 1, \ldots, 6$$

which we may obtain by following the lines in Figure B.1.

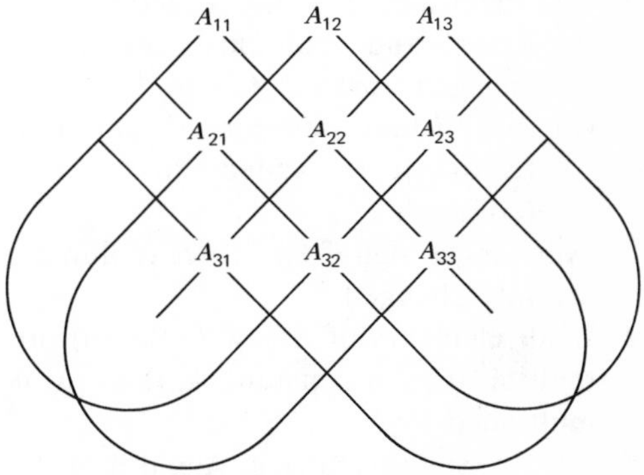

P_1: 123 requires 0 permutations $p = 0$.
P_2: 231 requires 2 permutations $p = 2$
P_3: 312 requires 2 permutations $p = 2$
P_4: 132 requires 1 permutation $p = 1$
P_5: 213 requires 1 permutation $p = 1$
P_6: 321 requires 1 permutation $p = 1$.

Likewise

P_1 has no line segments $p_1 = 0$
P_2 has no line segments $p_2 = 0$
P_3 has no line segments $p_3 = 0$
P_4 has one line segment $p_4 = 1$
P_5 has one line segment $p_5 = 1$
P_6 has three line segments $p_6 = 3$

$$\therefore |\underline{\mathbf{A}}| = +A_{11}A_{22}A_{33} + A_{12}A_{23}A_{31} + A_{13}A_{21}A_{32} - A_{11}A_{23}A_{32} - A_{12}A_{21}A_{33} - A_{13}A_{22}A_{31}$$

For later comparison we may write this

$$|\underline{\mathbf{A}}| = A_{11}(A_{22}A_{33} - A_{23}A_{32}) - A_{12}(A_{21}A_{33} - A_{23}A_{31}) + A_{13}(A_{21}A_{32} - A_{22}A_{31})$$

We may list some properties of determinants

1. $|\underline{\mathbf{A}}| = 0$ if

(a) All elements of a row (column) are zero, that is, $A_{ij} = 0$ for all i or $A_{ij} = 0$ for all j.

*Francis B. Hildebrand, *Methods of Applied Mathematics*, Prentice-Hall, 1965.

(b) All elements of a row (column) may be obtained from the elements of another row (column) by multiplication by a constant, that is, $A_{ij} = kA_{lj}$ for all j or $A_{ij} = kA_{il}$ for all i.

2. The determinant is unchanged if
   (a) Rows and columns are interchanged, that is, $|\underline{\mathbf{A}}| = |\underline{\mathbf{A}}^T|$.
   (b) A constant multiple of any row (column) is added to a row (column).
3. $|\underline{\mathbf{A}}|$ changes sign if two rows (columns) are interchanged.
4. If all elements of a row (column) are multiplied by a constant, k, then $|\underline{\mathbf{A}}|$ is multiplied by k.
5. For two square matrices $\underline{\mathbf{A}}$ and $\underline{\mathbf{B}}$ of the same rank

$$|\underline{\mathbf{A}} \cdot \underline{\mathbf{B}}| = |\underline{\mathbf{A}}||\underline{\mathbf{B}}|$$

but

$$|\underline{\mathbf{A}} + \underline{\mathbf{B}}| \neq |\underline{\mathbf{A}}| + |\underline{\mathbf{B}}|$$

6. The determinant of a diagonal square matrix is

$$|\underline{\mathbf{A}}(m \times m)| = \prod_{i=1}^{m} A_{ii} = A_{11} \cdot A_{22} \cdot \cdots \cdot A_{mm}$$

For a general matrix $\underline{\mathbf{A}}$, which is not necessarily square, we may form all possible square *submatrices.* If at least one determinant of order r is nonzero and if all determinants of order $r+1$ or higher vanish, the matrix $\underline{\mathbf{A}}$ is said to be of *rank r.* The rank of a *product* is never greater than the rank of a *factor.* The rank of a matrix $\underline{\mathbf{A}}(n \times m)$ is $r \leq n$ or m, whichever is less. We may also write this

$$\text{rank } r \leq \min(n, m)$$

18. COFACTORS

The *cofactor* of an element A_{ij}, denoted

$$\text{Cof } A_{ij}$$

is the product of $(-1)^{i+j}$ and the *determinant* of order $(m-1)$ of the matrix obtained from $\underline{\mathbf{A}}(m \times m)$ by *deleting* the ith row *and* jth column.

Example

$$\underline{\mathbf{A}} = \begin{pmatrix} A_{11} & A_{12} & A_{13} \\ A_{21} & A_{22} & A_{23} \\ A_{31} & A_{32} & A_{33} \end{pmatrix}$$

$$\text{Cof } A_{11} = (+1)\begin{vmatrix} A_{22} & A_{23} \\ A_{32} & A_{33} \end{vmatrix} = +(A_{22}A_{33} - A_{32}A_{23})$$

$$\text{Cof } A_{12} = (-1)\begin{vmatrix} A_{21} & A_{23} \\ A_{31} & A_{33} \end{vmatrix} = -(A_{21}A_{33} - A_{31}A_{23})$$

$$\text{Cof } A_{13} = (+1)\begin{vmatrix} A_{21} & A_{22} \\ A_{31} & A_{32} \end{vmatrix} = +(A_{21}A_{32} - A_{31}A_{22})$$

$$\text{Cof } A_{21} = (-1)\begin{vmatrix} A_{12} & A_{13} \\ A_{32} & A_{33} \end{vmatrix} = -(A_{12}A_{33} - A_{32}A_{13})$$

$$\text{Cof } A_{22} = (+1)\begin{vmatrix} A_{11} & A_{13} \\ A_{31} & A_{33} \end{vmatrix} = +(A_{11}A_{33} - A_{31}A_{13})$$

$$\text{Cof } A_{23} = (-1)\begin{vmatrix} A_{11} & A_{12} \\ A_{31} & A_{32} \end{vmatrix} = -(A_{11}A_{32} - A_{31}A_{12})$$

$$\text{Cof } A_{31} = (+1)\begin{vmatrix} A_{12} & A_{13} \\ A_{22} & A_{23} \end{vmatrix} = +(A_{12}A_{23} - A_{22}A_{13})$$

$$\text{Cof } A_{32} = (-1)\begin{vmatrix} A_{11} & A_{13} \\ A_{21} & A_{23} \end{vmatrix} = -(A_{11}A_{23} - A_{21}A_{13})$$

$$\text{Cof } A_{33} = (+1)\begin{vmatrix} A_{11} & A_{12} \\ A_{21} & A_{22} \end{vmatrix} = +(A_{11}A_{22} - A_{21}A_{12})$$

19. ADJOINT MATRIX

Given a matrix $\underline{\mathbf{A}}(m \times m)$, the *adjoint matrix* (i.e., the matrix that is *adjoint* to $\underline{\mathbf{A}}$) is denoted $\hat{\underline{\mathbf{A}}}$ with elements

$$(\hat{\underline{\mathbf{A}}})_{ij} = \text{Cof } A_{ji}$$

Example

$$\underline{\mathbf{A}} = \begin{pmatrix} A_{11} & A_{12} & A_{13} \\ A_{21} & A_{22} & A_{23} \\ A_{31} & A_{32} & A_{33} \end{pmatrix}$$

$$\underline{\hat{\mathbf{A}}} = \begin{bmatrix} \text{Cof } A_{11} & \text{Cof } A_{21} & \text{Cof } A_{31} \\ \text{Cof } A_{12} & \text{Cof } A_{22} & \text{Cof } A_{32} \\ \text{Cof } A_{13} & \text{Cof } A_{23} & \text{Cof } A_{33} \end{bmatrix}$$

$$= \begin{bmatrix} +(A_{22}A_{33} - A_{32}A_{23}) & -(A_{12}A_{33} - A_{32}A_{13}) & +(A_{12}A_{23} - A_{22}A_{13}) \\ -(A_{21}A_{33} - A_{31}A_{23}) & +(A_{11}A_{33} - A_{31}A_{13}) & -(A_{11}A_{23} - A_{21}A_{13}) \\ +(A_{21}A_{32} - A_{31}A_{22}) & -(A_{11}A_{32} - A_{31}A_{12}) & +(A_{11}A_{22} - A_{21}A_{12}) \end{bmatrix}$$

20. EXPANSION OF A DETERMINANT

The *Laplace development* of a determinant is the relation

$$|\underline{\mathbf{A}}(m \times m)| = \sum_{i=1}^{m} A_{ik} \text{ Cof } A_{ik} \qquad \text{for any } k$$

$$= \sum_{k=1}^{m} A_{ik} \text{ Cof } A_{ik} \qquad \text{for any } i$$

Example:

$$\underline{\mathbf{A}} = \begin{pmatrix} A_{11} & A_{12} & A_{13} \\ A_{21} & A_{22} & A_{23} \\ A_{31} & A_{32} & A_{33} \end{pmatrix}$$

$$\begin{aligned} |\underline{\mathbf{A}}| &= \sum_{i=1}^{3} A_{ik} \text{ Cof } A_{ik} = A_{11} \text{ Cof } A_{11} \\ &\quad + A_{21} \text{ Cof } A_{21} + A_{31} \text{ Cof } A_{31} \\ &= A_{11}(A_{22}A_{33} - A_{32}A_{23}) \\ &\quad - A_{21}(A_{12}A_{33} - A_{32}A_{13}) \\ &\quad + A_{31}(A_{12}A_{23} - A_{22}A_{13}) \end{aligned}$$

Also,

$$\begin{aligned} |\underline{\mathbf{A}}| &= \sum_{k=1}^{3} A_{ik} \text{ Cof } A_{ik} = A_{11} \text{ Cof } A_{11} \\ &\quad + A_{12} \text{ Cof } A_{12} + A_{13} \text{ Cof } A_{13} \\ &= A_{11}(A_{22}A_{33} - A_{32}A_{23}) \\ &\quad - A_{12}(A_{21}A_{33} - A_{31}A_{23}) \\ &\quad + A_{13}(A_{21}A_{32} - A_{31}A_{22}) \end{aligned}$$

These agree with each other and with the previous result (Section 17).

21. DERIVATIVE OF A DETERMINANT

For a square matrix $\underline{\mathbf{A}}$,

$$\frac{\partial |\underline{\mathbf{A}}|}{\partial A_{ij}} = \text{Cof } A_{ij}$$

22. SINGULAR MATRIX

If $|\underline{\mathbf{A}}| = 0$, the square matrix $\underline{\mathbf{A}}$ is said to be *singular*. All nonsquare matrices are *defined* to be singular. A square matrix, $\underline{\mathbf{A}}(n \times n)$, is singular if its *rank* $r < n$.

23. INVERSE MATRIX

For a nonsingular matrix $\underline{\mathbf{A}}$, there is one and only one *reciprocal* or *inverse* matrix denoted

$$\underline{\mathbf{A}}^{-1}$$

such that

$$\underline{\mathbf{A}} \cdot \underline{\mathbf{A}}^{-1} = \underline{\mathbf{A}}^{-1} \cdot \underline{\mathbf{A}} = \underline{\mathbf{I}}$$

The elements of the inverse matrix may be written

$$(\underline{\mathbf{A}}^{-1})_{ij} = \frac{\text{Cof } A_{ji}}{|\underline{\mathbf{A}}|} = \frac{(\underline{\hat{\mathbf{A}}})_{ij}}{|\underline{\mathbf{A}}|}$$

This relation is always true but does *not* usually provide the most convenient method for *inverting* a matrix.

If $\underline{\mathbf{A}}$ is a diagonal matrix, then $\underline{\mathbf{A}}^{-1}$ is also diagonal and

$$(\underline{\mathbf{A}}^{-1})_{ii} = \frac{1}{A_{ii}}$$

24. MULTIPLICATION AND INVERSION

If both $\underline{\mathbf{A}}$ and $\underline{\mathbf{B}}$ are nonsingular and conformable, then

$$(\underline{\mathbf{A}} \cdot \underline{\mathbf{B}})^{-1} = \underline{\mathbf{B}}^{-1} \cdot \underline{\mathbf{A}}^{-1}$$

25. MATRIX EQUATIONS

Suppose we have the relation

$$\underline{\mathbf{A}} \cdot \underline{\mathbf{X}} = \underline{\mathbf{B}}$$

and $\underline{\mathbf{A}}$ is nonsingular. We may obtain a *unique* solution for $\underline{\mathbf{X}}$ by multiplying *from the left* by $\underline{\mathbf{A}}^{-1}$:

$$\underline{\mathbf{A}}^{-1} \cdot \underline{\mathbf{A}} \cdot \underline{\mathbf{X}} = \underline{\mathbf{A}}^{-1} \cdot \underline{\mathbf{B}} = \underline{\mathbf{I}} \cdot \underline{\mathbf{X}} = \underline{\mathbf{X}}$$

Also, if

$$\underline{\mathbf{X}}' \cdot \underline{\mathbf{A}} = \underline{\mathbf{B}}$$

we may solve uniquely for $\underline{\mathbf{X}}'$ if $\underline{\mathbf{A}}$ is nonsingular by multiplying *from the right* by $\underline{\mathbf{A}}^{-1}$:

$$\underline{\mathbf{X}}' \cdot \underline{\mathbf{A}} \cdot \underline{\mathbf{A}}^{-1} = \underline{\mathbf{B}} \cdot \underline{\mathbf{A}}^{-1} = \underline{\mathbf{X}}'$$

We note that $\underline{\mathbf{X}}' \neq \underline{\mathbf{X}}$ in general.

26. ORTHOGONAL MATRIX

A square matrix is *orthogonal* if

$$\underline{\mathbf{A}}^T \cdot \underline{\mathbf{A}} = \underline{\mathbf{A}} \cdot \underline{\mathbf{A}}^T = \underline{\mathbf{I}}$$

that is,

$$\underline{\mathbf{A}}^T = \underline{\mathbf{A}}^{-1}$$

For an orthogonal matrix $\underline{\mathbf{A}}$, det $\underline{\mathbf{A}} = \pm 1$.

27. UNITARY MATRIX

For a matrix whose elements are *complex* we define a *unitary* matrix to be one for which

$$\underline{\mathbf{A}}^{-1} = \underline{\mathbf{A}}\dagger$$

28. HOMOGENEOUS FUNCTIONS; BILINEAR FORM

A function $f(x_1, x_2, \ldots, x_n)$ is **homogeneous of degree α** in all of its variables if, for any parameter t,

$$f(tx_1, tx_2, \ldots, tx_n) = t^\alpha f(x_1, x_2, \ldots, x_n)$$

Example

The function

$$\sum_{i=1}^{n} \sum_{j=1}^{n} A_{ij} x_i y_j$$

is an *homogeneous polynomial of the second degree,* since

$$\sum_{i=1}^{n} \sum_{j=1}^{n} A_{ij}(tx_i)(ty_j) = t^2 \sum_{i=1}^{n} \sum_{j=1}^{n} A_{ij} x_i y_j$$

This particular function is called a *bilinear form* and may be written

$$A(\mathbf{x}, \mathbf{y}) = (\mathbf{x}^T)\underline{\mathbf{A}}\mathbf{y} = \sum_{i=1}^{n} \sum_{j=1}^{n} A_{ij} x_i y_j$$

where $\underline{\mathbf{A}}$ is an $(n \times n)$ matrix and $\mathbf{x}$, $\mathbf{y}$ are both $(n \times 1)$ column vectors.

29. QUADRATIC FORM

A special case of a bilinear form is a *quadratic form*:

$$Q = \sum_{i=1}^{n} \sum_{j=1}^{n} A_{ij} x_i x_j \qquad \text{where} \qquad A_{ij} = A_{ji}$$

This may be written

$$A(\mathbf{x}, \mathbf{x}) = \mathbf{x}^T \underline{\mathbf{A}} \mathbf{x} = Q$$

where $\underline{\mathbf{A}} = \underline{\mathbf{A}}^T$ and $\mathbf{x}$ is a $(n \times 1)$ column vector.

If $A(\mathbf{x}, \mathbf{x})$ is positive for all values of the variables, it is called a *positive definite* quadratic form; if it is positive or zero, it is called *semidefinite.* The matrix $\underline{\mathbf{A}}$ is then said to be *positive definite* or *semidefinite,* respectively..

A quadratic form is a *scalar* quantity.

30. DERIVATIVE OF A QUADRATIC FORM

If we differentiate the scalar $A(\mathbf{x}, \mathbf{x})$ with respect to each of the n variables x_i, we get the components of an $(n \times 1)$ column vector:

$$\frac{\partial}{\partial \mathbf{x}} A(\mathbf{x}, \mathbf{x}) = \underline{\mathbf{A}}\mathbf{x} + \underline{\mathbf{A}}^T \mathbf{x} = 2\underline{\mathbf{A}}\mathbf{x}$$

since $\underline{\mathbf{A}} = \underline{\mathbf{A}}^T$.

31. LINEAR EQUATIONS

A set of n simultaneous linear equations of the form

$$\sum_{j=1}^{n} A_{ij} x_j = y_i \qquad i = 1, \ldots, n$$

where the A_{ij} are constants, is said to be *homogeneous* if $y_i = 0$ for all i, and *inhomogeneous* otherwise.

These equations may be written in matrix form

$$\underline{\mathbf{A}}\mathbf{x} = \mathbf{y}$$

where $\underline{\mathbf{A}}(n \times n)$ and $\mathbf{x}$ and $\mathbf{y}$ are $(n \times 1)$ vectors.

The homogeneous equation is

$$\underline{\mathbf{A}}\mathbf{x} = \underline{\mathbf{0}}$$

If we are given $\underline{\mathbf{A}}$ and $\mathbf{y}$ and asked to find $\mathbf{x}$ (i.e., the solutions x_i, $i = 1, \ldots, n$), we must specify that $\mathbf{y} \neq 0$ and $|\underline{\mathbf{A}}| \neq 0$. Then the inverse *exists*, and we get

$$\mathbf{x} = \underline{\mathbf{A}}^{-1}\mathbf{y} = \frac{\hat{\underline{\mathbf{A}}}\mathbf{y}}{|\underline{\mathbf{A}}|}$$

or

$$x_i = \frac{\sum_{j=1}^{n} \text{Cof } A_{ji} y_j}{|\underline{\mathbf{A}}|}$$

This is known as *Cramer's rule.*

The numerator is the expansion of a determinant formed by replacing the column of the coefficients of x_i by the column of the y's.

Example

$$\begin{aligned} A_{11}x_1 + A_{12}x_2 + \cdots + A_{1n}x_n &= y_1 \\ A_{21}x_1 + A_{22}x_2 + \cdots + A_{2n}x_n &= y_2 \\ \vdots \qquad & \quad \vdots \\ A_{n1}x_1 + \quad \cdots \quad + A_{nn}x_n &= y_n \end{aligned}$$

where $|\underline{\mathbf{A}}| \neq 0$ and $\mathbf{y} \neq 0$ (not *all* $y_i = 0$).

$$x_2 = \frac{\begin{vmatrix} A_{11} & y_1 & \cdots & A_{1n} \\ A_{21} & y_2 & \cdots & A_{2n} \\ \vdots & \vdots & & \vdots \\ A_{n1} & y_n & \cdots & A_{nn} \end{vmatrix}}{|\underline{\mathbf{A}}|}$$

Cramer's rule, while always correct for $\mathbf{y} \neq 0$ and $|\underline{\mathbf{A}}| \neq 0$, does not usually represent the most convenient method of solution. Often, explicit substitution is more efficient.

In the homogeneous case

$$\underline{\mathbf{A}}\mathbf{x} = \mathbf{0}$$

where $|\underline{\mathbf{A}}| \neq 0$, we have only the trivial solutions

$$\mathbf{x} = \mathbf{0}$$

that is, $x_i = 0$, $i = 1, \ldots, n$.

We get a nontrivial solution for the homogeneous equations *only* when $|\underline{\mathbf{A}}| = 0$, with Cof $A_{ij} \neq 0$ for at least one i and j. The solution is not determined uniquely in this case. There are infinitely many solutions of the form

$$x_i = c \text{ Cof } A_{ji}$$

where c is an arbitrary constant. In this case, the *ratios* of the unknowns, x_i, may be determined. Equivalently, the x_i are determined to *within a multiplication constant.*

32. LINEAR TRANSFORMATIONS

If we consider a set of linear, inhomogeneous equations relating n variables x_i and m variables y_i:

$$y_j = \sum_{i=1}^{n} A_{ji}x_i \qquad j = 1, \ldots, m$$

we say that a *linear transformation* relates $\mathbf{x}$ to $\mathbf{y}$:

$$\mathbf{y} = \underline{\mathbf{A}}\mathbf{x}$$

The *transformation matrix,* $\underline{\mathbf{A}}(m \times n)$, is square if $m = n$.

Two matrices $\underline{\mathbf{A}}$ and $\underline{\mathbf{B}}$ are said to be *equivalent* if they are related by

$$\underline{\mathbf{B}} = \underline{\mathbf{M}}\underline{\mathbf{A}}\underline{\mathbf{N}}$$

where $\underline{\mathbf{M}}$ and $\underline{\mathbf{N}}$ are nonsingular matrices.

If $\underline{\mathbf{B}} = \underline{\mathbf{M}}^{-1}\underline{\mathbf{A}}\underline{\mathbf{M}}$, we say that $\underline{\mathbf{A}}$ and $\underline{\mathbf{B}}$ are related by a *similarity* transformation.

If the matrix $\underline{\mathbf{M}}$ is orthogonal (i.e., $\underline{\mathbf{M}}^T = \underline{\mathbf{M}}^{-1}$), then

$$\underline{\mathbf{B}} = \underline{\mathbf{M}}^T\underline{\mathbf{A}}\underline{\mathbf{M}} = \underline{\mathbf{M}}^{-1}\underline{\mathbf{A}}\underline{\mathbf{M}}$$

represents a real *orthogonal* transformation.

For complex matrix elements and $\underline{\mathbf{M}}$ a unitary matrix (i.e., $\underline{\mathbf{M}}^\dagger = \underline{\mathbf{M}}^{-1}$), the relation

$$\underline{\mathbf{B}} = \underline{\mathbf{M}}^\dagger\underline{\mathbf{A}}\underline{\mathbf{M}} = \underline{\mathbf{M}}^{-1}\underline{\mathbf{A}}\underline{\mathbf{M}}$$

defines a *unitary* transformation.

In general, $|\underline{\mathbf{M}}^{-1}\underline{\mathbf{A}}\underline{\mathbf{M}}| = |\underline{\mathbf{A}}|$

33. LINEAR TRANSFORMATION OF QUADRATIC FORMS

Consider a linear transformation

$$\mathbf{x}(n \times 1) = \underline{\mathbf{R}}(n \times n)\,\mathbf{y}(n \times 1)$$

The quadratic form

$$\mathbf{x}^T \underline{\mathbf{A}}\mathbf{x}$$

becomes

$$\mathbf{x}^T \underline{\mathbf{A}}\mathbf{x} = \mathbf{y}^T(\underline{\mathbf{R}}^T \underline{\mathbf{A}}\underline{\mathbf{R}})\mathbf{y} = \mathbf{y}^T \underline{\mathbf{B}}\mathbf{y}$$

where

$$\underline{\mathbf{B}} = \underline{\mathbf{R}}^T \underline{\mathbf{A}}\underline{\mathbf{R}}$$

Note that $|\underline{\mathbf{B}}| = |\underline{\mathbf{A}}|$.

For an arbitrary *symmetric* matrix $\underline{\mathbf{A}}$ there always exists an *orthogonal* matrix $\underline{\mathbf{R}}$ such that

$$\underline{\mathbf{B}} = \underline{\mathbf{R}}^T \cdot \underline{\mathbf{A}} \cdot \underline{\mathbf{R}}$$

is *diagonal.* Here

$$|\underline{\mathbf{A}}| = |\underline{\mathbf{B}}| = \prod_{i=1}^{n} B_{ii}$$

34. TRACE

The *trace* of a square matrix $\underline{\mathbf{A}}(n \times n)$ is the *sum of the diagonal elements*:

$$Tr\,\underline{\mathbf{A}} = \sum_{i=1}^{n} A_{ii}$$

The *trace of the product* of two or more matrices is *independent of the order* of multiplication.

$$Tr(\underline{\mathbf{A}}_1 \cdot \underline{\mathbf{A}}_2 \cdot \cdots \cdot \underline{\mathbf{A}}_n) = Tr(\underline{\mathbf{A}}_n \cdot \underline{\mathbf{A}}_{n-1} \cdot \cdots \cdot \underline{\mathbf{A}}_1)$$

The trace is *invariant* under a similarity transformation of the matrix:

$$Tr\,(\underline{\mathbf{M}}^{-1}\underline{\mathbf{A}}\underline{\mathbf{M}}) = Tr\,\underline{\mathbf{A}}$$

Also,

$$Tr(\underline{\mathbf{A}} \pm \underline{\mathbf{B}}) = Tr\,\underline{\mathbf{A}} \pm Tr\,\underline{\mathbf{B}}$$

35. COVARIANCE MATRIX

Suppose we have r variates, x_i, $i = 1, \ldots, r$, which may be expressed as a vector $\mathbf{x}$. We define the $(r \times r)$ **covariance matrix**

$$[\underline{\mathbf{Covar}}(\mathbf{x})] = \begin{bmatrix} \mathrm{Var}(x_1) & \mathrm{Cov}(x_1, x_2) \cdots & \mathrm{Cov}(x_1, x_r) \\ \mathrm{Cov}(x_2, x_1) & \mathrm{Var}(x_1) \quad \cdots & \mathrm{Cov}(x_2, x_r) \\ \vdots & & \vdots \\ \mathrm{Cov}(x_r, x_1) & \cdots & \mathrm{Var}(x_r) \end{bmatrix}$$

Each element

$$\begin{aligned}[\underline{\mathbf{Covar}}(\mathbf{x})]_{ij} &= \mathrm{Cov}(x_i, x_j) \\ &= \langle (x_i - \langle x_i \rangle)(x_j - \langle x_j \rangle) \rangle \\ &= \langle (x_i x_j - \langle x_i \rangle \langle x_j \rangle) \rangle\end{aligned}$$

where $\langle\ \rangle$ denotes *expectation value.*

We may write this as a matrix relation

$$\begin{aligned}\underline{\mathbf{Covar}}(\mathbf{x}) &= \langle (\mathbf{x} - \langle \mathbf{x} \rangle)(\mathbf{x} - \langle \mathbf{x} \rangle)^T \rangle \\ &= E[(\mathbf{x} - \langle \mathbf{x} \rangle)(\mathbf{x} - \langle \mathbf{x} \rangle)^T]\end{aligned}$$

36. LAW OF PROPAGATION OF ERRORS FOR LINEAR TRANSFORMATIONS

Suppose we have the relation

$$\mathbf{y} = \underline{\mathbf{A}}\mathbf{x}$$

relating the r variates x_i, $i = 1, \ldots, r$ to the s variates y_j, $j = 1, \ldots, s$, where we are *given* the $(r \times r)$ covariance matrix of $\mathbf{x}$, $[\underline{\mathbf{Covar}}(\mathbf{x})]$. We wish to *find* the $(s \times s)$-covariance matrix of $\mathbf{y}$.

$$\begin{aligned}[\underline{\mathbf{Covar}}(\mathbf{y})]_{ij} &= \mathrm{Cov}(y_i, y_j) \\ &= \langle (y_i - \langle y_i \rangle)(y_j - \langle y_j \rangle) \rangle \\ &= \langle (y_i y_j - \langle y_i \rangle \langle y_j \rangle) \rangle\end{aligned}$$

by (20.11),

$$\begin{aligned}&= \left\langle \sum_{\rho=1}^{r} A_{i\rho} x_\rho \sum_{\sigma=1}^{r} A_{j\sigma} x_\sigma \right\rangle \\ &\quad - \left\langle \sum_{\rho=1}^{r} A_{i\rho} x_\rho \right\rangle \left\langle \sum_{\sigma=1}^{r} A_{i\sigma} x_\sigma \right\rangle \\ &= \sum_{\rho=1}^{r} \sum_{\sigma=1}^{r} A_{i\rho} A_{i\sigma} \{\langle x_\rho x_\sigma \rangle \\ &\qquad - \langle x_\rho \rangle \langle x_\sigma \rangle\}\end{aligned}$$

$$= \sum_{\rho=1}^{r} \sum_{\sigma=1}^{r} (\underline{\mathbf{A}})_{i\rho} (\underline{\mathbf{A}}^T)_{\sigma i} \operatorname{Cov}(x_\rho, x_\sigma)$$

$$= \sum_{\rho=1}^{r} \sum_{\sigma=1}^{r} [\underline{\mathbf{A}}(s \times r)]_{i\rho} \times [\underline{\mathbf{Covar}}(\mathbf{x})]_{\rho\sigma} \times [\underline{\mathbf{A}}^T(r \times s)]_{\sigma i}$$

i.e.,

$$[\underline{\mathbf{Covar}}(\mathbf{y})] = \underline{\mathbf{A}}[\underline{\mathbf{Covar}}(\mathbf{x})]\underline{\mathbf{A}}^T$$

This is called the "*law of propagation of errors.*"

37. EXPECTATION VALUE OF A QUADRATIC FORM

Given the quadratic form

$$A(\mathbf{x}, \mathbf{x}) = \mathbf{x}^T \underline{\mathbf{A}} \mathbf{x} = Q$$

where $\mathbf{x}$ is an $(r \times 1)$-vector of *uncorrelated* variates with constant variance, that is, the $r \times r$ *covariance matrix* is

$$[\underline{\mathbf{Covar}}(\mathbf{x})] = \sigma^2 \underline{\mathbf{I}}(r \times r)$$

and *expectation values*

$$\langle \mathbf{x} \rangle = E\mathbf{x}$$

then the **expectation value**

$$EQ = \sigma^2 \, Tr \, \underline{\mathbf{A}} + \langle \mathbf{x} \rangle^T \underline{\mathbf{A}} \langle \mathbf{x} \rangle$$

This is straightforwardly shown as follows.

$$EQ = \langle \mathbf{x}^T \underline{\mathbf{A}} \mathbf{x} \rangle = \sum_{\rho=1}^{r} \sum_{\sigma=1}^{r} A_{\rho\sigma} \langle x_\rho x_\sigma \rangle$$

$$= \sum_{\rho=1}^{r} \sum_{\sigma=1}^{r} A_{\rho\sigma} \langle (x_\rho x_\sigma - \langle x_\rho \rangle \langle x_\sigma \rangle) \rangle + \sum_{\rho=1}^{r} \sum_{\sigma=1}^{r} A_{\rho\sigma} \langle x_\rho \rangle \langle x_\sigma \rangle$$

$$= \sum_{\rho=1}^{r} \sum_{\sigma=1}^{r} A_{\rho\sigma} \operatorname{Cov}(x_\rho, x_\sigma) + \sum_{\rho=1}^{r} \sum_{\sigma=1}^{r} \langle x_\rho \rangle A_{\rho\sigma} \langle x_\sigma \rangle$$

The matrix $[\underline{\mathbf{Covar}}(\mathbf{x})]$ is *diagonal* by hypothesis, that is,

$$\operatorname{Cov}(x_\rho, x_\sigma) = \sigma^2 \delta_{\rho\sigma}$$

where the *Kronecker delta* is defined

$$\delta_{\rho\sigma} = 0 \qquad \rho \neq \sigma$$
$$= 1 \qquad \rho = \sigma$$

$$\therefore EQ = \sum_{t=1}^{r} A_{ii} \sigma^2 + \langle \mathbf{x} \rangle^T \underline{\mathbf{A}} \langle \mathbf{x} \rangle = \sigma^2 \, Tr \, \underline{\mathbf{A}} + \langle \mathbf{x} \rangle^T \underline{\mathbf{A}} \langle \mathbf{x} \rangle \qquad \text{Q.E.D.}$$

38. EIGENVALUE EQUATION OF A MATRIX

Given a square matrix $\underline{\mathbf{A}}(n \times n)$, the unit matrix $\underline{\mathbf{I}}(n \times n)$ and a scalar λ, we define the *eigenvalue equation* of the matrix to be

$$|\underline{\mathbf{A}} - \lambda \underline{\mathbf{I}}| = 0$$

This equation has n roots or solutions for λ called the *eigenvalues* λ_i, $i = 1, \ldots, n$.

If

$$\mathbf{B} = \underline{\mathbf{M}}^{-1} \underline{\mathbf{A}} \underline{\mathbf{M}}$$

then

$$\underline{\mathbf{B}} - \lambda \underline{\mathbf{I}} = \underline{\mathbf{M}}^{-1} \underline{\mathbf{A}} \underline{\mathbf{M}} - \lambda \underline{\mathbf{I}} = \underline{\mathbf{M}}^{-1} [\underline{\mathbf{A}} - \lambda \underline{\mathbf{I}}] \underline{\mathbf{M}}$$

and

$$|\underline{\mathbf{B}} - \lambda \underline{\mathbf{I}}| = |\underline{\mathbf{M}}^{-1}| \, |\underline{\mathbf{A}} - \lambda \underline{\mathbf{I}}| \, |\underline{\mathbf{M}}| = |\underline{\mathbf{A}} - \lambda \underline{\mathbf{I}}|$$

which shows that the eigenvalues of $\underline{\mathbf{B}}$ and $\underline{\mathbf{A}}$ are the *same* (solutions of the same eigenvalue equation).

If $\underline{\mathbf{D}} = \underline{\mathbf{M}}^{-1} \underline{\mathbf{A}} \underline{\mathbf{M}}$ is diagonal, the eigenvalue equation is especially simple,

$$\begin{vmatrix} (D_{11} - \lambda) & 0 & & \\ 0 & (D_{22} - \lambda) & & \\ & & \ddots & \\ & & & (D_{nn} - \lambda) \end{vmatrix} = (D_{11} - \lambda)(D_{22} - \lambda) \cdots (D_{nn} - \lambda) = 0$$

which has the roots

$$\lambda = D_{11}, D_{22}, \ldots, D_{nn}$$

39. EIGENVECTORS OF A MATRIX AND DIAGONALIZATION

The equation

$$\underline{\mathbf{A}} \mathbf{x} = \lambda \mathbf{x}$$

where λ is a scalar quantity, defines the *eigenvectors* $\mathbf{x}$ of the matrix $\underline{\mathbf{A}}$. There are n eigenvectors $\mathbf{x}$, n eigenvalues λ for

$\underline{\mathbf{A}}(n \times n)$. The only effect of the matrix $\underline{\mathbf{A}}$ on the eigenvector $\mathbf{x}$ is to multiply it by a scalar.

To solve for $\mathbf{x}$ we note that

$$(\underline{\mathbf{A}} - \lambda \underline{\mathbf{I}})\mathbf{x} = 0$$

is an homogeneous set of equations. We get a nontrivial solution only if

$$|\underline{\mathbf{A}} - \lambda \underline{\mathbf{I}}| = 0$$

which is the eigenvalue equation.

We find the λ_i, $i = 1, \ldots, n$, and construct a matrix $\underline{\mathbf{X}}$ such that

$$\underline{\mathbf{X}}^{-1}\underline{\mathbf{A}}\underline{\mathbf{X}} = \underline{\mathbf{D}} \qquad \text{where } \underline{\mathbf{D}} \text{ is diagonal}$$

When the eigenvalues λ_i are all different, we say they are *nondegenerate.* We select each λ_i in turn, obtaining each time the n linear homogeneous equations

$$(\underline{\mathbf{A}} - \lambda \underline{\mathbf{I}})\mathbf{x} = 0$$

which we may only solve for $\mathbf{x}$ to within a multiplicative constant (we may use this to *normalize* the eigenvectors). We thus have

$$\mathbf{x}_i = \begin{bmatrix} x_{1i} \\ x_{2i} \\ \cdot \\ \cdot \\ \cdot \\ x_{ni} \end{bmatrix}$$

and we form the matrix

$$(\underline{\mathbf{X}})_{ij} = x_{ij}$$

This satisfies the equation

$$\underline{\mathbf{X}}^{-1}\underline{\mathbf{A}}\underline{\mathbf{X}} = \underline{\mathbf{D}}$$

where

$$(\underline{\mathbf{D}})_{ii} = \lambda_i$$

40. NUMERICAL METHODS FOR MATRIX INVERSION

The problem of *inverting* an arbitrary matrix is one of the most important, and most difficult, in the whole area of computation. We have already pointed out that the use of the *Laplace development of a determinant* is *not* usually the best method to use for calculating the inverse of a matrix. The methods are many and varied, and no one method is uniformly best for all situations. The book of Hildebrand already cited contains a very good discussion of the *Crout method* which is widely used.

Most digital computer installations have a selection of matrix inversion routines based on various methods. The choice of which to use depends not only on the *particular* problem but the *availability* of routines and the *familiarity* that the user has with each.

Two good references for the user interested in understanding the *bases* of the techniques widely used are:

Householder, A. S., *Principles of Numerical Analysis*, McGraw-Hill, New York, 1953.

Ralston, Anthony and Herbert S. Wilf (eds.), *Mathematical Methods for Digital Computers*, Wiley, New York, 1960.

APPENDIX I

UNITS AND STANDARDS OF WEIGHTS AND MEASURES

Since all empirical science involves measurement, the manner of presenting the *results* of measurement is quite important.

The basic dimensions are **mass** (M), **length** (L) and **time** (T). We shall present a summary of the version of the metric system that promises to be the legal standard in all "developed" nations. The metric system was legalized by the United States in 1866, but its use will not be obligatory at least until 1984.

Since the "English" system of units remains the system of necessity in the English-speaking countries for years and will continue to have vestigial importance for decades, we summarize its features.

Finally, since measurements are used by *everyone*, we summarize some units that have *specific application* or *historical* or *literary* interest.

THE INTERNATIONAL SYSTEM OF UNITS (SYSTEME INTERNATIONALE, SI)

This is a rational system based on *decimal multiples* of the units meter, kilogram, second, ampere, kelvin, and candela. It is the current version of the *metric system.*

The *meter* is the base unit of metric **length** while the metric measures of *surface* and *volume* are the squares and cubes respectively of the meter and of its decimal fractions and multiples.

The *kilogram* is the base unit of metric **mass** with the derived units of **density** being decimally related to kilograms per cubic meter (kg/m^3).

The *second* is the base unit of metric **time** with derived units such as **velocity** and **acceleration** being expressed in meters per second (m/sec) and meters per second per second (m/sec^2), respectively.

We summarize the SI units below. Commonly used units are italicized with their alternative name and abbreviation in parentheses.

SI PREFIXES	MULT.	CORRES. ABBREVIATIONS
Tera	10^{12}	T
Giga	10^9	G
Mega	10^6	M
Myria	10^4	
Kilo	10^3	k
Hecto	10^2	h
Deca	10	da
Deci	10^{-1}	d
Centi	10^{-2}	c

(continued)

SI PREFIXES	MULT.	CORRES. ABBREVIATIONS
Milli	10^{-3}	m
Micro	10^{-6}	μ
Nano	10^{-9}	n
Pico	10^{-12}	p
Femto	10^{-15}	f
Atto	10^{-18}	a

In older usage, all prefixes denoting positive powers of 10 were abbreviated with upper-case letters: for example, kilo (K), hecto (H), and deca (D).

SI MEASURE OF LENGTH

Metric units of **length** are based on the *meter* (m).

Myriameter	$= 10^4$ m
Kilometer	$= 10^3$ m = (km)
Hectometer	$= 10^2$ m
Decameter	= 10 m
Meter	= 1 m = (m)
Decimeter	$= 10^{-1}$ m
Centimeter	$= 10^{-2}$ m = (cm)
Millimeter	$= 10^{-3}$ m = (mm)
Micrometer	$= 10^{-6}$ m = (μ, micron, μm)
Decinanometer	$= 10^{-10}$ m = (Å, angstrom)
Femtometer	$= 10^{-15}$ m = (F, *fermi*, fm)

SI MEASURE OF AREA

Metric units of **area** are based on the *square meter* [(meter)^2, m^2] and are sometimes referred to the *are* (100 m^2).

Square kilometer	$= 10^2$ ha $= 10^4$ a $= 10^6$ m^2 = (km^2)
Hectare	$= 10^2$ ares $= 10^4$ m^2 = (ha)
Are	$= 10^2$ m^2 = (a)
Square meter (*centiare*)	= 1 m^2 = (m^2, ca)
Square decimeter	$= 10^{-2}$ m^2 = (dm^2)
Square centimeter (cm^2)	$= 10^{-4}$ m^2
Square millimeter (mm^2)	$= 10^{-6}$ m^2
Barn (b)	$= 10^{-24}$ cm^2 $= 10^{-28}$ m^2
Millibarn (mb)	$= 10^{-27}$ cm^2 $= 10^{-31}$ m^2
Microbarn (μb)	$= 10^{-30}$ cm^2 $= 10^{-34}$ m^2

SI MEASURE OF VOLUME

Metric units of **volume** are based on the *cubic meter* [(meter)^3, m^3], sometimes called the *stere* when applied to **solid measures** such as timber. The principal unit of **capacity** is the *liter* although one can always employ cubic kilometers (km^3) or cubic fermis (F^3) if the context proves that to be convenient.

Decastere	= 10 m^3	= (10 steres)
Cubic meter	= 1 m^3	= (m^3, stere, *kiloliter*, kl or Kl)
Hectoliter	$= 10^{-1}$ m^3	= (decistere, hl or Hl)
Decaliter	$= 10^{-2}$ m^3	= (da 1 or D1, *dekaliter*)
Liter	$= 10^{-3}$ m^3	= (l)
Deciliter	$= 10^{-4}$ m^3	= (dl)
Centiliter	$= 10^{-5}$ m^3	= (cl)
Milliliter	$= 10^{-6}$ m^3	= (ml, *cubic centimeter*, cc, cm^3)

SI MEASURE OF MASS

Metric units of **mass** are called the *kilogram* (kg or Kg) or the *gram* (g). Although not now defined this way, the *gram* is very closely equal to the mass of 1 milliliter of pure water at 4°C temperature and 760 mm of mercury pressure, the *kilogram* is the **mass** of 1 liter of water, and the (metric) *ton* (T) is the mass of 1 cubic meter of water.

Ton $= 10^3$ kg = (T)
Quintal $= 10^2$ kg

Myriagram = 10 kg
Kilogram = 1 kg = (kg or Kg, *kilo*)
Hectogram $= 10^{-1}$ kg = (hg or Hg)
Decagram $= 10^{-2}$ kg = (da g or Dg)
Gram $= 10^{-3}$ kg = (g)
Decigram $= 10^{-4}$ kg = (dg)
Centigram $= 10^{-5}$ kg = (cg)
Milligram $= 10^{-6}$ kg = (mg)
Microgram $= 10^{-9}$ kg = (μg)

In addition, the *carat* is defined to be exactly 2 decigrams in international usage.

The *atomic mass unit* (amu) is defined to be 1/12 the mass of the atom ^{12}C.

$$1 \text{ amu} = (1.660531 \pm 0.000011) \times 10^{-27} \text{ kg}$$

SI MEASURE OF TIME

The standard unit of **time** is the *second* (sec) defined since 1964 as 9.192631770×10^9 cycles of the frequency associated with a transition between two hyperfine levels of the cesium 133 (Cs^{133}) isotope. As usual, there are 60 seconds in a *minute* and 60 minutes in an *hour*.

Until 1956 the second was defined in terms of the

$$\textit{Mean solar day} = 24 \text{ hours} = 8.64 \times 10^4 \text{ sec}$$

The solar day is tied to the earth's *rotation with respect to the sun.* The period between times when the sun is at its highest point in the sky is called the *apparent solar day.* Since the day is not always the same length, the *mean solar day* was defined to be the *average* length of the apparent solar days in a year. However, the rate of rotation of the earth is slowing; therefore it was decided in 1956 to specify the second in terms of the mean solar day *for the year* 1900. The *mean solar year* (also called the **equinoctial** year since measured between one vernal equinox and the next) defined in terms of the earth's position with respect to the sun is

$$\textit{Mean solar year} = 365 \text{ days, } 5 \text{ hours, } 48 \text{ minutes, } 45.5 \text{ seconds} = 3.15596255 \times 10^7 \text{ sec}$$

Since the earth is moving in its orbit while it is rotating on its axis, it takes a somewhat different amount of time to complete one rotation measured with respect to the fixed stars, that is, time measured by observing the successive passages of a star across a line running north and south that passes directly overhead. This interval, called a *sidereal day,* is slightly smaller than a solar day because the motion of the earth in its orbit makes it necessary for the earth to make a little more than one complete rotation between one passage of the sun directly overhead and the next such passage. The time interval with respect to the fixed stars is precisely defined as the time interval between two successive transits of the March equinox over the upper meridian of a place and is called the

$$\textit{Sidereal day} = 23 \text{ hours, } 56 \text{ minutes, } 4.09 \text{ seconds} = 8.616409 \times 10^4 \text{ seconds}$$

The time in which the earth completes one revolution in its orbit around the sun measured with respect to the fixed stars is called the

$$\textit{Sidereal year} = 365 \text{ days, } 6 \text{ hours, } 9 \text{ minutes, } 9.54 \text{ seconds} = 3.155814954 \times 10^7 \text{ seconds}$$

Commonly used units of time are the:

Millisecond $= 10^{-3}$ sec = (msec)
Microsecond $= 10^{-6}$ sec = (μ sec)
Nanosecond $= 10^{-9}$ sec = (ns, millimicro-second, mμ sec)
Picosecond $= 10^{-12}$ sec = (psec)

SI MEASURE OF FORCE

As mentioned in Part I, **force** is not a fundamental dimension but is related to *M*, *L*, *T* by $F \overset{D}{=} MLT^{-2}$. The SI unit of **force** is the *newton* (N), which is the force that, when applied to a mass of 1 kilogram for 1 second, will give to the mass a speed of 1 meter per second, that is, a force of 1 newton will produce on a mass of 1 kilogram an acceleration of 1 meter per second per second. Related to the newton is the CGS system unit, the *dyne*. The dyne is the force that produces an acceleration of 1 cm per second per second on a mass of 1 gram. Clearly,

$$10^5 \text{ dynes} = 1 \text{ newton} = 1 \text{ kg-m-sec}^{-2}$$

Weight is related to mass through the constant that expresses the *acceleration of gravity*. This varies from place to place but when used for units it is taken to mean

$$g \equiv 9.80665 \text{ m/sec}^2 = 980.665 \text{ cm/sec}^2$$

SI MEASURE OF PRESSURE

Pressure is a quantity derived from force and is thus a secondary (or even tertiary) quantity. Its dimensions are those of **force per unit area** and are *newtons*/m^2 or *dynes*/cm^2. It is convenient to describe a *normalized atmosphere of pressure* to be that which is exerted by a column of mercury (density 13.5951 g/cm^3) exactly 76 cm high where the acceleration of gravity is as above:

$$\begin{aligned} 1 \text{ atmosphere} &= 1.013246 \times 10^6 \text{ dyne/cm}^2 \\ &= 1.013246 \times 10^5 \text{ newtons/m}^2 \\ &= 760 \text{ mm of } Hg = 760 \textit{ torr} \\ &= 1 \textit{ bar} = 1 \textit{ atm} \end{aligned}$$

SI MEASURE OF TEMPERATURE

The *thermodynamic* or Kelvin scale of **temperature** used in SI has a defined fixed point at one atmosphere of pressure called the *triple point* of water, where the solid, liquid, and vapor states exist together in equilibrium. The temperature at this point is defined to be 273.16 K (273.16 *kelvins*) or 0.01 *degrees Celsius* (formerly called *Centigrade*) written 0.01°C. At normal pressure, the ice point of water is at 0°C. and the boiling point of water is at 100°C.

The **International Practical Temperature Scale of 1968** provides that the *definition* of the Kelvin and Celsius scales is in terms of the following fixed points in addition to the triple point of water:

Triple point of equilibrium hydrogen	13.81 K	(− 259.34°C.)
Equilibrium between liquid and vapor phases of equilibrium hydrogen at 33330.6 N/m^2	17.042 K	(− 256.108°C.)
Boiling point of equilibrium hydrogen	20.28 K	(− 252.87°C.)
Boiling point of neon	27.102 K	(− 246.048°C.)
Triple point of oxygen	54.361 K	(− 218.789°C.)
Boiling point of oxygen	90.188 K	(− 182.962°C.)
Boiling point of water	273.16 K	(100°C.)

Freezing point of zinc	692.73 K	(419.58°C.)
Freezing point of silver	1235.08 K	(961.93°C.)
Freezing point of gold	1337.58 K	(1064.43°C.)

Formally, the Celsius scale is defined in terms of the Kelvin scale and is related to it only by a displacement of the zero point:

$$\text{Temp (°C.)} = \text{Temp (K)} + 273.15$$

The Fahrenheit scale is defined in terms of the Celsius scale and is different in scale unit as well as zero point:

$$\text{Temp (°F)} = 1.8\ \text{Temp (°C.)} + 32$$

This places the triple point of water at 32.02°F.

SI MEASURE OF FREQUENCY

The unit of **frequency** is the *hertz* (Hz), which is one cycle per second.

SI MEASURE OF WORK AND ENERGY

Being a derived quantity, **work** and **energy** have dimensions

$$E \overset{D}{=} W \overset{D}{=} F \times L = MLT^{-2} = ML^2T^{-2}$$

The SI unit of energy is the

Joule (J) = 1 newton-meter = 1 N-m

Anticipating the definition of *watt*, we also note that

$$\text{Joule} = 2.77\dot{7} \times 10^{-7}\ \text{kilowatt-hours}$$

A common unit in the cgs system is the

Erg = 1 dyne-cm = 10^{-7} joules

Another unit in common usage is the

Electron-volt (eV) = $1.6021917 \times 10^{-12}$ erg
= $1.6021917 \times 10^{-19}$ joules

Related units are the
Kiloelectron volt = 10^3 eV = (keV)
Million electron volts = 10^6 eV = (MeV)
Billion electron volts = 10^9 eV = (BeV, *GeV*, gigaelectron volt)
Teraelectron volt = 10^{12} eV = (TeV)

HEAT UNIT

Since heat is a form of energy, it has the same dimensions.

1 *calorie* = 4.18400* joules = (gram-calorie, cal)
1 gross calorie = 10^3 cal = [calorie (dietetic), kilogram-calorie, k cal]

*This is the definition of the National Bureau of Standards. Formally, the calorie is defined as the quantity of heat that will raise the temperature of one gram of water (initially at 15°C.) one degree Celsius.

$$1\ \text{international calorie} = \frac{1}{859.858}\ \text{watt-hr} = 4.18674\ \text{joules}$$

SI MEASURE OF POWER

The unit of power has dimensions of energy per unit time and is the

Watt (W) = 1 joule per second

SI ELECTRICAL MEASURE

The primary SI unit is that of **current** called the *ampere* (A), defined as the magnitude of the current that, when flowing through each of 2 long parallel wires separated by one meter in free space, results in a force between the two wires of 2×10^{-7} N for each meter of length. A secondary unit is that of **potential:**

$$\textit{Volt} = 1 \text{ watt per ampere}$$
$$\text{V} = 1 \text{ W/A}$$

The unit of **resistance** is

$$\textit{Ohm}\ (\Omega) = 1 \text{ volt/ampere}$$

SI MEASURE OF LUMINOUS INTENSITY

The last *primary* SI unit is that of **luminous intensity** called the *candela* (cd). The candela is defined as the luminous intensity of 1/600,000 of a square meter of a radiating cavity at the temperature of freezing platinum (Pt at 2042 K) under a pressure of 101325 N/m^2. The candela refers to the intensity of a radiating source of energy.

SI MEASURE OF LIGHT FLUX

The SI unit of **light flux** (light energy per unit solid angle per unit time) is the *lumen* (lm) defined as follows:

A source with an intensity of 1 candela in all directions radiates a flux of 4π lumens.

Thus 1 candela = 1 lumen per steradian:

$$1 \text{ cd} = 1 \text{ lm/Sr}$$

The unit of **illumination** measures the light flux falling on a surface per square meter and is the *lux.*

$$1 \text{ lux} = 1 \text{ lumen/m}^2$$
$$\text{lx} = \text{lm/m}^2$$

At a wavelength of 5550 Å a luminous flux of 680 lm is equivalent to 1 watt.

SI STANDARD OF LENGTH

Since 1960, the Standard International definition of the *meter* is 1.65076373×10^6 wavelengths in vacuum of the red-orange radiation emitted by the krypton − 86 (Kr^{86}) atom undergoing a transition between the $5d_5$ and $2P_{10}$ levels where the krypton atoms are maintained at the triple point temperature of nitrogen (−210°C.).

In October, 1973 the Comité International des Poids et Mesures (CIPM) adopted recommended values for the wavelength of light from two stabilized lasers. This is preparatory to a possible later recommendation for the redefinition of the standard meter. A committee of the CIPM (the Comité Consultatif pour la Definition du Metre) recommended (and the CIPM adopted) the following values:

1. $3.39223140 \times 10^{-6}$ m for the wavelength emitted by a helium-neon laser stabilized to the $P(7)$ line in the $\nu 3$ band of the methane molecule.
2. $6.32991399 \times 10^{-7}$ m for the wavelength emitted by a helium-neon laser stabilized to the "i" component of the $R(127)$ line in the 11-5 band of I^{127}.

At the same meeting in October 1973, the following value for the speed of light was recommended and adopted

$$c = 2.99792458 \times 10^8 \text{ m sec}^{-1}.$$

SI STANDARD OF MASS

The standard is the mass of the International Prototype Kilogram No. 1 kept at the International Bureau of Weights and Measures at Sevres, France. This standard kilogram is measured to have the same mass as 1.000027×10^{-3} m^3 of pure water at 4°C and 1 atmosphere pressure.

ASTRONOMICAL UNITS OF LENGTH

1 astronomical unit = 1.49501201 $\times 10^8$ kilometers

This represents the mean distance between the earth and the sun.

1 light-second = 2.997924562 $\times 10^8$ meters (new value as of December 1972)

1 light-year = 9.460895 $\times 10^{15}$ meters (sidereal year)

1 parsec = 206,265 astronomical units
= 3.08369×10^{16} meters
= 3.25940 light year

This represents the distance at which the annual parallax of a star is 1″ (1 second of arc).

U.S. CUSTOMARY SYSTEM OF UNITS

The *U.S. Customary System* is in general usage in the United States and is likely to continue to be so at least until 1984. It is historically related to, but now different from, the *British Imperial System of Units*, the system in use in Great Britain and the Commonwealth countries until 1975, at which time conversion to the metric system is expected to be complete.

Since 1866 the metric system has been legal in the United States, but the "Law of 1866" was permissive and not obligatory, and a congeries of unit systems have continued to flourish, each legal in its restricted domain. As we shall see when we discuss United States standards, the legal definition of units is in terms of the international system; therefore we shall give metric equivalents wherever possible.

U.S. MEASURES OF LENGTH

The entire U.S. system is referred to the *yard* according to formal definition. We shall, however, be more interested in summarizing the whole system and shall start with the smallest unit in common usage and build up to the largest, referring to whichever unit is most common.

Here is the most general system in use in the United States.

1 mil	= 10^{-3} in.	= 2.54×10^{-5} m (exactly)
1 inch (in., ″)	= 1 in. (in., ″)	= 2.54×10^{-2} m (exactly)
1 foot (ft, ′)	= 12 in. (ft, ′)	= 3.048×10^{-1} m (exactly)
1 yard (yd)	= 3 ft	= 0.9144 m (exactly)
1 rod (rd)	= $5\frac{1}{2}$ yd	= 5.0292 m (exactly)
	= (perch, p; pole, p)	
1 furlong (fur)	= 40 rd = 1/8 mile	= 201.168 m (exactly)
1 statute mile (mi)	= 5280 ft	= 1.609344 km (exactly)
1 league	= 3 mi	= 4.828032 km (exactly)

NAUTICAL MEASURE OF LENGTH

A system in usage by geographers and those concerned with sea or air travel.

1 span	= 9 in.		
1 fathom (fm)	= 8 spans	= 6 ft	= 1.8288 m
1 cable's length (U.S.)	= 120 fms	= 720 ft	
1 cable's length (British)	= 100 fms	= 608 ft	
1 British nautical mile	= 10 cable's lengths (British)		
		= 6080 ft	
1 U.S. nautical mile	= 1.1516 mi	= 6080.20 ft	= 1853.248 m
1 nautical mile (U.S. Navy)	= 2000 yd	= 6000 ft	

1 league (U.S.) = 3 U.S. nautical miles = 3.45466 statute miles
1 degree (U.S.) = 60 U.S. nautical miles = 69.0931818 miles

In 1954, the United States officially adopted the International Nautical Mile, which is defined to be exactly 1852 meters. In 1954, this led to an equivalence of 1 nautical mile = $6076.1033\dot{3}$ (repeated 3's). This was later revised because of a change in the definition of the foot.

1 international nautical mile
= 6076.11549 ft
= 1852 meters (exactly)

A specialized system of linear measure used in surveying and a variation on this used in some engineering work are known as Gunter's chain and the Engineer's Chain, respectively.

SURVEYOR'S MEASURE OR GUNTER'S CHAIN

1 link (li)	= 7.92 in.		
1 rod (rd)	= 25 li = 16.5 ft = 5.5 yd		
1 chain (ch)	= 4 rd	= 100 li	= 66 ft
1 furlong (fur)	= 10 ch		= 660 ft
1 mile	= 8 fur		= 5280 ft

ENGINEER'S CHAIN

1 link (li)	= 1 ft	
1 chain (ch)	= 100 li	= 100 ft
1 mile (mi)	= 52.8 ch	= 5280 ft

U.S. MEASURE OF AREA

General system

1 square inch (sq in., in.^{2})		= 6.4516 cm^2
1 square foot (sq ft, ft^2)	= 144 sq in.	= 9.290304×10^{-2} m^2
1 square yard (sq yd, yd^2)	= 9 sq ft	= 8.3162736×10^{-1} m^2
1 square rod (sq rd, rd^2)	= $3\frac{1}{2}$ sq yd	= (1 pole)
		= 25.5018845 m^2
1 acre (A)	= 160 sq rd	= 0.40468564224 hectares
1 square mile (sq mi)	= 640 A	= 2.58998811 km^2

Surveyor's Measure of Area

1 square link (sq li)	= 62.73 sq in.	
1 square pole (sq p)	= 625 sq li	= 30.25 sq yd
1 square chain (sq ch)	= 16 sq p	= 484 sq yd
1 acre (A)	= 10 sq ch	= 4840 sq yd
1 section (sec)	= 640 A	= 1 sq mi
1 township (tp)	= 36 sec	= 36 sq mi

There are other *special* units of **area** measure.

Electrical wire

1 circular mil (cir mil) = 7.85×10^{-7} sq in.
(This is the area of a circular cross-section 1 mil in diameter.)
1 MCM = 10^3 cir mil = 7.85×10^{-4} sq in.
1 circular inch = 10^3 MCM = 0.785 sq in.

U.S. MEASURES OF VOLUME

General system

1 cubic inch (cu in., in.^{3})		= 16.387064 cm^3
1 cubic foot (cu ft, ft^3)	= 1728 cu in.	= $2.8316846592 \times 10^{-2}$ m^3
1 cubic yard (cu yd)	= 27 cu ft	= 0.764554857984 m^3

Although **volume** and **capacity** are the same quantity, the U.S. and British Customary systems distinguish two kinds of capacity, *liquid* and *dry*. The distinction is nonexistent in the metric system, and we shall give equivalents wherever convenient. **Exact** figures are in **bold face**.

U.S. LIQUID MEASURE

1 gill (gi)		= **7.21875** cu in.	= **0.11829411825** liters
1 pint (pt)	= 4 gi	= **28.875** cu in.	= **0.473176473** liters
1 quart (qt)	= 2 pt	= **57.75** cu in.	= **0.946352946** liters
1 gallon (gal)	= 4 qt	= **231** cu in.	= **3.785411784** liters
1 barrel (bbl)	= $31\frac{1}{2}$ gal	= **7276.5** cu in.	= 119.2404712 liters
1 barrel of petroleum (bbl)	= 42 gal	= **9702** cu in.	= 158.9872949 liters

British Imperial Liquid Measure

1 imperial quart = 1.2009 U.S. quart = 69.3185 cu in.
1 imperial gallon = 1.2009 U.S. gallon = 277.420 cu in.

U.S. APOTHECARIES' FLUID MEASURE

1 minim or drop (min)	= 3.75977×10^{-3} cu in.	= 0.06161152 milliliters
1 fluid dram (fl. dr)	= 60 min	= 3.6966912 milliliters
1 fluid ounce (fl. oz, liquid ounce)	= 8 fl. dr	= 1.8046875 cu in.
		= 29.5735296 milliliters
1 pint (O.) = 16 fl. oz.	= **28.875** cu in.	= **0.473176473** liters
1 gallon (C.) = 8 O.	= **231** cu in.	= **3.785411784** liters

U.S. COMMON CAPACITY MEASURE

1 teaspoon = 1/6 fluid ounce
1 tablespoon = 3 teaspoons = 1/2 fl. oz = 0.90234375 cu in. = 14.786765 milliliters
1 cup = 16 tablespoons = 8 fl. oz = 14.437500 cu in. = **0.2365882365** liters
1 pint = 2 cups = **28.875** cu in. = **0.473176473** liters
1 quart = 2 pints
1 gallon = 4 quarts

U.S. DRY CAPACITY MEASURE

1 pint (pt)	= 33.600312 cu in.		= 0.550610 liters
1 quart (qt)	= 2 pt	= **67.200625** cu in.	= 1.101221 liters
1 peck (pk)	= 8 quarts	= **537.605** cu in.	= 8.8098675 liters
1 bushel (bl)	= 4 pecks	= **2150.42** cu in.	= 35.23907 liters
1 dry barrel (bbl)	= 7056 cu in.		= 115.62 liters

U.S. SHIPPING MEASURE

1 barrel bulk = 5 cu ft = **8640** cu in.
1 freight ton = 8 barrels bulk = (shipping ton, measurement ton)
= 40 cu ft
1 displacement ton = 35 cu ft
1 register ton = 100 cu ft

U.S. WOOD MEASURE

1 board foot (bd ft) = 144 cu in. (in the form 1′ × 1′ × 1″ = 0.00236 m^3
1 cord foot (cd ft) = 16 cu ft (in the form 4′ × 4′ × 1′) = 0.4528 m^3
1 cord (cs) = 8 cu ft = 128 cu ft (in the form 4′ × 4′ × 8′) = 3.6224 m^3
= 3.6224 stere

U.S. MEASURES OF MASS

We shall ignore the distinction between mass and weight here, since it is not material to the comparison of the systems in usage in the United States: *avoirdupois*, *troy* and *apothecaries*. The important point of reference is that the *grain* is common to all three systems and is identical. **Exact** figures are in **boldface**.

Avoirdupois system

1 grain (gr) = **0.06479891** g
1 dram (dr) = 27.34375 gr = 1.7718452 g
1 ounce (oz) = 16 dr = 437.5 gr = **28.349523125** g
1 pound (lb) = 16 oz = 7000 gr = **0.45359237** kg
1 short hundredweight (cwt, hundredweight) = 100 lb = **45.359237** kg
1 short ton (s.t.) = 2000 lb = **907.18474** kg
= **0.90718474** T

British system

1 stone (st) = 14 lb	= 6.350293 kg
1 hundredweight (cwt, quintal) = 112 lb	= 50.802345 kg
1 long ton (l.t) = 20 cwt = 2240 lb	= 1016.0469088 kg
	= 1.0160469088 T

Troy system

1 grain (gr)	= **0.06479891** g
1 carat (c) = 3.086 gr	= **0.2** g
1 pennyweight (dwt) = 24 gr	= **1.55517384** g
1 ounce (oz t) = 20 dwt = 480 gr	= **31.1034768** g
1 pound (lb t) = 12 oz t = 5760 gr	= **373.2417216** g

Apothecaries' system

1 grain (gr)		= **0.06479891** g
1 scruple (s. ap.) = 20 gr		= **1.29597820**
1 dram (dr. ap.)	= 3 s. ap.	= **3.88793460** g
1 ounce (oz. ap.) = 8 dr. ap.		= **31.103476** g
1 pound (lb. ap.) = 12 oz. ap.		= **373.2417216** g

U.S. STANDARDS OF WEIGHTS AND MEASURES

In 1893 an order was issued by T. C. Mendenhall, then Superintendent of Weights and Measures of the United States, to the effect that the international meter and the international kilogram would henceforward be regarded as the fundamental standards of **length** and **mass** in the United States. The National Bureau of Standards was founded in 1901 and until 1959 used the following relations to define the principal units in the U.S. Customary System: 1 *yard* = 3600/3937 meter; 1 *pound avoirdupois* = 0.4535924277 kilograms; 1 *gallon* = 231 cubic inches; and 1 *bushel* = 2150.42 cubic inches.

Effective July 1, 1959, the National Bureau of Standards adopted the following exact equivalents in terms of the SI standards:

1 yard = 0.9144 meter
1 pound (avoirdupois) = 0.45359237 kilogram
1 gallon = 231 cubic inches
1 bushel = 2150.42 cubic inches

The only exception to this was the continued usage by the U.S. Coast and Geodetic Survey of the relation 1 foot = 1200/3937 meter, which now defines what is known as the *U.S. Survey Foot.* The standard of mass is referred to the International Prototype Kilogram No. 1 kept in France, but the United States reference standard *for most purposes* is the U.S. Prototype Kilogram No. 20, a cylinder of equal height and diameter made from platinum and iridium in the approximate weight ratio of 9 : 1 and received from the International Bureau of Weights and Measures in 1889.

MISCELLANEOUS UNITS USED IN THE U.S. AND BRITISH CUSTOMARY SYSTEMS

Counting measure

1 dozen (doz.) = 12 units	1 great gross = 12 gr = 1728 units
1 gross (gr.) = 12 doz = 144 units	1 score = 20 units

Paper measure

1 quire = 24 or 25 sheets
1 ream = 20 quires = 500 sheets
1 perfect ream = 516 sheets
1 bundle = 2 reams = 1000 sheets
1 bale = 5 bundles = 5000 sheets

Book measure

A book of sheets folded in:
2 leaves is a folio
4 leaves is a quarto
8 leaves is an octavo
12 leaves is a duodecimo
16 leaves is a 16mo.

Printer's measure

1 point ≈ 1/72 in. = 0.0139 in.
1 pica = 12 points ≈ 1/6 in. = 0.167 in.

Gunnery measure of arc

360° = 6400 mils

Linear measure

1 barleycorn = 1/3 inch (British only)
1 cubit = 18 inches (British only)
1 palm = 3 or 4 inches
1 hand = 4 inches (used to measure horses from the ground to the withers)
1 pace = $2\frac{1}{2}$ ft (U.S. Army quick time)
= 3 ft (U.S. Army double time)
1 geometrical pace = 5 ft
1 Roman pace = 58 in.
1 stadium = 607 ft (British)

Cloth measures

1 nail = 2.25 in.
1 ell = 45 in.
1 bolt = 120 ft
1 line = 1/40 in.
(used to measure buttons)
1 skein = 360 ft (used to measure yarn)

Area measure

1 square = 100 sq ft
(used for roofing material)
1 rood = 1/4 acre
(used for land measure)

Volume measure

Liquid measures

1 magnum = 50 fl oz (wine)
1 jeroboam = 104 fl oz (wine)
1 nebuchadnezzar = 4 gallons (wine)
1 pin = 4 1/2 gallons
1 kilderkin = 18 Imperial Gallons
1 tierce = 42 gallons
1 hogshead = 63 gallons (U.S.)
= 54 British Imperial Gallons
= 64.85 U.S. gallons
1 puncheon = 84 gallons
1 butt = 126 gallons
= 1 pipe (used for wine)
1 tun = 252 gallons

Solid measures

1 firkin = 9 Imperial Gallons (used for lard or butter and sometimes taken to be a unit of 56 pounds weight)
1 Imperial Bushel = 2219.36 cu in. (British)
1 chaldron = 36 bushels (British, used to measure coal)
1 perch = 24.75 cu ft (used for masonry)
1 load = 1 cubic yard (used for earth, gravel or other fill)
1 assay ton = 29.167 grams (used for testing ore)
1 catty ≈ 1 1/3 lb (used for tea)
1 quarter = 25 lb
1 slug = 32.1740 lb

Force measure

1 pound-force = 1 slug-ft-sec^{-2}
= 32.1740 lb-ft-sec^{-2}
= 4.4482 N
1 poundal = 1 lb-ft-sec^{-2}
= 0.138255 N

Energy measure

1 foot-pound (ft-lb) = 1.3558 joules

Heat measure

1 British thermal unit* (Btu) = 778.26 ft-lb
= 251.996 cal

Power measure

1 horsepower = 550 ft-lb/sec = 746 watt

Illumination measure

1foot-candle = 1 lumen incident per sq ft = 10.76 lux

*The British thermal unit is defined to be the quantity of heat that will raise the temperature of one pound-mass of water (initially at 39.1°F.) one degree Fahrenheit.

APPENDIX II

DIMENSIONAL ANALYSIS

Given a physical situation describable by physical variables $x_1, x_2, \ldots,$ we can sometimes deduce from *dimensional analysis* certain limitations on the *form* of any possible relationship among the variables. Dimensional analysis is *not* capable of completely determining the unknown functional relationship, but it can delimit the possibilities and in some simple cases, it can give the complete relationship to within a constant of proportionality.

A *primary dimension* is referred to a *primary standard*: for example, in mechanics, *mass, time,* and *length. Relative* lengths, *relative* times, and *relative* masses all have significance *independent* of the choice of the primary standard, for example, the ratio of the width and length of a given rectangle is independent of whether the lengths are measured in feet, inches, meters, or Biblical cubits! We *may* take a quantity such as force and relate it to a *primary* force standard, such as a standard spring maintained at a standard elongation under standard conditions in the *International Standards Laboratory of Lower Slobovia.* However, the **Newtonian relation** $F = ma$ enables force to be expressed *in terms of* the primary quantities M, L, and T and, hence, force may be regarded as a *secondary* quantity.

Here are some secondary quantities:

$$\text{Force} \overset{D}{=} MLT^{-2}$$

$$\text{Pressure} = \frac{\text{force}}{\text{area}} \overset{D}{=} ML^{-1}T^{-1}$$

$$\text{Energy} = \text{force} \times \text{distance}$$
$$\overset{D}{=} LMLT^{-2} = ML^2T^{-2}$$

$$\text{Momentum} = \text{mass} \times \text{velocity}$$
$$\overset{D}{=} MLT^{-1}$$

$$\text{Torque} = L = \text{force} \times \text{distance}$$
$$\overset{D}{=} ML^2T^{-2}$$

The dimensional formulas of all quantities have the form of products of powers.

A factor of proportionality that *varies* when the *units change* is called a **dimensional constant**.

An equation cast into a form that holds without change when the size of the fundamental units changes is called *a complete equation.* For example,

$$v = at$$
$$s = \tfrac{1}{2}at^2$$

Any mathematical relationship among the physical quantities $x_1, x_2, \ldots$ needed to describe a system may be written in the form

$$f(x_1, x_2, \ldots) = 0$$

For physical systems that are understood this may be written as a complete equation. Dimensional analysis is based on the restrictions that obtain on the mathematical form of any complete equation.

The fundamental theorem is that any functional relationship correlating physical quantities and dimensional constants can be cast into a form such that it contains *as arguments* only such products of powers of the physical parameters and dimensional constants as have *zero dimensions in the fundamental quantities.* This theorem is subject to the restriction that there be not more than one independent functional relationship among the quantities.

This theorem, called the Π theorem, restricts only the *arguments* of the functional relation, not the relation itself. It is frequently of value because the number of possible independent dimensionless arguments is generally less than the full number of physical parameters.

Consider $F(\Pi_1, \Pi_2, \ldots) = 0$ where the Π_i are dimensionless products formed from all the parameters. Solve for one of the Π_i and express it in terms of powers of the physical quantities

$$\Pi_1 = f(\Pi_2, \Pi_3 \cdots) = x_1^{\alpha} x_2^{\beta} x_3^{\gamma} \cdots$$

$$\therefore x_1^{\alpha} = x_2^{-\beta} x_3^{-\gamma} \cdots f(\Pi_2, \Pi_3 \cdots)$$

where the function $f(\Pi_2, \ldots)$ contributes only dimensionless terms.

$$x_1^{\alpha} \overset{D}{=} x_2^{-\beta} x_3^{-\gamma} \cdots$$

that is, the right-hand side has the same dimensions as the term on the left-hand side.

Therefore, the equation expressing this relationship is called *dimensionally homogeneous.*

Suppose there are *m fundamental units* whose values may be changed when referred to different primary standards and in terms of which the equation of relationship must be complete. An example is M, L, and T, that is, $m = 3$.

We wish to apply dimensional analysis to a physical system with n parameters including all dimensional constants.

The n parameters can usually be combined into only $n - m$ *independent dimensionless products.* Therefore, any possible functional relationship has only $n - m$ arguments instead of the full n.

Consider an idealized *simple pendulum.*

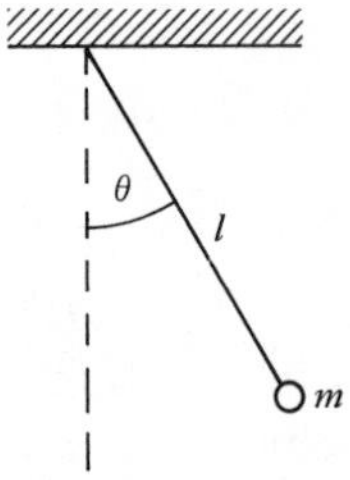

The parameters of this system and their dimensions are

m	M
l	L
θ	0
g	LT^{-2}
τ (period)	T

If we try to form all the independent dimensionless products, there will be 2, since

$$n = \text{number of parameters} = 5$$
$$m = \text{number of primary units } (M, L, T) = 3$$

and

$$n - m = 5 - 3 = 2$$

1. One product is clearly θ by itself.
2. One dimensionless product formed from the remaining four parameters is

$$m^{\alpha} l^{\beta} g^{\gamma} T^{\delta} = M^{\alpha} L^{\beta} (LT^{-2})^{\gamma} T^{\delta}$$

which must be dimensionless. This requires that

$$\alpha = 0$$
$$\beta + \gamma = 0$$
$$\delta - 2\gamma = 0$$

Here we have three equations and four unknowns, which means that we may set one arbitrarily.

Choose δ.

$$\therefore \alpha = 0; \quad \gamma = \frac{\delta}{2}; \quad \beta = -\gamma = -\frac{\delta}{2}$$

The dimensionless product is, therefore,

$$l^{-\delta/2} g^{\delta/2} \tau^{\delta}$$

The Π theorem states that the solution is some functional relation between the two dimensionless products.

$$F[(l^{-1/2} g^{1/2} \tau)^{\delta}, \theta] = 0$$

We may solve for the first in terms of the dimensionless second (therefore the $1/\delta$ root makes no difference):

$$l^{-1/2} g^{1/2} \tau = f(\theta)$$

$$\therefore \tau = \sqrt{\frac{l}{g}} f(\theta)$$

that is, dimensional analysis has yielded the result that τ is

1. Independent of mass.
2. Proportional to $\sqrt{l/g}$.

It may be any function of the amplitude. Detailed physical analysis will show that

$$f(\theta) = 2\pi\left\{1 + \frac{1}{2^2}\sin^2\frac{\theta}{2} + \frac{1}{2^2}\cdot\frac{3^2}{4^4}\sin^4\frac{\theta}{2} + \cdots\right\}$$

where θ is the maximum angular amplitude. For small θ, $f(\theta) \approx 2\pi$.

Example:

Uniform circular motion

Table **A.II.1**

Parameter	Dimension
a	LT^{-2}
m	M
v	LT^{-1}
R	L

In Table A.II.1 the number of *parameters* is $n = 4$ and the number of *fundamental units* is $m = 3$.

$$a^{\alpha} m^{\beta} v^{\gamma} R^{\delta} \overset{D}{=} (LT^{-2})^{\alpha} M^{\beta} (LT^{-1})^{\gamma} L^{\delta}$$

$$\beta = 0$$
$$\gamma + \alpha + \delta = 0$$
$$-2\alpha - \gamma = 0$$

$$\therefore \beta = 0 \quad \alpha = \frac{-\gamma}{2} \quad \frac{\gamma}{2} = -\delta$$

or

$$\gamma = -2\delta$$
$$\alpha = \delta$$

$$\therefore F(a^{\delta} v^{-2\delta} R^{\delta}) = 0$$

$$\left(\frac{a}{v^2}R\right)^{\delta} = K$$

$$\therefore a = \frac{v^2}{R}C$$

APPENDIX III

SOME COMMENTS ON THE FACTORIAL, GAMMA, AND ERROR FUNCTIONS

We have introduced the factorial function in (18.1) in terms of integer values: $n!$. These are summarized in Table A.III.1.

Table **A.III.1.** Factorials and Logarithms of Factorials

n	$n!$	$\log_{10} n!$	n	$n!$	$\log_{10} n!$	n	$\log n!$	n	$\log n!$
			50	3.0414×10^{64}	64.48307	100	157.97000	150	262.75689
1	1.0000	0.00000	51	1.5511×10^{66}	66.19065	101	159.97432	151	264.93587
2	2.0000	0.30103	52	8.0658×10^{67}	67.90665	102	161.98293	152	267.11771
3	6.0000	0.77815	53	4.2749×10^{69}	69.63092	103	163.99576	153	269.30241
4	2.4000×10	1.38021	54	2.3084×10^{71}	71.36332	104	166.01280	154	271.48993
5	1.2000×10^{2}	2.07918	55	1.2696×10^{73}	73.10368	105	168.03399	155	273.68026
6	7.2000×10^{2}	2.85733	56	7.1100×10^{74}	74.85187	106	170.05929	156	275.87338
7	5.0400×10^{3}	3.70243	57	4.0527×10^{76}	76.60774	107	172.08867	157	278.06928
8	4.0320×10^{4}	4.60552	58	2.3506×10^{78}	78.37117	108	174.12210	158	280.26794
9	3.6288×10^{5}	5.55976	59	1.3868×10^{80}	80.14202	109	176.15952	159	282.46934
10	3.6288×10^{6}	6.55976	60	8.3210×10^{81}	81.92017	110	178.20092	160	284.67346
11	3.9917×10^{7}	7.60116	61	5.0758×10^{83}	83.70550	111	180.24624	161	286.88028
12	4.7900×10^{8}	8.68034	62	3.1470×10^{85}	85.49790	112	182.29546	162	289.08980
13	6.2270×10^{9}	9.79428	63	1.9826×10^{87}	87.29724	113	184.34854	163	291.30198
14	8.7178×10^{10}	10.94041	64	1.2689×10^{89}	89.10342	114	186.40544	164	293.51683
15	1.3077×10^{12}	12.11650	65	8.2477×10^{90}	90.91633	115	188.46614	165	295.73431
16	2.0923×10^{13}	13.32062	66	5.4434×10^{92}	92.73587	116	190.53060	166	297.95442
17	3.5569×10^{14}	14.55107	67	3.6471×10^{94}	94.56195	117	192.59878	167	300.17714
18	6.4024×10^{15}	15.80634	68	2.4800×10^{96}	96.39446	118	194.67067	168	302.40245
19	1.2165×10^{17}	17.08509	69	1.7112×10^{98}	98.23331	119	196.74621	169	304.63033
20	2.4329×10^{18}	18.38612	70	1.1979×10^{100}	100.07841	120	198.82539	170	306.86078
21	5.1091×10^{19}	19.70834	71	8.5048×10^{101}	101.92966	121	200.90818	171	309.09378
22	1.1240×10^{21}	21.05077	72	6.1234×10^{103}	103.78700	122	202.99454	172	311.32931
23	2.5852×10^{22}	22.41249	73	4.4701×10^{105}	105.65032	123	205.08444	173	313.56735
24	6.2045×10^{23}	23.79271	74	3.3079×10^{107}	107.51955	124	207.17787	174	315.80790
25	1.5511×10^{25}	25.19065	75	2.4809×10^{109}	109.39461	125	209.27478	175	318.05094
26	4.0329×10^{26}	26.60562	76	1.8855×10^{111}	111.27543	126	211.37515	176	320.29645
27	1.0889×10^{28}	28.03698	77	1.4518×10^{113}	113.16192	127	213.47895	177	322.54443
28	3.0489×10^{29}	29.48414	78	1.1324×10^{115}	115.05401	128	215.58616	178	324.79485
29	8.8418×10^{30}	30.94654	79	8.9462×10^{116}	116.95164	129	217.69675	179	327.04770
30	2.6525×10^{32}	32.42366	80	7.1569×10^{118}	118.85473	130	219.81069	180	329.30297
31	8.2228×10^{33}	33.91502	81	5.7971×10^{120}	120.76321	131	221.92796	181	331.56065
32	2.6313×10^{35}	35.42017	82	4.7536×10^{122}	122.67703	132	224.04854	182	333.82072
33	8.6833×10^{36}	36.93869	83	3.9455×10^{124}	124.59610	133	226.17239	183	336.08317
34	2.9523×10^{38}	38.47016	84	3.3142×10^{126}	126.52038	134	228.29949	184	338.34799
35	1.0333×10^{40}	40.01423	85	2.8171×10^{128}	128.44980	135	230.42983	185	340.61516
36	3.7199×10^{41}	41.57054	86	2.4227×10^{130}	130.38430	136	232.56337	186	342.88467
37	1.3764×10^{43}	43.13874	87	2.1078×10^{132}	132.32382	137	234.70009	187	345.15652
38	5.2302×10^{44}	44.71852	88	1.8548×10^{134}	134.26830	138	236.83997	188	347.43067
39	2.0398×10^{46}	46.30959	89	1.6508×10^{136}	136.21769	139	238.98298	189	349.70714
40	8.1592×10^{47}	47.91165	90	1.4857×10^{138}	138.17194	140	241.12911	190	351.98589
41	3.3453×10^{49}	49.52443	91	1.3520×10^{140}	140.13098	141	243.27833	191	354.26692
42	1.4050×10^{51}	51.14768	92	1.2438×10^{142}	142.09477	142	245.43062	192	356.55022
43	6.0415×10^{52}	52.78115	93	1.1568×10^{144}	144.06325	143	247.58595	193	358.83578
44	2.6583×10^{54}	54.42460	94	1.0874×10^{146}	146.03638	144	249.74432	194	361.12358
45	1.1962×10^{56}	56.07781	95	1.0330×10^{148}	148.01410	145	251.90568	195	363.41362
46	5.5026×10^{57}	57.74057	96	9.9168×10^{149}	149.99637	146	254.07004	196	365.70587
47	2.5862×10^{59}	59.41267	97	9.6193×10^{151}	151.98314	147	256.23735	197	368.00034
48	1.2414×10^{61}	61.09391	98	9.4269×10^{153}	153.97437	148	258.40762	198	370.29701
49	6.0828×10^{62}	62.78410	99	9.3326×10^{155}	155.97000	149	260.58080	199	372.59586
			100	9.3326×10^{157}	157.97000			200	374.89689

Stirling's Formula

For the *factorial function* we may consider a more general function that reduces to $n!$ for n a *positive integer.* Such a function is the *gamma function* defined for **any** n

$$\Gamma(n+1) = \int_0^\infty z^n e^{-z}\, dz \qquad \text{(A.III.1)}$$

We may integrate this by parts for n a positive integer

$$\Gamma(n+1) = -z^n e^{-z}\Big|_0^\infty + n\int_0^\infty z^{n-1}e^{-z}\,dz$$

$$= n\int_0^\infty z^{n-1}e^{-z}\,dz$$

We thus obtain the *recursion relation* for the gamma function

$$\Gamma(n+1) = n\Gamma(n) \qquad \text{(A.III.2)}$$

We may integrate directly to obtain

$$\Gamma(1) = 1 \qquad \text{(A.III.3)}$$

We thus see that *for integer n*

$$\Gamma(n+1) = n(n-1)(n-2)\cdots 1 = n! \qquad \text{(A.III.4)}$$

This gives us an integral representation

$$n! = \int_0^\infty z^n e^{-z}\,dz \qquad \text{(A.III.5)}$$

We are interested when n is large. Make the change of variable $z = n + \sqrt{n}\,x$. Now

$$n! = \sqrt{n}\,e^{-n}\int_{-\sqrt{n}} e^{n\ln(n+\sqrt{n}x)-\sqrt{n}x}\,dx \qquad \text{(A.III.5a)}$$

We may rewrite the exponent since

$$\ln(n+\sqrt{n}x) = \ln\left\{n\left[1+\frac{x}{\sqrt{n}}\right]\right\}$$

$$= \ln n + \ln\left(1+\frac{x}{\sqrt{n}}\right)$$

$$\approx \ln n + \frac{x}{\sqrt{n}} - \frac{x^2}{2n} + \frac{1}{3}\left(\frac{x}{\sqrt{n}}\right)^3 + \cdots$$

Since n is assumed to be large, it suffices to stop at terms second order in x. We may substitute into (A.III.5a)

$$n! = \sqrt{n}\,e^{-n}\int_{-\sqrt{n}}^{\infty} e^{n\ln n+\sqrt{n}x-(nx^2/2n)-\sqrt{n}x}\,dx$$

$$= \sqrt{n}\,e^{-n}n^n\int_{-\sqrt{n}}^{\infty} e^{-x^2/2}\,dx \qquad \text{(A.III.6)}$$

The following integral defines the error function $E(x)$.

$$\sqrt{\frac{2}{\pi}}\int_x^\infty e^{-u^2/2}\,du = E(x) \qquad \text{(A.III.7)}$$

We show below in (A.III.17) that

$$\sqrt{\frac{2}{\pi}}\int_0^\infty e^{-u^2/2}\,du = 1 \qquad \text{(A.III.8a)}$$

Hence, $E(0) = 1$. Also

$$\int_{-\infty}^{\infty} e^{-u^2/2}\,du = \sqrt{2\pi} \qquad \text{(A.III.9)}$$

since $E(x)$ is an *even* function of x [i.e., $E(x) = E(-x)$]. The integrand, $e^{-u^2/2}$, plotted versus u is shown in Figure A.III.1.

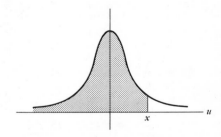

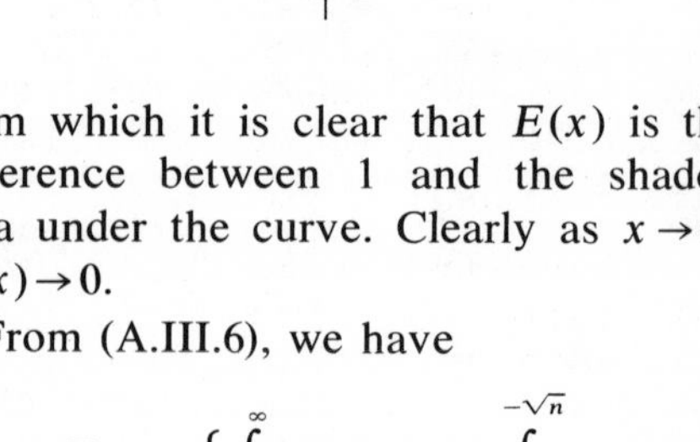

from which it is clear that $E(x)$ is the difference between 1 and the shaded area under the curve. Clearly as $x \to \infty$, $E(x) \to 0$.

From (A.III.6), we have

$$n! \approx \sqrt{n}e^{-n}n^n\left\{\int_{-\infty}^{\infty} e^{-u^2/2}\,du - \int_{-\infty}^{-\sqrt{n}} e^{-u^2/2}\,dn\right\}$$

The bracket is equivalent to

$$\{\ \} = \sqrt{2\pi} - \int_{-\infty}^{-\sqrt{n}} e^{-u^2/2}\,du$$

$$= \sqrt{2\pi} + \int_{\sqrt{n}}^{\infty} e^{-u^2/2}\,du$$

$$= \sqrt{2\pi} + \sqrt{\frac{\pi}{2}}E(\sqrt{n})$$

Since we are interested in the case when the argument, $\sqrt{n}$, is large, we neglect the term, $\sqrt{\pi/2}\,E(\sqrt{n})$.

We are thus left with the approximation

$$n! \approx \sqrt{n}e^{-n}n^n\sqrt{2\pi} \qquad \text{(A.III.10)}$$

which is just (18.9), *Stirling's approximation.*

A more accurate asymptotic expansion is

$$n! \approx \sqrt{2\pi n}e^{-n}n^n\left\{1 + \frac{1}{12n} + \frac{1}{288n^2} - \frac{139}{51840n^3} + \cdots\right\} \qquad \text{(A.III.11)}$$

but Stirling's approximation is quite good enough for many applications.

For example, $10! = 3628800 = 3.628800 \times 10^6$. Equation 18.9 gives $10! \approx 3.59869 \times 10^6 = (1-\epsilon)10!$ where $\epsilon = 0.008$.

Let us return to our assertions (A.III.8) and (A.III.9).

Consider the integral

$$I = \int_{-\infty}^{\infty} e^{-u^2}\,du$$

where the u is a dummy variable. We may introduce another dummy variable v and write

$$I^2 = \int_{\infty}^{\infty} e^{-u^2}\,du \int_{-\infty}^{\infty} e^{-v^2}\,dv \qquad \text{(A.III.12)}$$

This, however, can be interpreted as a double integral in the u and v plane.

$$I^2 = \int_{-\infty}^{\infty}\int_{-\infty}^{\infty} e^{-(u^2+v^2)}\,du\,dv \qquad \text{(A.III.13)}$$

We may perform this integration in polar coordinates such that $r^2 = u^2 + v^2$, and the differential element is $r\,dr\,d\theta$.

$$I^2 = \int_{r=0}^{\infty}\int_{0}^{2\pi} e^{-r^2}r\,dr\,d\theta$$

$$= 2\pi\int_{0}^{\infty} e^{-r^2}r\,dr = \pi \qquad \text{(A.III.14)}$$

since the remaining integral is easily shown to be

$$\int_0^{\infty} e^{-r^2}r\,dr = \frac{1}{2} \qquad \text{(A.III.15)}$$

Therefore,

$$I = \int_{-\infty}^{\infty} e^{-u^2}du = \sqrt{\pi} \qquad \text{(A.III.16)}$$

If we substitute $v/\sqrt{2} = u$, we get

$$\int_{-\infty}^{\infty} e^{-v^2/2}\,dv = \sqrt{2\pi} \qquad \text{(A.III.17)}$$

which is (A.III.9) and leads immediately to (A.III.8a) because the integrand is an even function.

Gamma and Beta Functions

We state and derive some properties of the *gamma function*

$$\int_0^{\infty} x^{p-1}e^{-\alpha x}dx = \frac{\Gamma(p)}{\alpha^p} \qquad \text{(A.III.18a)}$$

$$\int_0^1 x^{p-1}(1-x)^{q-1}dx = \frac{\Gamma(p)\Gamma(q)}{\Gamma(p+q)} \qquad \text{(A.III.19a)}$$

where $p, q > 0$

$$\Gamma^2\left(\frac{1}{2}\right) = \int_0^1 \frac{dx}{\sqrt{x(1-x)}}$$

$$= \int_{-1}^{1} \frac{dy}{\sqrt{1-y^2}} = \pi \qquad \text{(A.III.20a)}$$

$$\Gamma\left(\frac{1}{2}\right) = \sqrt{\pi} \qquad \text{(A.III.20b)}$$

$$\int_{-\infty}^{\infty} e^{-\alpha y^2} = \sqrt{\frac{\pi}{\alpha}}. \qquad \text{(A.III.21a)}$$

Equation A.III.18a follows from the defining equation A.III.1 written as

$$\Gamma(p) = \int_0^{\infty} z^{p-1}e^{-z}dz \qquad \text{(A.III.18b)}$$

sometimes called *Euler's integral of the second kind.* The substitution $z = \alpha x$ yields (A.III.18a).

The function

$$B(p, q) = \int_0^1 x^{p-1}(1-x)^{q-1}\, dx \quad \text{(A.III.19b)}$$

is the *Beta function* or *Euler's integral of the first kind.*

Substituting $1/t$ for x yields the alternative form

$$B(p, q) = \int_1^\infty t^{-p-q}(t-1)^{q-1}\, dt. \quad \text{(A.III.19c)}$$

The substitution $t = \sin^2\theta$ in (A.III.19b) yields another form

$$B(p, q) = 2\int_0^{\pi/2} \sin^{2p-1}\theta \cos^{2q-1}\theta\, d\theta \quad \text{(A.III.19d)}$$

The substitution $z = y^2$ in (A.III.18b) yields

$$\Gamma(a) = \int_0^\infty y^{2a-2}e^{-y^2}2y\,dy$$

Hence,

$$\Gamma(p)\Gamma(q) = 4\int_0^\infty e^{-y^2}y^{2p-1}dy\int_0^\infty e^{-x^2}x^{2q-1}dx$$

$$= 4\int_0^\infty\int_0^\infty e^{-(x^2+y^2)}x^{2q-1}y^{2p-1}dxdy$$

Transforming to polar coordinates via

$$x = r\cos\theta;\ y = r\sin\theta$$

We get

$$\Gamma(p)\Gamma(q) = \left\{\int_0^\infty dr^2 e^{-r^2}(r^2)^{p+q-1}\right\}$$

$$\times\left\{2\int_0^{\pi/2} \sin^{2p-1}\theta\cos^{2q-1}\theta\,d\theta\right\}.$$

The first bracketed term is $\Gamma(p+q)$ while the second is $B(p, q)$ by (A.III.19c). This proves relation A.III.19a and, incidentally, shows the *symmetry property* of the beta function

$$B(p, q) = B(q, p) \quad \text{(A.III.19e)}$$

If we substitute $p = q = 1/2$ and use (A.III.3) we get relation A.III.20a and, hence, (A.III.20b).

In relation A.III.9 we may substitute

$$\alpha y^2 = \frac{u^2}{2}$$

and get

$$\sqrt{2\pi} = \int_{-\infty}^\infty e^{-\alpha y^2}\frac{2\alpha y}{\sqrt{2\alpha y^2}}dy$$

$$= \sqrt{2\alpha}\int_{-\infty}^\infty e^{-\alpha y^2}\,dy$$

and, hence, (A.III.21a).

We now have formulas for the gamma function of *integer* and *half-integer* arguments:

$$\Gamma(n) = (n-1)! = (n-1)(n-2)\cdots(1) \quad \text{(A.III.22)}$$

for integer n.

$$\Gamma(m) = (m-1)(m-2)\cdots\left(\frac{3}{2}\right)\left(\frac{1}{2}\right)(\sqrt{\pi}) \quad \text{(A.III.23)}$$

for m = half-integer

$$= \frac{1}{2} \text{ or } \frac{3}{2} \text{ or } \frac{5}{2}\cdots$$

APPENDIX IV

EIGHT HUNDRED UNIFORMLY DISTRIBUTED RANDOM NUMBERS AND EIGHT HUNDRED RANDOM NORMAL DEVIATES

Table ***A.IV.1*** Four Hundred Random Numbers

400 UNIFORMLY RANDOM

.399	.906	.178	.811	.876
.794	.315	.751	.282	.269
.821	.518	.247	.239	.931
.900	.736	.859	.368	.900
.050	.015	.470	.127	.352
.643	.327	.572	.742	.680
.872	.906	.956	.385	.278
.885	.541	.622	.932	.661
.389	.521	.341	.708	.551
.009	.214	.346	.779	.167
.906	.700	.861	.080	.037
.422	.633	.728	.839	.633
.471	.313	.018	.168	.498
.121	.924	.047	.513	.615
.096	.178	.094	.123	.357
.515	.233	.532	.572	.109
.357	.833	.578	.752	.447
.761	.006	.133	.196	.959
.816	.333	.896	.026	.471
.389	.374	.174	.102	.439
.189	.850	.286	.303	.331
.770	.023	.434	.603	.741
.937	.306	.242	.878	.398
.461	.180	.270	.362	.110
.465	.079	.158	.757	.871
.553	.679	.730	.474	.351
.813	.287	.460	.635	.238
.955	.978	.256	.434	.252
.958	.198	.040	.596	.078
.962	.213	.426	.234	.574
.949	.568	.889	.164	.403
.482	.578	.126	.803	.490
.274	.457	.733	.999	.096
.570	.772	.354	.295	.909
.297	.392	.811	.291	.793
.495	.939	.312	.227	.835
.435	.265	.817	.115	.410
.827	.072	.147	.625	.765
.978	.463	.934	.542	.439
.911	.685	.362	.413	.874

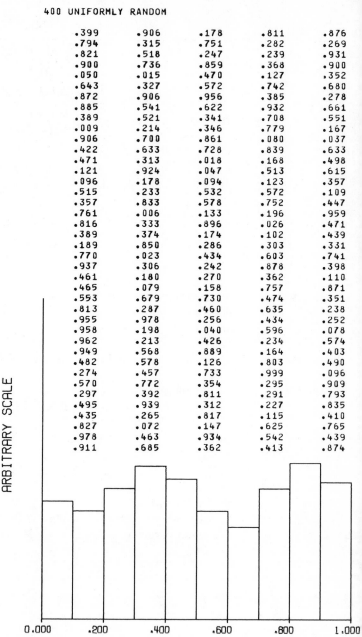

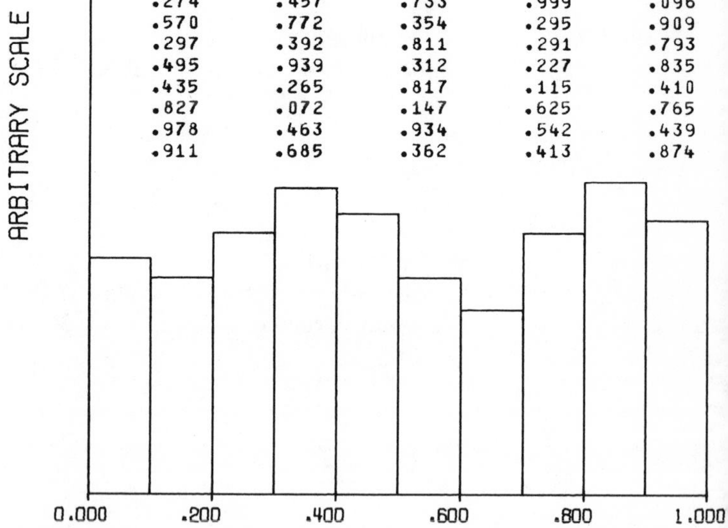

Figure A.IV.1
Histogram of Table A.IV.1.

with Calculated Mean and Variance.

.680	.439	.985	.517	.441
.125	.883	.788	.009	.239
.941	.447	.404	.570	.334
.370	.394	.094	.795	.145
.559	.013	.994	.538	.649
.428	.547	.633	.922	.006
.066	.496	.224	.964	.253
.318	.847	.041	.983	.476
.865	.212	.648	.894	.840
.775	.420	.084	.715	.812
.475	.627	.571	.562	.205
.196	.914	.372	.299	.638
.708	.277	.029	.118	.370
.091	.408	.113	.076	.823
.367	.393	.851	.245	.945
.069	.379	.588	.394	.410
.229	.145	.860	.632	.706
.360	.824	.382	.930	.293
.933	.238	.924	.686	.611
.956	.708	.151	.381	.932
.947	.789	.948	.498	.488
.813	.541	.675	.314	.816
.905	.438	.740	.857	.056
.921	.291	.817	.756	.376
.795	.865	.782	.896	.343
.962	.212	.104	.284	.957
.909	.754	.377	.598	.008
.789	.858	.957	.305	.488
.011	.786	.595	.541	.363
.086	.412	.775	.212	.305
.460	.018	.354	.674	.574
.612	.818	.622	.500	.781
.216	.756	.445	.178	.766
.910	.821	.070	.842	.017
.753	.217	.058	.105	.428
.836	.972	.738	.806	.874
.539	.602	.518	.019	.209
.720	.710	.544	.301	.394
.167	.896	.429	.726	.155
.289	.667	.892	.134	.641

```
  400  UNIFORMLY RANDOM
MEAN=        .511466
SD SQUARE=       .083923
SD=              .289695
```

Table A.IV.2 Four Hundred Random Number[s]

400 UNIFORMLY RANDOM

.140	.511	.808	.962	.394
.962	.570	.011	.056	.216
.909	.713	.642	.997	.517
.545	.569	.831	.606	.063
.501	.330	.132	.896	.033
.715	.704	.883	.990	.405
.415	.900	.002	.792	.933
.124	.986	.446	.606	.241
.068	.141	.178	.383	.328
.119	.455	.062	.649	.998
.349	.048	.231	.346	.792
.742	.017	.136	.393	.750
.860	.476	.253	.051	.485
.521	.519	.379	.982	.273
.595	.522	.848	.229	.114
.515	.236	.893	.685	.651
.885	.027	.795	.514	.463
.908	.860	.759	.579	.441
.005	.698	.170	.857	.028
.012	.640	.724	.200	.514
.259	.709	.540	.425	.761
.333	.201	.780	.629	.785
.941	.691	.114	.847	.464
.705	.574	.074	.241	.121
.596	.273	.572	.537	.558
.857	.218	.380	.982	.940
.512	.567	.443	.831	.997
.235	.394	.134	.464	.166
.685	.070	.330	.231	.371
.900	.759	.350	.805	.356
.460	.697	.645	.944	.026
.719	.864	.351	.263	.642
.775	.063	.592	.032	.388
.277	.356	.557	.068	.519
.337	.615	.552	.922	.843
.619	.921	.192	.204	.100
.919	.032	.349	.094	.438
.404	.608	.067	.297	.927
.202	.735	.620	.912	.453
.870	.439	.905	.254	.318

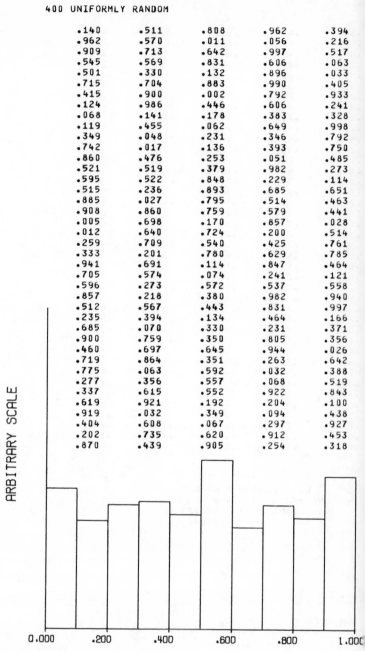

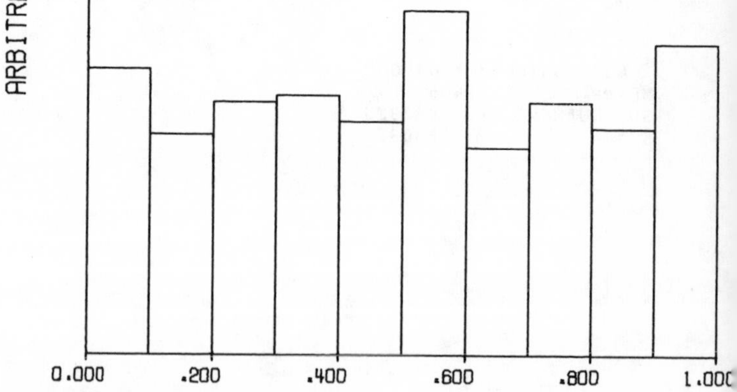

Figure A.IV.2
Histogram of Table A.IV.2.

with Calculated Mean and Variance.

.558	.402	.839	.666	.582
.026	.030	.658	.044	.845
.447	.809	.737	.058	.774
.709	.857	.540	.327	.299
.114	.994	.958	.581	.844
.162	.263	.165	.271	.767
.059	.726	.361	.475	.366
.436	.933	.286	.154	.482
.720	.336	.962	.522	.542
.812	.009	.957	.909	.461
.009	.189	.910	.543	.022
.238	.334	.018	.665	.562
.311	.511	.239	.696	.132
.477	.886	.493	.542	.114
.668	.719	.842	.065	.982
.915	.078	.239	.108	.136
.503	.846	.983	.454	.079
.265	.861	.691	.975	.236
.739	.610	.953	.973	.877
.401	.281	.826	.755	.599
.305	.156	.255	.435	.922
.148	.531	.232	.190	.586
.474	.941	.636	.660	.388
.520	.572	.752	.038	.078
.394	.555	.708	.919	.099
.959	.113	.719	.459	.720
.581	.766	.773	.685	.429
.368	.029	.945	.337	.791
.595	.578	.699	.937	.284
.880	.087	.962	.224	.726
.019	.724	.453	.961	.123
.490	.818	.515	.544	.995
.116	.525	.503	.327	.356
.107	.418	.909	.133	.519
.925	.384	.367	.668	.354
.580	.212	.978	.222	.207
.383	.014	.737	.441	.829
.568	.199	.305	.173	.276
.459	.671	.817	.033	.628
.383	.956	.891	.701	.252

400 UNIFORMLY RANDOM
MEAN= .502959
SD SQUARE= .086033
SD= .293314

800 UNIFORMLY RANDOM
MEAN= .507212
SD SQUARE= .084890
SD= .291359

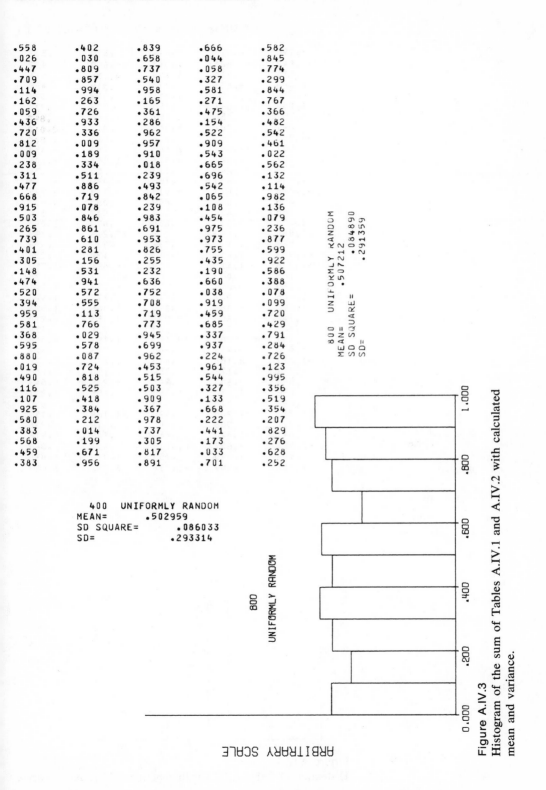

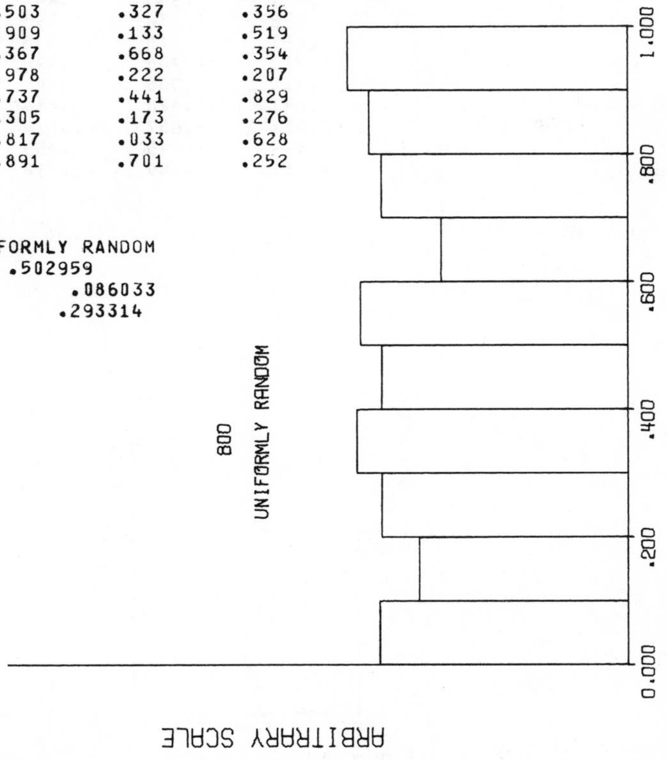

Figure A.IV.3
Histogram of the sum of Tables A.IV.1 and A.IV.2 with calculated mean and variance.

Table A.IV.3 Four Hundred Random Normal

400 RANDOM NORMAL DEVIATES: MEAN = 0.0047(2) ;

-.195	-.913	-.044	.529	.503
.642	-1.024	-.353	-.060	-.432
.625	-.210	-1.008	-.184	-.614
-1.080	-.006	-.737	.319	.488
-.374	.302	-1.861	.383	-.829
-1.054	-2.097	.153	.333	-.238
-1.751	-1.250	1.697	2.166	.852
.341	.707	-.368	-2.347	-2.012
-.258	.353	.014	.825	1.502
.031	-1.484	-2.685	.788	-.453
-.027	1.627	-.517	1.263	-.106
.799	-1.242	-.733	.089	.974
-.634	.663	1.755	-.099	-.459
-1.286	.473	-.618	-.249	-.935
.395	-.064	-.809	.087	-.497
.166	-1.438	.159	-.842	.314
-1.374	1.336	.268	-.073	1.056
-1.066	1.403	.206	-2.215	.153
.527	.117	-.337	.237	-.069
-1.903	1.412	-2.047	1.132	.232
-1.258	1.411	-.310	-.054	-1.713
.169	-.062	.261	-1.233	.075
-1.020	-.064	.232	-.187	-1.101
1.406	.207	.110	-.477	1.923
1.290	-.366	-.345	1.006	1.523
.742	-.663	2.201	1.351	-.238
.485	.715	1.103	-.058	.179
.257	-.306	-.657	1.151	-.556
-.636	-.948	2.600	-.215	.521
-1.317	2.165	.419	-1.270	.072
-.841	-.172	.154	-.110	-.608
.591	-.603	-.947	-.238	-.649
-.272	1.670	1.795	.258	1.391
1.801	.510	-1.076	.287	-.835
-.311	-1.206	-.470	1.188	.153
-.610	.966	.846	.410	-.336
-.041	-1.020	.454	.545	-.978
-2.057	-.667	-1.019	1.017	.863
.949	1.945	1.753	1.250	1.050
-.567	-.389	-1.218	1.562	1.633

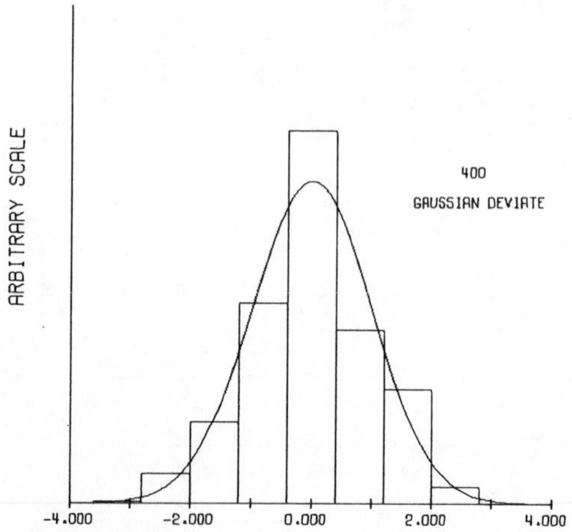

Figure A.IV.4
Histogram of Table A.IV.3 with superimposed gaussian curve.

Deviates with Calculated Mean and Variance.

VARIANCE = 0.9669; S.D. = 0.9833

-.312	.377	-.178	-.239	.211
1.047	.515	-.101	.367	-.474
-2.832	1.006	.279	-.902	1.473
.406	.520	.468	.156	.438
.204	.666	2.058	.805	1.069
1.728	.061	.092	.089	.281
-1.429	1.920	-.427	.924	-1.707
-.038	-.157	-1.486	-.263	.878
.006	-.120	-1.397	.544	-1.274
-.744	.856	1.221	.397	-.483
.238	.284	1.722	.085	.653
-.694	.376	-.044	-2.148	.168
-1.555	-.458	-.149	.193	-.120
.170	-.206	-1.029	-.406	.117
.953	-1.828	-.048	-2.000	-.399
.259	1.788	1.280	-.682	.940
-.267	.116	-.174	-1.167	.349
.125	-1.895	1.564	-1.318	-.768
1.161	.931	-.875	-.742	-.734
1.514	-2.434	.932	1.245	.169
.526	.687	.144	-.307	-.530
-.552	-2.306	.452	-.273	-1.129
-1.834	.556	-.996	-.026	.159
-1.524	.656	-.420	-1.605	-.774
.317	-.594	-1.833	1.643	1.290
-.042	-.567	-.348	-.415	1.649
1.748	-.458	-.845	2.311	.354
-1.064	-.874	.872	-.953	-.353
-.132	-.290	-.441	1.204	1.945
-.667	.187	.752	-.619	-1.323
-.309	-.643	-.657	.874	-.336
-.201	-1.862	1.246	2.142	.214
.602	.496	.177	-.435	-.236
-1.267	-1.263	-.556	-.555	.394
1.828	1.610	.149	1.160	.593
.131	.517	-2.118	1.266	.051
1.337	.137	-.160	.216	1.505
.514	-.384	-.451	-.397	.225
1.243	-.153	-.201	.359	.274
-1.110	.103	-.207	.175	-.353

Table ***A.IV.4*** Four Hundred Random Normal

400 RANDOM NORMAL DEVIATES: MEAN = -0.0162(0) ;

-1.677	-.353	-.552	-.368	.544
1.715	-.806	.802	-1.142	2.049
.302	-1.861	-.608	1.280	-.657
-1.095	1.226	-.333	.028	.419
1.508	-1.542	.083	-.142	1.276
.045	-.022	.523	-.495	.937
-2.434	-.263	.878	-.238	.319
-.206	1.680	.326	-.618	-.113
.711	-.787	-1.565	1.076	.729
.097	.423	-.002	.073	-.629
-.106	-.657	.131	-.475	-1.110
-1.095	-.362	-1.583	-1.524	-.005
-1.571	-.972	-.718	1.964	-.089
.631	-.336	-1.167	.504	-1.801
.085	-.863	-2.145	.314	.383
.498	-.722	-.557	.111	-.042
-.448	.335	-1.601	-.373	.099
-.913	2.245	1.264	-.245	-2.043
-.429	-.804	-.242	.505	-.831
.762	1.029	-.260	.216	.805
.863	1.008	.205	-.262	-.846
-1.056	.149	-.286	.789	-.113
1.567	.023	.419	-.916	-1.803
-.913	.814	-.788	-.310	-.236
-1.020	-1.874	-.330	.159	1.025
-.923	-.574	.937	1.511	-1.027
.283	-.994	-.383	1.025	-.644
-.211	-.011	.097	-1.080	-.366
-.634	.090	.457	-1.566	1.157
-.175	.005	.371	-.120	-.923
-1.326	1.502	-.330	1.015	1.331
.091	.775	-1.472	-.123	.417
.952	.419	-.383	1.076	-.994
-1.070	.312	.784	-.402	.504
-1.006	.704	1.020	.075	.341
1.585	-.362	-.798	-.239	1.326
.919	-.311	-1.071	-.642	-1.114
.085	1.094	-.437	-2.664	.762
-1.214	-1.024	.169	-.026	.485
.020	-1.293	.129	.581	-1.374

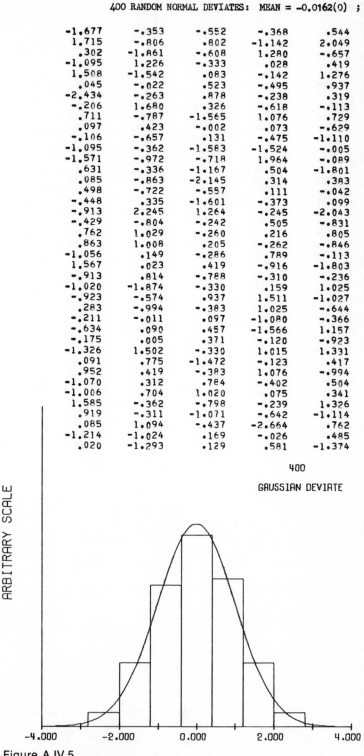

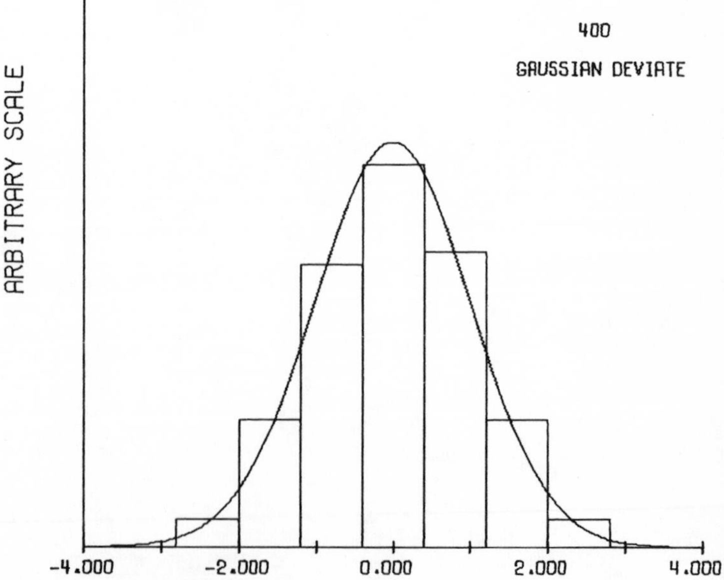

Figure A.IV.5
Histogram of Table A.IV.4 with superimposed gaussian curve.

Deviates with Calculated Mean and Variance.

VARIANCE = 1.0106; S.D. = 1.0053

.169	-.263	.422	-.310	.259
.066	-2.650	2.165	-.842	-.206
.130	-.846	-.264	.250	1.680
1.170	-.923	-.509	-.494	.462
.375	1.010	-1.214	-.831	-1.145
.618	-1.128	-.120	.663	1.246
-1.716	-.178	-.455	1.585	.956
2.043	.405	1.564	-.620	.680
.310	.460	-.242	-1.362	-.189
.588	.762	-1.080	.302	.149
-1.762	.474	.394	-.755	.821
.353	-.030	-1.442	-.209	1.299
-.884	2.125	1.378	-1.144	1.022
.557	.289	.805	-2.306	1.722
.790	1.320	1.211	-1.400	1.585
-1.623	.644	-.505	.936	.680
.265	-.259	-1.105	.717	0.000
1.057	.028	-1.024	-.394	.309
1.575	2.019	-1.801	-.958	.703
-.435	-1.581	1.020	1.403	-.353
.126	-2.066	.519	-1.477	-.296
1.458	.659	-1.141	.283	-.398
-2.092	1.163	.646	-1.980	-.864
1.643	.150	1.081	-1.707	1.050
1.440	-1.200	.434	-1.472	-.296
-.121	.417	.806	-1.155	.650
.475	.743	.802	-.911	-1.774
1.403	-.874	.139	-.449	2.600
1.489	-.178	-1.356	.972	-1.289
1.337	.952	.729	.950	-.787
-.675	1.263	-1.540	1.055	-1.074
-.905	1.564	1.511	-.175	-.866
.621	.154	-1.279	-.475	.893
-.557	-.662	.149	-.451	-1.397
-1.245	-.976	.967	2.089	.291
-.162	.541	-.814	1.457	.148
.630	.315	-1.699	1.992	2.019
-1.023	1.027	.169	1.643	-.400
-1.200	-.876	.129	.971	-.734
1.511	-.113	-.123	-2.198	.692

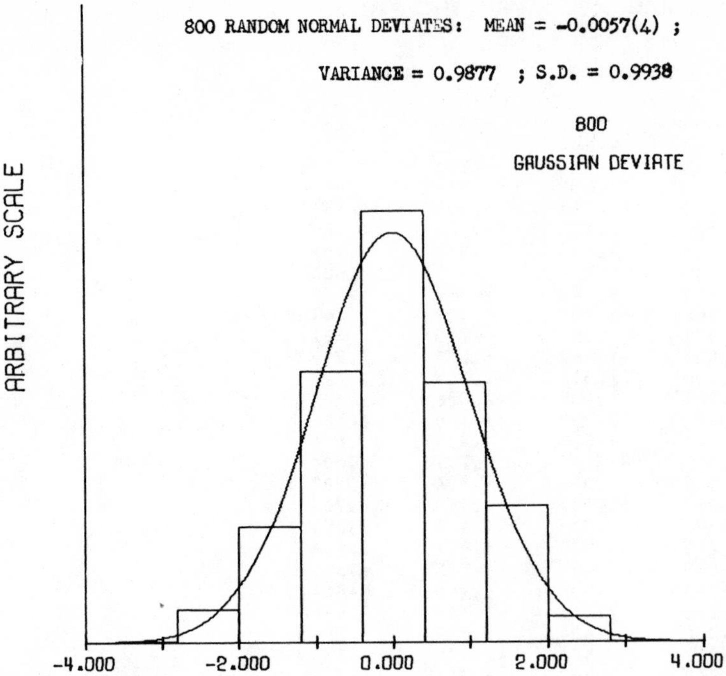

Figure A.IV.6
Histogram of the sum of Tables A.IV.3 and A.IV.4.

APPENDIX V

TABLES OF THE NEGATIVE EXPONENTIAL e^{-x}

Table A.V.1 e^{-x}: $x = 0.00$ to $x = 3.49$ in steps of 0.01.

Table A.V.2 e^{-x}: $x = 3.50$ to $x = 6.99$ in steps of 0.01.

Table A.V.3 e^{-x}: $x = 7.00$ to $x = 9.99$ in steps of 0.01. $x = 10.0$ to $x = 59.0$ in steps of 1.0.

Table **A.V.1** e^{-x}: $x = 0.00$ to

EXPONENTIAL TABLE

EXP(-X)

X	0.00	.01	.02	.03	.04
0.0	1.00000000	.99004983	.98019867	.97044553	.96078944
.1	.90483742	.89583414	.88692044	.87809543	.86935824
.2	.81873075	.81058425	.80251880	.79453360	.78662786
.3	.74081822	.73344696	.72614904	.71892373	.71177032
.4	.67032005	.66365025	.65704682	.65050909	.64403642
.5	.60653066	.60049558	.59452055	.58860497	.58274825
.6	.54881164	.54335087	.53794444	.53259180	.52729242
.7	.49658530	.49164420	.48675226	.48190899	.47711392
.8	.44932896	.44485807	.44043165	.43604929	.43171052
.9	.40656966	.40252422	.39851904	.39455371	.39062784
1.0	.36787944	.36421898	.36059494	.35700696	.35345468
1.1	.33287108	.32955896	.32627979	.32303326	.31981902
1.2	.30119421	.29819728	.29523017	.29229258	.28938422
1.3	.27253179	.26982006	.26713530	.26447726	.26184567
1.4	.24659696	.24414328	.24171402	.23930892	.23692776
1.5	.22313016	.22090998	.21871189	.21653567	.21438110
1.6	.20189652	.19988761	.19789870	.19592957	.19398004
1.7	.18268352	.18086579	.17906615	.17728441	.17552040
1.8	.16529889	.16365414	.16202575	.16041357	.15881743
1.9	.14956862	.14808039	.14660696	.14514820	.14370395
2.0	.13533528	.13398867	.13265547	.13133552	.13002871
2.1	.12245643	.12123797	.12003163	.11883729	.11765484
2.2	.11080316	.10970065	.10860911	.10752843	.10645850
2.3	.10025884	.09926125	.09827359	.09729575	.09632764
2.4	.09071795	.08981529	.08892162	.08803683	.08716085
2.5	.08208500	.08126824	.08045961	.07965902	.07886640
2.6	.07427358	.07353454	.07280286	.07207846	.07136127
2.7	.06720551	.06653681	.06587475	.06521929	.06457035
2.8	.06081006	.06020499	.05960594	.05901285	.05842567
2.9	.05502322	.05447573	.05393369	.05339704	.05286573
3.0	.04978707	.04929168	.04880122	.04831564	.04783489
3.1	.04504920	.04460096	.04415717	.04371780	.04328280
3.2	.04076220	.04035661	.03995506	.03955750	.03916390
3.3	.03688317	.03651617	.03615283	.03579311	.03543696
3.4	.03337327	.03304120	.03271243	.03238694	.03206469

$x = 3.49$ in steps of 0.01.

.05	.06	.07	.08	.09
.95122942	.94176453	.93239382	.92311635	.91393119
.86070798	.85214379	.84366482	.83527021	.82695913
.77880078	.77105159	.76337949	.75578374	.74826357
.70468809	.69767633	.69073433	.68386141	.67705687
.63762815	.63128365	.62500227	.61878339	.61262639
.57694981	.57120906	.56552544	.55989837	.55432728
.52204578	.51685133	.51170858	.50661699	.50157607
.47236655	.46766643	.46301307	.45840601	.45384480
.42741493	.42316208	.41895155	.41478291	.41065575
.38674102	.38289289	.37908304	.37531110	.37157669
.34993775	.34645581	.34300852	.33959553	.33621649
.31663677	.31348618	.31036694	.30727874	.30422126
.28650480	.28365403	.28083162	.27803730	.27527078
.25924026	.25666078	.25410696	.25157855	.24907530
.23457029	.23223627	.22992549	.22763769	.22537266
.21224797	.21013607	.20804518	.20597510	.20392561
.19204991	.19013898	.18824707	.18637398	.18451952
.17377394	.17204486	.17033299	.16863815	.16696017
.15723717	.15567263	.15412366	.15259011	.15107181
.14227407	.14085842	.13945686	.13806924	.13669543
.12873490	.12745397	.12618578	.12493021	.12368714
.11648416	.11532512	.11417762	.11304153	.11191675
.10539922	.10435048	.10331218	.10228421	.10126646
.09536916	.09442022	.09348073	.09255058	.09162968
.08629359	.08543495	.08458486	.08374323	.08290997
.07808167	.07730474	.07653555	.07577400	.07502004
.07065121	.06994822	.06925223	.06856315	.06788094
.06392786	.06329177	.06266200	.06203851	.06142121
.05784432	.05726876	.05669893	.05613476	.05557621
.05233971	.05181892	.05130331	.05079283	.05028744
.04735892	.04688770	.04642115	.04595926	.04550195
.04285213	.04242574	.04200360	.04158566	.04117187
.03877421	.03838840	.03800643	.03762826	.03725385
.03508435	.03473526	.03438964	.03404745	.03370868
.03174564	.03142976	.03111703	.03080741	.03050087

Table **A.V.2** e^{-x}: $x = 3.50$ to

EXPONENTIAL TABLE

EXP(-X)

X	0.00	.01	.02	.03	.04
3.5	.03019738	.02989691	.02959944	.02930492	.02901333
3.6	.02732372	.02705185	.02678268	.02651618	.02625234
3.7	.02472353	.02447752	.02423397	.02399284	.02375410
3.8	.02237077	.02214818	.02192780	.02170962	.02149360
3.9	.02024191	.02004050	.01984109	.01964367	.01944821
4.0	.01831564	.01813340	.01795296	.01777433	.01759747
4.1	.01657268	.01640777	.01624451	.01608288	.01592285
4.2	.01499558	.01484637	.01469864	.01455239	.01440759
4.3	.01356856	.01343355	.01329988	.01316755	.01303653
4.4	.01227734	.01215518	.01203423	.01191449	.01179594
4.5	.01110900	.01099846	.01088902	.01078068	.01067341
4.6	.01005184	.00995182	.00985280	.00975476	.00965770
4.7	.00909528	.00900478	.00891518	.00882647	.00873865
4.8	.00822975	.00814786	.00806679	.00798652	.00790705
4.9	.00744658	.00737249	.00729913	.00722650	.00715460
5.0	.00673795	.00667090	.00660453	.00653881	.00647375
5.1	.00609675	.00603608	.00597602	.00591656	.00585769
5.2	.00551656	.00546167	.00540733	.00535353	.00530026
5.3	.00499159	.00494193	.00489275	.00484407	.00479587
5.4	.00451658	.00447164	.00442715	.00438310	.00433948
5.5	.00408677	.00404611	.00400585	.00396599	.00392653
5.6	.00369786	.00366107	.00362464	.00358858	.00355287
5.7	.00334597	.00331267	.00327971	.00324708	.00321477
5.8	.00302755	.00299743	.00296761	.00293808	.00290884
5.9	.00273944	.00271219	.00268520	.00265848	.00263203
6.0	.00247875	.00245409	.00242967	.00240549	.00238156
6.1	.00224287	.00222055	.00219846	.00217658	.00215492
6.2	.00202943	.00200924	.00198925	.00196945	.00194986
6.3	.00183630	.00181803	.00179994	.00178203	.00176430
6.4	.00166156	.00164502	.00162866	.00161245	.00159641
6.5	.00150344	.00148848	.00147367	.00145901	.00144449
6.6	.00136037	.00134683	.00133343	.00132016	.00130703
6.7	.00123091	.00121866	.00120654	.00119453	.00118265
6.8	.00111378	.00110269	.00109172	.00108086	.00107010
6.9	.00100779	.00099776	.00098783	.00097800	.00096827

$x = 6.99$ in steps of 0.01.

.05	.06	.07	.08	.09
.02872464	.02843882	.02815585	.02787570	.02759833
.02599113	.02573251	.02547647	.02522297	.02497200
.02351775	.02328374	.02305206	.02282269	.02259560
.02127974	.02106800	.02085837	.02065083	.02044535
.01925470	.01906311	.01887343	.01868564	.01849971
.01742237	.01724902	.01707739	.01690747	.01673923
.01576442	.01560756	.01545226	.01529851	.01514628
.01426423	.01412230	.01398178	.01384266	.01370493
.01290681	.01277839	.01265124	.01252536	.01240073
.01167857	.01156236	.01144732	.01133341	.01122064
.01056720	.01046206	.01035796	.01025490	.01015286
.00956160	.00946646	.00937227	.00927901	.00918669
.00865170	.00856561	.00848038	.00839600	.00831246
.00782838	.00775048	.00767337	.00759701	.00752142
.00708341	.00701293	.00694315	.00687406	.00680566
.00640933	.00634556	.00628242	.00621991	.00615802
.00579940	.00574170	.00568457	.00562801	.00557201
.00524752	.00519530	.00514361	.00509243	.00504176
.00474815	.00470091	.00465413	.00460782	.00456197
.00429630	.00425356	.00421123	.00416933	.00412784
.00388746	.00384878	.00381048	.00377257	.00373503
.00351752	.00348252	.00344787	.00341356	.00337959
.00318278	.00315111	.00311976	.00308872	.00305798
.00287990	.00285124	.00282287	.00279479	.00276698
.00260584	.00257991	.00255424	.00252883	.00250366
.00235786	.00233440	.00231117	.00228818	.00226541
.00213348	.00211225	.00209124	.00207043	.00204983
.00193045	.00191125	.00189223	.00187340	.00185476
.00174675	.00172937	.00171216	.00169512	.00167826
.00158052	.00156480	.00154923	.00153381	.00151855
.00143012	.00141589	.00140180	.00138785	.00137404
.00129402	.00128115	.00126840	.00125578	.00124328
.00117088	.00115923	.00114769	.00113627	.00112497
.00105946	.00104891	.00103848	.00102814	.00101791
.00095864	.00094910	.00093965	.00093030	.00092105

Table **A.V.3** e^{-x}: $x = 7.00$ to $x = 9.99$ in steps of 0.01.

EXPONENTIAL TABLE

EXP(-X)

X	0.00	.01	.02	.03	.04
7.0	.00091188	.00090281	.00089383	.00088493	.00087613
7.1	.00082510	.00081689	.00080877	.00080072	.00079275
7.2	.00074659	.00073916	.00073180	.00072452	.00071731
7.3	.00067554	.00066882	.00066216	.00065557	.00064905
7.4	.00061125	.00060517	.00059915	.00059319	.00058729
7.5	.00055308	.00054758	.00054213	.00053674	.00053140
7.6	.00050045	.00049547	.00049054	.00048566	.00048083
7.7	.00045283	.00044832	.00044386	.00043944	.00043507
7.8	.00040973	.00040566	.00040162	.00039763	.00039367
7.9	.00037074	.00036705	.00036340	.00035979	.00035621
8.0	.00033546	.00033212	.00032882	.00032555	.00032231
8.1	.00030354	.00030052	.00029753	.00029457	.00029164
8.2	.00027465	.00027192	.00026922	.00026654	.00026388
8.3	.00024852	.00024604	.00024360	.00024117	.00023877
8.4	.00022487	.00022263	.00022041	.00021822	.00021605
8.5	.00020347	.00020144	.00019944	.00019745	.00019549
8.6	.00018411	.00018227	.00018046	.00017866	.00017689
8.7	.00016659	.00016493	.00016329	.00016166	.00016005
8.8	.00015073	.00014923	.00014775	.00014628	.00014482
8.9	.00013639	.00013503	.00013369	.00013236	.00013104
9.0	1.2341E-04	1.2218E-04	1.2097E-04	1.1976E-04	1.1857E-04
9.1	1.1167E-04	1.1055E-04	1.0945E-04	1.0837E-04	1.0729E-04
9.2	1.0104E-04	1.0003E-04	9.9039E-05	9.8053E-05	9.7078E-05
9.3	9.1424E-05	9.0515E-05	8.9614E-05	8.8722E-05	8.7839E-05
9.4	8.2724E-05	8.1901E-05	8.1086E-05	8.0279E-05	7.9480E-05
9.5	7.4852E-05	7.4107E-05	7.3370E-05	7.2640E-05	7.1917E-05
9.6	6.7729E-05	6.7055E-05	6.6388E-05	6.5727E-05	6.5073E-05
9.7	6.1283E-05	6.0674E-05	6.0070E-05	5.9472E-05	5.8881E-05
9.8	5.5452E-05	5.4900E-05	5.4354E-05	5.3813E-05	5.3277E-05
9.9	5.0175E-05	4.9675E-05	4.9181E-05	4.8692E-05	4.8207E-05

X	0.0	1.0	2.0	3.0	4.0
10.0	4.5400E-05	1.6702E-05	6.1442E-06	2.2603E-06	8.3153E-07
20.0	2.0612E-09	7.5826E-10	2.7895E-10	1.0262E-10	3.7751E-11
30.0	9.3576E-14	3.4425E-14	1.2664E-14	4.6589E-15	1.7139E-15
40.0	4.2484E-18	1.5629E-18	5.7495E-19	2.1151E-19	7.7811E-20
50.0	1.9287E-22	7.0955E-23	2.6103E-23	9.6027E-24	3.5326E-24

$x = 10.0$ to $x = 59.0$ in steps of 1.0.

.05	.06	.07	.08	.09
.00086741	.00085878	.00085023	.00084177	.00083340
.00078486	.00077705	.00076932	.00076167	.00075409
.00071017	.00070311	.00069611	.00068919	.00068233
.00064259	.00063620	.00062987	.00062360	.00061740
.00058144	.00057566	.00056993	.00056426	.00055864
.00052611	.00052088	.00051569	.00051056	.00050548
.00047604	.00047131	.00046662	.00046197	.00045738
.00043074	.00042646	.00042221	.00041801	.00041385
.00038975	.00038587	.00038203	.00037823	.00037447
.00035266	.00034915	.00034568	.00034224	.00033883
.00031910	.00031593	.00031278	.00030967	.00030659
.00028874	.00028586	.00028302	.00028020	.00027741
.00026126	.00025866	.00025609	.00025354	.00025101
.00023640	.00023404	.00023172	.00022941	.00022713
.00021390	.00021177	.00020966	.00020758	.00020551
.00019355	.00019162	.00018971	.00018782	.00018596
.00017513	.00017338	.00017166	.00016995	.00016826
.00015846	.00015688	.00015532	.00015378	.00015225
.00014338	.00014196	.00014054	.00013914	.00013776
.00012974	.00012845	.00012717	.00012590	.00012465
1.1739E-04	1.1622E-04	1.1507E-04	1.1392E-04	1.1279E-04
1.0622E-04	1.0516E-04	1.0412E-04	1.0308E-04	1.0205E-04
9.6112E-05	9.5155E-05	9.4209E-05	9.3271E-05	9.2343E-05
8.6965E-05	8.6100E-05	8.5243E-05	8.4395E-05	8.3555E-05
7.8690E-05	7.7907E-05	7.7131E-05	7.6364E-05	7.5604E-05
7.1201E-05	7.0493E-05	6.9791E-05	6.9097E-05	6.8409E-05
6.4426E-05	6.3785E-05	6.3150E-05	6.2522E-05	6.1899E-05
5.8295E-05	5.7715E-05	5.7140E-05	5.6572E-05	5.6009E-05
5.2747E-05	5.2222E-05	5.1703E-05	5.1188E-05	5.0679E-05
4.7728E-05	4.7253E-05	4.6783E-05	4.6317E-05	4.5856E-05

5.0	6.0	7.0	8.0	9.0
3.0590E-07	1.1254E-07	4.1399E-08	1.5230E-08	5.6028E-09
1.3888E-11	5.1091E-12	1.8795E-12	6.9144E-13	2.5437E-13
6.3051E-16	2.3195E-16	8.5330E-17	3.1391E-17	1.1548E-17
2.8625E-20	1.0531E-20	3.8740E-21	1.4252E-21	5.2429E-22
1.2996E-24	4.7809E-25	1.7588E-25	6.4702E-26	2.3803E-26

APPENDIX VI

TABLES OF THE GAUSSIAN (NORMAL) DISTRIBUTION

Table A.VI.1	The normal p.d.f. $\varphi(z)$ versus z; $z = 0.00$ to $z = 2.99$ in steps of 0.01, $z = 3.0$ to $z = 9.9$ in steps of 0.1.
Table A.VI.2	$A(z) = \int_{-z}^{+z} \varphi(t)\,dt$ versus z; $z = 0.00$ to $z = 2.99$ in steps of 0.01. $1 - A(z)$ versus z; $z = 3.0$ to $z = 9.9$ in steps of 0.1.
Table A.VI.3	The cumulative distribution function, $\Phi(z)$ versus z $P(z) = \Phi(z) = \int_{-\infty}^{z} \varphi(t)\,dt$ versus z; $z = 0.00$ to $z = 2.99$ in steps of 0.01. $1 - P(z)$ vs. z; $z = 3.0$ to $z = 9.9$ in steps of 0.01.

Table A.VI.1 The normal p.d.f. $\varphi(z)$ versus z; $z = 0.00$ to

GAUSSIAN PROBABILITY DISTRIBUTION. THE GAUSSIAN OR NORMAL

Z	0.00	.01	.02	.03	.04
0.0	.39894228	.39892233	.39886250	.39876280	.39862325
.1	.39695255	.39653597	.39608021	.39558542	.39505174
.2	.39104269	.39024188	.38940376	.38852859	.38761662
.3	.38138782	.38022635	.37903053	.37780068	.37653716
.4	.36827014	.36678166	.36526267	.36371360	.36213488
.5	.35206533	.35029188	.34849251	.34666772	.34481800
.6	.33322460	.33121468	.32918396	.32713298	.32506226
.7	.31225393	.31006029	.30785126	.30562741	.30338928
.8	.28969155	.28736890	.28503636	.28269448	.28034381
.9	.26608525	.26368804	.26128630	.25888055	.25647129
1.0	.24197072	.23955110	.23713195	.23471376	.23229700
1.1	.21785218	.21545816	.21306915	.21068555	.20830779
1.2	.19418606	.19186015	.18954316	.18723542	.18493728
1.3	.17136859	.16914676	.16693704	.16473972	.16255506
1.4	.14972747	.14763850	.14556413	.14350455	.14145997
1.5	.12951760	.12758295	.12566464	.12376279	.12187754
1.6	.11092083	.10915477	.10740608	.10567483	.10396110
1.7	.09404908	.09245913	.09088698	.08933262	.08779607
1.8	.07895016	.07753789	.07614327	.07476626	.07340681
1.9	.06561581	.06437766	.06315656	.06195242	.06076517
2.0	.05399097	.05291923	.05186358	.05082390	.04980009
2.1	.04398360	.04306742	.04216611	.04127953	.04040755
2.2	.03547459	.03470094	.03394076	.03319392	.03246027
2.3	.02832704	.02768157	.02704810	.02642649	.02581658
2.4	.02239453	.02186237	.02134071	.02082943	.02032836
2.5	.01752830	.01709467	.01667010	.01625445	.01584758
2.6	.01358297	.01323370	.01289213	.01255811	.01223153
2.7	.01042093	.01014283	.00987115	.00960580	.00934664
2.8	.00791545	.00769651	.00748287	.00727444	.00707110
2.9	.00595253	.00578210	.00561598	.00545410	.00529634

	0.00	.10	.20	.30	.40
3.0	4.4318E-03	3.2668E-03	2.3841E-03	1.7226E-03	1.2322E-03
4.0	1.3383E-04	8.9262E-05	5.8943E-05	3.8535E-05	2.4942E-05
5.0	1.4867E-06	8.9724E-07	5.3610E-07	3.1713E-07	1.8574E-07
6.0	6.0759E-09	3.3179E-09	1.7938E-09	9.6014E-10	5.0881E-10
7.0	9.1347E-12	4.5135E-12	2.2080E-12	1.0694E-12	5.1278E-13
8.0	5.0523E-15	2.2588E-15	9.9984E-16	4.3816E-16	1.9011E-16
9.0	1.0280E-18	4.1586E-19	1.6656E-19	6.6046E-20	2.5929E-20

$z = 2.99$ in steps of 0.01, $z = 3.9$ to $z = 9.9$ in steps of 0.1.

ERROR DISTRIBUTION φ VS. Z= (X-U)/SIGMA

.05	.06	.07	.08	.09
.39844391	.39822483	.39796607	.39766771	.39732983
.39447933	.39386836	.39321901	.39253148	.39180597
.38666812	.38568337	.38466266	.38360629	.38251457
.37524035	.37391061	.37254832	.37115388	.36972768
.36052696	.35889029	.35722533	.35553253	.35381237
.34294386	.34104579	.33912431	.33717994	.33521320
.32297236	.32086380	.31873714	.31659291	.31443166
.30113743	.29887241	.29659476	.29430503	.29200378
.27798489	.27561825	.27324443	.27086397	.26847740
.25405906	.25164434	.24922765	.24680949	.24439035
.22988214	.22746963	.22505994	.22265350	.22025077
.20593627	.20357139	.20121354	.19886312	.19652050
.18264909	.18037116	.17810384	.17584743	.17360225
.16038333	.15822479	.15607970	.15394829	.15183080
.13943057	.13741654	.13541806	.13343530	.13146843
.12000900	.11815730	.11632253	.11450480	.11270421
.10226492	.10058637	.09892547	.09728227	.09565680
.08627732	.08477636	.08329319	.08182778	.08038011
.07206487	.07074039	.06943331	.06814357	.06687109
.05959471	.05844094	.05730379	.05618314	.05507890
.04879202	.04779957	.04682264	.04586108	.04491477
.03955004	.03870686	.03787786	.03706291	.03626187
.03173965	.03103193	.03033696	.02965458	.02898466
.02521822	.02463127	.02405557	.02349099	.02293735
.01983735	.01935628	.01888498	.01842331	.01797113
.01544935	.01505962	.01467825	.01430511	.01394006
.01191224	.01160014	.01129507	.01099694	.01070560
.00909356	.00884645	.00860520	.00836969	.00813981
.00687277	.00667932	.00649068	.00630673	.00612738
.00514264	.00499290	.00484703	.00470496	.00456659

.50	.60	.70	.80	.90
8.7268E-04	6.1190E-04	4.2478E-04	2.9195E-04	1.9866E-04
1.5984E-05	1.0141E-05	6.3698E-06	3.9613E-06	2.4390E-06
1.0770E-07	6.1826E-08	3.5140E-08	1.9773E-08	1.1016E-08
2.6696E-10	1.3867E-10	7.1313E-11	3.6310E-11	1.8303E-11
2.4343E-13	1.1442E-13	5.3241E-14	2.4529E-14	1.1188E-14
8.1662E-17	3.4730E-17	1.4623E-17	6.0958E-18	2.5158E-18
1.0078E-20	3.8781E-21	1.4775E-21	5.5730E-22	2.0812E-22

Table **A.VI.2** $A(z) = \int_{-z}^{+z} \varphi(t)\,dt$ versus z; $z = 0.00$ to $z = 2.99$ in steps

THE INTEGRAL OF THE GAUSSIAN PROBABILITY DISTRIBUTION A(Z) VS.

Z	0.00	.01	.02	.03	.04
0.0	0.00000000	.00797871	.01595663	.02393295	.03190687
.1	.07965567	.08759063	.09551685	.10343357	.11134001
.2	.15851942	.16633233	.17412885	.18190823	.18966974
.3	.23582284	.24343904	.25103167	.25860004	.26614347
.4	.31084348	.31819405	.32551455	.33280436	.34006289
.5	.38292492	.38994854	.39693642	.40388807	.41080297
.6	.45149376	.45813819	.46474221	.47130542	.47782740
.7	.51607270	.52229586	.52847500	.53460982	.54070001
.8	.57628920	.58205982	.58778389	.59346122	.59909161
.9	.63187975	.63717749	.64242724	.64762892	.65278244
1.0	.68268949	.68750471	.69227154	.69698999	.70166010
1.1	.72866788	.73300097	.73728624	.74152378	.74571370
1.2	.76936066	.77372111	.77753513	.78130290	.78502461
1.3	.80639903	.80980416	.81316498	.81648173	.81975466
1.4	.83848668	.84146032	.84439232	.84728298	.85013260
1.5	.86638560	.86895658	.87148902	.87398327	.87643965
1.6	.89040142	.89260214	.89476772	.89689850	.89899483
1.7	.91036907	.91273413	.91456756	.91636972	.91814098
1.8	.92813936	.92970421	.93124100	.93275006	.93423176
1.9	.94256688	.94386679	.94514210	.94639316	.94762031
2.0	.95449974	.95556881	.95661661	.95764346	.95864967
2.1	.96427116	.96514164	.96599395	.96682839	.96764523
2.2	.97219310	.97289484	.97353123	.97425256	.97490903
2.3	.97855178	.97911185	.97965912	.98019385	.98071626
2.4	.98360493	.98404748	.98447949	.98490118	.98531274
2.5	.98758067	.98792688	.98826452	.98859375	.98891475
2.6	.99067762	.99094578	.99120702	.99146151	.99170940
2.7	.99306605	.99327168	.99347181	.99366657	.99385608
2.8	.99488974	.99504585	.99519764	.99534520	.99548865
2.9	.99626837	.99638571	.99649969	.99661038	.99671788

1-A

	0.00	.10	.20	.30	.40
3.0	2.6998E-03	1.9352E-03	1.3743E-03	9.6685E-04	6.7386E-04
4.0	6.3342E-05	4.1315E-05	2.6692E-05	1.7080E-05	1.0825E-05
5.0	5.7330E-07	3.3965E-07	1.9929E-07	1.1580E-07	6.6641E-08
6.0	1.9732E-09	1.0607E-09	5.6463E-10	2.9765E-10	1.5538E-10
7.0	2.5596E-12	1.2476E-12	6.0213E-13	2.8777E-13	1.3618E-13
8.0	1.2442E-15	5.4959E-16	2.4039E-16	1.0411E-16	4.4648E-17
9.0	2.2572E-19	9.0332E-20	3.5795E-20	1.4045E-20	5.4563E-21

of 0.01. $1-A(z)$ versus z; $z=3.0$ to $z=9.9$ in steps of 0.1.

Z = (X-U)/SIGMA

.05	.06	.07	.08	.09
.03987761	.04784437	.05580634	.06376274	.07171279
.11923538	.12711893	.13498986	.14284743	.15069087
.19741265	.20513623	.21283975	.22052250	.22818376
.27366130	.28115287	.28861751	.29605458	.30346345
.34728956	.35448378	.36164498	.36877261	.37586610
.41768063	.42452056	.43132230	.43808538	.44480935
.48430778	.49074617	.49714221	.50349554	.50980581
.54674530	.55274542	.55870011	.56460912	.57047223
.60467491	.61021096	.61569960	.62114069	.62653411
.65788775	.66294479	.66795351	.67291388	.67782588
.70628189	.71085540	.71538069	.71985782	.72428686
.74985613	.75395119	.75799903	.76199979	.76595361
.78870045	.79233064	.79591537	.79945486	.80294934
.82298402	.82617008	.82931310	.83241336	.83547112
.85294148	.85570993	.85843825	.86112675	.86377576
.87885848	.88124012	.88358489	.88589313	.88816519
.90105706	.90308555	.90508064	.90704268	.90897205
.91988169	.92159219	.92327286	.92492404	.92654609
.93568645	.93711447	.93851618	.93989192	.94124204
.94882388	.95000421	.95116163	.95229647	.95340906
.95963557	.96060146	.96154766	.96247447	.96338220
.96844479	.96922733	.96999315	.97074254	.97147576
.97555105	.97617875	.97679242	.97739231	.97797868
.98122659	.98172506	.98221191	.98268736	.98315163
.98571438	.98610630	.98648869	.98686176	.98722569
.98922771	.98953278	.98983015	.99011997	.99040241
.99195082	.99216593	.99241488	.99263778	.99285460
.99404047	.99421986	.99439437	.99456411	.99472920
.99562808	.99576359	.99589528	.99602325	.99614758
.99682226	.99692361	.99702200	.99711752	.99721023

.50	.60	.70	.80	.90
4.6526E-04	3.1822E-04	2.1560E-04	1.4470E-04	9.6193E-05
6.7953E-06	4.2249E-06	2.6016E-06	1.5867E-06	9.5837E-07
3.7979E-08	2.1435E-08	1.1981E-08	6.6315E-09	3.6350E-09
8.0320E-11	4.1116E-11	2.0842E-11	1.0462E-11	5.2003E-12
6.3818E-14	2.9613E-14	1.3607E-14	6.1907E-15	2.7890E-15
1.8959E-17	7.9716E-18	3.3188E-18	1.3682E-18	5.5847E-19
2.0989E-21	7.9944E-22	3.0150E-22	1.1259E-22	4.1628E-23

Table **A.VI.3** The cumulative distribution function, $\Phi(z)$ versus z. $1-P(z)$ versus z; $z=3.0$

THE GAUSSIAN CUMULATIVE DISTRIBUTION FUNCTION: P(Z) = Φ(Z) VS. Z =

Z	0.00	.01	.02	.03	.04
0.0	.50000000	.50398936	.50797831	.51196647	.51595344
.1	.53982784	.54379531	.54775843	.55171679	.55567000
.2	.57925971	.58316616	.58706442	.59095412	.59483487
.3	.61791142	.62171952	.62551583	.62930002	.63307174
.4	.65542174	.65909703	.66275727	.66640218	.67003145
.5	.69146246	.69497427	.69846821	.70194403	.70540148
.6	.72574688	.72906910	.73237111	.73565271	.73891370
.7	.75803635	.76114793	.76423750	.76730491	.77035000
.8	.78814460	.79102991	.79389195	.79673061	.79954581
.9	.81593987	.81858875	.82121362	.82381446	.82639122
1.0	.84134475	.84375235	.84613577	.84849500	.85083005
1.1	.86433394	.86650049	.86864312	.87076189	.87285685
1.2	.88493033	.88686055	.88876756	.89065145	.89251230
1.3	.90319952	.90490208	.90658249	.90824086	.90987733
1.4	.91924334	.92073016	.92219616	.92364149	.92506630
1.5	.93319280	.93447829	.93574451	.93699164	.93821982
1.6	.94520071	.94630107	.94738386	.94844925	.94949742
1.7	.95543454	.95636706	.95728378	.95818486	.95907049
1.8	.96406968	.96485211	.96562050	.96637503	.96711588
1.9	.97128344	.97193339	.97257105	.97319658	.97381016
2.0	.97724987	.97778441	.97830831	.97882173	.97932484
2.1	.98213558	.98257082	.98299698	.98341419	.98382262
2.2	.98609655	.98644742	.98679062	.98712628	.98745454
2.3	.98927589	.98955592	.98982956	.99009692	.99035813
2.4	.99180246	.99202374	.99223975	.99245059	.99265637
2.5	.99379033	.99396344	.99413226	.99429687	.99445738
2.6	.99533881	.99547289	.99560351	.99573076	.99585470
2.7	.99653303	.99663584	.99673590	.99683328	.99692804
2.8	.99744487	.99752293	.99759882	.99767260	.99774432
2.9	.99813419	.99819286	.99824984	.99830519	.99835894

1-P

	0.00	.10	.20	.30	.40
3.0	1.3499E-03	9.6760E-04	6.8714E-04	4.8342E-04	3.3693E-04
4.0	3.1671E-05	2.0658E-05	1.3346E-05	8.5399E-06	5.4125E-06
5.0	2.8665E-07	1.6983E-07	9.9644E-08	5.7901E-08	3.3320E-08
6.0	9.8659E-10	5.3034E-10	2.8232E-10	1.4882E-10	7.7688E-11
7.0	1.2798E-12	6.2378E-13	3.0106E-13	1.4388E-13	6.8092E-14
8.0	6.2210E-16	2.7480E-16	1.2019E-16	5.2056E-17	2.2324E-17
9.0	1.1286E-19	4.5166E-20	1.7897E-20	7.0223E-21	2.7282E-21

$P(z) = \Phi(z) = \int_{-\infty}^{z} \varphi(t)\, dt$ versus z; $z = 0.00$ to $z = 2.99$ in steps of 0.01. to $z = 9.9$ in steps of 0.1.

(X-U)/SIGMA

.05	.06	.07	.08	.09
.51993881	.52392218	.52790317	.53188137	.53585639
.55961769	.56355946	.56749493	.57142372	.57534543
.59870633	.60256811	.60641987	.61026125	.61409188
.63683065	.64057643	.64430875	.64802729	.65173173
.67364478	.67724189	.68082249	.68438630	.68793305
.70884031	.71226028	.71566115	.71904269	.72240468
.74215389	.74537309	.74857110	.75174777	.75490291
.77337265	.77637271	.77935005	.78230456	.78523612
.80233746	.80510548	.80784980	.81057035	.81326706
.82894387	.83147239	.83397675	.83645694	.83891294
.85314094	.85542770	.85769035	.85992891	.86214343
.87492806	.87697560	.87899952	.88099989	.88297680
.89435023	.89616532	.89795768	.89972743	.90147467
.91149201	.91308504	.91465655	.91620668	.91773556
.92647074	.92785496	.92921912	.93056338	.93188788
.93942924	.94062006	.94179244	.94294657	.94408260
.95052853	.95154277	.95254032	.95352134	.95448602
.95994084	.96079610	.96163643	.96246202	.96327304
.96784323	.96855724	.96925809	.96994596	.97062102
.97441194	.97500210	.97558081	.97614824	.97670453
.97981778	.98030073	.98077383	.98123723	.98169110
.98422239	.98461367	.98499658	.98537127	.98573788
.98777553	.98808937	.98839621	.98869616	.98898934
.99061329	.99086253	.99110596	.99134368	.99157581
.99285719	.99305315	.99324435	.99343088	.99361285
.99461385	.99476639	.99491507	.99505998	.99520120
.99597541	.99609297	.99620744	.99631889	.99642740
.99702024	.99710993	.99719719	.99728206	.99736460
.99781404	.99788179	.99794764	.99801162	.99807379
.99841113	.99846180	.99851100	.99855876	.99860511

.50	.60	.70	.80	.90
2.3263E-04	1.5911E-04	1.0780E-04	7.2348E-05	4.8096E-05
3.3977E-06	2.1125E-06	1.3008E-06	7.9333E-07	4.7919E-07
1.8990E-08	1.0718E-08	5.9904E-09	3.3157E-09	1.8175E-09
4.0160E-11	2.0558E-11	1.0421E-11	5.2310E-12	2.6001E-12
3.1909E-14	1.4807E-14	6.8033E-15	3.0954E-15	1.3945E-15
9.4795E-18	3.9858E-18	1.6594E-18	6.8408E-19	2.7923E-19
1.0495E-21	3.9972E-22	1.5075E-22	5.6293E-23	2.0813E-23

APPENDIX VII

TABLES AND GRAPHS OF THE CHI-SQUARE DISTRIBUTION

Tables of the upper-tail area function, $Q(\chi^2) = \Pr(X^2 > \chi^2)$ as a function of χ^2 and n (the number of degrees of freedom).

Table A.VII.1*a*	$Q(\chi^2)$ versus n, χ^2; $n = 1 - 100$, $\chi^2 = 1 - 10$ in steps of 1.
Table A.VII.1*b*	$Q(\chi^2)$ versus n, χ^2; $n = 1 - 100$, $\chi^2 = 11 - 20$ in steps of 1.
Table A.VII.1*c*	$Q(\chi^2)$ versus n, χ^2; $n = 1 - 100$, $\chi^2 = 21 - 30$ in steps of 1.

Tables of χ^2 as a function of $Q(\chi^2)$ and n.

Table A.VII.2*a*	χ^2 versus n, Q; $n = 1 - 200$, $Q = 0.99$ to $Q = 0.75$.
Table A.VII.2*b*	χ^2 versus n, Q; $n = 1 - 200$, $Q = 0.70$ to $Q = 0.35$.
Table A.VII.2*c*	χ^2 versus n, Q; $n = 1 - 200$, $Q = 0.30$ to $Q = 0.01$.
Table A.VII.2*d*	χ^2 versus n, Q; $n = 1 - 200$, $Q = 0.005$ to $Q = 10^{-8}$.
Figure A.VII.1*a*	Plot of ordinate Q versus abscissa χ^2 for various n, $Q = 1.0 - 10^{-5}$.
Figure A.VII.1*b*	Plot of ordinate Q versus abscissa χ^2 for various n, $Q = 10^{-5} - 10^{-26}$.

Table **A.VII.1a** $Q(\chi^2)$ versus n, χ^2;

n \ χ^2	1	2	3	4	5
1	.31731051	.60653066	.80125196	.90979599	.96256577
2	.15729921	.36787944	.57240670	.73575888	.84914504
3	.08326452	.22313016	.39162518	.55782540	.69998584
4	.04550026	.13533528	.26146413	.40600585	.54941595
5	.02534732	.08208500	.17179714	.28729750	.41588019
6	.01430588	.04978707	.11161022	.19914827	.30621892
7	.00815097	.03019738	.07189777	.13588823	.22064031
8	.00467773	.01831564	.04601171	.09157819	.15623563
9	.00269980	.01110900	.02929089	.06109948	.10906416
10	.00156540	.00673795	.01856614	.04042768	.07523525
11	.00091112	.00408677	.01172588	.02656401	.05137998
12	.00053201	.00247875	.00738316	.01735127	.03478778
13	.00031149	.00150344	.00463660	.01127579	.02337877
14	.00018281	.00091188	.00290515	.00729506	.01560942
15	.00010751	.00055308	.00181665	.00470122	.01036234
16	6.3342E-05	3.3546E-04	1.1340E-03	3.0192E-03	6.8441E-03
17	3.7300E-05	2.0347E-04	7.0674E-04	1.9329E-03	4.4998E-03
18	2.2000E-05	1.2341E-04	4.3985E-04	1.2341E-03	2.9464E-03
19	1.3000E-05	7.4852E-05	2.7340E-04	7.8594E-04	1.9221E-03
20	7.7400E-06	4.5400E-05	1.6974E-04	4.9940E-04	1.2497E-03
21	4.6000E-06	2.7536E-05	1.0528E-04	3.1667E-04	8.1006E-04
22	2.7000E-06	1.6702E-05	6.5231E-05	2.0042E-04	5.2360E-04
23	1.6000E-06	1.0130E-05	4.0382E-05	1.2663E-04	3.3757E-04
24	9.6000E-07	6.1442E-06	2.4979E-05	7.9875E-05	2.1711E-04
25	5.7000E-07	3.7267E-06	1.5440E-05	5.0310E-05	1.3933E-04
26	3.4000E-07	2.2603E-06	9.5369E-06	3.1645E-05	8.9235E-05
27	2.0000E-07	1.3710E-06	5.8868E-06	1.9879E-05	5.7042E-05
28	1.2000E-07	8.3153E-07	3.6315E-06	1.2473E-05	3.6398E-05
29	7.2000E-08	5.0435E-07	2.2389E-06	7.8174E-06	2.3187E-05
30	4.3000E-08	3.0590E-07	1.3795E-06	4.8944E-06	1.4748E-05
40	2.5000E-10	2.0612E-09	1.0122E-08	4.3284E-08	1.4880E-07
50	1.5000E-12	1.3888E-11	0.	3.6109E-10	0.
60	1.0000E-14	9.3576E-14	0.	2.9009E-12	0.
70	0.	6.3051E-16	0.	2.2698E-14	0.
80	0.	0.	0.	1.7418E-16	0.
90	0.	0.	0.	0.	0.
100	0.	0.	0.	0.	0.

$n = 1 - 100$, $\chi^2 = 1 - 10$ in steps of 1.

6	7	8	9	10
.98561232	.99482854	.99824838	.99943750	.99982788
.91969860	.95984037	.98101184	.99146761	.99634015
.80884683	.88500223	.93435755	.96429497	.98142406
.67667642	.77977741	.85712346	.91141253	.94734698
.54381312	.65996323	.75757613	.83430826	.89117802
.42319008	.53974935	.64723189	.73991829	.81526324
.32084720	.42887986	.53663267	.63711941	.72544495
.23810331	.33259390	.43347012	.53414622	.62883694
.17357807	.25265605	.34229596	.43727419	.53210358
.12465202	.18857347	.26502592	.35048521	.44049329
.08837643	.13861902	.20169920	.27570894	.35751800
.06196880	.10055887	.15120388	.21330930	.28505650
.04303595	.07210839	.11184961	.16260626	.22367182
.02963616	.05118135	.08176542	.12232523	.17299161
.02025672	.03599940	.05914546	.09093598	.13206186
1.3754E-02	2.5116E-02	4.2380E-02	6.6882E-02	9.9632E-02
9.2832E-03	1.7396E-02	3.0109E-02	4.8716E-02	7.4364E-02
6.2322E-03	1.1970E-02	2.1226E-02	3.5174E-02	5.4964E-02
4.1636E-03	8.1873E-03	1.4860E-02	2.5193E-02	4.0263E-02
2.7694E-03	5.5697E-03	1.0336E-02	1.7912E-02	2.9253E-02
1.8346E-03	3.7701E-03	7.1474E-03	1.2650E-02	2.1094E-02
1.2109E-03	2.5404E-03	4.9159E-03	8.8790E-03	1.5105E-02
7.9648E-04	1.7046E-03	3.3642E-03	6.1963E-03	1.0747E-02
5.2226E-04	1.1394E-03	2.2918E-03	4.3013E-03	7.6004E-03
3.4145E-04	7.5880E-04	1.5546E-03	2.9712E-03	5.3455E-03
2.2264E-04	5.0367E-04	1.0503E-03	2.0430E-03	3.7402E-03
1.4481E-04	3.3328E-04	7.0699E-04	1.3988E-03	2.6043E-03
9.3963E-05	2.1989E-04	4.7425E-04	9.5387E-04	1.8052E-03
6.0837E-05	1.4469E-04	3.1710E-04	6.4804E-04	1.2460E-03
3.9308E-05	9.4959E-05	2.1138E-04	4.3872E-04	8.5664E-04
4.5551E-07	1.2583E-06	3.2037E-06	7.5980E-06	1.6945E-05
4.7011E-09	1.3940E-08	4.0868E-08	1.0720E-07	2.6691E-07
4.5010E-11	0.	4.6610E-10	0.	3.6243E-09
4.0889E-13	0.	4.9144E-12	0.	4.4338E-11
3.5729E-15	0.	4.8889E-14	0.	5.0205E-13
3.0300E-17	0.	4.6504E-16	0.	5.3559E-15
0.	0.	0.	0.	5.4497E-17

Table **A.VII.1b** $Q(\chi^2)$ versus n, χ^2

χ^2 / n	11	12	13	14	15
1	.99994961	.99998584	.99999617	.99999900	.99999975
2	.99849588	.99940582	.99977375	.99991676	.99997035
3	.99072589	.99554402	.99793432	.99907401	.99959780
4	.96991702	.98343639	.99119139	.99546619	.99773734
5	.93116661	.95797896	.97519313	.98581269	.99212641
6	.87336425	.91608206	.94615296	.96649146	.97974775
7	.79908350	.85761355	.90215156	.93471190	.95764975
8	.71330383	.78513039	.84360028	.88932602	.92378270
9	.62189233	.70293043	.77294354	.83105058	.87751745
10	.53038715	.61596065	.69393437	.76218346	.81973992
11	.44326328	.52891869	.61081762	.68603598	.75259437
12	.36364322	.44567964	.52764386	.60630278	.67902906
13	.29332541	.36904068	.44781167	.52652362	.60229794
14	.23299348	.30070828	.37384398	.44971106	.52552913
15	.18249693	.24143645	.30735277	.37815469	.45141721
16	.14113088	.19123606	.24912985	.31337428	.38205166
17	.10787559	.14959731	.19930407	.25617786	.31886441
18	.08158061	.11569052	.15751946	.20678084	.26266556
19	.06109351	.08852845	.12310366	.16494924	.21373388
20	.04534067	.06708596	.09521026	.13014142	.17193269
21	.03337105	.05038045	.07292863	.10163250	.13682932
22	.02437324	.03751981	.05536178	.07861437	.10780391
23	.01767513	.02772594	.04167628	.06026972	.08413986
24	.01273320	.02034103	.03113006	.04582231	.06509349
25	.00911668	.01482287	.02308373	.03456739	.04994343
26	.00648992	.01073389	.01700084	.02588692	.03802268
27	.00459523	.00772720	.01244109	.01925362	.02873635
28	.00323734	.00553205	.00904983	.01422792	.02156903
29	.00226996	.00393999	.00654593	.01045036	.01608463
30	.00158459	.00279243	.00470970	.00763190	.01192150
40	3.5775E-05	7.1909E-05	1.3823E-04	2.5512E-04	4.5350E-04
50	6.2545E-07	1.3971E-06	2.9809E-06	6.1063E-06	1.2041E-05
60	0.	2.2573E-08	5.2047E-08	1.1732E-07	0.
70	0.	3.2030E-10	0.	1.9301E-09	0.
80	0.	4.1273E-12	0.	2.8296E-11	0.
90	0.	4.9374E-14	0.	3.7951E-13	0.
100	0.	5.5678E-16	0.	4.7424E-15	0.

$n = 1 - 100,\ \chi^2 = 11 - 20$ in steps of 1.

16	17	18	19	20
.99999994	.99999999	1.00000000	1.00000000	1.00000000
.99998975	.99999656	.99999887	.99999964	.99999989
.99983043	.99993050	.99997226	.99998921	.99999590
.99890328	.99948293	.99976255	.99989366	.99995350
.99575330	.99777084	.99885975	.99943096	.99972265
.98809550	.99318566	.99619701	.99792846	.99889751
.97326108	.98354890	.99012634	.99421326	.99668506
.94886638	.96654666	.97863657	.98667088	.99186776
.91341353	.94026180	.95974269	.97347940	.98290727
.86662833	.90361029	.93190637	.95294580	.96817194
.80948528	.85656399	.89435668	.92383845	.94622253
.74397976	.80013722	.84723749	.88562533	.91607598
.67275778	.73618604	.79157303	.83857105	.87738405
.59871384	.66710194	.72909127	.78369131	.83049594
.52463853	.59548165	.66196712	.72259733	.77640761
.45296081	.52383493	.59254734	.65727800	.71662426
.38559710	.45436611	.52310505	.58986782	.65297366
.32389696	.38884088	.45565260	.52243827	.58740824
.26866318	.32853216	.39182348	.45683613	.52182602
.22022065	.27422927	.33281968	.39457818	.45792971
.17851058	.22629029	.27941305	.33680090	.39713260
.14319153	.18471904	.23198513	.28425626	.34051064
.11373451	.14925069	.19059013	.23734180	.28879454
.08950450	.11943497	.15502778	.19615236	.24239216
.06982546	.09470961	.12491620	.16054222	.20143110
.05402825	.07446053	.09975791	.13018901	.16581188
.04148315	.05806781	.07899549	.10465307	.13526399
.03161966	.04493821	.06205520	.08342862	.10939937
.02393612	.03452612	.04837906	.06598513	.08775936
.01800219	.02634508	.03744649	.05179846	.06985366
7.7859E-04	1.2942E-03	2.0873E-03	3.2723E-03	4.9954E-03
2.2925E-05	4.2240E-05	7.5483E-05	1.3106E-04	2.2148E-04
5.2337E-07	0.	2.0461E-06	3.8705E-06	7.1218E-06
9.9790E-09	0.	4.5193E-08	0.	1.8214E-07
1.6640E-10	0.	8.5693E-10	0.	3.9259E-09
2.5018E-12	0.	1.4440E-11	0.	7.4129E-11
3.4640E-14	0.	2.2150E-13	0.	1.2596E-12

Table A.VII.1c $Q(\chi^2)$ versus n, χ^2;

n \ χ^2	21	22	23	24	25
1	1.00000000	1.00000000	1.00000000	1.00000000	1.00000000
2	.99999997	.99999999	1.00000000	1.00000000	1.00000000
3	.99999848	.99999945	.99999980	.99999993	.99999998
4	.99998013	.99999169	.99999660	.99999864	.99999946
5	.99986784	.99993837	.99997186	.99998740	.99999447
6	.99942618	.99970766	.99985410	.99992861	.99996573
7	.99814223	.99898061	.99945189	.99971101	.99985048
8	.99514424	.99716023	.99837218	.99908477	.99949494
9	.98921405	.99333133	.99595747	.99759572	.99859620
10	.97891186	.98630473	.99127665	.99454691	.99665264
11	.96278682	.97474875	.98318834	.98901186	.99294560
12	.93961783	.95737908	.97047068	.97990804	.98656782
13	.90862396	.93316121	.95199004	.96612044	.97650130
14	.86959927	.90147921	.92687124	.94665038	.96173244
15	.82295181	.86223798	.89463358	.92075869	.94138257
16	.76965111	.81588579	.85526872	.88807600	.91482880
17	.71110619	.76336198	.80925154	.84866205	.88179375
18	.64900423	.70598832	.75748933	.80300838	.84239072
19	.58514009	.64532843	.70122462	.75198961	.79712054
20	.52126125	.58303975	.64191179	.69677615	.74682531
21	.45894421	.52073813	.58108752	.63872522	.69260967
22	.39950989	.45988870	.52025178	.57926676	.63574403
23	.34397841	.40172961	.46077089	.51979809	.57756337
24	.29305853	.34722942	.40380844	.46159733	.51937357
25	.24716408	.29707474	.35028534	.40576069	.46237366
26	.20644905	.25168203	.30086623	.35316493	.40759869
27	.17085318	.21122648	.25596760	.30445316	.35588453
28	.14015134	.17568121	.21578162	.26003992	.30785327
29	.11400151	.14486079	.18030984	.22013086	.26391601
30	.09198801	.11846441	.14940165	.18475180	.22428900
40	7.4368E-03	1.0812E-02	1.5369E-02	2.1387E-02	2.9164E-02
50	3.6480E-04	5.8646E-04	9.2132E-04	1.4160E-03	2.1312E-03
60	1.2776E-05	2.2349E-05	3.8212E-05	6.3877E-05	1.0461E-04
70	0.	6.6144E-07	0.	2.1865E-06	0.
80	0.	1.6202E-08	0.	6.0842E-08	0.
90	0.	3.4273E-10	0.	1.4416E-09	0.
100	0.	6.4502E-12	0.	3.0044E-11	0.

$n = 1 - 100$, $\chi^2 = 21 - 30$ in steps of 1.

26	27	28	29	30
1.00000000	1.00000000	1.00000000	1.00000000	1.00000000
1.00000000	1.00000000	1.00000000	1.00000000	1.00000000
.99999999	1.00000000	1.00000000	1.00000000	1.00000000
.99999979	.99999992	.99999997	.99999999	1.00000000
.99999762	.99999899	.99999958	.99999983	.99999993
.99998385	.99999252	.99999660	.99999848	.99999933
.99992404	.99996209	.99998140	.99999102	.99999574
.99972628	.99985423	.99992367	.99996068	.99998007
.99919486	.99954614	.99974841	.99986279	.99992634
.99798115	.99880304	.99930201	.99959948	.99977375
.99554912	.99723879	.99831488	.99898786	.99940143
.99117252	.99429445	.99637151	.99772850	.99859965
.98397336	.98924716	.99289982	.99538406	.99704424
.97300023	.98125472	.98718861	.99137738	.99428280
.95733413	.96943196	.97843535	.98501495	.98973957
.93620280	.95294725	.96581930	.97553596	.98274301
.90908293	.93112246	.94858889	.96218127	.97257465
.87577343	.90351971	.92614923	.94427238	.95853367
.83642971	.87000144	.89813593	.92128800	.94000801
.79155648	.83075612	.86446442	.89292709	.91654153
.74196393	.78628827	.82534904	.85914941	.88788788
.68869665	.73737721	.78129117	.82018942	.85404401
.63294706	.68501245	.73304037	.77654315	.81525988
.57596525	.63031609	.68153563	.72893167	.77202453
.51897522	.57446199	.62783534	.67824747	.72503188
.46310475	.51860045	.57304456	.62549104	.67513153
.40933318	.46379482	.51824704	.57170510	.62327113
.35845842	.41097351	.46444756	.51791303	.57043671
.31108220	.36089916	.41252792	.46506624	.51759670
.26761103	.31415383	.36321784	.41400364	.46565371
3.9012E-02	5.1237E-02	6.6128E-02	8.3937E-02	1.0486E-01
3.1441E-03	4.5508E-03	6.4675E-03	9.0317E-03	1.2402E-02
1.6770E-04	2.6381E-04	4.0728E-04	6.1801E-04	9.2068E-04
6.6346E-06	0.	1.8610E-05	0.	4.8549E-05
2.0964E-07	0.	6.6749E-07	0.	1.9756E-06
5.5622E-09	0.	1.9826E-08	0.	6.5673E-08
1.2835E-10	0.	5.0645E-10	0.	1.8568E-09

Table **A.VII.2a** χ^2 versus *n*, *Q*; *n*

n \ Q	9.900E-01	9.800E-01	9.750E-01	9.500E-01
1	.00016	.00063	.00098	.00393
2	.02010	.04041	.05064	.10259
3	.11483	.18483	.21580	.35185
4	.29711	.42940	.48442	.71072
5	.55430	.75189	.83121	1.14548
6	.87209	1.13442	1.23734	1.63539
7	1.23904	1.56430	1.68987	2.16735
8	1.64650	2.03248	2.17973	2.73264
9	2.08790	2.53238	2.70039	3.32512
10	2.55821	3.05905	3.24697	3.94030
11	3.05349	3.60869	3.81575	4.57481
12	3.57058	4.17829	4.40379	5.22604
13	4.10691	4.76545	5.00875	5.89186
14	4.66043	5.36821	5.62873	6.57064
15	5.22936	5.98491	6.26214	7.26094
16	5.81222	6.61425	6.90766	7.96165
17	6.40776	7.25501	7.56418	8.67176
18	7.01492	7.90623	8.23075	9.39046
19	7.63274	8.56703	8.90651	10.11702
20	8.26040	9.23670	9.59078	10.85083
22	9.54250	10.60003	10.98232	12.33802
24	10.85638	11.99184	12.40115	13.84843
26	12.19815	13.40960	13.84391	15.37918
28	13.56475	14.84749	15.30786	16.92789
30	14.95350	16.30621	16.79077	18.49268
32	16.36224	17.78273	18.29076	20.07192
34	17.78916	19.27544	19.80625	21.66429
36	19.23269	20.78296	21.33588	23.26863
38	20.69151	22.30401	22.87848	24.88391
40	22.16426	23.83757	24.43304	26.50932
42	23.65010	25.38274	25.99866	28.14405
44	25.14809	26.93861	27.57457	29.78748
46	26.65732	28.50453	29.16005	31.43901
48	28.17701	30.07982	30.75451	33.09808
50	29.70669	31.66389	32.35736	34.76427
60	37.48488	39.69947	40.48175	43.18797
70	45.44173	47.89346	48.75756	51.73928
80	53.54012	56.21292	57.15317	60.39148
90	61.75410	64.63468	65.64662	69.12603
100	70.06502	73.14222	74.22193	77.92952
120	86.92345	90.35675	91.57264	95.70464
140	104.03443	107.81490	109.13687	113.65936
160	121.34571	125.44000	126.87005	131.75613
180	138.82047	143.20963	144.74126	149.96881
200	156.43203	161.10028	162.72798	168.27862

= 1 − 200, $Q = 0.99$ to $Q = 0.75$.

9.000E-01	8.500E-01	8.000E-01	7.500E-01
.01579	.03577	.06418	.10153
.21072	.32504	.44629	.57536
.58437	.79777	1.00517	1.21253
1.06362	1.36648	1.64878	1.92256
1.61031	1.99382	2.34253	2.67460
2.20413	2.66127	3.07009	3.45460
2.83311	3.35828	3.82232	4.25485
3.48954	4.07820	4.59357	5.07064
4.16816	4.81652	5.38005	5.89882
4.86519	5.57006	6.17908	6.73720
5.57778	6.33644	6.98867	7.58414
6.30380	7.11384	7.80733	8.43842
7.04150	7.90084	8.63386	9.29906
7.78954	8.69630	9.46733	10.16532
8.54675	9.49929	10.30696	11.03653
9.31224	10.30903	11.15212	11.91223
10.08519	11.12487	12.00226	12.79193
10.86494	11.94626	12.85695	13.67529
11.65091	12.77272	13.71579	14.56200
12.44261	13.60386	14.57844	15.45178
14.04150	15.27876	16.31405	17.23963
15.65870	16.96856	18.06181	19.03726
17.29189	18.67139	19.82020	20.84344
18.93925	20.38573	21.58797	22.65717
20.59924	22.11035	23.36412	24.47761
22.27059	23.84420	25.14779	26.30411
23.95226	25.58641	26.93827	28.13609
25.64331	27.33625	28.73496	29.97305
27.34295	29.09308	30.53735	31.81457
29.05053	30.85633	32.34496	33.66030
30.76542	32.62553	34.15741	35.50992
32.48714	34.40025	35.97436	37.36314
34.21517	36.18011	37.79549	39.21971
35.94913	37.96477	39.62052	41.07943
37.68866	39.75394	41.44921	42.94209
46.45891	48.75867	50.64062	52.29382
55.32895	57.84429	59.89783	61.69834
64.27786	66.99380	69.20694	71.14452
73.29110	75.19543	78.55844	80.62467
82.35814	85.44062	87.94535	90.13323
100.52363	104.03739	106.80561	109.21967
119.02929	122.74761	125.75806	128.38004
137.54569	141.54751	144.78338	147.59881
156.15263	160.42065	163.86822	166.86529
174.83531	179.35507	183.00279	186.17170

Table **A.VII.2b** χ^2 versus n, Q

n \ Q	7.000E-01	6.500E-01	6.000E-01	5.500E-01
1	.14847	.20590	.27500	.35732
2	.71335	.86157	1.02165	1.19567
3	1.42365	1.64157	1.86917	2.10946
4	2.19470	2.47009	2.75284	3.04695
5	2.99991	3.32510	3.65550	3.99594
6	3.82755	4.19727	4.57015	4.95187
7	4.67133	5.08164	5.49323	5.91251
8	5.52742	5.97529	6.42265	6.87663
9	6.39330	6.87626	7.35703	7.84341
10	7.26722	7.78324	8.29547	8.81239
11	8.14786	8.69523	9.23728	9.78306
12	9.03428	9.61152	10.18197	10.75527
13	9.92568	10.53150	11.12913	11.72877
14	10.82148	11.45475	12.07848	12.70339
15	11.72116	12.38088	13.02974	13.67900
16	12.62435	13.30961	13.98274	14.65550
17	13.53067	14.24064	14.93726	15.63277
18	14.43986	15.17381	15.89321	16.61078
19	15.35165	16.10890	16.85042	17.58942
20	16.26586	17.04577	17.80883	18.56868
22	18.10072	18.92427	19.72879	20.52878
24	19.94323	20.80842	21.65249	22.49076
26	21.79240	22.69748	23.57943	24.45439
28	23.64746	24.59088	25.50925	26.41946
30	25.50776	26.48815	27.44162	28.38581
32	27.37278	28.38889	29.37629	30.35331
34	29.24205	30.29278	31.31303	32.32185
36	31.11522	32.19952	33.25166	34.29134
38	32.99194	34.10889	35.19201	36.26169
40	34.87194	36.02066	37.13396	38.23284
42	36.75496	37.93466	39.07738	40.20472
44	38.64080	39.85073	41.02216	42.17729
46	40.52924	41.76873	42.96820	44.15048
48	42.42013	43.68852	44.91543	46.12427
50	44.31331	45.61000	46.86378	48.09861
60	53.80913	55.23941	56.62000	57.97759
70	63.34602	64.89903	66.39612	67.86640
80	72.91535	74.58251	76.18794	77.76307
90	82.51110	84.28542	85.99256	87.66609
100	92.12896	94.00457	95.80785	97.57443
120	111.41859	113.48254	115.46455	117.40408
140	130.76569	133.00275	135.14906	137.24754
160	150.15827	152.55638	154.85552	157.10191
180	169.58792	172.13726	174.57993	176.96519
200	189.04861	191.74096	194.31933	196.83591

$n = 1 - 200,\ Q = 0.70$ to $Q = 0.35.$

5.000E-01	4.500E-01	4.000E-01	3.500E-01
.45493	.57065	.70832	.87345
1.38629	1.59701	1.83258	2.09964
2.36597	2.64300	2.94616	3.28310
3.35669	3.68713	4.04463	4.43769
4.35145	4.72775	5.13186	5.57306
5.34812	5.76520	6.21076	6.69476
6.34580	6.79996	7.28320	7.80611
7.34412	7.83251	8.35053	8.90935
8.34283	8.86315	9.41363	10.00598
9.34182	9.89221	10.47324	11.09714
10.34099	10.91987	11.52982	12.18361
11.34032	11.94632	12.58384	13.26609
12.33975	12.97169	13.63556	14.34504
13.33927	13.99612	14.68529	15.42091
14.33885	15.01967	15.73321	16.49400
15.33850	16.04248	16.77954	17.56462
16.33817	17.06455	17.82437	18.63298
17.33790	18.08601	18.86790	19.69931
18.33764	19.10685	19.91019	20.76374
19.33743	20.12717	20.95137	21.82648
21.33704	22.16631	23.03066	23.94726
23.33673	24.20369	25.10635	26.06252
25.33646	26.23954	27.17888	28.17296
27.33623	28.27401	29.24861	30.27913
29.33603	30.30727	31.31586	32.38150
31.33586	32.33942	33.38086	34.48044
33.33570	34.37058	35.44382	36.57627
35.33557	36.40082	37.50494	38.66927
37.33545	38.43023	39.56435	40.75968
39.33534	40.45887	41.62219	42.84771
41.33525	42.48679	43.67859	44.93353
43.33516	44.51406	45.73363	47.01729
45.33508	46.54070	47.78743	49.09915
47.33500	48.56677	49.84005	51.17921
49.33494	50.59229	51.89158	53.25760
59.33467	60.71281	62.13483	63.62770
69.33447	70.82358	72.35834	73.96770
79.33433	80.92662	82.56625	84.28398
89.33421	91.02338	92.76141	94.58085
99.33413	101.11487	102.94594	104.86154
119.33399	121.28495	123.28899	125.38333
139.33390	141.44130	143.60429	145.86294
159.33383	161.58680	163.89769	166.30919
179.33377	181.72341	184.17317	186.72819
199.33372	201.85259	204.43367	207.12440

Table **A.VII.2c** χ^2 versus *n*, *Q*;

n \ Q	3.000E-01	2.500E-01	2.000E-01	1.500E-01
1	1.07419	1.32330	1.64237	2.07224
2	2.40795	2.77259	3.21888	3.79424
3	3.66486	4.10833	4.64161	5.31703
4	4.87843	5.38527	5.98862	6.74488
5	6.06442	6.62566	7.28926	8.11517
6	7.23114	7.84080	8.55806	9.44610
7	8.38342	9.03713	9.80323	10.74787
8	9.52446	10.21885	11.03009	12.02707
9	10.65636	11.38873	12.24212	13.28801
10	11.78072	12.54886	13.44196	14.53393
11	12.89865	13.70067	14.63140	15.76706
12	14.01110	14.84540	15.81198	16.98930
13	15.11871	15.98388	16.98477	18.20194
14	16.22210	17.11693	18.15077	19.40622
15	17.32168	18.24506	19.31063	20.60296
16	18.41789	19.36865	20.46508	21.79304
17	19.51100	20.48865	21.61453	22.97698
18	20.60135	21.60438	22.75954	24.15546
19	21.68910	22.71778	23.90039	25.32880
20	22.77454	23.82769	25.03751	26.49756
22	24.93901	26.03926	27.30145	28.82244
24	27.09596	28.24114	29.55331	31.13244
26	29.24633	30.43456	31.79461	33.42945
28	31.39088	32.62049	34.02656	35.71497
30	33.53023	34.79974	36.25018	37.99024
32	35.66491	36.97298	38.46631	40.25628
34	37.79537	39.14077	40.67564	42.51397
36	39.92198	41.30361	42.87880	44.76405
38	42.04504	43.46190	45.07627	47.00715
40	44.16487	45.61601	47.26853	49.24383
42	46.28167	47.76624	49.45597	51.47457
44	48.39569	49.91289	51.63891	53.69980
46	50.50711	52.05619	53.81770	55.91989
48	52.61609	54.19636	55.99257	58.13518
50	54.72279	56.33360	58.16379	60.34598
60	65.22651	66.98145	68.97206	71.34107
70	75.68927	77.57665	79.71464	82.25533
80	86.11971	88.13025	90.40535	93.10573
90	96.52376	98.64993	101.05371	103.90404
100	106.90576	109.14123	111.66670	114.65879
120	127.61590	130.05459	132.80628	136.06193
140	148.26862	150.89409	153.85373	157.35166
160	168.87594	171.67520	174.82830	178.55171
180	189.44618	192.40863	195.74343	199.67854
200	209.98541	213.10217	216.60878	220.74410

$n = 1 - 200$, $Q = 0.30$ to $Q = 0.01$.

1.000E-01	5.000E-02	2.000E-02	1.000E-02
2.70552	3.84141	5.41178	6.63465
4.60517	5.99146	7.82400	9.21034
6.25136	7.81467	9.83726	11.34457
7.77944	9.48773	11.66784	13.27670
9.23632	11.07043	13.38805	15.08594
10.64464	12.59159	15.03321	16.81200
12.01700	14.06706	16.62224	18.47495
13.36157	15.50731	18.16823	20.09023
14.68361	16.91890	19.67882	21.66561
15.98718	18.30704	21.16077	23.20925
17.27496	19.67505	22.61773	24.72456
18.54935	21.02607	24.05395	26.21697
19.81188	22.36194	25.47129	27.68782
21.06414	23.68479	26.87276	29.14123
22.30708	24.99569	28.25926	30.57746
23.54183	26.29622	29.63318	32.00000
24.76900	27.58701	30.99480	33.40820
25.98942	28.86930	32.34616	34.80530
27.20351	30.14341	33.68717	36.19039
28.41197	31.41042	35.01963	37.56623
30.81328	33.92444	37.65950	40.28935
33.19624	36.41502	40.27036	42.97982
35.56317	38.88513	42.85583	45.64167
37.91592	41.33713	45.41885	48.27823
40.25602	43.77297	47.96180	50.89216
42.58474	46.19424	50.48670	53.48575
44.90315	48.60234	52.99523	56.06091
47.21216	50.99844	55.43884	58.61920
49.51258	53.33354	57.96876	61.16205
51.80506	55.75848	60.43613	63.69073
54.09020	58.12404	62.89178	66.20622
56.36854	60.48086	65.33666	68.70948
58.64053	62.82962	67.77141	71.20140
60.90661	65.17076	70.19673	73.68264
63.16710	67.50480	72.61324	76.15389
74.39701	79.08194	84.57992	88.37942
85.52703	90.53120	96.38753	100.42513
96.57819	101.87947	108.06933	112.32879
107.56500	113.14526	119.64836	124.11632
118.49800	124.34211	131.14157	135.80672
140.23255	146.56736	153.91813	158.95001
151.82698	168.61295	176.47082	181.84031
133.31056	190.51638	198.84633	204.52995
204.70367	212.30391	221.07713	227.05581
226.02103	233.99427	243.18691	249.44502

Table A.VII.2d χ^2 versus n, Q;

n \ Q	5.000E-03	1.000E-03	5.000E-04	1.000E-04
1	7.87894	10.82502		
2	10.59663	13.81551	15.20180	18.42068
3	12.83757	16.26332		
4	14.86026	18.46683	19.99734	23.51274
5	16.74895	20.51183		
6	18.54758	22.45774	24.10278	27.85627
7	20.27703	24.31849		
8	21.95495	26.12448	27.86803	31.82755
9	23.58860	27.87358		
10	25.18818	29.58830	31.41979	35.56398
11	26.75606	31.26038		
12	28.29952	32.90949	34.82125	39.13435
13	29.81864	34.52427		
14	31.31935	36.12327	38.10938	42.57922
15	32.80046	37.69324		
16	34.26718	39.25235	41.30805	45.92482
17	35.71757	40.78602		
18	37.15645	42.31239	44.43374	49.18930
19	38.58133	43.81536		
20	39.99684	45.31474	47.49844	52.38587
22	42.79565	48.26794	50.51111	55.52448
24	45.55851	51.17858	53.47874	58.61285
26	48.28988	54.05136	56.40688	61.65714
28	50.99337	56.89228	59.30002	64.66231
30	53.67196	59.70306	62.16184	67.63249
32	56.32811	62.48721	64.99545	70.57110
34	58.96392	65.24722	67.80345	73.48105
36	61.58118	67.98517	70.58806	76.36482
38	64.18141	70.70288	73.35122	79.22455
40	66.76596	73.40196	76.09459	82.06212
42	69.33600	76.08375	78.81965	84.87916
44	71.89255	78.74951	81.52768	87.67710
46	74.43653	81.40031	84.21984	90.45724
48	76.96877	84.03712	86.89714	93.22072
50	79.48998	86.66080	89.56051	95.96856
60	91.95170	99.60723	102.69474	109.50288
70	104.21490	112.31693	115.57757	122.75460
80	116.32105	124.83922	128.26130	135.78246
90	128.29894	137.20834	140.78227	148.62723
100	140.16949	149.44925	153.16694	161.31860
120	163.64818	173.61744	177.60289	186.32573
140	186.84684	197.45077	201.68279	210.93196
160	209.82386	221.01897	225.48072	235.22040
180	232.61980	244.37045	249.04816	259.24902
200	255.26415	267.54051	272.42259	283.05998

$n = 1 - 200,\ Q = 0.005$ to $Q = 10^{-8}$.

1.000E-05	1.000E-06	1.000E-07	1.000E-08
23.02585	27.63085	32.23616	36.84136
28.47326	33.37673	38.23959	43.07157
33.10706	38.25825	43.33775	48.36263
37.33137	42.70085	47.97246	53.16948
41.29595	46.86299	52.30954	57.66396
45.07594	50.82520	56.43366	61.93423
48.71590	54.63526	60.39528	66.03281
52.24478	58.32435	64.22741	69.99467
55.68272	61.91419	67.95313	73.84379
59.04437	65.42065	71.58931	77.59794
62.34081	68.85574	75.14876	81.27060
65.58066	72.22883	78.64147	84.87233
68.77080	75.54738	82.07544	88.41155
71.91680	78.81747	85.45715	91.89515
75.02326	82.04412	88.79199	95.32880
78.09403	85.23153	92.08447	98.71729
81.13237	88.38326	95.33840	102.06467
84.14108	91.50237	98.55707	105.37443
87.12256	94.59150	101.74332	108.64956
90.07892	97.65294	104.89961	111.89270
93.01200	100.68872	108.02813	115.10617
95.92343	103.70062	111.13079	118.29200
98.81465	106.69022	114.20930	121.45201
101.68696	109.65895	117.26520	124.58782
104.54151	112.60808	120.29985	127.70091
118.58129	127.09635	135.19400	142.96760
132.29982	141.22945	149.70268	157.82135
145.76351	155.08051	163.90498	172.34661
159.01859	168.70054	177.85612	186.60240
172.09878	182.12677	191.59608	200.63191
197.83091	208.50436	218.56003	228.13582
223.10521	234.37398	244.97080	255.04525
248.01655	259.84066	270.94254	281.48263
272.63096	284.97758	296.55480	307.53332
296.99659	309.83809	321.86567	333.25970

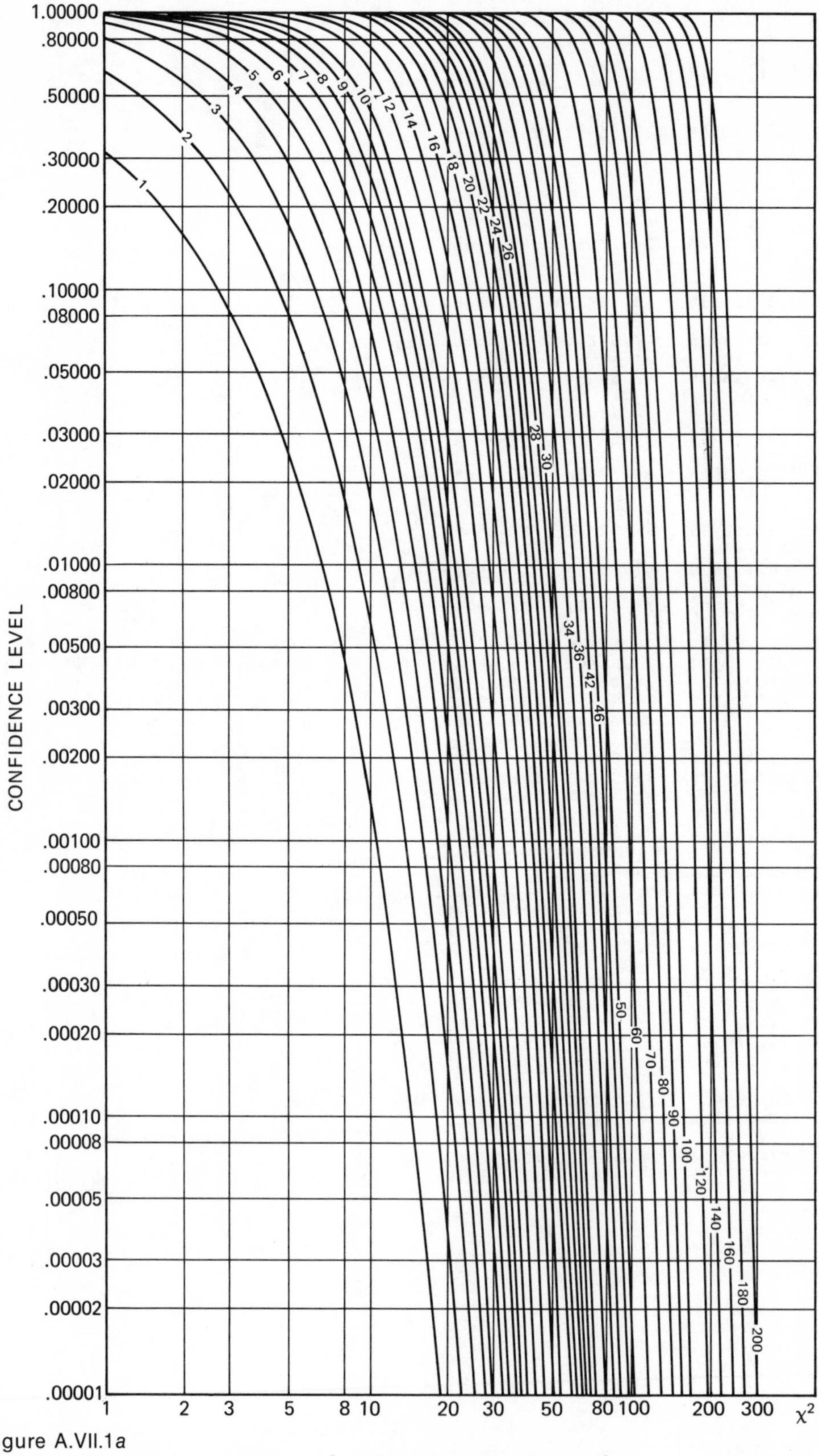

Figure A.VII.1*a*
Plot of ordinate Q versus abscissa χ^2 for various n, $Q = 1.0 - 10^{-5}$.

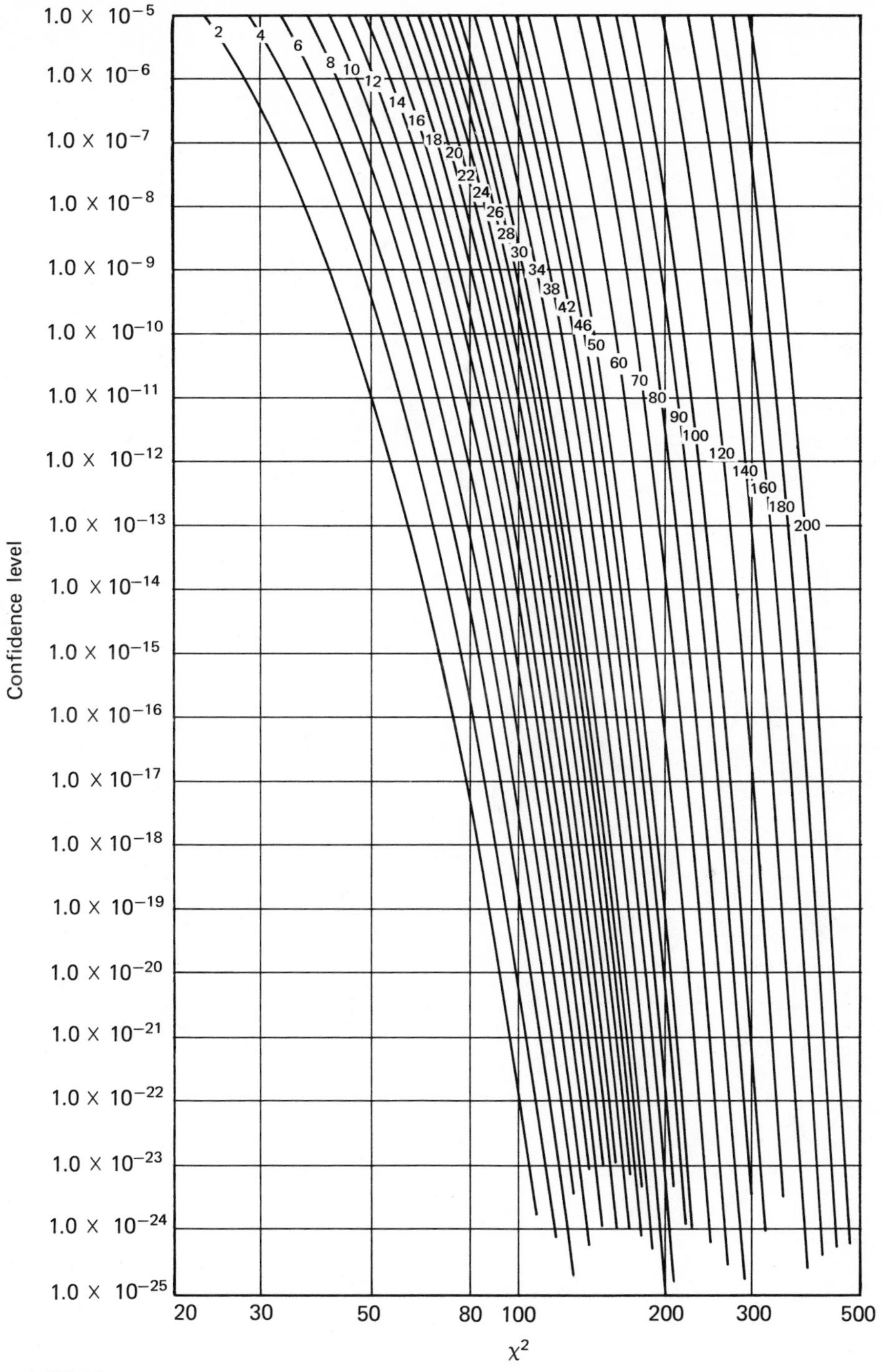

Figure A.VII.1*b*
Plot of ordinate Q versus abscissa χ^2 for various n, $Q = 10^{-5} - 10^{-26}$.

APPENDIX VIII

TABLES OF THE STUDENT'S t DISTRIBUTION

Tables of the cumulative Student's t distribution:

$$P(t;n) = \int_{-\infty}^{t} \mathrm{St}(x;n)\,dx = \Pr(x < t):$$

Table A.VIII.1*a*	$P(t;n)$ versus t and n; $n = 1-1000$, $t =$ 0.0 to 0.9 in steps of 0.1.
Table A.VIII.1*b*	$P(t;n)$ versus t and n; $n = 1-1000$, $t =$ 1.0 to 1.9 in steps of 0.1.
Table A.VIII.1*c*	$P(t;n)$ versus t and n; $n = 1-1000$, $t =$ 2.0 to 2.9 in steps of 0.1.
Table A.VIII.1*d*	$P(t;n)$ versus t and n; $n = 1-1000$, $t =$ 3.0 to 3.9 in steps of 0.1.
Table A.VIII.1*e*	$P(t;n)$ versus t and n; $n = 1-1000$, $t =$ 4.0 to 4.9 in steps of 0.1.
Table A.VIII.1*f*	$P(t;n)$ versus t and n; $n = 1-1000$, $t =$ 5.0 to 5.9 in steps of 0.1.

Table of the percentage points of the cumulative Student's t distribution: $P = \Pr(x < t)$:

Table A.VIII.2	t tabulated as a function of n and P; $n = 1-1000$, $P = 0.6$ to 0.9995.

Table **A. VIII.1a** $P(t; n)$ versus t and n;

TABLE OF CUMULATIVE

N \ T	0.00	.10	.20	.30	.40
1	.50000000	.53172552	.56283296	.59277358	.62111894
2	.50000000	.53526728	.57001400	.60375717	.63608276
3	.50000000	.53667383	.57286484	.60811835	.64203242
4	.50000000	.53742208	.57438149	.61043929	.64520137
5	.50000000	.53788493	.57531974	.61187548	.64716344
6	.50000000	.53819902	.57595650	.61285039	.64849588
7	.50000000	.53842597	.57641662	.61355497	.64945917
8	.50000000	.53859755	.57676450	.61408776	.65018775
9	.50000000	.53873178	.57703668	.61450465	.65075795
10	.50000000	.53883964	.57725540	.61483970	.65121630
11	.50000000	.53892820	.57743500	.61511482	.65159271
12	.50000000	.53900221	.57758508	.61534476	.65190735
13	.50000000	.53906499	.57771238	.61553980	.65217424
14	.50000000	.53911889	.57782170	.61570731	.65240349
15	.50000000	.53916569	.57791661	.61585273	.65260253
16	.50000000	.53920670	.57799977	.61598017	.65277696
17	.50000000	.53924292	.57807324	.61609275	.65293107
18	.50000000	.53927516	.57813862	.61619294	.65306822
19	.50000000	.53930403	.57819717	.61628267	.65319106
20	.50000000	.53933003	.57824992	.61636350	.65330171
22	.50000000	.53937499	.57834112	.61650327	.65349307
24	.50000000	.53941251	.57841721	.61661989	.65365275
26	.50000000	.53944429	.57848167	.61671868	.65378802
28	.50000000	.53947155	.57853696	.61680344	.65390408
30	.50000000	.53949519	.57858492	.61687695	.65400474
40	.50000000	.53957807	.57875304	.61713464	.65435765
50	.50000000	.53962789	.57885410	.61728956	.65456984
60	.50000000	.53966114	.57892155	.61739296	.65471147
70	.50000000	.53968491	.57896976	.61746688	.65481272
80	.50000000	.53970275	.57900595	.61752235	.65488871
90	.50000000	.53971663	.57903410	.61756552	.65494784
100	.50000000	.53972773	.57905663	.61760006	.65499516
120	.50000000	.53974440	.57909044	.61765189	.65506617
140	.50000000	.53975631	.57911460	.61768893	.65511692
160	.50000000	.53976524	.57913272	.61771672	.65515498
180	.50000000	.53977219	.57914682	.61773834	.65518460
200	.50000000	.53977775	.57915810	.61775563	.65520830
1000	.50000000	.53982784	.57925971	.61791142	.65542174

$n = 1 - 1000$, $t = 0.0$ to 0.9 in steps of 0.1.

STUDENT T DISTRIBUTION

.50	.60	.70	.80	.90
.64758362	.67202087	.69440011	.71477671	.73326229
.66666667	.69528337	.72180349	.74618298	.76844749
.67427602	.70459940	.73283650	.75890052	.78277448
.67833502	.70957942	.73874992	.76573643	.79049724
.68085056	.71266986	.74242553	.76999297	.79531440
.68256000	.71477186	.74492831	.77289482	.79860238
.68379643	.71629325	.74674122	.77499865	.80098824
.68473196	.71744497	.74811447	.77659333	.80279790
.68546435	.71834696	.74919048	.77784350	.80421734
.68605320	.71907241	.75005621	.77884979	.80536037
.68653689	.71966847	.75076777	.77967716	.80630049
.68694126	.72016688	.75136292	.78036938	.80708728
.68728433	.72058982	.75186806	.78095706	.80775541
.68757904	.72095321	.75230216	.78146220	.80832982
.68783494	.72126878	.75267921	.78190103	.80882893
.68805922	.72154540	.75300976	.78228581	.80926663
.68825740	.72178984	.75330190	.78262593	.80965359
.68843377	.72200742	.75356196	.78292874	.80999815
.68859175	.72220232	.75379496	.78320006	.81030691
.68873408	.72237792	.75400489	.78344456	.81058518
.68898022	.72268165	.75436805	.78386756	.81106666
.68918564	.72293515	.75467120	.78422073	.81146872
.68935966	.72314994	.75492808	.78452003	.81180950
.68950898	.72333424	.75514852	.78477691	.81210202
.68963850	.72349412	.75533978	.78499979	.81235586
.69009263	.72405480	.75601061	.78578174	.81324656
.69036572	.72439202	.75641418	.78625228	.81378269
.69054802	.72461717	.75668367	.78656654	.81414081
.69067836	.72477816	.75687637	.78679129	.81439696
.69077618	.72489899	.75702102	.78696000	.81458926
.69085230	.72499302	.75713359	.78709131	.81473894
.69091322	.72506827	.75722370	.78719642	.81485876
.69100464	.72518121	.75735892	.78735417	.81503859
.69106997	.72526192	.75745556	.78746692	.81516713
.69111898	.72532248	.75752808	.78755152	.81526359
.69115711	.72536959	.75758449	.78761734	.81533863
.69118762	.72540728	.75762963	.78767001	.81539869
.69146246	.72574688	.75803635	.78814460	.81593987

Table **A.VIII.1b** $P(t;n)$ versus t and n;

TABLE OF CUMULATIVE

N \ T	1.00	1.10	1.20	1.30	1.40
1	.75000000	.76514617	.77885794	.79128560	.80256846
2	.78867513	.80698006	.82349832	.83837648	.85176324
3	.80449889	.82415840	.84186894	.85776625	.87199634
4	.81304952	.83345818	.85182430	.86827420	.88294969
5	.81839127	.83927459	.85805447	.87484968	.88979806
6	.82204116	.84325193	.86231642	.87934741	.89447930
7	.82469167	.84614187	.86541403	.88261608	.89787948
8	.82670325	.84833612	.86776645	.88509819	.90046031
9	.82828180	.85005862	.86961340	.88704681	.90248570
10	.82955343	.85144659	.87110185	.88861709	.90411732
11	.83059965	.85258878	.87232686	.88990937	.90545973
12	.83147547	.85354513	.87335263	.89099141	.90658350
13	.83221936	.85435755	.87422410	.89191064	.90753797
14	.83285903	.85505625	.87497362	.89270120	.90835871
15	.83341493	.85566352	.87562511	.89338834	.90907196
16	.83390251	.85619621	.87619660	.89399109	.90969753
17	.83433362	.85666726	.87670199	.89452409	.91025065
18	.83471753	.85708677	.87715210	.89499879	.91074320
19	.83506160	.85746277	.87755554	.89542425	.91118462
20	.83537171	.85780169	.87791920	.89580775	.91158247
22	.83590837	.85838824	.87854860	.89647148	.91227094
24	.83635656	.85887816	.87907432	.89702585	.91284590
26	.83673649	.85929349	.87952003	.89749583	.91333328
28	.83706265	.85965007	.87990270	.89789933	.91375167
30	.83734569	.85995954	.88023482	.89824952	.91411475
40	.83833909	.86104582	.88140069	.89947874	.91538895
50	.83893718	.86169994	.88210277	.90021893	.91615603
60	.83933675	.86213699	.88257190	.90071350	.91666848
70	.83962257	.86244965	.88290751	.90106730	.91703504
80	.83983716	.86268440	.88315951	.90133294	.91731023
90	.84000421	.86286715	.88335568	.90153973	.91752444
100	.84013792	.86301344	.88351272	.90170527	.91769591
120	.84033864	.86323303	.88374845	.90195376	.91795329
140	.84048211	.86339001	.88391697	.90213139	.91813727
160	.84058977	.86350781	.88404342	.90226469	.91827532
180	.84067354	.86359946	.88414182	.90236840	.91838273
200	.84074058	.86367281	.88422057	.90245141	.91846869
1000	.84134475	.86433394	.88493033	.90319952	.91924334

$n = 1 - 1000$, $t = 1.0$ to 1.9 in steps of 0.1.

STUDENT T DISTRIBUTION

1.50	1.60	1.70	1.80	1.90
.81283296	.82219232	.83074697	.83858553	.84578589
.86380344	.87463432	.88438329	.89316683	.90109031
.88470807	.89604762	.90615468	.91516004	.92318423
.89600000	.90757543	.91782253	.92688081	.93488057
.90304816	.91475238	.92506161	.93412121	.94206835
.90785963	.91964199	.92998017	.93902379	.94691509
.91135076	.92318415	.93353555	.94255777	.95039699
.91399835	.92586695	.93622356	.94522350	.95301605
.91607467	.92796860	.93832617	.94730466	.95505596
.91774634	.92965912	.94001533	.94897388	.95668877
.91912099	.93104820	.94140179	.95034205	.95802478
.92027125	.93220975	.94256007	.95148366	.95913786
.92124785	.93319536	.94354210	.95245053	.96007931
.92208734	.93404215	.94438522	.95327984	.96088586
.92281667	.93477749	.94511689	.95399894	.96158449
.92345617	.93542200	.94575783	.95462838	.96219543
.92402148	.93599152	.94632390	.95518392	.96273419
.92452477	.93649841	.94682748	.95567784	.96321280
.92497573	.93695245	.94727837	.95611982	.96364080
.92538211	.93736149	.94768441	.95651765	.96402579
.92608519	.93806891	.94838630	.95720487	.96469026
.92667218	.93865928	.94897170	.95777757	.96524345
.92716964	.93915942	.94946738	.95826217	.96571112
.92759659	.93958854	.94989247	.95867752	.96611166
.92796704	.93996077	.95026106	.95903747	.96645854
.92926661	.94126580	.95155227	.96029699	.96767059
.93004859	.94205047	.95232779	.96105238	.96839616
.93057082	.94257424	.95284509	.96155579	.96887912
.93094429	.94294869	.95321473	.96191526	.96922369
.93122464	.94322969	.95349203	.96218480	.96948189
.93144283	.94344835	.95370774	.96239440	.96968258
.93161747	.94362334	.95388034	.96256205	.96984304
.93187958	.94388593	.95413927	.96281348	.97008359
.93206691	.94407357	.95432425	.96299304	.97025529
.93220747	.94421434	.95446299	.96312769	.97038400
.93231683	.94432386	.95457092	.96323240	.97048408
.93240434	.94441148	.95465726	.96331617	.97056411
.93319280	.94520071	.95543454	.96406968	.97128344

Table **A.VIII.1c** $P(t; n)$ versus t and n;

TABLE OF CUMULATIVE

N \ T	2.00	2.10	2.20	2.30	2.40
1	.85241638	.85853697	.86420025	.86945241	.87433408
2	.90824829	.91472507	.92059551	.92592593	.93077489
3	.93033702	.93671740	.94241402	.94750582	.95206276
4	.94194174	.94817336	.95367366	.95853048	.96282184
5	.94903026	.95512338	.96045305	.96511377	.96918960
6	.95378684	.95976116	.96494891	.96944977	.97335256
7	.95719034	.96306440	.96813449	.97250445	.97626692
8	.95974188	.96553124	.97050305	.97476458	.97841164
9	.96172359	.96744086	.97232971	.97650031	.98005106
10	.96330598	.96896138	.97377947	.97787284	.98134218
11	.96459802	.97019984	.97495694	.97898404	.98238372
12	.96567249	.97122753	.97593159	.97990121	.98324067
13	.96657982	.97209370	.97675124	.98067058	.98395749
14	.96735602	.97283343	.97744985	.98132486	.98456554
15	.96802750	.97347237	.97805221	.98188785	.98508754
16	.96861402	.97402971	.97857680	.98237725	.98554036
17	.96913070	.97452008	.97903767	.98280648	.98593676
18	.96958927	.97495480	.97944571	.98318593	.98628657
19	.96999898	.97534280	.97980945	.98352371	.98659747
20	.97036723	.97569121	.98013570	.98382629	.98687556
22	.97100215	.97629115	.98069668	.98434566	.98735197
24	.97153007	.97678925	.98116159	.98477521	.98774505
26	.97197591	.97720934	.98155310	.98513629	.98807480
28	.97235738	.97756839	.98188726	.98544399	.98835531
30	.97268748	.97787876	.98217578	.98570931	.98859678
40	.97383883	.97895899	.98317736	.98662755	.98942961
50	.97452647	.97960233	.98377186	.98717043	.98991974
60	.97498348	.98002912	.98416538	.98752884	.99024235
70	.97530919	.98033289	.98444503	.98778308	.99047069
80	.97555308	.98056013	.98465398	.98797278	.99064079
90	.97574252	.98073650	.98481601	.98811972	.99077239
100	.97589391	.98087737	.98494533	.98823690	.99087723
120	.97612074	.98108829	.98513880	.98841203	.99103374
140	.97628256	.98123865	.98527661	.98853665	.99114498
160	.97640381	.98135126	.98537975	.98862986	.99122810
180	.97649805	.98143875	.98545985	.98870220	.99129257
200	.97657341	.98150868	.98552385	.98875997	.99134402
1000	.97724987	.98213558	.98609655	.98927589	.99180246

$n = 1 - 1000$, $t = 2.0$ to 2.9 in steps of 0.1.

STUDENT T DISTRIBUTION

2.50	2.60	2.70	2.80	2.90
.87888106	.88312494	.88709368	.89081209	.89430219
.93519414	.93922930	.94292063	.94630370	.94940990
.95614668	.95981209	.96310698	.96607355	.96874888
.96661673	.96997593	.97295284	.97559422	.97794102
.97275495	.97587527	.97860797	.98100319	.98310465
.97673588	.97966886	.98221200	.98441803	.98633276
.97950389	.98228742	.98468049	.98673787	.98850704
.98152898	.98419109	.98646298	.98840108	.99005415
.98306909	.98563089	.98780315	.98964368	.99120231
.98427658	.98675425	.98884332	.99060273	.99208320
.98524681	.98765300	.98967163	.99136262	.99277744
.98604230	.98838705	.99034532	.99197788	.99333684
.98670561	.98899701	.99090302	.99248514	.99379604
.98726667	.98951133	.99137167	.99290984	.99417897
.98774710	.98995050	.99177061	.99327014	.99450266
.98816289	.99032960	.99211400	.99357932	.99477950
.98852610	.99065997	.99241248	.99384730	.99501871
.98884599	.99095031	.99267417	.99408164	.99522730
.98912979	.99120739	.99290536	.99428817	.99541066
.98938323	.99143653	.99311101	.99447146	.99557299
.98981643	.99182723	.99346070	.99478222	.99584730
.99017291	.99214779	.99374666	.99503541	.99606991
.99047126	.99241537	.99398467	.99524547	.99625395
.99072454	.99264200	.99418574	.99542243	.99640851
.99094218	.99283635	.99435777	.99557345	.99654004
.99168982	.99350100	.99494316	.99608449	.99698244
.99212752	.99388780	.99528156	.99637770	.99723417
.99241461	.99414049	.99550164	.99656745	.99739616
.99261731	.99431840	.99565610	.99670013	.99750898
.99276802	.99445040	.99577042	.99679806	.99759200
.99288446	.99455221	.99585842	.99687329	.99765562
.99297711	.99463311	.99592824	.99693288	.99770592
.99311523	.99475354	.99603201	.99702126	.99778035
.99321327	.99483888	.99610541	.99708365	.99783277
.99328645	.99490251	.99616006	.99713003	.99787168
.99334317	.99495178	.99620233	.99716587	.99790171
.99338841	.99499106	.99623600	.99719438	.99792557
.99379033	.99533881	.99653303	.99744487	.99813419

Table **A.VIII.1d** $P(t;n)$ versus t and n;

TABLE OF CUMULATIVE

N \ T	3.00	3.10	3.20	3.30	3.40
1	.89758362	.90067391	.90358875	.90634223	.90894700
2	.95226702	.95489963	.95732956	.95957615	.96165663
3	.97116556	.97335223	.97533408	.97713327	.97876934
4	.98002902	.98188944	.98354959	.98503329	.98636134
5	.98495038	.98657341	.98800241	.98926225	.99037448
6	.98799590	.98944176	.99069997	.99179606	.99275204
7	.99002894	.99133886	.99246709	.99343964	.99427876
8	.99146416	.99266712	.99369383	.99457057	.99531977
9	.99252182	.99363877	.99458435	.99538505	.99606334
10	.99332817	.99437467	.99525415	.99599325	.99661446
11	.99396008	.99494793	.99577266	.99646101	.99703545
12	.99446665	.99540500	.99618373	.99682962	.99736514
13	.99488055	.99577662	.99651621	.99712614	.99762884
14	.99522424	.99608380	.99678973	.99736886	.99784355
15	.99551363	.99634138	.99701808	.99757055	.99802111
16	.99576025	.99656006	.99721115	.99774035	.99816992
17	.99597265	.99674773	.99737622	.99788495	.99829611
18	.99615729	.99691034	.99751875	.99800935	.99840425
19	.99631914	.99705244	.99764291	.99811732	.99849777
20	.99646205	.99717756	.99775189	.99821181	.99857932
22	.99670269	.99738744	.99793396	.99836896	.99871434
24	.99689713	.99755624	.99807966	.99849405	.99882120
26	.99705727	.99769468	.99819863	.99859570	.99890760
28	.99719130	.99781013	.99829744	.99867977	.99897874
30	.99730502	.99790776	.99838070	.99875035	.99903820
40	.99768493	.99823155	.99865467	.99898059	.99923045
50	.99789915	.99841230	.99880597	.99910625	.99933405
60	.99803615	.99852713	.99890136	.99918485	.99939828
70	.99813114	.99860634	.99896682	.99923846	.99944182
80	.99820080	.99866422	.99901445	.99927729	.99947320
90	.99825404	.99870833	.99905063	.99930669	.99949686
100	.99829604	.99874304	.99907903	.99932969	.99951532
120	.99835805	.99879414	.99912071	.99936335	.99954223
140	.99840161	.99882994	.99914982	.99938677	.99956088
160	.99843388	.99885641	.99917129	.99940400	.99957457
180	.99845874	.99887676	.99918777	.99941720	.99958503
200	.99847848	.99889290	.99920082	.99942764	.99959328
1000	.99865010	.99903240	.99931286	.99951658	.99966307

$n = 1 - 1000$, $t = 3.0$ to 3.9 in steps of 0.1.

STUDENT T DISTRIBUTION

3.50	3.60	3.70	3.80	3.90
.91141447	.91375494	.91597774	.91809132	.92010336
.96358632	.96537892	.96704664	.96860044	.97005013
.98025948	.98161890	.98286101	.98399771	.98503951
.98755192	.98862093	.98958229	.99044817	.99122927
.99135778	.99222837	.99300030	.99368576	.99429535
.99358683	.99431675	.99495581	.99551610	.99600801
.99500348	.99563009	.99617250	.99664261	.99705056
.99596046	.99650885	.99697870	.99738167	.99772767
.99663824	.99712584	.99753972	.99789132	.99819029
.99713675	.99757605	.99794575	.99825710	.99851950
.99751485	.99791504	.99824920	.99852836	.99876171
.99780907	.99817705	.99848211	.99873509	.99894496
.99804298	.99838407	.99866498	.99889633	.99908691
.99823240	.99855076	.99881134	.99902460	.99919912
.99838823	.99868717	.99893046	.99912839	.99928939
.99851823	.99880039	.99902882	.99921363	.99936311
.99862798	.99889555	.99911108	.99928458	.99942415
.99872164	.99897640	.99918067	.99934430	.99947529
.99880233	.99904578	.99924013	.99939511	.99951860
.99887244	.99910583	.99929139	.99943874	.99955561
.99898795	.99920426	.99937496	.99950946	.99961527
.99907882	.99928120	.99943985	.99956398	.99966093
.99915189	.99934272	.99949142	.99960704	.99969676
.99921176	.99939287	.99953323	.99964175	.99972546
.99926160	.99943441	.99956770	.99967022	.99974887
.99942115	.99956606	.99967574	.99975843	.99982054
.99950596	.99963502	.99973145	.99980317	.99985626
.99955804	.99967695	.99976496	.99982977	.99987724
.99959310	.99970496	.99978717	.99984724	.99989089
.99961824	.99972493	.99980290	.99985954	.99990043
.99963711	.99973985	.99981460	.99986863	.99990745
.99965179	.99975141	.99982363	.99987562	.99991281
.99967310	.99976812	.99983662	.99988563	.99992046
.99968781	.99977961	.99984551	.99989244	.99992563
.99969856	.99978799	.99985196	.99989737	.99992935
.99970677	.99979435	.99985685	.99990109	.99993216
.99971323	.99979936	.99986069	.99990401	.99993435
.99976737	.99984089	.99989220	.99992765	.99995190

Table **A.VIII.1e** $P(t;n)$ versus t and n;

TABLE OF CUMULATIVE

N \ T	4.00	4.10	4.20	4.30	4.40
1	.92202087	.92385022	.92559723	.92726726	.92886519
2	.97140452	.97267155	.97385837	.97497142	.97601655
3	.98599577	.98687479	.98768396	.98842987	.98911839
4	.99193495	.99257348	.99315209	.99367719	.99415441
5	.99483829	.99532259	.99575522	.99614228	.99648907
6	.99644051	.99682132	.99715710	.99745360	.99771581
7	.99740504	.99771348	.99798222	.99821670	.99842157
8	.99802511	.99828112	.99850174	.99869212	.99885662
9	.99844479	.99866165	.99884665	.99900467	.99913981
10	.99874083	.99892771	.99908565	.99921928	.99933246
11	.99895690	.99912030	.99925721	.99937202	.99946840
12	.99911915	.99926382	.99938405	.99948405	.99956730
13	.99924396	.99937343	.99948022	.99956836	.99964117
14	.99934198	.99945893	.99955473	.99963323	.99969759
15	.99942034	.99952687	.99961355	.99968410	.99974155
16	.99948399	.99958172	.99966075	.99972467	.99977638
17	.99953640	.99962664	.99969919	.99975751	.99980441
18	.99958009	.99966389	.99973089	.99978445	.99982727
19	.99961690	.99969513	.99975734	.99980681	.99984614
20	.99964824	.99972159	.99977964	.99982557	.99986189
22	.99969843	.99976370	.99981490	.99985503	.99988646
24	.99973655	.99979544	.99984125	.99987685	.99990449
26	.99976625	.99981998	.99986148	.99989348	.99991813
28	.99978990	.99983940	.99987737	.99990645	.99992869
30	.99980908	.99985506	.99989011	.99991677	.99993703
40	.99986704	.99990174	.99992756	.99994671	.99996087
50	.99989540	.99992414	.99994516	.99996047	.99997159
60	.99991184	.99993694	.99995507	.99996811	.99997744
70	.99992243	.99994511	.99996133	.99997287	.99998105
80	.99992978	.99995073	.99996559	.99997609	.99998346
90	.99993515	.99995481	.99996867	.99997839	.99998517
100	.99993924	.99995790	.99997099	.99998012	.99998644
120	.99994503	.99996224	.99997423	.99998251	.99998820
140	.99994892	.99996515	.99997637	.99998409	.99998935
160	.99995171	.99996722	.99997790	.99998520	.99999015
180	.99995380	.99996877	.99997903	.99998602	.99999075
200	.99995543	.99996997	.99997991	.99998666	.99999120
1000	.99996833	.99997934	.99998665	.99999146	.99999459

$n = 1 - 1000$, $t = 4.0$ to 4.9 in steps of 0.1.

STUDENT T DISTRIBUTION

4.50	4.60	4.70	4.80	4.90
.93039551	.93186237	.93326956	.93462062	.93591877
.97699905	.97792369	.97879484	.97961646	.98039216
.98975479	.99034378	.99088956	.99139595	.99186633
.99458872	.99498455	.99534580	.99567592	.99597800
.99680023	.99707984	.99733145	.99755817	.99776277
.99794803	.99815399	.99833692	.99849963	.99864456
.99860084	.99875791	.99889574	.99901686	.99912344
.99899895	.99912227	.99922927	.99932223	.99940311
.99925553	.99935474	.99943992	.99951314	.99957618
.99942845	.99950994	.99957922	.99963818	.99968844
.99954940	.99961754	.99967493	.99972333	.99976419
.99963667	.99969453	.99974285	.99978323	.99981703
.99970136	.99975115	.99979239	.99982658	.99985495
.99975040	.99979377	.99982940	.99985872	.99988285
.99978835	.99982651	.99985764	.99988306	.99990383
.99981823	.99985212	.99987957	.99990184	.99991990
.99984212	.99987247	.99989690	.99991657	.99993242
.99986150	.99988888	.99991077	.99992830	.99994233
.99987741	.99990227	.99992204	.99993777	.99995028
.99989061	.99991333	.99993129	.99994550	.99995674
.99991106	.99993033	.99994540	.99995720	.99996643
.99992595	.99994258	.99995548	.99996547	.99997322
.99993710	.99995169	.99996290	.99997151	.99997812
.99994567	.99995863	.99996851	.99997604	.99998177
.99995240	.99996405	.99997286	.99997952	.99998455
.99997133	.99997903	.99998468	.99998883	.99999186
.99997964	.99998544	.99998962	.99999261	.99999476
.99998410	.99998883	.99999217	.99999453	.99999619
.99998681	.99999086	.99999368	.99999565	.99999702
.99998861	.99999219	.99999466	.99999637	.99999754
.99998987	.99999312	.99999534	.99999686	.99999790
.99999081	.99999380	.99999584	.99999722	.99999815
.99999209	.99999472	.99999650	.99999769	.99999848
.99999292	.99999532	.99999692	.99999799	.99999869
.99999349	.99999573	.99999721	.99999819	.99999883
.99999392	.99999603	.99999742	.99999834	.99999894
.99999424	.99999626	.99999758	.99999845	.99999901
.99999660	.99999789	.99999870	.99999921	.99999952

Table **A.VIII.1f** $P(t; n)$ versus t and n;

TABLE OF CUMULATIVE

N \ T	5.00	5.10	5.20	5.30	5.40
1	.93716704	.93836821	.93952485	.94063938	.94171402
2	.98112522	.98181867	.98247525	.98309749	.98368770
3	.99230378	.99271107	.99309071	.99344494	.99377582
4	.99625478	.99650870	.99674193	.99695641	.99715390
5	.99794764	.99811492	.99826648	.99840397	.99852887
6	.99877383	.99888930	.99899258	.99908509	.99916806
7	.99921736	.99930025	.99937350	.99943832	.99949576
8	.99947359	.99953507	.99958880	.99963581	.99967699
9	.99963052	.99967742	.99971797	.99975307	.99978350
10	.99973133	.99976799	.99979935	.99982623	.99984929
11	.99979874	.99982798	.99985276	.99987379	.99989167
12	.99984534	.99986910	.99988905	.99990583	.99991996
13	.99987852	.99989813	.99991445	.99992807	.99993943
14	.99990274	.99991915	.99993271	.99994391	.99995319
15	.99992082	.99993472	.99994612	.99995548	.99996316
16	.99993457	.99994649	.99995618	.99996408	.99997052
17	.99994521	.99995553	.99996387	.99997061	.99997606
18	.99995358	.99996259	.99996983	.99997564	.99998031
19	.99996025	.99996819	.99997452	.99997957	.99998360
20	.99996563	.99997268	.99997826	.99998269	.99998620
22	.99997366	.99997932	.99998375	.99998722	.99998994
24	.99997922	.99998386	.99998747	.99999026	.99999243
26	.99998320	.99998709	.99999008	.99999237	.99999413
28	.99998613	.99998945	.99999197	.99999389	.99999535
30	.99998835	.99999122	.99999338	.99999501	.99999624
40	.99999408	.99999570	.99999688	.99999773	.99999836
50	.99999628	.99999737	.99999814	.99999869	.99999908
60	.99999736	.99999817	.99999873	.99999913	.99999940
70	.99999796	.99999861	.99999906	.99999936	.99999957
80	.99999834	.99999888	.99999925	.99999950	.99999967
90	.99999859	.99999906	.99999938	.99999959	.99999973
100	.99999877	.99999919	.99999947	.99999965	.99999977
120	.99999901	.99999936	.99999958	.99999973	.99999983
140	.99999916	.99999946	.99999965	.99999978	.99999986
160	.99999925	.99999952	.99999970	.99999981	.99999988
180	.99999932	.99999957	.99999973	.99999983	.99999990
200	.99999937	.99999961	.99999975	.99999985	.99999991
1000	.99999971	.99999983	.99999990	.99999994	.99999997

$n = 1 - 1000$, $t = 5.0$ to 5.9 in steps of 0.1.

STUDENT T DISTRIBUTION

5.50	5.60	5.70	5.80	5.90
.94275085	.94375182	.94471874	.94565329	.94655707
.98424800	.98478035	.98528656	.98576828	.98622704
.99408520	.99437478	.99464608	.99490049	.99513929
.99733594	.99750393	.99765913	.99780267	.99793556
.99864246	.99874590	.99884020	.99892629	.99900496
.99924257	.99930957	.99936989	.99942428	.99947337
.99954673	.99959202	.99963233	.99966823	.99970026
.99971313	.99974488	.99977281	.99979742	.99981913
.99980991	.99983286	.99985285	.99987026	.99988546
.99986911	.99988616	.99990085	.99991352	.99992447
.99990688	.99991984	.99993090	.99994036	.99994844
.99993187	.99994193	.99995043	.99995763	.99996374
.99994893	.99995687	.99996354	.99996912	.99997382
.99996088	.99996727	.99997257	.99997698	.99998066
.99996947	.99997467	.99997896	.99998250	.99998542
.99997577	.99998006	.99998357	.99998645	.99998880
.99998048	.99998407	.99998697	.99998934	.99999126
.99998406	.99998709	.99998952	.99999149	.99999308
.99998683	.99998940	.99999147	.99999312	.99999445
.99998899	.99999120	.99999297	.99999437	.99999549
.99999208	.99999375	.99999507	.99999611	.99999692
.99999411	.99999541	.99999642	.99999721	.99999782
.99999549	.99999652	.99999732	.99999793	.99999841
.99999646	.99999730	.99999794	.99999843	.99999880
.99999716	.99999786	.99999838	.99999878	.99999908
.99999881	.99999914	.99999938	.99999955	.99999967
.99999935	.99999954	.99999968	.99999978	.99999984
.99999959	.99999972	.99999981	.99999987	.99999991
.99999971	.99999980	.99999987	.99999991	.99999994
.99999978	.99999985	.99999990	.99999994	.99999996
.99999982	.99999988	.99999992	.99999995	.99999997
.99999985	.99999991	.99999994	.99999996	.99999997
.99999989	.99999993	.99999996	.99999997	.99999998
.99999991	.99999994	.99999997	.99999998	.99999999
.99999993	.99999995	.99999997	.99999998	.99999999
.99999994	.99999996	.99999998	.99999999	.99999999
.99999994	.99999996	.99999998	.99999999	.99999999
.99999998	.99999999	.99999999	1.00000000	1.00000000

Table **A.VIII.2** *t* tabulated as a function

TABLE OF CUMULATIVE STUDENT T

N \ P	.6000	.7000	.8000	.9000	.9500
1	.32491969	.72654253	1.37638192	3.07768351	6.31375151
2	.28867511	.61721339	1.06066017	1.88561801	2.91998558
3	.27667065	.58438971	.97847230	1.63774432	2.35336336
4	.27072229	.56864904	.94096457	1.53320625	2.13184677
5	.26718085	.55942964	.91954378	1.47588404	2.01504837
6	.26483452	.55338092	.90570328	1.43975571	1.94318026
7	.26316685	.54910964	.89602964	1.41492392	1.89457848
8	.26192109	.54593374	.88888952	1.39681529	1.85954800
9	.26095533	.54348023	.88340386	1.38302874	1.83311292
10	.26018482	.54152802	.87905783	1.37218362	1.81246112
11	.25955585	.53993785	.87552998	1.36343032	1.79588481
12	.25903273	.53861766	.87260929	1.35621730	1.78228749
13	.25859084	.53750408	.87015153	1.35017127	1.77093339
14	.25821264	.53655217	.86805478	1.34503037	1.76131006
15	.25788528	.53572912	.86624497	1.34060558	1.75305035
16	.25759917	.53501044	.86466700	1.33675717	1.74588367
17	.25734698	.53437746	.86327902	1.33337939	1.73960673
18	.25712302	.53381573	.86204867	1.33039094	1.73406360
19	.25692280	.53331386	.86095055	1.32772816	1.72913274
20	.25674273	.53286277	.85996444	1.32534065	1.72471818
22	.25643201	.53208494	.85826605	1.32123674	1.71714437
24	.25617337	.53143803	.85685546	1.31783588	1.71088205
26	.25595474	.53089157	.85566523	1.31497186	1.70561792
28	.25576750	.53042384	.85464748	1.31252675	1.70113092
30	.25560536	.53001898	.85376726	1.31041501	1.69726084
40	.25503868	.52860567	.85069979	1.30307705	1.68385100
50	.25469934	.52776044	.84886924	1.29871365	1.67590497
60	.25447339	.52719812	.84765301	1.29582106	1.67064883
70	.25431213	.52679703	.84678628	1.29376285	1.66691442
80	.25419127	.52649652	.84613735	1.29222357	1.66412450
90	.25409731	.52626298	.84563327	1.29102888	1.66196099
100	.25402217	.52607626	.84523042	1.29007474	1.66023430
120	.25390952	.52579638	.84462684	1.28864620	1.65765087
140	.25382908	.52559661	.84419621	1.28762779	1.65581047
160	.25376877	.52544686	.84387351	1.28686504	1.65443286
180	.25372188	.52533044	.84362269	1.28627244	1.65336297
200	.25368437	.52523732	.84342213	1.28579875	1.65250806
1000	.25334710	.52440051	.84162123	1.28155157	1.64485361

of n and P; $n = 1 - 1000$, $P = 0.6$ to 0.9995.

DISTRIBUTION (PERCENTAGE POINT)

.9750	.9800	.9900	.9950	.9995
12.70620473	15.89454219	31.82051595	63.65674116	636.61900000
4.30265269	4.84873208	6.96455669	9.92484320	31.59905452
3.18244621	3.48190819	4.54070282	5.84090931	12.92393455
2.77644499	2.99852756	3.74694739	4.60409487	8.61030156
2.57058172	2.75650852	3.36493000	4.03214297	6.86882663
2.44691185	2.61224181	3.14266833	3.70742790	5.95881616
2.36462415	2.51675240	2.99795157	3.49948313	5.40787757
2.30600414	2.44898497	2.89645928	3.35538721	5.04130537
2.26215714	2.39844096	2.82143777	3.24983552	4.78091258
2.22813884	2.35931461	2.76376941	3.16927261	4.58689385
2.20098516	2.32813982	2.71807918	3.10580648	4.43697934
2.17881280	2.30272167	2.68099799	3.05453939	4.31779124
2.16036854	2.28160355	2.65030875	3.01227577	4.22083170
2.14478665	2.26378127	2.62449384	2.97684271	4.14045391
2.13144937	2.24854028	2.60248007	2.94671280	4.07276514
2.11990521	2.23535840	2.58348719	2.92078162	4.01499632
2.10981557	2.22384529	2.56693394	2.89823034	3.96512206
2.10092203	2.21370324	2.55237949	2.87844027	3.92164569
2.09302400	2.20470135	2.53948305	2.86093433	3.88340023
2.08596326	2.19665774	2.52797687	2.84533947	3.84951200
2.07387307	2.18289264	2.50832455	2.81875604	3.79213001
2.06389847	2.17154467	2.49215923	2.79693909	3.74539659
2.05552944	2.16202886	2.47862980	2.77871445	3.70660961
2.04840699	2.15393486	2.46714008	2.76326238	3.67390639
2.04227234	2.14696627	2.45726117	2.74999494	3.64595514
2.02107538	2.12290981	2.42325671	2.70445902	3.55096576
2.00855898	2.10872128	2.40327153	2.67779256	3.49600978
2.00029773	2.09936284	2.39011922	2.66028257	3.46019875
1.99443696	2.09272700	2.38080738	2.64790391	3.43501212
1.99006337	2.08777660	2.37386814	2.63869037	3.41633460
1.98667448	2.08394188	2.36849732	2.63156489	3.40193215
1.98397145	2.08088390	2.36421718	2.62589021	3.39048795
1.97993032	2.07631318	2.35782439	2.61742078	3.37345019
1.97705362	2.07306027	2.35327816	2.61140230	3.36137403
1.97490145	2.07062706	2.34987940	2.60690538	3.35236777
1.97323071	2.06873840	2.34724236	2.60341777	3.34539285
1.97189611	2.06722988	2.34513679	2.60063396	3.33983172
1.95996398	2.05374885	2.32634773	2.57582927	3.29052636

GUIDE FOR FURTHER READING AND BIBLIOGRAPHY

Guide for Further Reading

Following is a very personal choice of references for collateral and advanced reading. It represents my "guide for the consumer" to the enormous literature and is of necessity incomplete.

PART I

Aitken (1957), Baird (1962), Barford (1967), Beers (1953), Bulmer (1967), Croxton (1959), Fisher (1970), Langley (1971), Mack (1967), Parratt (1961), Shchigolev (1965), Topping (1960), Wilson (1952), Young (1962).

PART II

Acton (1959), Adams (1950), Davis (1955), Deming (1943), Douglass (1947), Karsten (1923), King (1971), Levens (1965), Worthing (1943).

PART III

Arley (1950), Cramer (1955), Cramer (1946), Dwass (1970), Feller (1957), Gnedenko (1954), Gnedenko (1962), Meyer (1965), Parratt (1961), Parzen (1960), Wilks (1948). *Geometrical Probability*: Coolidge (1925), O'Beirne (1965). *Monte Carlo Methods*: Cashwell (1959), Hammersley (1964), Meyer (1956).

PART IV

Dwass (1970), Feller (1957), Fry (1928), Lindsay (1941), Lippmann (1971), Meyer (1965), Peters (1940), Whittaker (1942).

PART V

Bevington (1969), Brandt (1970), Burford (1968), Cox (1958), Eadie (1971), Fisher (1971), Good (1965), Guttmann (1971), Hamilton (1964), Hoel (1947), Lindley (1965), Kempthorne (1971), Maritz (1970), Martin (1971), Plackett (1960), Snedecor (1967), Sprent (1969), Tribus (1969).

TABLES

Abramowitz (1965), Beyer (1968), Burington (1953), David (1938), Fisher (1970), Greenwood (1962), Hald (1952), Jahnke (1945), NBS (1949), Pearson (1970), Rand (1955).

HANDBOOKS OF FORMULAS

Beyer (1968), Burington (1953), Crow (1960), Quenouille (1959).

POPULAR AND INTRODUCTORY

Huff (1954), Langley (1971), Mosteller (1961), Tanur (1972), Wallis (1956), Weaver (1963).

GENERAL REFERENCE

Kendall (1967–1969).

SPECIALIZED TOPICS

Decision theory

Chernoff (1959), Lindgren (1962), Lindgren (1971), Raiffa (1961), Von Neumann (1944).

Dimensional analysis

Bridgman (1922), Huntley (1967).

Foundations and philosophy

Ayer (1972), Good (1950), Hogben (1968), Jeffreys (1948), Keynes (1962), Nagel (1955), Salmon (1967), Savage (1972).

History

David (1962), De Moivre (1967), Laplace (1951), Todhunter (1949).

Mathematical methods

Margenau (1956), Rektorys (1969), Whittaker (1942).

Bibliography

Abramowitz, Milton, and Irene Stegun (eds.), *Handbook of Mathematical Functions*, Dover, New York, 1965.

Acton, Forman S., *Analysis of Straight-Line Data*, Wiley, New York, 1959.

Adams, Douglas P., *An Index of Nomograms*, The M.I.T. Press, Cambridge, Mass., 1950. A compiled and categorized list of nomograms that appeared in print before 1950.

Aitken, A. C., *Statistical Mathematics*, Eighth Edition, Oliver and Boyd, London, 1957.

Arley, Niels, and K. Rander Buch, *Introduction to the Theory of Probability and Statistics*, Wiley, New York, 1950. A compact yet complete and readable book. It has the useful feature of describing equivalent terms in the various bodies of literature (e.g., French and German).

Ayer, A. J., *Probability and Evidence*, Columbia University Press, New York, 1972.

Baird, D. C., *Experimentation: An Introduction to Measurement Theory and Experimental Design*, Prentice-Hall, Englewood Cliffs, N.J., 1962.

Barford, N. C., *Experimental Measurements: Precision, Error and Truth*, Addison-Wesley, Reading, Mass., 1967. An attractively written book at an elementary level.

Beers, Yardley, *Introduction to the Theory of Error*, Addison-Wesley, Reading, 1953. A very clearly written little book about error analysis for beginning students.

Bevington, Philip R., *Data Reduction and Error Analysis for the Physical Sciences*, McGraw-Hill, New York, 1969. A nicely balanced book with broad coverage of data analysis. Contains many relevant FORTRAN programs.

Beyer, William H. (ed.), *Handbook of Tables for Probability and Statistics*, Second Edition, The Chemical Rubber Company, Cleveland, Ohio, 1968.

Brandt, Siegmund, *Statistical and Computational Methods in Data Analysis*, North-Holland, Amsterdam, 1970. A good description of methods used in modern physics research with emphasis on data analysis. Contains FORTRAN programs and sample computer printout with brief summary of optimization techniques.

Bridgman, P. W., *Dimensional Analysis*, Yale University Press, 1922. A classic. Still readable and useful.

Bulmer, M. G., *Principles of Statistics*, The M.I.T. Press, Cambridge, Mass., 1967.

Burford, Roger L., *Statistics: A Computer Approach*, Charles E. Merrill, Columbia, Ohio, 1968.

Burington, Richard Stevens, and Donald Curtis May, *Handbook of Probability and Statistics with Tables*, Handbook Publishers, Sandusky, Ohio, 1953. A useful desk reference well worth buying.

Cashwell, E. D., and C. J. Everett, *A Practical Manual on the Monte Carlo Method*, Pergamon Press, New York, 1959. Emphasis is on physics situations involving neutrons and photons.

Chernoff, Herman, and Lincoln E. Moses, *Elementary Decision Theory*, Wiley, New York, 1959. The clearest available introduction to decision theory.

Coolidge, Julian Lowell, *An Introduction to Mathematical Probability*, Oxford University Press, London, 1925. Contains much material not readily available elsewhere. Especially good on geometrical probability.

Cox, D. R., *Planning of Experiments*, Wiley, New York, 1958.

Cramer, Harald, *Mathematical Methods of Statistics*, Princeton University Press, Princeton, N. J., 1946. A classic. Good source for theorems and proofs.

Cramer, Harald, *The Elements of Probability Theory*, Wiley, New York, 1955. Introductory and clear with many insights.

Crow, Edwin, L., Frances A. Davis, and Margaret W. Maxfield, *Statistics Manual*, Dover, New York, 1960.

Croxton, Frederick E., *Elementary Statistics with Applications in Medicine and the Biological Sciences*, Dover, New York, 1959.

David, F. N., *Tables of the Correlation Coefficient*, Cambridge University Press, Cambridge, 1938.

David, F. N., *Games, Gods and Gambling*, Charles Griffin, London, 1962. Very interesting history of probability.

Davis, Dale S., *Nomography and Empirical Equations*, Reinhold, New York, 1955.

De Moivre, A., *The Doctrine of Chances*, Chelsea, New York, 1967. A photographic reprint of a classic. Little pedagogical value but an interesting slice of history.

Deming, W. E., *Statistical Adjustment of Data*, Wiley, New York, 1943.

Douglass, Raymond D., and Douglas P. Adams, *Elements of Nomography*, McGraw-Hill, New York, 1947.

Dwass, Meyer, *Probability: Theory and Applications*, W. A. Benjamin, New York, 1970. An excellent introduction for mathematics students.

Eadie, W. T., D. Dryard, F. E. James, M. Roos, and B. Sadoulet, *Statistical Methods in Experimental Physics*, North-Holland, Amsterdam, 1971. Presents in good detail the procedures useful in modern physics research.

Feller, W., *An Introduction to Probability Theory and Its Applications*, Vol. I, Second Edition, Wiley, New York, 1957. A classic well worth reading.

Fisher, R. A., *Statistical Methods for Research Workers*, 14th Edition, Hafner, Darien, Conn., 1970. Worth reading. A good example of the precept to study the masters.

Fisher, R. A., *The Design of Experiments*, Hafner, New York, 1971. Anything by R. A. Fisher is worth reading.

Fisher, Ronald A., and Frank Yates, *Statistical Tables*, Hafner, Darien, 1970.

Fletcher, R., (ed.), *Optimization*, Academic Press, New York, 1969.

Fry, T. C., *Probability and Its Engineering Uses*, Van Nostrand, New York, 1928. A classic.

Good, I. J., *Probability and the Weighing of Evidence*, Charles Griffin, London, 1950.

Good, I. J., *The Estimation of Probabilities*, The M.I.T. Press, Cambridge, Mass., 1965. A modern essay on "empirical Bayes' methods."

Gnedenko, B. V., and A. Ya. Khinchin, *An Elementary Introduction to the Theory of Probability*, English Edition, Dover, New York, 1962. A clear little book.

Gnedenko, B. V., and A. N. Kolmogorov, *Limit Distributions for Sums of Independent Random Variables*, English translation by K. L. Chung, Addison-Wesley, Reading, Mass., 1954. Contains a complete discussion not only of the normal convergence theorem but also other limit theorems for sums of independent random variables.

Greenwood, J. A., and H. O. Hartley, *Guide to Tables in Mathematical Statistics*, Princeton University Press, Princeton, N.J., 1962.

Guttman, Irwin, S. S. Wilks, and J. Stuart Hunter, *Introductory Engineering Statistics*, Second Edition, Wiley, New York, 1971. A comprehensive treatment at an introductory level.

Hald, A., *Statistical Tables and Formulas*, Wiley, New York, 1952.

Hamilton, Walter Clark, *Statistics in Physical Science: Estimation, Hypothesis Testing and Least Squares*, The Ronald Press, New York, 1964. Useful treatment for scientists with emphasis on chemistry.

Hammersley, J. M., and D. C. Handscomb, *Monte Carlo Methods*, Wiley, New York, 1964. An excellent introduction.

Hildebrand, Francis B., *Methods of Applied*

Mathematics, Second Edition, Prentice-Hall, Englewood Cliffs, N.J., 1965.

Hogben, Lancelot, *Statistical Theory*, Norton, New York, 1968. A philosophical examination stressing the "behaviorist" viewpoint.

Huff, Darrell, *How to Lie with Statistics*, Norton, New York, 1954. An amusing little book.

Hoel, Paul G., *Introduction to Mathematical Statistics*, Wiley, New York, 1947. A widely used text book.

Huntley, H. E., *Dimensional Analysis*, Dover, New York, 1967. Very readable and more available than Bridgman.

Jahnke, Eugene, and Fritz Emde, *Tables of Functions*, Dover, New York, 1945.

Jeffreys, Harold, *Theory of Probability*, Second Edition, Clarendon Press, Oxford, 1948. Actually more concerned with scientific method, this is a presentation of a philosophical position.

Karsten, Karl G., *Charts and Graphs*, Prentice-Hall, New York, 1923.

Kendall, Maurice G., and Alan Stuart, *The Advanced Theory of Statistics*,
Volume 1 (*Distribution Theory*), Third Edition, 1969;
Volume 2 (*Inference and Relationship*), Second Edition, 1967;
Volume 3 (*Design and Analysis, and Time Series*), Second Edition, 1968;
Charles Griffin & Company, London. The definitive reference. Encyclopedic.

Kempthorne, Oscar, and Leroy Folks, *Probability, Statistics and Data Analysis*, The Iowa State University Press, Ames, Iowa, 1971. A comprehensive introduction directed at data analysis.

Kenney, John F., *Mathematics of Statistics*, Van Nostrand, New York, 1939.

Keynes, J. M., *A Treatise on Probability*, reprinted Harper & Row, Evanston, Ill., 1962.

King, James R., *Probability Charts for Decision Making*, Industrial Press, New York, 1971. A very useful book.

Kowalik, J., and M. R. Osborne, *Methods for Unconstrained Optimization Problems*, American Elsevier, New York, 1968.

Laplace, P. S. de, *A Philosophical Essay on Probabilities*, Dover, New York, 1951. A reprint of a classic. Still readable.

Langley, Russell, *Practical Statistics Simply Explained*, Dover, New York, 1971. Largely nonmathematical but usable discussion of many statistical procedures.

Lavi, A., and T. P. Vogl (eds.), *Recent Advances in Optimization Techniques*, Wiley, New York, 1966.

Levens, A. S., *Graphical Methods in Research*, Wiley, New York, 1965.

Lindgren, B. W., *Elements of Decision Theory*, Macmillan, New York, 1971. Clear introduction to decision theory.

Lindgren, B. W., *Statistical Theory*, Macmillan, New York, 1962.

Lindley, D. V., *Introduction to Probability and Statistics*, Cambridge University Press, London, 1965. Bayesian point of view.

Lindsay, Robert Bruce, *Introduction to Physical Statistics*, Wiley, New York, 1941. Excellent discussion of ideas of probability and statistics in physics.

Lippmann, Steven A., *Elements of Probability and Statistics*, Holt, Rinehart & Winston, New York, 1971. Requires little background but contains some interesting problems with solutions by elementary methods.

Mack, C., *Essentials of Statistics for Scientists and Technologists*, Plenum Press, New York, 1967.

Margenau, Henry, and George Moseley Murphy, *The Mathematics of Physics and Chemistry*, Van Nostrand, New York, 1956. An old standby.

Maritz, J. S., *Empirical Bayes Methods*, Methuen, London, 1970. A clear statement of the modern version of Bayesian techniques. Readable but requires some background.

Martin, B. R., *Statistics for Physicists*, Academic Press, New York, 1971. A logical presentation of statistical methods used in physics research. Contains a good appendix on optimization techniques.

Meyer, Herbert A. (ed.), *Symposium on Monte Carlo Methods*, Wiley, New York, 1956.

Meyer, Paul L., *Introductory Probability and Statistical Applications*, Addison-Wesley, Reading, Mass., 1965. An excellent introductory text.

Mosteller, Frederick, Robert E. K. Rourke, and George B. Thomas, Jr., *Probability with Statistical Applications*, Addison-Wesley, Reading, Mass., 1961. A well-written, widely used text using elementary methods exclusively (e.g., no calculus).

Nagel, Ernest, *Principles of the Theory of*

Probability, in *International Encyclopedia of Unified Science*, University of Chicago Press, 1955. A learned survey of the interpretations of probability *circa* 1939. Makes a harsh judgment about the utility of Bayesian methods.

National Bureau of Standards, *Tables of the Binomial Probability Distribution*, Applied Mathematics Series: 6, United States Government Printing Office, Washington, 1949.

O'Beirne, T. H., *Puzzles and Paradoxes*, Oxford University Press, Oxford, 1965. Contains a good discussion of the Buffon Needle problem.

Parratt, Lyman G., *Probability and Experimental Errors in Science*, Wiley, New York, 1961. Much useful material clearly presented.

Parzen, Emanuel, *Modern Probability Theory and Its Applications*, Wiley, New York, 1960.

Pearson, E. S., and H. O. Hartley, *Biometrika Tables for Statisticians*, Cambridge University Press, Cambridge, 1970.

Peters, Charles C., and Walter R. Van Voorhis, *Statistical Procedures and Their Mathematical Bases*, McGraw-Hill, New York, 1940.

Plackett, R. L., *Principles of Regression Analysis*, Oxford University Press, Oxford, 1960. A brief but insightful book on the theory of least squares.

Quenouille, M. H., *Rapid Statistical Calculations*, Griffin, London, 1959. A unique vest-pocket compendium of useful formulas and approximate procedures.

Raiffa, Howard, and Robert Schlaifer, *Applied Statistical Decision Theory*, The M.I.T. Press, Cambridge, Mass., 1961. A classic in the area of managerial economics; it is for very mathematically oriented managers.

Ralston, Anthony, and Herbert S. Wilf (eds.), *Mathematical Methods for Digital Computers*, Wiley, New York, 1960.

Rand Corporation, *A Million Random Digits with* 100,000 *Normal Deviates*, Free Press Publishers, Glencoe, Illinois, 1955. Not much of a plot but surprisingly useful.

Rektorys, Karel (ed.), *Survey of Applicable Mathematics*, The M.I.T. Press, Cambridge, Mass., 1969. A very useful compendium.

Salmon, Wesley C., *The Foundations of Scientific Inference*, University of Pittsburgh Press, Pittsburgh, 1967. A readable survey of the philosophical problems of probability and inference.

Savage, Leonard J., *The Foundations of Statistics*, Second Edition, Dover, New York, 1972. A classic. A deep and penetrating analysis of foundations and interpretations for the serious student. Sometimes tough going but worth the effort.

Shchigolev, B. M., *Mathematical Analysis of Observations*, American Elsevier, New York, 1965. Comprehensive discussion of observations, especially in astronomy. Contains a bibliography with references to the extensive Russian literature.

Snedecor, George W., and William G. Cochran, *Statistical Methods*, Sixth Edition, Iowa State University Press, Ames, Iowa, 1967. A widely-used text.

Sprent, Peter, *Models in Regression*, Methuen, London, 1969. A fairly detailed treatment of curve-fitting methods.

Tanur, Judith M., Frederick Mosteller, William H. Kruskal, Richard F. Link, Richard S. Pieters, and Gerald R. Rising (eds.), *Statistics: A Guide to the Unknown*, Holden-Day, San Francisco, 1972. A collection of essays describing applications of probability and statistics to many areas. Nonmathematical.

Todhunter, I., *A History of the Mathematical Theory of Probability from the Time of Pascal to Laplace*, reprinted Chelsea, New York, 1949. A classical reference.

Topping, J., *Errors of Observation and Their Treatment*, Reinhold, New York, 1960.

Tribus, Myron, *Rational Descriptions, Decisions and Designs*, Pergamon Press, New York, 1969. A formal presentation of the use of statistical inference in engineering decisions. The use of computer methods is involved throughout.

Von Neumann, John, and Oskar Morgenstern, *Theory of Games and Economic Behavior*, A classic that is (perhaps) surprisingly readable.

Wallis, W. Allen, and Harry V. Roberts, *Statistics: A New Approach*, The Free Press, Glencoe, Illinois, 1956. Essentially nonmathematical but very good discussion.

Wallis, W. Allen, and Harry V. Roberts, *The Nature of Statistics*, Collier Books, New

York, 1962. A paperback reprint of the first sections of Wallis (1956).

Weaver, Warren, *Lady Luck*, Doubleday, Garden City, New York, 1963. A well-written book at the popular level.

Whittaker, E. T., and G. Robinson, *The Calculus of Observations*, Blackie & Son, London, 1942. Many useful numerical techniques.

Wilks, S. S., *Elementary Statistical Analysis*, Princeton University Press, Princeton, N.J., 1948. A clear introductory treatment.

Wilks, S. S., *Mathematical Statistics*, Wiley, New York, 1962. A definitive reference.

Wilson, E. Bright, Jr., *An Introduction to Scientific Research*, McGraw-Hill, New York, 1952. A classic that should be owned by every student of science.

Worthing, Archie G., and Joseph Geffner, *Treatment of Experimental Data*, Wiley, New York, 1943. Good section on graphical methods.

Young, Hugh D., *Statistical Treatment of Experimental Data*, McGraw-Hill, New York, 1962.

INDEX

Bold face numbers refer to the main references.